# Quantum Field Theory

This modern text combines fundamental principles with advanced topics and recent techniques in a rigorous and self-contained treatment of quantum field theory.

Key features include:
- a review of basic principles, starting with quantum mechanics and special relativity, and covering elementary aspects of quantum field theory and perturbation theory;
- standard results and tools relevant to many applications, including canonical quantization, path integrals, non-Abelian gauge theories and the renormalization group;
- advanced topics such as effective field theories, quantum anomalies, stable extended field configurations, lattice field theory and field theory at a finite temperature or in the strong field regime;
- two chapters dedicated to new methods for calculating scattering amplitudes (spinor-helicity, on-shell recursion and generalized unitarity), equipping students with practical skills for future research; and
- an accessible style featuring numerous worked examples, applications and end-of-chapter problems.

This is an essential text for graduate students in physics, and an equally excellent reference for researchers in the field.

# Quantum Field Theory
## From Basics to Modern Topics

**FRANÇOIS GELIS**  CEA, Saclay

# CAMBRIDGE
## UNIVERSITY PRESS

University Printing House, Cambridge CB2 8BS, United Kingdom

One Liberty Plaza, 20th Floor, New York, NY 10006, USA

477 Williamstown Road, Port Melbourne, VIC 3207, Australia

314–321, 3rd Floor, Plot 3, Splendor Forum, Jasola District Centre, New Delhi – 110025, India

79 Anson Road, #06–04/06, Singapore 079906

Cambridge University Press is part of the University of Cambridge.

It furthers the University's mission by disseminating knowledge in the pursuit of education, learning, and research at the highest international levels of excellence.

www.cambridge.org
Information on this title: www.cambridge.org/9781108480901
DOI: 10.1017/9781108691550

© Cambridge University Press 2019

This publication is in copyright. Subject to statutory exception and to the provisions of relevant collective licensing agreements, no reproduction of any part may take place without the written permission of Cambridge University Press.

First published 2019

Printed in the United Kingdom by TJ International Ltd, Padstow Cornwall

*A catalogue record for this publication is available from the British Library.*

*Library of Congress Cataloging-in-Publication Data*
Names: Gelis, François, 1972- author.
Title: Quantum field theory : from basics to modern topics / François Gelis (CEA, Saclay).
Description: Cambridge, United Kingdom ; New York, NY : Cambridge University Press, 2019. | Includes bibliographical references and index.
Identifiers: LCCN 2019005987 | ISBN 9781108480901 (hardback ; alk. paper) | ISBN 110848090X (hardback ; alk. paper)
Subjects: LCSH: Quantum field theory.
Classification: LCC QC174.45 .G45 2019 | DDC 530.14/3–dc23
LC record available at https://lccn.loc.gov/2019005987
ISBN 978-1-108-48090-1 Hardback

Additional resources for this publication at www.cambridge.org/gelis.

Cambridge University Press has no responsibility for the persistence or accuracy of URLs for external or third-party internet websites referred to in this publication and does not guarantee that any content on such websites is, or will remain, accurate or appropriate.

To Kanako and Nathan.

In order to improve the mind,
we ought less to learn,
than to contemplate.

RENÉ DESCARTES

# Contents

| | | |
|---|---|---|
| *List of Figures and Tables* | | page xii |
| *Preface* | | xv |
| *Acknowledgments* | | xvii |
| **1** | **Basics of Quantum Field Theory** | **1** |
| | 1.1 Why Quantum Field Theory? | 1 |
| | 1.2 Special Relativity | 3 |
| | 1.3 Free Scalar Fields, Mode Decomposition | 7 |
| | 1.4 Interacting Scalar Particles | 14 |
| | 1.5 From Field Correlations to Reaction Rates | 17 |
| | 1.6 Källén–Lehmann Spectral Representation | 24 |
| | 1.7 Generating Functional | 27 |
| | Exercises | 32 |
| **2** | **Perturbation Theory** | **35** |
| | 2.1 Perturbative Expansion and Feynman Rules | 35 |
| | 2.2 Calculation of Loop Integrals | 41 |
| | 2.3 Ultraviolet Divergences and Renormalization | 44 |
| | 2.4 Perturbative Unitarity | 53 |
| | Exercises | 61 |
| **3** | **Quantum Electrodynamics** | **63** |
| | 3.1 Spin-1/2 Fields | 63 |
| | 3.2 Spin-1 Fields | 70 |
| | 3.3 Quantum Electrodynamics | 74 |
| | 3.4 Charge Conservation, Ward–Takahashi Identities | 80 |
| | 3.5 Ultraviolet Renormalization | 82 |
| | 3.6 Cutting Rules in QED and Unitarity | 86 |

|   |     |                                                       |     |
|---|-----|-------------------------------------------------------|-----|
|   | 3.7 | Infrared Divergences                                  | 88  |
|   |     | Exercises                                             | 93  |
| **4** | **Spontaneous Symmetry Breaking**                         | **95**  |
|   | 4.1 | Potential Energy Landscape                            | 95  |
|   | 4.2 | Conserved Currents and Charges                        | 101 |
|   | 4.3 | Spectral Properties                                   | 103 |
|   | 4.4 | Coleman's Theorem                                     | 105 |
|   | 4.5 | Linear Sigma Model                                    | 107 |
|   | 4.6 | Heisenberg Model of Ferromagnetism                    | 111 |
|   |     | Exercises                                             | 114 |
| **5** | **Functional Quantization**                               | **116** |
|   | 5.1 | Path Integral in Quantum Mechanics                    | 116 |
|   | 5.2 | Functional Manipulations                              | 120 |
|   | 5.3 | Path Integral in Scalar Field Theory                  | 125 |
|   | 5.4 | Functional Determinants                               | 127 |
|   | 5.5 | Quantum Effective Action                              | 129 |
|   | 5.6 | Two-Particle Irreducible Effective Action             | 138 |
|   | 5.7 | Euclidean Path Integral and Statistical Mechanics     | 144 |
|   |     | Exercises                                             | 147 |
| **6** | **Path Integrals for Fermions and Photons**               | **150** |
|   | 6.1 | Grassmann Variables                                   | 150 |
|   | 6.2 | Path Integral for Fermions                            | 156 |
|   | 6.3 | Path Integral for Photons                             | 157 |
|   | 6.4 | Schwinger–Dyson Equations                             | 160 |
|   | 6.5 | Quantum Anomalies                                     | 162 |
|   |     | Exercises                                             | 170 |
| **7** | **Non-Abelian Gauge Symmetry**                            | **173** |
|   | 7.1 | Non-Abelian Lie Groups and Algebras                   | 173 |
|   | 7.2 | Yang–Mills Lagrangian                                 | 181 |
|   | 7.3 | Non-Abelian Gauge Theories                            | 185 |
|   | 7.4 | Spontaneous Gauge Symmetry Breaking                   | 190 |
|   | 7.5 | θ-term and Strong-CP Problem                          | 194 |
|   | 7.6 | Non-Local Gauge Invariant Operators                   | 201 |
|   |     | Exercises                                             | 209 |

## 8 Quantization of Yang–Mills Theory — 212
- 8.1 Naive Quantization of the Gauge Bosons — 212
- 8.2 Gauge Fixing — 214
- 8.3 Fadeev–Popov Quantization and Ghost Fields — 215
- 8.4 Feynman Rules for Non-Abelian Gauge Theories — 217
- 8.5 On-Shell Non-Abelian Ward–Takahashi Identities — 221
- 8.6 Ghosts and Unitarity — 223
- Exercises — 232

## 9 Renormalization of Gauge Theories — 234
- 9.1 Ultraviolet Power Counting — 234
- 9.2 Symmetries of the Quantum Effective Action — 236
- 9.3 Renormalizability — 241
- 9.4 Background Field Method — 245
- Exercises — 251

## 10 Renormalization Group — 252
- 10.1 Scale Dependence of Correlation Functions — 252
- 10.2 Correlators Containing Composite Operators — 256
- 10.3 Operator Product Expansion — 258
- 10.4 Example: QCD Corrections to Weak Decays — 261
- 10.5 Non-Perturbative Renormalization Group — 267
- Exercises — 276

## 11 Effective Field Theories — 278
- 11.1 General Principles of Effective Theories — 279
- 11.2 Example: Fermi Theory of Weak Decays — 282
- 11.3 The Standard Model as an Effective Field Theory — 285
- 11.4 Effective Theories in QCD — 291
- 11.5 EFT of Spontaneous Symmetry-Breaking — 300
- Exercises — 309

## 12 Quantum Anomalies — 312
- 12.1 Axial Anomalies in a Gauge Background — 312
- 12.2 Generalizations — 323
- 12.3 Wess–Zumino Consistency Conditions — 329
- 12.4 't Hooft Anomaly Matching — 333

12.5 Scale Anomalies — 334
Exercises — 341

## 13 Localized Field Configurations — 342
13.1 Domain Walls — 343
13.2 Monopoles — 346
13.3 Instantons — 355
13.4 Skyrmions — 366
Exercises — 370

## 14 Modern Tools for Tree Amplitudes — 372
14.1 Color Decomposition of Gluon Amplitudes — 373
14.2 Spinor-Helicity Formalism — 379
14.3 Britto–Cachazo–Feng–Witten On-Shell Recursion — 388
14.4 Tree-Level Gravitational Amplitudes — 398
14.5 Cachazo–Svrcek–Witten Rules — 406
Exercises — 415

## 15 Worldline Formalism — 418
15.1 Worldline Representation — 418
15.2 Quantum Electrodynamics — 423
15.3 Schwinger Mechanism — 427
15.4 Calculation of One-Loop Amplitudes — 430
Exercises — 438

## 16 Lattice Field Theory — 440
16.1 Discretization of Bosonic Actions — 441
16.2 Lattice Fermions — 445
16.3 Hadron Mass Determination on the Lattice — 449
16.4 Wilson Loops and Color Confinement — 451
16.5 Gauge Fixing on the Lattice — 454
16.6 Lattice Hamiltonian — 457
16.7 Lattice Worldline Formalism — 458
Exercises — 462

## 17 Quantum Field Theory at Finite Temperature — 465
17.1 Canonical Thermal Ensemble — 465
17.2 Finite-$T$ Perturbation Theory — 466

17.3 Large-Distance Effective Theories 482
17.4 Out-of-Equilibrium Systems 495
Exercises 501

**18 Strong Fields and Semi-Classical Methods** **504**
18.1 Situations Involving Strong Fields 505
18.2 Observables at Leading and Next-to-Leading Orders 512
18.3 Green's Formulas 517
18.4 Mode Functions 526
18.5 Multi-Point Correlation Functions at Tree Level 530
Exercises 543

**19 From Trees to Loops** **546**
19.1 Dualities Between Loops and Trees 546
19.2 Reduction of One-Loop Amplitudes 553
19.3 One-Loop Amplitudes from Unitarity Cuts 564
19.4 The Frontier: Multi-Loop Amplitudes 573
Exercises 581

*Further Reading* 584
*Index* 586

# Figures and Tables

## Figures

| | | |
|---|---|---|
| 1.1 | Geometry of a two-body cross-section in the target frame. | page 20 |
| 1.2 | Example of Källen–Lehmann spectral function. | 26 |
| 1.3 | Integration contours in the $z$ plane for the distribution $1/(x + i0^+)$. | 32 |
| 2.1 | A four-loop graph and its embedding on a torus. | 40 |
| 2.2 | Wick rotation in the complex $k_0$ plane. | 42 |
| 2.3 | Time contour in the Schwinger–Keldysh formalism. | 59 |
| 3.1 | Feynman rules of quantum electrodynamics in Coulomb gauge. | 77 |
| 3.2 | Spatial resolution of a virtual photon. | 81 |
| 3.3 | Potentially ultraviolet divergent correlators in QED. | 83 |
| 3.4 | Off-shell Ward–Takahashi identity between 3- and 2-point functions. | 85 |
| 3.5 | Soft photon corrections to a hard amplitude. | 89 |
| 4.1 | Scalar potential that exhibits spontaneous symmetry breaking. | 98 |
| 4.2 | Potential for the Hamiltonians $H_0$, $H_0 + V$ and $H_0 + V + \widetilde{V}$. | 99 |
| 4.3 | Graphical determination of $\sigma_*$ and $m_\pi$. | 110 |
| 5.1 | Illustration of eq. (5.7) with ten intermediate points. | 118 |
| 5.2 | Illustration of eq. (5.19). | 120 |
| 5.3 | Graphs in the one-loop effective potential in scalar QED. | 136 |
| 5.4 | Coleman-Weinberg effective potential. | 138 |
| 5.5 | Beginning of the diagrammatic expansion of the 2PI effective action. | 143 |
| 7.1 | Lie group and Lie algebra. | 176 |
| 7.2 | Geometrical interpretation of Trotter's formula. | 176 |
| 7.3 | Symmetry-breaking pattern for a potential with $O(3)$ symmetry. | 193 |
| 7.4 | Kinematics of the eikonal limit. | 208 |
| 8.1 | Illustration of the gauge fixing procedure. | 215 |
| 8.2 | Feynman rules of non-Abelian gauge theories in covariant gauge. | 219 |
| 8.3 | One-loop diagrams contributing to $q\bar{q} \to q\bar{q}$. | 225 |
| 8.4 | Illustration of the BRST cohomology. | 231 |
| 10.1 | Order-$g^2$ QCD corrections to the operator $\mathcal{O}_1$. | 263 |
| 10.2 | Kadanoff's block-spin renormalization. | 269 |
| 10.3 | Renormalization group flow in theory space. | 271 |
| 10.4 | Sliding cutoff in the functional renormalization group. | 274 |
| 11.1 | Main classes of higher dimensional operators. | 286 |

| | | |
|---|---|---|
| 11.2 | Diagrammatic illustration of the see-saw mechanism. | 289 |
| 11.3 | Cartoon of the fluctuations inside a nucleon. | 296 |
| 11.4 | Degrees of freedom in the color glass condensate. | 298 |
| 11.5 | Typical graph in a hadron–hadron collision. | 298 |
| 11.6 | Symmetry-breaking pattern for a model with $O(3)$ symmetry. | 301 |
| 11.7 | Perturbative expansion in the nonlinear sigma model. | 304 |
| 11.8 | $(\sigma, \boldsymbol{\pi})$ coordinates for an $O(3)$ model. | 306 |
| 11.9 | Successive stages of the Ricci flow on a 2-dimensional manifold. | 308 |
| 12.1 | Vector and axial currents in $1+1$ dimensions. | 314 |
| 12.2 | Axial anomaly in a gauge background. | 319 |
| 12.3 | Axial anomaly in a gravitational background. | 326 |
| 13.1 | Quartic potential exhibiting spontaneous symmetry breaking. | 343 |
| 13.2 | Domain wall profile corresponding to the potential of Figure 13.1. | 345 |
| 13.3 | Magnetic field lines of the Dirac monopole. | 346 |
| 13.4 | Cartoon of the hedgehog configuration. | 350 |
| 13.5 | Illustration of the symmetry-breaking pattern. | 353 |
| 13.6 | Hemispheres with gauge potentials $\mathbf{A}$ and $\mathbf{A}'$. | 354 |
| 13.7 | Cartoon of an instanton solution. | 356 |
| 13.8 | Stereographic projection that maps the plane $\mathbb{R}^2$ to the sphere $S_2$. | 367 |
| 14.1 | 4, 5 and 6-gluon color-ordered tree amplitudes. | 377 |
| 14.2 | Rules for color-ordered graphs in Yang–Mills theory. | 378 |
| 14.3 | Propagators affected by the shift. | 391 |
| 14.4 | BCFW recursion with shifts on the lines 1 and $n$. | 394 |
| 14.5 | BCFW recursion for MHV amplitudes. | 395 |
| 14.6 | Photon-scalar and graviton-scalar scattering amplitudes. | 399 |
| 14.7 | BCFW shift for the calculation of the $\gamma\gamma\phi\phi$ amplitude. | 401 |
| 14.8 | BCFW shift for the calculation of the $hh\phi\phi$ amplitude. | 405 |
| 14.9 | Propagators affected by the shift of eq. (14.157). | 408 |
| 14.10 | Propagators affected by the shift of eq. (14.176). | 413 |
| 15.1 | Typical worldloop in the effective action. | 423 |
| 15.2 | Tunneling in the Schwinger mechanism. | 428 |
| 15.3 | One-loop photon polarization tensor in scalar QED. | 436 |
| 16.1 | Discretization of Euclidean space-time on a hyper-cubic lattice. | 441 |
| 16.2 | Continuous and discrete bosonic dispersion relations. | 443 |
| 16.3 | Link variable and plaquette. | 444 |
| 16.4 | Continuous and discrete fermionic dispersion relations. | 447 |
| 16.5 | Discrete dispersion curve with a Wilson term. | 448 |
| 16.6 | Various types of fermionic contributions. | 449 |
| 16.7 | Hadron mass determination from lattice QCD. | 450 |
| 16.8 | Tiling of a closed loop by elementary plaquettes. | 452 |
| 16.9 | Rectangular Wilson loop in the $A_4 \equiv 0$ gauge. | 453 |
| 16.10 | Gauge orbit that intersects multiple times the gauge fixing manifold. | 455 |
| 16.11 | Worldlines on a cubic lattice. | 459 |
| 17.1 | Contour deformations to evaluate Matsubara sums. | 478 |

| | | |
|---|---|---|
| 17.2 | Free energy at non-zero temperature in the $\phi^4$ scalar theory. | 486 |
| 17.3 | Evolution of a scalar potential with increasing temperature. | 486 |
| 17.4 | Fermion and photon self-energies at one loop. | 487 |
| 17.5 | List of hard thermal loops in QCD. | 490 |
| 17.6 | Photon (left) and quark (right) dispersion relations. | 490 |
| 17.7 | Debye screening in QED. | 492 |
| 17.8 | Relevant distance scales in a relativistic plasma at high temperature. | 493 |
| 18.1 | Vertices for expansions in an initial coherent state. | 509 |
| 18.2 | Generic connected graph in the strong source regime. | 512 |
| 18.3 | Contributions to observables at NLO in the strong field regime. | 513 |
| 18.4 | Illustration of eq. (18.58). | 516 |
| 18.5 | Typical contribution to $\Phi(x)$. | 519 |
| 18.6 | Illustration of eq. (18.70). | 520 |
| 18.7 | Typical contribution to $a(x)$. | 521 |
| 18.8 | Schwinger–Keldysh rules for the generating functional. | 532 |
| 18.9 | Relationship between the fields $\phi_1$ and $\phi_2$. | 538 |
| 18.10 | Causal structure of the three-point correlation function. | 542 |
| 19.1 | Loop momentum labeling. | 549 |
| 19.2 | Tree-level diagram contributing to the response function. | 550 |
| 19.3 | One-loop diagram contributing to the response function. | 551 |
| 19.4 | Two-loop contribution containing the 111 vertex. | 552 |
| 19.5 | Pattern of reductions of one-loop tensor integrals. | 555 |
| 19.6 | Cutkosky's unitarity cuts for a one-loop diagram. | 565 |
| 19.7 | Number of solutions for several types of maximal cuts. | 578 |
| 19.8 | Two-loop skeleton topologies | 580 |

## Tables

| | | |
|---|---|---|
| 7.1 | Dimensions of a few common Lie algebras. | 177 |
| 12.1 | Quantum numbers of the Standard Model fermions. | 328 |
| 14.1 | Number of diagrams in tree-level Yang–Mills amplitudes. | 373 |
| 14.2 | Number of color-ordered diagrams in Yang–Mills theory. | 377 |

# Preface

This book started in the form of lecture notes for a quantum field theory course I taught at the Master's program on High Energy Physics at École Polytechnique (Palaiseau, France). Students enrolled in these courses have already had an introductory course to quantum field theory, and therefore I started directly with more advanced topics, such as path integral quantization, non-Abelian gauge theories and the renormalization group.

By virtue of the audience to which these lectures were delivered, a large portion of the material in these notes was directed toward applications in particle physics. However, in the last 25 years the traditional workflow (Lagrangian $\to$ Feynman rules $\to$ scattering amplitudes $\to$ cross-sections) has been shaken by a flood of new developments that has shattered how we view and calculate scattering amplitudes. More importantly, these new methods uncovered the great unity that exists across seemingly unrelated quantum field theories, something that was impossible to foresee from their respective Feynman rules. Thus, one of the goals I set for these lecture notes was to include – in addition to the traditional methods, not as a replacement of them – a detailed exposition of these new techniques, which are not addressed in most of the existing textbooks because they did not exist or were not mature enough when they were written.

At the same time, I wanted to expose students to a broader picture by presenting as much as possible quantum field theory as an agnostic tool that can also be used in other contexts, such as cosmology or condensed matter physics. To that effect, I decided to treat topics like the path integral in statistical mechanics, the Schwinger–Keldysh formalism and quantum field theory at finite temperature. Although each of these subjects is well-covered in specialized textbooks, they are usually missing in the more general ones, which may lead to an artificial impression of segmentation of the field.

When the idea to convert my lecture notes into a textbook started to take form, it became obvious that more introductory material was also needed. This led to the addition of the first four chapters, which provide a refresher on basic concepts and set the notations used in the rest of the book. A word of caution is in order regarding the intended role of these introductory chapters: They are meant to make the book sufficiently self-contained for readers who have had prior exposure to quantum field theory, but first-timers will probably find them too fast-paced.

Of course, many topics had to be left out due to a combination of limited space and a lack of familiarity by the author. One of the most prominent absences is supersymmetry, both in its use in extensions of the Standard Model and in toy theories used as playgrounds for formal QFT developments (e.g., supersymmetric Yang–Mills theory). Another topic not treated in this book is the AdS/CFT correspondence, despite it having been a very active subfield over the past 20 years and providing numerous insights about strongly coupled gauge theories.

Each chapter is supplemented by exercises (about 180 in total). Although these exercises range from straightforward (but not necessarily easy) verifications of some technical aspects of the text to more elaborate and time-consuming ones, a general guidance when choosing which exercises to include was the belief that fluency comes by first solving many "simple" exercises in order to master the manipulation of the objects and concepts discussed in the text. Some "recommended" (either because they provide a derivation of an important point of the text, or because they discuss interesting extensions) exercises are marked by a star.

Let me end with a note on the mathematics behind quantum field theory. The subject often carries the reputation of requiring a rather high level of mathematical sophistication. Consequently, some mathematical prerequisites are unavoidable for reading this text comfortably – most notably complex analysis, Fourier transforms, plus some basics of group theory and distribution theory. For completeness, when mathematical tools outside of these areas are necessary, I have included some discussion and sometimes a basic derivation of the tools and results needed to treat the subject at hand.

# Acknowledgments

This book would certainly not exist without Stéphane Munier and Pascal Paganini, who offered me the opportunity to teach quantum field theory in the Master's program they are in charge of. Early versions of this manuscript circulated in the form of lecture notes handed to the students, and benefited enormously from their questions, remarks and criticisms. Conversely, this book owes a lot to many talented teachers to whom I have had the pleasure of listening. In particular, Jean-Marie Mercier, Jean-Claude Le Guillou, François Delduc and Patrick Aurenche had a long-lasting impact on how I think about physics in general and quantum field theory in particular.

Moreover, over the years I have learned a lot from inspiring discussions with many colleagues and collaborators. There is not enough space to list them all here, but I would like to specially thank Ian Balitsky, Jean-Paul Blaizot, Yuri Dokshitzer, Thomas Epelbaum, Kenji Fukushima, Edmond Iancu, Sangyong Jeon, Keijo Kajantie, Tuomas Lappi, Larry McLerran, Guy Moore, Al Mueller, Jean-Yves Ollitrault, André Peshier, Gregory Soyez, Raju Venugopalan and Heribert Weigert.

I am also extremely grateful for the very professional help and advice received at all stages of the making of this book from Simon Capelin, Lisa Pinto, Heather Brolly and Rosie Crawley at Cambridge University Press, and Davide Napoletano for his careful reading of several chapters.

Last but not least, more than often the preparation of this book leaked out of the office. For this, for her patience and her constant encouragement, I would like to address my warmest thanks to my wife, Kanako, and to my son, Nathan, who did his best to remind me when to stop.

# CHAPTER 1

# Basics of Quantum Field Theory

## 1.1 Why Quantum Field Theory?

After the revolutions of relativity and quantum mechanics at the beginning of the twentieth century, theoretical physics seemed to rest on two pillars that had few connections with each other. On the one hand, one had a very successful non-quantum theory of electromagnetic radiation that had relativistic invariance built in, and on the other hand quantum mechanics provided a very effective way of predicting the energy spectrum of a particle in a given external potential but was obviously not relativistically covariant. This situation was not fully satisfactory but one could live with it, provided certain questions are not asked.

Simple estimates tell us that the electron in the hydrogen atom is not relativistic. Indeed, the energy of the electron in the ground state is $E_0 = m_e e^4/(32\pi^2 \hbar^2 \epsilon_0^2) \approx 13.6$ eV, while its mass is $m_e \approx 0.5$ meV/$c^2$. From this energy, we may estimate the ratio of the electron velocity to the speed of light by $(v_e/c)^2 \sim E_0/(m_e c^2) = \alpha^2/2$, where $\alpha = e^2/(4\pi\epsilon_0 \hbar c) \approx 1/137$ is a dimensionless constant – called the fine structure constant – that encodes the strength of the electromagnetic interactions that bind the electron to the hydrogen nucleus ($\alpha$ is proportional to the product of the electrical charges of the electron and of the hydrogen nucleus). If $\alpha$ was much larger,[1] the energy of the electron would get closer to $m_e c^2$, and we would expect relativistic corrections to become non-negligible. Moreover, one may view the non-relativistic framework of quantum mechanics as the zeroth-order approximation of a more general expansion in powers of $v_e/c$. Since the dimensionless constant $\alpha$ contains a factor $c^{-1}$, the expansion in $v_e/c$ is also an expansion in $\alpha$. Even if these corrections are very small, they should exist in principle (in fact, the fine structure of atomic spectra, known since 1887,

---

[1] Experimentally, this could occur with atoms of high Z that are highly ionized, i.e., stripped of most of their electrons. In this case, $\alpha$ would be replaced by $Z\alpha$.

has later been interpreted as such a relativistic correction), but their calculation certainly requires incorporating some ingredients of special relativity into the framework of quantum mechanics.

The quantum mechanical setup for calculating the spectrum of the hydrogen atom uses the Coulomb potential of the hydrogen nucleus. In this framework, this potential simply acts as a background in which the dynamics of the electron take place, but it does not play an active role in the problem. It would be much more satisfactory to have a theoretical framework in which matter and radiation are treated on the same footing, especially given the fact that quantum mechanics completely blurs the frontier between waves and particles. This unsatisfactory dichotomy arises in the excitation (respectively, de-excitation) of an atom by absorption (respectively, emission) of radiation. But since photons are massless, any framework that incorporates them on the same footing as the electron must be relativistic. Another place where this classical treatment of electromagnetic fields is lacking is in the very concept of a point-like electron. If one imagines reaching it as the zero radius limit of a spherical charge distribution, one can define the electrostatic energy contained in this sphere (defined as the energy necessary to bring these charges from infinity into the sphere). This energy becomes infinite when the radius goes to zero (and becomes comparable to the rest energy $m_e c^2$ of the electron at a finite radius known as the classical radius of the electron).

Quantum mechanics is all about measurements, whose expectations are calculated as averages of operators in the state vector of the system. In this context, one could imagine a "composite" measurement that consists of two local measurements performed at space-time points with a space-like separation. Special relativity tells us that, because no signal may propagate faster than the speed of light, these two measurements are not causally connected and their results should be independent. In quantum mechanics, the independence of two measurements is encoded in the fact that the corresponding operators commute. However, the notion that space-like separated local operators should commute is not naturally present in quantum mechanics. Therefore, we anticipate that it makes predictions that are not fully consistent with relativity in this type of situation.

Another difficulty with the usual formulation of quantum mechanics is that for each particle in the system under consideration, the wavefunction depends on the position of this particle. This becomes rapidly untractable in systems with more than a few particles. While this issue is just a technical difficulty for non-relativistic systems, it becomes an unsurmountable stumbling block in relativistic systems, for which even the number of particles can change. An obvious example is that of atomic transitions that are accompanied by the emission or absorption of a photon. It should be quite clear that a wavefunction that describes a predefined number of particles cannot accommodate this type of transition. This remark suggests that the framework that brings relativistic covariance into quantum mechanics has to be a somewhat radical departure from the usual formulation of quantum mechanics, at least on a technical level.

*Quantum field theory* is a theoretical framework that promotes quantum mechanics into a relativistic theory. Historically, the first developed quantum field theory was quantum electrodynamics. However, we shall not start with this example in order to avoid the extra complications related to the non-zero spin of photons and electrons, and to the redundancy due to the gauge invariance of classical electrodynamics. Instead, in the first two chapters, we will introduce the basic concepts of quantum field theory with the example of a *scalar* (i.e., zero spin) field. Although this example has less obvious applications in nature, it has considerable didactical virtues because it allows explanation of the structural aspects of quantum field theory without encumbering the exposition with the extra difficulties posed by spin and gauge invariance. These will be deferred until Chapter 3.

## 1.2 Special Relativity

### 1.2.1 Lorentz Transformations

Given these premises, special relativity obviously plays a crucial role in quantum field theory. A major requirement is that various observers in frames that are moving at a constant speed relative to each other should be able to describe physical phenomena using the same laws of physics. This does not imply that the equations governing these phenomena are independent of the observer's frame, but that these equations transform in a constrained fashion – depending on the nature of the objects they contain – under a change of reference frame. This property is called *relativistic covariance*, or *Lorentz covariance*. Let us consider two reference frames $\mathcal{F}$ and $\mathcal{F}'$, in which the coordinates of a given event are respectively $x^\mu$ and $x'^\mu$. A *Lorentz transformation* is a linear transformation such that the interval $ds^2 \equiv dt^2 - dx^2$ is the same in the two frames.[2] If we denote the coordinate transformation by

$$x'^\mu = \Lambda^\mu{}_\nu x^\nu, \tag{1.1}$$

the matrix $\Lambda$ of the transformation must obey

$$g_{\mu\nu} \Lambda^\mu{}_\rho \Lambda^\nu{}_\sigma = g_{\rho\sigma}, \tag{1.2}$$

where $g_{\mu\nu}$ is the Minkowski metric tensor $g_{\mu\nu} \equiv \text{diag}(1, -1, -1, -1)$. When equipped with the composition law, the set of Lorentz transformations becomes a group, the *Lorentz group*. From eq. (1.2), we can see that the inverse of a Lorentz transformation is given by

$$\Lambda_\mu{}^\nu = (\Lambda^{-1})_\nu{}^\mu. \tag{1.3}$$

Infinitesimal Lorentz transformations are those that relate reference frames that have a very small relative velocity. They can be written as a small deviation about the identical transformation,

$$\Lambda^\mu{}_\nu = \delta^\mu{}_\nu + \omega^\mu{}_\nu, \tag{1.4}$$

with all components of $\omega$ much smaller than unity. The definition of Lorentz transformations implies that $\omega_{\mu\nu} = -\omega_{\nu\mu}$ (with all indices down). Consequently, there are six independent Lorentz transformations, three of which are ordinary *rotations* and three are *boosts*. Note that the infinitesimal transformations (1.4) have a determinant[3] equal to $+1$ (they are called *proper* transformations), and do not change the direction of the time axis since $\Lambda^0{}_0 \approx 1 \geq 0$ (they are called *orthochronous*). Any combination of such infinitesimal transformations shares the same properties, and their set forms a subgroup of the Lorentz group.

---

[2] The physical premises of special relativity require that the speed of light be the same in all inertial frames, which implies solely that $ds^2 = 0$ be preserved in all inertial frames. The group of transformations that achieves this is called the *conformal group*. In four space-time dimensions, the conformal group is 15 dimensional, and in addition to the six orthochronous Lorentz transformations it contains translations, dilatations, as well as nonlinear transformations called *special conformal transformations* (see Exercise 12.9).

[3] From eq. (1.2), the determinant may be equal to $\pm 1$.

## 1.2.2 Lorentz and Poincaré Algebras

In the previous section, we introduced Lorentz transformations via their action on the coordinates $x^\mu$. But of course, coordinates are not the only objects that vary when changing the reference frame. For instance, any tensor transforms as

$$T'^{\mu\nu\cdots} = \Lambda^\mu{}_\alpha \Lambda^\nu{}_\beta \cdots T^{\alpha\beta\cdots}. \tag{1.5}$$

In addition, in a quantum system, a Lorentz transformation $\Lambda$ should also act on the states in the Hilbert space via a linear transformation $U(\Lambda)$,

$$|\alpha_{\mathcal{F}'}\rangle = U(\Lambda)|\alpha_{\mathcal{F}}\rangle, \tag{1.6}$$

that forms a representation of the Lorentz group, i.e.,

$$U(\Lambda'\Lambda) = U(\Lambda')U(\Lambda). \tag{1.7}$$

This property simply means that under a succession of two Lorentz transformations $\Lambda$ and $\Lambda'$, the resulting state can either be obtained in a single transformation corresponding to the product of the Lorentz transformations, or in a two-step process in which the two transformations are applied successively. For an infinitesimal Lorentz transformation, we can write

$$U(1+\omega) = 1 + \frac{i}{2}\omega_{\mu\nu}M^{\mu\nu} + \mathcal{O}(\omega^2). \tag{1.8}$$

(The prefactor $i/2$ in the second term of the right-hand side is conventional.) Since the $\omega_{\mu\nu}$ are antisymmetric, the $M^{\mu\nu}$ can also be chosen as antisymmetric. The $M^{\mu\nu}$ are called the *generators* of the Lorentz group in the representation $U$. By using eq. (1.7) for the Lorentz transformation $\Lambda^{-1}\Lambda'\Lambda$, we arrive at

$$U^{-1}(\Lambda)M^{\mu\nu}U(\Lambda) = \Lambda^\mu{}_\rho \Lambda^\nu{}_\sigma M^{\rho\sigma}, \tag{1.9}$$

indicating that $M^{\mu\nu}$ transforms as a rank-2 tensor. When used with an infinitesimal transformation $\Lambda = 1+\omega$, this identity leads (see Exercise 1.1) to the commutation relation that defines the Lie algebra of the Lorentz group,

$$[M^{\mu\nu}, M^{\rho\sigma}] = i(g^{\mu\rho}M^{\nu\sigma} - g^{\nu\rho}M^{\mu\sigma}) - i(g^{\mu\sigma}M^{\nu\rho} - g^{\nu\sigma}M^{\mu\rho}). \tag{1.10}$$

When necessary, it is possible to divide the six generators $M^{\mu\nu}$ into three generators $J^i$ for ordinary spatial rotations, and three generators $K^i$ for the Lorentz boosts along each of the spatial directions:

Rotations: $\quad J^i \equiv \tfrac{1}{2}\epsilon_{ijk}M^{jk},$

Lorentz boosts: $\quad K^i \equiv M^{i0},$ (1.11)

where $\epsilon_{ijk}$ is the three-dimensional Levi-Civita symbol[4] normalized by $\epsilon_{123} = +1$ (thus, $J^1 = M^{23}, J^2 = M^{31}, J^3 = M^{12}$).

---

[4]Throughout this book, the Levi-Civita symbol on a set of ordered indices $\{i_1, i_2, \cdots, i_n\}$ is consistently defined by $\epsilon_{i_1 i_2 \cdots i_n} = +1$, *with lowered indices*. In circumstances where it makes sense to raise the indices, this is done as usual by multiplication with the metric tensor.

## 1.2 SPECIAL RELATIVITY

**Poincaré group:** The group of Lorentz transformations can be extended by adding the translations, resulting in a larger group of transformations known as the *Poincaré group*. A translation is parameterized by the 4-vector $a^\mu$ by which all coordinates are shifted, $x^\mu \to x^\mu + a^\mu$. A generic transformation for the Poincaré group is thus a pair $(a, \Lambda)$ such that

$$x^\mu \to \Lambda^\mu{}_\nu x^\nu + a^\mu. \tag{1.12}$$

(In this definition, the Lorentz transformation is applied first.) As for the Lorentz transformations, the action of infinitesimal translations on states can be represented by

$$U(a) = 1 + i\, a_\mu P^\mu + \mathcal{O}(a^2), \tag{1.13}$$

where $P^\mu$ is the generator of translations (which is nothing but the 4-momentum operator). In a fashion similar to eq. (1.9), we obtain

$$U^{-1}(\Lambda) P^\mu U(\Lambda) = \Lambda^\mu{}_\rho P^\rho, \tag{1.14}$$

which leads to the following commutation relation (see Exercise 1.2) between $P^\mu$ and $M^{\mu\nu}$,

$$\begin{aligned}\left[P^\mu, M^{\rho\sigma}\right] &= i(g^{\mu\sigma}P^\rho - g^{\mu\rho}P^\sigma),\\ \left[P^\mu, P^\nu\right] &= 0.\end{aligned} \tag{1.15}$$

Let us illustrate on a simple example how to relate the measurements of the momentum in a certain state performed by two observers in different reference frames. The two observers measure the expectation values $\langle \alpha_\mathcal{F} | P^\mu | \alpha_\mathcal{F} \rangle$ and $\langle \alpha_{\mathcal{F}'} | P^\mu | \alpha_{\mathcal{F}'} \rangle$. These expectation values are related by

$$\langle \alpha_{\mathcal{F}'} | P^\mu | \alpha_{\mathcal{F}'} \rangle = \langle \alpha_\mathcal{F} | U^{-1}(\Lambda) P^\mu U(\Lambda) | \alpha_\mathcal{F} \rangle = \Lambda^\mu{}_\nu \langle \alpha_\mathcal{F} | P^\nu | \alpha_\mathcal{F} \rangle. \tag{1.16}$$

Unsurprisingly, since the two observers measure the 4-momentum of the same system in two different frames, the results of their measurements are related in a simple way by the Lorentz transformation of a vector.

### 1.2.3 One-Particle States

Let us denote $|\mathbf{p}, \sigma\rangle$ as a one-particle state, where $\mathbf{p}$ is the 3-momentum of that particle, and $\sigma$ denotes its other quantum numbers. Since this state contains a particle with a definite momentum, it is an eigenstate of the momentum operator $P^\mu$, namely

$$P^\mu |\mathbf{p}, \sigma\rangle = p^\mu |\mathbf{p}, \sigma\rangle, \quad \text{with } p^0 \equiv \sqrt{\mathbf{p}^2 + m^2}. \tag{1.17}$$

Let us now act on this state with a Lorentz transformation, to obtain $U(\Lambda) |\mathbf{p}, \sigma\rangle$. We have

$$P^\mu U(\Lambda) |\mathbf{p}, \sigma\rangle = U(\Lambda) \underbrace{U^{-1}(\Lambda) P^\mu U(\Lambda)}_{\Lambda^\mu{}_\nu P^\nu} |\mathbf{p}, \sigma\rangle = \Lambda^\mu{}_\nu p^\nu\, U(\Lambda) |\mathbf{p}, \sigma\rangle. \tag{1.18}$$

Therefore, $U(\Lambda)|\mathbf{p}, \sigma\rangle$ is an eigenstate of momentum with eigenvalue $(\Lambda p)^\mu$, and we may write it as a linear combination of all the states with momentum $\Lambda p$,

$$U(\Lambda) |\mathbf{p}, \sigma\rangle = \sum_{\sigma'} C_{\sigma\sigma'}(\Lambda; \mathbf{p}) |\Lambda \mathbf{p}, \sigma'\rangle. \tag{1.19}$$

## 1.2.4 Little Group

Consider a momentum $p^\mu$ such that $p^0 > 0$ and $p^2 = m^2$ (it is said to be *on-shell*). Any such vector can be obtained by applying an orthochronous Lorentz transformation to some reference momentum $q^\mu$ located on the same mass-shell (i.e., $q^0 > 0$, $q^2 = m^2$),

$$p^\mu \equiv L^\mu{}_\nu(\mathbf{p})\, q^\nu. \tag{1.20}$$

The choice of the reference 4-vector is not important, but depends on whether the particle under consideration is massive or not. Convenient choices are the following:

- $m > 0$: $q^\mu \equiv (m, 0, 0, 0)$, the 4-momentum of a massive particle at rest;
- $m = 0$: $q^\mu \equiv (\omega, 0, 0, \omega)$, the 4-momentum of a massless particle moving in the third direction of space.

Then, we may define generic one-particle states from those corresponding to the reference momentum as follows:

$$|\mathbf{p}, \sigma\rangle \equiv \mathcal{N}_\mathbf{p}\, U(L(\mathbf{p}))|\mathbf{q}, \sigma\rangle, \tag{1.21}$$

where $L(\mathbf{p})$ is the Lorentz transformation that transforms $q^\mu$ into $p^\mu$ and $\mathcal{N}_\mathbf{p}$ is a numerical prefactor that may be necessary to properly normalize the states. This definition leads to

$$U(\Lambda)|\mathbf{p}, \sigma\rangle = \mathcal{N}_\mathbf{p}\, U(L(\Lambda \mathbf{p}))\, U\big(\underbrace{L^{-1}(\Lambda \mathbf{p})\Lambda L(\mathbf{p})}_{\Sigma}\big)|\mathbf{q}, \sigma\rangle. \tag{1.22}$$

Note that the Lorentz transformation $\Sigma \equiv L^{-1}(\Lambda \mathbf{p})\Lambda L(\mathbf{p})$ maps $q^\mu$ to itself,

$$q^\mu \xrightarrow[L(\mathbf{p})]{} p^\mu \xrightarrow[\Lambda]{} (\Lambda p)^\mu \xrightarrow[L^{-1}(\Lambda \mathbf{p})]{} q^\mu, \tag{1.23}$$

and therefore belongs to the subgroup of the Lorentz group that leaves $q^\mu$ invariant, called the *little group* of $q^\mu$. Thus, when $U(\Sigma)$ acts on the reference state, the momentum remains unchanged and only the other quantum numbers may vary:

$$U(\Sigma)|\mathbf{q}, \sigma\rangle = \sum_{\sigma'} C_{\sigma\sigma'}(\Sigma)|\mathbf{q}, \sigma'\rangle. \tag{1.24}$$

Moreover, the coefficients $C_{\sigma\sigma'}(\Sigma)$ in the right-hand side of this formula define a representation of the little group,

$$C_{\sigma\sigma'}(\Sigma_2 \Sigma_1) = \sum_{\sigma''} C_{\sigma\sigma''}(\Sigma_2)\, C_{\sigma''\sigma'}(\Sigma_1). \tag{1.25}$$

**Massive particles:** In the case of a massive particles, the little group is made of the Lorentz transformations that leave the vector $q^\mu = (m, 0, 0, 0)$ invariant, which is the group of all rotations in three-dimensional space, $SO(3)$. The additional quantum number $\sigma$ is therefore a label that enumerates the possible states in a given representation of the rotation group. These representations correspond to the angular momentum, but since we are in the rest frame of the particle, this is also its spin. For a spin $s$, the dimension of the representation is $2s + 1$, and $\sigma$ takes the values $-s, 1-s, \cdots, +s$.

**Massless particles:** In the massless case, we look for Lorentz transformations $\Sigma^\mu{}_\nu$ that leave $q^\nu = (\omega, 0, 0, \omega)$ invariant. For an infinitesimal transformation, $\Sigma^\mu{}_\nu \approx \delta^\mu{}_\nu + \omega^\mu{}_\nu$, this gives the following general form:

$$\omega_{\mu\nu} = \begin{pmatrix} 0 & \alpha_1 & \alpha_2 & 0 \\ -\alpha_1 & 0 & -\theta & \alpha_1 \\ -\alpha_2 & \theta & 0 & \alpha_2 \\ 0 & -\alpha_1 & -\alpha_2 & 0 \end{pmatrix}, \tag{1.26}$$

where $\alpha_{1,2}, \theta$ are three real infinitesimal parameters. Therefore, an infinitesimal transformation $U(\Sigma)$ reads

$$U(\Sigma) \approx 1 - i\theta \underbrace{M^{12}}_{J^3} - i\alpha_1 \underbrace{(M^{10} + M^{31})}_{K^1 + J^2 \equiv B^1} - i\alpha_2 \underbrace{(M^{20} - M^{23})}_{K^2 - J^1 \equiv B^2}. \tag{1.27}$$

Thus, the little group for massless particles is three-dimensional, with generators $J^3$ (the projection of the angular momentum in the direction of the momentum) and the combinations $B^{1,2}$.[5] Using eq. (1.10), we have

$$[J^3, B^1] = i B^2, \quad [J^3, B^2] = -i B^1, \quad [B^1, B^2] = 0. \tag{1.28}$$

The last commutator implies that we may choose states that are simultaneous eigenstates of $B^1$ and $B^2$. However, non-zero eigenvalues for $B^{1,2}$ may be shown to lead to a continuum of states with the same momentum, which is not realized in nature. Therefore, the only eigenvalue that labels the massless states is that of $J^3$, that generates rotations about the direction of momentum,

$$J^3 |q, \sigma\rangle = \sigma |q, \sigma\rangle, \quad U(\Sigma) |q, \sigma\rangle \underset{\alpha_{1,2}=0}{=} e^{-i\sigma\theta} |q, \sigma\rangle. \tag{1.29}$$

The number $\sigma$ is called the *helicity* of the particle. After a rotation of angle $\theta = 2\pi$, the state must return to itself (bosons) or its opposite (fermions), implying that the helicity must be a half-integer,

$$\text{bosons: } \sigma = 0, \pm 1, \pm 2, \cdots, \quad \text{fermions: } \sigma = \pm \tfrac{1}{2}, \pm \tfrac{3}{2}, \cdots \tag{1.30}$$

## 1.3 Free Scalar Fields, Mode Decomposition

### 1.3.1 Scalars and Scalar Fields

In special relativity, a *scalar* quantity is any quantity invariant under a Lorentz transformation. One may think of a scalar as simply being a plain number. A *scalar field* extends this notion

---

[5]The generators $B^{1,2}$ are the generators of *Galilean boosts* in the $(x^1, x^2)$ plane transverse to the particle momentum, i.e., the transformations that shift the transverse velocity, $v^j \to v^j + \delta v^j$. The physical reason for their appearance in the discussion of massless particles is time dilation: in the observer's frame, the transverse dynamics of a particle moving at the speed of light is infinitely slowed down by time dilation, and is therefore non-relativistic (this intuitive idea can be further substantiated by light-cone quantization – see Exercise 1.10).

to functions of space–time: $\phi(x)$ is a scalar field operator if it is invariant under a Lorentz transformation, except for the change of coordinate induced by the transformation:

$$U^{-1}(\Lambda)\phi(x)U(\Lambda) = \phi(\Lambda^{-1}x). \tag{1.31}$$

This formula just reflects the fact that the point x where the transformed field is evaluated was located at the point $\Lambda^{-1}x$ before the transformation. The simplicity of this transformation law is the reason why we start our study of quantum field theory with scalar fields. As we shall see, scalar fields describe particles that have no other quantum number besides their momentum, i.e., spin-0 particles (also called scalar particles). The first derivative $\partial^\mu\phi$ of the field transforms as a 4-vector,

$$U^{-1}(\Lambda)\partial^\mu\phi(x)U(\Lambda) = \Lambda^\mu{}_\nu \partial^{\overline{\nu}}\phi(\Lambda^{-1}x), \tag{1.32}$$

where the bar in $\partial^{\overline{\nu}}$ indicates that we are differentiating with respect to the whole argument of $\phi$, i.e., $\Lambda^{-1}x$. Likewise, the second derivative $\partial^\mu\partial^\nu\phi$ transforms like a rank-2 tensor, but the d'Alembertian $\Box\phi$ transforms as a scalar.

### 1.3.2 Quantum Harmonic Oscillators

In order to introduce scalar fields, let us make a detour by a familiar problem in quantum mechanics, that of the harmonic oscillator. But instead of a single oscillator, let us consider a continuous collection of *independent* harmonic oscillators, each of them corresponding to particles with a given momentum **p**. Each of these harmonic oscillators can be described by a pair of creation and annihilation operators $a_\mathbf{p}^\dagger, a_\mathbf{p}$, where **p** is a 3-momentum that labels the corresponding mode. Note that the energy of the particles is fixed from their 3-momentum by the relativistic dispersion relation,

$$p^0 = E_\mathbf{p} \equiv \sqrt{\mathbf{p}^2 + m^2}. \tag{1.33}$$

**Harmonic oscillators and free particles:** A simple but essential remark is that *independent harmonic oscillators describe a collection of non-interacting particles*. In order to see this, recall that the operators creating or destroying particles with a given momentum **p** obey the usual commutation relations,

$$[a_\mathbf{p}, a_\mathbf{p}] = [a_\mathbf{p}^\dagger, a_\mathbf{p}^\dagger] = 0, \quad [a_\mathbf{p}, a_\mathbf{p}^\dagger] \neq 0. \tag{1.34}$$

In contrast, by our assumption that oscillators with different momenta are independent, operators acting on different momenta always commute,

$$[a_\mathbf{p}, a_\mathbf{q}] = [a_\mathbf{p}^\dagger, a_\mathbf{q}^\dagger] = [a_\mathbf{p}, a_\mathbf{q}^\dagger] = 0. \tag{1.35}$$

As we shall see shortly, the normalization of the only non-zero commutator can be chosen as follows:

$$[a_\mathbf{p}, a_\mathbf{q}^\dagger] = (2\pi)^3\, 2E_\mathbf{p}\, \delta(\mathbf{p}-\mathbf{q}). \tag{1.36}$$

The independence between the momenta also implies that the Hamiltonian of this system is a sum of the Hamiltonians of independent harmonic oscillators, which we choose to normalize as follows:[6]

$$\mathcal{H} = \int \frac{d^3\mathbf{p}}{(2\pi)^3 2E_\mathbf{p}} \, E_\mathbf{p} \, (a_\mathbf{p}^\dagger a_\mathbf{p} + V E_\mathbf{p}), \tag{1.37}$$

where $V$ is the volume of the system. The term $V E_\mathbf{p}$ simply shifts all the energy levels of the system by a constant, but energy differences are unaffected by this term. We have included it nevertheless in order to facilitate the contact with the usual form[7] of the Hamiltonian of harmonic oscillators in quantum mechanics. Note that once we have chosen the normalization of the creation and annihilation operators via eq. (1.36), the Hamiltonian is completely constrained. Indeed, we can now check that

$$[\mathcal{H}, a_\mathbf{p}^\dagger] = +E_\mathbf{p} a_\mathbf{p}^\dagger, \quad [\mathcal{H}, a_\mathbf{p}] = -E_\mathbf{p} a_\mathbf{p}. \tag{1.38}$$

The meaning of this equation is as follows: When $a_\mathbf{p}^\dagger$ acts on an energy eigenstate of energy E, the result is another energy eigenstate, of energy $E + E_\mathbf{p}$. This property is equivalent to the statement that particles are non-interacting, since it tells us that adding a particle of momentum $\mathbf{p}$ does not affect the rest of the system. In other words, a system with N particles has no binding energy.

**Occupation number:** In order to gain more intuition about the Hamiltonian (1.37), it is useful to introduce the *occupation number* $f_\mathbf{p}$ defined by

$$2E_\mathbf{p} V f_\mathbf{p} \equiv a_\mathbf{p}^\dagger a_\mathbf{p}. \tag{1.39}$$

In terms of $f_\mathbf{p}$, the above Hamiltonian reads

$$\mathcal{H} = V \int \frac{d^3\mathbf{p}}{(2\pi)^3} \, E_\mathbf{p} \, (f_\mathbf{p} + \tfrac{1}{2}). \tag{1.40}$$

The expectation value of $f_\mathbf{p}$ has the interpretation of the number of particles per unit of phase-space (i.e., per unit of volume in coordinate space and per unit of volume in momentum space), and the $1/2$ in $f_\mathbf{p} + \tfrac{1}{2}$ is the ground state occupation of each oscillator. This is

---

[6] The measure $d^3\mathbf{p}/(2\pi)^3 2E_\mathbf{p}$ is Lorentz invariant. Moreover, it emerges naturally from the four-dimensional momentum integration $d^4p/(2\pi)^4$ constrained by the positive energy mass-shell condition $2\pi \theta(p^0) \delta(p^2 - m^2)$.

[7] In relativistic quantum field theory, it is customary to use a system of units, called *natural units*, in which $\hbar = 1$, $c = 1$, $\epsilon_0 = 1$ (and also $k_{\scriptscriptstyle B} = 1$ when the Boltzmann constant is needed to relate energies and temperature). In this system of units, the action $\mathcal{S}$ is dimensionless. Mass, energy, momentum and temperature have the same dimension, which is the inverse of the dimension of length and duration:

$$[\text{mass}] = [\text{energy}] = [\text{momentum}] = [\text{temperature}] = [\text{length}^{-1}] = [\text{duration}^{-1}].$$

Moreover, in four dimensions, the creation and annihilation operators introduced in eq. (1.37) have the dimension of an inverse energy:

$$[a_\mathbf{p}] = [a_\mathbf{p}^\dagger] = [\text{energy}^{-1}]$$

(the occupation number $f_\mathbf{p}$ defined in eq. (1.39) is dimensionless).

reminiscent of the fact that the energy of the level $n$ of a quantized harmonic oscillator of base energy $\omega$ is $E_n = (n + \frac{1}{2})\omega$.

**Ground state:** The ground state of the Hamiltonian (1.37) is the empty state, in which the expectation values of number operators are zero, $\langle 0|a_\mathbf{p}^\dagger a_\mathbf{p}|0\rangle = 0$. This state, also called the *vacuum*, is usually denoted $|0\rangle$. The vacuum of a theory is invariant under Lorentz transformation, $U(\Lambda)|0\rangle = |0\rangle$. The physical meaning of this property is that the vacuum state appears empty to all observers that are moving at constant velocities[8] relative to each other. Let us now consider one-particle states, obtained by acting on the vacuum with a single creation operator,

$$|\mathbf{p}\rangle \equiv a_\mathbf{p}^\dagger |0\rangle. \tag{1.41}$$

The standard notation for states populated with a few particles is simply to list the momenta (and their other quantum numbers, if applicable) contained in the state. Since scalar states have no other quantum numbers, we have

$$a_{\Lambda\mathbf{p}}^\dagger |0\rangle = |\Lambda\mathbf{p}\rangle = U(\Lambda)|\mathbf{p}\rangle = \underbrace{U(\Lambda)\, a_\mathbf{p}^\dagger\, U^{-1}(\Lambda)}_{a_{\Lambda\mathbf{p}}^\dagger}\, \underbrace{U(\Lambda)|0\rangle}_{|0\rangle}, \tag{1.42}$$

from which we read off the action of $U(\Lambda)$ on the creation operators. For instance, if two observers in frames $\mathcal{F}$ and $\mathcal{F}'$ measure the occupation number in a state $\alpha$, their respective measurements will be

$$\begin{aligned} f_\mathbf{p}(\alpha_\mathcal{F}) &= \frac{\langle \alpha_\mathcal{F}|a_\mathbf{p}^\dagger a_\mathbf{p}|\alpha_\mathcal{F}\rangle}{2E_\mathbf{p}V}, \\ f_\mathbf{p}(\alpha_{\mathcal{F}'}) &= \frac{\langle \alpha_{\mathcal{F}'}|a_\mathbf{p}^\dagger a_\mathbf{p}|\alpha_{\mathcal{F}'}\rangle}{2E_\mathbf{p}V} = \frac{\langle \alpha_\mathcal{F}|U^{-1}(\Lambda)\, a_\mathbf{p}^\dagger a_\mathbf{p}\, U(\Lambda)|\alpha_\mathcal{F}\rangle}{2E_\mathbf{p}V} = f_{\Lambda^{-1}\mathbf{p}}(\alpha_\mathcal{F}). \end{aligned} \tag{1.43}$$

(The factor $E_\mathbf{p}V$ is Lorentz invariant.) As expected, the observer in $\mathcal{F}'$ measures at $\mathbf{p}$ the same occupation number as the observer in $\mathcal{F}$ at the momentum $\Lambda^{-1}\mathbf{p}$.

### 1.3.3 Scalar Field Operator

At this point, we have a collection of non-interacting quantum oscillators, one for each possible momentum $\mathbf{p}$, that describe a system of non-interacting particles. Before we turn to something more useful with interactions, let us first rephrase the description of this non-interacting system in a way that is more explicitly Lorentz covariant.

---

[8] For this to hold, it is important that there is no relative acceleration. A state that appears to be empty in one frame may appear populated in an accelerating frame, a phenomenon known as the *Unruh effect*.

## 1.3 FREE SCALAR FIELDS, MODE DECOMPOSITION

**Field operator:** Recall that in quantum mechanics, a particle with a well-defined momentum **p** is not localized at a specific point in space, due to the *uncertainty principle*. Thus, when we say that $a_{\mathbf{p}}^\dagger$ creates a particle of momentum **p**, this production process may happen anywhere in space and at any time since the energy is also well defined. Instead of using the momentum basis, one may introduce an operator that depends on space-time in order to give preeminence to the time and position at which a particle is created or destroyed. It is possible to encapsulate all the $a_{\mathbf{p}}, a_{\mathbf{p}}^\dagger$ into the following operator:[9]

$$\phi(x) \equiv \int \frac{d^3\mathbf{p}}{(2\pi)^3 2E_{\mathbf{p}}} \left[ a_{\mathbf{p}}^\dagger e^{+i p \cdot x} + a_{\mathbf{p}} e^{-i p \cdot x} \right], \tag{1.44}$$

where $p \cdot x \equiv p_\mu x^\mu$ with $p^0 = +E_{\mathbf{p}}$. Loosely speaking, one may view this as a Fourier transform, to which we add the Hermitian conjugate in order to obtain a Hermitian operator. In the following, we will also need the time derivative of this operator, denoted $\Pi(x)$,

$$\Pi(x) \equiv \partial_0 \phi(x) = i \int \frac{d^3\mathbf{p}}{(2\pi)^3 2E_{\mathbf{p}}} E_{\mathbf{p}} \left[ a_{\mathbf{p}}^\dagger e^{+i p \cdot x} - a_{\mathbf{p}} e^{-i p \cdot x} \right]. \tag{1.45}$$

**Canonical commutation relations:** Given the commutation relation (1.36), we obtain the following *equal-time*[10] commutation relations (see Exercise 1.3) for $\phi$ and $\Pi$:

$$\left[\phi(x), \phi(y)\right]_{x^0=y^0} = \left[\Pi(x), \Pi(y)\right]_{x^0=y^0} = 0, \quad \left[\phi(x), \Pi(y)\right]_{x^0=y^0} = i\delta(\mathbf{x}-\mathbf{y}). \tag{1.46}$$

These are called the *canonical field commutation relations*. In this approach (known as *canonical quantization*), the quantization of a field theory corresponds to promoting the classical Poisson bracket between a dynamical variable and its conjugate momentum to a commutator,

$$\{Q_i, P_j\} = \delta_{ij} \quad \rightarrow \quad [\hat{Q}_i, \hat{P}_j] = i\hbar \delta_{ij}. \tag{1.47}$$

In addition to these relations that hold for equal times, one may prove that $\phi(x)$ and $\Pi(y)$ commute for space-like intervals $(x-y)^2 < 0$. Physically, this is related to the absence of causal relations between two measurements performed at space-time points with a space-like separation.

---

[9] In four space-time dimensions, this field has the same dimension as energy,

$$[\phi(x)] = [\text{energy}].$$

[10] The time in special relativity is frame-dependent, and it is possible to define canonical quantization with respect to the time of any reference frame. Besides the formalism presented in this chapter, where commutators are defined on slices of constant $x^0$, another useful choice is to take slices of constant $x^+ \equiv (x^0 + x^3)/\sqrt{2}$, which is the time variable in a frame moving at the speed of light in the z direction. This alternative is known as *light-cone quantization*.

**Hamiltonian in terms of $\phi, \Pi$:** It is possible to invert eqs. (1.44) and (1.45) in order to obtain the creation and annihilation operators given the operators $\phi$ and $\Pi$. These inversion formulas read

$$a_p^\dagger = -i\int d^3x\, e^{-ip\cdot x}\left[\Pi(x) + iE_p\phi(x)\right] = -i\int d^3x\, e^{-ip\cdot x}\, \overleftrightarrow{\partial_0}\, \phi(x),$$

$$a_p = +i\int d^3x\, e^{+ip\cdot x}\left[\Pi(x) - iE_p\phi(x)\right] = +i\int d^3x\, e^{+ip\cdot x}\, \overleftrightarrow{\partial_0}\, \phi(x), \qquad (1.48)$$

where the operator $\overleftrightarrow{\partial_0}$ is defined as

$$A\, \overleftrightarrow{\partial_0}\, B \equiv A\left(\partial_0 B\right) - \left(\partial_0 A\right) B. \qquad (1.49)$$

Note that these expressions, although they appear to contain $x^0$, do not actually depend on time. Using these formulas, we can rewrite the Hamiltonian in terms of $\phi$ and $\Pi$,

$$\mathcal{H} = \int d^3x\, \left\{\tfrac{1}{2}\Pi^2(x) + \tfrac{1}{2}(\boldsymbol{\nabla}\phi(x))^2 + \tfrac{1}{2}m^2\phi^2(x)\right\}. \qquad (1.50)$$

From this Hamiltonian, one may obtain equations of motion in the form of Hamilton's equations. Formally, they read[11]

$$\partial_0 \phi(x) = \frac{\delta \mathcal{H}}{\delta \Pi(x)} = \Pi(x),$$

$$\partial_0 \Pi(x) = -\frac{\delta \mathcal{H}}{\delta \phi(x)} = \left(\boldsymbol{\nabla}^2 - m^2\right)\phi(x). \qquad (1.51)$$

**Lagrangian formulation:** One may also obtain a Lagrangian $\mathcal{L}(\phi, \partial_0\phi)$ that leads to the Hamiltonian (1.50) by the usual manipulations. First, the momentum canonically conjugated to $\phi(x)$ should be given by

$$\Pi(x) = \frac{\delta \mathcal{L}}{\delta \partial_0 \phi(x)}. \qquad (1.52)$$

For this to be consistent with the first Hamilton equation, the Lagrangian must contain the following kinetic term:

$$\mathcal{L} = \int d^3x\, \tfrac{1}{2}(\partial_0\phi(x))^2 + \cdots \qquad (1.53)$$

The missing potential term of the Lagrangian is obtained by requesting that we have

$$\mathcal{H} = \int d^3x\, \Pi(x)\partial_0\phi(x) - \mathcal{L}. \qquad (1.54)$$

---

[11]The symbol $\delta$ denotes a *functional derivative*, i.e., loosely speaking, a "derivative with respect to an object that depends on space-time." For now, it is sufficient to know that this operation obeys $\delta\phi(y)/\delta\phi(x) = \delta(x-y)$. This equation just means that the values taken by $\phi(x)$ at different points should be considered independent. All the other properties of ordinary derivatives extend straightforwardly to functional derivatives. See Section 5.2.2 for more details.

## 1.3 FREE SCALAR FIELDS, MODE DECOMPOSITION

This gives the following Lagrangian:[12]

$$L = \int d^3x \; \{\tfrac{1}{2}(\partial_\mu \phi(x))(\partial^\mu \phi(x)) - \tfrac{1}{2}m^2\phi^2(x)\}. \tag{1.55}$$

Note that the action, $S = \int dx^0 \, L$, is a Lorentz scalar (this is not true of the Hamiltonian, which may be considered as the time component of a 4-vector from the point of view of Lorentz transformations). The Lagrangian (1.55) leads to the following Euler–Lagrange equation of motion:

$$\left(\Box_x + m^2\right) \phi(x) = 0, \tag{1.56}$$

which is known as the *(free) Klein–Gordon equation*. This equation is of course equivalent to the pair of Hamilton equations derived earlier.

### 1.3.4 Noether's Theorem

Conservation laws in a physical theory are intimately related to the continuous symmetries of the system. This is well known in the Lagrangian formulation of classical mechanics, and can be extended to quantum field theory. Consider a generic Lagrangian density $\mathcal{L}(\phi, \partial_\mu \phi)$ that depends on some fields and their derivatives with respect to the space-time coordinates, and assume that the theory is invariant under the following variation of the field:

$$\phi(x) \quad \to \quad \phi(x) + \varepsilon \, \Psi(x). \tag{1.57}$$

Such an invariance is said to be continuous when it is valid for any value of the infinitesimal parameter $\varepsilon$. If the Lagrangian density is unchanged by this transformation, we can write

$$\begin{aligned}
0 = \delta \mathcal{L} &= \frac{\partial \mathcal{L}}{\partial \phi}\, \varepsilon \Psi + \frac{\partial \mathcal{L}}{\partial(\partial_\mu \phi)}\, \varepsilon \partial_\mu \Psi \\
&= \partial_\mu \left(\frac{\partial \mathcal{L}}{\partial(\partial_\mu \phi)}\right) \varepsilon \Psi + \frac{\partial \mathcal{L}}{\partial(\partial_\mu \phi)}\, \varepsilon \partial_\mu \Psi \\
&= \varepsilon \, \partial_\mu \big( \underbrace{\tfrac{\partial \mathcal{L}}{\partial(\partial_\mu \phi)} \Psi}_{J^\mu} \big).
\end{aligned} \tag{1.58}$$

In the second line, we have used the Euler–Lagrange equation obeyed by the field. The 4-vector $J^\mu$ is known as the *Noether current* associated to this symmetry. The fact that the variation of the Lagrangian density is zero implies the following continuity equation for this current:

$$\partial_\mu J^\mu = 0. \tag{1.59}$$

This is the simplest case of *Noether's theorem*, where the Lagrangian density itself is invariant. But for the theory to be unmodified by the transformation of eq. (1.57), it is only necessary

---

[12] One often uses the *Lagrangian density*, defined as the integrand in this expression. The symbol $\mathcal{L}$ is also used for the density (the context and the dimension of the expression makes it clear which one it is).

that the action be invariant, which is also realized if the Lagrangian density is modified by a total derivative, i.e., $\delta\mathcal{L} = \varepsilon\, \partial_\mu K^\mu$. (The proportionality to $\varepsilon$ follows from the fact that the variation must vanish when $\varepsilon \to 0$.) When the variation is a total derivative instead of zero, the continuity equation is modified into:

$$\partial_\mu \left( J^\mu - K^\mu \right) = 0, \tag{1.60}$$

where $J^\mu$ is the same current as before. As we shall see later, there are situations where a conservation equation such as (1.59) is violated by quantum effects, due to a delicate interplay between the symmetry responsible for the conservation law and the short-distance behavior of the theory.

## 1.4 Interacting Scalar Particles

### 1.4.1 Interaction Term

Until now, we have only considered non-interacting particles, which is of course of very limited use in practice. That the Hamiltonian (1.37) does not allow interactions follows from the fact that the only non-trivial term it contains is of the form $a_\mathbf{p}^\dagger a_\mathbf{p}$, which destroys a particle of momentum $\mathbf{p}$ and then creates a particle of momentum $\mathbf{p}$ (hence nothing changes in the state of the system under consideration). By momentum conservation, this is the only allowed Hermitian operator which is quadratic in the creation and annihilation operators. Therefore, in order to include interactions, we must include in the Hamiltonian terms of higher degree in the creation and annihilation operators. The additional term must be Hermitian, since $\mathcal{H}$ generates the time evolution, which must be unitary.

The simplest Hermitian addition to the Hamiltonian is a term of the form

$$\mathcal{H}_I = \frac{\lambda}{n!} \int d^3\mathbf{x}\, \phi^n(x), \tag{1.61}$$

where $n$ is a power larger than 2. The real constant $\lambda$ is called a *coupling constant* and controls the strength of the interactions, while the denominator $n!$ is a symmetry factor that will prove convenient later on. At this point, it seems that any degree $n$ may provide a reasonable interaction term. However, theories with an odd $n$ have an unstable vacuum, and theories with $n > 4$ have problems at short distance in four space-time dimensions (they are said to be non-renormalizable), as we shall see later. For these reasons, $n = 4$ is the only case which is widely studied in practice, and we will stick to this value in the rest of this chapter.

With this choice, the Hamiltonian and Lagrangian read

$$\begin{aligned}\mathcal{H} &= \int d^3\mathbf{x}\, \left\{ \tfrac{1}{2}\Pi^2(x) + \tfrac{1}{2}(\boldsymbol{\nabla}\phi(x))^2 + \tfrac{1}{2}m^2\phi^2(x) + \tfrac{\lambda}{4!}\phi^4(x) \right\}, \\ \mathcal{L} &= \int d^3\mathbf{x}\, \left\{ \tfrac{1}{2}(\partial_\mu\phi(x))(\partial^\mu\phi(x)) - \tfrac{1}{2}m^2\phi^2(x) - \tfrac{\lambda}{4!}\phi^4(x) \right\},\end{aligned} \tag{1.62}$$

and the Klein–Gordon equation is modified into

$$\left(\Box_x + m^2\right)\phi(x) + \frac{\lambda}{6}\phi^3(x) = 0, \tag{1.63}$$

called the *nonlinear Klein–Gordon equation*.

## 1.4.2 Interaction Representation

A field operator that obeys this nonlinear equation of motion can no longer be represented as a linear superposition of plane waves such as eq. (1.44). Let us assume that the coupling constant is very slowly time-dependent, in such a way that

$$\lim_{x^0 \to \pm\infty} \lambda = 0. \tag{1.64}$$

What we have in mind here is that $\lambda$ goes to zero *adiabatically* at asymptotic times, i.e., much slower than all the physically relevant timescales of the theory under consideration. Therefore, at $x^0 = \pm\infty$, the theory is a free theory whose spectrum is made of the eigenstates of the free Hamiltonian. Likewise, the field $\phi(x)$ should be in a certain sense "close to a free field" in these limits. Consider the limit $x^0 \to -\infty$, and denote[13]

$$\lim_{x^0 \to -\infty} \phi(x) = \phi_{in}(x), \tag{1.65}$$

where $\phi_{in}$ is a free field operator that admits a Fourier decomposition similar to eq. (1.44),

$$\phi_{in}(x) \equiv \int \frac{d^3\mathbf{p}}{(2\pi)^3 2E_\mathbf{p}} \left[ a^\dagger_{\mathbf{p},in} e^{+ip\cdot x} + a_{\mathbf{p},in} e^{-ip\cdot x} \right]. \tag{1.66}$$

**Time evolution operator:** Eq. (1.65) can be made more explicit by writing

$$\phi(x) = U(-\infty, x^0) \phi_{in}(x) U(x^0, -\infty), \tag{1.67}$$

where $U$ is a unitary *time evolution operator* defined by

$$\partial_t U(t, t') = \left[ i \int d^3\mathbf{x}\, \mathcal{L}_I(\phi_{in}(x)) \right] U(t, t'), \quad U(t, t) = 1, \tag{1.68}$$

where $\mathcal{L}_I(\phi(x)) \equiv -\frac{\lambda}{4!} \phi^4(x)$ is the interaction term in the Lagrangian density. This evolution operator is a bit special because it evolves only under the effect of the interactions. This is the essence of the *interaction representation*, which consists in splitting the time evolution into two parts: the trivial evolution of free fields (encoded in the time-dependence of $\phi_{in}(x)$), and a modification due to the interactions (encoded in $U(t, t')$). (Such a separation is common in ordinary quantum mechanics, but there the interaction and Heisenberg pictures are usually made to coincide at $t = 0$ instead of $t = -\infty$.)

Note that although eq. (1.68) is a linear first-order differential equation, its solution is not straightforward because in the right-hand side the spatial integral of the interaction term does not commute with itself at different times. In particular, the naive solution that would consist in exponentiating the square bracket is not correct. Instead, the correct answer is a *time-ordered exponential* of the interaction term,

$$U(t_2, t_1) \equiv T \exp\left( i \int_{t_1}^{t_2} dx^0 d^3\mathbf{x}\, \mathcal{L}_I(\phi_{in}(x)) \right), \tag{1.69}$$

---

[13] In this equation, we ignore for now the issue of field renormalization, which we shall come back to later (see Section 1.6).

obtained by writing the Taylor expansion of the exponential, and by reordering all the products in such a way that the operators are ordered from left to right in order of decreasing time arguments:

$$\mathrm{T}\,\exp\int_0^t d\tau\, A(\tau) \equiv 1 + \int_0^\tau d\tau_1\, A(\tau_1) + \int_0^\tau d\tau_1 \int_0^{\tau_1} d\tau_2\, A(\tau_1)A(\tau_2)$$
$$+ \int_0^\tau d\tau_1 \int_0^{\tau_1} d\tau_2 \int_0^{\tau_2} d\tau_3\, A(\tau_1)A(\tau_2)A(\tau_3) + \cdots \quad (1.70)$$

Note that the time-ordered exponential is equal to an ordinary exponential if $A(\tau)$ commutes with $A(\tau')$ for any $\tau,\tau'$ (the Taylor coefficient $1/n!$ comes from the fact that in the above equation we integrate only over one out of $n!$ possible orderings of the $\tau_i$s). In order to check that this series solves the differential equation that defines the time evolution operator, we simply differentiate it term by term with respect to $\tau$, which gives

$$\partial_t \left[\mathrm{T}\,\exp\int_0^t d\tau\, A(\tau)\right] = A(t) \left[\mathrm{T}\,\exp\int_0^t d\tau\, A(\tau)\right]. \quad (1.71)$$

The time evolution operator (1.69) satisfies the following properties:

$$U(t_3, t_1) = U(t_3, t_2)\,U(t_2, t_1) \quad \text{(for all } t_2\text{)}$$
$$U(t_1, t_2) = U^{-1}(t_2, t_1) = U^\dagger(t_2, t_1). \quad (1.72)$$

**Equation of motion in the interaction representation:** One can then prove the following identity (see Exercise 1.5) for the left-hand side of the nonlinear Klein–Gordon equation obeyed by the field $\phi(x)$:

$$(\Box_x + m^2)\phi(x) + \frac{\lambda}{6}\phi^3(x) = U(-\infty, x^0)\left[(\Box_x + m^2)\phi_{\mathrm{in}}(x)\right] U(x^0, -\infty). \quad (1.73)$$

This equation shows that $\phi_{\mathrm{in}}$ obeys the free Klein–Gordon equation if $\phi$ obeys the nonlinear interacting one, and justifies a posteriori our choice of the unitary operator $U$ that connects $\phi$ and $\phi_{\mathrm{in}}$. The main benefit of the interaction representation is organizational, since it splits the time evolution into a trivial part already present in the non-interacting theory, and a modification due to the interactions. This separation is an essential ingredient of perturbation theory.

### 1.4.3 In and Out States

The *in* creation and annihilation operators can be used to define a space of eigenstates of the free Hamiltonian, starting from a ground state (vacuum) denoted $|0_{\mathrm{in}}\rangle$. For instance, one-particle states would be defined as

$$|\mathbf{p}_{\mathrm{in}}\rangle = a^\dagger_{\mathbf{p},\mathrm{in}} |0_{\mathrm{in}}\rangle. \quad (1.74)$$

The physical interpretation of these states, called *Fock states*, is that they are states with a definite particle content at $x^0 = -\infty$, before the interactions are turned on.[14]

In the same way as we have constructed *in* field operators, creation and annihilation operators and states, we may construct *out* ones such that the field $\phi_{\text{out}}(x)$ is a free field that coincides with the interacting field $\phi(x)$ in the limit $x^0 \to +\infty$ (with the same caveat about field renormalization). Starting from a vacuum state $|0_{\text{out}}\rangle$, we may also define a full set of states, such as $|\mathbf{p}_{\text{out}}\rangle$, that have a definite particle content at $x^0 = +\infty$. It is crucial to observe that the in and out states are in general not identical:

$$|0_{\text{out}}\rangle \neq |0_{\text{in}}\rangle \quad \text{(they differ by the phase } \langle 0_{\text{out}}|0_{\text{in}}\rangle\text{)}, \quad |\mathbf{p}_{\text{out}}\rangle \neq |\mathbf{p}_{\text{in}}\rangle, \cdots \quad (1.75)$$

Taking the limit $x^0 \to +\infty$ in eq. (1.67), we first see that[15]

$$\begin{aligned} a_{\mathbf{p},\text{out}} &= U(-\infty,+\infty)\, a_{\mathbf{p},\text{in}}\, U(+\infty,-\infty), \\ a^\dagger_{\mathbf{p},\text{out}} &= U(-\infty,+\infty)\, a^\dagger_{\mathbf{p},\text{in}}\, U(+\infty,-\infty), \end{aligned} \quad (1.76)$$

from which we deduce that the in and out states must be related by

$$|\alpha_{\text{out}}\rangle = U(-\infty,+\infty)\,|\alpha_{\text{in}}\rangle. \quad (1.77)$$

The two sets of states are identical for a free theory, since the time evolution operator reduces to the identity in this case.

## 1.5 From Field Correlations to Reaction Rates

### 1.5.1 Invariant Cross-Sections

All experiments in particle physics amount to a measurement that answers the following question: Given a certain setup that defines an initial state, how many reactions of a certain type occur per unit of time? The concept of "reaction of a certain type" may vary widely depending on the number of criteria that are imposed on the final state for the reaction to be worth counting. For instance, one may consider the reaction $e^+e^- \to$ *anything*, the reaction $e^+e^- \to \mu^+\mu^-$, or even a reaction where, in addition, the final muons are required to have momenta in a certain range. In this section, we establish the link between these physical observables and correlation functions of the field operators that can in principle be calculated in quantum field theory, with transition amplitudes as an intermediate step.

**Definition of a cross-section:** In a scattering experiment such as those performed in a particle collider, the observed reaction rate results from a combination of some factors that depend on the accelerator design (the fluxes of particles in the colliding beams), and a factor that contains the genuine microscopic information about the reaction. In general, this microscopic input is

---

[14] For an interacting system, it is not possible to enumerate the particle content of states because of quantum fluctuations that may temporarily create additional virtual particles.

[15] The evolution operator from $x^0 = -\infty$ to $x^0 = +\infty$ is sometimes called the S-matrix, $S \equiv U(+\infty,-\infty)$.

expressed in terms of a quantity called a *cross-section*,[16] that has the dimension of an area. Consider two colliding beams containing particles of type 1 and 2, respectively. For simplicity, assume that the two beams have a uniform particle density, and let us denote $\mathcal{S}$ their transverse overlap area. If, during the experiment, $N_1$ particles of the first beam and $N_2$ particles of the second beam fly by the interaction zone, the cross-section for the process $1 + 2 \to F$, where F is some final state, is the quantity $\sigma_{12 \to F}$ defined by

$$\begin{pmatrix} \text{number of times F is} \\ \text{seen in the experiment} \end{pmatrix} = \frac{N_1 N_2}{\mathcal{S}} \, \sigma_{12 \to F}. \tag{1.78}$$

In this formula, the left-hand side is measured experimentally, while in the right-hand side the ratio $N_1 N_2 / \mathcal{S}$ depends only on the setup of the collider.[17] Therefore, the cross-section can be obtained as the ratio of two known quantities. Note that the cross-section in general depends on the momenta $\mathbf{p}_{1,2}$ of the particles participating in the collision (and on the momenta of the particles in the final state F), but in a Lorentz covariant way, i.e., only through Lorentz scalars such as $(p_1 + p_2)^2$.

**Normalization of one-particle states:** An important point in the calculation of the cross-section is the normalization of the one-particle states. We have

$$\langle \mathbf{p}_{\text{in}} | \mathbf{p}_{\text{in}} \rangle = \int d^3 \mathbf{x} \, |\langle \mathbf{x} | \mathbf{p}_{\text{in}} \rangle|^2 = 2 E_\mathbf{p} \, \underbrace{(2\pi)^3 \delta(0)}_{V}. \tag{1.79}$$

In the first equality, we have inserted a complete set of position eigenstates in order to highlight the interpretation of $\langle \mathbf{p}_{\text{in}} | \mathbf{p}_{\text{in}} \rangle$ as the integral of the square of some kind of wavefunction. The second equality follows from the canonical commutation relation between creation and annihilation operators. The proportionality of the result to the volume means that one should not view the state $|\mathbf{p}_{\text{in}}\rangle$ as containing a single particle of momentum $\mathbf{p}$. Instead, this state contains a uniform density of particles, all with momentum $\mathbf{p}$ (see Exercise 1.6). Eq. (1.79) means that our convention of normalization of the states corresponds to "$2 E_\mathbf{p}$ particles per unit volume." We are using quotes here because $2 E_\mathbf{p}$ does not have the correct dimension to be a proper density of particles. This is mostly an aesthetic problem: This convention of normalization will cancel out eventually, since cross-sections are defined in such a way that they do not depend on the incoming fluxes of particles.

**Example of a non-interacting theory:** These normalization issues can be clarified by considering the trivial example of a non-interacting theory. In this case, the exact result for the transition amplitude between two one-particle states is

$$\langle \mathbf{q}_{\text{out}} | \mathbf{p}_{\text{in}} \rangle = 2 E_\mathbf{p} \, (2\pi)^3 \delta(\mathbf{q} - \mathbf{p}). \tag{1.80}$$

---

[16] The terminology comes from the classical problem of a gas of hard spheres. Two such spheres collide if their impact parameter (i.e., their minimal approach distance) is smaller than their diameter. This happens if the trajectory of one of the spheres passes in a disk of area $\sigma = \pi(\text{diameter})^2$ centered at the other sphere ($\sigma$ is called the cross-section). This concept can be generalized to objects that have more complicated interactions, but then it loses its clear-cut geometrical meaning.

[17] In practice, the beam conditions are monitored by measuring in parallel the event rate of another reaction, whose cross-section is already accurately known.

## 1.5 FROM FIELD CORRELATIONS TO REACTION RATES

By squaring this amplitude, we obtain

$$|\langle \mathbf{q}_{\text{out}}|\mathbf{p}_{\text{in}}\rangle|^2 = 4E_\mathbf{p} E_\mathbf{q} V (2\pi)^3 \delta(\mathbf{q}-\mathbf{p}), \tag{1.81}$$

and integrating over $\mathbf{q}$ with the Lorentz invariant measure $d^3\mathbf{q}/((2\pi)^3 2E_\mathbf{q})$ gives

$$\int \frac{d^3\mathbf{q}}{(2\pi)^3 2E_\mathbf{q}} |\langle \mathbf{q}_{\text{out}}|\mathbf{p}_{\text{in}}\rangle|^2 = \langle \mathbf{p}_{\text{in}}|\mathbf{p}_{\text{in}}\rangle = \text{initial number of particles.} \tag{1.82}$$

Since we are considering a non-interacting theory, we know a priori that every particle in the initial state is present in the final state with the same momentum. Therefore, the integral in the left-hand side of the previous equation is also the number of particles in the final state, and the quantity

$$|\langle \mathbf{q}_{\text{out}}|\mathbf{p}_{\text{in}}\rangle|^2 \frac{d^3\mathbf{q}}{(2\pi)^3 2E_\mathbf{q}} \tag{1.83}$$

counts those that have their momentum in a volume $d^3\mathbf{q}$ centered around $\mathbf{q}$. More generally, for an $n$-particle final state and a generic initial state,

$$|\langle \mathbf{q}_1 \cdots \mathbf{q}_{n\,\text{out}}|\cdots_{\text{in}}\rangle|^2 \prod_{j=1}^n \frac{d^3\mathbf{q}_j}{(2\pi)^3 2E_{\mathbf{q}_j}} \tag{1.84}$$

is the number of events where the final state particles have their momenta in the volume $d^3\mathbf{q}_1 \cdots d^3\mathbf{q}_n$ centered on $(\mathbf{q}_1, \cdots, \mathbf{q}_n)$.

**General squared amplitude:** Consider now a transition amplitude from a two-particle state to a final state with $n$ particles, $\langle \mathbf{q}_1 \cdots \mathbf{q}_{n\,\text{out}}|\mathbf{p}_1\mathbf{p}_{2\,\text{in}}\rangle$. By momentum conservation, all the contributions to this amplitude are proportional to a delta function,

$$\langle \mathbf{q}_1 \cdots \mathbf{q}_{n\,\text{out}}|\mathbf{p}_1\mathbf{p}_{2\,\text{in}}\rangle \equiv (2\pi)^4 \delta(p_1 + p_2 - \sum_{j=1}^n q_j)\, \mathcal{T}(\mathbf{q}_{1,\cdots,n}|\mathbf{p}_{1,2}), \tag{1.85}$$

and its squared modulus reads

$$|\langle \mathbf{q}_1 \cdots \mathbf{q}_{n\,\text{out}}|\mathbf{p}_1\mathbf{p}_{2\,\text{in}}\rangle|^2 = (2\pi)^4 \delta(p_1 + p_2 - \sum_{j=1}^n q_j)$$
$$\times \underbrace{(2\pi)^4 \delta(0)}_{VT}\, |\mathcal{T}(\mathbf{q}_{1,\cdots,n}|\mathbf{p}_{1,2})|^2. \tag{1.86}$$

This expression contains the square of the delta function. One of these factors becomes a delta of zero, which has the interpretation of space-time volume $VT$ in which the process takes place. Since the initial state contains a fixed number of particles of each type (1 and 2) per unit volume in all space-time, we expect the total number of events to be extensive, because interactions may happen in all the volume at any time. This is the meaning of the factor $VT$ that appears in this square.

From the insight gained by studying the non-interacting theory, this square weighted by the Lorentz invariant phase-space measure of the final state counts the number of events in which the final state particles have momenta in the volume $d^3q_1 \cdots d^3q_n$ centered on $(q_1, \cdots, q_n)$:

Number of events

$$
\begin{aligned}
&= \left|\langle q_1 \cdots q_{n\,\text{out}}|p_1 p_{2\,\text{in}}\rangle\right|^2 \prod_{j=1}^{n} \frac{d^3 q_j}{(2\pi)^3 2 E_{q_j}} \\
&= VT \left|\mathcal{T}(q_1, \cdots, n|p_{1,2})\right|^2 \underbrace{(2\pi)^4 \delta(p_1 + p_2 - \sum_{j=1}^{n} q_j) \prod_{j=1}^{n} \frac{d^3 q_j}{(2\pi)^3 2 E_{q_j}}}_{d\Gamma_n(p_{1,2})}.
\end{aligned}
\tag{1.87}
$$

($d\Gamma_n(p_{1,2})$ is the invariant final state measure subject to the constraint of momentum conservation.)

**Cross-section in the target frame:** At this point, the relationship with the cross-section of this transition is most easily established in the rest frame of one of the initial state particles, e.g., the particle 2 (this frame is called the *target frame*). Consider a thin slice of this target, of transverse section $S$ and infinitesimal thickness $\ell$, as shown in Figure 1.1. The interaction volume is the volume of this target, i.e., $V = S\ell$. Given the normalization of the one-particle states, it contains $N_2 = 2m_2 S\ell$ particles of type 2. In the target frame, the particles 1 have a velocity $v_1 = p_1/E_{p_1}$. Therefore, within a time interval T, $N_1 = 2E_{p_1} S v_1 T = 2p_1 ST$ of them travel through the interaction zone. Using eqs. (1.78) and (1.87), we thus obtain the following expression for the cross-section in the target frame:[18]

$$
\sigma_{12 \to 1 \cdots n}\bigg|_{\substack{\text{target} \\ \text{frame}}} = \frac{1}{4 m_2 p_1} \int d\Gamma_n(p_{1,2}) \left|\mathcal{T}(q_1, \cdots, n|p_{1,2})\right|^2. \tag{1.88}
$$

Note that this is the total cross-section for a final state with $n$ particles, since we have integrated over all the final state momenta. By undoing some of these integrations, we may obtain

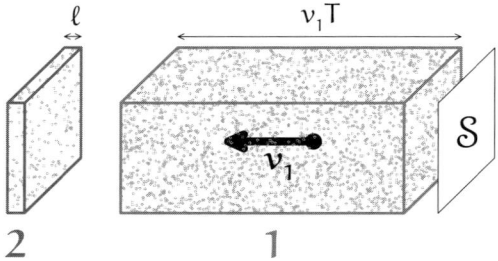

FIGURE 1.1: Geometry of a two-body cross-section in the target frame. The two volumes represent the particles that can take part in the reaction in duration T.

---

[18]Regarding dimensions, $\langle q_1 \cdots q_{n\,\text{out}}|p_1 p_{2\,\text{in}}\rangle \sim (\text{mass})^{-(2+n)}$, $\mathcal{T}(q_1, \cdots, n|p_{1,2}) \sim (\text{mass})^{2-n}$ and $d\Gamma_n \sim (\text{mass})^{2n-4}$, and therefore this formula indeed gives an area.

a cross-section which is differential with respect to some of the kinematical variables that characterize the final state (e.g., the angle at which a final state particle is scattered).

**Cross-section in the center of momentum frame:** Another important frame in view of the setup of many experiments in particle physics is the *center of momentum* frame. This is the observer's frame in experiments where two beams of same-mass particles and equal energies collide head-on. In this frame, we have

$$\mathbf{p}_1 + \mathbf{p}_2 = 0, \quad E_1 = E_2, \quad s \equiv (p_1 + p_2)^2 = 4E_1^2. \tag{1.89}$$

Since s is Lorentz invariant, its expression in the rest frame of the particle 2 is $s = (m_2 + E_1')^2 - \mathbf{p}_1'^2$ (in this paragraph, the primes indicate kinematical variables in the target frame). Moreover, simple kinematics show that the combination $m_2 p_1'$ in the target frame becomes $m_2 p_1' = \sqrt{s}\, p_1$ in the center of momentum frame. Therefore, the expression of the cross-section in this frame reads

$$\sigma_{12 \to 1 \cdots n} \bigg|_{\substack{\text{center of}\\ \text{momentum}}} = \frac{1}{4\sqrt{s}\, p_1} \int d\Gamma_n(p_{1,2}) \left| \mathcal{T}(q_1, \cdots, n | p_{1,2}) \right|^2. \tag{1.90}$$

Likewise, obtaining the expression of a cross-section in a frame where the two beams have different momenta is a simple matter of relativistic kinematics (this is useful when the detector apparatus is neither the rest frame of one of the particles, nor the center of momentum frame, and one counts events in terms of some kinematical variable measured in this frame – alternatively, one may boost all the measured final state momenta in order to convert them to momenta in one of the above two frames).

### 1.5.2 Decay Rates

Another very common type of observable is the decay rate $\Gamma$ of an unstable particle, defined so that $\Gamma T$ is the decay probability of a particle at rest in the infinitesimal time interval T. The decay rate can be obtained from matrix elements with a one-particle initial state,

$$\langle q_1 \cdots q_{n\,\text{out}} | p_{1\,\text{in}} \rangle \equiv (2\pi)^4 \delta\!\left(p_1 - \sum\nolimits_{j=1}^n q_j\right) \mathcal{T}(q_1, \cdots, n | p_1). \tag{1.91}$$

Squaring this matrix element again produces a space-time volume factor VT, and integrating over the invariant phase space of the final state particles gives

$$\underbrace{\int \prod_{j=1}^n \frac{d^3 q_j}{(2\pi)^3 2 E_{q_j}} \left| \langle q_1 \cdots q_{n\,\text{out}} | p_{1\,\text{in}} \rangle \right|^2}_{\text{Total number of decays}} = VT \underbrace{\int d\Gamma_n(p_1) \left| \mathcal{T}(q_1, \cdots, n | p_1) \right|^2}_{\text{Decays per unit of time and volume}}. \tag{1.92}$$

Given the normalization of the one-particles states, a sample of volume V contains $N_1 = 2E_{p_1} V$ particles, and the average number of decays in the time interval T is therefore $2E_{p_1} \Gamma VT$. From this, we get the following expression for the decay rate:

$$\Gamma = \frac{1}{2E_{p_1}} \int d\Gamma_n(p_1) \left| \mathcal{T}(q_1, \cdots, n | p_1) \right|^2. \tag{1.93}$$

A differential decay rate can be obtained by leaving some of the final state kinematical variables unintegrated.

### 1.5.3 Lehmann–Symanzik–Zimmermann (LSZ) Reduction Formulas

In the previous two subsections we have seen that both cross-section and decay rates can be expressed in terms of transition amplitudes such as $\langle \mathbf{q}_1\mathbf{q}_2\cdots_{\text{out}}|\mathbf{p}_1\mathbf{p}_2\cdots_{\text{in}}\rangle$. We now need to relate them to more elementary objects that can be calculated in quantum field theory. A first step in view of calculating transition amplitudes is to relate them to expectation values involving the field operator $\phi(x)$. In order to illustrate the main steps in deriving such a relationship, let us consider the simple case of the transition amplitude between two one-particle states, $\langle \mathbf{q}_{\text{out}}|\mathbf{p}_{\text{in}}\rangle$. First, we write the state $|\mathbf{p}_{\text{in}}\rangle$ as the action of a creation operator on the corresponding vacuum state, and we replace the creation operation by its expression in terms of $\phi_{\text{in}}$,

$$\langle \mathbf{q}_{\text{out}}|\mathbf{p}_{\text{in}}\rangle = \langle \mathbf{q}_{\text{out}}|a^\dagger_{\mathbf{p},\text{in}}|0_{\text{in}}\rangle$$
$$= -i\int d^3x\, e^{-ip\cdot x}\langle \mathbf{q}_{\text{out}}|\Pi_{\text{in}}(x) + iE_{\mathbf{p}}\phi_{\text{in}}(x)|0_{\text{in}}\rangle. \tag{1.94}$$

Next, we use the fact that $\phi_{\text{in}}, \Pi_{\text{in}}$ are the limits when $x^0 \to -\infty$ of the interacting fields $\phi, \Pi$, and we express this limit by means of the following trick:

$$\lim_{x^0\to-\infty} F(x^0) = \lim_{x^0\to+\infty} F(x^0) - \int_{-\infty}^{+\infty} dx^0\, \partial_{x^0} F(x^0). \tag{1.95}$$

The term with the limit $x^0 \to +\infty$ produces a term identical to the right-hand side of the first line of eq. (1.94), but with an $a^\dagger_{\mathbf{p},\text{out}}$ instead of $a^\dagger_{\mathbf{p},\text{in}}$. At this stage we have

$$\langle \mathbf{q}_{\text{out}}|\mathbf{p}_{\text{in}}\rangle = \langle 0_{\text{out}}|a_{\mathbf{q},\text{out}}a^\dagger_{\mathbf{p},\text{out}}|0_{\text{in}}\rangle$$
$$+ i\int d^4x\, \partial_{x^0}\, e^{-ip\cdot x}\langle \mathbf{q}_{\text{out}}|\Pi(x) + iE_{\mathbf{p}}\phi(x)|0_{\text{in}}\rangle. \tag{1.96}$$

In the first line, we use the commutation relation between creation and annihilation operators to obtain

$$\langle 0_{\text{out}}|a_{\mathbf{q},\text{out}}a^\dagger_{\mathbf{p},\text{out}}|0_{\text{in}}\rangle = (2\pi)^3 2E_{\mathbf{p}}\,\delta(\mathbf{p}-\mathbf{q}). \tag{1.97}$$

This term does not involve any interaction, since the initial state particle simply goes through to the final state (in other words, this particle just acts as a spectator). Such trivial terms always appear when expressing transition amplitudes in terms of the field operator, and they are usually dropped since they do not carry any interesting physical information. We can then perform explicitly the time derivative in the second line to obtain[19]

$$\langle \mathbf{q}_{\text{out}}|\mathbf{p}_{\text{in}}\rangle \doteq i\int d^4x\, e^{-ip\cdot x}(\Box_x + m^2)\langle \mathbf{q}_{\text{out}}|\phi(x)|0_{\text{in}}\rangle, \tag{1.98}$$

---

[19]We use here the dispersion relation $p_0^2 - \mathbf{p}^2 = m^2$ of the incoming particle to arrive at this expression. The mass that should enter in this formula is the physical mass of the particles. This remark will become important when we discuss renormalization.

## 1.5 FROM FIELD CORRELATIONS TO REACTION RATES

where we use the symbol $\doteq$ to indicate that the trivial terms involving spectators have been dropped.

**Time-ordered products of fields:** Next, we repeat the same procedure for the final state particle: (1) replace the annihilation operator $a_{q,\text{out}}$ by its expression in terms of $\phi_{\text{out}}$; (2) write $\phi_{\text{out}}$ as a limit of $\phi$ when $x^0 \to +\infty$; (3) write this limit as an integral of a time derivative plus a term at $x^0 \to -\infty$, that we rewrite as the annihilation operator $a_{q,\text{in}}$,

$$\langle q_{\text{out}}|p_{\text{in}}\rangle \doteq i\int d^4x\, e^{-ip\cdot x}\,(\square_x + m^2)\,\Big\{\langle 0_{\text{out}}|a_{q,\text{in}}\phi(x)|0_{\text{in}}\rangle$$
$$+ i\int d^4y\, \partial_{y^0}\, e^{iq\cdot y}\,\langle 0_{\text{out}}|(\Pi(y) - iE_q\phi(y))\,\phi(x)|0_{\text{in}}\rangle\Big\}. \quad (1.99)$$

However, we are stuck at this point because in the first term we would like to bring the $a_{q,\text{in}}$ to the right where it would annihilate $|0_{\text{in}}\rangle$, but we do not know the commutator between $a_{q,\text{in}}$ and the *interacting* field operator $\phi(x)$. The remedy is to go one step back and use a special trick – known as a *time-ordered product* – to order the operators in a more suitable way. Given two time-dependent (they may also depend on space, but this does not play any role in the definition) operators $A(t)$ and $B(t)$, their time-ordered product (or T-product) is defined by

$$T(A(t)B(t')) \equiv \theta(t-t')\,A(t)B(t') + \theta(t'-t)\,B(t')A(t). \quad (1.100)$$

In other words, the time-ordering consists in always placing on the left the operator that carries the largest time. Likewise, the T-product of $n$ operators is defined as a sum over all the permutations of the time variables,

$$T(A_1(t_1)\cdots A_n(t_n)) = \sum_{\sigma\in\mathfrak{S}_n}\theta(t_{\sigma_n}-t_{\sigma_{n-1}})\cdots\theta(t_{\sigma_2}-t_{\sigma_1})\,A_{\sigma_n}(t_{\sigma_n})\cdots A_{\sigma_1}(t_{\sigma_1}).$$
$$(1.101)$$

The time-ordered exponential introduced earlier simply consists in applying the time-ordering operator to every term in the Taylor series of the ordinary exponential. Returning now to the evaluation of $\langle q_{\text{out}}|p_{\text{in}}\rangle$, note that we are free to insert a T-product in

$$(\Pi_{\text{out}}(y) - iE_q\phi_{\text{out}}(y))\,\phi(x) \underset{y^0\to+\infty}{=} T\left((\Pi(y) - iE_q\phi(y))\,\phi(x)\right) \quad (1.102)$$

since the time $y^0 \to +\infty$ is obviously larger than $x^0$. Then the boundary term at $y^0 \to -\infty$ will automatically lead to the desired ordering $\phi(x)\,a_{q,\text{in}}$,

$$\langle q_{\text{out}}|p_{\text{in}}\rangle \doteq i\int d^4x\, e^{-ip\cdot x}\,(\square_x + m^2)\,\Big\{\langle 0_{\text{out}}|\phi(x)\underbrace{a_{q,\text{in}}|0_{\text{in}}\rangle}_{0}$$
$$+ i\int d^4y\, \partial_{y^0}\, e^{iq\cdot y}\,\langle 0_{\text{out}}|T\left(\Pi(y) - iE_q\phi(y)\right)\phi(x)|0_{\text{in}}\rangle\Big\}. \quad (1.103)$$

Performing the derivative with respect to $y^0$, we finally arrive at

$$\langle q_{out}|p_{in}\rangle \doteq i^2 \int d^4x\, d^4y\, e^{i(q\cdot y - p\cdot x)}\, (\Box_x + m^2) \\ \times (\Box_y + m^2)\, \langle 0_{out}|T\, \phi(x)\phi(y)|0_{in}\rangle. \tag{1.104}$$

Such a formula is known as a (Lehmann–Symanzik–Zimmermann) *reduction formula*.

The method that we have exposed above on a simple case can easily be extended to a general transition amplitude, with the following result for the part of the amplitude that does not involve any spectator:

$$\langle q_1 \cdots q_{n\,out}|p_1 \cdots p_{m\,in}\rangle \doteq i^{m+n} \int \prod_{i=1}^{m} d^4x_i\, e^{-ip_i\cdot x_i}\, (\Box_{x_i} + m^2) \\ \times \int \prod_{j=1}^{n} d^4y_j\, e^{iq_j\cdot x_j}\, (\Box_{y_j} + m^2) \\ \times \langle 0_{out}|T\, \phi(x_1)\cdots\phi(x_m)\phi(y_1)\cdots\phi(y_n)|0_{in}\rangle. \tag{1.105}$$

The bottom line is that an amplitude with $m+n$ particles is related to the vacuum expectation value of a time-ordered product of $m + n$ interacting field operators (a slight but important modification to this formula will be introduced in Section 1.6, in order to account for field renormalization). Note that the vacuum states on the left and on the right of the expectation value are respectively the *out* and the *in* vacua.

## 1.6 Källen–Lehmann Spectral Representation

As we shall see now the limit in eq. (1.65) that relates the interacting field $\phi$ and the free field of the interaction picture $\phi_{in}$ is too naive. One of the consequences is that we will have to make a slight modification to the reduction formula (1.105). Consider the time-ordered two-point function,

$$\langle 0_{out}|T\, \phi(x)\phi(y)|0_{in}\rangle = \theta(x^0 - y^0)\, \langle 0_{out}|\phi(x)\phi(y)|0_{in}\rangle \\ + \theta(y^0 - x^0)\, \langle 0_{out}|\phi(y)\phi(x)|0_{in}\rangle. \tag{1.106}$$

For each of the expectation values on the right-hand side, let us insert an identity operator between the two field operators, written in the form of a sum over all the possible physical states,

$$1 = \sum_{\text{states } \lambda} |\lambda\rangle\langle\lambda|. \tag{1.107}$$

The states $\lambda$ can be arranged into classes inside which the states differ only by a boost. A class of states, which we will denote $\alpha$, is characterized by its particle content and by the relative momenta of these particles. Within a class, the total momentum of the state can be varied by applying a Lorentz boost. For a class $\alpha$, we will denote $|\alpha_p\rangle$ the state of total momentum **p**. Each class of states has an *invariant mass* $m_\alpha$, such that the total energy $p^0$ and total

## 1.6 KÄLLEN–LEHMANN SPECTRAL REPRESENTATION

momentum $\mathbf{p}$ of the states in this class obey $p_0^2 - \mathbf{p}^2 = m_\alpha^2$. In addition, it is useful to isolate the vacuum in the sum over the states. Therefore, the identity operator can be rewritten as

$$1 = |0\rangle\langle 0| + \sum_{\text{classes } \alpha} \int \frac{d^3\mathbf{p}}{(2\pi)^3 2\sqrt{\mathbf{p}^2 + m_\alpha^2}} |\alpha_\mathbf{p}\rangle\langle\alpha_\mathbf{p}|, \tag{1.108}$$

where we have written the integral over the total momentum of the states in a Lorentz invariant fashion. (We need not specify if we are using *in* or *out* states here.)

When we insert this identity operator between the two field operators, the vacuum does not contribute. For instance,

$$\langle 0_{\text{out}}|\phi(x)|0\rangle = 0. \tag{1.109}$$

($\phi$ creates or destroys a particle, and therefore has a vanishing matrix element between vacuum states.) Using the momentum operator $\hat{P}$, we can write

$$\begin{aligned}\langle 0_{\text{out}}|\phi(x)|\alpha_\mathbf{p}\rangle &= \langle 0_{\text{out}}|e^{i\hat{P}\cdot x}\phi(0)e^{-i\hat{P}\cdot x}|\alpha_\mathbf{p}\rangle \\ &= \langle 0_{\text{out}}|\phi(0)|\alpha_\mathbf{p}\rangle\, e^{-ip\cdot x} \\ &= \langle 0_{\text{out}}|\phi(0)|\alpha_\mathbf{0}\rangle\, e^{-ip\cdot x}.\end{aligned} \tag{1.110}$$

The second line uses the fact that the total momentum in the vacuum state is zero, and is $\mathbf{p}$ for the state $\alpha_\mathbf{p}$. In the last equality, we have applied a boost that cancels the total momentum $\mathbf{p}$, and used the fact that the vacuum is invariant, as well as the scalar field $\phi(0)$. Therefore, we obtain the following representation for the time-ordered two-point function,

$$\langle 0_{\text{out}}|T\,\phi(x)\phi(y)|0_{\text{in}}\rangle = \sum_{\text{classes } \alpha} \langle 0_{\text{out}}|\phi(0)|\alpha_\mathbf{0}\rangle\langle\alpha_\mathbf{0}|\phi(0)|0_{\text{in}}\rangle$$

$$\times \underbrace{\int \frac{d^3\mathbf{p}}{(2\pi)^3 2\sqrt{\mathbf{p}^2 + m_\alpha^2}} \left\{\theta(x^0 - y^0)e^{-ip\cdot(x-y)} + \theta(y^0 - x^0)e^{ip\cdot(x-y)}\right\}}_{G_F^0(x,y;m_\alpha^2)}, \tag{1.111}$$

where the underlined integral, $G_F^0(x, y; m_\alpha^2)$, is the value of $\langle 0_{\text{in}}|T\phi_{\text{in}}(x)\phi_{\text{in}}(y)|0_{\text{in}}\rangle$ (this function is called the *Feynman propagator* – see Section 1.7.4) for a hypothetical free scalar field of mass $m_\alpha$ (compare this integral with eq. (1.136)). It is customary to rewrite the above representation as

$$\langle 0_{\text{out}}|T\,\phi(x)\phi(y)|0_{\text{in}}\rangle = \int_0^\infty \frac{dM^2}{2\pi} \rho(M^2)\, G_F^0(x, y; M^2), \tag{1.112}$$

where $\rho(M^2)$ is the *spectral function* defined as

$$\rho(M^2) \equiv 2\pi \sum_{\text{classes } \alpha} \delta(M^2 - m_\alpha^2)\, \langle 0_{\text{out}}|\phi(0)|\alpha_\mathbf{0}\rangle\langle\alpha_\mathbf{0}|\phi(0)|0_{\text{in}}\rangle. \tag{1.113}$$

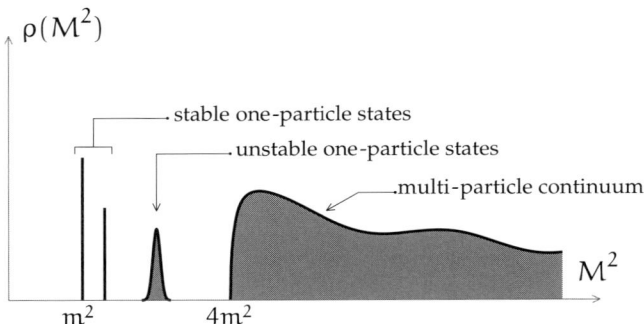

FIGURE 1.2: Example of Källen–Lehmann spectral function.

This function describes the invariant mass distribution of the non-empty states of the theory under consideration, and the exact two-point correlation function is a sum of free correlation functions with varying masses, weighted by this mass distribution.

In a theory with massive particles, the spectral function has a delta function corresponding to states containing a single particle of mass m, and a continuum distribution[20] that starts at the minimal invariant mass (2m) of a two-particle state,

$$\rho(M^2) = 2\pi Z \delta(M^2 - m^2) + \text{ continuum for } M^2 \geq 4m^2, \tag{1.114}$$

where Z is the product of matrix elements that appear in eq. (1.113), in the case of one-particle states. In a theory with interactions, Z in general differs from unity (in fact, it may be infinite). Note that in this equation, m must be the *physical* mass of the particles, as it would be inferred from the simultaneous measurement of their energy and momentum. In Section 2.3.2, we will see that this may not be the same as the parameter we denoted m in the Lagrangian.

As we shall see shortly, by taking the Fourier transform of eq. (1.112) and using eq. (1.114) for the spectral function, we obtain the following pole structure for the exact Feynman propagator:

$$\widetilde{G}_F(p) = \frac{iZ}{p^2 - m^2 + i0^+} + \text{ terms without poles.} \tag{1.115}$$

Therefore, the parameter Z that appears in the spectral function has also the interpretation of the residue of the single particle pole in the exact Feynman propagator.

The fact that $Z \neq 1$ calls for a modification of the LSZ reduction formulas. Eq. (1.114) implies that a factor $\sqrt{Z}$ appears in the overlap between the state $\phi(x)|0_{in}\rangle$ and the one-particle state $|p_{in}\rangle$. In other words, $\phi(x)$ creates a particle with probability Z rather than 1. Or, equivalently,

$$\lim_{x^0 \to -\infty} \phi(x) = \sqrt{Z}\, \phi_{in}(x), \quad \lim_{x^0 \to +\infty} \phi(x) = \sqrt{Z}\, \phi_{out}(x). \tag{1.116}$$

---

[20] Between the one-particle delta function and the two-particle continuum, there may be additional peaks corresponding to multi-particle bound states (to have a stable bound state, the binding energy should decrease the mass of the state compared to the mass 2m of two free particles at rest). The width of these peaks is related to the lifetime of the corresponding states (zero width = infinite lifetime = stable). See Figure 1.2.

Therefore, there should be a factor $Z^{-1/2}$ for each incoming and outgoing particle in the LSZ reduction formulas that relate transition amplitudes to time-ordered products of fields $\phi$,

$$\langle q_1 \cdots q_{n\ \text{out}} | p_1 \cdots p_{m\ \text{in}} \rangle \doteq \left(\frac{i}{\sqrt{Z}}\right)^{m+n}$$
$$\times \int \prod_{i=1}^{m} d^4 x_j\ e^{-ip_i \cdot x_i}\ (\Box_{x_i} + m^2) \prod_{j=1}^{n} d^4 y_j\ e^{iq_j \cdot x_j}\ (\Box_{y_j} + m^2)$$
$$\times \langle 0_{\text{out}} | T\ \phi(x_1) \cdots \phi(x_m) \phi(y_1) \cdots \phi(y_n) | 0_{\text{in}} \rangle. \tag{1.117}$$

## 1.7 Generating Functional

### 1.7.1 Definition

To facilitate the bookkeeping, it is useful to introduce a *generating functional* that encapsulates all the expectation values by defining[21]

$$Z[j] \equiv \sum_{n=0}^{\infty} \frac{i^n}{n!} \int d^4 x_1 \cdots d^4 x_n\ j(x_1) \cdots j(x_n) \langle 0_{\text{out}} | T\ \phi(x_1) \cdots \phi(x_n) | 0_{\text{in}} \rangle$$
$$= \langle 0_{\text{out}} | T\ \exp i \int d^4 x\ j(x) \phi(x) | 0_{\text{in}} \rangle. \tag{1.119}$$

Roughly speaking, a *functional* is a "function of a function." If we know $Z[j]$ for any test function $j(x)$ defined over space-time, we can extract the vacuum expectation values of T-products of fields by taking functional derivatives,

$$\langle 0_{\text{out}} | T\ \phi(x_1) \cdots \phi(x_n) | 0_{\text{in}} \rangle = \left. \frac{\delta^n Z[j]}{i\delta j(x_1) \cdots i\delta j(x_n)} \right|_{j=0}. \tag{1.120}$$

The knowledge of $Z[j]$ would therefore give access to all the transition amplitudes. However, it is in general not possible to derive $Z[j]$ in closed form, and we need to resort to *perturbation theory*, in which the answer is obtained as an expansion in powers of the coupling constant.

### 1.7.2 Factorization of the Interactions

The generating functional can be brought to a more useful form by first writing

$$\phi(x_1) \cdots \phi(x_n) = U(-\infty, x_1^0)\ \phi_{\text{in}}(x_1)\ U(x_1^0, x_2^0)\ \phi_{\text{in}}(x_2) \cdots \phi_{\text{in}}(x_n)\ U(x_n^0, \infty). \tag{1.121}$$

---

[21] Note that

$$Z[0] = \langle 0_{\text{out}} | 0_{\text{in}} \rangle \neq 1 \tag{1.118}$$

in an interacting theory (but if the vacuum state is stable, then this vacuum-to-vacuum transition amplitude must be a pure phase whose squared modulus is 1).

For convenience, we split the leftmost evolution operator as

$$U(-\infty, x_1^0) = U(-\infty, +\infty) \, U(+\infty, x_1^0). \qquad (1.122)$$

Noticing that the formula (1.121) is true for any ordering of the times $x_i^0$ and using the expression of the Us as a time-ordered exponential, we have

$$T\, \phi(x_1) \cdots \phi(x_n) = U(-\infty, +\infty)\, T\, \phi_{in}(x_1) \cdots \phi_{in}(x_n)\, \exp i \int d^4x\, \mathcal{L}_I(\phi_{in}(x)), \qquad (1.123)$$

where the time-ordering on the right-hand side applies to all the operators on its right. This leads to the following representation of the generating functional:[22]

$$Z[j] = \underbrace{\langle 0_{out}|U(-\infty, +\infty)}_{\langle 0_{in}|}\, T\, \exp i \int d^4x\, \Big[j(x)\phi_{in}(x) + \mathcal{L}_I(\phi_{in}(x))\Big] |0_{in}\rangle$$

$$= \exp i \int d^4x\, \mathcal{L}_I\left(\frac{\delta}{i\delta j(x)}\right) \underbrace{\langle 0_{in}|T\, \exp i \int d^4x\, j(x)\phi_{in}(x)|0_{in}\rangle}_{Z_0[j]}. \qquad (1.124)$$

This expression of $Z[j]$ is the most useful, since it factorizes the interactions into a (functional) differential operator acting on $Z_0[j]$, the generating functional for the non-interacting theory.

### 1.7.3 Free Generating Functional

It turns out that $Z_0[j]$ is calculable analytically. The main difficulty in evaluating it is to deal with the non-commuting objects contained in the exponential. A central mathematical result that we shall need is a particular case of the *Baker–Campbell–Hausdorff formula* (see Exercise 1.7, and Section 7.1.5 for a generalization under less restrictive assumptions):

$$\text{if}\quad [A,[A,B]] = [B,[A,B]] = 0, \qquad e^A\, e^B = e^{A+B}\, e^{\frac{1}{2}[A,B]}. \qquad (1.125)$$

This formula is useful here because the commutators $[a, a^\dagger]$ are themselves commuting objects. In order to apply it, we first slice the time axis into an infinite number of small intervals by writing

$$T \exp \int_{-\infty}^{+\infty} d^4x\, O(x) = \prod_{i=-\infty}^{+\infty} T \exp \int_{x_i^0}^{x_{i+1}^0} d^4x\, O(x), \qquad (1.126)$$

where the intermediate times are ordered according to $\cdots x_i^0 < x_{i+1}^0 < \cdots$. The product on the right-hand side should be understood with the convention that the factors are ordered

---

[22]In the last step, we use the functional analogue of $e^{F(x)+jx} = e^{F(\partial_j)} e^{jx}$.

## 1.7 GENERATING FUNCTIONAL

from left to right when the index i decreases. When the size $\Delta \equiv x^0_{i+1} - x^0_i$ of these intervals goes to zero, the time-ordering can be removed in the individual factors:[23]

$$T \exp \int_{-\infty}^{+\infty} d^4x\, O(x) = \lim_{\Delta \to 0^+} \prod_{i=-\infty}^{+\infty} \exp \int_{x^0_i}^{x^0_{i+1}} d^4x\, O(x). \tag{1.127}$$

A first application of the Baker–Campbell–Hausdorff formula leads to

$$T \exp\left(i \int d^4x\, j(x)\phi_{in}(x)\right) = \exp\left\{i \int d^4x\, j(x)\phi_{in}(x)\right\}$$
$$\times \exp\left\{-\frac{1}{2} \int d^4x d^4y\, \theta(x^0 - y^0)\, j(x)j(y)\, [\phi_{in}(x), \phi_{in}(y)]\right\}. \tag{1.128}$$

Note that the exponential in the second line is a commuting number. In the end, we will need to evaluate the expectation value of this operator in the $|0_{in}\rangle$ vacuum state. Therefore, it is desirable to transform it in such a way that the annihilation operators are on the right and the annihilation operators are on the left. This can be achieved by writing

$$\phi_{in}(x) = \phi^{(+)}_{in}(x) + \phi^{(-)}_{in}(x),$$
$$\phi^{(+)}_{in}(x) \equiv \int \frac{d^3p}{(2\pi)^3 2E_p}\, a^\dagger_{p,in}\, e^{+ip\cdot x},$$
$$\phi^{(-)}_{in}(x) \equiv \int \frac{d^3p}{(2\pi)^3 2E_p}\, a_{p,in}\, e^{-ip\cdot x}, \tag{1.129}$$

and by using again the Baker–Campbell–Hausdorff formula. We obtain

$$T \exp\left(i \int d^4x\, j(x)\phi_{in}(x)\right)$$
$$= \exp\left\{i \int d^4x\, j(x)\phi^{(+)}_{in}(x)\right\} \exp\left\{i \int d^4x\, j(x)\phi^{(-)}_{in}(x)\right\}$$
$$\times \exp\left\{\frac{1}{2} \int d^4x d^4y\, j(x)j(y)\, [\phi^{(+)}_{in}(x), \phi^{(-)}_{in}(y)]\right\}$$
$$\times \exp\left\{-\frac{1}{2} \int d^4x d^4y\, \theta(x^0 - y^0)\, j(x)j(y)\, [\phi_{in}(x), \phi_{in}(y)]\right\}. \tag{1.130}$$

The operator that appears in the first line of the right-hand side is called a *normal-ordered exponential*, and is denoted by bracketing the exponential between a pair of colons (: ⋯ :),

$$:\exp i\left(\int d^4x\, j(x)\phi_{in}(x)\right): \equiv \exp\left\{i \int d^4x\, j(x)\phi^{(+)}_{in}(x)\right\} \exp\left\{i \int d^4x\, j(x)\phi^{(-)}_{in}(x)\right\}. \tag{1.131}$$

---

[23] Field operators commute for space-like intervals,

$$[O(x), O(y)] = 0 \quad \text{if } (x - y)^2 < 0.$$

Moreover, when $\Delta \to 0$, the separation between any pair of points x, y with $x^0_i < x^0, y^0 < x^0_{i+1}$ is always space-like.

A crucial property of the normal-ordered exponential is that its *in*-vacuum expectation value is equal to unity:

$$\langle 0_{in}|:\exp\left(i\int d^4x\, j(x)\phi_{in}(x)\right):|0_{in}\rangle = 1. \tag{1.132}$$

Therefore, the generating functional of the free theory is Gaussian in $j(x)$,

$$Z_0[j] = \exp\left\{-\frac{1}{2}\int d^4x d^4y\, j(x)j(y)\, G_F^0(x,y)\right\}, \tag{1.133}$$

where $G_F^0(x,y)$ is a two-point function called the *free Feynman propagator* and defined as

$$G_F^0(x,y) = \theta(x^0-y^0)[\phi_{in}(x),\phi_{in}(y)] - [\phi_{in}^{(+)}(x),\phi_{in}^{(-)}(y)]. \tag{1.134}$$

### 1.7.4 Feynman Propagator

Since the commutators in the right-hand side of eq. (1.134) are commuting numbers, we can also write them as a vacuum expectation value and then rearrange the operators,

$$\begin{aligned}G_F^0(x,y) &= \langle 0_{in}|\theta(x^0-y^0)[\phi_{in}(x),\phi_{in}(y)] - [\phi_{in}^{(+)}(x),\phi_{in}^{(-)}(y)]|0_{in}\rangle \\ &= \langle 0_{in}|T\,\phi_{in}(x)\phi_{in}(y)|0_{in}\rangle.\end{aligned} \tag{1.135}$$

In other words, the free Feynman propagator is the *in*-vacuum expectation value of the time-ordered product of two free fields. Using the Fourier mode decomposition of $\phi_{in}$ and the commutation relation between creation and annihilation operators, the Feynman propagator can be rewritten as follows:

$$G_F^0(x,y) = \int\frac{d^3p}{(2\pi)^3 2E_p}\left\{\theta(x^0-y^0)\,e^{-ip\cdot(x-y)} + \theta(y^0-x^0)\,e^{+ip\cdot(x-y)}\right\}. \tag{1.136}$$

In the following, we will also make extensive use of the Fourier transform of this propagator (with respect to the difference of coordinates $x^\mu - y^\mu$, since it is translation invariant),

$$\begin{aligned}\widetilde{G}_F^0(k) &\equiv \int d^4(x-y)\,e^{ik\cdot(x-y)}\,G_F^0(x,y) \\ &= \frac{1}{2E_k}\left\{\int_0^{+\infty}dz^0\,e^{i(k^0-E_k)z^0} + \int_{-\infty}^0 dz^0\,e^{i(k^0+E_k)z^0}\right\}.\end{aligned} \tag{1.137}$$

The remaining Fourier integrals over $z^0$ are not defined as ordinary functions. Instead, they are distributions, which can also be viewed as the limiting value of a family of ordinary functions. In order to see this, let us write

$$\int_0^{+\infty}dz^0\,e^{iaz^0} = \lim_{\epsilon\to 0^+}\int_0^{+\infty}dz^0\,e^{i(a+i\epsilon)z^0} = \frac{i}{a+i0^+}. \tag{1.138}$$

## 1.7 GENERATING FUNCTIONAL

Likewise

$$\int_{-\infty}^{0} dz^0 \, e^{iaz^0} = \lim_{\epsilon \to 0^+} \int_{+\infty}^{0} dz^0 \, e^{i(a-i\epsilon)z^0} = -\frac{i}{a - i0^+}. \tag{1.139}$$

Therefore, the Fourier space Feynman propagator reads:

$$\widetilde{G}_F^0(k) = \frac{i}{k^2 - m^2 + i0^+}. \tag{1.140}$$

Note that $\widetilde{G}_F^0(k)$ is Lorentz invariant. Henceforth, $G_F^0(x, y)$ is also Lorentz invariant.[24] It is sometimes useful to have a representation of eq. (1.140) in terms of distributions. This is provided by the following identity:

$$\frac{i}{z + i0^+} = i P\left(\frac{1}{z}\right) + \pi \delta(z), \tag{1.141}$$

where $P(1/z)$ is the *principal value* of $1/z$, i.e., the distribution obtained by cutting out – symmetrically – an infinitesimal interval around $z = 0$:

$$\int_{-\infty}^{+\infty} dz \, P\left(\frac{1}{z}\right) f(z) = \lim_{\epsilon \to 0^+} \left\{ \int_{-\infty}^{-\epsilon} \frac{dz}{z} f(z) + \int_{+\epsilon}^{+\infty} \frac{dz}{z} f(z) \right\}. \tag{1.142}$$

(The limit is well defined if the test function $f(z)$ is continuous at $z = 0$.) As far as integration over the variable $z$ is concerned, this $i0^+$ prescription in eq. (1.140) amounts to shifting the pole slightly below the real axis, or equivalently to deform the contour around the pole at $z = 0$ by going slightly above (the term in $\pi\delta(z)$ can be viewed as the result of the integral on the infinitesimal half-circle around the pole), as shown in Figure 1.3. From eq. (1.140), it is trivial to check that $G_F^0(x, y)$ is a Green's function of the operator $\Box_x + m^2$ (up to a normalization factor $-i$),

$$(\Box_x + m^2) \, G_F^0(x, y) = -i\delta(x - y). \tag{1.143}$$

Strictly speaking, the operator $\Box_x + m^2$ is not invertible, since it admits as zero modes all the plane waves $\exp(\pm i k \cdot x)$ with an on-shell momentum $k_0^2 = \mathbf{k}^2 + m^2$. The $i0^+$ prescription in the denominator of eq. (1.140) amounts to shifting infinitesimally the zeroes of $k_0^2 = \mathbf{k}^2 + m^2$ in the complex $k_0$ plane, in order to have a well-defined inverse. The regularization used in eq. (1.140) is specific to the time-ordered propagator, but it is not the only possibility.

---

[24]This is somewhat obfuscated by the fact that the step functions $\theta(\pm(x^0 - y^0))$ that enter in the definition of the time-ordered product are not Lorentz invariant. The Lorentz invariance of time-ordered products follows from the following properties:

- if $(x - y)^2 < 0$, then the two fields commute and the time ordering is irrelevant;
- if $(x - y)^2 \geq 0$, then the sign of $x^0 - y^0$ is Lorentz invariant.

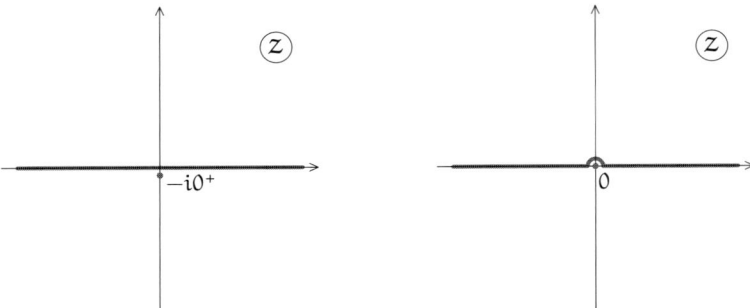

FIGURE 1.3: Integration contours in the z plane for the distribution $1/(x+i0^+)$.

Other regularizations would provide different Green's functions; for instance, the free *retarded propagator* is given by

$$\widetilde{G}^0_R(k) = \frac{i}{(k_0 + i0^+)^2 - (\mathbf{k}^2 + m^2)}. \tag{1.144}$$

One can easily check that its inverse Fourier transform is a function $G^0_R(x,y)$ that satisfies

$$(\Box_x + m^2)\, G^0_R(x,y) = -i\delta(x-y), \quad G^0_R(x,y) = 0 \text{ if } x^0 < y^0. \tag{1.145}$$

In other words, $G^0_R$ is also a Green's function of the operator $\Box_x + m^2$, but with boundary conditions that differ from those of $G^0_F$.

## Exercises

**1.1** Prove eq. (1.9) and derive the commutator $[M^{\mu\nu}, M^{\rho\sigma}]$. *Hints:*
- Denote $\Lambda'^\mu{}_\nu \approx \delta^\mu{}_\nu + \chi^\mu{}_\nu$. Check $(\Lambda^{-1}\Lambda'\Lambda)_{\rho\sigma} \approx \delta_{\rho\sigma} + \chi_{\mu\nu}\Lambda^\mu{}_\rho\Lambda^\nu{}_\sigma$.
- Denote $\Lambda^\mu{}_\nu \approx \delta^\mu{}_\nu + \omega^\mu{}_\nu$. Check $U^{-1}(\Lambda)M^{\mu\nu}U(\Lambda) \approx M^{\mu\nu} + \tfrac{i}{2}\omega_{\rho\sigma}[M^{\mu\nu}, M^{\rho\sigma}]$.
- Identify this expression with $\Lambda^\mu{}_\rho\Lambda^\nu{}_\sigma M^{\rho\sigma}$ (when simplifying by $\omega_{\rho\sigma}$, one needs to enforce the antisymmetry of the residual factors).

**1.2** Prove eq. (1.14) and derive the commutators $[M^{\mu\nu}, P^\rho]$ and $[P^\mu, P^\nu]$. *Hints:*
- Note that $\Lambda^{-1}a\Lambda$ is a translation by $(\Lambda^{-1}a)_\rho = a_\mu\Lambda^\mu{}_\rho$.
- Denote $\Lambda^\mu{}_\nu \approx \delta^\mu{}_\nu + \omega^\mu{}_\nu$. Check that $U^{-1}(\Lambda)P^\mu U(\Lambda) \approx P^\mu + \tfrac{i}{2}\omega_{\rho\sigma}[P^\mu, M^{\rho\sigma}]$.
- Identify this with $\Lambda^\mu{}_\rho P^\rho$.
- What elementary property of translations implies $[P^\mu, P^\nu] = 0$?

**1.3** Show that $[\phi(x), \Pi(y)] = 0$ for any space-like separation (in the massless case, for simplicity). *Hints:*
- Show that $[\phi(x), \Pi(y)] = \frac{i}{2\pi^2 r}\int_0^\infty dp\, p\, \cos(pr^0)\sin(pr)$, where $r^0 \equiv x^0 - y^0$ and $r \equiv |\mathbf{x} - \mathbf{y}|$.
- Perform the integral, to obtain $[\phi(x), \Pi(y)] = -\frac{i}{4\pi r}\frac{\partial}{\partial r}[\delta(r+r^0) + \delta(r-r^0)]$.

**1.4** Consider a *complex* scalar field $\phi$, with the Lagrangian $\mathcal{L} \equiv (\partial_\mu \phi^*)(\partial^\mu \phi) - m^2 \phi^* \phi - V(\phi^* \phi)$, where $V$ is an interaction potential that we need not specify (other than the fact that it depends on the product $\phi^* \phi$).

- Determine the equation of motion associated with this Lagrangian.
- Show that this Lagrangian is invariant under the transformation $\phi \to e^{i\vartheta} \phi$, where $\vartheta$ is a constant. Write the infinitesimal form of this transformation.
- Derive the corresponding Noether current $J^\mu$.
- Check that this current is conserved when the equation of motion is satisfied.
- Write the density $J^0$ in terms of creation and annihilation operators (Note: One needs to introduce two sets of creation and annihilation operators, one for the real part of $\phi$ and one for its imaginary part).

**1.5** Prove eq. (1.73). *Hints:*

- *Start from eq. (1.67). Note that spatial derivatives do not act on the Us.*
- *Show that* $\partial_0 \Big[ U(-\infty, x^0) \phi_{in}(x) U(x^0, -\infty) \Big] = U(-\infty, x^0) \Pi_{in}(x) U(x^0, -\infty)$. *How did the terms coming from the time derivative of the Us cancel?*
- *Apply a second time derivative to this result, to obtain*

$$\partial_0^2 \Big[ U(-\infty, x^0) \phi_{in}(x) U(x^0, -\infty) \Big]$$
$$= U(-\infty, x^0) \Big[ \partial_0^2 \phi_{in}(x) + i \int d^3 y \, [\Pi_{in}(x), \mathcal{L}_I(\phi_{in}(x^0, y))] \Big] U(x^0, -\infty).$$

- *Calculate the commutator on the right-hand side (one may prove that if $[A, B]$ is an object that commutes with all other operators, then $[A, f(B)] = f'(B)[A, B]$).*

**1.6** Show that the expectation value of the occupation number in a one-particle Fock state is given by $\langle p_{in} | f_k | p_{in} \rangle = (2\pi)^3 \, 2E_k \, \delta(p - k)$. *This result, independent of space-time, confirms the fact that Fock states represent a uniform density of particles spread out over all space, rather than a fixed total number of particles.*

**1.7** Prove the Baker–Campbell–Hausdorff formula (1.125). *Hints:*

- *Define $F(t) \equiv e^A e^{tB} e^{-\frac{t}{2}[A,B]}$ and $G(t) \equiv e^{A+tB}$. Check that $F(0) = G(0) = e^A$.*
- *Prove that $F'(t) = F(t)\left(B - \frac{1}{2}[A, B]\right)$.*
- *Prove that $\left[B, (A + tB)^p\right] = -p(A + tB)^{p-1}[A, B]$.*
- *Using the Taylor series of the exponential, prove that $G'(t) = G(t)\left(B - \frac{1}{2}[A, B]\right)$.*

**1.8** Calculate the expression in coordinate space of the retarded propagator given in eq. (1.144) (for $m = 0$). *Hint: Perform the $k_0$ integral in the complex plane with the theorem of residues. The remaining integrals are elementary.*

**1.9** Consider a hypothetical quantum field theory with a kinetic term that involves higher derivatives of the form

$$\mathcal{L}_0 \equiv -\frac{1}{2\mu^2} \phi \, (\Box + m^2)^2 \, \phi,$$

with $\mu$ a constant having the dimension of a mass.

- Give the expression of the retarded propagator in this theory, and relate it to the free propagator of the scalar field theory with a standard kinetic term.
- What would be the Källen–Lehman spectral function for this theory? Is it positive definite?
- Answer the same questions for a kinetic term of the form $-\frac{1}{2\mu^2}\phi\,(\Box+m_1^2)(\Box+m_2^2)\,\phi$.

**1.10** For any 4-vector $a^\mu$, define its light-cone components as $a^\pm \equiv (a^0 \pm a^3)/\sqrt{2}$, and denote also $\mathbf{a}_\perp \equiv (a^1, a^2)$ its transverse component. In light-cone quantization, a field theory is quantized by imposing that the canonical commutation relations be defined at equal $x^+$.

- Rewrite the Lagrangian of a $\phi^4$ scalar field theory in terms of light-cone coordinates, and obtain the conjugate momentum $\Pi$ of $\phi$.
- Rewrite the integration measure $d^4p\,\theta(p^0)\delta(p^2-m^2)$ in terms of the light-cone components of $p^\mu$, and adapt the Fourier decomposition of the free field operator to use these coordinates (denote $\alpha^\dagger_{\vec{p}}, \alpha_{\vec{p}}$, with $\vec{p} \equiv (p^+, \mathbf{p}_\perp)$, the creation and annihilation operators introduced in this decomposition).
- Starting from the canonical commutation relation $\big[\phi(x), \Pi(y)\big]_{x^+=y^+} = i\delta(x^- - y^-)\delta(\mathbf{x}_\perp - \mathbf{y}_\perp)$, derive the commutation relations between these creation and annihilation operators.
- Derive the expression of the free Hamiltonian, and its commutation relations with the creation and annihilation operators.

**1.11** This exercise is a continuation of **Exercise 1.10**. Denote $P^\pm, P^{1,2}, M^{12} \equiv J^3, M^{+-} \equiv K^3, M^{1+} \equiv S^1, M^{2+} \equiv S^2, M^{1-} \equiv B^1, M^{-2} \equiv S^2$ the non-zero generators of the Poincaré algebra in light-cone coordinates.

- Justify the relation $M^{+-} = K^3$, and show that

$$B^1 = \tfrac{K^1+J^2}{\sqrt{2}}, \quad B^2 = \tfrac{K^2-J^1}{\sqrt{2}}, \quad S^1 = \tfrac{K^1-J^2}{\sqrt{2}}, \quad S^2 = \tfrac{K^2+J^1}{\sqrt{2}}.$$

- Show that $P^+, J^3, P^{1,2}, B^{1,2}$ do not change $x^+$. Note that $P^-$ is the generator of translations in $x^+$.
- Show that the set $\{P^-, P^+, J^3, P^{1,2}, B^{1,2}\}$ defines a subalgebra of the Poincaré algebra. *Hint: Calculate their commutation relations and check that they are closed.*
- Show that this subalgebra is isomorphic to the algebra of *Galilean transformations* in the two-dimensional transverse plane, provided one makes the following identification

$$\begin{aligned}
P^+ &\longleftrightarrow & \text{mass,} \\
P^- &\longleftrightarrow & \text{Hamiltonian,} \\
J^3 &\longleftrightarrow & \text{rotations in the 1, 2 plane,} \\
P^{1,2} &\longleftrightarrow & \text{translations in the 1, 2 plane,} \\
B^{1,2} &\longleftrightarrow & \text{Galilean velocity shift in the 1, 2 plane.}
\end{aligned}$$

What is the physical interpretation of this isomorphism?

# CHAPTER 2

# Perturbation Theory

In the previous chapter, using the example of a simple scalar field theory, we have seen how observable quantities such as cross-sections and decay rates may be related to the vacuum expectation values of time-ordered products of field operators. All these correlation functions can be encapsulated in a generating functional Z[j]. However, except in non-interacting theories, this functional is usually not known analytically in closed form. Instead, it is given indirectly by eq. (1.124) as the action of a functional differential operator that acts on the generating functional of the free theory. The latter is a Gaussian in j, whose variance is given by the free Feynman propagator $G_F^0$. Although not fully explicit, this formula provides in principle a straightforward method for obtaining vacuum expectation values of T-products of fields to any given order in the coupling constant $\lambda$. Indeed, by expanding the exponential that contains the interactions, we naturally obtain a series in powers of the coupling constant. In this chapter, we discuss various aspects of this perturbative expansion.

## 2.1 Perturbative Expansion and Feynman Rules

### 2.1.1 Example of a Two-Point Function

Let us assume we want to calculate a $1 \to 1$ transition amplitude such as $\langle \mathbf{q}_{\text{out}} | \mathbf{p}_{\text{in}} \rangle$. The Lehmann–Symanzik–Zimmermann (LSZ) reduction formula tells us that we need to evaluate the vacuum expectation value of the time-ordered product of two fields, $\langle 0_{\text{out}} | T \, \phi(x)\phi(y) | 0_{\text{in}} \rangle$. For a reason that will become clear shortly, it is useful to calculate also the constant $\langle 0_{\text{out}} | 0_{\text{in}} \rangle$. In this example, we will calculate them to order 1 in the coupling constant. In order to make the notations a bit more compact, we denote $G^0_{xy} \equiv G^0_F(x,y)$. At order 1 in $\lambda$, we have

$$\langle 0_{out}|0_{in}\rangle = Z[0] = \left[1 - i\frac{\lambda}{4!}\int d^4z \left(\frac{\delta}{i\delta j(z)}\right)^4 + \mathcal{O}(\lambda^2)\right] Z_0[j]\bigg|_{j=0}$$
$$= 1 - i\frac{\lambda}{8}\int d^4z\, G^{0\,2}_{zz} + \mathcal{O}(\lambda^2). \tag{2.1}$$

Note that the fourth derivative of $Z_0[j]$ produces many terms, but most of them vanish when we set $j = 0$ at the end of the calculation, yielding a much simpler result. However, it would have been necessary to keep all the intermediate terms if we had wanted to go to the next order, for which we would need to take four more derivatives. A similar but a bit lengthier calculation gives

$$\langle 0_{out}|T\,\phi(x)\phi(y)|0_{in}\rangle$$
$$= \left[1 - i\frac{\lambda}{4!}\int d^4z \left(\frac{\delta}{i\delta j(z)}\right)^4 + \mathcal{O}(\lambda^2)\right] \frac{\delta^2 Z_0[j]}{i^2 \delta j(x)\delta j(y)}\bigg|_{j=0}$$
$$= G^0_{xy} - i\,G^0_{xy}\frac{\lambda}{8}\int d^4z\, G^{0\,2}_{zz} - i\frac{\lambda}{2}\int d^4z\, G^0_{xz}G^0_{zz}G^0_{zy} + \mathcal{O}(\lambda^2)$$
$$= \underbrace{\left[1 - i\frac{\lambda}{8}\int d^4z\, G^{0\,2}_{zz} + \mathcal{O}(\lambda^2)\right]}_{Z[0]}$$
$$\times \left[G^0_{xy} - i\frac{\lambda}{2}\int d^4z\, G^0_{xz}G^0_{zz}G^0_{zy} + \mathcal{O}(\lambda^2)\right]. \tag{2.2}$$

Here also, the final expression is much simpler than the intermediate results obtained before setting $j = 0$. Moreover, the expression of the two-point function $\langle 0_{out}|T\,\phi(x)\phi(y)|0_{in}\rangle$ becomes simpler after we notice that one can factor out $Z[0]$. This property is in fact completely general; all transition amplitudes contain a factor $Z[0]$. From the remark made after eq. (1.118), this factor is a pure phase and its squared modulus is 1 and will have no effect in transition probabilities. Therefore, it would be desirable to identify from the start the terms that lead to this prefactor, in order to avoid unnecessary calculations.

## 2.1.2 Diagrammatic Representation

As should be clear from the previous example, the explicit calculation of functional derivatives of $Z_0[j]$ does not seem to be a practical method for contributions of moderately high order in $\lambda$, since we would need to compute $4p + n$ (the factor 4 is specific to an interaction term which is the fourth power of the field) successive derivatives for a term of order $\lambda^p$ in a $n$-point function. But the simplifications that we observed on this example suggest that there may be a better way of obtaining the answer, which bypasses the naive method. In fact, these simplifications follows a rather transparent pattern if we represent the above expressions diagrammatically, by introducing the following notations:

$$G^0_{xy} \equiv x\!-\!\!-\!\!-\!y, \quad -i\lambda\int d^4z \equiv \underset{z}{\times}. \tag{2.3}$$

## 2.1 PERTURBATIVE EXPANSION AND FEYNMAN RULES

The functions that we have calculated above can then be represented as follows:

$$\langle 0_{out}|0_{in}\rangle = 1 + \tfrac{1}{8}\left[\infty_z\right] + \mathcal{O}(\lambda^2)$$

$$\langle 0_{out}|T\,\phi(x)\phi(y)|0_{in}\rangle$$

$$= x\!-\!\!-\!\!y + \tfrac{1}{8}\left[x\!-\!\!-\!\!y\right]\times\left[\infty_z\right] + \tfrac{1}{2}\left[x\!-\!\!\!\bigcirc_z\!\!\!-\!\!y\right] + \mathcal{O}(\lambda^2)$$

$$= \left\{1 + \tfrac{1}{8}\left[\infty_z\right]\right\} \times \left\{x\!-\!\!-\!\!y + \tfrac{1}{2}\left[x\!-\!\!\!\bigcirc_z\!\!\!-\!\!y\right]\right\} + \mathcal{O}(\lambda^2). \tag{2.4}$$

The graphs that appear in the right-hand sides of these equations are called *Feynman diagrams*. With the two rules of eq. (2.3), these graphs are in one-to-one correspondence with the expressions one may obtain from the explicit calculation of derivatives. The vacuum-to-vacuum transition amplitude is made of graphs that have no open-ended line (i.e., all lines have their two endpoints connected to a vertex of the graph). Such graphs are called *vacuum graphs*. The expectation value of the T-product of two fields is given by graphs that have two open ends (these are called *external lines*), the T-product of three fields by graphs with three open ends, etc. The coordinates carried by the fields in the T-product are assigned to the open ends of the graph. From these examples, it appears that one should include all the possible graphs with the appropriate number of external lines, and a number of vertices equal to the desired order in $\lambda$. As we have already emphasized, a factor $Z[0]$ appears in the perturbative expansion of any of the T-products. By a close examination of the second part of eq. (2.4), it appears that this comes from *disconnected* graphs, i.e., graphs that contain a subgraph not connected to any of the external points. Since their only role is to produce the factor $Z[0]$, which is a pure phase, the disconnected graphs are usually discarded from the start. In eq. (2.4), we have simply copied the numerical prefactors $(1/8, 1/2,...)$ that we had obtained via the explicit calculation of derivatives. But as we shall see, they can in fact be recovered simply from the symmetries of the graphs.

### 2.1.3 Feynman Rules

The diagrammatic representation of eq. (2.4) can in fact be used to completely bypass the explicit calculation of the functional derivatives of $Z_0[j]$. The rules that govern this construction are called *Feynman rules*. All the contributions of order $\lambda^p$ to an $n$-point time-ordered product of fields $\langle 0_{out}|T\phi(x_1)\cdots\phi(x_n)|0_{in}\rangle$ can be obtained as follows:

1. Draw all the graphs with only vertices of valence four that connect the $n$ points $x_1$ to $x_n$ and have exactly $p$ vertices. Graphs that contain a subgraph which is not connected to any $x_i$ should be ignored.
2. Each line of a graph represents a free Feynman propagator $G_F^0$ whose two arguments are the coordinates assigned to the endpoints of the line.
3. Each vertex represents a factor $-i\lambda$ and an integral over the coordinate is assigned to this vertex.
4. The numerical prefactor for a given graph is the inverse of the order of its discrete symmetry group. As an illustration, we indicate below the generators of these discrete symmetry groups and the corresponding prefactors in eq. (2.4):

$$\frac{1}{2^3} = \frac{1}{8}, \qquad \frac{1}{2}.$$

(See also Exercise 2.1.) Note that this rule for obtaining the *symmetry factor* associated to a given graph is correct only if the corresponding term in the Lagrangian has been properly symmetrized. For instance, the operator $\phi^4$ should appear in the Lagrangian with a prefactor $1/4!$.

## 2.1.4 Connected Graphs

At step 1, graphs made of several disconnected subgraphs can usually appear in certain functions, provided that each subgraph is connected to at least one of the points $x_i$. For instance, a four-point function contains a piece which is simply made of the product of two two-point functions. In addition, it contains terms that correspond to a genuine four-point function, not factorizable in a product of two-point functions. The factorizable pieces are usually less interesting because they can be recovered from already calculated simpler building blocks.[1] For this reason, it is sometimes useful to introduce the *generating functional of the connected graphs*, denoted $W[j]$. This functional is simply the logarithm of $Z[j]$, $W[j] = \log\bigl(Z[j]/Z[0]\bigr)$. In order to sketch the proof of this identity (see Exercise 2.6), let us write

$$W[j] \equiv \sum_{n=1}^{\infty} \frac{1}{n!} \int d^4x_1 \cdots d^4x_n \, C_n(x_1,\cdots,x_n)\, j(x_1)\cdots j(x_n), \qquad (2.5)$$

where $C_n(x_1,\cdots,x_n)$ are n-point functions whose diagrammatic representation contain only connected graphs. If we expand $Z[j] = Z[0]\exp W[j]$, we obtain

$$\frac{Z[j]}{Z[0]} = 1 + \int d^4x\, C_1(x)\, j(x) + \frac{1}{2!}\int d^4x d^4y\, \underbrace{\bigl[C_2(x,y) + C_1(x)C_1(y)\bigr]}_{\langle 0_{out}|T\phi(x)\phi(y)|0_{in}\rangle} j(x)j(y)$$

$$+ \frac{1}{3!}\int d^4x d^4y d^4z\, \Bigl[C_3(x,y,z) + C_2(x,y)C_1(z) + C_2(y,z)C_1(x)$$

$$\underbrace{+ C_2(z,x)C_1(y) + C_1(x)C_1(y)C_1(z)\Bigr] j(x)j(y)j(z)}_{\langle 0_{out}|T\phi(x)\phi(y)\phi(z)|0_{in}\rangle} + \cdots$$

(2.6)

---

[1] Moreover, in scattering amplitudes, these disconnected contributions are not physically interesting. For instance, if an $m+n \to p+q$ amplitude factorizes into the product of two sub-amplitudes ($m \to p$ and $n \to q$, respectively), then the corresponding subprocesses can happen at very distant locations in space-time, which is usually not what one wants. This can be seen in the derivation of the connection between cross-sections and scattering amplitudes. If an amplitude factorizes into k disconnected sub-amplitudes, its square contains a factor $(VT)^k$ instead of $VT$, because each sub-amplitude has its own delta function of momentum conservation. The factor $(VT)^k$ clearly indicates that we have k independent processes, each happening anywhere in space-time.

## 2.1.5 Feynman Rules in Momentum Space

Until now, we have obtained Feynman rules in terms of objects that depend on space-time coordinates, leading to expressions for the perturbative expansion of the vacuum expectation value of time-ordered products of fields. However, in most practical applications, we need subsequently to use the LSZ reduction formula (1.105) to turn these expectation values into transition amplitudes. This involves the application of the operator $i(\Box + m^2)$ and a Fourier transform, for each external leg. First, note that thanks to eq. (1.143), the application of $i(\Box + m^2)$ simply removes the external line to which it is applied:

$$i(\Box_x + m^2) \left[ x \underset{z}{\rule{2em}{0.4pt}} \bullet \right] = x \rule{2em}{0.4pt} \bullet \,. \tag{2.7}$$

Thus, these operators just produce Feynman graphs that are *amputated* of all their external lines. Then, the Fourier transform can be propagated to all the internal lines of the graph, leading to an expression that involves propagators and vertices that depend only on momenta. The Feynman rules for obtaining directly these momentum space expressions are the following:

1'. The graph topologies that must be considered are of course unchanged. The momenta of the initial state particles are entering into the graph, and the momenta of the final state particles are going out of the graph.
2'. Each line of a graph represents a free Feynman propagator in momentum space $\widetilde{G}^0_F(k)$.
3'. Each vertex represents a factor $-i\lambda(2\pi)^4\delta(k_1 + \cdots + k_4)$, where the $k_i$ are the four momenta *entering* into this vertex.
3''. All the internal momenta that are not constrained by these delta functions should be integrated over with a measure $d^4k/(2\pi)^4$. As we shall see, there is an unconstrained momentum for each loop of a graph.
4'. Symmetry factors are calculated as before.

For instance, these rules lead to

$$\begin{aligned} &= -i\frac{\lambda}{2} \int \frac{d^4k}{(2\pi)^4} \frac{i}{k^2 - m^2 + i0^+} \\ &= \frac{(-i\lambda)^2}{2} \int \frac{d^4k}{(2\pi)^4} \frac{i}{k^2 - m^2 + i0^+} \frac{i}{(p_1+p_2-k)^2 - m^2 + i0^+}. \end{aligned} \tag{2.8}$$

The first one, and more generally all loop corrections to the amputated two-point function, is called a *self-energy*. The terminology comes from the fact that these corrections modify the mass of the particle, thereby changing its rest energy. We will defer the very important question of the calculation of these integrals to later in this chapter. Let us mention right away a major difficulty with these rules: With loops and/or many external legs, a very large number of graphs may contribute and this combinatorial complexity is a huge obstacle in practical applications. In the last 25 years, novel methods have been elaborated to calculate more efficiently scattering amplitudes, often bypassing the representation in terms of Feynman diagrams. We will expose them in Chapters 14 and 19.

### 2.1.6 Counting the Powers of $\lambda$ and $\hbar$

**Order in $\lambda$:** The order in $\lambda$ of a (connected) graph $\mathcal{G}$ is of course related to the number of vertices $n_V$ in the graph,

$$\mathcal{G} \sim \lambda^{n_V}. \tag{2.9}$$

This can also be related to the number of loops of the graph, which is a better measure of its complexity since it also determines how many momentum integrals it contains. Let us denote $n_E$ the number of external lines, $n_I$ the number of internal lines and $n_L$ the number of loops. These parameters are related by the following two identities:

$$4n_V = 2n_I + n_E, \quad n_L = n_I - n_V + 1. \tag{2.10}$$

The first of these identities equates the number of "handles" carried by the vertices, and the number of propagator endpoints that must be attached to them. Strictly speaking, the right-hand side of the second equation counts the number of internal momenta that are not constrained by the delta functions of momentum conservation carried by the vertices (the $+1$ comes from the fact that not all these delta functions are independent – a linear combination of them simply tells that the sum of the external momenta must be zero, and therefore does not constrain the internal ones in any way). That this is also the number of loops in the graph follows from

$$\underbrace{\#(\text{nodes})}_{n_V} - \underbrace{\#(\text{edges})}_{n_I} + \underbrace{\#(\text{faces})}_{n_L+1} + 2 \times (\text{genus}) = 2. \tag{2.11}$$

This equation defines the Euler characteristic of the graph amputated of its external lines. The genus is the number of holes (i.e., tori) of the minimal surface on which the graph can be embedded without edges crossing (see Figure 2.1 for an example). From the two identities (2.10), one obtains

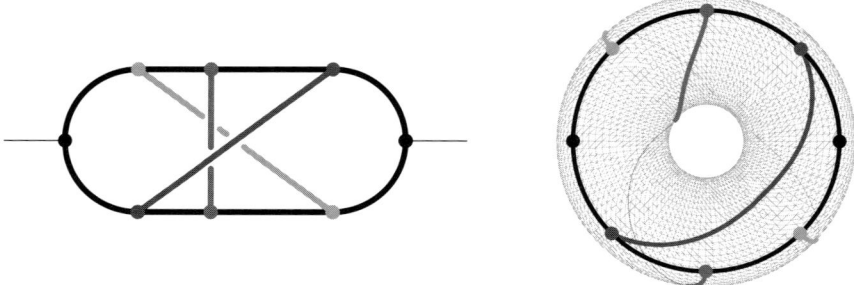

FIGURE 2.1: Illustration of eq. (2.11): a four-loop graph and its embedding on a torus (three faces, genus $= 1$, $n_V = 8$, $n_I = 11$, $n_L = 4$). Thin lines are on the opposite face of the torus.

$$n_V = n_L - 1 + \frac{n_E}{2}, \tag{2.12}$$

and the order in $\lambda$ of the graph is also

$$\mathcal{G} \sim \lambda^{n_L - 1 + n_E/2}. \tag{2.13}$$

According to this formula, the order of a graph depends only on the number of external lines $n_E$ (i.e., on the number of particles involved in the transition amplitude under consideration), and on the number of loops. Thus, the perturbative expansion is also a *loop expansion*, with the leading order being given by *tree diagrams* (i.e., graphs with no loop), the first correction in $\lambda$ by *one-loop graphs*, etc.

**Order in $\hbar$:** It turns out that the number of loops also counts the order in the Planck constant $\hbar$ of a graph. Although we have been using a system of units in which $\hbar = 1$, the original dimension of $\hbar$ is that of an action, and it is easy to reinstate $\hbar$ by the substitution

$$S \quad \to \quad \frac{S}{\hbar} = -\int d^4x \left\{ \frac{1}{2}\phi(x)\frac{\Box_x + m^2}{\hbar}\phi(x) + \frac{\lambda}{4!\,\hbar}\phi^4(x) \right\}. \tag{2.14}$$

From this, we see that $\hbar$ enters in the Feynman rules as follows:

$$\text{Propagator}: \quad \frac{i\hbar}{p^2 - m^2 + i0^+}, \qquad \text{Vertex}: \quad -i\frac{\lambda}{\hbar},$$

and the order in $\hbar$ of a graph is given by

$$\mathcal{G} \sim \hbar^{n_I - n_V} \sim \hbar^{n_L - 1}. \tag{2.15}$$

Therefore, each additional loop brings a power of $\hbar$, and the loop expansion can also be viewed as an expansion in powers of $\hbar$.

## 2.2 Calculation of Loop Integrals

### 2.2.1 Wick's Rotation

Let us consider the first of the examples given in eq. (2.8) and define

$$-i\Sigma(P) \equiv -i\frac{\lambda}{2}\int \frac{d^4k}{(2\pi)^4} \frac{i}{k_0^2 - \mathbf{k}^2 - m^2 + i0^+}, \tag{2.16}$$

where we have expanded the norm $k^2 = k_0^2 - \mathbf{k}^2$ in order to make explicit the dependence on the energy and the 3-momentum. For a simple integral like this example, it may be possible to perform first the $k_0$ integration, and then the integral over the spatial components. However, one may also see that the fact that $k_0$ and $\mathbf{k}$ appear on different footings is entirely due to the signs in the Minkowski metric tensor. The integrand would become significantly simpler if

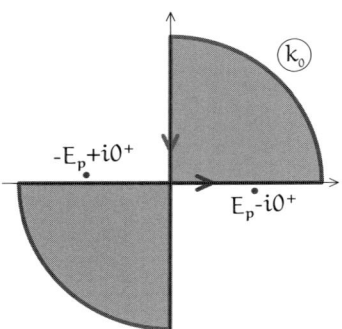

FIGURE 2.2: Wick rotation in the complex $k_0$ plane.

we could rewrite it in terms of a momentum whose square is given by the Euclidean metric. This transformation is called a *Wick rotation*, and consists in rotating the $k_0$ integration axis by 90 degrees in order to bring it along the imaginary axis, as illustrated in Figure 2.2. The integrals along the horizontal and vertical axis are opposite because the shaded domain does not contain any of the poles of the Feynman propagator, and because the propagator vanishes as $k_0^{-2}$ when $|k_0| \to \infty$. The integral along the vertical axis amounts to writing $k_0 = -i\kappa$ with $\kappa$ varying from $-\infty$ to $+\infty$. After this transformation, the integral of eq. (2.16) becomes

$$\Sigma(P) = \frac{\lambda}{2} \int \frac{d^4 k_E}{(2\pi)^4} \frac{1}{k_E^2 + m^2}, \qquad (2.17)$$

where $k_E$ is the *Euclidean* 4-vector defined by $k_E^i = \mathbf{k}$ ($i = 1, 2, 3$) and $k_E^4 = \kappa$, with squared norm $k_E^2 = \mathbf{k}^2 + \kappa^2$. After this transformation, the integrand has become rotationally invariant in four dimensions. Therefore, if we calculate the integral in four-dimensional spherical coordinates, only the radial integral will be non-trivial.

## 2.2.2 Angular Integration in D Dimensions

The four-dimensional volume element for a spherically symmetric integral can be found easily and reads $d^4 k_E = 2\pi^2\, k_E^3\, dk_E$. However, in view of applications to *dimension regularization*, it is useful to derive the analogue of this formula for a D-dimensional Euclidean space. This can be obtained as follows. Let us denote $V_D(k_E)$, the volume of the D-dimensional ball of radius $k_E$. Obviously, it scales like $V_D(k_E) = V_D(1)\, k_E^D$. The radial volume element that we are looking for is the differential of $V_D(k_E)$,

$$d^D k_E = D\, V_D(1)\, k_E^{D-1}\, dk_E. \qquad (2.18)$$

These volumes can be determined recursively by viewing the D-dimensional ball as a stack of $D - 1$ dimensional ones,

$$V_D(k_E) = k_E \int_0^\pi d\theta\, \sin\theta\, V_{D-1}(k_E \sin\theta), \quad V_1(k_E) = 2k_E. \qquad (2.19)$$

(The starting point of the recursion, $V_1(k_E)$, is the length of the line segment $[-k_E, +k_E]$.) Therefore, for low values of D, we obtain

$$V_2(k_E) = \pi k_E^2, \quad V_3(k_E) = \frac{4\pi}{3} k_E^3, \quad V_4(k_E) = \frac{\pi^2}{2} k_E^4. \tag{2.20}$$

More generally, we have

$$V_{D+1}(1) = V_D(1)\pi^{1/2} \frac{\Gamma(\frac{D}{2}+1)}{\Gamma(\frac{D}{2}+\frac{3}{2})} \quad \text{and} \quad V_D(1) = \frac{2\pi^{D/2}}{D\Gamma(\frac{D}{2})}, \tag{2.21}$$

where $\Gamma(z)$ is Euler's gamma function (for positive integer arguments, $\Gamma(n) = (n-1)!$). Some useful properties of the gamma function are:

$$\Gamma(z+1) = z\Gamma(z), \quad \Gamma(\tfrac{1}{2}) = \sqrt{\pi}, \quad \Gamma(z)\Gamma(1-z) = \frac{\pi}{\sin(\pi z)}. \tag{2.22}$$

The last formula tells us that the gamma function has poles in the complex plane at all the negative integers, $z = -n$, with residues $(-1)^n/n!$.

### 2.2.3 Feynman Parametrization of Denominators

Let us now consider the second diagram of eq. (2.8) (with the notation $P \equiv p_1 + p_2$),

$$-i\Gamma_4(P) \equiv \frac{(-i\lambda)^2}{2} \int \frac{d^4k}{(2\pi)^4} \frac{i}{k^2 - m^2 + i0^+} \frac{i}{(P-k)^2 - m^2 + i0^+}. \tag{2.23}$$

In this more complicated example, an extra difficulty is that the integrand is not rotationally invariant. The following trick, known as *Feynman parametrization*, can be used to rearrange the denominators:[2]

$$\frac{1}{AB} = \int_0^1 \frac{dx}{[xA + (1-x)B]^2}. \tag{2.24}$$

The denominator resulting from this transformation is

$$x(k^2 - m^2 + i0^+) + (1-x)((P-k)^2 - m^2 + i0^+) = l^2 - m^2 - \Delta(x, P) + i0^+, \tag{2.25}$$

where we denote $l \equiv k - (1-x)P$ and $\Delta(x, P) \equiv -x(1-x)P^2$. At this point, we can apply a Wick rotation[3] to the shifted integration variable $l$, in order to obtain

$$\Gamma_4(P) = -\frac{\lambda^2}{2} \int_0^1 dx \int \frac{d^4l_E}{(2\pi)^4} \frac{1}{[l_E^2 + m^2 + \Delta(x, P)]^2}, \tag{2.26}$$

where the integrand is again invariant by rotation in four-dimensional Euclidean space.

---

[2]For n denominators, this formula can be generalized into

$$\frac{1}{A_1 A_2 \cdots A_n} = \Gamma(n) \int_0^1 dx_1 \cdots dx_n \, \delta(1 - \sum_i x_i) \frac{1}{[x_1 A_1 + \cdots + x_n A_n]^n}.$$

[3]It is allowed because the integration axis can be rotated counterclockwise without passing through any of the poles in the variable $l_0$.

### 2.2.4 Integration by Parts and Differential Equations

Wick's rotation and Feynman parametrization are sufficient for simple Feynman integrals, but more complicated cases require other tools. Consider for instance the following two-loop example:

$$\Sigma_2(P) \equiv -\frac{\lambda^2}{6} \int \frac{d^4k_E \, d^4\ell_E}{(2\pi)^8} \frac{1}{k_E^2 \ell_E^2 (k_E + \ell_E - P)^2}. \tag{2.27}$$

(We have already performed the Wick rotation of the two loop momenta, and we are considering the massless case to simplify the subsequent manipulations in this example – see Exercise 2.7.) Here, a useful trick is to insert under the integral the following trivial identity:

$$1 = \frac{1}{8}\left(\frac{\partial k_E^\mu}{\partial k_E^\mu} + \frac{\partial \ell_E^\mu}{\partial \ell_E^\mu}\right). \tag{2.28}$$

Ignoring for the moment that we may be manipulating divergent integrals, this leads to

$$\begin{aligned}
\Sigma_2(P) &= \frac{\lambda^2}{48} \int \frac{d^4k_E \, d^4\ell_E}{(2\pi)^8} \underbrace{\left(\frac{\partial k_E^\mu}{\partial k_E^\mu} + \frac{\partial \ell_E^\mu}{\partial \ell_E^\mu}\right)}_{-k_E^\mu \frac{\partial}{\partial k_E^\mu} - \ell_E^\mu \frac{\partial}{\partial \ell_E^\mu}} \frac{1}{k_E^2 \ell_E^2 (k_E + \ell_E - P)^2} \\
&= \frac{3}{4}\Sigma_2(P) + \frac{\lambda^2}{24} \int \frac{d^4k_E \, d^4\ell_E}{(2\pi)^8} \frac{P \cdot (k_E + \ell_E - P)}{k_E^2 \ell_E^2 (k_E + \ell_E - P)^4} \\
&= \frac{3}{4}\Sigma_2(P) + \frac{P^\mu}{8}\frac{\partial \Sigma_2(P)}{\partial P^\mu} = \frac{3}{4}\Sigma_2(P) + \frac{P^2}{4}\frac{\partial \Sigma_2(P)}{\partial P^2}.
\end{aligned} \tag{2.29}$$

The last equality uses the fact that $\Sigma_2(P)$ is in fact a function of the Lorentz invariant quantity $P^2$. Thus, we have obtained a homogeneous differential equation for $\Sigma_2$. If all integrals were finite and our manipulations safe, this equation would simply tell us that $\Sigma_2$ is proportional to $P^2$. Variants and extensions of this technique, known as the *integration by parts method*, are routinely used to derive relationships between various loop integrals. They can be automated with the *Laporta algorithm*.

## 2.3 Ultraviolet Divergences and Renormalization

Until now, we have not really attempted to calculate explicitly the Euclidean integrals encountered in eqs. (2.17), (2.26) and (2.27). In fact, these integrals do not converge when $|k_E, \ell_E| \to \infty$, and as such they are therefore infinite. These infinities are called *ultraviolet divergences*.

### 2.3.1 Regularization of Divergent Integrals

As we shall see shortly, this has very deep implications for how we should interpret the theory. However, before we can discuss this, it is crucial to make all the loop integrals temporarily finite in order to secure all subsequent manipulations. For instance, it should be clear that

## 2.3 ULTRAVIOLET DIVERGENCES AND RENORMALIZATION

the integration by parts method cannot make sense as it is an integral that diverges. This procedure, called *regularization*, amounts to altering the theory to make all the integrals finite. There is no unique method for achieving this, and the most common ones are the following.

**Pauli–Villars method:** In this approach, we modify the free propagator according to

$$\frac{i}{k^2 - m^2 + i0^+} \rightarrow \frac{i}{k^2 - m^2 + i0^+} - \frac{i}{k^2 - M^2 + i0^+}. \tag{2.30}$$

When $|k_E| \gg M$, this modified propagator decreases as $|k_E|^{-4}$ instead of $|k_E|^{-2}$ for the unmodified propagator, which is usually sufficient to render the integrals convergent. The original theory (and its ultraviolet divergences) are recovered in the limit $M \to \infty$.

**Lattice regularization:** With this regularization, the continuous space-time is replaced by a discrete lattice of points, for instance a cubic lattice with a spacing $a$ between the nearest-neighbor sites. On such a lattice, the momenta are themselves discrete (if the lattice is finite), and bounded by a maximal momentum of order $a^{-1}$. Therefore, the momentum integrals are replaced by discrete sums that are all finite. The original theory is recovered when $a \to 0$ and the volume becomes infinite. A shortcoming of lattice regularization is that the discrete momentum sums are usually much more difficult to evaluate than continuum integrals, and that it breaks the usual space-time symmetries such as translation and rotation invariance. This is, however, the basis of a numerical Monte-Carlo method known as *lattice field theory* (see Chapter 16).

**Cutoff regularization:** one may also cut the integration over the norm of the Euclidean momentum by imposing $|k_E| \leq \Lambda$. The underlying theory is recovered in the limit $\Lambda \to \infty$. This is a commonly used regularization in scalar theories at one-loop, due to its simplicity and because it preserves all the symmetries of the theory. But it becomes complicated to implement beyond one-loop.

**Dimensional regularization:** This method is based on the observation that the integral

$$\int_0^\infty dk_E \frac{k_E^{D-1}}{[k_E^2 + \Delta]^n} = \frac{1}{2} \int_0^\infty du \frac{u^{\frac{D}{2}-1}}{[u+\Delta]^n} = \frac{\Delta^{\frac{D}{2}-n}}{2} \underbrace{\int_0^1 dx\, x^{n-\frac{D}{2}-1}(1-x)^{\frac{D}{2}-1}}_{\frac{\Gamma(n-\frac{D}{2})\Gamma(\frac{D}{2})}{\Gamma(n)}} \tag{2.31}$$

is well defined for almost any D except for $D = 2n, 2n+2, 2n+4, \cdots$ (and $D = 0, -2, -4, \cdots$) thanks to the analytical properties of the gamma function. Dimensional regularization keeps the number of space-time dimensions D arbitrary in all the intermediate calculations, and at the end one usually writes $D = 4 - 2\epsilon$ with $\epsilon \ll 1$. This regularization does not break any of the symmetries of the theory, including gauge invariance (which is not the case of cutoff regularization). Note that an extra ingredient is necessary: The coupling constant $\lambda$ is dimensionless only when $D = 4$. In order to keep the dimension of $\lambda$ unchanged, we must introduce a parameter $\mu$ that has the dimension of a mass, and replace $\lambda$ by $\lambda\mu^{4-D}$. The field $\phi(x)$ now has the dimension of a mass to the power $(D-2)/2$. Setting $D = 4 - 2\epsilon$, the singular behavior of the integrals $\Sigma(P)$ and $\Gamma_4(P)$ introduced earlier as examples appears in the form of poles at $\epsilon = 0$,

$$\Sigma(P) = -\frac{\lambda}{2} \frac{m^2}{(4\pi)^2} \frac{1}{\epsilon} + \mathcal{O}(1), \quad \Gamma_4(P) = -\frac{\lambda^2}{2} \frac{1}{(4\pi)^2} \frac{1}{\epsilon} + \mathcal{O}(1). \quad (2.32)$$

For reference, we list here the result in dimensional regularization of some integrals that are frequently encountered in the calculation of loop integrals (see Exercises 2.2 and 2.8):

$$\int \frac{d^D k_E}{(2\pi)^D} \frac{1}{(k_E^2 + \Delta)^n} = \frac{\Delta^{\frac{D}{2}-n}}{(4\pi)^{\frac{D}{2}}} \frac{\Gamma(n - \frac{D}{2})}{\Gamma(n)},$$

$$\int \frac{d^D k_E}{(2\pi)^D} \frac{k_E^\mu}{(k_E^2 + \Delta)^n} = 0,$$

$$\int \frac{d^D k_E}{(2\pi)^D} \frac{k_E^\mu k_E^\nu}{(k_E^2 + \Delta)^n} = \frac{g^{\mu\nu}}{2} \frac{\Delta^{\frac{D}{2}+1-n}}{(4\pi)^{\frac{D}{2}}} \frac{\Gamma(n - 1 - \frac{D}{2})}{\Gamma(n)},$$

$$\int \frac{d^D k_E}{(2\pi)^D} \frac{k_E^\mu k_E^\nu k_E^\rho k_E^\sigma}{(k_E^2 + \Delta)^n} = \frac{g^{\mu\nu}g^{\rho\sigma} + g^{\mu\rho}g^{\nu\sigma} + g^{\mu\sigma}g^{\nu\rho}}{4} \frac{\Delta^{\frac{D}{2}+2-n}}{(4\pi)^{\frac{D}{2}}} \frac{\Gamma(n - 2 - \frac{D}{2})}{\Gamma(n)}. \quad (2.33)$$

### 2.3.2 Mass Renormalization

The results in eq. (2.32) are finite as long as we keep $\epsilon$ non-zero. However, nature is four-dimensional, and we would like to take the limit $\epsilon \to 0$ at some point. What this means is that, although it makes all integrals finite and thus provides a safe way to manipulate them, regularization does not solve the problem of infinite perturbative corrections. In order to progress toward a physical understanding of these divergences and how to dispose of them, let us make a few observations:

- the above divergent terms are momentum independent;[4]
- these divergences appear in two-point and four-point functions only.

Moreover, it is important to note that the parameters ($m^2$ and $\lambda$) in the Lagrangian are not directly observable quantities by themselves.[5] In fact, the physical substance of a theory is not contained in how an observable quantity depends on $m^2$ and $\lambda$, but in how various physical quantities relate to one another. For instance, one may consider several observables $\mathcal{O}_{1,2,3,...}$. For a theorist, they are given by certain functions of the parameters of the Lagrangian,

$$\mathcal{O}_i = f_i(m^2, \lambda, \ldots), \quad (2.34)$$

---

[4]These examples are not completely general. As we shall see later, divergent terms proportional to $P^2$ also appear in the two-point function.

[5]In this regard, it is important to realize that the renormalization of the parameters of the Lagrangian would be necessary even in a theory that has no divergent loop integrals.

## 2.3 ULTRAVIOLET DIVERGENCES AND RENORMALIZATION

but what matters for an experimentalist are the relationships between observables. Renormalization consists in changing the parameters of the Lagrangian (possibly making them infinite) in such a way that the relationships between the $\mathcal{O}_i$ remain well defined.

Consider, for instance, the physical (as opposed to the parameter we call $m$ in the Lagrangian) mass of a particle. It is a measurable property of the particle that can be obtained for instance by measuring both its energy and its momentum, via $p_0^2 - \mathbf{p}^2$. In quantum field theory, this definition of the mass corresponds to the location of the poles of the propagator in the complex $p_0$ plane. However, as we shall see, loop corrections modify substantially the propagator, and it turns out that the parameter $m$ in the free propagator has in fact little to do with the physical mass. If we *dress* the propagator by summing the repeated insertions of the one-loop correction $-i\Sigma$,

$$\widetilde{G}_F(P) \equiv \underset{P}{\longrightarrow} + \underset{P}{\longrightarrow}\!\!\bigcirc\!\!\underset{}{\longrightarrow} + \underset{P}{\longrightarrow}\!\!\bigcirc\!\!\bigcirc\!\!\underset{}{\longrightarrow} + \underset{P}{\longrightarrow}\!\!\bigcirc\!\!\bigcirc\!\!\bigcirc\!\!\underset{}{\longrightarrow} + \cdots, \qquad (2.35)$$

we obtain a geometrical series whose sum is easy to calculate,

$$\widetilde{G}_F(P) = \frac{i}{p_0^2 - \mathbf{p}^2 - m^2 - \Sigma + i0^+}, \qquad (2.36)$$

from which it is immediate to see that the location of the pole has changed. It is now given by

$$p_0^2 - \mathbf{p}^2 = \underbrace{m^2 + \Sigma}_{\text{new squared mass}}. \qquad (2.37)$$

Since the propagator given in eq. (2.36) includes some corrections due to the interactions, its poles ought to give a value of the mass closer to the physical one. Therefore, it is tempting to write

$$m_{\text{phys}}^2 = m^2 + \Sigma + \mathcal{O}(\lambda^2). \qquad (2.38)$$

Of course, since $\Sigma$ is infinite, the only way this can be satisfied is that the parameter $m^2$ that appears in the Lagrangian be itself infinite, with an opposite sign in order to cancel the infinity from $\Sigma$. To further distinguish it from the physical mass, the parameter $m$ in the Lagrangian is usually called the *bare mass*, while $m_{\text{phys}}$ is the physical – or *renormalized* – mass.

### 2.3.3 Field Renormalization

The one-loop function $\Sigma$ in a theory with a $\phi^4$ interaction is somewhat special because at this order, it is independent of the momentum $P$. Being a constant, the infinity it contains can be absorbed entirely into a redefinition of the bare mass, but the residue of the pole remains equal to 1. However, starting at two loops, the two-point functions that correct the propagator are usually momentum dependent, as is the case of the two-loop correction in eq. (2.27). Note that when the two-point function $\Sigma(P)$ depends on the momentum $P^\mu$ in a scalar field theory, Lorentz invariance requires that it depends only on the norm $P^2$. It is convenient to expand $\Sigma(P^2)$ around the physical mass,

$$\Sigma(P^2) = \Sigma(m_{\text{phys}}^2) + (P^2 - m_{\text{phys}}^2)\Sigma'(m_{\text{phys}}^2) + \tfrac{1}{2}(P^2 - m_{\text{phys}}^2)^2 \Sigma''(m_{\text{phys}}^2) + \cdots \qquad (2.39)$$

For the resumed propagator $\widetilde{G}_F$ to have a pole at $P^2 = m_{phys}^2$, we must impose

$$m_{phys}^2 = m^2 + \Sigma(m_{phys}^2), \tag{2.40}$$

that generalizes eq. (2.38) to a momentum-dependent $\Sigma$. Then, in the vicinity of the pole, the dressed propagator behaves as

$$\widetilde{G}_F(P) \underset{P^2 \to m_{phys}^2}{\approx} \frac{i}{(1 - \Sigma'(m_{phys}^2))(P^2 - m_{phys}^2) + i0^+}. \tag{2.41}$$

This indicates that the field renormalization factor Z introduced in (1.115) cannot be equal to 1 when the propagator is corrected by a momentum-dependent $\Sigma$. Instead, we have

$$Z = \frac{1}{1 - \Sigma'(m_{phys}^2)}. \tag{2.42}$$

Moreover, *Weinbergs's theorem* implies that the ultraviolet divergences of the two-point function $\Sigma(P^2)$ arise only in $\Sigma(m_{phys}^2)$ and in the first derivative $\Sigma'(m_{phys}^2)$, while higher derivatives are all finite. Eqs. (2.40) and (2.42) therefore indicate that these infinities can be "hidden" in the bare mass $m^2$ and in a multiplicative field renormalization factor Z.

### 2.3.4 Ultraviolet Power Counting

Likewise, the divergence in the four-point function $\Gamma_4$ can also be absorbed into an infinite redefinition of the coupling constant $\lambda$ in the Lagrangian. But for this redefined $\lambda$ to truly be a constant, it is important that the divergence in $\Gamma_4$ (the residue of the pole in $\epsilon^{-1}$) be independent of momentum. From these examples, it appears crucial that $\Sigma$ has divergences only in its zeroth- and first-order Taylor coefficients and $\Gamma_4$ only in the zeroth order, in order to be able to absorb the divergences by a proper definition of $m^2$, Z and $\lambda$. A simple dimensional argument gives plausibility to this assertion (of which *Weinberg's theorem* provides a rigorous justification). Let us assume that we scale up all the loop momenta of a graph[6] by some factor $\xi$. In doing this, a graph $\mathcal{G}$ with $n_L$ loops and $n_I$ internal lines will scale as

$$\mathcal{G} \sim \xi^{D n_L - 2n_I}, \tag{2.43}$$

assuming D space-time dimensions for more generality. The exponent $\omega(\mathcal{G}) \equiv D n_L - 2n_I$ is called the *superficial degree of divergence* of the graph. This exponent characterizes how the graph diverges when all its internal momenta are rescaled uniformly:

- $\omega(\mathcal{G}) \geq 0$: The graph may have an intrinsic divergence.
- $\omega(\mathcal{G}) < 0$: The graph may be finite, or may contain a divergent subgraph. More precisely, the *convergence theorem* states that a graph $\mathcal{G}$ is finite if $\omega(\mathcal{G}) < 0$, and the degrees of divergence of all its subgraphs are negative as well. Of course, subgraphs do not always satisfy this condition. But in the renormalization process, the divergent subgraphs will have been dealt with at an earlier stage since they occur at a lower order of the perturbative expansion.

---

[6]This is possible without changing the external momenta, precisely because the loop momenta are those that are unconstrained by momentum conservation.

## 2.3 ULTRAVIOLET DIVERGENCES AND RENORMALIZATION

Therefore, the superficial degree of divergence indicates which n-point functions may have ultraviolet divergences of their own (as opposed to being divergent because of a divergent subgraph). Using eq. (2.10), the exponent $\omega(\mathcal{G})$ can be rewritten as follows:

$$\omega(\mathcal{G}) = 4 - n_E + (D-4)n_L. \qquad (2.44)$$

An important consequence of this formula is that in four dimensions the superficial degree of divergence of a graph does not depend on the number of loops, but only on the number of external lines. When $D = 4$, the only functions that have a non-negative $\omega$ are the two-point function and the four-point function.[7] It is important to realize that this does not mean that a six-point cannot be divergent. However, it can diverge only if it contains a divergent two-point or four-point subgraph. Moreover, the value of the superficial degree of divergence indicates the maximal power of the ultraviolet cutoff $\Lambda$ (assuming momentarily regularization by a cutoff) that may appear in these functions:

- two-point: up to $\Lambda^2$;
- four-point: up to $\log(\Lambda)$.

(The two-point function is said to have a quadratic divergence, while the four-point function has a logarithmic divergence.) Note also that if we differentiate a graph with respect to the invariant norm $P^2$ of one of its external momenta, we get

$$\omega\left(\frac{\partial \mathcal{G}}{\partial P^2}\right) = 2 - n_E + (D-4)n_L. \qquad (2.45)$$

($\omega$ further decreases by two units with each additional derivative with respect to $P^2$.) Therefore, the momentum derivative $\Sigma'(P^2)$ of the two-point function has $\omega = 0$ in $D = 4$, and its higher derivatives all have $\omega < 0$. The fact that only $\Gamma_4(m_{\text{phys}}^2)$, $\Sigma(m_{\text{phys}}^2)$ and $\Sigma'(m_{\text{phys}}^2)$ have $\omega \geq 0$ is the reason why it is possible to get rid of all the divergences of this theory (in four dimensions) by a redefinition of the parameters of the Lagrangian. This theory is said to be *renormalizable*.

### 2.3.5 Ultraviolet Classification of Quantum Field Theories

In dimensions lower than four, the $\omega(\mathcal{G})$ given in eq. (2.44) is a strictly decreasing function of the number of loops, which indicates that graphs with a given $n_E$ do not develop new divergences beyond a certain loop order. Such theories are said to be *super renormalizable* because they only have a finite number of divergent graphs. Conversely, in dimensions higher than four, $\omega(\mathcal{G})$ increases with the number of loops, and any function will eventually become divergent at some loop order. These theories are usually[8] *non-renormalizable*. One may think of introducing, as they become necessary, additional operators in the Lagrangian with a coupling constant adjusted to cancel the new divergences that arise at a given loop order.

---

[7]Functions with an odd number of external lines vanish in the theory under consideration. Note also that zero-point functions (vacuum graphs) have a superficial degree of divergence equal to 4, indicating that they may contain up to quartic divergences $\sim \Lambda^4$.

[8]It may happen that an internal symmetry, such as a gauge symmetry, renders a function finite while its superficial degree of divergence is non-negative.

However, an infinite number of such parameters would need to be introduced, thereby reducing drastically the predictive power of this type of theory.[9]

As we have seen, the renormalizability of a field theory depends both on the interaction terms it contains, and on the dimension of space-time. In fact, a simpler equivalent criterion is the mass dimension of the coupling constant in front of the interaction term:

- dim $> 0$: super-renormalizable;
- dim $= 0$: renormalizable;
- dim $< 0$: non-renormalizable.

For instance, the "coupling constant" $m^2$ in front of the mass term always has a mass dimension equal to two, and this term is therefore super-renormalizable. In contrast, the coupling constant $\lambda$ in front of a $\phi^4$ interaction has a mass dimension of $4 - D$, and is (super)renormalizable in dimensions less than or equal to four.

### 2.3.6 Renormalization in Perturbation Theory: Counterterms

A convenient setup for casting the renormalization procedure within perturbation theory is to start from the bare Lagrangian,

$$\mathcal{L} = \frac{1}{2}(\partial_\mu \phi_b)(\partial^\mu \phi_b) - \frac{1}{2}m_b^2 \phi_b^2 - \frac{\lambda_b}{4!}\phi_b^4, \qquad (2.46)$$

where we denote $\phi_b$, $m_b$ and $\lambda_b$ the bare field, mass and coupling, to stress that they are not the physical ones. This bare Lagrangian is then decomposed as the sum of a renormalized Lagrangian and a correction,

$$\mathcal{L} = \mathcal{L}_r + \Delta\mathcal{L}$$
$$\mathcal{L}_r \equiv \frac{1}{2}(\partial_\mu \phi_r)(\partial^\mu \phi_r) - \frac{1}{2}m_r^2 \phi_r^2 - \frac{\lambda_r}{4!}\phi_r^4$$
$$\Delta\mathcal{L} \equiv \frac{1}{2}\Delta_z(\partial_\mu \phi_r)(\partial^\mu \phi_r) - \frac{1}{2}\Delta_m \phi_r^2 - \frac{1}{4!}\Delta_\lambda \phi_r^4. \qquad (2.47)$$

$\mathcal{L}_r$ contains the renormalized (i.e., physical) mass $m_r$ and coupling constant $\lambda_r$ (the latter may be defined from the measurement of some cross-section chosen as a reference). The relationship between the bare and renormalized fields is defined to be

$$\phi_b \equiv \sqrt{Z}\,\phi_r. \qquad (2.48)$$

This choice ensures that the propagator defined in terms of $\phi_r$ has a residue equal to 1 at the pole, instead of $Z$ for the propagator defined from the bare field. In $\Delta\mathcal{L}$, the coefficients $\Delta_z, \Delta_m, \Delta_\lambda$ are called *counterterms*. The bare and physical parameters and the counterterms must be related by

$$\Delta_z = Z - 1, \quad \Delta_m = Zm_b^2 - m_r^2, \quad \Delta_\lambda = Z^2 \lambda_b - \lambda_r. \qquad (2.49)$$

---

[9]Non-renormalizable field theories may nevertheless be used as low-energy effective field theories, where they approximate below a certain cutoff a more fundamental – possibly unknown – theory supposedly valid above the cutoff.

## 2.3 ULTRAVIOLET DIVERGENCES AND RENORMALIZATION

The terms in $\Delta\mathcal{L}$ are treated as a perturbation to $\mathcal{L}_r$, and one may introduce extra Feynman rules for the various terms it contains,

$$\frac{1}{2}\Delta_z(\partial_\mu\phi_r)(\partial^\mu\phi_r) - \frac{1}{2}\Delta_m\phi_r^2 \quad \rightarrow \quad \xrightarrow{P}\!\!\otimes\!\!- = -i\left(\Delta_z P^2 + \Delta_m\right)$$

$$-\frac{1}{4!}\Delta_\lambda\phi_r^4 \quad \rightarrow \quad \times = -i\Delta_\lambda. \tag{2.50}$$

At tree level, one uses only $\mathcal{L}_r$ since $\Delta\mathcal{L}$ is formally of higher order in the coupling, and by construction the physical quantities computed at this order will depend directly on physical parameters. Higher orders contain divergent loop corrections, but also contributions coming from $\Delta\mathcal{L}$. The counterterms $\Delta_z, \Delta_m, \Delta_\lambda$ should be adjusted at every order to cancel the new divergences that arise at this order. In particular, after having included the contribution of the counterterms, the self-energy $\Sigma(P^2)$ is usually required to satisfy the following conditions:[10]

$$\Sigma(m_r^2) = 0, \quad \Sigma'(m_r^2) = 0. \tag{2.51}$$

**Self-energy corrections in the LSZ formula:** With the renormalization condition (2.51), the self-energy vanishes like $(P^2 - m_r^2)^2$ in the vicinity of the mass-shell, and it is not necessary to dress the external lines with the self-energy in the LSZ reduction formulas for transition amplitudes. Indeed, we have

$$i(\square + m_r^2)\, G_F = 1, \quad \lim_{p^2 \to m_r^2}(-i\Sigma)G_F = 0. \tag{2.52}$$

For each external line, the reduction formula contains an operator $i(\square_x + m_r^2)$ acting on the corresponding external propagator. If this propagator is dressed with $\Sigma$, this gives

$$i(\square + m_r^2)\underbrace{\left\{G_F + G_F(-i\Sigma)G_F + G_F(-i\Sigma)G_F(-i\Sigma)G_F + \cdots\right\}}_{\text{dressed propagator}} = 1. \tag{2.53}$$

Therefore, all the terms are zero except the first one, and we can ignore self-energy corrections on the external lines.

### 2.3.7 BPHZ Renormalization

The actual proof of renormalizability is more complicated than this superficial discussion based on power counting may suggest. Indeed, a crucial aspect is to show that the divergences can be removed via the subtraction of local terms only, i.e., that the divergences are polynomial in the external momenta. While this is trivial in all the one-loop graphs

---

[10]Strictly speaking, the only requirement is that the counterterms cancel the infinities, which does not fix uniquely their finite part. Various *renormalization schemes* are possible, which differ in how these finite parts are chosen. This freedom usually involves the choice of a momentum scale, since one should specify the external momenta at which the renormalization condition is satisfied.

we have considered, it is not obviously true beyond one loop. As an illustration of the difficulty, let us consider the following example of a two-loop contribution to the four-point function,

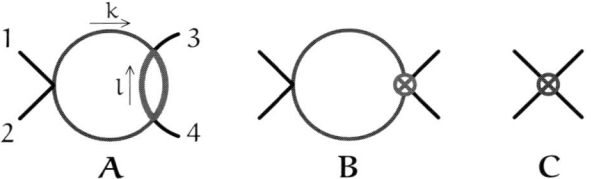

In graph A, the loop we have represented with a thicker line is divergent, and is multiplied by a non-polynomial function of $P^2 \equiv (p_1 + p_2)^2$ coming from the rest of the graph. The Feynman rules give the following integrand for this graph:

$$\mathcal{I}_A = \frac{(-i\lambda)^3}{2} G_F^0(k) G_F^0(k-P) \, G_F^0(l) G_F^0(l+k+p_3). \tag{2.54}$$

The superficial degree of divergence of the integration over $l$ is $\omega(A;l) = 0$ (at fixed $k$), and therefore the loop carrying the momentum $l$ is logarithmically divergent. Diagram B is obtained by *subtracting from this loop a polynomial in its external momenta, whose degree is precisely equal to its superficial degree of divergence*. Since $\omega(A;l) = 0$, the subtraction is the zeroth order of the Taylor expansion of that loop (underlined in the following equation):

$$\mathcal{I}_{A+B} = \frac{(-i\lambda)^3}{2} G_F^0(k) G_F^0(k-P) \left[ G_F^0(l) G_F^0(l+k+p_3) - \underline{G_F^0(l) G_F^0(l)} \right]. \tag{2.55}$$

Now, the degree of divergence in $l$ of the combination inside the square brackets is $\omega(A+B;l) = -1$, and the integration over $l$ is therefore convergent in four dimensions. After the momentum $l$ has been integrated out, we are left with a function of $k$ whose behavior is $k^0$, up to logarithms, whose integral is thus divergent. Since the degree of divergence in $k$ is $\omega(A+B;k) = 0$, this overall divergence can again be removed by subtracting the zeroth order of the Taylor expansion with respect to the external momenta, i.e.,

$$\mathcal{I}_{A+B+C} = \frac{(-i\lambda)^3}{2} \left\{ G_F^0(k) G_F^0(k-P) \left[ G_F^0(l) G_F^0(l+k+p_3) - G_F^0(l) G_F^0(l) \right] \right.$$
$$\left. - G_F^0(k) G_F^0(k) \left[ G_F^0(l) G_F^0(l+k) - G_F^0(l) G_F^0(l) \right] \right\}. \tag{2.56}$$

After these two successive subtractions, we have obtained a function whose integral on both $k$ and $l$ is finite. Moreover, at each step, we have subtracted only quantities that are polynomial in the external momenta of the corresponding loop (with a degree equal to the superficial degree of divergence of the loop). This recursive procedure for constructing a subtracted integrand is known as *Bogoliubov–Parasiuk–Hepp–Zimmermann renormalization*. Note that this approach performs renormalization without the need for regularization since the subtractions that make the result finite are performed at the level of the integrand.

## 2.4 Perturbative Unitarity

Another important aspect of any quantum field theory is unitarity, which is one of the pillars of quantum mechanics since it is closely related to the "conservation of probability." Although unitarity is manifest in the classical theory since the Lagrangian is Hermitian, one should ensure that this property is preserved at all orders in the perturbative expansion. As we shall see now, a consequence of unitarity is the *optical theorem*, an identity that has a concrete realization in perturbation theory in the form of the so-called *Cutkosky's cutting rules*.

### 2.4.1 Optical Theorem

In canonical quantization, the *in* and *out* states are related by the S-matrix, i.e., the time evolution operator from $-\infty$ to $+\infty$,

$$\langle \alpha_{\text{out}}| \equiv \langle \alpha_{\text{in}}| \, S. \tag{2.57}$$

In a unitary field theory, the S-matrix is a unitary operator *on the space of physical states*,

$$SS^\dagger = S^\dagger S = 1. \tag{2.58}$$

This property means that for a properly normalized initial physical state $|\alpha_{\text{in}}\rangle$, we have

$$\sum_{\text{states } \beta} |\langle \beta_{\text{out}}|\alpha_{\text{in}}\rangle|^2 = 1, \tag{2.59}$$

where the sum includes only physical states. In other words, a physical state $\alpha$ must evolve with probability 1 into other physical states. In general, one subtracts from the S-matrix the identity operator that corresponds to the absence of interactions, by writing $S \equiv 1 + iT$. Therefore, one has

$$1 = (1 + iT)(1 - iT^\dagger) = 1 + iT - iT^\dagger + TT^\dagger, \tag{2.60}$$

or equivalently $-i(T - T^\dagger) = TT^\dagger$. Let us now take the expectation value of this identity in the state $|\alpha_{\text{in}}\rangle$, and insert the identity operator written as a complete sum over physical states between $T$ and $T^\dagger$ on the right-hand side. This leads to

$$-i\langle \alpha_{\text{in}}|T - T^\dagger|\alpha_{\text{in}}\rangle = \sum_{\text{states } \beta} |\langle \alpha_{\text{in}}|T|\beta_{\text{in}}\rangle|^2. \tag{2.61}$$

Equivalently, this identity reads

$$\text{Im}\,\langle \alpha_{\text{in}}|T|\alpha_{\text{in}}\rangle = \frac{1}{2} \sum_{\text{states } \beta} |\langle \alpha_{\text{in}}|T|\beta_{\text{in}}\rangle|^2. \tag{2.62}$$

This identity is known as the *optical theorem*. It implies that the total probability to scatter from the state $\alpha$ to any state $\beta$ (with at least one interaction) equals twice the imaginary part of the forward-scattering amplitude $\alpha \to \alpha$.

## 2.4.2 Cutkosky's Cutting Rules

Eq. (2.62) is valid to all orders in the interactions. But as we shall see, it also manifests itself in some properties of the perturbative expansion. In all the examples of this section, we will consider a scalar field theory with a cubic interaction in $-\frac{\lambda}{3!}\phi^3$. Compared to the field theory with a $\phi^4$ interaction term, the only difference in perturbation theory is that the Feynman graphs have trivalent vertices, instead of tetravalent ones.

**Usual Feynman rules:** First, decompose the free Feynman propagator in two terms, depending on the ordering between the times at the two endpoints,

$$G_F^0(x,y) \equiv \theta(x^0 - y^0)G_{-+}^0(x,y) + \theta(y^0 - x^0)G_{+-}^0(x,y). \tag{2.63}$$

The two-point functions $G_{-+}^0$ and $G_{+-}^0$ are therefore defined as

$$G_{-+}^0(x,y) \equiv \langle 0_{in}|\phi_{in}(x)\phi_{in}(y)|0_{in}\rangle, \quad G_{+-}^0(x,y) \equiv \langle 0_{in}|\phi_{in}(y)\phi_{in}(x)|0_{in}\rangle. \tag{2.64}$$

In order to make the notations more uniform, it is convenient to rename $G_F^0$ by $G_{++}^0$. The usual Feynman rules in coordinate space amount to connecting a vertex at $x$ and a vertex at $y$ by the propagator $G_{++}^0(x,y)$. The coordinate $x$ of each vertex is integrated out over all space-time, and a factor $-i\lambda$ is attached to each vertex. We will call $+$ this type of vertex. Thus, the Feynman rules for calculating scattering amplitudes involve only the $+$ vertex and the $G_{++}^0$ propagator. The integrand corresponding to a given Feynman graph $\mathcal{G}$ is a function $\mathcal{G}(x_1, x_2, \cdots)$ of the coordinates $x_i$ at the vertices.

**Generalized rules:** Let us now introduce another vertex, of type $-$, to which we assign a factor $+i\lambda$ (instead of $-i\lambda$ for the vertex of type $+$). We also introduce another propagator, denoted $G_{--}^0$, defined with a reversed time ordering,

$$G_{--}^0(x,y) \equiv \theta(x^0 - y^0)G_{+-}^0(x,y) + \theta(y^0 - x^0)G_{-+}^0(x,y). \tag{2.65}$$

By combining the $+$ and the $-$ vertices, we may construct generalized diagrams that have a mixture of vertices of the two kinds. Thus, the integrand $\mathcal{G}(x_1, x_2, \cdots)$ is replaced as follows:

$$\mathcal{G}(x_1, x_2, \cdots) \quad \to \quad \mathcal{G}_{\epsilon_1 \epsilon_2 \cdots}(x_1, x_2, \cdots), \tag{2.66}$$

where the indices $\epsilon_i = \pm$ indicate which is the type of the $i$th vertex. The usual Feynman rules thus correspond to the function $\mathcal{G}_{++\cdots}$. These generalized integrands are constructed according to the following rules:

$$+ \text{ vertex:} \quad -i\lambda,$$
$$- \text{ vertex:} \quad +i\lambda,$$
$$\text{Propagator from } \epsilon \text{ to } \epsilon': \quad G_{\epsilon\epsilon'}^0(x,y). \tag{2.67}$$

**Largest time equation:** Let us assume that the $i$th vertex carries the largest time among all the vertices of the graph. Since $x_i^0$ is larger than all the other times, the propagator that connects this vertex to an adjacent vertex of type $\epsilon$ at position $x$ is given by

## 2.4 PERTURBATIVE UNITARITY

$$G^0_{\pm\epsilon}(x_i, x) = G^0_{-\epsilon\,\epsilon}(x_i, x). \tag{2.68}$$

In other words, this propagator depends only on the type $\epsilon$ of the neighboring vertex, but not on the type of the ith vertex. Therefore, we have

$$\mathcal{G}_{\ldots[i+]\ldots}(x_1, x_2, \cdots) + \mathcal{G}_{\ldots[i-]\ldots}(x_1, x_2, \cdots) = 0, \tag{2.69}$$

where the notation $[i^\pm]$ indicates that the ith vertex has type $+$ or $-$ (the types of the vertices not written explicitly are the same in the two terms, but otherwise arbitrary). This identity, known as the *largest time equation*, follows from eq. (2.68) and from the sign change when a vertex changes from $+$ to $-$.

A similar identity also applies to the sum extended to all the possible assignments of the plus and minus indices,

$$\sum_{\{\epsilon_i = \pm\}} \mathcal{G}_{\epsilon_1 \epsilon_2 \ldots}(x_1, x_2, \cdots) = 0. \tag{2.70}$$

This is obtained by pairing the terms and using eq. (2.69). It is crucial to observe that this identity is now valid for any ordering of the times at the vertices of the graph. Therefore, it is also valid in momentum space after a Fourier transform. If we isolate the two terms where all the vertices are of type $+$ or all of type $-$, this also reads

$$\mathcal{G}_{++\cdots} + \mathcal{G}_{--\cdots} = -\sum_{\{\epsilon_i = \pm\}'} \mathcal{G}_{\epsilon_1 \epsilon_2 \ldots}, \tag{2.71}$$

where the symbol $\{\epsilon_i = \pm\}'$ indicates the set of all the vertex assignments, except $++\cdots$ and $--\cdots$.

**Generalized rules in momentum space:** Using eq. (1.136),

$$G^0_{++}(x,y) = \int \frac{d^3p}{(2\pi)^3 2E_p} \left\{ \theta(x^0 - y^0)\, e^{-ip\cdot(x-y)} + \theta(y^0 - x^0)\, e^{+ip\cdot(x-y)} \right\}, \tag{2.72}$$

and comparing with eq. (2.65), we can read off the expressions of $G^0_{-+}$ and $G^0_{+-}$,

$$G^0_{-+}(x,y) = \int \frac{d^3p}{(2\pi)^3 2E_p}\, e^{-ip\cdot(x-y)}, \quad G^0_{+-}(x,y) = \int \frac{d^3p}{(2\pi)^3 2E_p}\, e^{+ip\cdot(x-y)}. \tag{2.73}$$

Likewise, we obtain

$$G^0_{--}(x,y) = \int \frac{d^3p}{(2\pi)^3 2E_p} \left\{ \theta(x^0 - y^0)\, e^{+ip\cdot(x-y)} + \theta(y^0 - x^0)\, e^{-ip\cdot(x-y)} \right\}. \tag{2.74}$$

Note that $G^0_{++} + G^0_{--} = G^0_{-+} + G^0_{+-}$ (see Exercises 2.10 and 2.11). With the following representation for the step function,

$$\theta(x^0 - y^0) = \int \frac{dp_0}{2\pi}\, \frac{-i}{p_0 - i0^+}\, e^{ip_0(x^0 - y^0)}, \tag{2.75}$$

we can derive the momentum space expressions of all the propagators:

$$G^0_{++}(p) = \frac{i}{p^2 - m^2 + i0^+},$$
$$G^0_{--}(p) = \frac{-i}{p^2 - m^2 - i0^+} = [G^0_{++}(p)]^*,$$
$$G^0_{-+}(p) = 2\pi\,\theta(+p_0)\delta(p^2 - m^2),$$
$$G_{+-}(p) = 2\pi\,\theta(-p_0)\delta(p^2 - m^2). \qquad (2.76)$$

Therefore, the momentum space Feynman rules for the $-$ sector are the complex conjugate of those for the $+$ sector, since we also have $+i\lambda = (-i\lambda)^*$. Note that for this assertion to be true, it is crucial that the coupling constant $\lambda$ be real, which is a condition for unitarity.

**Imaginary part of scattering amplitudes:** The Fourier transform of an amputated Feynman graph $\mathcal{G}$ gives a contribution to a transition amplitude (recall the LSZ reduction formula), i.e., a matrix element of the S-matrix, and $\Gamma \equiv i\mathcal{G}$ gives a matrix element of the operator T. Therefore, after Fourier transform, eq. (2.71) becomes

$$\mathrm{Im}\,\Gamma_{++\cdots} = \frac{1}{2} \sum_{\{\epsilon_i = \pm\}'} [i\Gamma]_{\epsilon_1 \epsilon_2 \cdots}. \qquad (2.77)$$

If the graph contains N vertices, there are a priori $2^N - 2$ terms on the right-hand side of this equation. However, this number is considerably reduced if we notice that the $+-$ and $-+$ propagators can carry energy only in one direction because of the factors $\theta(\pm p^0)$. This constraint on energy flow forbids "islands" of vertices of type $+$ surrounded by only type $-$ vertices, or the reverse. In fact, from the LSZ reduction formula (1.104) and the definition (1.137) of the Fourier transformed propagators, we see that the notation $G_{-+}(p)$ implies a momentum p defined as flowing from the $+$ endpoint to the $-$ endpoint:

$$G_{-+}(p) = \underset{+\qquad\;-}{\xrightarrow{p}}. \qquad (2.78)$$

Thus, the proportionality $G_{-+}(p) \propto \theta(p^0)$ indicates that the energy flows from the $+$ endpoint to the $-$ endpoint.

**Cutting rules:** Consider the example of a very simple one-loop two-point function $\Gamma(p)$,

$$-i\Gamma(p) = \xrightarrow{p}\!\!\bigcirc\!\!\text{---}. \qquad (2.79)$$

Because of the constrained energy flow in $G_{-+}, G_{+-}$, if the energy enters into the graph from the left, then the only assignments that mix $+$ and $-$ vertices must divide the graph into two connected subgraphs: a connected part made only of $+$ vertices that comprises the vertex

## 2.4 PERTURBATIVE UNITARITY

where $p^0 > 0$ enters in the graph, and a connected part containing only $-$ vertices comprising the vertex where the energy leaves the graph. For the topology shown in eq. (2.79), there is only one possibility,

$$-i\Gamma_{+-}(p) = \quad \text{[diagram]} \quad , \qquad (2.80)$$

where the vertex of type $-$ is indicated by a circle. The division of the graph into these two subgraphs may be materialized by drawing a line (shown in gray above) through the graph. This line is called a *cut*, and the rules for calculating the value of a graph with a given assignment of $+$ and $-$ vertices are called *Cutkosky's cutting rules*. For instance, in the case of the above example, they lead immediately to the following expression[11] for the imaginary part of $\Gamma_{++}$:

$$\text{Im}\,\Gamma_{++}(p) = \frac{\lambda^2}{2}\frac{1}{2}\int \frac{d^4k}{(2\pi)^4} G_{-+}(k) G_{-+}(p-k), \qquad (2.81)$$

which can be rewritten as

$$\text{Im}\,\Gamma_{++}(p) = \frac{\lambda^2}{4}\int \frac{d^4k_1 d^4k_2}{(2\pi)^8}\,(2\pi)^4\delta(p-k_1-k_2)$$
$$\times\,2\pi\,\theta(k_1^0)\delta(k_1^2-m^2)\,2\pi\,\theta(k_2^0)\delta(k_2^2-m^2). \qquad (2.82)$$

In the right-hand side of this equation, we recognize the square of the transition amplitude $\langle k_1 k_{2\,\text{out}}|p_{\text{in}}\rangle$ (whose value at tree level is simply $\lambda$), integrated over the (symmetrized) accessible phase-space for a two-particle final state. We can therefore view this equation as a perturbative realization of the optical theorem at order $\lambda^2$. Indeed, at this order, the only states $\beta$ that may be included in the sum over final states are two-particle states.[12]

The considerations developed on this example can be generalized to the two-point function at any loop order. We can write

$$\text{Im}\,\Gamma_{++}(p) = \frac{1}{2}\sum_{\text{cuts }\gamma}(i\Gamma_\gamma(p)), \qquad (2.83)$$

where the sum is now limited to a sum over all the possible cuts (with the $+$ vertices on the left of the cut and the $-$ vertices on the right of the cut). As an illustration, let us consider the following two-loop example, for which three cuts are possible:

---

[11] The first factor $1/2$ comes from eq. (2.77), and the second $1/2$ is the symmetry factor of the graph for a scalar loop. In the formula for $\text{Im}\,\Gamma_{++}$, it has the interpretation of the factor that symmetrizes a two-particle final state.

[12] This result is consistent with eq. (1.93) for a decay rate, if we note that the decay rate $\Gamma$ of a particle is related to the imaginary part of the corresponding self-energy by $\Gamma = (\text{Im}\,\Gamma_{++}(p))/E_p$. This can be seen as follows: after resumming the self-energy $\Gamma_{++}(p)$ on the propagator, the imaginary part makes it decay as $G_{++}(x,y) \sim \exp(-(\text{Im}\,\Gamma_{++})|x^0-y^0|/2E_p)$, and the particle density, quadratic in the field operator, decays as the square of the propagator.

$$\text{Im}\,\Gamma_{++}(p) = \quad\text{[diagram]}\quad . \tag{2.84}$$

At this order there are various types of contributions in the right-hand side of eq. (2.62): The central cut corresponds to a three-body final state, while the other two cuts correspond to an interference between the tree level and the one-loop correction to a two-body decay.

### 2.4.3 Schwinger–Keldysh Formalism

The cutting rules that give the imaginary part of Feynman diagrams are also closely related (in fact, formally identical) to an extension of the usual perturbation theory, known as the *Schwinger–Keldysh formalism*, that plays an important role for instance in applications to condensed matter physics. Ordinary perturbation theory is tailored for the calculation of scattering amplitudes that are amenable via the LSZ reduction formulas to the expectation value of time-ordered products of field operators, between the in and out vacuum states, such as $\langle 0_{\text{out}}|T\,\phi(x_1)\phi(x_2)\phi(x_3)\phi(x_4)|0_{\text{in}}\rangle$. Since the *out* Fock states provide a basis for the space of all possible final states, any observable can in principle be reduced to a calculation of such scattering amplitudes. For instance, the expectation value of the number operator that counts the particles of momentum $\mathbf{p}$ in the final state given an initially empty state, $\langle 0_{\text{in}}|a^\dagger_{\mathbf{p},\text{out}} a_{\mathbf{p},\text{out}}|0_{\text{in}}\rangle$, can be written as

$$\langle 0_{\text{in}}|a^\dagger_{\mathbf{p},\text{out}} a_{\mathbf{p},\text{out}}|0_{\text{in}}\rangle = \sum_\beta \langle 0_{\text{in}}|a^\dagger_{\mathbf{p},\text{out}}|\beta_{\text{out}}\rangle\langle \beta_{\text{out}}|a_{\mathbf{p},\text{out}}|0_{\text{in}}\rangle$$
$$= \sum_\beta \langle 0_{\text{in}}|\mathbf{p}+\beta_{\text{out}}\rangle\langle \mathbf{p}+\beta_{\text{out}}|0_{\text{in}}\rangle, \tag{2.85}$$

where $|\mathbf{p}+\beta_{\text{out}}\rangle$ denotes the final state with the content of $\beta$ plus an extra particle of momentum $\mathbf{p}$. Alternatively, one may mimic the derivation of the LSZ reduction to relate this observable to the correlation function $\langle 0_{\text{in}}|\phi(x)\phi(y)|0_{\text{in}}\rangle$, which has no final state sum. If we were able to calculate such a correlation function, we would therefore completely bypass the final state sum present in eq. (2.85).

**Schwinger–Keldysh perturbation theory:** The main feature of $\langle 0_{\text{in}}|\phi(x)\phi(y)|0_{\text{in}}\rangle$ is that the two field operators are not time-ordered. In fact, it is the simplest of a more general family of correlation functions:[13]

$$\langle 0_{\text{in}}|\overline{T}\left(\phi(x_1)\cdots\phi(x_n)\right) T\left(\phi(y_1)\cdots\phi(y_p)\right)|0_{\text{in}}\rangle, \tag{2.86}$$

where $\overline{T}$ denotes the reversed time-ordering (note that the usual time-ordered functions are a special case). As we did in the derivation of ordinary perturbation theory, let us first replace

---

[13] Here, the initial state is the vacuum (i.e., a pure state). In Chapters 17 and 18 we will generalize this formalism to situations where the initial state is a mixed state such as a thermal state or a coherent state.

## 2.4 PERTURBATIVE UNITARITY

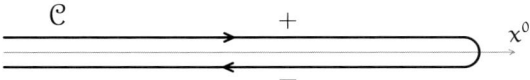

FIGURE 2.3: Time contour in the Schwinger–Keldysh formalism.

each field operator $\phi$ by its counterpart $\phi_{in}$ in the interaction representation, using eq. (1.67). After some rearrangement of the evolution operators, we get

$$\langle 0_{in}|T\,\phi(x_1)\cdots\phi(x_n)\,\overline{T}\phi(y_1)\cdots\phi(y_p)|0_{in}\rangle$$
$$= \langle 0_{in}|\overline{T}\left[\phi_{in}(x_1)\cdots\phi_{in}(x_n)\exp i\int_{-\infty}^{+\infty}d^4x\,\mathcal{L}_I(\phi_{in}(x))\right]$$
$$\times T\left[\phi_{in}(y_1)\cdots\phi_{in}(y_p)\exp i\int_{-\infty}^{+\infty}d^4x\,\mathcal{L}_I(\phi_{in}(x))\right]|0_{in}\rangle. \qquad(2.87)$$

Here, we have exploited the fact that the factor $U(-\infty,+\infty)$ that appears in these manipulations is the anti-time-ordered exponential of the interaction term, in order to write this formula in a more symmetric way. To go further, it is useful to imagine that the time axis is in fact a contour $\mathcal{C}$ made of two branches labeled $+$ and $-$ running parallel to the real axis, as illustrated in Figure 2.3. This contour is oriented with the $+$ branch running in the direction of increasing time, followed by the $-$ branch running in the direction of decreasing time. Then, it is convenient to introduce a *path ordering*, denoted by P and defined as a standard ordering along the contour $\mathcal{C}$. In detail, one has

$$P\,A(x)B(y) = \begin{cases} T\,A(x)B(y) & \text{if} \quad x^0,y^0 \in \mathcal{C}_+, \\ \overline{T}\,A(x)B(y) & \text{if} \quad x^0,y^0 \in \mathcal{C}_-, \\ A(x)B(y) & \text{if} \quad x^0 \in \mathcal{C}_-,\,y^0 \in \mathcal{C}_+, \\ B(y)A(x) & \text{if} \quad x^0 \in \mathcal{C}_+,\,y^0 \in \mathcal{C}_-. \end{cases} \qquad(2.88)$$

One can use this contour ordering to write the previous equations in a much more compact way. In fact, eq. (2.87) can be generalized into

$$\langle 0_{in}|P\,\phi^-(x_1)\cdots\phi^-(x_n)\phi^+(y_1)\cdots\phi^+(y_p)|0_{in}\rangle$$
$$= \langle 0_{in}|P\,\phi_{in}^-(x_1)\cdots\phi_{in}^-(x_n)\phi_{in}^+(y_1)\cdots\phi_{in}^+(y_p)\exp i\int_{\mathcal{C}}d^4x\,\mathcal{L}_I(\phi_{in}(x))|0_{in}\rangle. \qquad(2.89)$$

The differences compared to eq. (2.87) are threefold:

1. A single overall path ordering automatically takes care of both the time ordering and the anti-time ordering contained in the original formula.
2. For this trick to work, one must assume that the fields on the upper and lower branch of the contour $\mathcal{C}$ are independent: we denote them $\phi^+$ and $\phi^-$, respectively.
3. The time integration in the exponential is now running over both branches of the contour $\mathcal{C}$.

The advantage of having introduced this more complicated time contour is that it leads to expressions that are formally identical to those of ordinary perturbation theory, provided one replaces the time ordering by the path ordering and provided one extends the time integration from $\mathbb{R}$ to $\mathcal{C}$. In particular, one can first define a generating functional,

$$Z^{\text{SK}}[j] \equiv \langle 0_{\text{in}} | T \exp i \int_{\mathcal{C}} d^4 x \, j(x) \phi(x) | 0_{\text{in}} \rangle, \tag{2.90}$$

that encodes all the correlators considered in this section, provided the external source j has distinct values $j_+$ and $j_-$ on the two branches of the contour (the superscript SK is used to distinguish this generating functional from the standard one). As in the case of Feynman perturbation theory, one can write this generating functional as

$$Z^{\text{SK}}[j] = \exp\left( i \int_{\mathcal{C}} d^4 x \, \mathcal{L}_{\text{I}}\left( \frac{\delta}{i \delta j(x)} \right) \right) \underbrace{\langle 0_{\text{in}} | T \exp i \int_{\mathcal{C}} d^4 x \, j(x) \phi_{\text{in}}(x) | 0_{\text{in}} \rangle}_{Z_0^{\text{SK}}[j]}, \tag{2.91}$$

with

$$Z_0^{\text{SK}}[j] = \exp\left\{ -\frac{1}{2} \int_{\mathcal{C}} d^4 x \, d^4 y \, j(x) j(y) \, G_e^0(x, y) \right\}$$

$$G_e^0(x, y) \equiv \langle 0_{\text{in}} | P \, \phi_{\text{in}}(x) \phi_{\text{in}}(y) | 0_{\text{in}} \rangle. \tag{2.92}$$

The free propagator $G_e^0$, defined on the contour $\mathcal{C}$, is a natural extension of the Feynman propagator (in particular, it coincides with the Feynman propagator if the two time arguments are on the branch $\mathcal{C}_+$ of the contour). Besides the propagator, the other change to the perturbative expansion in the *Schwinger–Keldysh* formalism is that the time integration at the vertices of a diagram must run over the contour $\mathcal{C}$ instead of the real axis.

The connection with Cutkosky's cutting rules appears when we break down the propagator into four components $G_{\pm\pm}^0(x, y)$, depending on whether the times $x^0, y^0$ are on the upper or lower branch of the contour. An explicit calculation of these free propagators shows that they are in fact identical to those of eq. (2.76). Moreover, the time integration on the contour $\mathcal{C}$ can be split into two terms, the upper branch corresponding to a vertex + (with Feynman rule $-i\lambda$) and the lower branch to a vertex − (with Feynman rule $+i\lambda$, because of the minus sign due to integrating from $+\infty$ to $-\infty$). One may understand why the formalism to calculate the type of observables we started the section with turns out to be identical to Cutkosky's rules by writing them generically as follows:

$$\langle 0_{\text{in}} | \mathcal{O}_{\text{out}} | 0_{\text{in}} \rangle = \sum_{\beta',\beta} \underbrace{\langle 0_{\text{in}} | \beta'_{\text{out}} \rangle}_{\text{only } -} \underbrace{\langle \beta'_{\text{out}} | \mathcal{O}_{\text{out}} | \beta_{\text{out}} \rangle}_{\text{mixed } -+} \underbrace{\langle \beta_{\text{out}} | 0_{\text{in}} \rangle}_{\text{only } +}. \tag{2.93}$$

(The double final state sum over $\beta', \beta$ simplifies into a single one, if the observable is diagonal in the Fock state basis – this is for instance the case with the number operator $a_{\mathbf{p},\text{out}}^\dagger a_{\mathbf{p},\text{out}}$.) The physical meaning of this formula is quite transparent: In order to measure something in the final state, one has to evolve the initial state (here, the vacuum) into some final state, evaluate the observable in that final state (in general the observable may not be diagonal in the basis of final states), and sum over all possible final states. The fact that we need the − rules in the first factor in the right-hand side of eq. (2.93) comes from the fact that $\langle 0_{in}|\beta'_{out}\rangle = (\langle \beta'_{out}|0_{in}\rangle)^*$.

Note that in the Schwinger–Keldysh formalism, the vacuum–vacuum diagrams are simpler than in conventional perturbation theory. Here, one has

$$Z^{SK}[0] = \langle 0_{in}|0_{in}\rangle = 1, \qquad (2.94)$$

which means that all the connected vacuum–vacuum diagrams are zero. This is due to the fact that in this formalism one is calculating correlators that have the in-vacuum on both sides. This property, which follows from eq. (2.70), results from a cancellation between the various ways of assigning the $+$ and $-$ indices to the vertices of a diagram (a vacuum–vacuum diagram with a fixed assignment of $+$ and $-$ vertices is not zero in general) and works for each individual topology.

**Relation between the functionals Z[j] and $Z^{SK}$[j]:** There is a compact functional relation between the generating functional of conventional perturbation theory Z[j], and that of the *Schwinger–Keldysh* formalism:

$$Z^{SK}[j_+, j_-] = \exp\left[\int d^4x d^4y \; G^0_{+-}(x,y) \Box_x \Box_y \frac{\delta^2}{\delta j_+(x) \delta j_-(y)}\right] Z[j_+] Z^*[j_-]. \qquad (2.95)$$

(Here, in order to avoid any confusion, we write explicitly the two components $+$ and $-$ of the source j in the Schwinger–Keldysh generating functional – see Exercise 5.9 for a proof.) Thanks to this formula, one can construct diagrams in the Schwinger–Keldysh formalism by *stitching* an ordinary Feynman diagram and the complex conjugate of another Feynman diagram. In order to prove this relation, it is sufficient to establish it for the free theory, since the interactions are always trivially factorizable (see eqs. (1.124) and (2.91)).

## Exercises

**2.1** Give the symmetry factors of the following graphs:

**\*2.2** Check the formulas listed in eq. (2.33). *Hint: For the integrals that have Lorentz indices, construct the most general tensor that can carry these indices, and determine the prefactor by appropriate contractions. A crucial point of being in D dimensions is that the trace of the metric tensor is $g_{\mu\nu}g^{\mu\nu} = D$.*

**2.3** Consider a scalar field theory with both cubic and quartic interaction terms: $\mathcal{L}_I = -\frac{g}{3!}\phi^3 - \frac{\lambda}{4!}\phi^4$. Derive the identities that relate the numbers of external points, propagators, vertices and loops. What is the dimension of the coupling constant g? Is this theory renormalizable in four dimensions? List the necessary counterterms. What is the main difference compared to a pure $\phi^4$ interaction, and what is the origin of this difference? What would be a reasonable renormalization condition to set the value of the new counterterm?

**2.4** Draw all the one-loop and two-loop graphs for the six-point function in a scalar theory with a $\phi^4$ interaction. Write the corresponding Feynman integrals.

***2.5** Calculate the one-loop integral of the previous exercise:

$$\text{[diagram]} = \int \frac{d^D\ell}{(2\pi)^D} \frac{(-i\lambda)^3 i^3}{(\ell^2 - m^2 + i0^+)((\ell+p)^2 - m^2 + i0^+)((\ell-q)^2 - m^2 + i0^+)}.$$

*Hints:*
- *Rearrange the denominators into a single one using Feynman parametrization, and shift the momentum to obtain a simpler expression.*
- *Perform a Wick rotation to convert it to an Euclidean integral. Does the integral exist in four dimensions? If yes, set $D = 4$ from now on.*
- *Calculate the momentum integral.*

**2.6** Prove to all orders the formula $W[j] = \log Z[j]$ for the generating functional of connected graphs. *Hint: Consider a graph $\mathcal{G}$ that factorizes as $\prod_i \mathcal{C}_i^{n_i}$ where the $\mathcal{C}_i$ are connected graphs. What is the symmetry factor associated with this factorization?*

**2.7** Reproduce the integration by parts applied to eq. (2.27) in dimensional regularization, and with a non-zero mass.

**2.8** Denote $I_n(\Delta)$ the first integral of eq. (2.33). Without calculating it, show that

$$(2n - D)I_n(\Delta) = 2n\Delta I_{n+1}(\Delta).$$

*Hint: Use integration by parts.*

**2.9** Consider the following loop integrals:

$$I_{m,n} \equiv \int \frac{d^D k_E}{(2\pi)^D} \frac{1}{k_E^{2m}(k_E + q)^{2n}}.$$

Using integration by parts, find recurrence relations that relate these integrals at various $m, n$. What is the value of the integral if $m = 0$ or $n = 0$?

**2.10** Show that the Schwinger–Keldysh propagators obey $G^0_{++} + G^0_{--} = G^0_{+-} + G^0_{-+}$. Why is this property in fact true to all orders in the coupling? Consider now a self-energy $\Sigma$, which in this formalism is in fact four functions $\Sigma_{++}, \Sigma_{--}, \Sigma_{+-}, \Sigma_{-+}$. How are these functions related? *Hint: For the last question, recall that a self-energy inserted between a pair of free propagators is a contribution to the dressed propagator.*

***2.11** Check that $G_{++} - G_{+-}$ is the retarded propagator. *Hints: This may be seen explicitly for the free propagators in momentum space. For an all-orders proof, one needs the definition of the various propagators as expectation values of products of fields.*

# CHAPTER 3

# Quantum Electrodynamics

The scalar field theory that we have considered as an example in the first two chapters has a great didactical value, but is rather limited in its applications to concrete physical systems that may be studied experimentally. Indeed, most of the presently known elementary particles carry some conserved charges, and/or have a non-zero intrinsic angular momentum (spin), both of which require fields that have additional structure. *Quantum electrodynamics*, the quantized theory of the interactions between electrically charged particles, is a very important example of such a rich theory. First, since all the elementary charged particles in nature are spin-1/2 fermions (e.g., the electron), we will need to extend the concept of quantum field in order to describe them. The main consequence of enforcing Lorentz covariance on them will be that any such particle must have an anti-particle of identical mass but opposite charge. In addition to the charge carriers, the electromagnetic field itself must be quantized, and its elementary excitations may be viewed as a new (spin-1) particle, the photon.

## 3.1 Spin-1/2 Fields

### 3.1.1 Two-Dimensional Representation of the Rotation Group

In ordinary quantum mechanics, the spin $s$ is related to the dimension $n$ of representations of the rotation group $SO(3)$ by $n = 2s + 1$. Thus, spin-1/2 corresponds to representations of dimension 2. Such a representation is based on the (Hermitian) Pauli matrices,

$$\sigma^1 = \begin{pmatrix} 0 & 1 \\ 1 & 0 \end{pmatrix}, \quad \sigma^2 = \begin{pmatrix} 0 & -i \\ i & 0 \end{pmatrix}, \quad \sigma^3 = \begin{pmatrix} 1 & 0 \\ 0 & -1 \end{pmatrix}, \tag{3.1}$$

from which we can construct the following unitary $2 \times 2$ matrices:

$$U \equiv \exp\left(-\tfrac{i}{2}\theta^i \sigma^i\right). \tag{3.2}$$

That the Pauli matrices divided by two are generators of the Lie algebra[1] of rotations can be seen from

$$[J^i, J^j] = i\,\epsilon_{ijk}\, J^k \quad \text{with} \quad J^i \equiv \frac{\sigma^i}{2}. \tag{3.3}$$

### 3.1.2 Spinor Representation of the Lorentz Group

This idea can be extended to quantum field theory in order to encompass all the Lorentz transformations rather than just the spatial rotations. We are therefore seeking a dimension-2 representation of the commutation relations (1.10). First, let us assume that we know a set of four $n \times n$ matrices $\gamma^\mu$ that satisfy the following anti-commutation relation:

$$\{\gamma^\mu, \gamma^\nu\} = 2\,g^{\mu\nu}\, 1_{n \times n}. \tag{3.4}$$

Such matrices are called *Dirac matrices*. From these matrices, it is easy to check (see Exercise 2.2) that the matrices

$$M^{\mu\nu} \equiv \frac{i}{4}[\gamma^\mu, \gamma^\nu] \tag{3.5}$$

form an $n$-dimensional representation of the Lorentz algebra. However, an exhaustive search indicates that the smallest matrices that fulfill eq. (3.4) in four space-time dimensions are $4 \times 4$ (see Exercise 3.1). Several unitarily equivalent choices exist for these matrices. A possible representation (known as the Weyl – or chiral – representation) is the following:[2]

$$\gamma^0 \equiv \begin{pmatrix} 0 & 1 \\ 1 & 0 \end{pmatrix}, \quad \gamma^i \equiv \begin{pmatrix} 0 & \sigma^i \\ -\sigma^i & 0 \end{pmatrix}. \tag{3.6}$$

In this representation, the generators for the boosts and for the rotations are

$$M^{0i} = -\frac{i}{2}\begin{pmatrix} \sigma^i & 0 \\ 0 & -\sigma^i \end{pmatrix}, \quad M^{ij} = \frac{1}{2}\epsilon_{ijk}\begin{pmatrix} \sigma^k & 0 \\ 0 & \sigma^k \end{pmatrix}. \tag{3.7}$$

Given a Lorentz transformation $\Lambda$ defined by the parameters $\omega_{\mu\nu}$, let us define

$$U_{1/2}(\Lambda) \equiv \exp\left(-\tfrac{i}{2}\omega_{\mu\nu} M^{\mu\nu}\right). \tag{3.8}$$

A *Dirac spinor* is defined to be a four-component field $\psi(x)$ that transforms as follows:

$$\psi(x) \quad \to \quad U_{1/2}(\Lambda)\,\psi(\Lambda^{-1}x). \tag{3.9}$$

---

[1]The notions of *Lie group* and *Lie algebra* will be studied more thoroughly in Chapter 7.
[2]Although it is sometimes convenient to have an explicit representation of the Dirac matrices, most manipulations only rely on the fact that they obey the anti-commutation relations (3.4).

In other words, the matrix $U_{1/2}$ defines how the four components of this field transform under a Lorentz transformation (since these four components mix, $\psi(x)$ is not the juxtaposition of four scalar fields). The fact that the lowest dimension for the Dirac matrices is four suggests that the spinor $\psi(x)$ may describe two spin-1/2 particles instead of just one: a particle and its *anti-particle*, that are distinct from each other.

### 3.1.3 Dirac Equation and Lagrangian

Let us now determine an equation of motion obeyed by this field, such that it is invariant under Lorentz transformations. Since the $M^{\mu\nu}$s act only on the Dirac indices, a trivial answer is the Klein–Gordon equation, $(\Box_x + m^2)\psi(x) = 0$. But there is in fact a stronger equation that remains invariant when $\psi$ is transformed according to eq. (3.9). Notice first that

$$U_{1/2}^{-1}(\Lambda)\gamma^\mu U_{1/2}(\Lambda) = \Lambda^\mu{}_\nu \gamma^\nu. \tag{3.10}$$

This equation indicates that rotating the Dirac indices of $\gamma^\mu$ with $U_{1/2}$ is equivalent to transforming the $\mu$ index as one would do for a normal 4-vector. Using this identity, we can check that under the same Lorentz transformation we have

$$(i\gamma^\mu \partial_\mu - m)\psi(x) \quad \to \quad U_{1/2}(\Lambda)(i\gamma^\mu \partial'_\mu - m)\psi(\Lambda^{-1}x), \tag{3.11}$$

where the prime indicates a derivative with respect to $\Lambda^{-1}x$. Therefore, the *Dirac equation*,

$$(i\gamma^\mu \partial_\mu - m)\psi(x) = 0, \tag{3.12}$$

is Lorentz invariant. This equation implies the Klein–Gordon equation (to see it, apply the operator $i\gamma^\mu \partial_\mu + m$ on the left), and is therefore stronger. In fact, each of the four components of the spinor $\psi(x)$ obeys individually the Klein–Gordon equation, but in addition the Dirac equation implies certain relationships among the components.

The Dirac matrices are not Hermitian. Instead, they can be chosen to satisfy $(\gamma^\mu)^\dagger = \gamma^0 \gamma^\mu \gamma^0$. Therefore, $U_{1/2}(\Lambda)$ is not unitary, and we have instead

$$U_{1/2}^\dagger(\Lambda) = \exp\left(\tfrac{i}{2}\omega_{\mu\nu}(M^{\mu\nu})^\dagger\right) = \gamma^0 \exp\left(\tfrac{i}{2}\omega_{\mu\nu}M^{\mu\nu}\right)\gamma^0 = \gamma^0 U_{1/2}^{-1}(\Lambda)\gamma^0. \tag{3.13}$$

Because of this, the simplest Lorentz scalar bilinear combination of $\psi$s is $\psi^\dagger\gamma^0\psi$ (instead of the naive $\psi^\dagger\psi$). It is common to denote $\overline{\psi} \equiv \psi^\dagger\gamma^0$. From this, we conclude that the Lorentz scalar Lagrangian density that leads to the Dirac equation reads:

$$\mathcal{L} = \overline{\psi}\left(i\gamma^\mu\partial_\mu - m\right)\psi(x). \tag{3.14}$$

Note that since the Lagrangian has a mass dimension equal to four (in four space-time dimensions), $\psi(x)$ must have a mass dimension 3/2.

### 3.1.4 Basis of Free Spinors

Before quantizing the spinor field in a similar fashion as the scalar field, we need to find the plane wave solutions of the Dirac equation. Since the Dirac equation is a first-order equation, the positive and negative frequency solutions are proportional to different spinor coefficients:

$$\psi(x) = u(\mathbf{p})\, e^{-i p \cdot x} \quad \text{with} \quad (p_\mu \gamma^\mu - m)\, u(\mathbf{p}) = 0,$$
$$\psi(x) = v(\mathbf{p})\, e^{+i p \cdot x} \quad \text{with} \quad (p_\mu \gamma^\mu + m)\, v(\mathbf{p}) = 0. \qquad (3.15)$$

The solutions $u(\mathbf{p})$ and $v(\mathbf{p})$ each form a two-dimensional linear space, and it is customary to denote a basis by $u_s(\mathbf{p})$ and $v_s(\mathbf{p})$ (the index s, which takes two values $s = \pm$, is interpreted as the two spin states of a spin-1/2 particle). A convenient normalization of these spinors is the one that leads to

$$\bar{u}_r(\mathbf{p}) u_s(\mathbf{p}) = 2m\delta_{rs}, \quad u_r^\dagger(\mathbf{p}) u_s(\mathbf{p}) = 2E_\mathbf{p} \delta_{rs},$$
$$\bar{v}_r(\mathbf{p}) v_s(\mathbf{p}) = -2m\delta_{rs}, \quad v_r^\dagger(\mathbf{p}) v_s(\mathbf{p}) = 2E_\mathbf{p} \delta_{rs},$$
$$\bar{u}_r(\mathbf{p}) v_s(\mathbf{p}) = \bar{v}_r(\mathbf{p}) u_s(\mathbf{p}) = 0, \quad u_r^\dagger(\mathbf{p}) v_s(-\mathbf{p}) = v_r^\dagger(\mathbf{p}) u_s(-\mathbf{p}) = 0. \qquad (3.16)$$

When summing over the spin states, we have

$$\sum_{s=\pm} u_s(\mathbf{p}) \bar{u}_s(\mathbf{p}) = \slashed{p} + m, \quad \sum_{s=\pm} v_s(\mathbf{p}) \bar{v}_s(\mathbf{p}) = \slashed{p} - m, \qquad (3.17)$$

where we have introduced the notation $\slashed{p} \equiv p_\mu \gamma^\mu$. With this normalization, the spinors $u_s(\mathbf{p}), v_s(\mathbf{p})$ have a mass dimension equal to $1/2$.

The most general free (i.e., solution of the vacuum Dirac equation) spinor $\psi(x)$ is a linear combination of the positive and negative solutions obtained above,

$$\psi(x) \equiv \sum_{s=\pm} \int \frac{d^3\mathbf{p}}{(2\pi)^3 2E_\mathbf{p}} \left\{ a_{s\mathbf{p}}^\dagger v_s(\mathbf{p}) e^{+i p \cdot x} + b_{s\mathbf{p}} u_s(\mathbf{p}) e^{-i p \cdot x} \right\}. \qquad (3.18)$$

Note that, since $u_s(\mathbf{p})$ and $v_s(\mathbf{p})$ are not related by Hermitian conjugation, there is no reason why the creation and annihilation operators ($b_{s\mathbf{p}}$ and $a_{s\mathbf{p}}^\dagger$) weighting them should be mutual conjugates. Therefore, the spinor $\psi(x)$ encodes two types of particles, created and destroyed respectively by $a_{s\mathbf{p}}^\dagger, a_{s\mathbf{p}}$ and $b_{s\mathbf{p}}^\dagger, b_{s\mathbf{p}}$. Thanks to our normalization of the spinors $u_s(\mathbf{p}), v_s(\mathbf{p})$, the creation and annihilation operators introduced in $\psi(x)$ have the same dimension as the scalar ones.

### 3.1.5 Canonical Quantization

From the Lagrangian (3.14), the momentum canonically conjugated to $\psi(x)$ is $\Pi(x) = i\psi^\dagger(x)$, and the Hamiltonian is

$$\mathcal{H} = \int d^3\mathbf{x}\, \bar{\psi}(x)(m - i\gamma^i \partial_i)\psi(x). \qquad (3.19)$$

A naive generalization of the canonical commutation relation used for scalar fields would amount to defining

$$[a_{r\mathbf{p}}, a_{s\mathbf{q}}^\dagger] = (2\pi)^3\, 2E_\mathbf{p}\, \delta_{rs} \delta(\mathbf{p} - \mathbf{q}), \quad [b_{r\mathbf{p}}, b_{s\mathbf{q}}^\dagger] = (2\pi)^3\, 2E_\mathbf{p}\, \delta_{rs} \delta(\mathbf{p} - \mathbf{q}), \qquad (3.20)$$

## 3.1 SPIN-1/2 FIELDS

with all other commutators zero. However, these commutation relations lead to

$$\mathcal{H} = \sum_{s=\pm} \int \frac{d^3p}{(2\pi)^3 2E_p} \, E_p \left( b_{sp}^\dagger b_{sp} - a_{sp}^\dagger a_{sp} - 4E_p V \right), \tag{3.21}$$

where $V$ is the volume of the system (this comes from $V = (2\pi)^3 \delta(\mathbf{p} - \mathbf{p}))$. Ignoring the last term that does not depend on creation or annihilation operators, the first two terms are the number operators counting the two kinds of particles encoded in $\psi$. However, this would-be Hamiltonian has the unpleasant feature that one of these number operators appears with a negative sign in front. Therefore, if correct, this would imply that the energy of the system decreases when more particles of this kind are added to it. Another bad consequence is that this Hamiltonian has no ground state, which casts serious doubts on the stability of the system it would describe.

If we trace back the origin of the term $-a_{sp}^\dagger a_{sp} - 4E_p V$, we see it was originally a term $-a_{sp} a_{sp}^\dagger$, before we used the commutation relation (3.20) to swap the creation and annihilation operators. This suggests a way of solving this problem: If we use *anti-commutation relations* instead,

$$\{a_{rp}, a_{sq}^\dagger\} = (2\pi)^3 \, 2E_p \, \delta_{rs} \delta(\mathbf{p} - \mathbf{q}), \quad \{b_{rp}, b_{sq}^\dagger\} = (2\pi)^3 \, 2E_p \, \delta_{rs} \delta(\mathbf{p} - \mathbf{q}) \tag{3.22}$$

(we denote $\{A, B\} \equiv AB + BA$), then the Hamiltonian would instead be

$$\mathcal{H} = \sum_{s=\pm} \int \frac{d^3p}{(2\pi)^3 2E_p} \, E_p \left( b_{sp}^\dagger b_{sp} + a_{sp}^\dagger a_{sp} - 4E_p V \right), \tag{3.23}$$

(see Exercise 3.6). Now the energy is perfectly bounded from below, since it grows when particles of any kind are added into the system. The anti-commutation relations (3.22) imply that the square of creation operators is zero, which means that it is not possible to have two particles with the same momentum and spin in a quantum state. This is nothing but the *Pauli exclusion principle*. This is the simplest example of the *spin-statistics theorem*, which states that half-integer spin particles must obey Fermi statistics. Given the anti-commutation relations obeyed by the creation and annihilation operators, the spinor and its conjugate momentum satisfy the following equal time anti-commutation relation:

$$\{\psi_a(x), \psi_b^\dagger(y)\}_{x^0=y^0} = \delta_{ab} \, \delta(\mathbf{x} - \mathbf{y}). \tag{3.24}$$

(For clarity, we have indicated explicitly the Dirac indices $a, b$ that label the four components of the spinors.) Equivalently, we may write it as

$$\{\psi_a(x), \overline{\psi}_b(y)\}_{x^0=y^0} = \gamma^0_{ab} \, \delta(\mathbf{x} - \mathbf{y}). \tag{3.25}$$

(Here, the indices are somewhat superfluous because the presence of $\gamma^0$ on the right-hand side makes it clear that the result is a $4 \times 4$ matrix.)

## 3.1.6 LSZ Reduction Formula for Spin-1/2

The Lehmann–Symanzik–Zimmermann (LSZ) reduction formula for scattering amplitudes with fermions and/or anti-fermions in the initial and final states reads

$$\langle \underbrace{q_r \bar{q}_{\bar{r}} \cdots}_{n\text{ particles}} {}_{\text{out}} | \underbrace{p_s \bar{p}_{\bar{s}} \cdots}_{m\text{ particles}} {}_{\text{in}} \rangle \doteq \left( \frac{i}{Z^{1/2}} \right)^{m+n} \int d^4x\, d^4\bar{x}\, d^4y\, d^4\bar{y} \cdots$$

$$\times \cdots \left[ e^{-i\bar{p}\cdot\bar{x}} \bar{v}_{\bar{s}}(\bar{p})(-i\vec{\partial}_{\bar{x}} + m) \right] \left[ e^{+i q\cdot y} \bar{u}_r(q)(-i\vec{\partial}_y + m) \right]$$

$$\times \langle 0_{\text{out}} | T \psi(y)\psi(\bar{x}) \cdots \bar{\psi}(\bar{y})\bar{\psi}(x) | 0_{\text{in}} \rangle$$

$$\times \left[ (i\overleftarrow{\partial}_x + m) u_s(p) e^{-i p\cdot x} \right] \left[ (i\overleftarrow{\partial}_{\bar{y}} + m) v_{\bar{r}}(\bar{q}) e^{+i\bar{q}\cdot\bar{y}} \right] \cdots , \tag{3.26}$$

where we give examples for fermions and anti-fermions (the latter are indicated by a bar over the momentum and spin), both for the initial and final states. The derivation closely mimics the scalar case (see Exercise 3.7 for an example), and uses the following inversion formulas that give the creation/annihilation operators in terms of the field $\psi$:

$$a_{sp}^{\dagger} = \int d^3x\, e^{-i p\cdot x} v_s^{\dagger}(p)\psi(x), \quad b_{sp} = \int d^3x\, e^{i p\cdot x} u_s^{\dagger}(p)\psi(x). \tag{3.27}$$

A time-ordering operator is necessary in eq. (3.26), for the same reason as in the scalar case. We have also anticipated the need to include field renormalization factors $Z^{-1/2}$ in the reduction formula. Besides the requirement that the external lines of the Feynman graphs should be amputated, this formula leads to the following prescriptions for the amputated external fermionic lines:

In-fermion : $= u(p)$, In-anti-fermion : $= \bar{v}(p)$,

Out-fermion : $= \bar{u}(p)$, Out-anti-fermion : $= v(p)$.

Note that when writing the expression corresponding to a given Feynman graph, the fermion lines it contains must be read in the direction opposite to the arrow carried by the lines.

## 3.1.7 Free Spin-1/2 Propagator

Since it is quadratic in the spinors, the Dirac Lagrangian (3.14) describes non-interacting fermions, which makes the field expectation values that appear in the above LSZ formula trivial: There must be an equal number of $\psi$s and $\bar{\psi}$s, and the result of the expectation value of the T-product factorizes into products of the following free propagator:

$$S^0_{F\,ab}(x,y) \equiv \langle 0_{\text{in}} | \underbrace{\theta(x^0 - y^0)\psi_{\text{in},a}(x)\bar{\psi}_{\text{in},b}(y) - \theta(y^0 - x^0)\bar{\psi}_{\text{in},b}(y)\psi_{\text{in},a}(x)}_{T(\psi_{\text{in},a}(x)\bar{\psi}_{\text{in},b}(y))} | 0_{\text{in}} \rangle.$$

$$\tag{3.28}$$

## 3.1 SPIN-1/2 FIELDS

Using the Fourier decomposition of $\psi_{in}$, we get

$$\langle 0_{in}|\psi_{in,a}(x)\overline{\psi}_{in,b}(y)|0_{in}\rangle = (i\slashed{\partial}_x + m)_{ab} \int \frac{d^3\mathbf{p}}{(2\pi)^3 2E_\mathbf{p}} \, e^{-ip\cdot(x-y)},$$

$$\langle 0_{in}|\overline{\psi}_{in,b}(y)\psi_{in,a}(x)|0_{in}\rangle = -(i\slashed{\partial}_x + m)_{ab} \int \frac{d^3\mathbf{p}}{(2\pi)^3 2E_\mathbf{p}} \, e^{ip\cdot(x-y)}, \quad (3.29)$$

and the fermion free propagator is related to the scalar one by

$$S_F^0(x,y) = (i\slashed{\partial}_x + m) \, G_F^0(x,y). \quad (3.30)$$

Since $(i\slashed{\partial} - m)(i\slashed{\partial} + m) = -(\Box + m^2)$, this formula makes obvious the fact that $S_F^0(x,y)$ is a Green's function of the Dirac operator $i\slashed{\partial}_x - m$ since $G_F^0(x,y)$ is a Green's function of $\Box_x + m^2$. In momentum space, the fermion propagator reads

$$S_F^0(p) = \xrightarrow{p} = \frac{i}{\slashed{p} - m + i0^+} = \frac{i(\slashed{p} + m)}{p^2 - m^2 + i0^+}. \quad (3.31)$$

Now, we can explain the minus sign introduced in eq. (3.28). The derivation shows that this sign is necessary in order to obtain the relationship (3.30), and therefore a Green's function of the Dirac operator. Moreover, without this minus sign, the fermion propagator would not be Lorentz invariant. First note that both parts of eq. (3.29) are Lorentz invariant, but the step functions by which they are multiplied in eq. (3.28) are not in general (except for a time-like interval $(x-y)^2 > 0$, for which the sign of $x^0 - y^0$ is Lorentz invariant). When the interval is space-like, $(x-y)^2 < 0$, the sign of $x^0 - y^0$ is not Lorentz invariant, but with the minus sign in the T-product, the two terms seamlessly combine into a Lorentz invariant formula thanks to the fact that $\psi(x)$ and $\overline{\psi}(y)$ anti-commute for space-like intervals.

In order to obtain expectation values of T-products with more than one $\psi$ and one $\overline{\psi}$, we can introduce the following generating functional:

$$Z_0[\overline{\eta}, \eta] \equiv \langle 0_{in}|T \exp i \int d^4x \Big(\overline{\eta}(x)\psi_{in}(x) + \overline{\psi}_{in}(x)\eta(x)\Big)|0_{in}\rangle. \quad (3.32)$$

In order to be able to pull down either a $\psi$ or a $\overline{\psi}$, we need to couple them to independent fictitious sources $\eta, \overline{\eta}$. Moreover, in order to simplify the expressions, it is useful to assume that these sources are anti-commuting quantities called *Grassmann numbers* (see Chapter 6 for a detailed discussion of their properties). Here, it is sufficient to know that they anti-commute among themselves, and with the spinors. Since we are so far considering a theory of free fermions, this generating functional is simply the exponential of the quadratic term in the expansion in powers of $\eta, \overline{\eta}$:

$$Z_0[\overline{\eta}, \eta] = \exp\left(-\int d^4x d^4y \, \overline{\eta}(x) \, S_F^0(x,y) \, \eta(y)\right). \quad (3.33)$$

Alternatively, this formula can be obtained from eq. (3.32) by a method similar to the scalar case, using the Baker–Campbell–Hausdorff formula.[3]

---

[3] For this, note that $[\overline{\eta}\psi, \overline{\psi}\eta] = \overline{\eta}\{\psi, \overline{\psi}\}\eta$, thanks to the anti-commuting nature of the sources.

## 3.2 Spin-1 Fields

### 3.2.1 Classical Electrodynamics

A prime example of a spin-1 particle is the photon, the particle associated to the quantized electromagnetic field. In *classical electrodynamics*, the electric field **E** and magnetic field **B** obey *Maxwell's equations*,

$$\nabla \cdot \mathbf{E} = \rho, \quad \nabla \times \mathbf{B} - \partial_t \mathbf{E} = \mathbf{J}, \quad \nabla \times \mathbf{E} + \partial_t \mathbf{B} = 0, \quad \nabla \cdot \mathbf{B} = 0, \tag{3.34}$$

written here in terms of charge density $\rho$ and current **J**. The local conservation of electrical charge implies the continuity equation $\partial_t \rho + \nabla \cdot \mathbf{J} = 0$. The last two Maxwell's equations are automatically satisfied if we write the **E**, **B** fields in terms of potentials $V$ and **A**,

$$\mathbf{E} \equiv \partial_t \mathbf{A} + \nabla V, \quad \mathbf{B} \equiv -\nabla \times \mathbf{A}. \tag{3.35}$$

This representation is not unique, since **E** and **B** are unchanged if we transform the potentials as follows:

$$V \to V + \partial_t \chi, \quad \mathbf{A} \to \mathbf{A} - \nabla \chi, \tag{3.36}$$

where $\chi$ is an arbitrary function of space and time. Eq. (3.36) is called a(n) (Abelian) *gauge transformation*. Quantities that do not change under eq. (3.36) are said to be *gauge invariant*. In particular, the electrical and magnetic fields are gauge invariant.

### 3.2.2 Classical Electrodynamics in Lorentz Covariant Form

In order to highlight the properties of Maxwell's equations under Lorentz transformations, let us first rewrite them in covariant form. Introduce a 4-vector $A^\mu$ and a rank-2 tensor $F^{\mu\nu}$,

$$A^\mu \equiv (V, \mathbf{A}), \quad F^{\mu\nu} \equiv \partial^\mu A^\nu - \partial^\nu A^\mu. \tag{3.37}$$

($F^{\mu\nu}$ is called the *field strength*.) Recalling that $\partial^\mu = (\partial_t, -\nabla)$, gauge transformations take the following form

$$A^\mu \to A^\mu + \partial^\mu \chi, \tag{3.38}$$

and $F^{\mu\nu}$ is gauge invariant. Moreover, we see that

$$E^i = F^{0i}, \quad B^i = \tfrac{1}{2} \epsilon_{ijk} F^{jk}. \tag{3.39}$$

If we also encapsulate $\rho$ and **J** in a 4-vector, $J^\mu \equiv (\rho, \mathbf{J})$, the first two Maxwell's equations and the continuity equation read

$$\partial_\mu F^{\mu\nu} = -J^\nu, \quad \partial_\mu J^\mu = 0. \tag{3.40}$$

The last two Maxwell's equations become $\epsilon_{\mu\nu\rho\sigma} \partial^\nu F^{\rho\sigma} = 0$, which is automatically satisfied thanks to the antisymmetry of $F^{\mu\nu}$.

## 3.2 SPIN-1 FIELDS

A Lorentz scalar Lagrangian density whose Euler–Lagrange equations of motion are the Maxwell's equations is

$$\mathcal{L} \equiv -\frac{1}{4} F_{\mu\nu} F^{\mu\nu} + J^\mu A_\mu. \tag{3.41}$$

Because of the term $J^\mu A_\mu$ that couples the potential to the sources, this Lagrangian density is not gauge invariant, but the action (integral of $\mathcal{L}$ over all space-time) is, provided that the current is conserved (i.e., satisfies the continuity equation). Indeed, we have

$$\int d^4x\, J^\mu A_\mu \to \int d^4x\, J^\mu (A_\mu + \partial_\mu \chi) = \int d^4x\, J^\mu A_\mu - \int d^4x\, \chi \underbrace{\partial_\mu J^\mu}_{0} + \text{boundary term}. \tag{3.42}$$

(The boundary term is zero if we assume that there are no sources at infinity.)

### 3.2.3 Canonical Quantization in Coulomb Gauge

Although it leads to Maxwell's equations, the above Lagrangian has an unusual property, related to gauge invariance: The conjugate momentum of the potential $A^0$ is identically zero:

$$\Pi^0(x) \equiv \frac{\partial \mathcal{L}}{\partial(\partial^0 A^0(x))} = 0. \tag{3.43}$$

Therefore, we cannot quantize electrodynamics simply by promoting the Poisson bracket between $A^0$ and its conjugate momentum to a commutator. However, this problem is not specific to quantum mechanics: The very same issue arises when trying to formulate classical electrodynamics in Hamiltonian form. The resolution of this problem is to fix the gauge, i.e., to impose an extra condition on the potential $A^\mu$ such that a unique $A^\mu$ corresponds to given $\mathbf{E}$ and $\mathbf{B}$ fields. Possible *gauge conditions* are

$$\begin{aligned}
&\text{Axial gauge:} && n^\mu A_\mu = 0 && (n^\mu \text{ is a fixed 4-vector}), \\
&\text{Lorenz gauge:} && \partial^\mu A_\mu = 0, \\
&\text{Coulomb gauge:} && \boldsymbol{\nabla} \cdot \mathbf{A} = 0.
\end{aligned} \tag{3.44}$$

Let us illustrate this procedure in Coulomb gauge.[4] First, let us decompose the vector potential $\mathbf{A}^i$ into longitudinal and transverse components,

$$\mathbf{A}^i = \mathbf{A}^i_\parallel + \mathbf{A}^i_\perp, \quad \text{with } \mathbf{A}^i_\parallel \equiv \frac{\partial^i \partial^j}{\partial^2} \mathbf{A}^j, \quad \mathbf{A}^i_\perp \equiv \left(\delta^{ij} - \frac{\partial^i \partial^j}{\partial^2}\right) \mathbf{A}^j. \tag{3.45}$$

---

[4]One may start from another gauge condition, and follow a similar line of reasoning in order to derive a quantized theory of the photon field in another gauge. However, as we shall see later, we can make the gauge fixing much more straightforward by using functional quantization.

The Coulomb gauge condition is equivalent to $\mathbf{A}_\parallel^i = 0$. The remaining components of $A^\mu$ are therefore $A^0$ and the two components of $\mathbf{A}_\perp^i$, in terms of which the Lagrangian reads

$$\mathcal{L} = \frac{1}{2}(\partial_t \mathbf{A}_\perp^i)(\partial_t \mathbf{A}_\perp^i) - \frac{1}{2}(\partial_j \mathbf{A}_\perp^i)(\partial_j \mathbf{A}_\perp^i) + \frac{1}{2}(\partial_i A^0)(\partial_i A^0)$$
$$+ \underline{(\partial_t \mathbf{A}_\perp^i)(\partial_i A^0)} + \underline{\frac{1}{2}(\partial_i \mathbf{A}_\perp^j)(\partial_j \mathbf{A}_\perp^i)} + J^0 A^0 - J^i \mathbf{A}_\perp^i. \tag{3.46}$$

The two underlined terms will vanish in the action, after an integration by parts (thanks to the transversality of $\mathbf{A}_\perp^i$). The Euler–Lagrange equation for the field $A^0$ is $\partial^2 A^0 = J^0$, i.e., the Poisson equation with source term $J^0$. Note that this equation has no time derivative, and is thus a constraint rather than a dynamical equation. Therefore, $A^0$ reflects instantaneously the changes of the charge density $J^0$ (this does not violate causality since $A^0$ is not an observable – only $\mathbf{E}$ and $\mathbf{B}$ are[5]). Ignoring all the terms that vanish in the action upon integration by parts, we may thus rewrite the Lagrangian as

$$\mathcal{L} = \frac{1}{2}(\partial_t \mathbf{A}_\perp^i)(\partial_t \mathbf{A}_\perp^i) - \frac{1}{2}(\partial_j \mathbf{A}_\perp^i)(\partial_j \mathbf{A}_\perp^i) + \frac{1}{2}(\partial_i A^0)(\partial_i A^0) + J^0 A^0 - J^i \mathbf{A}_\perp^i, \tag{3.47}$$

and obtain the following *Euler–Lagrange* equation of motion for the field $\mathbf{A}_\perp^i$,

$$\Box \mathbf{A}_\perp^i = -\left(\delta^{ij} - \frac{\partial^i \partial^j}{\partial^2}\right) J^j, \tag{3.48}$$

i.e., a massless *Klein–Gordon* equation with the transverse projection of the charge current as a source term.

In this form, electrodynamics has no redundant degrees of freedom, and can now be quantized in the vacuum ($J^0 = J^i = 0$) in the canonical way. First, we define the momentum conjugated to $\mathbf{A}_\perp^i$,

$$\Pi_\perp^i(x) \equiv \frac{\delta \mathcal{L}}{\delta \partial_t \mathbf{A}_\perp^i(x)} = \partial_t \mathbf{A}_\perp^i(x). \tag{3.49}$$

Then, we promote $\mathbf{A}_\perp^i$ and $\Pi_\perp^i$ to quantum operators, and we impose on them the following canonical equal-time commutation relations,

$$\left[\mathbf{A}_\perp^i(x), \Pi_\perp^j(y)\right]_{x^0=y^0} = i\left(\delta^{ij} - \frac{\partial^i \partial^j}{\partial^2}\right)\delta(\mathbf{x}-\mathbf{y}),$$
$$\left[\mathbf{A}_\perp^i(x), \mathbf{A}_\perp^j(y)\right]_{x^0=y^0} = \left[\Pi_\perp^i(x), \Pi_\perp^j(y)\right]_{x^0=y^0} = 0. \tag{3.50}$$

(In the first of these relations, the transverse projector on the right-hand side follows from the fact that $\mathbf{A}_\perp^i$ and $\Pi_\perp^j$ are both transverse.) These commutation relations can be realized by decomposing $\mathbf{A}_\perp^i$ on a basis of plane wave solutions of the Klein–Gordon equation,

---

[5]For instance, we have $\mathbf{E}^i = \partial^0 \mathbf{A}_\perp^i - \frac{\partial^i}{\partial^2} J^0$, and div $\mathbf{E} = -\partial^i \mathbf{E}^i = J^0$, in agreement with Maxwell's equations despite the supraluminal dependence of $A^0$ on $J^0$. This constraint on $A^0$ is equivalent to Gauss' law.

## 3.2 SPIN-1 FIELDS

$$\mathbf{A}_\perp^i(x) \equiv \sum_{\lambda=1,2} \int \frac{d^3\mathbf{p}}{(2\pi)^3 2|\mathbf{p}|} \left[ \epsilon_\lambda^i(\mathbf{p})\, a_{\lambda\mathbf{p}}^\dagger\, e^{+ip\cdot x} + \epsilon_\lambda^{i*}(\mathbf{p})\, a_{\lambda\mathbf{p}}\, e^{-ip\cdot x} \right], \tag{3.51}$$

where the two vectors $\epsilon_{1,2}^i(\mathbf{p})$ are *polarization vectors* orthogonal to $\mathbf{p}$ ($\mathbf{p}\cdot\boldsymbol{\epsilon}_\lambda(\mathbf{p})=0$), so that $\partial_i A_\perp^i = 0$. In three spatial dimensions, a basis of such vectors has two elements, that we have labeled with $\lambda = 1,2$. In addition, it is convenient to normalize them as follows:

$$\boldsymbol{\epsilon}_\lambda(\mathbf{p}) \cdot \boldsymbol{\epsilon}_{\lambda'}^*(\mathbf{p}) = \delta_{\lambda\lambda'}, \qquad \sum_{\lambda=1,2} \epsilon_\lambda^{i*}(\mathbf{p}) \epsilon_\lambda^j(\mathbf{p}) = \delta^{ij} - \frac{p^i p^j}{\mathbf{p}^2}. \tag{3.52}$$

(Thus, these two polarization vectors form an orthonormal basis of the plane orthogonal to the momentum $\mathbf{p}$.) With this choice, the commutation relations of eq. (3.50) are equivalent to the following commutation relations between creation and annihilation operators:

$$[a_{\lambda\mathbf{p}}, a_{\lambda'\mathbf{q}}] = [a_{\lambda\mathbf{p}}^\dagger, a_{\lambda'\mathbf{q}}^\dagger] = 0, \quad [a_{\lambda\mathbf{p}}, a_{\lambda'\mathbf{q}}^\dagger] = (2\pi)^3\, 2|\mathbf{p}|\, \delta_{\lambda\lambda'}\, \delta(\mathbf{p}-\mathbf{q}). \tag{3.53}$$

### 3.2.4 Feynman Rules for Photons

Eq. (3.51) can be inverted to obtain the creation and annihilation operators as

$$a_{\lambda\mathbf{p}}^\dagger = -i\, \epsilon_\lambda^{i*}(\mathbf{p}) \int d^3\mathbf{x}\, e^{-ip\cdot x}\, \overleftrightarrow{\partial_0}\, A_\perp^i(x),$$

$$a_{\lambda\mathbf{p}} = +i\, \epsilon_\lambda^i(\mathbf{p}) \int d^3\mathbf{x}\, e^{+ip\cdot x}\, \overleftrightarrow{\partial_0}\, A_\perp^i(x). \tag{3.54}$$

With these formulas, it is easy to derive the LSZ reduction formulas for photons in the initial and final states:

$$\langle \underbrace{q_{\lambda'}\cdots}_{n\text{ photons}} {}_{\text{out}} | \underbrace{p_\lambda\cdots}_{m\text{ photons}} {}_{\text{in}}\rangle \doteq \left(\frac{i}{Z^{1/2}}\right)^{m+n} \int d^4x\, e^{-ip\cdot x}\, \epsilon_\lambda^{i*}(\mathbf{p})\, \square_x \cdots$$

$$\times \int d^4y\, e^{+iq\cdot y}\, \epsilon_{\lambda'}^j(\mathbf{q})\, \square_y \cdots \langle 0_{\text{out}} | T\left( A_\perp^i(x) A_\perp^j(y)\cdots \right) | 0_{\text{in}}\rangle. \tag{3.55}$$

The operator $\epsilon_\lambda^i(\mathbf{p})\, \square_x$ in the reduction formula merely amputates the external photon line to which it is applied.[6] Scattering amplitudes with incoming and outgoing photons are therefore given by amputated graphs, with a polarization vector contracted to the Lorentz index of each external photon. The Feynman propagator of the photon in Coulomb gauge can be read off the Lagrangian (3.46). In momentum space, it reads

$$G_F^{0\,00}(p) = \frac{i}{\mathbf{p}^2}, \quad G_F^{0\,ij}(p) = \frac{i\left(\delta^{ij} - \frac{p^i p^j}{\mathbf{p}^2}\right)}{p^2 + i0^+}, \quad G_F^{0\,0i}(p) = G_F^{0\,i0}(p) = 0. \tag{3.56}$$

---

[6] Note that
$$\left(\delta^{ij} - \frac{p^i p^j}{\mathbf{p}^2}\right) \epsilon_\lambda^j(\mathbf{p}) = \epsilon_\lambda^i(\mathbf{p}).$$
Therefore, the transverse projectors attached to the external photon lines can be dropped.

(The 00 component of the propagator enters in the internal lines of Feynman graphs, even though $A^0$ does not carry any physical polarization.) For compactness, we may combine all the components as follows:

$$G_F^{0\,\mu\nu}(p) = \frac{i\,C^{\mu\nu}(p)}{p^2 + i0^+}, \quad \text{with } C^{\mu\nu}(p) \equiv \begin{pmatrix} \frac{p^2}{\mathbf{p}^2} & 0 \\ 0 & \delta^{ij} - \frac{p^i p^j}{\mathbf{p}^2} \end{pmatrix}. \tag{3.57}$$

The T-products that appear in eq. (3.55) can all be encapsulated in a generating functional, which is a Gaussian since the photons are non-interacting:

$$\begin{aligned} Z_0[J] &\equiv \langle 0_{\text{in}} | T \exp i \int d^4x \left( J_0(x) A^0(x) + J_i(x) A_\perp^i(x) \right) | 0_{\text{in}} \rangle \\ &= \exp\left( -\frac{1}{2} \int d^4x\, d^4y\, J_\mu(x)\, G_F^{0\,\mu\nu}(x,y)\, J_\nu(y) \right). \end{aligned} \tag{3.58}$$

## 3.3 Quantum Electrodynamics

So far, we have derived a quantized field theory of free fermions and a quantized field theory of free photons (in the absence of charged sources), but they appear as unrelated constructions of little use since there are no interactions. The next step is to combine the two into a quantum theory of charged fermions that interact electromagnetically via photon exchanges.

### 3.3.1 Global U(1) Symmetry of the Dirac Lagrangian

The fermion Lagrangian is invariant under the transformation $\psi \to \Omega^\dagger \psi$, where $\Omega$ is a phase (i.e., an element of the group $U(1)$), provided that we consider only rigid transformations (i.e., independent of the space-time point x). By *Noether's theorem* (see Section 1.3.4), this continuous symmetry corresponds to the existence of a conserved current,

$$J^\mu = \overline{\psi} \gamma^\mu \psi. \tag{3.59}$$

It is indeed straightforward to check from Dirac's equation that $\partial_\mu J^\mu = 0$. The physical interpretation of this current emerges from the spatial integral of the time component $J^0$,

$$Q \equiv \int d^3x\, J^0(x). \tag{3.60}$$

Using the Fourier mode decomposition (3.18) of the spinor $\psi(x)$, we obtain the following expression:

$$\begin{aligned} Q &= \sum_{s=\pm} \int \frac{d^3\mathbf{p}}{(2\pi)^3 2E_\mathbf{p}} \left\{ a_{s\mathbf{p}} a_{s\mathbf{p}}^\dagger + b_{s\mathbf{p}}^\dagger b_{s\mathbf{p}} \right\} \\ &= \sum_{s=\pm} \int \frac{d^3\mathbf{p}}{(2\pi)^3 2E_\mathbf{p}} \left\{ b_{s\mathbf{p}}^\dagger b_{s\mathbf{p}} - a_{s\mathbf{p}}^\dagger a_{s\mathbf{p}} \right\} + \text{(infinite) constant}. \end{aligned} \tag{3.61}$$

## 3.3.2 Minimal Coupling to a Spin-1 Field

The gauge transformation of the potential $A^\mu$ given in eq. (3.38) can also be written in the following form:[7]

$$A^\mu \;\to\; \Omega^\dagger A^\mu \Omega + i\Omega^\dagger \partial^\mu \Omega, \quad \Omega(x) \equiv e^{-i\chi(x)}. \tag{3.62}$$

When written in this form, the gauge transformation of the photon field appears to be also related to the group $U(1)$. Unlike the quantum field theory for fermions, the photon Lagrangian is in fact invariant under *local gauge transformations*, i.e., where $\Omega$ depends on $x$ in an arbitrary fashion. Therefore, at this point we have two disjoint quantum field theories: a theory of non-interacting charged fermions that has a global $U(1)$ invariance; and a theory of non-interacting photons that has a local $U(1)$ invariance.

Let us determine what minimal modification would promote the $U(1)$ symmetry of the fermion sector into a local symmetry. An immediate obstacle is that $\Omega(x)\,\partial^\mu \Omega^\dagger(x) \neq \partial^\mu$. Equivalently, the problem comes from the fact that the derivative $\partial_\mu \psi$ does not transform in the same way as $\psi$ itself when $\Omega$ depends on $x$. Instead, we have

$$\partial_\mu \psi \;\to\; \partial_\mu \Omega^\dagger \psi = \Omega^\dagger \partial_\mu \psi + (\partial_\mu \Omega^\dagger)\psi. \tag{3.63}$$

But we also see that the second term can be connected to the variation of a photon field under the same transformation. This suggests that the combination $(\partial_\mu - iA_\mu)\psi$ has a simpler transformation law. Indeed, we have

$$\begin{aligned}
(\partial_\mu - iA_\mu)\psi \;&\to\; \Big(\partial_\mu - i(\Omega^\dagger A_\mu \Omega + i\Omega^\dagger \partial_\mu \Omega)\Big)\Omega^\dagger \psi \\
&= \Omega^\dagger(\partial_\mu - iA_\mu)\psi + \Omega^\dagger \Big(\underbrace{\Omega(\partial_\mu \Omega^\dagger) + (\partial_\mu \Omega)\Omega^\dagger}_{\partial_\mu(\Omega\Omega^\dagger)=0}\Big)\psi.
\end{aligned} \tag{3.64}$$

The operator $D_\mu \equiv \partial_\mu - iA_\mu$ is called a *covariant derivative*. The above calculation shows that $D_\mu \psi$ transforms like $\psi$, and that $\overline{\psi} D_\mu \psi$ is invariant under local gauge transformations.

## 3.3.3 Abelian Gauge Theories

This observation is the basis of Abelian gauge theories: the minimal change to the Dirac Lagrangian that makes it locally gauge invariant introduces a coupling $\overline{\psi} A_\mu \psi$ between two fermion fields and a spin-1 field such as the photon. The complete Lagrangian of this theory therefore reads

$$\mathcal{L} = -\frac{1}{4} F_{\mu\nu} F^{\mu\nu} + \overline{\psi}(i\slashed{D} - m)\psi. \tag{3.65}$$

---

[7]Naturally, $\Omega^\dagger A^\mu \Omega = A^\mu$. We have used this somewhat more complicated form to highlight the analogy with the non-Abelian gauge theories that we will study later.

Quantum electrodynamics (QED) is the quantum field theory that describes the interactions between electromagnetic radiation (photons) and charged particles (electrons and positrons, for instance), whose Lagrangian is of the form of eq. (3.65). The only necessary generalization compared to the previous discussion is to introduce a parameter[8] $e$ that represents the (bare) electrical charge of the electron, which leads to the following changes,

Covariant derivative: $\quad D_\mu \equiv \partial_\mu - ieA_\mu,$

Gauge transformation of the photon: $\quad A_\mu \to \Omega^\dagger A^\mu \Omega + \frac{i}{e} \Omega^\dagger \partial^\mu \Omega,$

Electrical current: $\quad e\overline{\psi}\gamma^\mu\psi.$ (3.66)

The generating functional for time-ordered products of fermion and photon fields can be written as follows:

$$Z[\eta, \overline{\eta}, J] = \exp\left(-ie\gamma^\mu_{ab} \int d^4x \, \frac{\delta}{i\delta J_\mu(x)} \, \frac{\overrightarrow{\delta}}{i\delta\overline{\eta}_b(x)} \, \frac{\overleftarrow{\delta}}{i\delta\eta_a(x)}\right) Z_0[J] \, Z_0[\overline{\eta}, \eta], \quad (3.67)$$

where the arrows on the derivatives with respect to the fermionic sources indicate on which side of $Z_0[\overline{\eta}, \eta]$ they should act (in order to avoid unwanted signs due to the fact that these derivatives anti-commute). We already know the Feynman rules for the photon and fermion propagators, and the prescriptions for external photons and fermions, so that the only missing ingredient is the vertex that couples a photon, a fermion and an anti-fermion. From the previous formula, we see that this vertex is $-ie\gamma^\mu$. In addition, each fermion loop provides a minus sign to the Feynman rules because spinors are anti-commuting objects. In order to see this, consider a loop with $n$ photons attached to it. Such a contribution contains $n$ interaction terms that we can schematically write as

$$[\overline{\psi}_1 A\!\!\!/_1 \psi_1][\overline{\psi}_2 A\!\!\!/_2 \psi_2][\overline{\psi}_3 A\!\!\!/_3 \psi_3] \cdots [\overline{\psi}_n A\!\!\!/_n \psi_n]$$
$$= -[\psi_n \overline{\psi}_1] A\!\!\!/_1 [\psi_1 \overline{\psi}_2] A\!\!\!/_2 \cdots [\psi_{n-1} \overline{\psi}_n] A\!\!\!/_n. \quad (3.68)$$

On the right-hand side, we have moved the spinor $\psi_n$ to the leftmost position, where it can be directly paired with $\overline{\psi}_1$ to form a propagator (we have reorganized the brackets to indicate how the spinors are paired to make the successive propagators around the loop). While doing so, $\psi_n$ moves through $2n-1$ other spinors, producing an overall minus sign. Note that fermion loops should be included with only one of the two possible orientations of the arrow and that this orientation modifies the symmetry factor of the graph, compared to a real scalar loop, which has no orientation. The Feynman rules of QED are summarized in Figure 3.1.

### 3.3.4 Example: Electron–Muon Scattering

Consider as an example (see Exercises 3.9 and 3.10 for more examples) the elastic scattering of an electron and a muon at leading order. The amplitude $\mathcal{M}_{e\mu}$ receives the contribution of a single Feynman diagram,

---

[8] Without this parameter, the theory would be too rigid to be renormalizable. Indeed, loop corrections modify the coupling strength of the photon to the fermions.

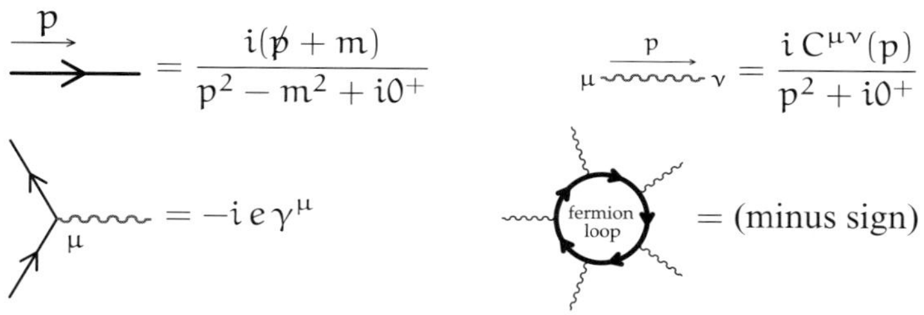

## 3.3 QUANTUM ELECTRODYNAMICS

$$\xrightarrow{p} = \frac{i(\not{p}+m)}{p^2-m^2+i0^+} \qquad \xrightarrow{p}_{\mu\phantom{xx}\nu} = \frac{iC^{\mu\nu}(p)}{p^2+i0^+}$$

$$\underset{\mu}{\diagup\!\!\!\!\diagdown}\!\!\sim = -ie\gamma^\mu \qquad \sim\!\!\bigcirc\!\!\text{fermion loop}\!\!\sim = (\text{minus sign})$$

FIGURE 3.1: Feynman rules of quantum electrodynamics in Coulomb gauge.

$$\mathcal{M}_{e\mu} = \underset{q\phantom{xx}q'}{\overset{p\phantom{xx}p'}{\diagdown\!\!\!\!\diagup}} = -i\frac{[\bar{u}_{s'}(\mathbf{p}')(-ie\gamma_\mu)u_s(\mathbf{p})][\bar{u}_{r'}(\mathbf{q}')(-ie\gamma^\mu)u_r(\mathbf{q})]}{(p-p')^2+i0^+},$$

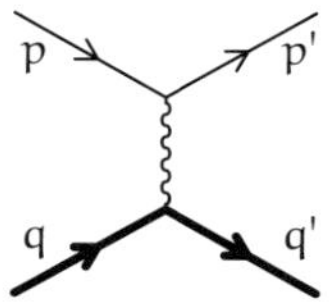

where we have expressed this diagram using the photon propagator in Feynman gauge (see Section 6.3.3) where it reads simply $-ig^{\mu\nu}/(k^2+i0^+)$ (but the result does not depend on this choice, as shown in Exercise 3.8). When squaring an amplitude containing spinors, we use

$$[\bar{u}_{s'}(\mathbf{p}')(-ie\gamma_\mu)u_s(\mathbf{p})]^* = [\bar{u}_{s'}(\mathbf{p}')(-ie\gamma_\mu)u_s(\mathbf{p})]^\dagger = [u_s^\dagger(\mathbf{p})(ie\gamma_\mu^\dagger)\bar{u}_{s'}^\dagger(\mathbf{p}')]$$
$$= [u_s^\dagger(\mathbf{p})(ie\gamma^0\gamma_\mu\gamma^0)\gamma^{0\dagger}u_{s'}(\mathbf{p}')] = [\bar{u}_s(\mathbf{p})(ie\gamma_\mu)u_{s'}(\mathbf{p}')].$$

Therefore, the complex conjugate of such a string of spinors and Dirac matrices is the reversed string with all factors $-i$ changed into $+i$. At this stage, the spin states $r, s, r', s'$ of the initial and final particles are still fixed. In most experiments, one uses in fact beams in which all spin states are mixed (such beams are called unpolarized), and the final state spins are not measured. This is achieved by averaging over the initial spins $r, s$ and by summing over the final spins $r', s'$. Besides the fact that this corresponds to a common experimental setup, these spin sums also simplify considerably the expressions, thanks to identities such as $\sum_s u_s(\mathbf{p})\bar{u}_s(\mathbf{p}) = \not{p}+m$. This results in the following expression for the squared amplitude:

$$\tfrac{1}{4}\sum_{rr'ss'}|\mathcal{M}_{e\mu}|^2 = \frac{e^4}{4}\frac{\text{tr}\left((\not{p}'+m)\gamma_\mu(\not{p}+m)\gamma_\nu\right)\text{tr}\left((\not{q}'+M)\gamma^\mu(\not{q}+M)\gamma^\nu\right)}{(p-p')^4},$$

where we denote $m$ the electron mass and $M$ the muon mass. In the denominator, we have dropped the $i0^+$ prescription since in physical scatterings the kinematics is such that the denominator does not vanish. Up to a relabeling, the two traces are identical. For instance, we have

$$\text{tr}\left((\not{p}'+m)\gamma_\mu(\not{p}+m)\gamma_\nu\right) = 4[p'_\mu p_\nu + p'_\nu p_\mu + 4g_{\mu\nu}(m^2-p\cdot p')].$$

The contraction of the Lorentz indices between the two traces will produce various kind of invariant scalar products: $p^2 = p'^2 = m^2$, $q^2 = q'^2 = M^2$, $p\cdot q, p\cdot p', p\cdot q'$, etc. For $2\to 2$ scatterings, the latter ones are all expressible in terms of three Lorentz invariants, known as the Mandelstam variables,

$$s \equiv (p+q)^2, \quad t \equiv (p-p')^2, \quad u \equiv (p-q')^2.$$

Note that s, t, u are not independent, but obey the constraint $s + t + u = 2(m^2 + M^2)$. (In the more general case of four distinct external masses, the right-hand side is the sum of these four masses.) The scalar products appearing in the square can be rewritten as

$$p \cdot q = p' \cdot q' = \frac{(p+q)^2 - p^2 - q^2}{2} = \frac{s - m^2 - M^2}{2},$$

$$p \cdot p' = q \cdot q' = \frac{p^2 - p'^2 - (p-p')^2}{2} = \frac{2m^2 - t}{2},$$

$$q \cdot q' = p' \cdot q = \frac{p^2 + q'^2 - (p-q')^2}{2} = \frac{m^2 + M^2 - u}{2},$$

and the squared amplitude reads

$$\tfrac{1}{4} \sum_{rr'ss'} |\mathcal{M}_{e\mu}|^2 = \frac{2e^4}{t^2}\left[s^2 + u^2 + 4t(m^2 + M^2) - 2(m^2 + M^2)^2\right].$$

**Cross-section in center-of-momentum frame:** Let us consider now the cross-section in the center of momentum of the incoming electron and positron. This implies that the momenta have the following form:

$$p^\mu = (E_p, \mathbf{p}), \quad q^\mu = (E_q, -\mathbf{p}), \quad p'^\mu = (E'_p, \mathbf{p}'), \quad q'^\mu = (E'_q, \mathbf{q}').$$

(We also have $\mathbf{q}' = -\mathbf{p}'$ from momentum conservation.) In this frame, the differential cross-section reads

$$d\sigma = \frac{1}{4\sqrt{s}|\mathbf{p}|}\left[\tfrac{1}{4}\sum_{rr'ss'}|\mathcal{M}_{e\mu}|^2\right]\frac{d^3\mathbf{p}'d^3\mathbf{q}'}{(2\pi)^6 4E'_p E'_q}(2\pi)^4\delta(p+q-p'-q')$$

$$= \frac{1}{4\sqrt{s}|\mathbf{p}|}\left[\tfrac{1}{4}\sum_{rr'ss'}|\mathcal{M}_{e\mu}|^2\right]\frac{\mathbf{p}'^2 d\mathbf{p}' d\Omega}{(2\pi)^3 4E'_p E'_q}\underbrace{2\pi\delta(2(E-E'))}_{\frac{2\pi E'_p E'_q}{|\mathbf{p}'|\sqrt{s}}\delta(p'-\cdots)}$$

$$= \frac{d\Omega}{64\pi^2 s}\frac{|\mathbf{p}'|}{|\mathbf{p}|}\left[\tfrac{1}{4}\sum_{rr'ss'}|\mathcal{M}_{e\mu}|^2\right],$$

where $d\Omega$ is the solid angle element in which the electron is deflected in the final state (in the center-of-momentum frame). In the ultrarelativistic limit, $u, s \gg m^2, M^2$, we have

$$\mathbf{p} = \mathbf{p}', \quad s = 4\mathbf{p}^2, \quad t = -4\mathbf{p}^2 \sin^2 \tfrac{\theta}{2}, \quad u = -4\mathbf{p}^2 \cos^2 \tfrac{\theta}{2},$$

where $\theta$ is the deflection angle of the electron, and the cross-section reads

$$\frac{d\sigma}{d\Omega} = \frac{e^4}{32\pi^2 s}\frac{1+\cos^4\tfrac{\theta}{2}}{\sin^4\tfrac{\theta}{2}}.$$

**Cross-section in the muon rest frame:** Consider now this scattering in the muon rest frame, where we have

$$p^\mu = (E_p, \mathbf{p}), \quad q^\mu = (M, \mathbf{0}), \quad p'^\mu = (E'_p, \mathbf{p}'), \quad q'^\mu \approx (M, \mathbf{q}').$$

## 3.3 QUANTUM ELECTRODYNAMICS

By eliminating from the final state phase space the variables that are constrained by momentum conservation, we get

$$\frac{d\sigma}{d\Omega} = \frac{1}{64\pi^2 M^2} \left[ \frac{1}{4} \sum_{rr'ss'} |\mathcal{M}_{e\mu}|^2 \right] = \frac{e^4}{64\pi^2 p^4} \frac{m^2 + p^2 \cos^2 \frac{\theta}{2}}{\sin^4 \frac{\theta}{2}},$$

where $d\Omega$ is the solid angle element of the scattered electron in the rest frame of the muon, and $\theta$ is its deflection angle.

### 3.3.5 Classical Equations of Motion

From the Lagrangian (3.65), one can also derive the corresponding Euler–Lagrange equations. For the fermions, we obtain a generalization of the Dirac equation,

$$\left( i \slashed{D} - m \right) \psi = 0, \tag{3.69}$$

in which the ordinary derivative has been replaced by a covariant derivative. Therefore, this equation describes how the fermion field evolves in the presence of an external electromagnetic field. The derivation of the equation of motion for the gauge potential $A^\mu$ requires the following derivatives:

$$\partial_\mu \frac{\partial \mathcal{L}}{\partial (\partial_\mu A_\nu)} = -\partial_\mu F^{\mu\nu}, \qquad \frac{\partial \mathcal{L}}{\partial A_\nu} = e \bar{\psi} \gamma^\nu \psi, \tag{3.70}$$

which lead to

$$\partial_\mu F^{\mu\nu} + \underbrace{e \bar{\psi} \gamma^\nu \psi}_{J^\nu} = 0. \tag{3.71}$$

If we note that the second term is the electromagnetic current induced by the charged particles described by $\psi$, this equation is nothing but the Maxwell equation that drives the electromagnetic field in the presence of charged sources.

### 3.3.6 Charge Conjugation

The *charge conjugation operator*, denoted $\mathbf{C}$, is an operator that transforms $e$ into $-e$, in the theory. In order to construct it, start from the complex conjugate of the Dirac equation. Then, inserting $\gamma^{0T} \gamma^{0T} = 1$ just before $\psi^*$ and multiplying on the left by $\gamma^{0T}$ leads to

$$\left( \gamma^{\mu T} (-i \partial_\mu + e A_\mu) - m \right) \bar{\psi}^T = 0. \tag{3.72}$$

The charge conjugation operator $\mathbf{C}$ is defined so that $\gamma^{\mu T} = -\mathbf{C}^{-1} \gamma^\mu \mathbf{C}$. Therefore, the previous equation becomes

$$\left( i \slashed{\partial} - e \slashed{A} - m \right) \underbrace{\mathbf{C} \bar{\psi}^T}_{-\psi^c} = 0, \tag{3.73}$$

where we have introduced the charge conjugated spinor $\psi^c$. Note that the precise matrix form of $\mathbf{C}$ depends on the representation chosen for the Dirac matrices, but it exists in any representation. In many applications, the explicit form is not needed, and it is sufficient to know how it relates the Dirac matrices and their transpose. The existence of the charge conjugation operator, which simply flips the sign of the electrical charge in the Dirac equation while preserving the mass, is another way to see that it describes both particles and their antiparticles. When applied to the fermionic time-ordered propagator, it transforms it as follows:

$$\mathbf{C}^{-1}\, S_F(x-y)\, \mathbf{C} = S_F^T(y-x), \quad \mathbf{C}^{-1}\, S_F(p)\, \mathbf{C} = S_F^T(-p). \tag{3.74}$$

## 3.4 Charge Conservation, Ward–Takahashi Identities

### 3.4.1 Charge of One-Particle States

The charge operator $Q$ defined in eq. (3.60) is invariant by translation in time (because $J^\mu$ is a conserved current) and in space (because it is integrated over all space). Since the current $J^\mu$ is a 4-vector, $Q$ is also invariant under Lorentz transformations. Therefore, $Q$ conserves the energy and momentum of the states on which it acts (this can also be seen in eq. (3.61)). When acting on the vacuum state, one has $Q\,|0\rangle = 0$. When acting on a one-particle state $|\alpha_\mathbf{p}\rangle$, $Q$ gives another state with the same 4-momentum. But since stable single particle states are isolated delta peaks in the spectral function of the theory, $Q\,|\alpha_\mathbf{p}\rangle$ must in fact be proportional to $|\alpha_\mathbf{p}\rangle$ itself, i.e., $Q\,|\alpha_\mathbf{p}\rangle = q_{\alpha,\mathbf{p}}\,|\alpha_\mathbf{p}\rangle$. In other words, one-particle states are eigenvectors of the charge operator. Since $Q$ is Lorentz invariant, the eigenvalue $q_{\alpha,\mathbf{p}}$ cannot depend on the momentum $\mathbf{p}$ (nor on the spin state of the particle), and it can only depend on the species of particle $\alpha$. We will thus denote it $q_\alpha$, and call it the electrical charge of the particle $\alpha$.

In theories with one-particle states that do not correspond to the fundamental fields of the Lagrangian (e.g., composite bound states made of several elementary particles), one may go a bit further. The canonical anti-commutation relations imply

$$\left[J^0(x), \psi(y)\right]_{x^0=y^0} = -e\,\psi(x)\,\delta(\mathbf{x}-\mathbf{y}), \quad \left[Q, \psi(y)\right] = -e\,\psi(y). \tag{3.75}$$

More generally, for any local function $F(\psi(x), \psi^\dagger(x))$, we have

$$\left[Q, F(\psi(y), \psi^\dagger(y))\right] = -e\,(n_+ - n_-)\,F(\psi(y), \psi^\dagger(y)), \tag{3.76}$$

where $n_+$ is the number of $\psi$ in $F$ and $n_-$ the number of $\psi^\dagger$. If we evaluate this identity between the vacuum and a one-particle state $|\alpha_\mathbf{p}\rangle$, we obtain

$$\langle 0|F(\psi(y), \psi^\dagger(y))|\alpha_\mathbf{p}\rangle\,(q_\alpha - (n_+ - n_-)e) = 0. \tag{3.77}$$

Therefore, if the operator $F(\psi, \psi^\dagger)$ can create the particle $\alpha$ from the vacuum (i.e., the matrix element on the left-hand side is non-zero), then we must have

$$q_\alpha = (n_+ - n_-)\,e. \tag{3.78}$$

In other words, the charge of the particle $\alpha$ is the number of $\psi$ it contains, minus the number of $\psi^\dagger$, times the electrical charge $e$ of the field $\psi$ (as it appears in the Lagrangian). The non-trivial aspect of this assertion comes from the fact that it does not depend on the (usually complicated and non-perturbative) interactions that produce the binding. As we shall see in the next section, the renormalized and bare electron charges are related by $e_r = \sqrt{Z_3}\,e_b$.

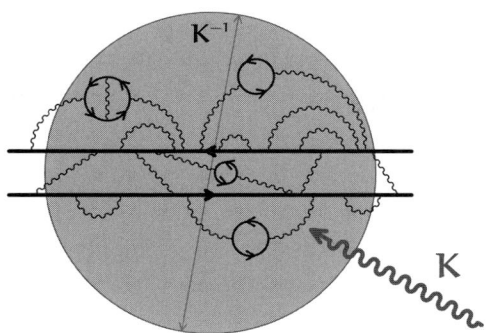

FIGURE 3.2: Illustration of the fact that the complicated internal details of a bound state are not resolved by a long wavelength (represented by the shaded disk) photon. Such a photon sees only the total charge of the valence constituents (shown by thick horizontal lines).

In combination with eq. (3.78), this means that the electrical charges of all one-particle states are renormalized by the same factor $\sqrt{Z_3}$, regardless of the internals of the bound state. For this to work, cancellations between various Feynman graphs are necessary. These cancellations are a consequence of the local gauge invariance of the theory, and in their simplest form they can be encapsulated in the *Ward–Takahashi identities*, that we shall derive now. Note that although non-trivial on a diagrammatic level, eq. (3.78) has a fairly intuitive interpretation once we realize that the operator Q measures the charge as seen by a photon of very long wavelength (due to the integration over all space in its definition). Such a soft photon cannot resolve the internal details of the bound state, and is only sensitive to the total charge carried by the "valence" constituents, as illustrated in Figure 3.2. However, these details would of course matter if one looks at the momentum dependence of the vertex by which a photon of non-zero momentum couples to that bound state. Note that even the electron develops such a cloud of virtual photons and $e^+e^-$ pairs. As a consequence, beyond tree level, the coupling of a photon to an electron gets a non-trivial momentum-dependent structure (see Exercise 3.11).

### 3.4.2 Ward–Takahashi Identities

On-shell amplitudes with amputated external photon lines can be obtained as follows:

$$M^{\mu_1\mu_2\cdots}(q_1, q_2, \cdots) = \int d^4x_1\, d^4x_2 \cdots e^{-iq_1\cdot x_1}\, e^{-iq_2\cdot x_2} \cdots$$
$$\times \langle \beta_{\text{out}} | T\{J^{\mu_1}(x_1) J^{\mu_2}(x_2) \cdots\} | \alpha_{\text{in}} \rangle, \qquad (3.79)$$

where only electromagnetic currents appear inside the T-product, and all the external charged particles are kept in the initial and final states $\alpha$ and $\beta$ (and are therefore on-shell). Let us now contract the Lorentz index $\mu_1$ with the momentum $q_1^{\mu_1}$ of the first photon. After an integration by parts, this reads

$$q_{1\mu_1} M^{\mu_1\mu_2\cdots}(q_1, q_2, \cdots) = -i \int d^4x_1\, d^4x_2 \cdots e^{-iq_1\cdot x_1}\, e^{-iq_2\cdot x_2} \cdots$$
$$\times \langle 0_{\text{out}} | \partial_{\mu_1} T\{J^{\mu_1}(x_1) J^{\mu_2}(x_2) \cdots\} | 0_{\text{in}} \rangle. \qquad (3.80)$$

The derivative of the T-product involves two types of terms: terms where the derivative acts directly on the current $J^{\mu_1}(x_1)$ that are zero thanks to current conservation, and terms where it acts on the step functions that order the operators inside the T-product. With two currents, the latter term reads:[9]

$$\frac{\partial}{\partial x^{\mu}} \, T\left\{J^{\mu}(x) J^{\nu}(y)\right\} = \delta(x^0 - y^0) \left[J^0(x), J^{\nu}(y)\right] = 0. \tag{3.81}$$

This generalizes to more than two currents, and we therefore have quite generally

$$q_{1\,\mu_1} \, M^{\mu_1 \mu_2 \cdots}(q_1, q_2, \cdots) = 0. \tag{3.82}$$

The same property would hold for all the external photon lines of the amplitude. This equation is known as the *Ward–Takahashi identity*.

An important consequence of eq. (3.82) is that QED scattering amplitudes are unchanged if the photon propagators or polarization vectors are modified by terms proportional to the momentum $p^{\mu}$,

$$G_F^{0\,\mu\nu}(p) \to G_F^{0\,\mu\nu}(p) + a^{\mu} \, p^{\nu} + b^{\nu} \, p^{\mu}, \quad \epsilon_{\lambda}^{\mu}(p) \to \epsilon_{\lambda}^{\mu}(p) + c \, p^{\mu}. \tag{3.83}$$

This is precisely the modification of the Feynman rules one would encounter by using a different gauge fixing in the quantization of the theory. Thus, the Ward–Takahashi identities imply the gauge invariance of the transitions amplitudes in QED.

## 3.5 Ultraviolet Renormalization

### 3.5.1 Power Counting

In order to determine which correlators may have intrinsic ultraviolet divergences, let us calculate the superficial degree of divergence of a generic graph. Consider a Feynman graph $\mathcal{G}$ made of fermions and photons, with $n_{\gamma_I}$ internal photons, $n_{\gamma_E}$ external ones, $n_{\psi_I}$ internal fermions, $n_{\psi_E}$ external ones, $n_L$ loops and $n_V$ vertices. These six parameters are related by

$$2\,n_V = 2\,n_{\psi_I} + n_{\psi_E}, \quad n_V = 2\,n_{\gamma_I} + n_{\gamma_E}, \quad n_L = n_{\psi_I} + n_{\gamma_I} - n_V + 1. \tag{3.84}$$

The ultraviolet superficial degree of divergence of this graph is then given (in four space-time dimensions) by

$$\omega(\mathcal{G}) \equiv 4\,n_L - 2\,n_{\gamma_I} - n_{\psi_I} = 4 - n_{\gamma_E} - \tfrac{3}{2}\,n_{\psi_E}. \tag{3.85}$$

As in the case of the four-dimensional scalar theory with quartic interaction, $\omega(\mathcal{G})$ depends only on the number of external photons and fermions (and not on the internal details of the graph), indicating that quantum electrodynamics is a renormalizable theory. This is consistent

---

[9]This step of the argument would fail if we had kept charged field operators inside the T-product, because their equal-time commutator with $J^0$ is non-zero. Therefore, the Ward–Takahashi identities are valid provided all the external charged particles are on-shell, but there is no such requirement for the neutral external particles (e.g., the photons).

## 3.5 ULTRAVIOLET RENORMALIZATION

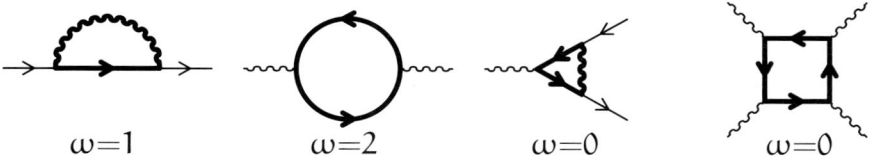

FIGURE 3.3: Non-zero one-loop diagrams with $\omega(\mathcal{G}) \geq 0$. The corresponding superficial degrees of divergence are indicated below each graph.

with the fact that, in four dimensions, all the couplings in the QED Lagrangian have zero or positive mass dimension. There are four non-zero correlators for which $\omega(\mathcal{G}) \geq 0$, indicated in Figure 3.3 (we are not considering the three-photon function, despite its superficial degree of divergence $\omega = 1$, because it is identically zero thanks to *Furry's theorem*, which states that a fermion loop with an odd number of photons attached to it is identically zero – see Exercise 3.4).

### 3.5.2 Renormalized Lagrangian

The divergences contained in the first three functions (respectively, the fermion self-energy, the photon self-energy and the photon–fermion vertex) can be absorbed into a redefinition of the coefficients already present in the QED Lagrangian (including the field normalization for $\psi$ and $A^\mu$). It is customary to parametrize the ultraviolet counterterms of QED by writing $\mathcal{L}_b = \mathcal{L}_r + \Delta \mathcal{L}$, with

$$\Delta \mathcal{L} \equiv -\tfrac{1}{4}(Z_3 - 1)\, F_{r\,\mu\nu} F_r^{\mu\nu} + (Z_2 - 1)(i\overline{\psi}_r \slashed{\partial}\psi_r - m_r \overline{\psi}_r \psi_r)$$
$$+ Z_2 \Delta m\, \overline{\psi}_r \psi_r + e_r (Z_1 - 1)\, \overline{\psi}_r \slashed{A}_r \psi_r, \qquad (3.86)$$

where the subscript r denotes renormalized fields. The bare and renormalized quantities are related by

$$A_b^\mu = \sqrt{Z_3}\, A_r^\mu, \quad \psi_b = \sqrt{Z_2}\, \psi_r, \quad m_r = m_b + \Delta m, \quad Z_1 e_r = Z_2 \sqrt{Z_3}\, e_b. \qquad (3.87)$$

For the correction $\Delta \mathcal{L}$ to have the same structure as the bare Lagrangian, it is necessary that $Z_1 = Z_2$. Otherwise, the renormalized theory would have the same operators as the bare theory, but would not be gauge invariant anymore.[10] As we shall see later, gauge invariance is crucial for unitarity, therefore the loss of gauge symmetry would make the theory non-physical. In fact, gauge symmetry enforces $Z_1 = Z_2$ to all orders, which also implies that the bare and renormalized electron charge are related by $e_r = \sqrt{Z_3}\, e_b$.

---

[10]The operator $\overline{\psi}_r \slashed{D}_r \psi_r$ is invariant under the simultaneous transformation $\psi_r \to e^{i e_r \alpha} \psi_r$, $A_r^\mu \to A_r^\mu + \partial^\mu \alpha$. For this $\Omega$ to leave invariant the bare operator $\overline{\psi}_b \slashed{D}_b \psi_b$, the bare gauge field must transform as $A_b^\mu \to A_b^\mu + (e_r/e_b)\partial^\mu \alpha = \sqrt{Z_3}(A_r^\mu + (Z_2/Z_1)\partial^\mu \alpha)$, from which we can read off an alternate form of the transformation law of $A_r^\mu$. For consistency, we should have $Z_1 = Z_2$.

### 3.5.3 Four-Photon Vertex

Another concern for the renormalizability of QED is the rightmost graph in Figure 3.3. Indeed, if the one-loop four-photon function is ultraviolet divergent, the only way out would be to introduce in the Lagrangian a counterterm for a four-photon local operator that was not present in the bare Lagrangian. Let us denote $\Gamma^{\mu\nu\rho\sigma}(k_1, \cdots, k_4)$ the amputated four-photon function. Thanks to the Ward–Takahashi identity, this function vanishes when contracted with any of the external photon momenta,

$$k_{1\mu}\Gamma^{\mu\nu\rho\sigma} = k_{2\nu}\Gamma^{\mu\nu\rho\sigma} = k_{3\rho}\Gamma^{\mu\nu\rho\sigma} = k_{4\sigma}\Gamma^{\mu\nu\rho\sigma} = 0. \tag{3.88}$$

This is also true for the ultraviolet divergent part of the function (the term with the $\epsilon^{-1}$ pole in dimensional regularization). Therefore, the counterterm one obtains by contracting the divergent part of this function with four photon fields, $\Gamma^{\mu\nu\rho\sigma}_{\text{divergent}} A_\mu A_\nu A_\rho A_\sigma$, is gauge invariant. This means that it can in fact be rewritten with four powers of the field strength $F_{\alpha\beta}$. But for this to be possible, $\Gamma^{\mu\nu\rho\sigma}_{\text{divergent}}(k_1, \cdots, k_4)$ must contain four powers of the external momenta, implying that the genuine ultraviolet degree of divergence of the loop is $-4$ instead of 0 and that the four-photon vertex is therefore ultraviolet finite. Note that this cancellation of ultraviolet divergences does not work graph by graph, but only among gauge invariant sets of graphs. In the present case, one needs to sum over the six possible orderings of the external lines 1, 2, 3, 4 around the loop.

### 3.5.4 Proof of $Z_1 = Z_2$

Consider the renormalized photon–fermion–anti-fermion correlator, amputated of its external photon propagator (but where we keep the external fermion propagators):

$$F^\mu_{r\,ab}(k, p, q) \equiv -iZ_1 \int d^4x\, d^4y\, d^4z\, e^{i(k\cdot x + p\cdot y + q\cdot z)} \langle 0_{\text{out}} | T\left(J^\mu_r(x)\psi_{r\,a}(y)\overline{\psi}_{r\,b}(z)\right) | 0_{\text{in}} \rangle, \tag{3.89}$$

where all the momenta are defined to be incoming. All the operators inside the T-product are the renormalized ones. Now, contract this function with $k_\mu$. The calculation is similar to our earlier proof of the Ward–Takahashi identity, but now we obtain in addition commutators of $J^0_r$ with either $\psi_r$ or $\overline{\psi}_r$,

$$k_\mu F^\mu_{r\,ab}(k, p, q) = Z_1 \int d^4x\, d^4y\, d^4z\, e^{i(k\cdot x + p\cdot y + q\cdot z)}$$
$$\times \Big( \delta(x^0 - y^0) \langle 0_{\text{out}} | T\left([J^0_r(x), \psi_{r\,a}(y)]\overline{\psi}_{r\,b}(z)\right) | 0_{\text{in}} \rangle$$
$$+ \delta(x^0 - z^0) \langle 0_{\text{out}} | T\left(\psi_{r\,a}(y)[J^0_r(x), \overline{\psi}_{r\,b}(z)]\right) | 0_{\text{in}} \rangle \Big). \tag{3.90}$$

The equal-time commutators can be obtained easily from their bare counterparts, by introducing the appropriate renormalization factors:

$$Z_2 \left[J^0_r(x), \psi_{r\,a}(y)\right]_{x^0=y^0} = -e_r\, \delta(\mathbf{x} - \mathbf{y})\, \psi_{r\,a}(y),$$
$$Z_2 \left[J^0_r(x), \overline{\psi}_{r\,b}(z)\right]_{x^0=z^0} = +e_r\, \delta(\mathbf{x} - \mathbf{z})\, \overline{\psi}_{r\,b}(z). \tag{3.91}$$

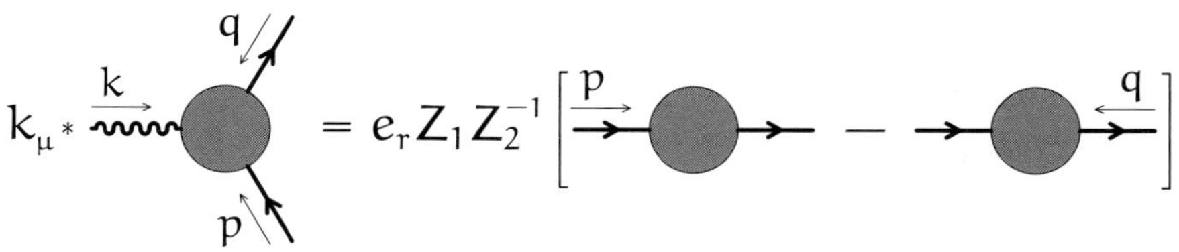

## 3.5 ULTRAVIOLET RENORMALIZATION

FIGURE 3.4: Illustration of the off-shell Ward–Takahashi identity that relates the vertex and the fermion propagator.

Therefore, we have

$$k_\mu F^\mu_{r\,ab}(k,p,q) = e_r\, Z_1 Z_2^{-1}\, (2\pi)^4 \delta(k+p+q) \left(S_{r\,ab}(p) - S_{r\,ab}(-q)\right), \quad (3.92)$$

where $S_r$ denotes the renormalized fermion propagator. This identity, which may be viewed as an off-shell extension of the Ward–Takahashi identity, is illustrated graphically in Figure 3.4. If we write $F^\mu_r(k,p,q) \equiv S_r(-q)\left(-i\Gamma^\mu(k,p,q)\right)S_r(p)$, this takes also the form

$$k_\mu \Gamma^\mu_{r\,ab}(k,p,q) = i\, e_r\, Z_1 Z_2^{-1}\, (2\pi)^4 \delta(k+p+q) \left(S^{-1}_{r\,ab}(-q) - S^{-1}_{r\,ab}(p)\right), \quad (3.93)$$

Because this identity relates renormalized functions, that are by definition finite, it implies that the ratio $Z_1/Z_2$ is itself finite. Let us adopt the following renormalization conditions:

$$\Gamma^\mu_r(k,p,q)\Big|_{\substack{p^2=q^2=m_r^2 \\ p+q\to 0}} = e_r\,(2\pi)^4 \delta(p+q)\,\gamma^\mu,$$

$$S_r^{-1}(p) = -i\left(\not{p} - m_r - \Sigma_r(p)\right) \quad \text{with} \quad \Sigma_r(p)\Big|_{p^2=m_r^2} = 0. \quad (3.94)$$

The first one means that a long-wavelength photon should couple to the renormalized on-shell electron as if it were a point-like object of charge $e_r$, and the second one states that the renormalized inverse propagator vanishes on the renormalized mass-shell. Evaluating eq. (3.93) with on-shell fermions in the limit $k \to 0$, and using these renormalization conditions, we obtain $Z_1 = Z_2$. As stated earlier, this implies that $e_b A^\mu_b = e_r A^\mu_r$, i.e., that the covariant derivative is not renormalized.

### 3.5.5 Dressed Photon Propagator

The Ward–Takahashi identities also constrain the form of the dressed photon propagator. Indeed, the photon self-energy (also called *photon polarization tensor*) $\Pi^{\mu\nu}(p)$ is symmetric and satisfies $p_\mu \Pi^{\mu\nu}(p) = 0$. Therefore, it must have the following form:[11]

$$\Pi^{\mu\nu}(p) = \left(g^{\mu\nu} p^2 - p^\mu p^\nu\right) \Pi(p^2). \quad (3.95)$$

---

[11] There is a subtlety here, since this decomposition implicitly assumes Lorentz covariance, while the Coulomb gauge Feynman rules are not covariant. But thanks to the Ward–Takahashi identities, the gauge dependence of the internal photon propagators cancels out, and we may use instead a covariant photon propagator such as the Feynman gauge one. Another way to see that $\Pi^{\mu\nu}$ is gauge invariant is to notice that $\Pi^{\mu\nu}(x,y)$ is the vacuum expectation value of the time-ordered product of two currents $T\left(J^\mu(x)J^\nu(y)\right)$ (combined with the fact that $J^\mu$ is gauge invariant).

Consider now the summation of this self-energy correction to all orders on the photon propagator,

$$G_F^{\mu\nu} = G_F^{0\ \mu\rho} \sum_{n=0}^{\infty} \left[(\Pi G_F^0)^n\right]_\rho^{\ \nu}. \tag{3.96}$$

Using the explicit form of the tensor $C^{\mu\nu}(p)$ that enters in the numerator of the Coulomb gauge photon propagator, we obtain

$$G_F^{\mu\nu}(p) = \frac{1}{1 + \Pi(p^2)} G_F^{0\ \mu\nu}(p). \tag{3.97}$$

Assuming that the function $p^2 \Pi(p^2)$ goes to zero when $p^2 \to 0$, this resummed photon propagator has a pole at $p^2 = 0$. In other words, the photon remains massless. The factor $(1 + \Pi(p^2))^{-1}$ provides the wavefunction renormalization $Z_3$ of the photon, and its square root is the renormalization factor of the electron charge (since $Z_1 = Z_2$).

## 3.6 Cutting Rules in QED and Unitarity

### 3.6.1 Fermions

Using eqs. (2.63) and (3.30), we have

$$\begin{aligned}(i\slashed{\partial}_x + m)G_F^0(x,y) &= \theta(x^0 - y^0)(i\slashed{\partial}_x + m)G_{-+}^0(x,y) \\ &+ \theta(y^0 - x^0)(i\slashed{\partial}_x + m)G_{+-}^0(x,y) \\ &+ i\gamma^0 \delta(x^0 - y^0)\left(G_{-+}^0(x,y) - G_{+-}^0(x,y)\right),\end{aligned} \tag{3.98}$$

but the last term is in fact zero because $G_{-+}^0$ and $G_{+-}^0$ coincide at equal times. This makes the extension of cutting rules to fermions straightforward, with the following rules:

$$S_{++}^0(p) = \frac{i(\slashed{p} + m)}{p^2 - m^2 + i0^+}, \quad S_{--}^0(p) = \frac{-i(\slashed{p} + m)}{p^2 - m^2 - i0^+},$$
$$S_{\mp\pm}^0(p) = 2\pi (\slashed{p} + m)\theta(\pm p_0)\delta(p^2 - m^2). \tag{3.99}$$

The cutting rules for fermions are therefore similar to those for scalar particles. The possibility to interpret the cut fermion propagators in terms of on-shell final state fermions is a consequence of the following identities:

$$\slashed{p} + m = \sum_{\text{spin } s} u_s(p)\bar{u}_s(p), \quad \slashed{p} - m = \sum_{\text{spin } s} v_s(p)\bar{v}_s(p), \tag{3.100}$$

which are valid when $p_0 = \sqrt{\mathbf{p}^2 + m^2} > 0$. In the case of the propagator $S_{-+}^0(p)$, we may attach the spinor $u_s(p)$ to the amplitude on the right of the cut, and the spinor $\bar{u}_s(p)$ to the amplitude on the left, which are precisely the spinors required by the LSZ formula for a fermion of momentum $\mathbf{p}$ in the final state. In the case of $S^0_{+-}(p)$, for which $p^0 < 0$, we should first write

$$S^0_{+-}(p) = -2\pi(-\not{p} - m)\theta(-p_0)\delta(p^2 - m^2)$$
$$= -2\pi \sum_{\text{spin } s} v_s(-\mathbf{p})\bar{v}_s(-\mathbf{p})\theta(-p_0)\delta(p^2 - m^2), \quad (3.101)$$

in order to see that it corresponds to an anti-fermion in the final state.

### 3.6.2 Photons

**Coulomb gauge:** For photons in Coulomb gauge, the reasoning is similar to the case of fermions. Note that the propagator $G^{00}_F$ is a bit special since it is proportional to $\delta(x^0 - y^0)$ in coordinate space. Since the field $A^0$ does not appear on the external lines of physical amplitudes, its only effect is to produce a four-fermion interaction, which is local in time but non-local in space. The time locality of this interaction is sufficient to extend the cutting rules without having to consider the $G^{00}_F$ propagator (in other words, this propagator will never cross the cut). Therefore, the propagators that matter in the cutting rules are

$$G^{0\,ij}_{++}(p) = \frac{i\left(\delta^{ij} - \frac{p^i p^j}{\mathbf{p}^2}\right)}{p^2 + i0^+}, \quad G^{0\,ij}_{--}(p) = \frac{-i\left(\delta^{ij} - \frac{p^i p^j}{\mathbf{p}^2}\right)}{p^2 - i0^+},$$
$$G^{0\,ij}_{\mp\pm}(p) = 2\pi\,\theta(\pm p^0)\left(\delta^{ij} - \frac{p^i p^j}{\mathbf{p}^2}\right)\delta(p^2). \quad (3.102)$$

Recalling also that

$$\sum_{\lambda=1,2} \epsilon^{i*}_\lambda(\mathbf{p})\epsilon^j_\lambda(\mathbf{p}) = \delta^{ij} - \frac{p^i p^j}{\mathbf{p}^2}, \quad (3.103)$$

we see that the projector that appears in the cut propagators can be interpreted as the polarization vectors that should be attached to amplitudes for each final state photon. Therefore, the cutting rules in Coulomb gauge have a direct interpretation in terms of the optical theorem. This simplicity follows from the fact that, in Coulomb gauge, the only propagating modes are physical modes.

**Feynman gauge:** This interpretation is not so direct in covariant gauges, such as the Feynman gauge, which we will discuss further in Chapter 6. For now, it is sufficient to know that the free photon propagator in this gauge is given by

$$G^{0\,\mu\nu}_{++}(p) = -g^{\mu\nu}\frac{i}{p^2 + i0^+}. \quad (3.104)$$

The factor $-g^{\mu\nu}$ does not change anything in the cutting rules, and simply appears as a prefactor in all the other propagators:

$$G^{0\,\mu\nu}_{--}(p) = -g^{\mu\nu}\frac{-i}{p^2 - i0^+},$$
$$G^{0\,\mu\nu}_{-+}(p) = -2\pi g^{\mu\nu}\theta(+p_0)\delta(p^2), \quad G^{0\,\mu\nu}_{+-}(p) = -2\pi g^{\mu\nu}\theta(-p_0)\delta(p^2). \quad (3.105)$$

The two physical polarizations vectors, $\epsilon^\mu_{1,2}$, can be chosen to form an orthonormal basis of the plane orthogonal to $\mathbf{p}$. They also obey $p_\mu \epsilon^\mu_{1,2}(\mathbf{p}) = 0$. However, the tensor $-g^{\mu\nu}$ that appears in the cut photon propagators cannot be written as a sum over physical polarizations,

$$-g^{\mu\nu} \neq \sum_{\lambda=1,2} \epsilon^\mu_\lambda(\mathbf{p}) \epsilon^{\nu*}_\lambda(\mathbf{p}). \tag{3.106}$$

As a consequence, Cutkosky's cutting rules in Feynman gauge seemingly lead to terms that we cannot interpret as physical photon final states, which would violate the optical theorem. If this was the case, then perturbation theory would not be consistent with unitarity. To see how this disastrous conclusion is avoided, let us introduce two more (unphysical) polarization vectors,

$$\epsilon^\mu_+(\mathbf{p}) \equiv \frac{1}{\sqrt{2}|\mathbf{p}|}(p_0, \mathbf{p}) = \frac{p^\mu}{\sqrt{2}|\mathbf{p}|}, \quad \epsilon^\mu_-(\mathbf{p}) \equiv \frac{1}{\sqrt{2}|\mathbf{p}|}(p_0, -\mathbf{p}). \tag{3.107}$$

Thanks to these additional vectors, we may now write

$$g^{\mu\nu} = \epsilon^\mu_+(\mathbf{p}) \epsilon^{\nu*}_-(\mathbf{p}) + \epsilon^\mu_-(\mathbf{p}) \epsilon^{\nu*}_+(\mathbf{p}) - \sum_{\lambda=1,2} \epsilon^\mu_\lambda(\mathbf{p}) \epsilon^{\nu*}_\lambda(\mathbf{p}). \tag{3.108}$$

In other words, the physical polarization sum in the right-hand side of eq. (3.106) is equal to $-g^{\mu\nu}$, plus some extra terms that are proportional to $p^\mu$ or $p^\nu$.

When we use Cutkosky's cutting rules in order to calculate the imaginary part of a graph, a cut photon line carrying the momentum $p^\mu$ leads to an expression of the form $M^\mu_L(p)[-g^{\mu\nu}]M^{\nu*}_R(p)$, where $M^\mu_L$ and $M^\nu_R$ are the amplitudes on the left and on the right of the cut, respectively. Here, we are focusing only on one of the cut photons, and the other cut lines have not been written explicitly since they do not play any role in the argument. Thanks to eq. (3.108), this quantity can be rewritten as

$$M^\mu_L(p)[-g^{\mu\nu}]M^{\nu*}_R(p) = M^\mu_L(p) \left[ \sum_{\lambda=1,2} \epsilon^\mu_\lambda(\mathbf{p}) \epsilon^{\nu*}_\lambda(\mathbf{p}) \right] M^{\nu*}_R(p). \tag{3.109}$$

Indeed, since $\epsilon^\mu_+(\mathbf{p}) \propto p^\mu$, the two unphysical terms cancel thanks to the Ward–Takahashi identity satisfied[12] by the amplitudes $M^\mu_{L,R}$. Therefore, the cancellation of the non-physical photon degrees from the final state sum (i.e., the unitarity of the theory) appears closely related to gauge symmetry.

## 3.7 Infrared Divergences

Quantum electrodynamics has the peculiarity of containing a massless particle – the photon – the production of which does not cost any energy (one says that there is no "mass gap", i.e.,

---

[12] When an amplitude has external charged particles, the Ward–Takahashi identity is satisfied only if these particles are on-shell. This is indeed the case here, because all the cut lines are on-shell, as well as all the external lines.

## 3.7 INFRARED DIVERGENCES

a minimal energy). As we shall see, this appears on the surface to be a disaster for the theory since it seems that any QED process receives arbitrarily large corrections due to low-energy photons. However, this is resolved after one takes into account the unobservability of these soft photons, by a cancellation between two types of corrections.

### 3.7.1 Attaching Soft Photons to Hard Lines

Consider a hard scattering amplitude, i.e., one in which all the external particles have an energy above a certain value $\Lambda$, such as the one shown in Figure 3.5. The external lines may be fermions or photons. The initial state is collectively denoted $\alpha$ and the final state $\beta$. We denote $p_i$ the external momenta, and $e_i$ the corresponding electrical charge. The hard amplitude is denoted $M^\Lambda_{\beta\alpha}$, with a superscript $\Lambda$ to recall that all external particles are hard. Consider now attaching a soft photon to one of the external fermion lines, as shown in the middle graph of Figure 3.5. If this line is an outgoing fermion, the insertion of the photon amounts to the following substitution for the free spinor that ends the line:

$$\overline{u}_r(\mathbf{p}_i) \to -ie_i\,\overline{u}_r(\mathbf{p}_i)\gamma^\mu \frac{i(\slashed{p}_i + \slashed{q} + m)}{(p_i+q)^2 - m^2 + i0^+} \underset{q\to 0}{\approx} e_i\,\overline{u}_r(\mathbf{p}_i)\gamma^\mu \frac{\slashed{p}_i + m}{2p_i\cdot q + i0^+}$$

$$= e_i\,\overline{u}_r(\mathbf{p}_i)\gamma^\mu \frac{\sum_{\text{spin }s} u_s(\mathbf{p}_i)\overline{u}_s(\mathbf{p}_i)}{2p_i\cdot q + i0^+} = \overline{u}_r(\mathbf{p}_i)\frac{e_i\,p_i^\mu}{p_i\cdot q + i0^+}. \qquad (3.110)$$

(The last equality follows from $\overline{u}_r(\mathbf{p}_i)\gamma^\mu u_s(\mathbf{p}_i) = 2\delta_{rs}\,p_i^\mu$.) Thus, the insertion of a soft photon leads to a very simple modification, since the spinor is just multiplied by a simple factor that does not contain Dirac matrices. Likewise, if the photon is attached to an incoming fermion line, the substitution reads

$$u_r(\mathbf{p}_i) \to \frac{-e_i\,p_i^\mu}{p_i\cdot q - i0^+}\,u(\mathbf{p}_i). \qquad (3.111)$$

The substitution rules for anti-fermions are identical, and also for charged scalar particles (see Exercise 3.13). If we sum over all external fermion lines where a photon may be attached, the corrected scattering amplitude becomes

$$M^\mu_{\beta\alpha}(q) \underset{q\to 0}{=} M^\Lambda_{\beta\alpha} \sum_{i\in\mathcal{E}} \frac{\eta_i e_i p_i^\mu}{p_i\cdot q + i\eta_i 0^+}, \qquad (3.112)$$

where $e_i$ is the electrical charge of the external line $i$, and where $\eta_i = +1$ if the momentum $p_i$ is outgoing and $-1$ if it is incoming. ($\mathcal{E}$ denotes the set of all the external charged lines

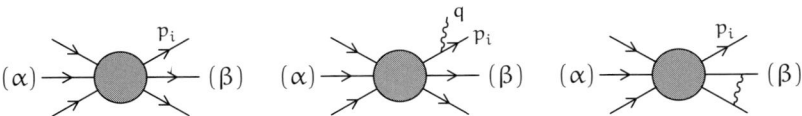

FIGURE 3.5: Left: generic hard scattering amplitude from a state $\alpha$ to a state $\beta$. Middle: real soft photon emission. Right: virtual correction by a soft photon.

of the hard amplitude.) Interestingly, this result is valid regardless of the spin of the charged external lines. Note that eq. (3.112) obeys the Ward–Takahashi identity. Indeed, we have

$$q_\mu M^\mu_{\beta\alpha}(q) = M^\Lambda_{\beta\alpha} \sum_{i\in\mathcal{E}} \frac{\eta_i e_i p_i \cdot q}{p_i \cdot q + i\eta_i 0^+} = M^\Lambda_{\beta\alpha} \underbrace{\sum_{i\in\mathcal{E}} \eta_i e_i}_{=0}. \tag{3.113}$$

The final zero comes from charge conservation: The incoming and outgoing charges in the hard process must have equal sums. Thus, in the limit of soft photons, this formula provides a very clear link between charge conservation and the Ward–Takahashi identities. Another property of eq. (3.112) is that it diverges when $q \to 0$. From the calculation, we can trace this back to the fact that the momentum $p_i$ is on-shell, $p_i^2 = m^2$. Therefore, generically, this divergence can only occur when the photon is attached to an external line. Soft photons attached to internal off-shell lines do not produce this kind of divergence.

Consider now the attachment of two soft photons with momenta $q_1$ and $q_2$. If they are inserted on two distinct external lines, $i \neq j$, the amplitude $M^\Lambda_{\beta\alpha}$ is simply multiplied by two factors of the type that appears in the right-hand side of eq. (3.112). The situation is less simple when the two photons are attached to the same external line. For an outgoing fermion, the substitution is

$$\overline{u}_r(p_i) \to (-ie_i)^2 \overline{u}_r(p_i)\gamma^{\mu_1} \frac{i(\slashed{p}_i + \slashed{q}_1 + m)}{(p+q_1)^2 - m^2 + i0^+} \gamma^{\mu_2} \frac{i(\slashed{p}_i + \slashed{q}_1 + \slashed{q}_2 + m)}{(p+q_1+q_2)^2 - m^2 + i0^+}$$

$$= \overline{u}_r(p_i) \frac{e_i p_i^{\mu_1}}{p_i \cdot q_1 + i0^+} \frac{e_i p_i^{\mu_2}}{p_i \cdot (q_1+q_2) + i0^+}. \tag{3.114}$$

This is the result for one of the two possible orderings of the photon insertions along the line of momentum $p_i$. If we add the term where the photon of momentum $q_2$ is the closest to the end of the line, we have

$$\overline{u}_r(p_i) \to \overline{u}_r(p_i) \left[ \frac{e_i p_i^{\mu_1}}{p_i \cdot q_1 + i0^+} \frac{e_i p_i^{\mu_2}}{p_i \cdot (q_1+q_2) + i0^+} \right.$$

$$\left. + \frac{e_i p_i^{\mu_2}}{p_i \cdot q_2 + i0^+} \frac{e_i p_i^{\mu_1}}{p_i \cdot (q_1+q_2) + i0^+} \right] = \overline{u}(p_i) \frac{e_i p_i^{\mu_1}}{p_i \cdot q_1 + i0^+} \frac{e_i p_i^{\mu_2}}{p_i \cdot q_2 + i0^+}. \tag{3.115}$$

Quite remarkably, this result has the same structure as if the photons were attached to different lines. Thus, in all cases, the hard amplitude corrected by two soft photons on the lines $i, j$ (with $i$ possibly equal to $j$) is

$$M^{\mu_1\mu_2}_{\beta\alpha}(q_1 q_2) \underset{q\to 0}{=} M^\Lambda_{\beta\alpha} \sum_{i,j\in\mathcal{E}} \frac{\eta_i e_i p_i^{\mu_1}}{p_i \cdot q_1 + i\eta_i 0^+} \frac{\eta_j e_j p_j^{\mu_2}}{p_j \cdot q_2 + i\eta_j 0^+}. \tag{3.116}$$

More generally, with $n$ soft photons attached to the same line, we must sum over all the orderings of the photon insertions along the line, so that eq. (3.115) becomes

$$\overline{u}_r(p_i) \to \overline{u}_r(p_i) \sum_{\sigma\in\mathfrak{S}_n} \frac{e_i p_i^{\mu_{\sigma(1)}}}{p_i \cdot q_{\sigma(1)} + i0^+} \frac{e_i p_i^{\mu_{\sigma(2)}}}{p_i \cdot (q_{\sigma(1)}+q_{\sigma(2)}) + i0^+} \cdots$$

$$\times \cdots \frac{e_i p_i^{\mu_{\sigma(n)}}}{p_i \cdot (q_{\sigma(1)} + \cdots + q_{\sigma(n)}) + i0^+}, \tag{3.117}$$

where $\mathfrak{S}_n$ is the group of permutations of $n$ elements. The sum over permutations can be done easily thanks to

$$\sum_{\sigma\in\mathfrak{S}_n}\frac{1}{a_{\sigma(1)}}\frac{1}{a_{\sigma(1)}+a_{\sigma(2)}}\cdots\frac{1}{a_{\sigma(1)}+\cdots+a_{\sigma(n)}}=\prod_{j=1}^n\frac{1}{a_j}, \qquad (3.118)$$

which may be proved by recursion on $n$, starting from the trivial case $n=1$. Thus, we conclude that the case of multiple photons attached to the same line is no different from the case in which they are attached to distinct lines: Each photon of momentum $q_a$ attached to an external line of momentum $p_i$ brings a factor $\eta_i e_i p_i^\mu/(p_i\cdot q_a+i0^+)$, and the corrected hard amplitude reads

$$M_{\beta\alpha}^{\mu_1\cdots\mu_n}(q_1\cdots q_n)\underset{q\to 0}{=}M_{\beta\alpha}^\Lambda\prod_{a=1}^n\left(\sum_{i\in\mathcal{E}}\frac{\eta_i e_i p_i^{\mu_a}}{p_i\cdot q_a+i\eta_i 0^+}\right). \qquad (3.119)$$

### 3.7.2 Virtual Corrections

From eq. (3.119), we can make photon loops such as the one shown in the graph on the right of Figure 3.5 by connecting pairwise the Lorentz indices with photon propagators. In Feynman gauge, it reads $G_F^{0\ \mu\mu'}(q)=-ig^{\mu\mu'}/(q^2+i0^+)$. In addition, we need a factor $1/2$ since when doing this we are connecting two photons of identical momenta (in the derivation of eq. (3.119), it has been implicitly assumed that all the photon momenta are distinct – when some have equal momenta, extra symmetry factors are necessary). Thus, the hard amplitude corrected by a loop with a soft photon is given by

$$M_{\beta\alpha}^{\text{virtual}}=M_{\beta\alpha}^\Lambda\frac{1}{2}\sum_{i,j\in\mathcal{E}}\eta_i\eta_j e_i e_j J_{ij}, \qquad (3.120)$$

where we denote

$$J_{ij}\equiv -i\int\frac{d^4q}{(2\pi)^4}\frac{p_i\cdot p_j}{(q^2+i0^+)(p_i\cdot q+i\eta_i 0^+)(-p_j\cdot q+i\eta_j\epsilon)}. \qquad (3.121)$$

If we perform $n$ such pairings in order to make $n$ loops, we need in addition a factor $1/n!$ since the $n$-photon integration domain is symmetric (without this factor, one would count the same contributions multiple times). Therefore, if we sum over all values of $n$, the amplitude fully corrected by soft virtual photon loops reads

$$M_{\beta\alpha}^{\text{virtual}}=M_{\beta\alpha}^\Lambda\,\exp\left\{\frac{1}{2}\left[\sum_{i,j\in\mathcal{E}}\eta_i\eta_j e_i e_j J_{ij}\right]\right\}. \qquad (3.122)$$

In this expression, the imaginary part of $J_{ij}$ modifies the amplitude by an irrelevant phase. Thus, we need only study the real part. The integral on $q$ diverges at small $q$ (this is called an *infrared divergence*). For the time being, we may regularize this integral by cutting out the soft region, by imposing $\lambda<|q|$. We should also impose $|q|<\Lambda$, because photons with momenta above $\Lambda$ are already included in the hard amplitude $M_{\beta\alpha}^\Lambda$. The integration over $q^0$ can be done in the complex plane with the theorem of residues. The integrand has four poles, $q_0 = \pm(|\mathbf{q}| - i0^+)$, $q_0 = \mathbf{p}_i \cdot \mathbf{q}/p_i^0 \pm i\eta_i\epsilon$, and an explicit calculation shows that

$$\text{Re } J_{ij} = \frac{1}{2}(p_i \cdot p_j) \int_{\lambda < q < \Lambda} \frac{d^3q}{(2\pi)^3} \frac{1}{q^3} \frac{1}{(p_i^0 - \mathbf{p}_i \cdot \hat{\mathbf{q}})(p_j^0 - \mathbf{p}_j \cdot \hat{\mathbf{q}})}, \tag{3.123}$$

where we denote $\hat{\mathbf{q}} \equiv \mathbf{q}/q$. The radial integral over $|\mathbf{q}|$ is trivial and gives a logarithm. The angular integral over the orientation of $\hat{\mathbf{q}}$ can be done simply by noting that the result is Lorentz invariant and should depend only on the invariants $p_i^2 = m_i^2$, $p_j^2 = m_j^2$ and $p_i \cdot p_j$. With this in mind, we may evaluate it in the frame where $\mathbf{p}_j = 0$, which renders the integral trivial, leading to

$$\text{Re } J_{ij} = \frac{1}{8\pi^2 \beta_{ij}} \ln\left(\frac{1 + \beta_{ij}}{1 - \beta_{ij}}\right) \ln\left(\frac{\Lambda}{\lambda}\right), \quad \text{with } \beta_{ij} \equiv \sqrt{1 - \frac{m_i^2 m_j^2}{(p_i \cdot p_j)^2}}. \tag{3.124}$$

($\beta_{ij}$ is the relative velocity of the particles of momenta $\mathbf{p}_{i,j}$.) Therefore, summing all virtual corrections with photon momenta in the range $\lambda \leq |\mathbf{q}| \leq \Lambda$, the squared amplitude becomes[13]

$$|M_{\beta\alpha}^{\text{virtual}}|^2 = |M_{\beta\alpha}^\Lambda|^2 \left(\frac{\lambda}{\Lambda}\right)^{A_{\beta\alpha}}, \quad \text{with } A_{\beta\alpha} \equiv -\frac{1}{8\pi^2} \sum_{i,j \in \mathcal{E}} \frac{\eta_i \eta_j e_i e_j}{\beta_{ij}} \ln\left(\frac{1 + \beta_{ij}}{1 - \beta_{ij}}\right). \tag{3.125}$$

### 3.7.3 Real Corrections and Cancellations

Consider now the emission of real soft photons in the final state, whose phase-spaces are integrated out since we are not interested in measuring them. Each emitted soft photon of momentum $\mathbf{q}$ brings a factor $\epsilon_\mu(\mathbf{q})\epsilon_{\mu'}^*(\mathbf{q})$ to the square of the amplitude. Recall that

$$\sum_{\text{pol.}} \epsilon_\mu(\mathbf{q})\epsilon_{\mu'}^*(\mathbf{q}) = -g_{\mu\mu'} + \text{terms in } q_\mu \text{ or } q_{\mu'}, \tag{3.126}$$

but the terms in $q_\mu, q_{\mu'}$ do not contribute thanks to the Ward–Takahashi identities. Thus, the squared amplitude with $n$ real photons emitted reads

$$|M_{\beta\alpha}(q_1 \cdots q_n)|^2 = |M_{\beta\alpha}^{\text{virtual}}|^2 \frac{1}{n!} \prod_{a=1}^n \left(-\sum_{i,j \in \mathcal{E}} \frac{\eta_i \eta_j e_i e_j (p_i \cdot p_j)}{(p_i \cdot q_a)(p_j \cdot q_a)}\right), \tag{3.127}$$

where the prefactor $1/n!$ is necessary in order to symmetrize the $n$-photon final state. Let us now integrate over the phase-space of these real photons. The lower bound of this integration

---

[13]Note how corrections that were additive when considered order by order exponentiate and become multiplicative when all orders are summed. Such multiplicative corrections are known as *Sudakov factors*.

should also be $\lambda < |\mathbf{q}|$ in order to avoid infrared divergences. The choice of the upper bound is dictated by the energy resolution of the detector apparatus, $E_{res}$. Indeed, since the emission of photons with $|\mathbf{q}| < E_{res}$ cannot be detected, we should allow the possibility of producing an arbitrary number of them. Summing on $n$ and integrating over the phase-space of the photons, we get

$$|M_{\beta\alpha}|^2 = |M_{\beta\alpha}^{virtual}|^2 \exp\left(-\sum_{ij\in\mathcal{E}} \eta_i\eta_j e_i e_j K_{ij}\right), \tag{3.128}$$

where $K_{ij}$ is the following integral

$$K_{ij} = \int_{\lambda < q < E_{res}} \frac{d^3\mathbf{q}}{(2\pi)^3 2q} \frac{1}{q^2} \frac{p_i \cdot p_j}{(p_i^0 - \mathbf{p}_i\cdot\widehat{\mathbf{q}})(p_j^0 - \mathbf{p}_j\cdot\widehat{\mathbf{q}})} = \frac{1}{8\pi^2\beta_{ij}} \ln\left(\frac{1+\beta_{ij}}{1-\beta_{ij}}\right) \ln\left(\frac{E_{res}}{\lambda}\right). \tag{3.129}$$

(It is identical to $\mathrm{Re}\,J_{ij}$, except for the integration range.) Therefore,

$$|M_{\beta\alpha}|^2 = \left(\frac{E_{res}}{\lambda}\right)^{A_{\beta\alpha}} \left(\frac{\lambda}{\Lambda}\right)^{A_{\beta\alpha}} |M_{\beta\alpha}^{\Lambda}|^2 = \left(\frac{E_{res}}{\Lambda}\right)^{A_{\beta\alpha}} |M_{\beta\alpha}^{\Lambda}|^2, \tag{3.130}$$

which has a finite limit when the infrared cutoff $\lambda$ goes to zero. What really happens is a cancellation of infrared divergent contributions between the virtual and real corrections. We can note here a profound aspect of infrared divergences, namely the fact that they are the result of asking a non-observable question. In the case of QED, the transition probability to a fully specified state (including the soft, unobservable, photons) is a rather pathological quantity, because it does not cost any energy to produce an extra zero-momentum photon. In fact, eq. (3.130) obtained after a careful summation tells us that this probability is zero if we take the limit $E_{res} \to 0$ (this limit corresponds to a perfect detector, a necessary condition to be able to fully specify the final state). This result can be understood from unitarity: The sum of the probabilities to go to a final state with some hard particles and $0, 1, 2, 3, \cdots$ soft photons must be less than 1. Since there is no limit on the number of soft photons, each probability in this sum must be zero.

# Exercises

**3.1** Check that the anti-commutation relations of Dirac matrices cannot be satisfied with $2 \times 2$ or $3 \times 3$ matrices in four space-time dimensions. *Hints:*

- *Assume matrices of size $n \times n$, and show that $(1-(-1)^n)\det(\gamma^\mu\gamma^\nu) = 0$ if $\mu \neq \nu$.*
- *For the $2 \times 2$ case, assume first that $\gamma^0$ is diagonalizable. What is its diagonal form? Show that the constraints on $\gamma^{1,2,3}$ have no solution.*
- *If $\gamma^0$ is not diagonalizable, what is its Jordan normal form?*

Extra: show that there are at most three non-singular anti-commuting $2 \times 2$ matrices.

**3.2** Check that the $M^{\mu\nu}$ defined in eq. (3.5) are generators of the Lorentz algebra. *Hint: Prove first the following intermediate results:*

$$\gamma^\mu \gamma^\nu = g^{\mu\nu} - 2i\, M^{\mu\nu}, \quad [M^{\mu\nu}, \gamma^\rho] = i(\gamma^\mu g^{\nu\rho} - \gamma^\nu g^{\mu\rho}).$$

**\*3.3** Calculate the traces of products of 1, 2, 3, 4 and 5 Dirac matrices. *Hint: For an even number of matrices, proceed by recursion, starting with 2. For an odd number of matrices, define* $\gamma^5 \equiv i\gamma^0 \gamma^1 \gamma^2 \gamma^3$, *and prove that* $(\gamma^5)^2 = 1$ *and* $\{\gamma^5, \gamma^\mu\} = 0$. *If G is the product of an odd number of Dirac matrices, show that* $\{\gamma^5, G\} = 0$ *and* $2G = [\gamma^5, \gamma^5 G]$.
Extra: calculate $\mathrm{tr}\,(\gamma^5 \gamma^\mu)$, $\mathrm{tr}\,(\gamma^5 \gamma^\mu \gamma^\nu)$, $\mathrm{tr}\,(\gamma^5 \gamma^\mu \gamma^\nu \gamma^\rho)$, and $\mathrm{tr}\,(\gamma^5 \gamma^\mu \gamma^\nu \gamma^\rho \gamma^\sigma)$.

**\*3.4** Show that a fermion loop with an odd number of photons attached is zero (this result is known as Furry's theorem). *Hint: Use the charge conjugation operator.*

**3.5** Prove the following identities,

$$\gamma^\mu \gamma^\nu \gamma_\mu = -2\gamma^\nu, \quad \gamma^\mu \gamma^\nu \gamma^\rho \gamma_\mu = 4g^{\nu\rho}, \quad \gamma^\mu \gamma^\nu \gamma^\rho \gamma^\sigma \gamma_\mu = -2\gamma^\sigma \gamma^\rho \gamma^\nu,$$

$$\gamma^\mu \gamma^\nu \gamma^\rho = g^{\mu\nu}\gamma^\rho + g^{\nu\rho}\gamma^\mu - g^{\mu\rho}\gamma^\nu - i\epsilon^{\mu\nu\rho\sigma}\gamma_\sigma \gamma^5.$$

**3.6** Determine the commutation relations of the free Hamiltonian with $a_{sp,in}$, $a^\dagger_{sp,in}$, $b_{sp,in}$ and $b^\dagger_{sp,in}$.

**3.7** Derive the reduction formula for the scattering amplitude $\langle q_{r\,out} | p_{s\,in} \rangle$ between two states populated by a fermion.

**3.8** Check that the cross-section calculated in Section 3.3.4 is gauge invariant.

**\*3.9** Calculate the electron–electron scattering cross-section.

**\*3.10** Calculate the total cross-section for the process $e^+ e^- \to \mu^+ \mu^-$.

**\*3.11** Consider the vertex function $\Gamma_r^\mu(k, p, q)$ of eq. (3.93). Show that its contribution to electron scattering off an external field $\mathcal{A}^\mu$ can be parametrized as follows:

$$\mathcal{A}_\mu(k)\bar{u}(-q)\Gamma_r^\mu(k,p,q)u(p) = e_r \mathcal{A}_\mu(k)\bar{u}(-q)\left[F_1(k^2)\gamma^\mu + i F_2(k^2)\frac{M^{\mu\nu}k_\nu}{m_r}\right]u(p).$$

Approximate this formula for a constant magnetic field. Calculate $F_2(0)$ at one loop.

**3.12** Consider the following Lagrangian for a complex scalar field:

$$\mathcal{L} \equiv (\partial_\mu \phi^*)(\partial^\mu \phi) - m^2 \phi^* \phi.$$

- Show that it has a global $U(1)$ invariance.
- What minimal coupling to a photon would turn this into a local symmetry?
- List the Feynman rules of the resulting theory.
- Calculate the photon self-energy at one loop. Check the Ward–Takahashi identity.

**3.13** Derive the rule for attaching a soft photon to a charged scalar particle described by the theory studied in the previous exercise. How does it differ from the emission of a soft photon by a spin-1/2 fermion?

# CHAPTER 4

# Spontaneous Symmetry Breaking

Until now, our discussion of the symmetry of a theory has been limited to a study of its Lagrangian or Hamiltonian, and we have tacitly assumed that the symmetry of the Lagrangian implies that the physics of this system exhibits the same symmetry to its full extent. However, strictly speaking, a symmetric Lagrangian only implies that the corresponding equations of motion are symmetric, i.e., that a symmetry transformation applied to a solution of the equations of motion gives another solution. In other words, the symmetry of the Lagrangian implies that the set of the solutions of the equations of motion is symmetric, not that every individual solution is symmetric. A *spontaneously broken symmetry* is a symmetry of the Lagrangian which is not realized by the ground state. In this chapter, we discuss this phenomenon, and its most important consequence, the Goldstone theorem.

## 4.1 Potential Energy Landscape

At lowest order in $\hbar$, the ground state of a system of fields is determined by the minima of their energy, which is the sum of a kinetic term (the sum of the squares of the derivatives of the field), and of the potential energy. Minimal energy configurations are obtained with fields that are spatially homogeneous and constant in time, which minimize the potential energy. In this section, we first consider some examples of potentials that have a non-trivial pattern of minima, and then discuss how degenerate classical minimal energy configurations are promoted into a quantum ground state.

## 4.1.1 $\phi^4$ Example

The scalar field theory we have considered as our main example in Chapters 1 and 2 is particularly simple because it has a single field and the potential energy,

$$V(\phi) = \frac{1}{2}m^2\phi^2 + \frac{\lambda}{4!}\phi^4, \tag{4.1}$$

is convex with a unique minimum at $\phi = 0$. Therefore, classically, a minimal energy density is obtained with a constant and homogeneous field whose value is $\phi = 0$ everywhere. Since this is the absolute minimum of the potential, such a configuration is classically stable, and moreover the stability cannot be upset by quantum fluctuations. Note also that the perturbative expansion, in which we assume $\lambda \ll 1$ and expand in powers of $\lambda$, may also be viewed as an expansion in small field amplitudes around $\phi = 0$, since the quartic term in the above potential becomes comparable to the quadratic term only for field amplitudes of order $m/\sqrt{\lambda} \gg m$. Moreover, in such a theory, the identification of the degrees of freedom is straightforward: It contains one scalar particle of (bare) mass m.

One may also consider the same theory with a quadratic term of the opposite sign,

$$V(\phi) = -\frac{1}{2}m^2\phi^2 + \frac{\lambda}{4!}\phi^4. \tag{4.2}$$

($\lambda$ should remain positive in order to prevent the fields from running away to infinity.) Now, $\phi = 0$ is a local maximum, and the potential has two minima at $\phi_* = \pm(6m^2/\lambda)^{1/2}$. Since $\phi = 0$ is not even stable classically, it is most certainly not a good point to start a perturbative expansion. In fact, even the identification of the degrees of freedom is not clear in this case; indeed, the quadratic term cannot be interpreted as a mass term since it has an incorrect sign for this. Intuitively, it seems one could choose one of the minima $\phi_*$, shift the field according to $\phi \equiv \phi_* + \varphi$, and rewrite the potential as

$$V(\phi) = -\frac{3m^4}{2\lambda} + m^2\varphi^2 + \frac{\lambda\phi_*}{6}\varphi^3 + \frac{\lambda}{4!}\varphi^4. \tag{4.3}$$

(The term linear in $\varphi$ is zero since $\phi_*$ is an extremum.) When quantizing this theory, $\phi_*$ is a c-number while $\varphi$ becomes the field operator, to which one applies the usual canonical formalism. This is all fine, except for the unjustified premise where we "choose" one of the two minima as the origin of the expansion (as opposed to a linear superposition of the two minima). We shall return to this point in Section 4.1.4.

## 4.1.2 Mass Matrix

The situation becomes more complicated in theories with multiple fields. Let us assume that the potential energy is a smooth function $V(\phi_1, \cdots, \phi_n)$ that depends on n real fields $\boldsymbol{\phi} \equiv (\phi_1, \cdots, \phi_n)$, and that this potential becomes infinite if $\boldsymbol{\phi}$ goes to infinity in any direction. Generically, such a potential may have multiple local extrema, determined by

$$\frac{\partial V(\boldsymbol{\phi})}{\partial \phi_i} = 0. \tag{4.4}$$

## 4.1 POTENTIAL ENERGY LANDSCAPE

The precise nature of these extrema depends on the mass matrix obtained from the second derivatives of the potential:

$$m_{ij}^2 \equiv \frac{\partial^2 V(\phi)}{\partial \phi_i \partial \phi_j}, \qquad (4.5)$$

evaluated at the extremum. If the eigenvalues of this matrix are all negative or zero,[1] then we have a local maximum, and a local minimum[2] if they are all positive or zero. The case of eigenvalues with mixed signs corresponds to a saddle point.

Classically, minima of the energy are obtained with constant fields whose value is one of the minima, and these minimal energy configurations are stable against small perturbations (provided that all the eigenvalues of the mass matrix are strictly positive). Quantum mechanically, local minima that are not the absolute minimum are unstable because quantum fluctuations allow the system to tunnel through the energy barrier. However, the tunneling rate is exponentially suppressed for field configurations that occupy a macroscopic region of space, implying that their lifetime may be extremely large. For this reason, such configurations may nevertheless be relevant in some physical situations (they are called *meta-stable*).

### 4.1.3 Symmetric Potentials

Let us now consider the important case where the theory is invariant under a local continuous transformation of the fields,

$$\phi_i(x) \quad \to \quad \Omega_{ij}\,\phi_j(x), \qquad (4.6)$$

where the matrix $\Omega_{ij}$ belongs to an orthogonal representation of some group of transformations $\mathcal{G}$. In such a theory, the potential energy must also be invariant under this transformation, as illustrated in Figure 4.1 in the case of a two-component field with an O(2) invariance. If $\phi_*$ is a local extremum of this potential, then all the fields obtained from $\phi_*$ by a symmetry transformation, $\phi_*^\Omega \equiv \Omega\phi_*$, are also extrema with the same potential energy. Although this is quite obvious, it can also be seen from the fact that the mass matrices at these two points are related by

$$m_{ij}^2(\phi_*^\Omega) = \Omega_{ik}^{-1}\Omega_{jl}^{-1}\, m_{kl}^2(\phi_*), \qquad (4.7)$$

and therefore have the same eigenvalues.

If we are at a local minimum $\phi_*$ of the potential, all the eigenvalues of the mass matrix at $\phi_*$ are positive or zero. Consider now the variations of the fields due to an infinitesimal symmetry transformation,

$$\phi_i - \phi_{*i} = i\,\epsilon^a\, t_{ij}^a\,\phi_{*j}. \qquad (4.8)$$

(The matrices $t_{ij}^a$ are called the generators of the transformations and the $\epsilon^a$ are real infinitesimal parameters.) For each generator such that $t^a\phi_* \neq 0$, let us define a vector

$$V_i^a \equiv i\,t_{ij}^a\,\phi_{*j}, \qquad (4.9)$$

---

[1] A vanishing eigenvalue indicates a direction along which the curvature of the potential is zero.
[2] If the potential is smooth and goes to $+\infty$ for large fields, then it has at least one minimum.

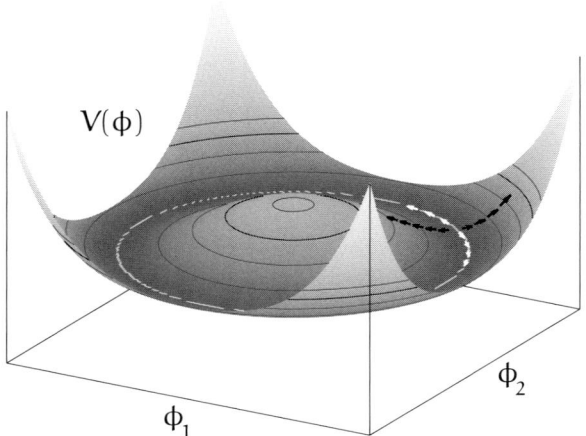

FIGURE 4.1: Simplest potential that exhibits the spontaneous breakdown of a continuous symmetry. A few contour lines are shown to guide the eye. The white arrows show the angular (massless) excitations, while the black arrows show the radial (massive) excitations.

and denote by $\mathcal{E}$ the subspace of $\mathbb{R}^n$ spanned by these vectors. The dimension of this subspace is the number of independent symmetry transformations that do not leave $\phi_*$ invariant. Since the potential energy is invariant under all the transformations (4.8), we have

$$m_{ij}^2(\phi_*) v_i w_j = 0 \quad \text{for all } v, w \in \mathcal{E}. \tag{4.10}$$

(This quantity is the second variation of the potential, and the first variation is zero since $\phi_*$ is an extremum.) The sum of the k smallest eigenvalues of a symmetric matrix A is given by (see Exercise 4.1)[3]

$$\sum \binom{\text{k smallest}}{\text{eigenvalues}} = \inf_{\substack{V \subset \mathbb{R}^n \\ \dim(V)=k}} \Big( \sum_{a=1}^{k} A_{ij} X_i^a X_j^a \Big), \tag{4.11}$$

where the vectors $\{X^a\}$ form an orthonormal basis of the subspace V (the sum on the right-hand side is the partial trace of A on the subspace V). Combining this result with eq. (4.10) and the fact that all eigenvalues of the mass matrix at $\phi_*$ are positive or zero, we conclude that there is a number of zero eigenvalues equal to the dimension of $\mathcal{E}$ (i.e., the number of symmetry transformations that change $\phi_*$). Here also, it is tempting to choose a specific minimum $\phi_*$ and expand the Lagrangian around this value of the field before quantization. From the above analysis, we would then conclude that the resulting quantum field theory has a number of massless excitations equal to the number of symmetries broken by the choice of that particular $\phi_*$. However, we should first clarify why it makes sense to choose one minimum out of many.

---

[3]This formula follows from Courant–Fischer's min–max theorem.

## 4.1 POTENTIAL ENERGY LANDSCAPE

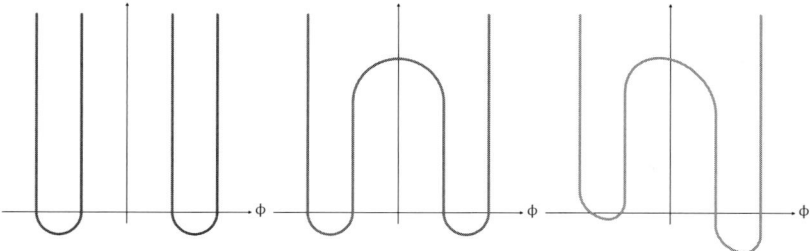

FIGURE 4.2: Left to right: potential for the Hamiltonians $H_0$, $H_0 + V$ and $H_0 + V + \tilde{V}$.

### 4.1.4 Degenerate Minima with Discrete Symmetries

Consider again the potential of eq. (4.2), which is invariant under a discrete symmetry R such that $R^2 = 1$ (a mirror symmetry $\phi \to -\phi$ in this case). The idea of spontaneous symmetry breaking is that the true ground state of the system should be one of the two minima of the potential. However, there is a standard result of quantum mechanics that – on the surface – seems to forbid the possibility of such a non-symmetric ground state. Indeed, the Hamiltonian H of this system commutes with the symmetry generator,

$$[R, H] = 0, \tag{4.12}$$

implying that H and R are diagonalizable simultaneously. Since $R^2 = 1$, the eigenvalues of R are $\pm 1$, and all eigenstates of H (including the ground state) are either symmetric or antisymmetric under R.

In order to see how to evade this obstruction in quantum field theory, let us consider a simple toy model of this situation with a potential made of two infinite wells centered at $\phi = \pm \phi_*$, mirror symmetric with respect to $\phi = 0$ (see left of Figure 4.2). Let us denote $H_0$ the Hamiltonian of this system. This Hamiltonian is a sum of two terms that do not talk to each other (because of the infinite energy barrier between the two wells), and each of the wells has its own ground state, which we denote $|0_+\rangle$ and $|0_-\rangle$, respectively. They are degenerate in energy, transform into one another by the action of R and have a vanishing overlap,

$$R|0_+\rangle = |0_-\rangle, \quad \langle 0_+|0_-\rangle = 0. \tag{4.13}$$

Then, we introduce a perturbation V, also mirror symmetric, such that the energy barrier between the two wells becomes finite (this interaction acts as a kind of coupling between the two wells, as shown in the middle of Figure 4.2). With this perturbation, we have

$$\langle 0_+|H_0 + V|0_+\rangle = \langle 0_-|H_0 + V|0_-\rangle = a,$$
$$\langle 0_+|H_0 + V|0_-\rangle = \langle 0_+|V|0_-\rangle = b \neq 0,$$
$$\langle 0_-|H_0 + V|0_+\rangle = \langle 0_-|V|0_+\rangle = b, \tag{4.14}$$

with $a, b$ real. When $b \neq 0$, the lowest energy eigenstates of the full Hamiltonian $H \equiv H_0 + V$ are no longer the states $|0_+\rangle$ and $|0_-\rangle$, but the combinations[4] $|0_+\rangle \pm |0_-\rangle$, whose eigenvalues

---

[4]For this conclusion to hold, the matrix elements of H between $|0_\pm\rangle$ and the higher-lying excited states should be negligible. Otherwise, the ground state of the perturbed Hamiltonian will be a more general linear combination of the eigenstates of $H_0$.

are $a \pm b$. These eigenstates are symmetric (with eigenvalue $a + b$) or antisymmetric (with eigenvalue $a - b$)[5] under R.

Note now that b is suppressed by the exponential of a quantity proportional to the spatial volume, since the field must go from $+\phi_*$ to $-\phi_*$ in the entire volume during the transition encoded in b (see Exercises 4.2 and 13.9 for illustrations of this property). In fact, b is zero if the volume is infinite, and the two eigenstates $|0_+\rangle \pm |0_-\rangle$ are then degenerate. If this system is then perturbed by an antisymmetric term $\widetilde{V}$ (see the right plot in Figure 4.2), no matter how small, the diagonal matrix elements[6] $\langle 0_\pm | \widetilde{V} | 0_\pm \rangle$ of the perturbation are much larger than the exponentially suppressed splitting 2b between the eigenvalues of H. Therefore, under the effect of this perturbation, the ground state (now unique since the perturbation $\widetilde{V}$ breaks the symmetry R) of the Hamiltonian is very close to one of the original states $|0_\pm\rangle$.

### 4.1.5 Degenerate Vacua with a Continuous Symmetry

Let us now return to the case of a continuous symmetry. When the volume is infinite, a ground state $|v\rangle$ is characterized by the fact that it is an eigenstate of the momentum $\mathbf{P}^i$ with a null eigenvalue:[7]

$$\mathbf{P}|v\rangle = 0. \tag{4.15}$$

There is in general a set of such states, that we may choose to be orthogonal,

$$\langle u|v\rangle = \delta_{uv}. \tag{4.16}$$

For any matrix element $\langle u|A(\mathbf{x})B(0)|v\rangle$ of the equal-time product of two local operators, we may insert a complete basis of states in order to get

$$\langle u|A(\mathbf{x})B(0)|v\rangle = \sum_{\text{vacua } w} \langle u|A(0)|w\rangle\langle w|B(0)|v\rangle$$
$$+ \int \frac{d^3\mathbf{p}}{(2\pi)^3} \sum_N \langle u|A(0)|N,\mathbf{p}\rangle\langle N,\mathbf{p}|B(0)|v\rangle\, e^{-i\mathbf{p}\cdot\mathbf{x}}, \tag{4.17}$$

where we have separated the ground states $|w\rangle$ from the continuum of populated states $|N,\mathbf{p}\rangle$ (the label N – possibly continuous – distinguishes all those states that have the same total momentum $\mathbf{p}$). To obtain this relationship, we have used the translation invariance of the ground states, and the fact that $\mathbf{P}$ is the generator of spatial translations. Since the states $|N,\mathbf{p}\rangle$ belong to a continuum of states, the integrand in the second line depends smoothly on $\mathbf{p}$ and the integral vanishes when $|\mathbf{x}| \to \infty$ by Riemann's lemma. Therefore, we have:

$$\lim_{|\mathbf{x}|\to+\infty} \langle u|A(\mathbf{x})B(0)|v\rangle = \sum_{\text{vacua } w} \langle u|A(0)|w\rangle\langle w|B(0)|v\rangle. \tag{4.18}$$

---

[5] The potential V that leads to the potential in the middle of Figure 4.2 is negative, and the off-diagonal matrix element b is therefore also negative. Thus, the ground state is the symmetric one.

[6] The non-diagonal matrix elements of $\widetilde{V}$ are zero per our assumption that $\widetilde{V}$ is odd under R.

[7] Multi-particle states whose total momentum is zero can be excluded by the fact that they are separated from ground states by a finite threshold.

Likewise, we may prove:

$$\lim_{|x|\to+\infty} \langle u|B(0)A(x)|v\rangle = \sum_{\text{vacua } w} \langle u|B(0)|w\rangle\langle w|A(0)|v\rangle. \tag{4.19}$$

Causality implies that $A(x)B(0) = B(0)A(x)$ since the separation between the two points is space-like. Therefore, the matrix elements $\langle u|A(0)|v\rangle$ and $\langle u|B(0)|v\rangle$ may be viewed as commuting Hermitian matrices, that we can diagonalize simultaneously.

Moreover, since A and B are arbitrary local Hermitian operators, this property is in fact true for all such operators. By choosing properly the basis of the vacua when the volume is infinite, all the local Hermitian operators have vanishing matrix elements between distinct vacua,

$$\langle u|A(0)|v\rangle = \delta_{uv} a_v. \tag{4.20}$$

In particular, any local interaction term that breaks the symmetry responsible for the degeneracy of these vacua is diagonal in this basis. Therefore, it lifts the degeneracy and promotes one of the states $|v\rangle$ to the status of true ground state of the system (instead of a symmetric linear combination of the $|v\rangle$s).

## 4.2 Conserved Currents and Charges

The fact that the Lagrangian is invariant under a continuous symmetry implies the existence of conserved currents $J_a^\mu(x)$ such that

$$\partial_\mu J_a^\mu(x) = 0. \tag{4.21}$$

An infinitesimal symmetry transformation of the fields is of the form

$$\phi_i(x) \;\to\; \phi_i(x) + i\,\epsilon^a\, t_{ij}^a \phi_j(x), \tag{4.22}$$

where the $t_{ij}^a$ are the generators of the Lie algebra of the group or transformations (see Chapter 7), in the representation where the fields $\phi_i$ live. For the fields $\phi_i$ to be Hermitian, the numbers $t_{ij}^a$ must be purely imaginary (this would be the case if the $\phi_i$ transform under the adjoint representation of the Lie group). From Noether's theorem, the conserved currents read

$$J_a^\mu(x) = \sum_i \frac{\partial \mathcal{L}}{\partial(\partial_\mu \phi_i(x))} \frac{\delta \phi_i(x)}{\delta \epsilon_a}. \tag{4.23}$$

By integrating over all space, we obtain conserved charges

$$Q_a(x^0) \equiv \int d^3x\, J_a^0(x^0, \mathbf{x}). \tag{4.24}$$

The time component of the currents has the form

$$J_a^0(x) = i\,\Pi_i(x)\, t_{ij}^a\, \phi_j(x), \tag{4.25}$$

where $\Pi_i(x)$ is the canonical momentum associated with $\phi_i(x)$. Since the matrices $t^a$ have imaginary components, these currents are Hermitian, as well as the charges $Q_a$. Using the canonical commutation relations,

$$[\phi_i(x), \phi_j(y)]_{x^0=y^0} = 0, \quad [\Pi_i(x), \Pi_j(y)]_{x^0=y^0} = 0,$$
$$[\phi_i(x), \Pi_j(y)]_{x^0=y^0} = i\delta_{ij}\delta(x-y), \tag{4.26}$$

we find the following equal-time commutator between components of the conserved currents:

$$[J_a^0(x), J_b^0(y)]_{x^0=y^0} = t_{ij}^a t_{kl}^b [\Pi_i(x)\phi_j(x), \Pi_k(y)\phi_l(y)]$$
$$= i\delta(x-y)[t^a, t^b]_{ij} \Pi_i(x)\phi_j(x). \tag{4.27}$$

Using also the commutation relation that defines the Lie bracket,

$$[t^a, t^b] = if^{abc} t^c \tag{4.28}$$

(the $f^{abc}$ are real numbers), we get

$$[J_a^0(x), J_b^0(y)]_{x^0=y^0} = \delta(x-y) f^{abc} J_c^0(x). \tag{4.29}$$

By integrating over the positions $x$ and $y$, this becomes a commutator between the conserved charges,

$$[Q_a(x^0), Q_b(x^0)] = f^{abc} Q_c(x^0). \tag{4.30}$$

In other words, the charges $Q_a(x^0)$ form a real representation of the Lie algebra. Moreover, the commutator between the conserved charges and the field operators is given by[8]

$$[Q_a(x^0), \phi_i(x)] = i\int d^3y \, [\Pi_k(y) t_{kl}^a \phi_l(y), \phi_i(x)]_{x^0=y^0}$$
$$= i\int d^3y (-i)\delta(x-y)\delta_{ki} t_{kl}^a \phi_l(x) = t_{ij}^a \phi_j(x). \tag{4.31}$$

Note that the above commutation relations are not affected by the spontaneous breaking of symmetry, since they follow from the properties of the field operators, regardless of the nature of the ground state of the system.

The ground state of the system is characterized by the expectation values of the field operators:

$$\langle \phi_i \rangle \equiv \langle 0|\phi_i(x)|0\rangle. \tag{4.32}$$

In order to see whether the ground state is invariant under the action of the symmetry transformations, let us study the variation of the quantity $\langle \phi_i \rangle$. It reads

$$\delta\langle \phi_i \rangle = \langle 0|\delta\phi_i(x)|0\rangle = i\epsilon_a t_{ij}^a \langle 0|\phi_j(x)|0\rangle$$
$$= i\epsilon_a \langle 0|[Q_a(x^0), \phi_i(x)]|0\rangle = i\epsilon_a \langle 0|Q_a\phi_i(x) - \phi_i(x)Q_a|0\rangle. \tag{4.33}$$

Thus, it is clear that these expectation values are invariant if the ground state is annihilated by all the generators of the Lie algebra (i.e., if $Q_a|0\rangle = 0$ for all $a$).

---

[8]Since the charges are conserved, we are free to evaluate them at the same time as the field $\phi_i$.

## 4.3 Spectral Properties

Consider now the expectation value in the ground state of the commutator between the conserved currents and the field operators,

$$\langle 0|[J_a^\mu(x),\phi_i(y)]|0\rangle = \langle 0|[J_a^\mu(x-y),\phi_i(0)]|0\rangle$$
$$= \sum_N \int d^4p\, \delta(p-p_N)\left[\langle 0|J_a^\mu(x-y)|N\rangle\langle N|\phi_i(0)|0\rangle\right.$$
$$\left. - \langle 0|\phi_i(0)|N\rangle\langle N|J_a^\mu(x-y)|0\rangle\right]$$
$$= \sum_N \int d^4p\, \delta(p-p_N)\left[\langle 0|J_a^\mu(0)|N\rangle\langle N|\phi_i(0)|0\rangle\, e^{ip\cdot(x-y)}\right.$$
$$\left. - \langle 0|\phi_i(0)|N\rangle\langle N|J_a^\mu(0)|0\rangle\, e^{ip\cdot(y-x)}\right]. \tag{4.34}$$

In the second line, we have summed over a complete set of states $|N\rangle$, arranged according to their 4-momentum $p_N$. We have also used the translation invariance of the ground state, and the properties of states with a definite momentum under translations. If we define

$$\mathcal{F}_{a,i}^\mu(p) \equiv (2\pi)^4 \sum_N \delta(p-p_N)\langle 0|J_a^\mu(0)|N\rangle\langle N|\phi_i(0)|0\rangle,$$
$$\widetilde{\mathcal{F}}_{a,i}^\mu(p) \equiv (2\pi)^4 \sum_N \delta(p-p_N)\langle 0|\phi_i(0)|N\rangle\langle N|J_a^\mu(0)|0\rangle, \tag{4.35}$$

we have

$$\mathcal{F}_{a,i}^\mu(p) = \left(\widetilde{\mathcal{F}}_{a,i}^\mu(p)\right)^*, \tag{4.36}$$

since $J_a^\mu$ and $\phi_i$ are Hermitian. Moreover, Lorentz invariance implies that these objects have the following form:

$$\mathcal{F}_{a,i}^\mu(p) = p^\mu\, \theta(p^0)\, \rho_{a,i}(p^2), \quad \widetilde{\mathcal{F}}_{a,i}^\mu(p) = p^\mu\, \theta(p^0)\, \widetilde{\rho}_{a,i}(p^2), \tag{4.37}$$

where $\rho_{a,i}$ and $\widetilde{\rho}_{a,i}$ are functions (so far unspecified) depending only on the invariant $p^2$. The factor $\theta(p^0)$ follows from the fact that the physical states $|N\rangle$ have a positive energy. Then, by inserting a unit factor expressed as

$$1 = \int ds\, \delta(p^2 - s), \tag{4.38}$$

we obtain

$$\langle 0|[J_a^\mu(x),\phi_i(y)]|0\rangle = \partial_x^\mu \int \frac{ds}{2\pi i}\left[\rho_{a,i}(s)\Delta(x-y;s) + \widetilde{\rho}_{a,i}(s)\Delta(y-x;s)\right], \tag{4.39}$$

where we denote

$$\Delta(x-y;s) \equiv \int \frac{d^4p}{(2\pi)^4}\, 2\pi\, \theta(p^0)\, \delta(p^2 - s)\, e^{ip\cdot(x-y)}. \tag{4.40}$$

This function obeys the Klein–Gordon equation with the "mass" s,

$$(\Box_x + s)\Delta(x - y; s) = 0, \tag{4.41}$$

and is Lorentz invariant. When the interval $x - y$ is space-like, it cannot depend separately on $x^0 - y^0$ since the sign of $x^0 - y^0$ is not invariant for a space-like separation. Therefore, for such an interval, it can depend only on $(x - y)^2$ and s, and we thus have

$$\Delta(x - y; s) = \Delta(y - x; s) \quad \text{if } (x - y)^2 < 0. \tag{4.42}$$

Therefore,

$$\langle 0|[J_a^\mu(x), \phi_i(y)]|0\rangle = \partial_x^\mu \int \frac{ds}{2\pi i} \left[\rho_{a,i}(s) + \tilde{\rho}_{a,i}(s)\right] \Delta(y - x; s) \quad \text{if } (x - y)^2 < 0. \tag{4.43}$$

Since the commutator on the left-hand side vanishes for local operators with a space-like separation, we get[9]

$$\rho_{a,i}(s) + \tilde{\rho}_{a,i}(s) = 0. \tag{4.44}$$

Returning to the case of a generic interval $x - y$, we thus have

$$\langle 0|[J_a^\mu(x), \phi_i(y)]|0\rangle = \partial_x^\mu \int \frac{ds}{2\pi i} \rho_{a,i}(s) \left[\Delta(x - y; s) - \Delta(y - x; s)\right]. \tag{4.45}$$

By applying the derivative $\partial_{x\mu}$ to both sides of this equation, and using the Klein–Gordon equation and the fact that the current $J_a^\mu(x)$ is conserved, we get

$$0 = \int \frac{ds}{2\pi i} \, s \, \rho_{a,i}(s) \left[\Delta(x - y; s) - \Delta(y - x; s)\right], \tag{4.46}$$

which implies

$$s \, \rho_{a,i}(s) = 0. \tag{4.47}$$

Therefore, $\rho_{a,i}(s) = 0$ for all $s \neq 0$, and the only possible support of $\rho_{a,i}(s)$ is localized at $s = 0$ (in the form of a delta function $\delta(s)$).

Let us now show that it is not possible that $\rho_{a,i}(s)$ be identically zero everywhere (including at $s = 0$) when the symmetry is spontaneously broken. By setting $\mu = 0$ and $x^0 = y^0$, and using eq. (4.40), we obtain

$$\langle 0|[J_a^0(x), \phi_i(y)]|0\rangle \Big|_{x^0=y^0} = 2i \int \frac{ds}{2\pi i} d^4p \sqrt{\mathbf{p}^2+s} \, \rho_{a,i}(s) \, e^{i\mathbf{p}\cdot(\mathbf{x}-\mathbf{y})} \delta(p^2 - s)$$

$$= i\delta(\mathbf{x} - \mathbf{y}) \int \frac{ds}{2\pi i} \rho_{a,i}(s). \tag{4.48}$$

---

[9]This property, combined with eq. (4.36), implies that $\rho_{a,i}(p^2)$ is real.

Then, we can integrate over $x$ and use the commutation relation (4.31) in order to get

$$t_{ij}^a \langle \phi_j \rangle = i \int \frac{ds}{2\pi i} \, \rho_{a,i}(s). \tag{4.49}$$

Thus, the functions $\rho_{a,i}(s)$ that have a non-zero integral are in one-to-one correspondence with the non-zero $t_{ij}^a \langle \phi_j \rangle$, i.e., with the fact that the ground state is non-invariant under the action of some of the symmetry generators. When this happens, we must have

$$i \rho_{a,i}(s) = 2\pi \, \delta(s) \, t_{ij}^a \langle \phi_j \rangle. \tag{4.50}$$

This equation is the essence of *Goldstone's theorem*. Note now that $\rho_{a,i}$ is a spectral function similar to the one defined in Section 1.6. Therefore, the presence of a $\delta(s)$ in this function signals the existence of a one-particle state with zero mass in the sum of eq. (4.35). Moreover, this result indicates that there are as many such massless particles (called Nambu–Goldstone modes) as there are broken symmetries by the ground state. Intuitively, these massless modes correspond to excitations along the bottom of the potential energy valley (illustrated with white arrows in Figure 4.1), which does not cost any energy since this is a contour line of minima of the potential energy. In contrast, the "radial" excitations have to climb up from the minimum and are thus massive (see the black arrows in Figure 4.1).

Finally, let us note that the state $\phi_i(0)|0\rangle$ is invariant under rotations, which implies that the matrix element $\langle N|\phi_i(0)|0\rangle$ is zero unless the state $\langle N|$ has a vanishing angular momentum. Thus, only spin 0 particles can contribute to the $\delta(s)$ in the non-zero spectral functions. Moreover, $\langle 0|J_a^0(0)|N\rangle$ vanishes for any state $|N\rangle$ whose quantum numbers differ from those of $J_a^0$. Thus, *the Nambu–Goldstone modes are spin-0 particles that have the same internal quantum numbers as $J_a^0$*.

## 4.4 Coleman's Theorem

In $1 + 1$ space-time dimensions, some of the above manipulations in the derivation of Goldstone's theorem are flawed because $\Delta(x - y; s = 0)$ is not defined. Indeed, we have

$$\Delta(x - y; 0) = \int \frac{d^2p}{(2\pi)^2} \, 2\pi \, \theta(p^0) \, \delta(p^2) \, e^{ip \cdot (x-y)}$$

$$= \int_0^{+\infty} \frac{dp_1}{\pi |p_1|} \, e^{i|p_1|(x^0 - y^0)} \, \cos(p_1(x^1 - y^1)) = \infty. \tag{4.51}$$

(The divergence of the integral comes from $p^1 \to 0$, i.e., from long-distance physics.) In order to ascertain what happens in $1 + 1$ dimensions, we need an alternative derivation that does not use the object $\Delta(x - y; s)$. Let us start from the two-dimensional version of the quantity $\mathcal{F}_{a,i}^\mu$ introduced earlier,

$$\mathcal{F}_{a,i}^\mu(p) \equiv \int d^2x \, e^{-ip \cdot x} \, \langle 0|J_a^\mu(x)\phi_i(0)|0\rangle, \tag{4.52}$$

complemented by the following two objects:

$$\mathcal{F}_{j,i}(p) \equiv \int d^2x \, e^{-ip \cdot x} \, \langle 0|\phi_j(x)\phi_i(0)|0\rangle,$$
$$\mathcal{F}_{a,b}^{\mu\nu}(p) \equiv \int d^2x \, e^{-ip \cdot x} \, \langle 0|J_a^\mu(x)J_b^\nu(0)|0\rangle. \tag{4.53}$$

Current conservation implies that $p_\mu \mathcal{F}_{a,i}^\mu(p) = 0$, which in two dimensions admits the following general solution:

$$\mathcal{F}_{a,i}^\mu(p) = \theta(p^0)\left[i\,\sigma_{a,i}\, p^\mu\, \delta(p^2) + \rho_{a,i}(p^2)\, \epsilon^{\mu\nu} p_\nu\right]. \tag{4.54}$$

(In two dimensions, there are only two possible linearly independent vectors.) The integral performed in the last step of the derivation of Goldstone's theorem, proportional to the infinitesimal variation of the field vacuum expectation value, is now given by (see Exercise 4.3)

$$t_{ij}^a \langle \phi_j \rangle = \int dx^1 \, \langle 0|[J_a^0(x^0, x^1), \phi_i(0)]|0\rangle = \frac{i\,\sigma_{a,i}}{2\pi}. \tag{4.55}$$

(The term in $\epsilon^{\mu\nu}$ does not contribute to this integral.) Given some arbitrary function $h(x)$ and constants $\alpha$ and $\beta$, let us now consider the state

$$|a, i; h\alpha\beta\rangle \equiv \int d^2x \, h(x)\left[\alpha J_a^0(x) + \beta \phi_i(x)\right]|0\rangle. \tag{4.56}$$

The norm of this state may be written as follows

$$\langle a, i; h\alpha\beta | a, i; h\alpha\beta\rangle = (\alpha^* \quad \beta^*) \begin{pmatrix} \mathbb{F}_{a,a}^{00} & \mathbb{F}_{a,i}^0 \\ \mathbb{F}_{a,i}^{0\,*} & \mathbb{F}_{i,i} \end{pmatrix}\begin{pmatrix} \alpha \\ \beta \end{pmatrix}, \tag{4.57}$$

where we have defined

$$\mathbb{F}_{a,a}^{00} \equiv \int \frac{d^2p}{(2\pi)^2}\, |\tilde{h}(p)|^2\, \mathcal{F}_{a,a}^{00}(p),$$
$$\mathbb{F}_{a,i}^{0} \equiv \int \frac{d^2p}{(2\pi)^2}\, |\tilde{h}(p)|^2\, \mathcal{F}_{a,i}^{0}(p),$$
$$\mathbb{F}_{i,i} \equiv \int \frac{d^2p}{(2\pi)^2}\, |\tilde{h}(p)|^2\, \mathcal{F}_{i,i}(p). \tag{4.58}$$

($\tilde{h}(p)$ is the Fourier transform of $h(x)$.) For this norm to be positive for all choices of $h, \alpha, \beta$, the quantities $\mathbb{F}_{a,a}^{00}$ and $\mathbb{F}_{i,i}$ must be positive, and the following inequality must also hold true:

$$\mathbb{F}_{i,i}\, \mathbb{F}_{a,a}^{00} \geq \left|\mathbb{F}_{a,i}^0\right|^2. \tag{4.59}$$

Let us now use the following special test function

$$\widetilde{h}(p) \equiv f(\lambda p^+)g(p^-) + f(\lambda p^-)g(p^+), \tag{4.60}$$

where $p^\pm \equiv p^0 \pm p^1$, $\lambda$ is a constant, $f$ is an even positive function peaked at the origin, and $g$ is a function that vanishes at the origin. With this function $\widetilde{h}$, we have (see Exercise 4.4)

$$\mathbb{F}^0_{a,i} = i\sigma_{a,i} \underbrace{|f(0)|^2 \int \frac{dp^-}{(2\pi)^2} |g(p^-)|^2}_{\text{constant indep. of } \lambda}. \tag{4.61}$$

On the other hand, given our assumptions about the functions $f$ and $g$, the left-hand side of the inequality (4.59) vanishes when $\lambda \to +\infty$. Therefore, $\sigma_{a,i} = 0$ for all $a$ and $i$, implying that spontaneous symmetry breaking cannot happen in $1+1$ dimensions. Physically, this is related to the divergence of the two-dimensional two-point function in eq. (4.51), which leads to very large quantum fluctuations. Any state initially centered on one of the degenerate vacua will subsequently develop large quantum fluctuations in the flat directions of the potential, such that the resulting state is a symmetric superposition of all the degenerate configurations.

## 4.5 Linear Sigma Model

### 4.5.1 Definition

The linear sigma model is a phenomenological description of the interactions between pions and nucleons, consistent with the symmetries of quantum chromodynamics (the theory of strong nuclear interactions between quarks and gluons, that we shall study later) at low energy. It is defined by the following Lagrangian:

$$\mathcal{L} \equiv \mathcal{L}_0 + \epsilon \mathcal{L}_1, \tag{4.62}$$

with

$$\mathcal{L}_0 \equiv \overline{\psi}_i \left( i\delta_{ij}\slashed{\partial} - g(\sigma\delta_{ij} + i\pi^a \sigma^a_{ij}\gamma^5) \right) \psi_j + \frac{1}{2}(\partial_\mu \sigma \partial^\mu \sigma + \partial_\mu \pi^a \partial^\mu \pi^a)$$
$$- \frac{\mu^2}{2}(\sigma^2 + \pi^a \pi^a) - \frac{\lambda^2}{4}(\sigma^2 + \pi^a \pi^a)^2,$$
$$\mathcal{L}_1 \equiv \sigma. \tag{4.63}$$

The matrix $\gamma^5$ is defined as $\gamma^5 \equiv i\gamma^0\gamma^1\gamma^2\gamma^3$ (see Exercise 3.3). The spinors $\psi_i(x)$ ($i = 1, 2$) represent the proton and the neutron. The fields $\sigma$ and $\pi^a$ ($a = 1, 2, 3$) are scalar fields, with the $\pi^a$ representing the three pions. $\lambda$ and $\epsilon$ are positive constants, but $\mu^2$ may have either sign. (In this section, be careful not to confuse the Pauli matrices $\sigma^a$ with the field $\sigma$.) It is useful to recall the following identity:

$$\sigma^a \sigma^b = \delta^{ab} + i\epsilon_{abc}\sigma^c. \tag{4.64}$$

The indices i and a can be omitted by combining the spinors, scalar fields and Pauli matrices into multiplets:

$$\psi \equiv \begin{pmatrix} \psi_1 \\ \psi_2 \end{pmatrix}, \quad \pi \equiv \begin{pmatrix} \pi^1 \\ \pi^2 \\ \pi^3 \end{pmatrix}, \quad \text{and} \quad \sigma \equiv \begin{pmatrix} \sigma^1 \\ \sigma^2 \\ \sigma^3 \end{pmatrix}. \quad (4.65)$$

With these compact notations, the Lagrangian $\mathcal{L}_0$ reads

$$\mathcal{L}_0 = \overline{\psi}[i\slashed{\partial} - g(\sigma + i\pi \cdot \sigma \gamma^5)]\psi$$
$$+ \frac{1}{2}(\partial_\mu \sigma \partial^\mu \sigma + \partial_\mu \pi \cdot \partial^\mu \pi) - \frac{\mu^2}{2}(\sigma^2 + \pi^2) - \frac{\lambda^2}{4}(\sigma^2 + \pi^2)^2. \quad (4.66)$$

### 4.5.2 Symmetries of the Model

**Invariance under $SU(2) \times SU(2)$:** The kinetic term of the fermions may be written as

$$\overline{\psi}\slashed{\partial}\psi = \overline{\psi}_R \gamma^\mu \partial_\mu \psi_R + \overline{\psi}_L \gamma^\mu \partial_\mu \psi_L, \quad (4.67)$$

where we have introduced the left- and right-handed projections of the spinors (see Exercise 4.5),

$$\psi_R \equiv \frac{1+\gamma^5}{2}\psi, \quad \psi_L \equiv \frac{1-\gamma^5}{2}\psi. \quad (4.68)$$

Likewise, the interaction term between the fermions and the scalar fields takes the form

$$\overline{\psi}(\sigma + i\pi \cdot \sigma \gamma^5)\psi = \overline{\psi}_L \Sigma \psi_R + \overline{\psi}_R \Sigma^\dagger \psi_L, \quad (4.69)$$

with the compact notation $\Sigma \equiv \sigma + i\pi \cdot \sigma$. Therefore, given any pair $U, V$ of matrices in the fundamental representation of $SU(2)$, the part of the Lagrangian that contains fermions is invariant under the following transformation:

$$\psi_R \to U\psi_R, \quad \psi_L \to V\psi_L, \quad \Sigma \to V\Sigma U^\dagger. \quad (4.70)$$

Moreover, since $\sigma^2 + \pi^2 = \det(\Sigma)$ is invariant in this transformation ($U, V$ have a determinant equal to 1), the other terms of $\mathcal{L}_0$ are also invariant (provided the transformation is a global one, so that the matrices $U, V$ commute with spatial derivatives). In contrast, $\mathcal{L}_1$ is not invariant under this transformation, because it mixes the $\sigma$ field with the pions. Likewise, an explicit Dirac mass term for the fermions would also explicitly break this invariance. Therefore, the symmetries of this model seem to forbid nucleon masses.

**Axial and vector symmetries:** Any matrix in the fundamental representation of $SU(2)$ can be parametrized as $\exp(i\theta \cdot \sigma/2)$, where $\theta$ is a vector with three real components. Thus, the matrices $U, V$ that define the transformation (4.70) can be written as

$$U \equiv e^{\frac{i}{2}(\alpha-\beta)\cdot\sigma}, \quad V \equiv e^{\frac{i}{2}(\alpha+\beta)\cdot\sigma}. \quad (4.71)$$

From this, we obtain the following infinitesimal form for the transformation of the various fields of the model:

$$\psi_L \to \psi_L + \tfrac{i}{2}(\alpha+\beta)\cdot\sigma\,\psi_L, \quad \psi_R \to \psi_R + \tfrac{i}{2}(\alpha-\beta)\cdot\sigma\,\psi_R,$$
$$\sigma \to \sigma - \beta\cdot\pi, \quad \pi \to \pi - \alpha\times\pi + \beta\sigma. \tag{4.72}$$

We see that the $\sigma$ field (i.e., the term $\mathcal{L}_1$ in the Lagrangian) is invariant under the subgroup of transformations for which $\beta = 0$ (this is the "diagonal" subgroup where $U = V$, i.e., the left- and right-handed parts of the spinors are rotated in the same way). It is this residual invariance that justifies a posteriori our choice of parametrizing $U$ and $V$ as in eq. (4.71). The invariance under the transformations induced by $\alpha$ is called *vector symmetry*, while the one under the transformations induced by $\beta$ is called *axial symmetry* (or *chiral symmetry*). Note also that the configuration $\sigma = \pi = 0$ is invariant under both vector and axial symmetries, while the configurations $\sigma \neq 0, \pi = 0$ are invariant under vector symmetry but break axial symmetry.

### 4.5.3 Symmetry Breaking

The self-interaction potential for the scalar fields, at $\pi = 0$, is the following function of $\sigma$

$$V(\sigma) = \frac{\lambda^2}{4}\sigma^4 + \frac{\mu^2}{2}\sigma^2 - \epsilon\sigma. \tag{4.73}$$

A value $\sigma_*$ of the sigma field that minimizes this potential must therefore obey

$$(\mu^2 + \lambda^2 \sigma_*^2)\sigma_* = \epsilon. \tag{4.74}$$

Consequently, when $\epsilon \neq 0$, the ground state of the system cannot be symmetric (since $\sigma_* = 0$ is not a solution of the above equation). This was of course perfectly expected since the term $\mathcal{L}_1$ explicitly breaks the axial symmetry.

More interesting is the fact that there can be non-symmetric minima even when $\epsilon = 0$, if $\mu^2 < 0$, despite the fact that the Lagrangian is itself perfectly symmetric under both vector and axial symmetries. This is a case of spontaneous symmetry breaking. By defining $\sigma \equiv \sigma_* + s$, the Lagrangian can be rewritten as

$$\mathcal{L} = \overline{\psi}[i\partial\!\!\!/ - g(s + i\pi\cdot\sigma\gamma^5)]\psi - g\sigma_*\overline{\psi}\psi + \frac{1}{2}(\partial_\mu s \partial^\mu s + \partial_\mu\pi\cdot\partial^\mu\pi)$$
$$- \frac{\mu^2}{2}(s^2 + \pi^2) - \frac{\lambda^2}{4}(4\sigma_*^2 s^2 + (s^2+\pi^2)(s^2+\pi^2+4\sigma_* s + 2\sigma_*^2)). \tag{4.75}$$

The term in $\overline{\psi}\psi$ gives a mass $m_N \equiv g\sigma_*$ to the nucleons. Likewise, we may read the mass of the $\sigma$ and $\pi$ fields from the corresponding quadratic terms:

$$m_\sigma^2 = \mu^2 + 3\lambda^2\sigma_*^2, \quad m_\pi^2 = \mu^2 + \lambda^2\sigma_*^2. \tag{4.76}$$

Note that the condition that defines the non-trivial minima can also be written as

$$m_\pi^2 = \frac{\epsilon}{\sigma_*}. \tag{4.77}$$

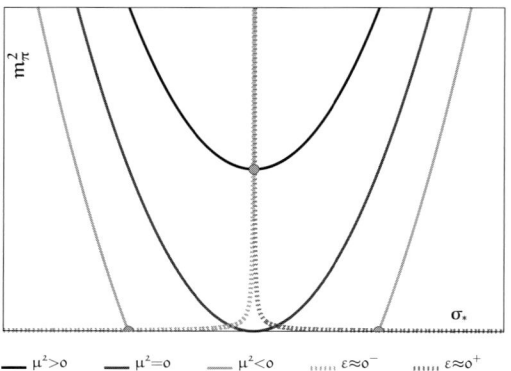

FIGURE 4.3: Graphical determination of $\sigma_*$ and $m_\pi$ (see the text for details). Solid: lines of constant $\mu^2$. Dashed: lines of constant $\epsilon$.

The various solutions may be visualized by considering the plane of the quantities $(m_\pi^2, \sigma_*)$, and by plotting the curves of constant $\mu^2$ and the lines of constant $\epsilon$ (see Figure 4.3). The intersection between the curve of constant $\mu^2$ and that of constant $\epsilon$ gives directly the value of $\sigma_*$ and the pion mass for this choice of parameters. In the limit $\epsilon \to 0$ of a symmetric Lagrangian, we have the following:

- If $\mu^2 > 0$, then $\sigma_* = 0$. Therefore, the ground state is symmetric, and the masses of the various fields are

$$m_N = 0, \quad m_\pi = m_\sigma = \mu. \qquad (4.78)$$

- If $\mu^2 < 0$, then the intersection happens at $\sigma_* \neq 0$, and the axial symmetry is spontaneously broken. Note also how changing $\epsilon$ from a positive infinitesimal quantity to a negative one changes the sign of $\sigma_*$. Moreover, the pion mass is zero, and the other masses are

$$m_N = g\sigma_*, \quad m_\sigma^2 = \mu^2 + 3\lambda^2 \sigma_*^2. \qquad (4.79)$$

The sigma model may be viewed as an effective description of the strong interactions in the very low energy regime, which reproduces the symmetries of quantum chromodynamics, in particular its $SU(2) \times SU(2)$ chiral symmetry due to the nearly zero masses of the two lightest quarks. In the sigma model, the term linear in $\sigma$ (that breaks explicitly the axial symmetry) mimics the effect of a non-zero mass for the light quarks. The fact that the pions are much lighter than the typical energy scales of strong interactions is naturally explained by the fact that they are the (quasi) Nambu–Goldstone bosons generated in the spontaneous breakdown of the axial symmetry, and the fact that there are three light pions is in line with the fact that three symmetries are broken, from $SU(2) \times SU(2)$ to a residual $SU(2)$ symmetry. In this mechanism, the nucleon mass is also related to spontaneous symmetry breaking, since it is proportional to the vacuum expectation value $\sigma_*$.

## 4.6 Heisenberg Model of Ferromagnetism

Spontaneous symmetry breaking also plays a central role in condensed matter systems, because many phase transitions amount to a reduction of symmetry when some external control parameters are changed (e.g., the temperature). Let us illustrate this by the example of the Heisenberg model, a phenomenological model of ferromagnets whose Hamiltonian reads

$$\mathcal{H} \equiv -J \sum_{\langle i,j \rangle} \mathbf{s}_i \cdot \mathbf{s}_j, \tag{4.80}$$

where the summation is extended to neighboring sites of a cubic lattice, on the nodes of which live spin operators $\mathbf{s}_i$. For a spin 1/2, $\mathbf{s} \equiv \boldsymbol{\sigma}/2$, where $\sigma^1, \sigma^2, \sigma^3$ are the three Pauli matrices. Since $\mathbf{s} \cdot \mathbf{s}$ and $\mathbf{s}^3$ commute, we may choose spin states that are common eigenstates $|m\rangle$ of these two operators, such that $\mathbf{s}^3|m\rangle = m|m\rangle$ ($m = \pm\frac{1}{2}$). The quantum number $m$ may be raised or lowered with $\mathbf{s}^\pm \equiv \mathbf{s}^1 \pm i\mathbf{s}^2$:

$$\mathbf{s}^\pm |m\rangle = \sqrt{\tfrac{3}{4} - m(m \pm 1)}\, |m \pm 1\rangle. \tag{4.81}$$

The constant J controls the strength of the interactions between neighboring spins. Assuming that $J > 0$, this Hamiltonian describes the fact that unpaired electrons on the atoms at each site of a lattice tend to align their spins in order to minimize the energy. The above Hamiltonian is invariant when all the spins are rotated by the same angle. This transformation is generated by the following unitary operator:[10]

$$U(\theta_1, \theta_2, \theta_3) \equiv e^{i\boldsymbol{\theta} \cdot \sum_i \mathbf{s}_i}, \tag{4.82}$$

where $\theta_{1,2,3}$ are the three angles defining the rotation. The invariance under rotations corresponds to the conservation of the total spin,

$$\mathbf{S} \equiv \sum_i \mathbf{s}_i. \tag{4.83}$$

(One may indeed check that $[\mathcal{H}, \mathbf{S}] = 0$ – see Exercise 4.6.)

Although the above Hamiltonian is invariant under rotations, it admits eigenstates that do break rotation symmetry. Consider for instance the state in which all the spins of the lattice are aligned in the $+z$ direction, which we denote by $|all \uparrow \rangle$. By rewriting the Hamiltonian as

$$\mathcal{H} = -J \sum_{\langle i,j \rangle} \left( \tfrac{1}{2}(s_i^+ s_j^- + s_i^- s_j^+) + s_i^3 s_j^3 \right), \tag{4.84}$$

we may check that

$$\mathcal{H}|all \uparrow \rangle = -\frac{3NJ}{4}|all \uparrow \rangle, \tag{4.85}$$

---

[10] Note that the exponential of the sum can be rewritten as a product of exponentials, because the spin operators on different lattice sites commute.

where N is the number of nodes of the cubic lattice (therefore, this lattice has 3N edges, if we assume periodic boundary conditions). Flipping any spin in this configuration will raise the energy since this spin will be surrounded by spins with the opposite orientation. Consider now the state obtained by applying the rotation of angle θ to all spins,

$$|all \nearrow\rangle_\theta \equiv U(\theta) |all \uparrow\rangle. \tag{4.86}$$

Given the rotational invariance of the Hamiltonian, this state has the same energy as the state with all spins up. Using the explicit form of the rotation matrix for a single spin 1/2,

$$U(\theta) = \cos\tfrac{\theta}{2} + 2i\, s^a \widehat{\theta}_a \sin\tfrac{\theta}{2}, \tag{4.87}$$

where $\theta \equiv \sqrt{\theta_a \theta_a}$ and $\widehat{\theta}_a \equiv \theta_a/\theta$, we obtain (see Exercise 4.7)

$$\langle all \uparrow | all \nearrow\rangle_\theta = \left(\cos\tfrac{\theta}{2} + i\widehat{\theta}_3 \sin\tfrac{\theta}{2}\right)^N. \tag{4.88}$$

Unless $\theta = (0, 0, \theta_3)$ (i.e., the rotation is around the z axis, which just changes the state by a phase), the number under the exponent on the right-hand side has a modulus strictly smaller than 1, and its Nth exponent vanishes when $N \to \infty$. Thus, in the limit of an infinite lattice, the states obtained by rotating the all-up state are all mutually orthogonal, and there is a continuous infinity of ground states related by rotations. Any perturbation that breaks rotational symmetry, such as the coupling to an external magnetic field, no matter how small, will lift this degeneracy and promote one of these states to the unique ground state of the system. When this happens, the rotation symmetry is spontaneously broken and the system exhibits a macroscopic magnetization.

In the case of the Heisenberg model, it is possible to identify explicitly the massless Nambu–Goldstone bosons, i.e., the excitations whose energy goes to zero as the momentum goes to zero. Let us introduce a state defined by

$$|\mathbf{k}\rangle \equiv \sum_{\mathbf{i}} e^{i a \mathbf{k} \cdot \mathbf{i}} |\uparrow \cdots \uparrow \underset{\mathbf{i}}{\downarrow} \uparrow \cdots \uparrow\rangle, \tag{4.89}$$

where on the right-hand side all spins are up except for a downward-oriented spin on the lattice site **i**. $a$ is the distance between nearest-neighbor nodes and the indices **i** that label the lattice nodes run over all triplets of integers. Thanks to the phase modulation introduced in the sum, this state has a definite momentum **k**, and an explicit calculation shows that

$$\mathcal{H}|\mathbf{k}\rangle = -J \left(\tfrac{3N}{4} + \underline{\cos(k_1 a) + \cos(k_2 a) + \cos(k_3 a) - 3}\right)|\mathbf{k}\rangle. \tag{4.90}$$

The underlined term provides a positive (recall that $-J < 0$) energy shift over the ground state energy, that goes to zero quadratically as the momentum goes to zero. Therefore, this gapless excitation is the massless Nambu–Goldstone mode (in this context, it is called a spin wave).

**Mermin–Wagner's theorem:** Coleman's theorem has a very important counterpart in the context of the Heisenberg model, obtained by Mermin and Wagner, which states that *the Heisen-*

berg model[11] *with finite range interactions cannot have a ferromagnetic phase at any non-zero temperature in one or two dimensions.* In this case, the spontaneous breaking of symmetry is prevented by thermal fluctuations, which are large enough in low dimensions to cause the angular variables to explore the entire range. In order to prove this result, let us first add to the Heisenberg model a term that couples the spins to an external magnetic field h in the z direction,

$$\mathcal{H} \equiv -J \sum_{\langle i,j \rangle} \mathbf{s}_i \cdot \mathbf{s}_j - h \sum_i s_i^3. \tag{4.91}$$

The spontaneous magnetization at temperature T is defined by

$$M(T) \equiv \lim_{h \to 0} \lim_{N \to \infty} \underbrace{\omega_T \left( \sum_i s_i^3 \right)}_{M(T,N,h)}, \tag{4.92}$$

where $\omega_T$ is the canonical thermal average at the temperature T,

$$\omega_T(A) \equiv \frac{\operatorname{tr}\left(e^{-\mathcal{H}/T} A\right)}{\operatorname{tr}\left(e^{-\mathcal{H}/T}\right)}. \tag{4.93}$$

(We use a system of units where the Boltzmann constant is $k_B = 1$.) Note that in the definition of spontaneous magnetization, we first take the infinite volume limit (in order to have a continuous set of degenerate ground states) and then the external field to zero. The proof of the Mermin–Wagner theorem relies on the *Bogoliubov inequality* (see Exercise 4.8),

$$\frac{1}{2T} \omega_T \left( A^\dagger A + A A^\dagger \right) \omega_T \left( [B, [\mathcal{H}, B^\dagger]] \right) \geq \left| \omega_T \left( [A, B^\dagger] \right) \right|^2, \tag{4.94}$$

with the following special choices for the operators $A$ and $B$,

$$A \equiv \underbrace{\sum_i e^{-i\mathbf{k}\cdot\mathbf{x}_i} s_i^2}_{\tilde{s}_\mathbf{k}^2}, \quad B \equiv \sum_i e^{-i\mathbf{k}\cdot\mathbf{x}_i} s_i^1, \tag{4.95}$$

where $\mathbf{k}$ is some fixed momentum. First, note that

$$\omega_T \left( A^\dagger A + A A^\dagger \right) = 2\, \omega_T \left( \tilde{s}_\mathbf{k}^{2\dagger} \tilde{s}_\mathbf{k}^2 \right), \quad \omega_T \left( [A, B^\dagger] \right) = N \left| M(T, N, h) \right|. \tag{4.96}$$

For the third thermal average that enters in the Bogoliubov inequality, one may obtain the following upper bound:

$$\omega_T \left( [B, [\mathcal{H}, B^\dagger]] \right) \leq \tfrac{3}{2} N J\, \mathbf{k}^2 a^2 + N h \left| M(T, N, h) \right|, \tag{4.97}$$

---

[11] For this theorem to apply, the system must have a continuous symmetry. For instance, it is not valid for the Ising model (a toy model of ferromagnetism, in which each lattice site carries a number that can take the values $\pm 1$), whose symmetries are discrete. Indeed, it is well known that the Ising model in two dimensions exhibits spontaneous magnetization below a certain critical temperature.

where $a$ is the distance between nearest neighbors on the lattice. Therefore, the Bogoliubov inequality gives

$$\frac{1}{NT} \omega_T(\tilde{s}_k^{2\dagger} \tilde{s}_k^2) \geq \frac{|\mathcal{M}(T, N, h)|^2}{\frac{3}{2} J k^2 a^2 + h|\mathcal{M}(T, N, h)|}. \tag{4.98}$$

Summing over all the momenta $\mathbf{k}$, we get

$$\frac{3}{4T} \geq |\mathcal{M}(T, N, h)|^2 \underbrace{\frac{1}{N}\sum_{\mathbf{k}} \frac{1}{\frac{3}{2} J k^2 a^2 + h|\mathcal{M}(T, N, h)|}}_{= \infty \text{ for } N = \infty, h = 0, d \leq 2}. \tag{4.99}$$

In the limit of an infinite lattice, the discrete sum becomes a continuous integral (in a compact domain since the lattice spacing provides an ultraviolet cutoff). When we set $h = 0$, this integral is divergent in dimensions $d \leq 2$ (this divergence is of the same nature as the one in eq. (4.51)), which implies that $\mathcal{M}(T) = 0$. The above derivation may be extended to couplings between non-nearest neighbors. In that case, $Ja^2$ is replaced by the average of $J(\mathbf{x}-\mathbf{y})(\mathbf{x}-\mathbf{y})^2$ over the intervals $\mathbf{x} - \mathbf{y}$. As long as this average is finite (i.e., the interaction strength decays fast enough with the distance), the same conclusion holds. Note also that because of the factor $1/T$ on the left-hand side, the inequality is inconclusive at zero temperature.

## Exercises

**4.1** Prove eq. (4.11). *Hints: Express the basis vectors as linear combinations of the eigenvectors of A. When looking for minima of the right-hand side, it is necessary to introduce Lagrange multipliers to enforce the orthonormality of the basis vectors.*

**\*4.2** This exercise provides an explicit example of the fact that the overlap of states centered at distinct values of the field is exponentially suppressed, with a factor proportional to the spatial volume inside the exponential. Let us consider a *coherent state* defined by

$$|\chi_{\text{in}}\rangle \equiv \mathcal{N}_\chi \exp\left\{\int \frac{d^3 \mathbf{k}}{(2\pi)^3 2E_\mathbf{k}} \chi(\mathbf{k}) a^\dagger_{\mathbf{k}, \text{in}}\right\} |0_{\text{in}}\rangle,$$

where $\chi(\mathbf{k})$ is a function of 3-momentum and $\mathcal{N}_\chi$ a normalization constant.
- Show that the coherent state is the ground state of a quadratic field theory with a shifted field.
- Consider two coherent states $|\chi_{\text{in}}\rangle, |\vartheta_{\text{in}}\rangle$, with $\chi(\mathbf{k}), \vartheta(\mathbf{k}) \propto \delta(\mathbf{k})$, and calculate $\langle\vartheta|\chi\rangle$.

**4.3** Check eq. (4.55).

**4.4** Check eq. (4.61) and the arguments developed afterwards to conclude about Coleman's theorem.

**\*4.5** Check that the operators $(1 \pm \gamma^5)/2$ introduced in eq. (4.68) are projectors. Check the left–right decompositions of eqs. (4.67) and (4.69).

**4.6** Check that the Hamiltonian of the Heisenberg model (defined in eq. (4.80)) commutes with the total spin operator $S \equiv \sum_i s_i$. Check eqs. (4.85) and (4.90).

**4.7** Prove eqs. (4.87) and (4.88).

**4.8** Prove the Bogoliubov inequality, eq. (4.94). *Hints:*

- *Given two operators* $U, V$, *define*

$$(U, V) \equiv -\sum_{i \neq j} \langle i|U|j\rangle^* \langle i|V|j\rangle \frac{w_i - w_j}{E_i - E_j},$$

*where the sum runs over the eigenstates of the Hamiltonian and where we denote* $w_i \equiv e^{-\beta E_i}/\mathrm{tr}\,(e^{-\beta \mathcal{H}})$. *Show that* $(U, V)$ *is a positive Hermitian scalar product.*

- *Prove the following inequalities:*

$$0 \leq \frac{w_j - w_i}{E_i - E_j} \leq \frac{w_i + w_j}{2T}, \quad 0 \leq (A, A) \leq \frac{1}{2T} \omega_T(\{A, A^\dagger\}).$$

- *Take* $\mathcal{C} \equiv [\mathcal{B}, \mathcal{H}]$. *Check* $(A, \mathcal{C}) = \omega_T([\mathcal{B}, A^\dagger])$ *and* $(\mathcal{C}, \mathcal{C}) = \omega_T([\mathcal{B}, [\mathcal{H}, \mathcal{B}^\dagger]])$.
- *Use the Cauchy–Schwarz inequality,* $(A, A)(\mathcal{C}, \mathcal{C}) \geq |(A, \mathcal{C})|^2$, *to conclude.*

# CHAPTER 5

# Functional Quantization

Until now, our approach to quantum field theory has been based on the canonical formalism, in which a field is an operator-valued distribution. This form of quantization in certain respects mimics the quantization procedure most commonly used in ordinary quantum mechanics: promote the classical variables of a Hamiltonian system to operators and promote the Poisson brackets to commutators. In this approach, the derivation of the generating functional from which we obtained the perturbative rules was arguably rather cumbersome, even in the simple setting of a scalar field theory. Moreover, this form of quantization makes it difficult to understand in which sense the quantum theory is a deformation of the underlying classical theory, since it appears to be a radically different mathematical construction. *Functional quantization*, also known as the path integral formalism, is an alternative quantization procedure that considerably simplifies the algebraic manipulations, and provides some rather intuitive insights into what makes a theory quantum.

## 5.1 Path Integral in Quantum Mechanics

Let us consider a non-relativistic quantum mechanical system with a single degree of freedom, whose Hamiltonian is

$$\mathcal{H} \equiv \frac{P^2}{2m} + V(Q). \tag{5.1}$$

The position and momentum operators Q and P obey the following commutation relation: $[Q, P] = i$. Consider the probability for the system starting at position $q_i$ at time $t_i$ and ending at position $q_f$ at time $t_f$. It may be obtained as $|\psi(q_f, t_f)|^2$ by solving Schrödinger's equation with an initial wavefunction localized at $q_i$,

$$i\partial_t \psi(q, t) = \mathcal{H} \psi(q, t), \qquad \psi(q, t_i) \equiv \delta(q - q_i). \tag{5.2}$$

More formally, in the Schrödinger picture, this probability is given by the squared modulus of the following transition amplitude:

$$\langle q_f | e^{-i\mathcal{H}(t_f - t_i)} | q_i \rangle, \tag{5.3}$$

where $|q\rangle$ denotes the eigenstate of the position operator with eigenvalue $q$. Let us subdivide the time interval $[t_i, t_f]$ into N equal sub-intervals by introducing

$$\Delta \equiv \frac{t_f - t_i}{N}, \qquad t_n \equiv t_i + n\Delta. \tag{5.4}$$

(Therefore, we have $t_0 = t_i$ and $t_N = t_f$.) The time evolution operator can be factorized as

$$e^{-i\mathcal{H}(t_f - t_i)} = e^{-i\mathcal{H}(t_N - t_{N-1})} \times e^{-i\mathcal{H}(t_{N-1} - t_{N-2})} \times \cdots \times e^{-i\mathcal{H}(t_1 - t_0)}. \tag{5.5}$$

Between the successive factors on the right-hand side, we can insert the identity operator written as a complete sum over the position eigenstates,

$$1 = \int_{-\infty}^{+\infty} dq \, |q\rangle\langle q|, \tag{5.6}$$

and the transition amplitude (5.3) becomes

$$\langle q_f | e^{-i\mathcal{H}(t_f - t_i)} | q_i \rangle = \int \prod_{j=1}^{N-1} dq_j \, \langle q_f | e^{-i\Delta\mathcal{H}} | q_{N-1} \rangle \langle q_{N-1} | e^{-i\Delta\mathcal{H}} | q_{N-2} \rangle \cdots$$
$$\cdots \langle q_1 | e^{-i\Delta\mathcal{H}} | q_i \rangle. \tag{5.7}$$

Note that this formula, illustrated in Figure 5.1, is exact for any value of N. In the Hamiltonian (5.1), the kinetic energy and potential energy terms do not commute, which complicates the evaluation of its exponential. We can remedy this situation by using the *Baker–Campbell–Hausdorff formula*, that we shall write here as follows:

$$e^{\Delta(A+B)} = e^{\Delta A} e^{\Delta B} e^{-\frac{\Delta^2}{2}[A,B] + \mathcal{O}(\Delta^3)}. \tag{5.8}$$

In the limit $\Delta \to 0$ (i.e., $N \to \infty$), we may neglect the last factor since the product of all such factors goes to unity[1] when $N \to \infty$. Therefore, each elementary factor of eq. (5.7) is rewritten as

---

[1] We use
$$\lim_{N \to \infty} e^{\alpha_1/N^2} e^{\alpha_2/N^2} \cdots e^{\alpha_N/N^2} = 1,$$
valid provided that the sum $\sum_i \alpha_i$s does not diverge too quickly.

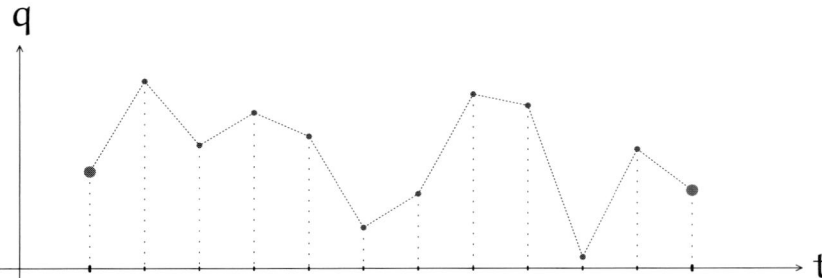

FIGURE 5.1: Illustration of eq. (5.7) with ten intermediate points. The endpoints are fixed, while the intermediate points are integrated over. The line segments connecting the points are to guide the eye, but there is no "path" at this stage.

$$\begin{aligned}\langle q_{i+1}|e^{-i\Delta\mathcal{H}}|q_i\rangle &\approx \langle q_{i+1}|e^{-i\Delta\frac{P^2}{2m}} e^{-i\Delta V(Q)}|q_i\rangle \\ &= \int \frac{dp_i}{2\pi} \langle q_{i+1}|e^{-i\Delta\frac{P^2}{2m}}|p_i\rangle \langle p_i|e^{-i\Delta V(Q)}|q_i\rangle,\end{aligned} \quad (5.9)$$

where we have inserted another identity operator, written this time as a complete sum over momentum eigenstates,

$$1 \equiv \int \frac{dp}{2\pi} |p\rangle\langle p|. \quad (5.10)$$

In the two factors, the exponential operator depends only on P or Q, and the matrix elements are trivial to evaluate by using the fact that the operators are enclosed between momentum and position eigenstates:

$$\begin{aligned}\langle q_{i+1}|e^{-i\Delta\frac{P^2}{2m}}|p_i\rangle &= e^{-i\Delta\frac{p_i^2}{2m}} \langle q_{i+1}|p_i\rangle, \\ \langle p_i|e^{-i\Delta V(Q)}|q_i\rangle &= e^{-i\Delta V(q_i)} \langle p_i|q_i\rangle.\end{aligned} \quad (5.11)$$

Using

$$\langle q|p\rangle = e^{ipq}, \quad (5.12)$$

we arrive at the formula[2]

$$\langle q_{i+1}|e^{-i\Delta\mathcal{H}}|q_i\rangle = \int \frac{dp_i}{2\pi} e^{-i\Delta\mathcal{H}(p_i,q_i)} e^{ip_i(q_{i+1}-q_i)} \left(1+\mathcal{O}(\Delta^2)\right). \quad (5.13)$$

---

[2] A bit more care is necessary for Hamiltonians that are not separable into a sum of a P-dependent term and a Q-dependent term. A proper treatment should use Weyl's prescription (see Exercise 5.1) for defining the quantum Hamiltonian from the classical Hamiltonian. In eq. (5.13), one would obtain $\mathcal{H}(p_i, \frac{1}{2}(q_i + q_{i+1}))$ instead of $\mathcal{H}(p_i, q_i)$.

## 5.1 PATH INTEGRAL IN QUANTUM MECHANICS

If we define $\dot{q}_i \equiv (q_{i+1} - q_i)/\Delta$ as the slope of the line segments in Figure 5.1, and we take the limit $N \to \infty$, we may write the transition amplitude as a *path integral*,

$$\langle q_f | e^{-i\mathcal{H}(t_f - t_i)} | q_i \rangle = \int_{\substack{q(t_i)=q_i \\ q(t_f)=q_f}} [Dp(t)Dq(t)]$$

$$\times \exp\left\{ i \int_{t_i}^{t_f} dt \, (p(t)\dot{q}(t) - \mathcal{H}(p(t), q(t))) \right\}. \quad (5.14)$$

One should be aware of the fact that the functional measure $[Dq(t)Dp(t)]$ in general lacks solid mathematical foundations, although it allows for some powerful manipulations that would be extremely cumbersome to perform at the level of quantum operators. Note also that at the boundaries $t_{i,f}$ the position is well defined, and therefore the momentum is not constrained (by the uncertainty principle). A crucial aspect of eq. (5.14) is that all the objects that appear on the right-hand side are ordinary commuting numbers, while the left-hand side is made of quantum operators and states. In this section, we have started from the conventional formulation of transition amplitudes in quantum mechanics, in order to arrive at eq. (5.14). However, one may now "forget" the canonical formalism and view the path integral expression of transition amplitudes as another way of going from a classical Hamiltonian $\mathcal{H}$ to a quantized theory.

For a Hamiltonian where the P dependence has no powers higher than quadratic, as in the example of eq. (5.1), it is possible to perform exactly the integral over $p(t)$. This type of integral is called a *Gaussian path integral*. Gaussian path integrals can be evaluated in the same way as their ordinary counterparts, using the following formulas:

$$\int_{-\infty}^{+\infty} dx \, e^{-x^2/(2\sigma)} = \sqrt{2\pi\sigma}, \quad \int_{-\infty}^{+\infty} dx \, e^{\pm ix^2/(2\sigma)} = e^{\pm i\frac{\pi}{4}} \sqrt{2\pi\sigma}, \quad (5.15)$$

and treating each $p(t)$ as an independent variable. In the present case, we need the integral

$$\int dp \, e^{i\Delta(p\dot{q} - \frac{p^2}{2m})} = \underbrace{e^{-i\frac{\pi}{4}} \sqrt{\frac{2\pi m}{\Delta}}}_{\substack{\text{prefactor} \\ \text{independent of } q, \dot{q}}} e^{i\Delta \frac{m\dot{q}^2}{2}}. \quad (5.16)$$

The (infinite in the limit $\Delta \to 0$) prefactors can be hidden in the measure $[Dq(t)]$ since they do not depend on the path, and we are therefore led to the following formula:

$$\langle q_f | e^{-i\mathcal{H}(t_f - t_i)} | q_i \rangle = \int_{\substack{q(t_i)=q_i \\ q(t_f)=q_f}} [Dq(t)] \, \exp\left\{ i \int_{t_i}^{t_f} dt \, \mathcal{L}(q(t)) \right\}$$

$$= \int_{\substack{q(t_i)=q_i \\ q(t_f)=q_f}} [Dq(t)] \, e^{i\mathcal{S}[q(t)]}, \quad (5.17)$$

where $\mathcal{L}(q)$ is the classical Lagrangian,

$$\mathcal{L}(q) \equiv \frac{m\dot{q}^2}{2} - V(q) \quad (5.18)$$

and $\mathcal{S}[q]$ is the corresponding classical action.

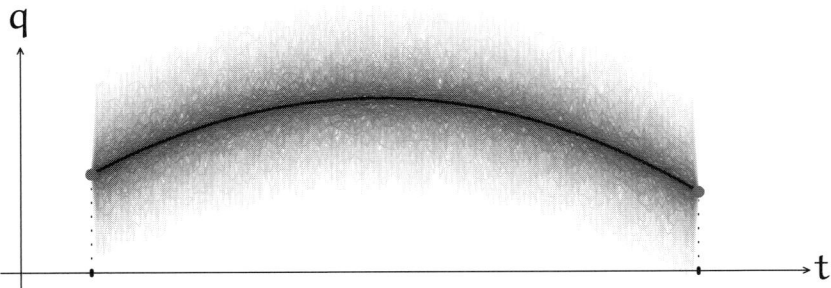

FIGURE 5.2: Illustration of eq. (5.19). The paths whose action is far apart from the classical extremum are plotted in fainter shades. The solid black line is the classical trajectory.

**Classical limit, least action principle:** In the previous section, we have written all the formulas with $\hbar = 1$. Had we kept the Planck constant, the final formula would have been

$$\langle q_f | e^{-i\mathcal{H}(t_f - t_i)} | q_i \rangle = \int_{\substack{q(t_i) = q_i \\ q(t_f) = q_f}} [Dq(t)] \, e^{\frac{i}{\hbar} S[q(t)]}. \tag{5.19}$$

(This can be guessed a posteriori based on the fact that $\hbar$ has the dimension of an action.) Because of the factor $i$ inside the exponential, this integral is wildly oscillating, except in the immediate vicinity of the function $q_c(t)$ that realizes the extremum of the action. Note that this function is precisely the solution of the classical Euler–Lagrange equations of motion. Roughly speaking, the phase oscillations become significant when

$$\left| S[q(t)] - S[q_c(t)] \right| \geq 2\pi \hbar, \tag{5.20}$$

and paths that fulfill this inequality do not contribute to the path integral. Therefore, in the limit $\hbar \to 0$, the path integral is dominated by the unique path $q_c(t)$, i.e., by the classical trajectory of the system (see Figure 5.2). The path integral formalism thus provides a very intuitive way of connecting smoothly quantum and classical mechanics.

## 5.2 Functional Manipulations

### 5.2.1 Time-Ordered Products

Consider now the matrix element

$$\langle q_f | e^{-i\mathcal{H}(t_f - t_1)} Q \, e^{-i\mathcal{H}(t_1 - t_i)} | q_i \rangle, \tag{5.21}$$

which measures the expectation value of the position at time $t_1$, in an evolution that starts at $q_i$ and ends at $q_f$. In order to evaluate this object, we need to insert on either side of the position operator $Q$ an identity operator written as a complete sum over position eigenstates,

## 5.2 FUNCTIONAL MANIPULATIONS

i.e.,

$$Q \to \int dq\, dq'\, |q\rangle \underbrace{\langle q|Q|q'\rangle}_{q\,\delta(q-q')} \langle q'| = \int dq\, q\, |q\rangle\langle q|. \tag{5.22}$$

This leads immediately to the following path integral representation:

$$\langle q_f|e^{-i\mathcal{H}(t_f-t_1)}\, Q\, e^{-i\mathcal{H}(t_1-t_i)}|q_i\rangle = \int_{\substack{q(t_i)=q_i \\ q(t_f)=q_f}} [Dq(t)]\, q(t_1)\, e^{iS[q(t)]}. \tag{5.23}$$

Likewise, if $t_2 > t_1$, we have

$$\langle q_f|e^{-i\mathcal{H}(t_f-t_2)}\, Q\, e^{-i\mathcal{H}(t_2-t_1)}\, Q\, e^{-i\mathcal{H}(t_1-t_i)}|q_i\rangle$$
$$\int_{\substack{q(t_i)=q_i \\ q(t_f)=q_f}} [Dq(t)]\, q(t_1)\, q(t_2)\, e^{iS[q(t)]}. \tag{5.24}$$

If we introduce a time-dependent position operator

$$Q(t) \equiv e^{i\mathcal{H}t}\, Q\, e^{-i\mathcal{H}t}, \tag{5.25}$$

and its eigenstates

$$|q,t\rangle \equiv e^{i\mathcal{H}t}|q\rangle, \tag{5.26}$$

the previous equation takes a much more compact form

$$\langle q_f, t_f | Q(t_2) Q(t_1) | q_i, t_i \rangle \underset{t_2 > t_1}{=} \int_{\substack{q(t_i)=q_i \\ q(t_f)=q_f}} [Dq(t)]\, q(t_1)\, q(t_2)\, e^{iS[q(t)]}. \tag{5.27}$$

The condition $t_2 > t_1$ is crucial here, because the left-hand side would be quite different if the times are ordered differently. In contrast, the objects $q(t_1)$ and $q(t_2)$ in the right-hand side are ordinary numbers that commute. One may render this formula true for any ordering between $t_1$ and $t_2$ by introducing a T-product, that ensures that the operator with the largest time is always on the left:

$$\langle q_f, t_f | T\left(Q(t_1) Q(t_2)\right) | q_i, t_i \rangle = \int_{\substack{q(t_i)=q_i \\ q(t_f)=q_f}} [Dq(t)]\, q(t_1)\, q(t_2)\, e^{iS[q(t)]}. \tag{5.28}$$

This formula generalizes to $n$ factors as follows:

$$\langle q_f, t_f | T\left(Q(t_1) \cdots Q(t_n)\right) | q_i, t_i \rangle = \int_{\substack{q(t_i)=q_i \\ q(t_f)=q_f}} [Dq(t)]\, q(t_1) \cdots q(t_n)\, e^{iS[q(t)]}. \tag{5.29}$$

This result is extremely important in applications to quantum field theory, since time-ordered products of field operators are the central objects that appear in the Lehmann–Symanzik–Zimmermann (LSZ) reduction formulas. One may also apply differential operators containing time derivatives on this equation. For instance,

$$\frac{\partial}{\partial t_1}\langle q_f, t_f | T\left(Q(t_1)\cdots Q(t_n)\right) | q_i, t_i \rangle = \int_{\substack{q(t_i)=q_i \\ q(t_f)=q_f}} [Dq(t)]\, \dot{q}(t_1) \cdots q(t_n)\, e^{iS[q(t)]}. \quad (5.30)$$

Note how a time derivative in the integrand of the path integral also applies to the step functions that enforce the time ordering on the left-hand side.

## 5.2.2 Functional Sources and Derivatives

The amplitudes of the form (5.29) can all be encapsulated into the following *generating functional*,

$$Z_{fi}[j(t)] \equiv \langle q_f, t_f | T \exp i \int_{t_i}^{t_f} dt\, j(t)\, Q(t) | q_i, t_i \rangle, \quad (5.31)$$

where $j(t)$ is some arbitrary function of time. From $Z_{fi}[j]$, the amplitudes can be recovered by *functional differentiation*,

$$\langle q_f, t_f | T\left(Q(t_1)\cdots Q(t_n)\right)| q_i, t_i \rangle = \left.\frac{\delta^n Z_{fi}[j]}{i^n \delta j(t_1)\cdots \delta j(t_n)}\right|_{j\equiv 0}. \quad (5.32)$$

Functional derivatives obey the usual rules of differentiation, with the additional property that the values of the function $j(t)$ at different times should be viewed as independent variables, i.e.,

$$\frac{\delta j(t)}{\delta j(t')} = \delta(t - t'). \quad (5.33)$$

From this formula, one may also read the dimension of a functional derivative,

$$\dim\left[\frac{\delta}{\delta j(t)}\right] = -\dim[j(t)] - \dim[t]. \quad (5.34)$$

From eq. (5.29), we can derive an expression of the generating functional $Z_{fi}$ as a path integral,

$$Z_{fi}[j(t)] = \int_{\substack{q(t_i)=q_i \\ q(t_f)=q_f}} [Dq(t)]\, e^{iS[q(t)] + i\int_{t_i}^{t_f} dt\, j(t) q(t)}, \quad (5.35)$$

## 5.2 FUNCTIONAL MANIPULATIONS

which involves only the commuting number q(t) and no time ordering. Note also that there is an Hamiltonian version of this path integral,

$$Z_{fi}[j(t)] = \int_{\substack{q(t_i)=q_i\\q(t_f)=q_f}} [Dp(t)Dq(t)]$$
$$\times \exp\left\{i\int_{t_i}^{t_f} dt\, \left(p(t)\dot{q}(t) - \mathcal{H}(p(t),q(t)) + j(t)q(t)\right)\right\}. \quad (5.36)$$

### 5.2.3 Projection on the Ground State at Asymptotic Times

So far in this section, we have considered amplitudes where the initial and final states are position eigenstates. However, the path integral formalism is not limited to this situation. Let us assume for instance that the system is in state $|\psi_i\rangle$ at time $t_i$ and in state $|\psi_f\rangle$ at time $t_f$. For any operator $\mathcal{O}$, the expectation value between these two states can be related to transitions between position eigenstates by writing

$$\langle \psi_f, t_f | \mathcal{O} | \psi_i, t_i \rangle = \int dq_i dq_f\, \psi_f^*(q_f)\psi_i(q_i)\, \langle q_f, t_f | \mathcal{O} | q_i, t_i \rangle, \quad (5.37)$$

where $\psi(q) \equiv \langle q|\psi\rangle$ is the position representation of the wavefunction of the state $|\psi\rangle$. However, the use of this formula is cumbersome in practice, because of the integrations over $q_i$ and $q_f$.

In the special case where the initial and final states are the ground state of the Hamiltonian, $|0\rangle$, and the initial and final times are $-\infty$ and $+\infty$, there is a trick to circumvent this difficulty. Let us introduce the eigenstates $|n\rangle$ of the Hamiltonian, with eigenvalue $E_n$ and eigenfunction $\psi_n(q) \equiv \langle q|n\rangle$, and write

$$|q_i, t_i\rangle = e^{i\mathcal{H}t_i}|q_i\rangle = \sum_{n=0}^{\infty} e^{i\mathcal{H}t_i}|n\rangle\langle n|q_i\rangle = \sum_{n=0}^{\infty} \psi_n^*(q_i)\, e^{iE_n t_i}|n\rangle. \quad (5.38)$$

We will assume that the Hamiltonian is shifted by a constant so that the energy of the ground state $|0\rangle$ is $E_0 = 0$. Now, we multiply the Hamiltonian by $1 - i0^+$, where $0^+$ denotes some positive infinitesimal number. All the factors $\exp(i(1-i0^+)E_n t_i)$ go to zero when $t_i \to -\infty$, except for $n=0$. Therefore, after this alteration of the Hamiltonian, we have

$$\lim_{t_i \to -\infty} |q_i, t_i\rangle = \psi_0^*(q_i)|0\rangle. \quad (5.39)$$

We can then weight this equation by a function $\varphi(q_i)$,

$$\lim_{t_i \to -\infty} \int dq_i\, \varphi(q_i)|q_i, t_i\rangle = \underbrace{\int dq_i\, \varphi(q_i)\psi_0^*(q_i)}_{\langle 0|\varphi\rangle}|0\rangle, \quad (5.40)$$

i.e.,

$$|0\rangle = \lim_{t_i \to -\infty} \frac{1}{\langle 0|\varphi\rangle} \int dq_i \, \varphi(q_i) \, |q_i, t_i\rangle. \tag{5.41}$$

Any function $\varphi(q)$ such that the state $|\varphi\rangle$ has a non-zero overlap with the ground state $\langle 0|$ is appropriate in this role, but the simplest expressions are obtained with the constant function $\varphi(q) = 1$, corresponding to the momentum eigenstate $p = 0$. Likewise, changing $\mathcal{H} \to (1 - i0^+)\mathcal{H}$ has a similar effect on the final state in the limit $t_f \to +\infty$,

$$\lim_{t_f \to +\infty} \langle q_f, t_f| = \psi_0(q_f) \langle 0|. \tag{5.42}$$

From these considerations, when the initial and final states at $\pm\infty$ are the ground state, we can write the generating functional in the following simple path integral form,

$$Z[j(t)] = \int [Dp(t)Dq(t)]$$
$$\times \exp\left\{i \int dt \, (p(t)\dot{q}(t) - (1 - i0^+)\mathcal{H}(p(t), q(t)) + j(t)q(t))\right\}. \tag{5.43}$$

From the discussion after eq. (5.41), we see that the boundary conditions on the paths are not important. They only affect an overall prefactor that can be adjusted by hand in such a way that $Z[0] = 1$. After performing the Gaussian functional integral over $p(t)$, we can rewrite this expression in Lagrangian form:

$$Z[j(t)] = \int [Dq(t)]$$
$$\times \exp\left\{i \int dt \, \left((1 + i0^+)\frac{m\dot{q}^2(t)}{2} - (1 - i0^+)V(q(t)) + j(t)q(t)\right)\right\}. \tag{5.44}$$

The term in $(i0^+)\dot{q}^2$ may be viewed as contributing to the convergence of the integral at large velocities. Likewise, for a confining potential such that $V(q) \to +\infty$ when $|q| \to \infty$, the term in $(i0^+)V(q)$ contributes to the convergence at large coordinates.

### 5.2.4 Functional Fourier Transform

Given a functional $F[q(t)]$, its *functional Fourier transform* is defined by

$$\widetilde{F}[p(t)] \equiv \int [Dq(t)] \, F[q(t)] \, \exp\left\{i \int dt \, p(t)q(t)\right\}. \tag{5.45}$$

In this definition, the function $p(t)$ plays the role of the Fourier conjugate of the integration variable $q(t)$. Eq. (5.45) may be inverted by

$$F[q(t)] \equiv \int [Dp(t)] \, \widetilde{F}[p(t)] \, \exp\left\{-i \int dt \, p(t)q(t)\right\}. \tag{5.46}$$

The usual properties of ordinary Fourier transforms extend to the functional case, e.g.,

- the Fourier transform of a constant is a delta function;
- the Fourier transform of a Gaussian is another Gaussian;
- the Fourier transform of a product is the convolution product of the Fourier transforms.

### 5.2.5 Functional Translation Operator

The functional derivative $\delta/\delta j(t)$ may be viewed as the generator of translations in the space of the functions $j(t)$. Its exponential provides a translation operator,

$$\exp\left\{\int dt\, a(t)\, \frac{\delta}{\delta j(t)}\right\} F[j(t)] = F[j(t) + a(t)], \qquad (5.47)$$

for any functional $F[j(t)]$. Another extremely important formula is

$$\underbrace{\exp\left\{\lambda \int dt \left(\frac{\delta}{\delta j(t)}\right)^n\right\} \exp\left\{\int dt\, j(t) q(t)\right\}}_{A[j,q;\lambda]} = \underbrace{\exp\left\{\int dt\, \left(j(t) q(t) + \lambda\, q^n(t)\right)\right\}}_{B[j,q;\lambda]}. \qquad (5.48)$$

The proof of this formula amounts to noticing that $A[j, q; \lambda = 0] = B[j, q; \lambda = 0]$, and comparing their (ordinary) derivatives with respect to $\lambda$:

$$\partial_\lambda A[j, q; \lambda] = \lambda \int dt \left(\frac{\delta}{\delta j(t)}\right)^n A[j, q; \lambda] = \lambda \int dt\, q^n(t)\, A[j, q; \lambda],$$

$$\partial_\lambda B[j, q; \lambda] = \lambda \int dt\, q^n(t)\, B[j, q; \lambda]. \qquad (5.49)$$

Since $A[j, q; \lambda]$ and $B[j, q; \lambda]$ are equal at $\lambda = 0$ and obey the same differential equation, they are equal at any $\lambda$.

## 5.3 Path Integral in Scalar Field Theory

The functional formalism that we have exposed in the context of quantum mechanics can easily be extended to quantum field theory. The main change is that the functions over which one integrates are functions of time and space (as opposed to functions of time only in quantum mechanics). All the results of the previous section can be translated into analogous formulas in quantum field theory, thanks to the following correspondence:

$$q(t) \to \phi(x), \quad p(t) \to \Pi(x), \quad j(t) \to j(x). \qquad (5.50)$$

The main results of the previous section, namely that time-ordered products of operators in the canonical formalism become ordinary products of functions in the path integral representation, and that the ground state at $\pm\infty$ can be obtained by relaxing the boundary conditions and multiplying the Hamiltonian by $1 - i0^+$, remain true in this new context. Thus, the analogue of eq. (5.43) in a real scalar field theory is

$$Z[j] = \int [D\Pi(x) D\phi(x)] \\ \times \exp\left\{i \int d^4x \, (\Pi(x)\dot{\phi}(x) - (1 - i0^+)\mathcal{H}(\Pi, \phi) + j(x)\phi(x))\right\}. \tag{5.51}$$

Since the Hamiltonian is quadratic in $\Pi$,

$$\mathcal{H} = \frac{1}{2}\Pi^2 + \frac{1}{2}(\nabla\phi)\cdot(\nabla\phi) + \frac{1}{2}m^2\phi^2 + V(\phi), \tag{5.52}$$

it is easy to perform the (Gaussian) functional integration on $\Pi$, to obtain

$$Z[j] = \int [D\phi(x)] \, \exp\left\{i \int d^4x \, (\mathcal{L}(\phi) + j(x)\phi(x))\right\}, \tag{5.53}$$

where

$$\mathcal{L}(\phi) \equiv \frac{1}{2}(1 + i0^+)\dot{\phi}^2 - \frac{1}{2}(1 - i0^+)((\nabla\phi)\cdot(\nabla\phi) + m^2\phi^2) - (1 - i0^+)V(\phi). \tag{5.54}$$

(Note that the $1 - i0^+$ in front of the interaction potential plays no role if we turn off adiabatically the coupling constant when $|x^0| \to \infty$.) Using the analogue of eq. (5.48), we can separate the interactions as follows:

$$Z[j] = \exp\left\{-i \int d^4x \, V\left(\frac{\delta}{i\delta j(x)}\right)\right\} Z_0[j], \tag{5.55}$$

with

$$Z_0[j] \equiv \int [D\phi(x)] \, \exp\left\{i \int d^4x \, (\mathcal{L}_0(\phi) + j(x)\phi(x))\right\},$$
$$\mathcal{L}_0(\phi) = \frac{1}{2}(1 + i0^+)\dot{\phi}^2 - \frac{1}{2}(1 - i0^+)((\nabla\phi)\cdot(\nabla\phi) + m^2\phi^2). \tag{5.56}$$

The functional integral that gives $Z_0[j]$ in eq. (5.56) is Gaussian in $\phi$ and can be performed in a straightforward manner, giving

$$Z_0[j] = \exp\left\{-\frac{1}{2} \int d^4x \, d^4y \, j(x) j(y) \, G_F^0(x, y)\right\}, \tag{5.57}$$

where $G_F^0(x, y)$ is the inverse of the operator

$$i\left[(1 + i0^+)\partial_0^2 - (1 - i0^+)(\nabla^2 - m^2)\right]. \tag{5.58}$$

Note that the terms in $i0^+$ ensure the existence of this inverse. Going to momentum space, we see that the Fourier transform of this inverse is

$$\widetilde{G}_F^0(k) = \frac{i}{(1+i0^+)k_0^2 - (1-i0^+)(\mathbf{k}^2 + m^2)}, \tag{5.59}$$

which after some rearrangement of the $i0^+$s appears to be nothing but eq. (1.140). Although the canonical quantization of a scalar field theory was tractable, we see on this example that the path integral approach provides a much quicker way of obtaining the expression of the free generating functional, with the correct pole prescription for the free time-ordered propagator.

## 5.4 Functional Determinants

In the earlier sections of this chapter, we have been a bit cavalier with Gaussian integrations, since we have disregarded the constant prefactors they produce. This was legitimate in the problems we were considering, since the normalization of the generating functional can be fixed by hand. However, in certain situations these prefactors depend crucially on quantities that have a physical significance, e.g., on a background field.

In order to compute this prefactor, let us start from a simple one-dimensional Gaussian integral,

$$\int_{-\infty}^{+\infty} dx\, e^{-\frac{1}{2}ax^2} = \sqrt{\frac{2\pi}{a}}. \tag{5.60}$$

The first stage of generalization is to replace $x$ by an $n$-component vector $\mathbf{x} \equiv (x_1, \cdots, x_n)$, and the positive number $a$ by a positive definite symmetric matrix $\mathbf{A}$, and to consider the integral

$$I(\mathbf{A}) \equiv \int \prod_{i=1}^{n} dx_i\, e^{-\frac{1}{2}\mathbf{x}^T \mathbf{A}\mathbf{x}}. \tag{5.61}$$

This integral can be calculated by representing the vector $\mathbf{x}$ in the orthonormal basis made of the eigenvectors of $\mathbf{A}$ (such a basis exists, since $\mathbf{A}$ is symmetric). The measure $\prod_i dx_i$ is unchanged, because the diagonalization of the matrix can be done by an orthogonal transformation. Therefore, the above integral also reads

$$I(\mathbf{A}) = \int \prod_{i=1}^{n} dy_i\, e^{-\frac{1}{2}\sum_i a_i y_i^2} = \prod_{i=1}^{n} \sqrt{\frac{2\pi}{a_i}}, \tag{5.62}$$

where the numbers $a_i$ are the eigenvalues of $\mathbf{A}$. This result can be written in a much more compact form,

$$I(\mathbf{A}) = \frac{(2\pi)^{n/2}}{\sqrt{\det \mathbf{A}}}. \tag{5.63}$$

This reasoning can be generalized to the functional case by writing

$$\int [D\phi(x)] \ \exp\left\{ -\frac{1}{2} \int d^4x d^4y \ \phi(x) A(x,y) \phi(y) \right\} = \left[\det(A)\right]^{-1/2}, \tag{5.64}$$

where $A(x,y)$ is a symmetric operator. In this formula, we have still disregarded some truly constant (and infinite) prefactors, made of powers of $2\pi$. One can also generalize this Gaussian integral to the case where the vector $x$ is complex,

$$J(\mathbf{A}) \equiv \int \prod_{i=1}^{n} dx_i dx_i^* \ e^{-x^\dagger \mathbf{A} x} = \frac{(2\pi)^n}{\det \mathbf{A}}, \tag{5.65}$$

where $\mathbf{A}$ is a Hermitian matrix. The functional analogue of this integral is

$$\int [D\phi(x) D\phi^*(x)] \ \exp\left\{ -\int d^4x d^4y \ \phi^*(x) A(x,y) \phi(y) \right\} = \left[\det(A)\right]^{-1}, \tag{5.66}$$

**Zeta function regularization:** Despite the elegance of these formulas, one should keep in mind that the functional determinant $\det A$ is most often infinite, because the spectrum of the operator extends to infinity. A common regularization technique for functional determinants is based on a generalization of Riemann's $\zeta$ function. Let the $\lambda_n$ be the eigenvalues of $A$, and define

$$\zeta_A(s) \equiv \text{tr}\left(A^{-s}\right) = \sum_n \frac{1}{\lambda_n^s}. \tag{5.67}$$

(The function $\zeta_A$ is called the zeta function of the operator $A$.) The determinant of $A$ is related to this function by

$$\det(A) = \exp\left(-\zeta_A'(0)\right). \tag{5.68}$$

The sum over $n$ in the definition of $\zeta_A$ usually converges only if $\text{Re}(s)$ is large enough (how large depends on the distribution of eigenvalues at large $n$), but not for $s = 0$. However, as in the case of Riemann's zeta function, $\zeta_A(s)$ can be analytically continued to most of the complex $s$-plane, which provides a regularized definition of the determinant.

**Diagrammatic interpretation:** Let us consider as an example the operator $A_\varphi \equiv \Box + \frac{\lambda}{2}\varphi^2$, where $\varphi(x)$ is a function of space-time (called a background field). The inverse of this operator is the propagator of a scalar particle (with a $\phi^4$ interaction) over the background field $\varphi$. We can skip the regularization step if we make a ratio with the determinant of the similar operator with no background field:

$$R \equiv \frac{\det(\Box)}{\det\left(\Box + \frac{\lambda}{2}\varphi^2\right)}. \tag{5.69}$$

A very useful formula relates the determinant of an operator to the trace of its logarithm,

$$\det(A) = \exp\left(\text{tr}\log(A)\right). \tag{5.70}$$

This formula can be proven (heuristically, since the objects we are manipulating may not be finite) by writing both sides of the equation in terms of the eigenvalues of A,

$$\det(A) = \prod_n \lambda_n = \exp \sum_n \log \lambda_n = \exp\left(\operatorname{tr}\log(A)\right). \tag{5.71}$$

Therefore, the ratio defined in eq. (5.69) can be rewritten as

$$R = \exp\left(-\operatorname{tr}\log\left(1 + \tfrac{\lambda \varphi^2}{2}\square^{-1}\right)\right). \tag{5.72}$$

Writing $\square^{-1} = iG_F^0$, and expanding the logarithm gives

$$R = \exp\left\{\sum_{n=1}^{\infty} \frac{1}{n} \operatorname{tr}\left(\left[-i\tfrac{\lambda\varphi^2}{2}G_F^0\right]^n\right)\right\}. \tag{5.73}$$

The argument of the exponential has a simple interpretation as a sum of one-loop diagrams made of a line dressed with insertions of the background field, the index $n$ being the number of such insertions:

$$\frac{1}{n}\operatorname{tr}\left(\left[-i\tfrac{\lambda\varphi^2}{2}G_F^0\right]^n\right) = \underbrace{\text{(diagram)}}_{n \text{ insertions}}. \tag{5.74}$$

Each of the insertions of the background field (shown by lines terminated by a dot in the above diagram) corresponds to a factor $-i\tfrac{\lambda}{2}\varphi^2$. The prefactor $1/n$ is the symmetry factor for the cyclic permutations of the $n$ insertions. All these diagrams are simply connected. Taking the exponential to obtain the ratio R produces all the multiply connected graphs made of products of such one-loop diagrams.

## 5.5 Quantum Effective Action

### 5.5.1 Definition

The action $S[\phi]$ that enters in the path integral representation of the generating functional $Z[j]$ is the *classical action*. Its parameters reflect the interactions among the constituents of the system at tree level, but in order to express higher-order corrections, loops are necessary. The *quantum effective action*, denoted $\Gamma[\phi]$, is defined as the functional that would produce the all-orders value of observables solely from tree-level contributions. $\Gamma[\phi]$ should coincide with the classical action at lowest order of perturbation theory, but also encapsulates all the higher-order corrections. One may formally write $\Gamma[\phi]$ as

$$\Gamma[\phi] \equiv \sum_{n=2}^{\infty} \frac{1}{n!} \int d^4x_1 \cdots d^4x_n\, \phi(x_1)\cdots\phi(x_n)\, \Gamma_n(x_1,\cdots,x_n). \tag{5.75}$$

$\Gamma_2(x_1, x_2)$ is therefore the inverse of the exact propagator, $\Gamma_4(x_1, \cdots, x_4)$ is the exact four-point function (in coordinate space), etc.

### 5.5.2 Relation between $\Gamma[\phi]$ and $W[j]$

Until now, we have introduced the generating functional of the vacuum expectation value of time-ordered products of fields, $Z[j]$, as well as the functional $W[j] \equiv \log Z[j]$ that generates the subset made of connected Feynman graphs. Recall that in terms of path integrals,

$$Z[j] = e^{W[j]} = \int [D\phi(x)] \, \exp\left[iS[\phi(x)] + i \int d^4x \, j(x)\phi(x)\right]. \tag{5.76}$$

Let us replace the classical action $S[\phi]$ by the quantum effective action $\Gamma[\phi]$ in the previous formula, in order to define

$$Z_\Gamma[j] \equiv e^{W_\Gamma[j]} \equiv \int [D\phi(x)] \, \exp\left[i\Gamma[\phi(x)] + i \int d^4x \, j(x)\phi(x)\right]. \tag{5.77}$$

This functional generates graphs whose building blocks are the exact propagator ($\Gamma_2^{-1}$), and the exact vertices ($\Gamma_3, \Gamma_4, \cdots$). From the definition of $\Gamma[\phi]$ as the "action" that would generate the exact theory with only tree-level contributions, we must have

$$W_\Gamma[j]\big|_{\text{tree}} = W[j]. \tag{5.78}$$

In other words, the tree diagrams of $W_\Gamma[j]$ should be equal to the all-orders $W[j]$. The tree diagrams may be isolated by introducing a small parameter $\varepsilon$ in the definition of $Z_\Gamma[j]$ as follows:

$$Z_\Gamma[j; \varepsilon] = e^{W_\Gamma[j; \varepsilon]} = \int [D\phi(x)] \, \exp\left[\frac{i}{\varepsilon}\left(\Gamma[\phi(x)] + \int d^4x \, j(x)\phi(x)\right)\right]. \tag{5.79}$$

By a counting similar to that of Section 2.1.6, we see that the order in $\varepsilon$ of a connected graph without external lines is $\varepsilon^{n_L - 1}$, where $n_L$ is the number of loops of the graph *built with the propagator and vertices from $\Gamma[\phi]$ as building blocks* (in other words, when counting the loops, we do not look inside the Taylor coefficients of $\Gamma[\phi]$ to see how they are themselves made in terms of bare propagator and vertices). The parameter $\varepsilon$ plays a very similar role to Planck's constant in ordering the graphs according to their number of loops, but here we should treat it as distinct from $\hbar$ since the Taylor coefficients of $\Gamma[\phi]$ still depend on $\hbar$ ($\hbar$ counts the number of loops of graphs expressed in terms of the bare propagator and vertices). Therefore, the functional $W_\Gamma[j; \varepsilon]$ has the following loop expansion:

$$W_\Gamma[j; \varepsilon] = \sum_{n_L = 0}^{\infty} \varepsilon^{n_L - 1} \underbrace{W_{\Gamma, n_L}[j]}_{n_L \text{ loops}}, \tag{5.80}$$

## 5.5 QUANTUM EFFECTIVE ACTION

and the tree-level contributions in $W_r[j]$ are the terms that survive in the formal limit $\varepsilon \to 0$:

$$W_r[j]\big|_{\text{tree}} = \lim_{\varepsilon \to 0} \left(\varepsilon\, W_r[j;\varepsilon]\right). \tag{5.81}$$

But from our discussion of the classical limit of path integrals in Section 5.1, we know that the limit $\varepsilon \to 0$ in eq. (5.79) corresponds to the extremum of the argument of the exponential, i.e.,

$$\frac{\delta \Gamma[\phi]}{\delta \phi(x)} + j(x) = 0. \tag{5.82}$$

Note that this equation is the analogue of the usual Euler–Lagrange equation of motion, with the quantum effective action instead of the classical action. This equation implicitly defines $\phi$ as a function of $j$, which we will denote $\phi_j$, in terms of which we can write

$$e^{W_r[j;\varepsilon]} \underset{\varepsilon \to 0}{=} \exp\left[\frac{i}{\varepsilon}\left(\Gamma[\phi_j(x)] + \int d^4x\, j(x)\phi_j(x)\right)\right]. \tag{5.83}$$

This leads to the following relationship between the quantum effective action and the generating functional of connected graphs:

$$\Gamma[\phi_j] = -i\, W[j] - \int d^4x\, j(x)\phi_j(x). \tag{5.84}$$

Therefore, $\Gamma[\phi]$ can be obtained as the Legendre transform of the generating functional $W[j]$ of the connected graphs (see Exercise 5.10 for a diagrammatic interpretation of the Legendre transform).

The "quantum equation of motion" (5.82) may also be viewed as defining $j$ in terms of $\phi$, that we shall denote $j_\phi$. Eq. (5.84) may therefore also be written as

$$\Gamma[\phi] = -i\, W[j_\phi] - \int d^4x\, j_\phi(x)\phi(x). \tag{5.85}$$

Taking a functional derivative of this equation with respect to $\phi(y)$ and using the chain rule, we obtain

$$\underbrace{\frac{\delta \Gamma[\phi]}{\delta \phi(y)}}_{-j_\phi(y)} = -i \int d^4x\, \frac{\delta W[j]}{\delta j(x)}\bigg|_{j=j_\phi} \frac{\delta j_\phi(x)}{\delta \phi(y)} - j_\phi(y) - \int d^4x\, \frac{\delta j_\phi(x)}{\delta \phi(y)}\phi(x). \tag{5.86}$$

This leads to

$$\phi(x) = -i\, \frac{\delta W[j]}{\delta j(x)}\bigg|_{j=j_\phi}, \quad \text{or equivalently} \quad \phi_j(x) = -i\, \frac{\delta W[j]}{\delta j(x)} = \langle \phi(x) \rangle_j. \tag{5.87}$$

In other words, $\phi_j$ is the connected one-point function (i.e., the vacuum expectation value of the field) in the presence of source $j$.

### 5.5.3 Second Derivative of the Effective Action

Differentiating eq. (5.82) with respect to j(y) gives

$$\delta(x-y) = -\frac{\delta}{\delta j(y)} \frac{\delta \Gamma[\phi_j]}{\delta \phi_j(x)} = -\int d^4z \, \frac{\delta \phi_j(z)}{\delta j(y)} \frac{\delta^2 \Gamma[\phi_j]}{\delta \phi_j(x) \delta \phi_j(z)}$$

$$= i \int d^4z \, \underbrace{\frac{\delta^2 W[j]}{\delta j(y) \delta j(z)}}_{-G(y,z)_{\text{connected}}} \underbrace{\frac{\delta^2 \Gamma[\phi_j]}{\delta \phi_j(z) \delta \phi_j(x)}}_{\Gamma_2(z,x)}. \tag{5.88}$$

This formula shows a posteriori that (up to a factor i) the coefficient $\Gamma_2$ in the expansion (5.75) is indeed the inverse of the exact connected two-point function, as was expected from our request that the effective action $\Gamma[\phi]$ reproduces the full content of the theory (see Exercise 5.3 for a more complicated example). By parametrizing the inverse propagator in terms of a self-energy $\Sigma$ as follows:

$$G^{-1} = G_0^{-1} + i\Sigma, \tag{5.89}$$

we see that, except for the bare inverse propagator $G_0^{-1}$ (i.e., the D'Alembertian that gives the kinetic term in the action), the second derivative of the quantum effective action is nothing but the self-energy. An important class of diagrams in this discussion are the *one-particle irreducible (1PI)* diagrams, defined as the graphs that remain connected if one cuts any one of their internal propagators. For instance, the first of these diagrams is 1PI while the second is not:

1PI :  , Non-1PI :

The concept of 1PI diagrams is crucial in the summation of a self-energy to all orders. Indeed, repeated insertions of a non-1PI self-energy would lead to the erroneous multiple countings of identical graphs. To avoid this, a self-energy should only contain 1PI graphs, and we conclude that the second derivative of the quantum effective action is one-particle irreducible.

### 5.5.4 One-Particle Irreducibility

The quantum effective action $\Gamma[\phi]$ is in fact one-particle irreducible at all orders in $\phi$, not just at quadratic order in $\phi$ as the above argument suggests. By exponentiating eq. (5.85) and using the path integral definition of $\exp(W[j])$, we first obtain

$$e^{i\Gamma[\phi]} = \int [D\varphi] \, \exp i \Big( S[\varphi] + \int d^4x \, j(x)(\varphi(x) - \phi(x)) \Big)_{j=j_\phi}$$

$$= \int [D\varphi] \, \exp i \Big( S[\varphi + \phi] + \int d^4x \, j(x)\varphi(x) \Big)_{j=j_\phi}. \tag{5.90}$$

(In the second line, we have shifted by $\phi$ the integration variable $\varphi$.) Thus, the quantum effective action can be obtained from a shifted classical action, to which is added a source $j_\phi$ that implicitly depends on $\phi$ via the quantum equation of motion (5.82). The expansion of the shifted classical action $S[\varphi+\phi]$ leads to a number of vertices, some of which are $\phi$-dependent. Thus, $\Gamma[\phi]$ is the sum of the connected (because we must take the logarithm in order to extract $\Gamma[\phi]$) vacuum graphs built with these $\phi$-dependent vertices and the $\phi$-dependent source $j_\phi$. To every line of such a graph is associated a free propagator $G_0$, determined from the quadratic term in the action.

A very important property is the fact that the expectation value of $\varphi(x)$ with this shifted action vanishes:

$$\begin{aligned}\langle\varphi(x)\rangle &\equiv \int[D\varphi]\,\varphi(x)\,\exp i\Big(S[\varphi+\phi]+\int j\varphi\Big)_{j=j_\phi}\\ &=\int[D\varphi]\,(\varphi(x)-\phi(x))\,\exp i\Big(S[\varphi]+\int d^4x\,j(\varphi-\phi)\Big)_{j=j_\phi}\\ &=-\phi(x)\,e^{i\Gamma[\phi]}+e^{-i\int j\phi}\,\frac{\delta}{i\delta j(x)}\,e^{W[j]}\Big|_{j=j_\phi}\\ &=e^{i\Gamma[\phi]}\underbrace{\Big(\frac{\delta W[j]}{i\delta j(x)}-\phi(x)\Big)_{j=j_\phi}}_{0}=0.\end{aligned} \qquad (5.91)$$

Note that in order to obtain the final zero, it is crucial that $j$ be set to $j_\phi$ at the end. Let us now consider a one-particle reducible vacuum graph $\mathcal{G}$ that may possibly contribute to $\Gamma[\phi]$. Because it is reducible, this graph contains at least one bare propagator that connects two subgraphs A and B, such that the two subgraphs would become disconnected when removing this propagator:

$$\mathcal{G}_{AB}\equiv\int_{x,y}\;\; \text{A} \overset{x\;\;y}{\bullet\!\!-\!\!\bullet} \text{B}. \qquad (5.92)$$

When summing over all graphs that may enter in B, we get

$$\sum_B \mathcal{G}_{AB}=\int d^4x\,A(x)\,\langle\varphi(x)\rangle=0, \qquad (5.93)$$

thanks to the previous result on the expectation value of $\varphi$. Therefore, the one-particle reducible graphs do not contribute to $\Gamma[\phi]$, which generalizes to all orders in $\phi$ what we have already seen for the quadratic terms.

### 5.5.5 One-Loop Effective Action

At one loop, one may obtain a closed expression for the quantum effective action. For this, write the Lagrangian as a renormalized Lagrangian plus counterterms,

$$\mathcal{L}\equiv\mathcal{L}_r(\phi_r)+\Delta\mathcal{L}(\phi_r), \qquad (5.94)$$

both depending on the renormalized field $\phi_r$. We will denote $\mathcal{S}_r$ and $\Delta\mathcal{S}$ the corresponding actions. Likewise, we write the external source as $j = j_r + \Delta j$, where $j_r$ is the current that solves the following equation:

$$\left.\frac{\delta \mathcal{S}_r[\phi_r]}{\delta \phi_r(x)}\right|_\varphi + j_r(x) = 0, \tag{5.95}$$

i.e., the current that solves at lowest order the defining equation of the effective action. The correction $\Delta j$ is then adjusted order by order so that the expectation value of the field remains equal to $\varphi$ at all orders:

$$\varphi(x) = \langle \phi_r(x) \rangle_{j_r + \Delta j}. \tag{5.96}$$

In the path integral representation of the generating functional $Z[j]$, we write the field as $\phi_r = \varphi + \eta$:

$$Z[j] = \int [D\eta(x)]\, e^{i\{\mathcal{S}_r[\varphi+\eta] + \Delta\mathcal{S}[\varphi+\eta] + \int d^4x\, (j_r + \Delta j)(\varphi+\eta)\}}, \tag{5.97}$$

and we expand the argument of the exponential in powers of $\eta$ up to quadratic order,

$$\begin{aligned}
\mathcal{S}_r[\varphi+\eta] + \int d^4x\, j_r(\varphi+\eta) &= \mathcal{S}_r[\varphi] + \int d^4x\, j_r(x)\varphi(x) \\
&\quad + \int d^4x\, \left(\left.\frac{\delta \mathcal{S}_r[\phi_r]}{\delta \phi_r(x)}\right|_\varphi + j_r\right) \eta(x) \\
&\quad + \frac{1}{2}\int d^4x\, d^4y\, \eta(x)\left(\left.\frac{\delta^2 \mathcal{S}_r[\phi_r]}{\delta \phi_r(x)\delta \phi_r(y)}\right|_\varphi\right)\eta(y) \\
&\quad + \cdots
\end{aligned} \tag{5.98}$$

Note that the linear term in $\eta$ is zero by virtue of eq. (5.95). Therefore, we may rewrite $Z[j]$ as follows:

$$Z[j] = e^{i\{\mathcal{S}_r[\varphi] + \Delta\mathcal{S}[\varphi] + \int d^4x\, j\varphi\}} \int [D\eta(x)]\, e^{i\{\mathcal{S}_\varphi[\eta] + \Delta\mathcal{S}_\varphi[\eta]\}}, \tag{5.99}$$

where we denote

$$\mathcal{S}_\varphi[\eta] \equiv \frac{1}{2}\int d^4x\, d^4y\, \eta(x)\left(\left.\frac{\delta^2 \mathcal{S}_r[\phi_r]}{\delta \phi_r(x)\delta \phi_r(y)}\right|_\varphi\right)\eta(y) + \cdots \tag{5.100}$$

(Likewise, $\Delta\mathcal{S}_\varphi[\eta]$ results from the expansion in powers of $\eta$ of the counterterms.) At one loop, it is sufficient to keep only the quadratic terms in $\eta$, and the path integral gives a determinant,

$$\left[\det\left(-i\left.\frac{\delta^2 \mathcal{S}_r[\phi_r]}{\delta \phi_r(x)\delta \phi_r(y)}\right|_\varphi\right)\right]^{-1/2} = \exp\left[-\frac{1}{2}\mathrm{tr}\,\ln\left(-i\left.\frac{\delta^2 \mathcal{S}_r[\phi_r]}{\delta \phi_r(x)\delta \phi_r(y)}\right|_\varphi\right)\right]. \tag{5.101}$$

At this order, the generating functional of connected graphs reads[3]

$$W[j] = i\{S_r[\varphi] + \Delta S[\varphi] + \int d^4x \, j\varphi\} - \frac{1}{2} \text{tr} \ln \left( \left. \frac{\delta^2 S_r[\phi_r]}{\delta \phi_r(x) \delta \phi_r(y)} \right|_\varphi \right) + \cdots \qquad (5.102)$$

from which we obtain the following quantum effective action

$$\Gamma[\varphi] = S_r[\varphi] + \Delta S[\varphi] + \frac{i}{2} \text{tr} \ln \left( \left. \frac{\delta^2 S_r[\phi_r]}{\delta \phi_r(x) \delta \phi_r(y)} \right|_\varphi \right) + \cdots \qquad (5.103)$$

Note that the object inside the logarithm is the inverse of the propagator dressed by the background field $\varphi$.

### 5.5.6 Example: Coleman–Weinberg Effective Potential

As a concrete application of the quantum effective action, let us study the effect of quantum fluctuations on a system close to a spontaneous breakdown of symmetry. Consider to that effect a complex scalar field coupled to an Abelian gauge potential,

$$\begin{aligned}\mathcal{L} &\equiv -\tfrac{1}{4} F^{\mu\nu} F_{\mu\nu} + (D_\mu \phi)^* (D^\mu \phi) - m^2 \phi^* \phi - \tfrac{\lambda}{6}(\phi^* \phi)^2 \\ &= -\tfrac{1}{4} F^{\mu\nu} F_{\mu\nu} + \tfrac{1}{2}(\partial_\mu \phi_1)(\partial^\mu \phi_1) + \tfrac{1}{2}(\partial_\mu \phi_2)(\partial^\mu \phi_2) + e A_\mu(\phi_2 \partial^\mu \phi_1 - \phi_1 \partial^\mu \phi_2) \\ &\quad + \tfrac{e^2}{2} A_\mu A^\mu (\phi_1^2 + \phi_2^2) - \tfrac{m^2}{2}(\phi_1^2 + \phi_2^2) - \tfrac{\lambda}{4!}(\phi_1^4 + \phi_2^4 + 2\phi_1^2 \phi_2^2), \end{aligned} \qquad (5.104)$$

where in the second equality we have written the Lagrangian explicitly in terms of the two components of the complex scalar field $\phi \equiv \frac{1}{\sqrt{2}}(\phi_1 + i\phi_2)$. The quartic coupling constant $\lambda$ must be positive for the system to have a well-defined ground state. Classically, the vacuum expectation value of the scalar field is zero if $m^2 \geq 0$, and differs from zero if $m^2 < 0$. When quantum corrections are taken into account, these conclusions are unchanged if the renormalized value of the parameter $m^2$ is strictly positive or strictly negative. The goal of the subsequent calculation is to use the quantum effective action formalism in order to show that the system may undergo spontaneous breakdown of symmetry also in the case where the renormalized $m^2$ is zero. In other words, when we attempt to renormalize the theory in such a way that it has no dimensionful parameter, the system nevertheless acquires a scale through a non-zero vacuum expectation value.

For this study, it is sufficient to evaluate the quantum effective action at a null value of the photon field, and for a spatially homogeneous scalar field (i.e., we assume that the vacuum remains Lorentz invariant even if spontaneous symmetry breaking occurs). Moreover, since the Lagrangian is $U(1)$ invariant, we can without loss of generality decide that spontaneous symmetry breaking (if it occurs) is along the scalar field component $\phi_1$. For this reason, it is sufficient to calculate the effective potential for a constant scalar field of the form $\phi \equiv \frac{1}{\sqrt{2}}(\phi_1 + i0)$. In this special configuration, the quantum effective action can be parametrized as follows:

$$\Gamma[\phi, A^\mu] = -V(\phi_1), \qquad (5.105)$$

---

[3] We have dropped a factor $-i$ inside the tr ln, since it only produces an additive constant.

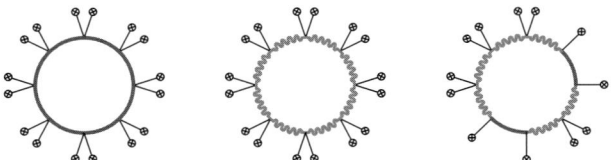

FIGURE 5.3: One-loop graphs contributing to the scalar effective potential in scalar QED.

where $V(\phi_1)$ is a local function of the scalar field called the *effective potential*. At tree level, the effective potential can simply be read off the classical action,

$$V_{\text{tree}}(\phi_1) = \tfrac{m^2}{2}\phi_1^2 + \tfrac{\lambda}{4!}\phi_1^4. \tag{5.106}$$

At one loop, the effective potential is given by the graphs in Figure 5.3, where the lines terminated by a cross denote the constant field $\phi_1$. The graphs in this figure are just templates for each topology, and we must sum over graphs with arbitrary numbers of insertions of the external field.

Let us start by the topologies represented in the right part of the figure, where the loop mixes photons and scalar particles. Recall that the vertex coupling a single photon to two scalars is proportional to $(p-q)_\mu$, where $p, q$ are the incoming momenta of the scalars. In this vertex, one of the scalars has a null momentum since it corresponds to the constant external field. Therefore, $(p-q)_\mu$ is also the momentum of the photon attached to this vertex. If we use Landau gauge (note that $A^\mu = 0$ is consistent with Landau gauge condition, $\partial_\mu A^\mu = 0$), in which the free photon propagator is transverse ($k_\mu G_0^{\mu\nu}(k) = 0$), these graphs are all zero.

Consider now the first graph, with a scalar loop. Under free propagation, the scalar components $\phi_{1,2}$ do not mix, and the loop propagators are thus entirely made of $\phi_1$s or entirely made of $\phi_2$s. Consider the first case. We have

$$V_{\phi_1 \text{ loop}}(\phi_1) = i\sum_{n=1}^{\infty}\int\frac{d^4k}{(2\pi)^4}\frac{1}{2n}\left(-i\tfrac{\lambda}{2}\phi_1^2\frac{i}{k^2+i0^+}\right)^n$$
$$= -\frac{i}{2}\int\frac{d^4k}{(2\pi)^4}\ln\left(1-\frac{\lambda\phi_1^2}{2(k^2+i0^+)}\right) = \frac{1}{2}\int\frac{d^4k_E}{(2\pi)^4}\ln\left(\frac{k_E^2+\frac{\lambda\phi_1^2}{2}}{k_E^2}\right). \tag{5.107}$$

(Thereafter, we assume that $m^2 = 0$.) In the first equality, the prefactor $1/(2^n(2n))$ is the symmetry factor of the graph. Observe that $\tfrac{\lambda}{2}\phi_1^2$ acts as a squared mass for the field running in the loop. A Wick's rotation has been performed in the last equality, leading to an ultraviolet divergent integral. Regularizing it with a cutoff $k_E^2 \leq \Lambda^2$, we get

$$V_{\phi_1 \text{ loop}}(\phi_1) = \frac{1}{64\pi^2}\left(\lambda\Lambda^2\phi_1^2 + \frac{\lambda^2\phi_1^4}{4}\left(\ln\left(\tfrac{\lambda\phi_1^2}{2\Lambda^2}\right) - \tfrac{1}{2}\right)\right). \tag{5.108}$$

When the loop carries the field $\phi_2$, the only change is the coupling strength to $\phi_1$. The simplest is to note that $\tfrac{\lambda}{6}\phi_1^2$ is a squared mass for the field $\phi_2$ in the loop. This observation leads immediately to

$$V_{\phi_2 \text{ loop}}(\phi_1) = \frac{1}{64\pi^2}\left(\frac{\lambda\Lambda^2\phi_1^2}{3} + \frac{\lambda^2\phi_1^4}{36}\left(\ln\left(\tfrac{\lambda\phi_1^2}{6\Lambda^2}\right) - \tfrac{1}{2}\right)\right). \tag{5.109}$$

## 5.5 QUANTUM EFFECTIVE ACTION

Consider now the second graph, with a photon loop. The calculation is almost the same as in the case of a scalar loop, with two differences: now, $-e^2\phi_1^2$ acts as an effective squared mass for the photon; and the photon propagator in Landau gauge reads

$$G_0^{\mu\nu}(k) = -\frac{i}{k^2 + i0^+}\left(g^{\mu\nu} - \frac{k^\mu k^\nu}{k^2}\right). \tag{5.110}$$

We obtain

$$V_{A^\mu \text{ loop}}(\phi_1) = \frac{3}{64\pi^2}\left(2e^2\Lambda^2\phi_1^2 + e^4\phi_1^4\left(\ln\left(\frac{e^2\phi_1^2}{\Lambda^2}\right) - \frac{1}{2}\right)\right). \tag{5.111}$$

(The prefactor 3 is the trace of the tensor $g^{\mu\nu} - \frac{k^\mu k^\nu}{k^2}$.) To these contributions, we must add ultraviolet counterterms,

$$V_{CT}(\phi_1) \equiv \frac{b_1}{2}\phi_1^2 + \frac{b_4}{4!}\phi_1^4. \tag{5.112}$$

(Since we are considering only a constant $\phi_1$, we do not need the wavefunction renormalization counterterm.) These two counterterms can be set by the following conditions:

$$\left.\frac{\partial^2 V(\phi_1)}{\partial\phi_1^2}\right|_{\phi_1=0} = 0, \quad \left.\frac{\partial^4 V(\phi_1)}{\partial\phi_1^4}\right|_{\phi_1=M} = \lambda. \tag{5.113}$$

The first condition is simply the statement that the parameter $m^2$ should remain zero at one loop, and the second condition sets the value of the coupling at the scale M (a natural choice would be to take M as the vacuum expectation value of $\phi_1$). Combining the tree-level potential, the one-loop contributions and the counterterms determined from these conditions, we get

$$V_{\text{tree + one-loop}}(\phi_1) = \frac{\lambda}{4!}\phi_1^4 + \frac{\phi_1^4}{64\pi^2}\left(\frac{5}{18}\lambda^2 + 3e^4\right)\left(\ln\left(\frac{\phi_1^2}{M^2}\right) - \frac{25}{6}\right). \tag{5.114}$$

The derivative of this potential reads

$$\frac{\partial V_{\text{tree + one-loop}}(\phi_1)}{\partial\phi_1} = \frac{\lambda}{6}\phi_1^3 + \frac{\phi_1^3}{64\pi^2}\left(\frac{5}{18}\lambda^2 + 3e^4\right)\left(\ln\left(\frac{\phi_1^2}{M^2}\right) - \frac{44}{3}\right). \tag{5.115}$$

$\phi_1 = 0$ is always an extremum of the potential. Let us assume that this derivative vanishes at some $\phi_{1*} \neq 0$. To simplify the expressions, it is convenient to set the renormalization scale M to be this extremum. We see that such a non-trivial extremum exists provided that the coupling constants are related as follows:

$$\lambda = \frac{11}{8\pi^2}\left(3e^4 + \frac{5}{18}\lambda^2\right). \tag{5.116}$$

If this condition is satisfied, the effective potential reads, at leading order in $e^4$,

$$V_{\text{tree + one-loop}}(\phi_1) = \frac{3e^4}{64\pi^2}\phi_1^4\left(\ln\left(\frac{\phi_1^2}{\langle\phi_1\rangle^2}\right) - \frac{1}{2}\right). \tag{5.117}$$

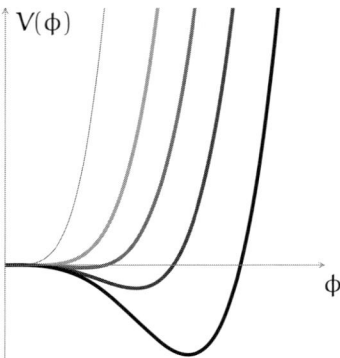

FIGURE 5.4: Illustration of the Coleman–Weinberg effective potential. Thin line: classical potential with $m^2 = 0$. Thick lines: one-loop effective potential (5.117), for varying values of $\langle \phi_1 \rangle$.

This quantity, illustrated in Figure 5.4, is known as the *Coleman–Weinberg effective potential*. First, let us comment on the number of parameters that this family of symmetry-broken theories depend on. The starting point was a theory that depends on two dimensionless couplings $\lambda, e$ and on the dimensionful renormalization scale M. The condition (5.116) that allows spontaneous symmetry breaking makes one of the dimensionless couplings depend on the other one, but leads to the appearance of the new dimensionful scale $\langle \phi_1 \rangle$, and a new dimensionless ratio $\langle \phi_1 \rangle / M$ (the latter does not appear explicitly in eq. (5.117) because of our special choice of the renormalization scale, $M = \langle \phi_1 \rangle$). Another interesting aspect of the potential (5.117) is that it is non-analytic at $\phi_1 = 0$. This suggests that finite-order truncations in $\phi_1$ lead to an incorrect behavior, and that resummations to all orders in $\phi_1$ are necessary. In the above calculation, this can be seen in the fact that truncations at finite $n$ give extremely infrared divergent contributions, that are regularized by $\phi_1^2$ when the insertions of the external field are summed to all orders. In other words, this summation dresses the loop propagators by a $\phi_1$-dependent mass. For instance, the propagator of the field $\phi_1$ becomes

$$\frac{i}{k^2 + i0^+} \to \frac{i}{k^2 - \frac{\lambda \phi_1^2}{2} + i0^+}. \tag{5.118}$$

## 5.6 Two-Particle Irreducible Effective Action

### 5.6.1 Definition and Equations of Motion

The quantum effective action $\Gamma[\phi]$ studied in the previous section can be generalized into a functional $\Gamma[\phi, G]$ that depends on a field $\phi$ and a propagator G. The starting point of this derivation is to introduce a second source $k(x, y)$ that couples to a pair of fields $\varphi(x)\varphi(y)$. The corresponding generating functional $W[j, k]$ for connected graphs is given by

$$e^{W[j,k]} = \int [D\varphi] \ \exp i \Big( S[\varphi] + \int_x j(x)\varphi(x) + \frac{1}{2} \int_{x,y} k(x,y) \, \varphi(x)\varphi(y) \Big). \tag{5.119}$$

In terms of graphs, $W[j, k]$ is the sum of the connected vacuum graphs built with the bare propagator and the vertices defined by the classical action $S[\varphi]$, with the external source $j$, and with a kind of non-local mass term $k(x, y)$. Let us denote

$$\frac{\delta W[j, k]}{i\delta j(x)} \equiv \phi_{j,k}(x),$$
$$\frac{\delta W[j, k]}{i\delta k(x, y)} \equiv \frac{1}{2}\left[\phi_{j,k}(x)\phi_{j,k}(y) + G_{j,k}(x, y)\right]. \quad (5.120)$$

In the second equation, we have separated a disconnected part $\phi_{j,k}(x)\phi_{j,k}(y)$ and a connected two-point function $G_{j,k}(x, y)$. Both the field $\phi_{j,k}$ and the propagator $G_{j,k}$ depend on the sources, which we indicate by the subscript $j, k$. Conversely, we may formally invert these equations to define $(\phi, G)$-dependent sources, $j_{\phi,G}$ and $k_{\phi,G}$.

Then, the Legendre transform that defined $\Gamma[\phi]$ from $W[j]$ can be generalized into

$$\Gamma[\phi, G] \equiv -i W[j_{\phi,G}, k_{\phi,G}] - \int d^4x\, j_{\phi,G}(x)\phi(x)$$
$$- \frac{1}{2}\int d^4x d^4y\, k_{\phi,G}(x, y)\left[\phi(x)\phi(y) + G(x, y)\right]. \quad (5.121)$$

By taking derivatives with respect to $\phi(x)$ or $G(x, y)$, we obtain the following equations:

$$\frac{\Gamma[\phi, G]}{\delta\phi(x)} + j_{\phi,G}(x) + \int d^4y\, k_{\phi,G}(x, y)\,\phi(y) = 0,$$
$$\frac{\Gamma[\phi, G]}{\delta G(x, y)} + \frac{1}{2}k_{\phi,G}(x, y) = 0. \quad (5.122)$$

Note that the first of these equations generalizes the quantum equation of motion (5.82) with the adjunction of a self-energy $k_{\phi,G}(x, y)$.

### 5.6.2 Two-Particle Irreducibility

From the Legendre transform of eq. (5.121), we obtain the following path integral representation of the functional $\Gamma[\phi, G]$:

$$e^{i\Gamma[\phi, G]} = \int [D\varphi]\, \exp i\Big(S[\varphi] + \int_x j(x)(\varphi(x) - \phi(x))$$
$$+ \frac{1}{2}\int_{x,y} k(x, y)\left[\varphi(x)\varphi(y) - \phi(x)\phi(y) - G(x, y)\right]\Big)$$
$$= \int [D\varphi]\, \exp i\Big(S[\varphi+\phi] + \int_x j(x)\varphi(x)$$
$$+ \frac{1}{2}\int_{x,y} k(x, y)\left[\varphi(x)\varphi(y) + 2\varphi(x)\phi(y) - G(x, y)\right]\Big), \quad (5.123)$$

where we have omitted the subscript $\phi, G$ on the sources $j, k$ for the sake of brevity. From the second equation, we first obtain

$$\begin{aligned}\langle \varphi(x) \rangle &= \int [D\varphi]\; \varphi(x)\; \exp i\Big(\mathcal{S}[\varphi+\phi] + \int j\varphi + \tfrac{k}{2}\left[\varphi\varphi + 2\varphi\phi - G\right]\Big)\\ &= e^{i\Gamma[\phi,G]}\left(\frac{\delta W[j,k]}{i\delta j(x)} - \phi(x)\right)_{\substack{j=j_{\phi,G}\\k=k_{\phi,G}}} = 0.\end{aligned} \quad (5.124)$$

As in the case of the 1PI functional $\Gamma[\phi]$, this identity ensures that one-particle reducible graphs do not contribute to $\Gamma[\phi, G]$. But as we shall see now, the functional $\Gamma[\phi, G]$ is limited to a much more restricted set of graphs, since only the *two-particle irreducible graphs* contribute, i.e., the graphs that cannot be disconnected by removing two arbitrary propagators. Consider a two-particle reducible graph,

$$\mathcal{G}_{AB} \equiv \int_{x,y} \;\; \begin{array}{c}\text{[diagram: blob A connected to blob B via two propagators labeled }x\text{ and }y\text{]}\end{array}, \quad (5.125)$$

in which we have exhibited the two bare propagators that would disconnect the graph if removed. Summing over the graphs that can contribute to B, we may write this as

$$\sum_B \mathcal{G}_{AB} = \int d^4x\, d^4y\; A(x,y)\; \langle \varphi(x)\varphi(y)\rangle_c \quad (5.126)$$

(the subscript c indicates that we keep only the connected part of the two-point function), with

$$\begin{aligned}\langle \varphi(x)\varphi(y)\rangle_c &\equiv e^{-i\Gamma[\phi,G]}\int [D\varphi]\; \varphi(x)\varphi(y)\; e^{i\left(\mathcal{S}[\varphi+\phi] + \int j\varphi + \tfrac{k}{2}[\varphi\varphi+2\varphi\phi-G]\right)}\\ &= -\phi(x)\phi(y) + 2\; \underbrace{e^{-i\Gamma[\phi,G]}e^{-i\int j\phi + \tfrac{k}{2}(G-\phi\phi)}}_{e^{-W[j,k]}}\; \frac{\delta e^{W[j,k]}}{i\delta k(x,y)}\\ &= G(x,y).\end{aligned} \quad (5.127)$$

In the second equality, we have ignored some terms that have already been shown to vanish when studying $\langle \varphi(x)\rangle$, and we have extracted the combination $\varphi(x)\varphi(y)$ by differentiating with respect to $k(x, y)$. From this identity, we obtain

$$\sum_B \int_{x,y} \;\begin{array}{c}\text{[diagram: A--B with two propagators }x,y\text{]}\end{array} = \int_{x,y}\;\begin{array}{c}\text{[diagram: A with loop }G(x,y)\text{]}\end{array}. \quad (5.128)$$

In other words, when summing over all the possible graphs contributing to B, the two-particle reducible block is replaced by a single propagator $G(x, y)$. Thus, the functional $\Gamma[\phi, G]$, when expressed in terms of the two-point function $G$, is made only of two-particle irreducible graphs, and its derivatives with respect to the field $\phi$ are the two-particle irreducible n-point functions. Thus, $\Gamma[\phi, G]$ is the generating functional in $\phi$ of the 2PI correlation functions.

## 5.6.3 Loop Expansion of $\Gamma[\phi, G]$

**2PI functional at null field:** Consider first the 2PI effective action at null field, $\Gamma[0, G]$. At $\phi \equiv 0$, we have

$$-2\frac{\delta \Gamma[0, G]}{\delta G} = k_{0, G}. \tag{5.129}$$

Using the path integral representation (5.123), and replacing $k_{0,G}$ by the above equality, we get:[4]

$$e^{i\Gamma[0,G]} = \int [D\varphi]\, e^{i\left(S[\varphi] + \mathrm{tr}\left([G - \varphi\varphi]\frac{\delta \Gamma[0,G]}{\delta G}\right) + \int j_{0,G}\varphi\right)}$$

$$= \sum \left\{ \begin{array}{c} \text{2PI vacuum diagrams built with the} \\ \text{propagator G and the vertices from } S[\varphi] \end{array} \right\}. \tag{5.130}$$

The first equality is as an implicit identity obeyed by $\Gamma[0, G]$. The diagrammatic interpretation in the second line follows from the discussion at the end of the previous subsection. On the right-hand side, the term in $j_{0,G}\varphi$ cancels the one-particle reducible tadpole contributions, while the term in $G - \varphi\varphi$ eliminates the two-particle reducible ones (this term replaces chains like $G_0 \Sigma G_0 \Sigma \cdots \Sigma G_0$ by $G$). Note that the bare propagator $G_0$ defined by the quadratic part of the classical action $S[\varphi]$ does not appear in the final result for $\Gamma[0, G]$, since it is replaced systematically by $G$. Only the interaction terms of the classical action matter, because they define the vertices that connect the Gs in the diagrammatic representation of $\Gamma[0, G]$.

**Legendre transform at fixed k:** When $\phi \neq 0$, it is useful to start with the first Legendre transform of $W[j, k]$, i.e., the 1PI effective action at fixed $k$,

$$\Gamma_k[\phi] \equiv -i\, W[j_{\phi,k}, k] - \int d^4x\, j_{\phi,k}(x)\phi(x), \tag{5.131}$$

where the source $j_{\phi,k}$ now has an implicit dependence on the field $\phi$ and on the second source $k$. This functional obeys

$$\frac{\delta \Gamma_k[\phi]}{\delta k(x,y)} = \frac{\delta W}{i\delta k(x,y)} + \int d^4z \underbrace{\frac{\delta W}{i\delta j(z)}\Big|_{j = j_{\phi,k}}}_{\phi(z)} \frac{\delta j_{\phi,k}(z)}{\delta k(x,y)} - \int d^4z\, \frac{\delta j_{\phi,k}(z)}{\delta k(x,y)} \phi(z)$$

$$= \frac{\delta W}{i\delta k(x,y)}. \tag{5.132}$$

Given this identity, it is natural to define $G(x, y)$ as follows:

$$\frac{\delta \Gamma_k[\phi]}{\delta k(x,y)} \equiv \frac{1}{2}\left(\phi(x)\phi(y) + G(x,y)\right). \tag{5.133}$$

---

[4] We use the compact notation $\int_{x,y} A(x,y) B(x,y) = \mathrm{tr}(AB)$.

In this equation, we may either view G as a function of k, ϕ or k as a function (that we denote $k_{\phi,G}$) of ϕ, G. Adopting the latter point of view, we may perform a second Legendre transform to obtain a functional of ϕ and G:

$$\Gamma[\phi, G] \equiv \Gamma_{k_{\phi,G}}[\phi] - \frac{1}{2}\operatorname{tr}\left(k_{\phi,G}[G + \phi\phi]\right). \tag{5.134}$$

(We use the same notation $\Gamma[\phi, G]$ because we shall prove shortly that this functional is identical to the one we have defined earlier by a double Legendre transform.) This definition leads to

$$\frac{\delta\Gamma[\phi, G]}{\delta G(x,y)} = -\frac{1}{2}k(x,y) + \underbrace{\operatorname{tr}\left(\frac{\delta\Gamma_k}{\delta k}\bigg|_{k_{\phi,G}} \frac{\delta k_{\phi,G}}{\delta G}\right) - \frac{1}{2}\operatorname{tr}\left(\frac{\delta k_{\phi,G}}{\delta G}[G + \phi\phi]\right)}_{0},$$

$$\frac{\delta\Gamma[\phi, G]}{\delta\phi(x)} = -j_{\phi,k} - \int_y k_{\phi,G}(x,y)\phi(y)$$
$$+ \underbrace{\operatorname{tr}\left(\frac{\delta\Gamma_k}{\delta k}\bigg|_{k_{\phi,G}} \frac{\delta k_{\phi,G}}{\delta\phi}\right) - \frac{1}{2}\operatorname{tr}\left(\frac{\delta k_{\phi,G}}{\delta G}[G + \phi\phi]\right)}_{0}, \tag{5.135}$$

which are identical to eqs. (5.122). This proves that the definition (5.134) of the 2PI effective action is equivalent to the original definition (5.121).

**Path integral representation of $\Gamma_k[\phi]$:** Since we have

$$e^{W[j,k]} = \int [D\varphi]\, e^{i\left(S[\varphi]+\int j\varphi+\frac{1}{2}\int \varphi k\varphi\right)}, \tag{5.136}$$

$\Gamma_k[\phi]$ is the 1PI effective action of the modified classical action

$$S_k[\varphi] \equiv S[\varphi] + \frac{1}{2}\int d^4x\, d^4y\, \varphi(x)\, k(x,y)\, \varphi(y), \tag{5.137}$$

and it admits the following path integral representation:

$$e^{i\Gamma_k[\phi]} = \int [D\varphi]\, e^{i\left(S_k[\phi+\varphi] - \int \varphi \frac{\delta\Gamma_k[\phi]}{\delta\phi}\right)}. \tag{5.138}$$

Thus, if we denote $\Gamma_{k,1}[\phi] \equiv \Gamma_k[\phi] - S_k[\phi]$ the terms with at least one loop, we have

$$e^{i\Gamma_{k,1}[\phi]} = \int [D\varphi]\, e^{i\left(S_k[\phi+\varphi] - S_k[\phi] - \int \varphi \frac{\delta(S_k[\phi]+\Gamma_{k,1}[\phi])}{\delta\phi}\right)}. \tag{5.139}$$

Let us now expand the shifted action $S_k[\phi + \varphi]$:

$$S_k[\phi + \varphi] \equiv S_k[\phi] + \int \varphi \frac{\delta S_k[\phi]}{\delta\phi} + \frac{1}{2}\int \varphi \underbrace{\frac{\delta^2 S_k[\phi]}{\delta\phi\delta\phi}}_{k+i\mathcal{G}_\phi^{-1}}\varphi + S_{\text{int}}[\phi;\varphi], \tag{5.140}$$

## 5.6 TWO-PARTICLE IRREDUCIBLE EFFECTIVE ACTION

FIGURE 5.5: Beginning of the diagrammatic expansion of $\Gamma_2[\phi, G]$ in a scalar field theory with quartic interaction. The lines terminated by a cross are the field $\phi$ and the black lines represent the propagator G.

where $\mathcal{S}_{\text{int}}[\phi; \varphi]$ denotes the terms of degree at least three in $\varphi$ in the Taylor expansion of $\mathcal{S}_k[\phi + \varphi]$, and $\mathcal{G}_\phi^{-1}$ is the inverse of the tree-level propagator in the background field $\phi$. Therefore, $\Gamma_{k,1}[\phi]$ can also be written as

$$e^{i\Gamma_{k,1}[\phi]} = \int [D\varphi]\, e^{i\left(\frac{1}{2}\int \varphi[k+i\mathcal{G}_\phi^{-1}]\varphi + \mathcal{S}_{\text{int}}[\phi;\varphi] - \int \varphi \frac{\delta\Gamma_{k,1}[\phi]}{\delta\phi}\right)}. \tag{5.141}$$

**Diagrammatic interpretation of $\Gamma[\phi, G]$:** Let us now write $\Gamma[\phi, G]$ as follows:

$$\Gamma[\phi, G] = \text{const} + \mathcal{S}[\phi] + \tfrac{i}{2}\operatorname{tr}\ln(G^{-1}) + \tfrac{i}{2}\operatorname{tr}(\mathcal{G}_\phi^{-1} G) + \Gamma_2[\phi, G]. \tag{5.142}$$

This equation should be understood as a definition of $\Gamma_2[\phi, G]$ so that the left-hand side is indeed $\Gamma[\phi, G]$. Then, combining eqs. (5.134), (5.141) and (5.142), we must have

$$\Gamma_2[\phi, G] = \text{const} - \tfrac{i}{2}\operatorname{tr}\left([k_{\phi,G} + i\mathcal{G}_\phi^{-1}]G\right) - \tfrac{i}{2}\operatorname{tr}\ln(G^{-1}) + \Gamma_{k,1}[\phi]. \tag{5.143}$$

In order to eliminate $k_{\phi,G}$, we use

$$\frac{\delta\Gamma_2[\phi, G]}{\delta G} + \frac{1}{2}k_{\phi,G} + \frac{i}{2}\left(\mathcal{G}_\phi^{-1} - G^{-1}\right) = 0, \tag{5.144}$$

that follows from eqs. (5.122) and (5.142). We thus obtain

$$e^{i\Gamma_2[\phi,G]} = \int [D\varphi]\, e^{i\left(\mathcal{S}_{\phi,G}[\varphi] + \operatorname{tr}\left([G-\varphi\varphi]\frac{\delta\Gamma_2}{\delta G}\right) - \int \varphi \frac{\delta\Gamma_{k,1}[\phi]}{\delta\phi}\right)}, \tag{5.145}$$

with

$$\mathcal{S}_{\phi,G}[\varphi] \equiv \frac{i}{2}\int \varphi\, G^{-1}\varphi + \mathcal{S}_{\text{int}}[\phi; \varphi]. \tag{5.146}$$

Finally, using the fact that $-j_{\phi,G} = \delta\Gamma_{k,1}/\delta\phi$ and comparing with eq. (5.130), we see that $\Gamma_2[\phi, G]$ is the sum of the 2PI vacuum graphs built with the propagator G and the vertices obtained from the expansion of $\mathcal{S}[\phi + \varphi]$. The first four terms of the expansion of $\Gamma_2[\phi, G]$ in a scalar field theory with $\phi^4$ interaction are shown in Figure 5.5.

### 5.6.4 Dyson Equation

After setting the source $k = 0$, the equation of motion for the propagator (5.144) becomes

$$-i\, G^{-1} = -i\,\mathcal{G}_\phi^{-1} - \underbrace{2\,\frac{\delta\Gamma_2[\phi, G]}{\delta G}}_{-\Sigma}, \tag{5.147}$$

which is known as the Dyson equation, that re-sums the self-energy $\Sigma$ on the propagator. Convoluting this equation by G on the right gives

$$-i\mathcal{G}_\phi^{-1} G + \Sigma G = -i. \tag{5.148}$$

Recall that $\mathcal{G}_\phi$ is the tree-level propagator in a background field $\phi$, i.e.,

$$i\mathcal{G}_\phi^{-1} = \frac{\delta^2 S[\phi]}{\delta\phi\delta\phi} = -(\Box + m^2) - V''(\phi), \tag{5.149}$$

where $V(\phi)$ is the interaction potential in the Lagrangian. Therefore, the equation of motion has the following more explicit form:

$$\left[\Box + m^2 + V''(\phi) + \Sigma\right] G = -i. \tag{5.150}$$

A closed system of equations is obtained by combining it with the equation of motion of $\phi$, obtained as $\delta\Gamma/\delta\phi = 0$ (in a system with symmetry $\phi \to -\phi$, and assuming that there is no spontaneous breaking of this symmetry, the field expectation value is zero and we may simply set $\phi = 0$ in the above equation for G). The self-energy $\Sigma$ is itself a functional of $\phi$ and G, obtained as the derivative of $\Gamma_2[\phi, G]$ with respect to the propagator,

$$\Sigma = -2 \frac{\delta \Gamma_2[\phi, G]}{\delta G}. \tag{5.151}$$

Diagrammatically, this derivative amounts to opening one internal line of the graphs that contribute to $\Gamma_2$. For instance, the graphs in Figure 5.5 give the following topologies in $\Sigma$:

Note that the two-particle irreducibility of $\Gamma_2$ is equivalent to the one-particle irreducibility of $\Sigma$, which is crucial in order to avoid including multiple times the same contributions when summing the self-energy to all orders. In practice, a truncation is necessary in order to obtain equations that have a finite number of terms. For instance, keeping only the first diagram in $\Gamma_2$ gives the tadpole diagram in $\Sigma$ (but bear in mind that since this tadpole contains the full propagator G, the solution of the corresponding Dyson equation already re-sums an infinite set of Feynman graphs).

## 5.7 Euclidean Path Integral and Statistical Mechanics

### 5.7.1 Statistical Mechanics in Path Integral Form

A path integral formalism also exists for statistical mechanics. In order to illustrate this, let us consider again the quantum mechanical system described by the Hamiltonian of eq. (5.1). Our goal is to calculate the *partition function* in the canonical ensemble,[5]

$$Z_\beta \equiv \text{tr}\left(e^{-\beta \mathcal{H}}\right), \tag{5.152}$$

---

[5] In theories with a conserved quantity, it is also possible to study the grand canonical ensemble. In this case, one should substitute $\mathcal{H} \to \mathcal{H} - \mu Q$ in the definition of the partition function, where Q is the operator of the conserved charge and $\mu$ the associated chemical potential.

## 5.7 EUCLIDEAN PATH INTEGRAL AND STATISTICAL MECHANICS

where $\beta$ is the inverse temperature (it is customary to use a system of units in which Boltzmann's constant $k_B$ is equal to unity – therefore temperature has the same dimension as energy). More generally, one may want to calculate the following canonical ensemble expectation values:

$$\langle \mathcal{O} \rangle_\beta \equiv Z_\beta^{-1} \, \mathrm{tr}\left(e^{-\beta \mathcal{H}} \mathcal{O}\right). \tag{5.153}$$

The cyclicity of the trace leads to an important identity for expectation values of products of operators,

$$\begin{aligned}
\langle \mathcal{O}_1(t)\mathcal{O}_2(t') \rangle_\beta &\equiv Z_\beta^{-1} \, \mathrm{tr}\left(e^{-\beta \mathcal{H}} \mathcal{O}_1(t)\mathcal{O}_2(t')\right) \\
&= Z_\beta^{-1} \, \mathrm{tr}\left(e^{-\beta \mathcal{H}} \mathcal{O}_1(t) \underbrace{e^{+\beta \mathcal{H}} e^{-\beta \mathcal{H}}}_{1} \mathcal{O}_2(t')\right) \\
&= Z_\beta^{-1} \, \mathrm{tr}\left(e^{-\beta \mathcal{H}} \mathcal{O}_2(t') \underbrace{e^{-\beta \mathcal{H}} \mathcal{O}_1(t) e^{+\beta \mathcal{H}}}_{\mathcal{O}_1(t+i\beta)}\right) \\
&= \langle \mathcal{O}_2(t') \mathcal{O}_1(t+i\beta) \rangle_\beta,
\end{aligned} \tag{5.154}$$

where we have formally identified the density operator $\exp(-\beta \mathcal{H})$ with a time evolution operator for an imaginary time $-i\beta$. This relationship is called the *Kubo–Martin–Schwinger (KMS) identity*. Although we have established it for an expectation value of a product of two operators, it is completely general.

The identification of the density operator with an imaginary time evolution operator is at the heart of the formalism to evaluate canonical ensemble expectation values. If we represent the trace that appears in the partition function in the coordinate basis,

$$Z_\beta = \int dq \, \langle q | e^{-\beta \mathcal{H}} | q \rangle, \tag{5.155}$$

the integrand on the right-hand side is a transition amplitude similar to eq. (5.3), except that initial and final coordinates are identical, and the time interval is imaginary. We can nevertheless formally reproduce all the manipulations of Section 5.1, with an initial time $t_i \equiv 0$ and a final time $t_f \equiv -i\beta$. It is common to introduce the *Euclidean time* $\tau \equiv it$, with $\tau$ varying from $0$ to $\beta$. The only change to our original derivation of the path integral is that the path $q(t)$ must be replaced by a path $q(\tau)$ whose time derivative is the *Euclidean velocity* $\dot{q}_E$, related to the usual velocity $\dot{q}$ by

$$\dot{q} \equiv \frac{dq}{dt} = i \underbrace{\frac{dq}{d\tau}}_{\dot{q}_E}. \tag{5.156}$$

We obtain the following path integral representation of the partition function,

$$\begin{aligned}
Z_\beta &= \int dq \int_{\substack{q(0)=q \\ q(\beta)=q}} [Dp(\tau)Dq(\tau)] \, \exp\left\{\int_0^\beta d\tau \, (i\, p(\tau)\dot{q}_E(\tau) - \mathcal{H}(p(\tau), q(\tau)))\right\} \\
&= \int_{q(0)=q(\beta)} [Dp(\tau)Dq(\tau)] \, \exp\left\{\int_0^\beta d\tau \, (i\, p(\tau)\dot{q}_E(\tau) - \mathcal{H}(p(\tau), q(\tau)))\right\}.
\end{aligned} \tag{5.157}$$

In the second line, we have simplified the boundary conditions of the path $q(\tau)$, since the only constraint it must obey is to be $\beta$-periodic in the imaginary time. The integration over the momentum $p(\tau)$ is again Gaussian, and after performing it we obtain the following expression:

$$Z_\beta = \int_{q(0)=q(\beta)} [Dq(\tau)] \exp\left\{-\underbrace{\int_0^\beta d\tau \left(\frac{m}{2}\dot{q}_E^2(\tau) + V(q(\tau))\right)}_{S_E[q(\tau)]}\right\}. \tag{5.158}$$

The quantity $S_E[q]$ is called the *Euclidean action*.

Then, we can generalize this formalism to calculate ensemble averages of time-ordered (in imaginary time) products of position operators. For instance, the analogue of eq. (5.28) is

$$\mathrm{tr}\left(e^{-\beta\mathcal{H}} T_\tau (Q(\tau_1)Q(\tau_2))\right) = \int_{q(0)=q(\beta)} [Dq(\tau)]\, e^{-S_E[q(\tau)]}\, q(\tau_1)q(\tau_2), \tag{5.159}$$

where the symbol $T_\tau$ denotes the time-ordering in the imaginary time $\tau$. Likewise, we may define a generating functional for these expectation values:

$$\mathrm{tr}\left(e^{-\beta\mathcal{H}} T_\tau \exp\int_0^\beta d\tau\, j(\tau)Q(\tau)\right) = \int_{q(0)=q(\beta)} [Dq(\tau)]\, e^{-S_E[q(\tau)]+\int_0^\beta d\tau\, j(\tau)q(\tau)}. \tag{5.160}$$

### 5.7.2 Statistical Field Theory

This formalism can be extended readily to a quantum field theory. In this context, it can be used to calculate canonical ensemble expectation values of operators for a system of relativistic particles at non-zero temperature. One can write directly the following generalization of eq. (5.160):

$$\underbrace{\mathrm{tr}\left(e^{-\beta\mathcal{H}} T_\tau \exp\int_0^\beta d^4x_E\, j(x)\phi(x)\right)}_{Z[j;\beta]} = \int_{\phi(0,\mathbf{x})=\phi(\beta,\mathbf{x})} [D\phi(x)]\, e^{-S_E[\phi(x)]+\int_0^\beta d^4x_E\, j(x)\phi(x)}, \tag{5.161}$$

where the measure $d^4x_E$ stands for $d\tau\, d^3\mathbf{x}$. As in the case of ordinary quantum field theory in Minkowski space-time, we can isolate the interactions by writing

$$Z[j;\beta] = \exp\left\{-\int d^4x_E\, \mathcal{L}_{E,I}\left(\frac{\delta}{\delta j(x)}\right)\right\} Z_0[j;\beta], \tag{5.162}$$

where $\mathcal{L}_{E,I}$ is the interaction term in the Euclidean Lagrangian density, and $Z_0[j;\beta]$ is the generating functional of the non-interacting theory,

$$Z_0[j;\beta] = \int_{\phi(0,\mathbf{x})=\phi(\beta,\mathbf{x})} [D\phi(x)]\, \exp\left[-\int_0^\beta d^4x_E\, \left(\frac{1}{2}((\partial_\tau\phi)^2+(\boldsymbol{\nabla}\phi)^2+m^2\phi^2)-j\phi\right)\right]. \tag{5.163}$$

The Gaussian path integral in this expression leads to

$$Z_0[j;\beta] = \exp\left\{\frac{1}{2}\int_0^\beta d^4x_E\, d^4y_E\; j(x)\, G_E^0(x,y)\, j(y)\right\}, \qquad (5.164)$$

where the free Euclidean propagator $G_E^0(x,y)$ is the inverse of the operator $m^2 - \partial_\tau^2 - \nabla^2$ over the space of functions that are $\beta$-periodic in the imaginary time variable. Because of this periodicity, the "energy" variable, conjugate to the Euclidean time, is discrete:

$$\omega_n \equiv \frac{2\pi n}{\beta} \qquad (n \in \mathbb{Z}). \qquad (5.165)$$

In terms of these energies, called *Matsubara frequencies*, the free Euclidean propagator in momentum space reads:

$$\widetilde{G}_E^0(\omega_n, \mathbf{p}) = \frac{1}{\omega_n^2 + \mathbf{p}^2 + m^2}. \qquad (5.166)$$

Note that the denominator cannot vanish, and therefore this propagator does not need an $i0^+$ prescription for being fully defined. Eqs. (5.162) and (5.164) lead to a perturbative expansion that can be cast into an expansion in terms of Feynman diagrams. The Feynman rules associated with these graphs are very similar to those already encountered when calculating scattering amplitudes, with only a few modifications:

Propagators: $\qquad\qquad \dfrac{1}{\omega_n^2 + \mathbf{p}^2 + m^2}, \qquad (5.167)$

Vertices: $\qquad\qquad -\lambda\, \beta\, \delta_{n_1+n_2+\cdots}\, (2\pi)^3 \delta\!\left(\sum_i \mathbf{p}_i\right), \qquad (5.168)$

Loops: $\qquad\qquad \dfrac{1}{\beta}\sum_{n\in\mathbb{Z}} \int \dfrac{d^3\mathbf{p}}{(2\pi)^3}. \qquad (5.169)$

In words, the main difference with the usual perturbative expansion is that the energies are replaced by the discrete Matsubara frequencies, and that the loop integration on $p^0$ is replaced by a discrete sum over these frequencies (see Exercise 5.8 for a simple example).

## Exercises

**5.1** Weyl's prescription for the quantization of a generic classical quantity $f(q,p)$ is

$$F(Q,P) \equiv \int \frac{dp\, dq\, d\mu\, d\nu}{(2\pi)^2}\, f(q,p)\, e^{i(\mu(Q-q)+\nu(P-p))},$$

where $Q, P$ are the position and momentum operators, while $q, p$ are the corresponding classical variables. (This is known as Weyl mapping.)

- Show that for a separable quantity, $f(q,p) = a(q) + b(p)$, this definition simply leads to $F(Q,P) = a(Q) + b(P)$.

- Show that $(\alpha q + \beta p)^n$ is mapped to $(\alpha Q + \beta P)^n$.
- Calculate the quantum operator corresponding to qp.
- Show that the inverse of Weyl's mapping is the Wigner transform:

$$f(q, p) \equiv \int ds\, e^{ips} \langle q - \tfrac{s}{2} | F(Q, P) | q + \tfrac{s}{2} \rangle.$$

- Derive the path integral for a non-separable Hamiltonian.

**5.2** Show that

$$\exp\left\{ \int dt\, \frac{\sigma(t)}{2} \left( \frac{\delta}{\delta j(t)} \right)^2 \right\} F[j] = \int [Da(t)]\, \exp\left\{ -\int dt\, \frac{a^2(t)}{2\sigma(t)} \right\} F[j + a].$$

(This is a functional version of the *Weierstrass transform*.)

**\*5.3** Express the four-point time-ordered connected correlator in terms of derivatives of the quantum effective action. Check that the answer is consistent with the fact that the quantum effective action generates all orders from tree diagrams only.

**\*5.4** Consider the free generating functional,

$$Z_0[j] = \exp\left( -\tfrac{1}{2} \int d^4x\, d^4y\, j(x)\, G_F^0(x, y)\, j(y) \right).$$

The odd derivatives are zero at $j = 0$. Show that the even derivatives are given by the following combinatorial formula:

$$\frac{\delta^{2n} Z_0[j]}{i\delta j(x_1) \cdots i\delta j(x_{2n})}\bigg|_{j=0} = \sum_{\text{pairs } \{(\alpha_i, \beta_i)\}_{1 \leq i \leq n}} \prod_{i=1}^{n} G_F^0(x_{\alpha_i}, x_{\beta_i}),$$

where the sum is over all ways of pairing the $2n$ coordinates (for each pairing, we have $\cup_{i=1}^{n} \{\alpha_i, \beta_i\} = \{1, \cdots, 2n\}$). How many terms are there in the sum? (This is known as *Wick's theorem* in quantum field theory, although an earlier derivation is due to Isserlis.)

**\*5.5** In the case of a scalar field theory, use functional methods to prove the following identity:

$$(\Box_x + m^2) \frac{\delta Z[j]}{i\delta j(x)} - \mathcal{L}_I'\left( \frac{\delta}{i\delta j(x)} \right) Z[j] = j(x)\, Z[j].$$

**5.6** The goal of this exercise if to derive the LSZ reduction formula in functional form. First, show that the Heisenberg, and the in, out fields are related by

$$\phi(x) = \sqrt{Z}\, \phi_{in}(x) + \int d^4y\, G_R^0(x, y)\, \frac{\partial \mathcal{L}_I}{\partial \phi(y)},$$

$$\phi(x) = \sqrt{Z}\, \phi_{out}(x) + \int d^4y\, G_A^0(x, y)\, \frac{\partial \mathcal{L}_I}{\partial \phi(y)},$$

where $G^0_{R,A}$ are the free retarded and advanced propagators. Introducing the S-matrix defined by $S \equiv U(+\infty, -\infty)$ and defining

$$\mathcal{E}[j; \phi] \equiv T \exp\left(i \int d^4x\, j(x)\phi(x)\right),$$

prove the following identity:

$$\sqrt{Z}\left[\phi_{in}(x), S\mathcal{E}\right] = \int d^4y\, [\phi_{in}(x), \phi_{in}(y)](\Box_y + m^2) \frac{\delta S\mathcal{E}}{\delta j(y)}.$$

Then, show that this equation is solved by

$$S\mathcal{E} = \exp\left(\int d^4y\, Z^{-1/2}\phi_{in}(y)(\Box_y + m^2) \frac{\delta}{\delta j(y)}\right) F[j],$$

where $F[j]$ is any functional of $j$. Finally, evaluate the expectation value of this equation in the state $|0_{in}\rangle$, and conclude that the S-matrix is given by

$$S = \,:\exp\left(\int d^4y\, Z^{-1/2}\phi_{in}(y)(\Box_y + m^2) \frac{\delta}{\delta j(y)}\right): Z[j]\bigg|_{j=0}.$$

**5.7** Give the symmetry factors of the 2PI graphs listed in Figure 5.5. Check that when these graphs are differentiated with respect to the propagator in order to obtain a self-energy, one obtains the correct symmetry factor for the latter.

**\*5.8** Give the expression of the one-loop self-energy in a $\phi^4$ theory in the Matsubara formalism. Calculate it in the limit $\beta m \ll 1$.

**5.9** Use functional methods to prove eq. (2.95) that relates the generating functionals in the usual time-ordered perturbation theory and in the Schwinger–Keldysh formalism. *Hint: Establish it only for the free theory, since the interactions can be trivially factorized.* What is the diagrammatic interpretation of this functional relationship?

**5.10** This exercise provides a combinatorial approach to the Legendre transform, that does not require series to converge in order to make sense. Consider a formal power series $S(X) \equiv \frac{1}{2} S^{(2)}_{ab} X_a X_b + \sum_{n>2} \frac{1}{n!} S^{(n)}_{a_1 \cdots a_n} X_{a_1} \cdots X_{a_n}$, with no constant and linear terms and with $S^{(2)}_{ab}$ invertible. Viewing this series as an "action," denote $T(J)$ the formal series of trees constructed with the "Feynman rules" defined by this action and a source $J_a$ coupled to the field $X_a$. We wish to prove combinatorially that $S(X(J)) + JX(J) = T(J)$, where $X(J)$ and $J$ are linked by $J_a + \partial S/\partial X_a = 0$. (The derivative $\partial S/\partial X$ is a well-defined operation in the ring of formal series, which does not require any notion of convergence.)

- Given a tree diagram $\mathcal{G}$, show that $n_V(\mathcal{G}) - n_E(\mathcal{G}) = 1$, where $n_V$ and $n_E$ are the number of vertices (including the sources $J$) and edges of $\mathcal{G}$, respectively.
- Consider the mth derivative $\partial^m T(J)/\partial J_{a_1} \cdots \partial J_{a_m}\big|_{J=0}$. What is its diagrammatic representation? How many times does each graph $\mathcal{G}$ occur?
- Consider now the mth derivative of $S(X(J)) + JX(J) = T(J)$. Show that it is represented by the same graphs as the derivative of $T(J)$, each of them occurring $n_V(\mathcal{G}) - n_E(\mathcal{G})$ times.

# CHAPTER 6

# Path Integrals for Fermions and Photons

In the previous chapter we learned how the quantization of a scalar field may be performed by means of the path integral representation. This leads to a much more concise derivation of the generating functional and of the free propagator, compared to the canonical approach. In this chapter, we seek a similar path integral formalism for other types of fields (fermions, gauge bosons), in view of the functional quantization of a gauge theory such as QED (and later, of non-Abelian gauge theories, for which a canonical approach would be extremely difficult to implement).

## 6.1 Grassmann Variables

### 6.1.1 Definition

In the functional formulation of a scalar field theory, we saw that time-ordered products of field operators correspond to the ordinary product of the integration variable in the integrand of the path integral (see eq. (5.29)). Ultimately, a path integral representation of the time-ordered product of fermion field operators should allow the same, but with a catch: The T-product for fermions involves a minus sign when two operators are exchanged (see eq. (3.28)), which we need to be able to generate in the integrand of a would-be fermionic path integral. This can be achieved with *Grassmann numbers*,[1] that are anti-commuting variables.

---

[1] Although we call them "numbers," they are not representable as scalar (e.g., real or complex) variables. A Grassmann number may be represented by a nilpotent matrix (the smallest being $2\times 2$), and the Grassmann algebra with N generators admits a representation in terms of $2^N \times 2^N$ matrices, which may be viewed as operators acting on the Hilbert space of N identical fermions of spin 1/2 (of dimension $2^N$ since each spin

## 6.1 GRASSMANN VARIABLES

In a sense, Grassmann numbers are the classical analogue of anti-commuting quantum operators. For a set of Grassmann variables $\psi_i$ ($i = 1 \cdots N$), we have

$$\{\psi_i, \psi_j\} = 0. \tag{6.1}$$

The linear space spanned by the $\psi_i$s is called a *Grassmann algebra*. A trivial consequence of this definition is that the square of Grassmann numbers is zero, $\psi_i^2 = 0$. This property is reminiscent of Fermi statistics, which forbids that two or more fermions occupy the same quantum state (in the canonical formalism, this translates into the fact that the square of creation operators is zero).

### 6.1.2 Functions of a Single Grassmann Variable

Consider first the case $N = 1$. The square of a Grassmann number $\psi$ is therefore zero, $\psi^2 = 0$, and by induction all higher powers of $\psi$ are also zero. The Taylor expansion of a function of $\psi$ is therefore limited to the first two terms,

$$f(\psi) = a + \psi b. \tag{6.2}$$

Often, we need functions $f(\psi)$ that are themselves commuting objects. Therefore, the coefficient $a$ is an ordinary number, while $b$ is another Grassmann number, so that $\{b, b\} = \{b, \psi\} = 0$. This implies that $f(\psi) = a - b\psi$. Because of the non-commuting nature of $b$ and $\psi$, we can define *left* and *right* derivatives, denoted by:

$$\overrightarrow{\partial}_\psi f(\psi) = b, \quad f(\psi) \overleftarrow{\partial}_\psi = -b. \tag{6.3}$$

In order to calculate these derivatives without making mistakes, it is useful to first move the variable which is going to be differentiated on the side where the derivative will hit. Otherwise, we have to move the derivative through the expression, keeping in mind that it is itself an anti-commuting object (for instance, $\partial_\psi$ anti-commutes with $b$).

Then, one may define a linear mapping acting on functions of a Grassmann variable, which behaves for most purposes like an integration (although it is not an integral in the Lebesgue sense), called the *Berezin integral*. We require two basic axioms:

- linearity:

$$\int d\psi \, \alpha \, f(\psi) = \alpha \int d\psi \, f(\psi); \tag{6.4}$$

- the integral of a total derivative must be zero:

$$\int d\psi \, \partial_\psi f(\psi) = 0. \tag{6.5}$$

---

has two states). For instance, when $N = 2$, one may represent the Grassmann numbers $\psi_{1,2}$ as

$$\psi_1 = \begin{pmatrix} 0 & 0 & 0 & 0 \\ 1 & 0 & 0 & 0 \\ 0 & 0 & 0 & 0 \\ 0 & 0 & 1 & 0 \end{pmatrix}, \quad \psi_2 = \begin{pmatrix} 0 & 0 & 0 & 0 \\ 0 & 0 & 0 & 0 \\ 1 & 0 & 0 & 0 \\ 0 & -1 & 0 & 0 \end{pmatrix}.$$

The only definition consistent with these requirements is

$$\int d\psi \, f(\psi) = b, \tag{6.6}$$

up to an overall factor that should be the same for all functions. Thus, integration and differentiation of functions of a Grassmann variable are essentially the same thing. In particular, the Berezin integral satisfies:

$$\int d\psi \, 1 = 0, \quad \int d\psi \, \psi = 1. \tag{6.7}$$

### 6.1.3 Functions of N Grassmann Variables

**Taylor expansion:** We will denote collectively $\psi \equiv (\psi_1, \cdots, \psi_N)$. The most general function of N Grassmann variables can be written as

$$f(\psi) = \sum_{p=0}^{N} \frac{1}{p!} \psi_{i_1} \psi_{i_2} \cdots \psi_{i_p} \, C_{i_1 i_2 \cdots i_p}, \tag{6.8}$$

with implicit summations on the indices $i_n$. Terms of degree higher than N cannot exist because they would contain the square of at least one of the $\psi_i$s, and therefore be zero. We have chosen to write the Grassmann variables on the left of the coefficients in order to simplify the calculation of the left derivatives $\vec{\partial}_\psi$. Note that the last coefficient $C_{i_1 \cdots i_N}$ must be proportional to the Levi-Civita symbol,

$$C_{i_1 \cdots i_N} \equiv \gamma \, \epsilon_{i_1 \cdots i_N}, \tag{6.9}$$

since the product $\psi_1 \psi_2 \cdots \psi_N$ is fully antisymmetric. Moreover, this last term can also be written as

$$\tfrac{1}{N!} \psi_{i_1} \cdots \psi_{i_N} \, \gamma \, \epsilon_{i_1 \cdots i_N} = \psi_1 \cdots \psi_N \, \gamma \tag{6.10}$$

by rearranging each term of the sum in such a way that the $\psi_i$s are in order of increasing index.

**Integration:** In order to be consistent with eq. (6.7), the integral of $f(\psi)$ over the N Grassmann variables $\psi_1, \cdots, \psi_N$ must be given by

$$\int d^N \psi \, f(\psi) = \gamma. \tag{6.11}$$

The terms of degree 0 through $N-1$ in the "Taylor expansion" of $f(\psi)$ cannot contribute to the integral, since at least one of the $\psi_i$ is absent in these terms, and the integral over this $\psi_i$ will therefore give zero. A somewhat more explicit formulation of an integral over N Grassmann variables is to write the measure as $d^N \psi \equiv d\psi_N \, d\psi_{N-1} \cdots d\psi_1$ (in this order), and to

## 6.1 GRASSMANN VARIABLES

perform the N integrals successively, starting with the innermost one (i.e., $d\psi_1$). Therefore

$$\int d^N\psi\, \psi_1 \cdots \psi_N = \int d\psi_N \cdots \left( \int d\psi_2 \underbrace{\left( \underbrace{\int d\psi_1\, \psi_1}_{1} \right) \psi_2}_{1} \right) \cdots \psi_N = 1. \qquad (6.12)$$

**Change of variables:** Let us now consider a linear change of variables,

$$\psi_i \equiv J_{ij}\, \theta_j, \qquad (6.13)$$

where $\theta_1 \cdots \theta_N$ are N Grassmann variables and the coefficients $J_{ij}$ are ordinary numbers. The last term of the expansion of $f(\psi)$, the only one relevant for integration, can be rewritten as

$$\psi_{i_1} \cdots \psi_{i_N}\, \epsilon_{i_1 \cdots i_N}\, \gamma = \left( J_{i_1 j_1} \theta_{j_1} \right) \cdots \left( J_{i_N j_N} \theta_{j_N} \right) \epsilon_{i_1 \cdots i_N}\, \gamma$$
$$= \det(J)\, \underbrace{\theta_{j_1} \cdots \theta_{j_N}\, \epsilon_{j_1 \cdots j_N}\, \gamma}_{\text{last term of } f(\theta)}. \qquad (6.14)$$

From this relationship, we have

$$\int d^N\psi\, \det(J)\, f(\theta) = \int d^N\psi\, f(\psi) = \int d^N\theta\, f(\theta). \qquad (6.15)$$

(The last equality is just a dummy replacement $\psi \to \theta$.) The comparison of the first and last integrals thus implies that

$$d^N\psi\, \det(J) = d^N\theta. \qquad (6.16)$$

In other words, a change of variables in a Grassmann integral involves the inverse of the Jacobian that would normally appear in the same change of variables for a scalar integral.

**Gaussian integrals:** Let $\psi_1, \cdots, \psi_N$ be N Grassmann variables, and consider the following integral

$$I(\mathbf{M}) \equiv \int d^N\psi\, \exp\left( \tfrac{1}{2} \psi_i M_{ij} \psi_j \right), \qquad (6.17)$$

where $\mathbf{M}$ is an antisymmetric $N \times N$ matrix made of commuting numbers (real or complex). First, note that such an integral is non-zero only if N is even. Indeed, when N is odd, there is always a mismatch between the number of integration variables and the number of factors in the terms of the Taylor expansion of the exponential. For $N = 2$, this matrix is of the form

$$\mathbf{M} = \begin{pmatrix} 0 & \mu \\ -\mu & 0 \end{pmatrix}, \qquad (6.18)$$

and the exponential in the integrand reads

$$\exp\left( \tfrac{1}{2} \psi_i M_{ij} \psi_j \right) = 1 + \mu\, \psi_1 \psi_2. \qquad (6.19)$$

(Recall that functions of two Grassmann variables are in fact polynomials of degree two.) Therefore, in the case N = 2, the Gaussian integral (6.17) reads:[2]

$$I(\mathbf{M}) = \mu = \left[\det(\mathbf{M})\right]^{1/2}. \tag{6.20}$$

In the case of a general even N, the matrix $\mathbf{M}$ may be written in the following block diagonal form:

$$\mathbf{M} = \mathbf{Q} \underbrace{\begin{pmatrix} 0 & \mu_1 & & & \\ -\mu_1 & 0 & & & \\ & & 0 & \mu_2 & \\ & & -\mu_2 & 0 & \\ & & & & \ddots \end{pmatrix}}_{\mathbf{D}} \mathbf{Q}^T, \tag{6.21}$$

where $\mathbf{Q}$ is a special[3] orthogonal matrix. Defining $\mathbf{Q}^T \psi \equiv \theta$, we have

$$I(\mathbf{M}) = \det(\mathbf{Q}^T) \underbrace{\int d^N\theta \, \exp\left(\tfrac{1}{2}\theta^T \mathbf{D}\theta\right)}_{\mu_1 \mu_2 \cdots = [\det(\mathbf{D})]^{1/2}}. \tag{6.22}$$

But since $\det(\mathbf{Q}) = +1$, this becomes

$$I(\mathbf{M}) = \left[\det(\mathbf{D})\right]^{1/2} = \left[\det(\mathbf{M})\right]^{1/2}. \tag{6.23}$$

Contrast this with the result of a Gaussian integral in the case of ordinary real variables, eq. (5.63), where the square root of the determinant appeared in the denominator.

It is often necessary to perform a Gaussian integral in the presence of a source that shifts the extremum of the quadratic form in the exponential,

$$I(\mathbf{M}, \eta) \equiv \int d^N\psi \, \exp\left(\tfrac{1}{2}\psi_i M_{ij} \psi_j + \eta_i \psi_i\right), \tag{6.24}$$

where $\eta$ is a set of N Grassmann sources. By introducing the new Grassmann variable $\psi'_i \equiv \psi_i - M^{-1}_{ij}\eta_j$, this integral falls back to the previous type, and we obtain

$$I(\mathbf{M}, \eta) = \left[\det(\mathbf{M})\right]^{1/2} \exp\left(\tfrac{1}{2}\eta^T \mathbf{M}^{-1} \eta\right). \tag{6.25}$$

**Gaussian integral with 2N variables:** Another useful type of Gaussian integral is

$$J(\mathbf{M}) \equiv \int d^N\xi \, d^N\psi \, \exp\left(\psi_i M_{ij} \xi_j\right), \tag{6.26}$$

---

[2]The determinant of a real antisymmetric matrix of even size is the square of its *Pfaffian* and is therefore positive.

[3]Orthogonal matrices have determinant +1 or −1. The special orthogonal matrices are the subgroup of those that have determinant +1.

## 6.1 GRASSMANN VARIABLES

where $\mathbf{M}$ is an $N \times N$ matrix of commuting numbers, and $\psi$ and $\xi$ are independent Grassmann variables. The only non-zero contribution to this integral comes from the term of order N in the Taylor expansion of the exponential,

$$
\begin{aligned}
J(\mathbf{M}) &= \frac{1}{N!} \int d^N \xi d^N \psi \, (\psi_{i_1} M_{i_1 j_1} \xi_{j_1}) \cdots (\psi_{i_N} M_{i_N j_N} \xi_{j_N}) \\
&= \frac{(-1)^{\frac{N(N-1)}{2}}}{N!} \int d^N \xi d^N \psi \, (\psi_{i_1} \cdots \psi_{i_N})(\xi_{j_1} \cdots \xi_{j_N}) M_{i_1 j_1} \cdots M_{i_N j_N} \\
&= \frac{(-1)^{\frac{N(N-1)}{2}}}{N!} \epsilon_{i_1 \cdots i_N} \epsilon_{j_1 \cdots j_N} M_{i_1 j_1} \cdots M_{i_N j_N}.
\end{aligned}
\tag{6.27}
$$

In the second line, we have reordered the Grassmann variables in order to bring all the $\psi_i$s on the left, and the sign in the prefactor keeps track of the number of permutations that are necessary to achieve this. To give a non-zero result, the indices $\{i_n\}$ and $\{j_n\}$ must be permutations of $[1 \cdots N]$, i.e.,

$$
\begin{aligned}
J(\mathbf{M}) &= \frac{(-1)^{\frac{N(N-1)}{2}}}{N!} \sum_{\sigma, \rho \in \mathfrak{S}_n} \epsilon(\sigma)\epsilon(\rho) \, M_{\sigma(1)\rho(1)} \cdots M_{\sigma(N)\rho(N)} \\
&= \frac{(-1)^{\frac{N(N-1)}{2}}}{N!} \sum_{\sigma, \tau \in \mathfrak{S}_n} \epsilon(\sigma)\epsilon(\tau\sigma) \, M_{1\tau(1)} \cdots M_{N\tau(N)}
\end{aligned}
\tag{6.28}
$$

where $\epsilon(\sigma)$ is the signature of the permutation $\sigma$, and with $\tau \equiv \rho\sigma^{-1}$ in the second line. Using $\epsilon(\sigma)\epsilon(\tau\sigma) = \epsilon(\tau)$, this becomes

$$
J(\mathbf{M}) = (-1)^{\frac{N(N-1)}{2}} \underbrace{\left(\frac{1}{N!} \sum_{\sigma \in \mathfrak{S}_n} 1\right)}_{1} \underbrace{\sum_{\tau \in \mathfrak{S}_n} \epsilon(\tau) \, M_{1\tau(1)} \cdots M_{N\tau(N)}}_{\det(\mathbf{M})}.
\tag{6.29}
$$

Note that this overall sign may be absorbed into a reordering of the measure, since:

$$
d^N \xi d^N \psi = (-1)^{\frac{N(N-1)}{2}} d\xi_N d\psi_N \cdots d\xi_1 d\psi_1.
\tag{6.30}
$$

Therefore, we have

$$
\int d\xi_N d\psi_N \cdots d\xi_1 d\psi_1 \, \exp(\psi_i M_{ij} \xi_j) = \det(\mathbf{M}).
\tag{6.31}
$$

### 6.1.4 Complex Grassmann Variables

Now, let us define *complex Grassmann variables* from two of the previously defined Grassmann variables $\psi$ and $\xi$:

$$
\chi \equiv \frac{\psi + i\xi}{\sqrt{2}}, \qquad \bar{\chi} \equiv \frac{\psi - i\xi}{\sqrt{2}}.
\tag{6.32}
$$

Conversely, we have

$$\psi = \frac{\overline{\chi}+\chi}{\sqrt{2}}, \qquad \xi = \frac{i(\overline{\chi}-\chi)}{\sqrt{2}}, \tag{6.33}$$

and the integrations over these variables are related by

$$d\xi d\psi = i\, d\chi d\overline{\chi}, \quad \psi\xi = -i\overline{\chi}\chi, \quad \int d\chi d\overline{\chi}\ \overline{\chi}\chi = \int d\xi d\psi\ \psi\xi = 1. \tag{6.34}$$

From this, we obtain

$$\int d\chi d\overline{\chi}\ \exp(\mu\overline{\chi}\chi) = \mu, \tag{6.35}$$

that can be generalized into

$$\int d\chi_N\, d\overline{\chi}_N \cdots d\chi_1\, d\overline{\chi}_1\ \exp(\overline{\chi}^{\scriptscriptstyle T} \mathbf{M}\chi) = \det(\mathbf{M}). \tag{6.36}$$

In the presence of sources $\eta$ and $\overline{\eta}$, we obtain the following Gaussian integral:

$$\int d\chi_N\, d\overline{\chi}_N \cdots d\chi_1\, d\overline{\chi}_1\ \exp\left(\overline{\chi}^{\scriptscriptstyle T}\mathbf{M}\chi + \overline{\eta}^{\scriptscriptstyle T}\chi + \overline{\chi}^{\scriptscriptstyle T}\eta\right) = \det(\mathbf{M})\ \exp\left(-\overline{\eta}^{\scriptscriptstyle T}\mathbf{M}^{-1}\eta\right). \tag{6.37}$$

## 6.2 Path Integral for Fermions

We now have all the ingredients for building a path integral for spin-1/2 fermions. Let us work our way backwards, starting from a generating functional that generates the free time-ordered products of spinors,

$$Z_0[\overline{\eta},\eta] \equiv \exp\left\{-\int d^4x d^4y\ \overline{\eta}(x) S_F^0(x,y)\eta(y)\right\}, \tag{6.38}$$

where $S_F^0(x,y)$ is the free Dirac time-ordered propagator and $\overline{\eta}$ and $\eta$ are a pair of complex Grassmann-valued sources. Indeed, we have

$$\frac{\overrightarrow{\delta}}{i\delta\overline{\eta}(x)}\, Z_0[\overline{\eta},\eta]\, \frac{\overleftarrow{\delta}}{i\delta\eta(y)}\bigg|_{\overline{\eta}=\eta=0} = S_F^0(x,y). \tag{6.39}$$

Taking more than two derivatives (but with an equal number of derivatives with respect to $\overline{\eta}$ and with respect to $\eta$) will lead to all the contributions in the free time-ordered product of spinors, with the correct signs to account for their anti-commuting nature. Note that using Grassmann-valued sources was necessary in order to get these signs.

Then, by comparing eqs. (6.37) and (6.38), we can represent this free generating function as a path integral over Grassmann variables:

$$Z_0[\overline{\eta},\eta] = \int [D\psi(x) D\overline{\psi}(x)]\ \exp\Big\{i\int d^4x\, (\overline{\psi}(x)(i\slashed{\partial}-m)\psi(x)$$
$$+\overline{\eta}(x)\psi(x) + \overline{\psi}(x)\eta(x))\Big\}. \tag{6.40}$$

## 6.3 Path Integral for Photons

### 6.3.1 Problems with the Naive Path Integral

In the case of photons, the difficulties encountered in the path integral formulation are of a different nature. Since photons are bosons, we expect that they can be represented by a functional integration over commuting functions $A_\mu(x)$. But the gauge invariance of the theory implies that there is an unavoidable redundancy in this representation: The naive path integral over $[DA_\mu(x)]$ would integrate over infinitely many copies of the same physical configurations. Therefore, we need a way to cut through this redundancy, which is achieved by *gauge fixing*.

In order to better see the nature of this difficulty, let us assume that we can treat $A_\mu(x)$ as a collection of four scalar fields, and write the following path integral:

$$Z_0[j^\mu] \equiv \int [DA_\mu(x)] \; \exp\left\{i\int d^4x \left(-\tfrac{1}{4} F^{\mu\nu}F_{\mu\nu} + j^\mu A_\mu\right)\right\}. \tag{6.41}$$

This is a Gaussian integral, since $F^{\mu\nu}F_{\mu\nu}$ is quadratic in the field $A_\mu$:

$$-\frac{1}{4}\int d^4x \, F^{\mu\nu}F_{\mu\nu} = -\frac{1}{4}\int d^4x \left(\partial^\mu A^\nu - \partial^\nu A^\mu\right)\left(\partial_\mu A_\nu - \partial_\nu A_\mu\right)$$

$$= +\frac{1}{2}\int d^4x \, A^\mu \left(g_{\mu\nu}\Box - \partial_\mu \partial_\nu\right) A^\nu$$

$$= -\frac{1}{2}\int \frac{d^4k}{(2\pi)^4} \, \widetilde{A}^\mu(k)\left(g_{\mu\nu}k^2 - k_\mu k_\nu\right)\widetilde{A}^\nu(-k). \tag{6.42}$$

Performing this Gaussian integral requires the inverse of the object $g_{\mu\nu}k^2 - k_\mu k_\nu$, which one may seek as a linear combination of the metric tensor $g^{\mu\nu}$ and $k^\mu k^\nu/k^2$. Thus, we are looking for coefficients $\alpha$ and $\beta$ such that

$$\underbrace{\left(g_{\mu\nu}k^2 - k_\mu k_\nu\right)\left(\alpha \, g^{\nu\rho} + \beta \, \tfrac{k^\nu k^\rho}{k^2}\right)}_{\alpha \, k^2 \delta^\rho_\mu - \alpha \, k_\mu k^\rho} = \delta^\rho_\mu. \tag{6.43}$$

This equation clearly has no solution, and therefore it is impossible to invert $g_{\mu\nu}k^2 - k_\mu k_\nu$. This means that some eigenvalues of this operator are zero, and that the quadratic form $\widetilde{A}^\mu(k)\left(g_{\mu\nu}k^2 - k_\mu k_\nu\right)\widetilde{A}^\nu(-k)$ has flat directions. Along these flat directions, the exponential in the path integral (6.41) does not oscillate, which spoils its convergence. In fact, these flat directions correspond to the projection of $\widetilde{A}^\mu(k)$ along $k^\mu$. Note that the longitudinal part of $A_\mu$ also does not contribute to the linear term $j^\mu A_\mu$, for a conserved current that satisfies $\partial_\mu j^\mu = 0$. Therefore, the simplest option would be not integrating over these components of $A^\mu$ in eq. (6.41).

## 6.3.2 Path Integral in Landau Gauge

A simple way out is to decompose $A^\mu$ into transverse and longitudinal components:

$$A^\mu = A^\mu_\perp + A^\mu_\parallel,$$

$$\tilde{A}^\mu_\perp(k) \equiv \left(g^{\mu\nu} - \frac{k^\mu k^\nu}{k^2}\right) \tilde{A}_\nu(k), \quad \tilde{A}^\mu_\parallel(k) \equiv \left(\frac{k^\mu k^\nu}{k^2}\right) \tilde{A}_\nu(k). \tag{6.44}$$

The functional measure can be factorized as follows:

$$[DA^\mu] = [DA^\mu_\perp][DA^\mu_\parallel], \tag{6.45}$$

and since nothing depends on $A^\mu_\parallel$ in the photon kinetic term, we can write

$$Z_0[j^\mu] \equiv \int [DA^\mu_\parallel(x)] \, \exp\left\{i \int d^4x \, j_\mu A^\mu_\parallel\right\}$$

$$\times \int [DA^\mu_\perp(x)] \, \exp\left\{i \int d^4x \, \left(-\tfrac{1}{4}F^{\mu\nu}F_{\mu\nu} + j_\mu A^\mu_\perp\right)\right\}. \tag{6.46}$$

By integrating by parts the argument of the exponential in the integral on $A^\mu_\parallel$, we obtain a delta function of $\partial_\mu j^\mu$. Thus, for external currents that obey the continuity equation $\partial_\mu j^\mu = 0$, this prefactor is an infinite constant that can be ignored. When restricted to the subspace of $A^\mu_\perp$, the operator $g_{\mu\nu}k^2 - k_\mu k_\nu$ is invertible, and we can now perform the Gaussian integral to obtain

$$Z_0[j^\mu] = \exp\left\{-\frac{1}{2}\int d^4x d^4y \, j_\mu(x) \, G_F^{0\,\mu\nu}(x,y) \, j_\nu(y)\right\}, \tag{6.47}$$

with the free photon propagator in momentum space given by

$$G_F^{0\,\mu\nu}(p) \equiv \frac{-i}{p^2 + i0^+}\left(g^{\mu\nu} - \frac{p^\mu p^\nu}{p^2}\right). \tag{6.48}$$

(We have introduced the $i0^+$ prescription that selects the ground state at $x^0 \to \pm\infty$, using the same argument as in Section 5.2.3.) The procedure used here is equivalent to imposing the gauge fixing condition $\partial_\mu A^\mu = 0$, called *Lorenz gauge* or *Landau gauge*. As one can see, the resulting propagator (6.48) differs from the Coulomb gauge propagator given in eq. (3.56).

## 6.3.3 General Covariant Gauges

All gauge fixings amount to constraining in some way the quantity $\partial_\mu A^\mu$, since it does not appear in the integrand of the photon path integral. Instead of imposing $\partial_\mu A^\mu = 0$, one may instead impose a more general condition,

$$\partial_\mu A^\mu(x) = \omega(x), \tag{6.49}$$

where $\omega(x)$ is some arbitrary function of space-time. This can be done by introducing a functional delta function, $\delta[\partial_\mu A^\mu - \omega]$, inside the path integral. However, the presence of

## 6.3 PATH INTEGRAL FOR PHOTONS

the function $\omega(x)$ breaks Lorentz invariance. To mitigate this problem, one usually integrates over all the functions $\omega(x)$, with a Gaussian weight. This amounts to defining the generating functional as follows:[4]

$$Z_0[j^\mu] \equiv \int [D\omega(x)] \, \exp\left\{-i\frac{\xi}{2}\int d^4x \, \omega^2(x)\right\}$$
$$\times \int [DA_\mu(x)] \, \delta[\partial_\mu A^\mu - \omega] \, \exp\left\{i\int d^4x \, \left(-\tfrac{1}{4}F^{\mu\nu}F_{\mu\nu} + j^\mu A_\mu\right)\right\}, \quad (6.50)$$

where $\xi$ is an arbitrary constant (the resulting gauge is also known as the $R_\xi$ gauge). Performing the integration over $\omega(x)$ thanks to the delta functional and integrating by parts, this becomes

$$Z_0[j^\mu] = \int [DA_\mu(x)] \, \exp\left\{i\int d^4x \, \left(\tfrac{1}{2}A^\mu(g_{\mu\nu}\Box - (1-\xi)\partial_\mu\partial_\nu)A^\nu + j^\mu A_\mu\right)\right\}. \quad (6.51)$$

From this formula, a standard Gaussian integration tells us that the corresponding photon propagator in momentum space should be the inverse of

$$i(g_{\mu\nu}p^2 - (1-\xi)p_\mu p_\nu). \quad (6.52)$$

Looking for an inverse of the form $\alpha \, g^{\nu\rho} + \beta \, \frac{p^\nu p^\rho}{p^2}$, we find

$$G_F^{0\,\mu\nu}(p) = \frac{-i g^{\mu\nu}}{p^2 + i0^+} + \frac{i}{p^2 + i0^+}\left(1 - \frac{1}{\xi}\right)\frac{p^\mu p^\nu}{p^2}. \quad (6.53)$$

The gauge fixing parameter $\xi$ appears in the propagator, but only in the term proportional to $p^\mu p^\nu$. Thanks to the Ward–Takahashi identities, it does not have any incidence on physical results, provided that all the external charged particles are on mass-shell. The Landau gauge of the previous subsection corresponds to $\xi \to \infty$. Another popular choice is the *Feynman gauge*, obtained for $\xi = 1$:

$$G_F^{0\,\mu\nu}(p)\Big|_{\xi=1} = \frac{-i g^{\mu\nu}}{p^2 + i0^+}. \quad (6.54)$$

Note that one could also introduce a non-Lorentz covariant condition inside the delta function, such as $\delta[\partial_i A^i - \omega]$, in order to derive the photon propagator in Coulomb gauge via the path integral.

---

[4]Since the argument of the delta function is linear in the variable $A^\mu_\parallel$, the Jacobian is a constant. It is possible to impose nonlinear gauge conditions of the form $\delta[F(\partial_\mu A^\mu) - \omega]$, but this should be done by writing the path integral as follows:

$$\int [D\omega(x)] \, \exp\left\{-i\frac{\xi}{2}\int d^4x \, \omega^2(x)\right\} \int [DA_\mu(x)] \, \underbrace{F'(\partial_\mu A^\mu)}_{\text{Jacobian}} \, \delta[F(\partial_\mu A^\mu) - \omega] \cdots$$

In general, the Jacobian cannot be ignored since it depends on the gauge field, but it can be expressed in terms of *ghost fields*. Doing this would be an unnecessary complication in QED, but is an essential step in the quantization of non-Abelian gauge theories.

## 6.4 Schwinger–Dyson Equations

### 6.4.1 Functional Derivation

Consider a Lagrangian density $\mathcal{L}(\phi, \partial_\mu \phi)$ ($\phi$ may be a collection of fields, but we do not write any index on it to keep the notation light), and $\mathcal{S} \equiv \int d^4x\, \mathcal{L}$ the corresponding action. The generating functional of time-ordered products of fields has the following path integral representation:

$$Z[j] = \int [D\phi(x)]\, e^{i\mathcal{S}[\phi] + i\int j\phi}. \tag{6.55}$$

On the right-hand side, $\phi(x)$ should be viewed as a dummy integration variable, and the result of the integral should be unmodified if we change $\phi(x) \to \phi(x) + \delta\phi(x)$. This translates into

$$0 = \delta Z[j] = i \int [D\phi(x)]\, e^{i\mathcal{S}[\phi] + i\int j\phi} \left\{ \int d^4x\, \delta\phi(x) \left( j(x) + \frac{\delta \mathcal{S}}{\delta \phi(x)} \right) \right\}. \tag{6.56}$$

Taking $n$ functional derivatives of this identity with respect to $ij(x_1), \ldots, ij(x_n)$ and setting $j$ to zero gives

$$0 = \int [D\phi(x)]\, e^{i\mathcal{S}[\phi]} \int d^4x\, \delta\phi(x) \Big\{ i\, \phi(x_1) \cdots \phi(x_n) \frac{\delta \mathcal{S}}{\delta \phi(x)}$$
$$+ \sum_{i=1}^{n} \delta(x - x_i) \prod_{j \neq i} \phi(x_j) \Big\}. \tag{6.57}$$

Since in this discussion the variation $\delta\phi(x)$ is arbitrary, this implies the following identities:

$$0 = \int [D\phi(x)]\, e^{i\mathcal{S}[\phi]} \left\{ i\, \phi(x_1) \cdots \phi(x_n) \frac{\delta \mathcal{S}}{\delta \phi(x)} + \sum_{i=1}^{n} \delta(x - x_i) \prod_{j \neq i} \phi(x_j) \right\}, \tag{6.58}$$

known as the *Schwinger–Dyson equations* (here written in functional form). For instance, in the case of a scalar field theory with a $\phi^4$ interaction term, this leads to

$$(\Box_x + m^2)\langle 0_{\text{out}} | T\, \phi(x_1) \cdots \phi(x_n) \phi(x) | 0_{\text{in}} \rangle + \tfrac{\lambda}{3!} \langle 0_{\text{out}} | T\, \phi(x_1) \cdots \phi(x_n) \phi^3(x) | 0_{\text{in}} \rangle$$
$$= -i \sum_{i=1}^{n} \delta(x - x_i)\, \langle 0_{\text{out}} | T \prod_{j \neq i} \phi(x_j) | 0_{\text{in}} \rangle. \tag{6.59}$$

(We have used the remark following eq. (5.30) in order to let the operator $\Box + m^2$ act also on the step functions that order the operators in the time-ordered product.) If we convolute this equation with the free Feynman propagator (i.e., the inverse of the operator $\Box_x + m^2$), the above Schwinger–Dyson equation can be represented diagrammatically as follows:

$$\underbrace{\begin{array}{c}\text{diagram}\end{array}}_{} \;+\; \underbrace{\begin{array}{c}\text{diagram}\end{array}}_{} \;=\; \sum_{i=1}^{n} \underbrace{\begin{array}{c}\text{diagram}\end{array}}_{\text{contact terms}} \tag{6.60}$$

The Schwinger–Dyson equations have several simple consequences. When applied to a free theory ($\lambda = 0$) in the case $n = 1$, we get

$$(\Box_x + m^2)\langle 0_{\text{out}}|T\,\phi(x_1)\phi(x)|0_{\text{in}}\rangle = -i\delta(x - x_1), \tag{6.61}$$

which is nothing but the equation of motion satisfied by the Feynman propagator. In the general case, if $x$ differs from all the $x_i$s, we obtain

$$\begin{aligned}(\Box_x + m^2)&\langle 0_{\text{out}}|T\,\phi(x_1)\cdots\phi(x_n)\phi(x)|0_{\text{in}}\rangle \\ &+ \tfrac{\lambda}{3!}\langle 0_{\text{out}}|T\,\phi(x_1)\cdots\phi(x_n)\phi^3(x)|0_{\text{in}}\rangle = 0.\end{aligned} \tag{6.62}$$

Thus, in a certain sense,[5] we can say that time-ordered products of fields satisfy the Euler–Lagrange equation of motion.

### 6.4.2 Schwinger–Dyson Equations and Conserved Currents

The functional derivative of the action $S$ with respect to $\phi(x)$ is given by

$$\frac{\delta S}{\delta \phi(x)} = \frac{\partial \mathcal{L}}{\partial \phi(x)} - \partial_\mu \frac{\partial \mathcal{L}}{\partial(\partial_\mu \phi(x))}, \tag{6.63}$$

and by equating this to zero we recover the Euler–Lagrange equation of motion. Under an infinitesimal variation $\delta\phi(x)$ of the field, the Lagrangian density varies by

$$\begin{aligned}\delta\mathcal{L} &= \frac{\partial \mathcal{L}}{\partial \phi(x)}\delta\phi(x) + \frac{\partial \mathcal{L}}{\partial(\partial_\mu \phi(x))}\partial_\mu(\delta\phi(x)) \\ &= \partial_\mu\left(\frac{\partial \mathcal{L}}{\partial(\partial_\mu \phi(x))}\delta\phi(x)\right) + \frac{\delta S}{\delta\phi(x)}\delta\phi(x).\end{aligned} \tag{6.64}$$

When the variation $\delta\phi(x)$ corresponds to a symmetry of the Lagrangian, we have $\delta\mathcal{L} = 0$ and therefore

$$\frac{\delta S}{\delta\phi(x)}\delta\phi(x) = -\partial_\mu \underbrace{\left(\frac{\partial \mathcal{L}}{\partial(\partial_\mu \phi(x))}\delta\phi(x)\right)}_{J^\mu(x)}, \tag{6.65}$$

where $J^\mu$ is the *Noether current* associated to this continuous symmetry. In the classical theory, this current is conserved, i.e., $\partial_\mu J^\mu = 0$, if the fields obey the Euler–Lagrange equation of motion. The Schwinger–Dyson equations provide a quantum analogue of this conservation law at the level of the expectation values of time-ordered products of fields. This can be achieved by replacing in eq. (6.57), $\delta\phi(\delta S/\delta\phi)$ by $-\partial_\mu J^\mu$. When the resulting identity is

---

[5]That is, up to the terms in $\delta(x - x_i)$ that may appear on the right-hand side, called *contact terms*. These contact terms in fact take care of the action of the time derivative on the step functions of the time-ordering operator T.

rewritten in terms of operators, the derivative $\partial_\mu$ should stay outside the time ordering, and we obtain

$$\partial_\mu \langle 0_{\text{out}} | T J^\mu(x) \phi(x_1) \cdots \phi(x_n) | 0_{\text{in}} \rangle$$
$$+ i \sum_{i=1}^n \delta(x - x_i) \langle 0_{\text{out}} | T \delta\phi(x) \prod_{j \neq i} \phi(x_j) | 0_{\text{in}} \rangle = 0. \tag{6.66}$$

Therefore, when a Noether current operator is inserted inside a time-ordered product, it satisfies the continuity equation up to contact terms (coming from the action of $\partial_0$ on the theta functions of the T product). Eq. (6.66) is a generalization of the Ward–Takahashi identities, already discussed in the context of electric charge conservation.

## 6.5 Quantum Anomalies

### 6.5.1 General Considerations

It may happen that some symmetries of the Lagrangian (i.e., symmetries of the classical theory) are broken by quantum corrections. This phenomenon is called a *quantum anomaly*. One way this may appear is through the introduction of a regularization (e.g., a cutoff), whose effect leaves an imprint on physical results even after the cutoff has been taken to infinity. Here we will adopt a functional point of view on this issue. In the previous section, a crucial point in the derivation of the Schwinger–Dyson equations was that the functional measure must be invariant under the symmetry under consideration. Quantum anomalies may be viewed as an obstruction in defining a functional measure which is invariant under certain symmetries, e.g., axial symmetry, as we shall see now.

Let us consider a set of fermion fields $\psi_n(x)$, which we encapsulate into a multiplet denoted $\psi(x)$, and assume that they interact with a gauge potential $A_\mu(x)$ in a non-chiral way (i.e., via couplings that are identical for the right-handed and left-handed projections of the spinors – this is the case of electromagnetic interactions and of strong interactions). Consider now the following transformation of the fermion fields:

$$\psi(x) \to U(x)\psi(x). \tag{6.67}$$

The Hermitic conjugate of $\psi$ transforms as $\psi^\dagger(x) \to \psi^\dagger(x) U^\dagger(x)$, so that we have

$$\overline{\psi}(x) \equiv \psi^\dagger(x)\gamma^0 \to \psi^\dagger(x) U^\dagger(x) \gamma^0 = \overline{\psi}(x) \gamma^0 U^\dagger(x) \gamma^0. \tag{6.68}$$

Since they are Grassmann variables, the measure is transformed with the inverse of the determinant of the transformation,

$$[D\psi D\overline{\psi}] \to \frac{1}{\det(\mathcal{U}) \det(\overline{\mathcal{U}})} [D\psi D\overline{\psi}], \tag{6.69}$$

where the matrices $\mathcal{U}$ and $\overline{\mathcal{U}}$ carry both indices for the fermion species and space-time indices:

$$\mathcal{U}_{xm,yn} \equiv U_{mn}(x)\, \delta(x-y),$$
$$\overline{\mathcal{U}}_{xm,yn} \equiv (\gamma^0 U^\dagger(x) \gamma^0)_{mn}\, \delta(x-y). \tag{6.70}$$

## 6.5.2 Non-Chiral Transformations

Let us consider the following transformation:
$$U(x) = e^{i\alpha(x)t}, \qquad (6.71)$$

where $\alpha(x) \in \mathbb{R}$ and $t$ is a Hermitian matrix that does not contain any Dirac matrices. (Thus, this transformation acts identically on right- and left-handed fermions.) Therefore, $U^\dagger(x) = e^{-i\alpha(x)t}$, and

$$\begin{aligned}(\overline{U}U)_{xm,yn} &= \int d^4 z \sum_p \overline{U}_{xm,zp}\, U_{zp,yn} \\ &= \int d^4 z\, \delta(x-z)\delta(z-y) \sum_p \left[e^{-i\alpha(z)t}\right]_{mp} \left[e^{i\alpha(z)t}\right]_{pn} \\ &= \delta_{mn}\delta(x-y). \end{aligned} \qquad (6.72)$$

Thus, $\overline{U}U = 1$, which implies $\det U \det \overline{U} = 1$, and the fermion measure is invariant under these kind of transformations. This means that this symmetry does not exhibit quantum anomalies.

## 6.5.3 Chiral Transformations

Let us now split the spinor into right-handed and left-handed projections,

$$\psi_R \equiv \left(\frac{1+\gamma^5}{2}\right)\psi, \quad \psi_L \equiv \left(\frac{1-\gamma^5}{2}\right)\psi, \qquad (6.73)$$

and consider a transformation that acts differently on these two components,

$$U(x) = e^{i\alpha(x)\gamma^5 t}, \qquad (6.74)$$

where $t$ is a Hermitian matrix that does not contain any Dirac matrices. Such transformations are called *chiral transformations*. From the discussion of the linear sigma model in Section 4.5, recall that a massless Dirac Lagrangian is invariant under this type of transformation. We recall here a few properties of the matrix $\gamma^5 \equiv i\gamma^0\gamma^1\gamma^2\gamma^3$:

$$(\gamma^5)^2 = 1, \quad \gamma^{5\dagger} = \gamma^5, \quad \{\gamma^5, \gamma^0\} = 0. \qquad (6.75)$$

These imply

$$\gamma^0 U^\dagger(x)\gamma^0 = \gamma^0 e^{-i\alpha(x)\gamma^5 t}\gamma^0 = e^{i\alpha(x)\gamma^5 t} = U(x). \qquad (6.76)$$

Thus, $\overline{U} = U$, and $\det U = \det \overline{U}$. Unless this determinant is equal to 1, the measure is not invariant and transforms according to

$$[D\psi D\overline{\psi}] \to \frac{1}{(\det U)^2}[D\psi D\overline{\psi}]. \qquad (6.77)$$

In order to calculate the determinant, we use the following formula:[6]

$$(\det \mathcal{U})^{-2} = \exp\left(-2\,\mathrm{tr}\,\ln \mathcal{U}\right) \tag{6.78}$$

$$= \exp\left(i \int d^4x\, \alpha(x) \underbrace{\left(-2\delta(x-x)\,\mathrm{tr}\,(\gamma^5 t)\right)}_{\equiv \mathcal{A}(x)}\right). \tag{6.79}$$

The function $\mathcal{A}(x)$ is called the *anomaly function*. In this equation, the trace denotes both a trace on the indices carried by the Dirac matrices and a trace on the fermion species. In terms of this function, the measure transforms as

$$[D\psi D\overline{\psi}] \to e^{i\int d^4x\, \alpha(x)\mathcal{A}(x)}\, [D\psi D\overline{\psi}]. \tag{6.80}$$

The fact that this measure may vary under the transformation (6.74) implies that there could exist fermion loop corrections that break the invariance under chiral transformations, even if the Dirac Lagrangian itself is invariant (this is the case when one considers a global transformation, i.e., a constant $\alpha(x)$, and the fermions are massless). The prefactor that alters the measure can be absorbed into a redefinition of the Lagrangian,

$$\mathcal{L}(x) \to \mathcal{L}(x) + \alpha(x)\mathcal{A}(x). \tag{6.81}$$

All happens as if the Lagrangian itself was not invariant under this transformation. In particular, if one integrates out the fermion fields in order to obtain an effective theory for the remaining fields, the term in $\alpha(x)\mathcal{A}(x)$ must be included in the Lagrangian of this effective theory in order to correctly account for the fermion sector that was integrated out.

### 6.5.4 Calculation of $\mathcal{A}(x)$

At first sight, the expression (6.79) of the anomaly function $\mathcal{A}(x)$ is very poorly defined: The trace may be zero, but is multiplied by an infinite $\delta(0)$. In order to manipulate finite expressions, we must first regularize the delta function. This can be done by writing

$$\mathcal{A}(x) = -2 \lim_{y\to x, M\to+\infty} \mathrm{tr}\left\{\gamma^5\, t\, \mathcal{F}\left(-\frac{\slashed{D}_x^2}{M^2}\right)\right\} \delta(x-y), \tag{6.82}$$

where $\slashed{D}_x$ is the *Dirac operator*[7]

$$\slashed{D}_x \equiv \gamma^\mu \left(\partial_\mu - i e\, A_\mu(x)\right), \tag{6.83}$$

---

[6] If the $\lambda_i$ are the eigenvalues of $\mathcal{U}$, we have:

$$\det \mathcal{U} = \prod_i \lambda_i = \exp\left(\sum_i \ln \lambda_i\right) = e^{\mathrm{tr}\,\ln\mathcal{U}}.$$

[7] The derivation of the anomaly function is also valid in the case where the fermions are coupled to a non-Abelian gauge field. See the next chapter for a thorough discussion of non-Abelian gauge symmetry.

## 6.5 QUANTUM ANOMALIES

and where $\mathcal{F}(s)$ is a function such that

$$\mathcal{F}(0) = 1, \quad \mathcal{F}(+\infty) = 0, \quad s\mathcal{F}'(s) = 0 \text{ at } s = 0, +\infty. \tag{6.84}$$

A covariant derivative is mandatory in eq. (6.82), since an ordinary derivative would break gauge invariance. Then, we replace the delta function by its Fourier representation,

$$\delta(x - y) = \int \frac{d^4k}{(2\pi)^4} e^{ik(x-y)}, \tag{6.85}$$

which leads to

$$\begin{aligned}
\mathcal{A}(x) &= -2 \int \frac{d^4k}{(2\pi)^4} \lim_{y \to x, M \to +\infty} \operatorname{tr}\left\{\gamma^5 t \mathcal{F}\left(-\frac{\slashed{D}_x^2}{M^2}\right)\right\} e^{ik(x-y)} \\
&= -2 \int \frac{d^4k}{(2\pi)^4} \lim_{M \to +\infty} \operatorname{tr}\left\{\gamma^5 t \mathcal{F}\left(-\frac{(i\slashed{k}+\slashed{D}_x)^2}{M^2}\right)\right\}.
\end{aligned} \tag{6.86}$$

The second equality follows from

$$\lim_{y \to x} \mathcal{F}(\partial_x) e^{ik\cdot(x-y)} = \mathcal{F}(ik + \partial_x). \tag{6.87}$$

The function $\mathcal{A}(x)$ can then be rewritten as follows:

$$\mathcal{A}(x) = -2 \lim_{M \to +\infty} M^4 \int \frac{d^4k}{(2\pi)^4} \operatorname{tr}\left\{\gamma^5 t \mathcal{F}\left(-\left[i\slashed{k} + \frac{\slashed{D}_x}{M}\right]^2\right)\right\}, \tag{6.88}$$

by redefining the integration variable, $k \to Mk$. Then, we can write

$$-\left[i\slashed{k} + \frac{\slashed{D}_x}{M}\right]^2 = k^2 - 2i\frac{k \cdot D_x}{M} - \left(\frac{\slashed{D}_x}{M}\right)^2, \tag{6.89}$$

and expand the function $\mathcal{F}(\cdot)$ in powers of $1/M$. The only terms that give a non-zero contribution to $\mathcal{A}(x)$ should not go to zero too quickly when $M \to +\infty$: only the terms decreasing at most as $1/M^4$ should be kept. Moreover, the Dirac trace should be non-zero, which implies that the matrix $\gamma^5$ must be accompanied by at least four ordinary $\gamma^\mu$ matrices. The matrices $\gamma^\mu$ come from the term $\slashed{D}_x^2$ in eq. (6.89), which brings two of them,[8] and we therefore need to go to the second order in the Taylor expansion of the function $\mathcal{F}(\cdot)$. In fact, a single term fulfills all these constraints:

$$\mathcal{A}(x) = -\int \frac{d^4k}{(2\pi)^4} \mathcal{F}''(k^2) \operatorname{tr}\left(\gamma^5 t \slashed{D}_x^4\right). \tag{6.90}$$

By a Wick's rotation ($k^0 \to i\kappa^0, k^2 \to -\kappa^2$), we obtain[9]

$$\int d^4k \, \mathcal{F}''(k^2) = 2i\pi^2 \int_0^{+\infty} d\kappa \, \kappa^3 \, \mathcal{F}''(-\kappa^2) = i\pi^2. \tag{6.91}$$

---

[8] In this counting, we assume that the matrix t does not contain Dirac matrices.
[9] Recall that the rotationally invariant measure in four-dimensional Euclidean space is $2\pi^2 \kappa^3 d\kappa$.

The last equality is obtained by two successive integrations by parts. We also have

$$\begin{aligned}\slashed{D}_x^2 &= D_x^\mu D_x^\nu \gamma_\mu \gamma_\nu = \frac{1}{2} D_x^\mu D_x^\nu (\{\gamma_\mu, \gamma_\nu\} + [\gamma_\mu, \gamma_\nu])\\ &= D_x^2 + \frac{1}{4}[D_x^\mu, D_x^\nu][\gamma_\mu, \gamma_\nu] = D_x^2 - \frac{ie}{4} F^{\mu\nu}[\gamma_\mu, \gamma_\nu].\end{aligned} \qquad (6.92)$$

In the last equality, we have used the fact that the field strength is proportional to the commutator of two covariant derivatives, $[D^\mu, D^\nu] = -ie F^{\mu\nu}$. It turns out that this property is also true in non-Abelian gauge theories, so that the results of this chapter regarding anomalies are also valid in that case. Using $\text{tr}(\gamma^5 \gamma_\mu \gamma_\nu \gamma_\rho \gamma_\sigma) = 4i\epsilon_{\mu\nu\rho\sigma}$ (see Exercise 3.3), we obtain

$$\mathcal{A}(x) = -\frac{e^2}{16\pi^2} \epsilon_{\mu\nu\rho\sigma} \, \text{tr}\left(t\, F^{\mu\nu}(x) F^{\rho\sigma}(x)\right), \qquad (6.93)$$

where the trace is now on the fermion species in case the matrix t acts on several flavors, and on internal group indices if the gauge theory under consideration is non-Abelian (in this case, the field strength is matrix-valued). For QED, the field strengths are commuting numbers that can be taken out of the trace. When t is the identity matrix, the integral of $\mathcal{A}(x)$ is called the *Chern–Pontryagin index*.

### 6.5.5 Anomaly of the Axial Current

The exponential of $i \int \alpha \mathcal{A}$ that arises from the non-invariance of the fermionic functional measure acts as if the action itself was not invariant under chiral transformations. Recall from the derivation of Noether's theorem that under an infinitesimal global transformation of the fields, $\psi \to e^{i\alpha \gamma^5 t}\psi$, the action varies by

$$\delta S = \int d^4x \, \alpha \, \partial_\mu J_5^\mu(x),$$

where $J_5^\mu \equiv -\overline{\psi}\gamma^\mu \gamma^5 t \psi$ is the Noether current (called the *axial current* in this context) associated to this transformation. Thus, identifying this variation with the exponential coming from the measure leads to

$$\partial_\mu \langle J_5^\mu(x)\rangle_A = -\frac{e^2}{16\pi^2} \epsilon_{\mu\nu\rho\sigma} \, \text{tr}\left(t\, F^{\mu\nu}(x) F^{\rho\sigma}(x)\right), \qquad (6.94)$$

where $\langle \cdot \rangle_A$ is an average over the fermion fields in a fixed gauge field configuration.

### 6.5.6 Axial Anomaly in the u and d Quarks Sector

Consider the sector of the two lightest quark flavors, u and d. If one neglects their mass, the corresponding action is invariant under the following infinitesimal chiral transformations:

$$\delta u = i\alpha\gamma^5 u, \quad \delta d = -i\alpha\gamma^5 d. \qquad (6.95)$$

## 6.5 QUANTUM ANOMALIES

The matrix t in quark flavor space that corresponds to this transformation is

$$t = \begin{pmatrix} 1 & 0 \\ 0 & -1 \end{pmatrix}. \tag{6.96}$$

**Strong interaction:** Through the strong interactions, all quark flavors couple identically with the gluons (i.e., all quarks belong to the same representation of the $\mathfrak{su}(3)$ algebra). In other words, $F^{\mu\nu}$ is the same for all quark flavors (equivalently, one may say that it is proportional to the identity in quark flavor space). The trace that appears in the anomaly function can be factored out into a trace over the color indices and a trace over the flavor,

$$\text{tr}\left(t\, F^{\mu\nu} F^{\rho\sigma}\right) = \text{tr}_{\text{color}} \left(F^{\mu\nu} F^{\rho\sigma}\right) \times \underbrace{\text{tr}_{\text{flavor}}(t)}_{1-1=0} = 0. \tag{6.97}$$

This means that the anomalies that may occur in the gluon sector cancel between the u and d flavors of quarks.

**Electromagnetic interaction:** The situation is different in the photon sector, because the u and d quarks have unequal (fractional) electrical charges. If we denote $\psi$ the fermion doublet that comprises the spinors for the u and d quarks, the covariant derivative acting on it reads:

$$\left(\partial^\mu - ie \underbrace{\begin{pmatrix} \tfrac{2}{3} & 0 \\ 0 & -\tfrac{1}{3} \end{pmatrix}}_{Q} A^\mu\right) \psi, \tag{6.98}$$

which leads to replacing $eF^{\mu\nu}$ by $eQ\, F^{\mu\nu}$ in the result for the anomaly function. Therefore, the trace it contains now reads:

$$\text{tr}_{\text{flavor}}\left(Q^2 t\right) \times \text{tr}_{\text{color}}\left(\mathbf{1}_{\text{color}}\right) = \frac{N_c}{3}, \tag{6.99}$$

where $N_c = 3$ is the number of colors. This leads to

$$\mathcal{A}(x) = -\frac{e^2 N_c}{48\pi^2} \epsilon_{\mu\nu\rho\sigma} F^{\mu\nu}(x)\, F^{\rho\sigma}(x), \tag{6.100}$$

where $F^{\mu\nu}$ is the electromagnetic field strength.

**Decay of the neutral pion in two photons:** At low energy, the strong interactions may be described by an effective theory that couples a doublet of fermions $\psi$ (the u and d quarks), the three pions $\pi$ and a field $\sigma$ (see Section 4.5 on the linear sigma model). The interaction term in this model is $\mathcal{L}_I \equiv \lambda\, \overline{\psi}(\sigma + i\pi \cdot \sigma\gamma^5)\psi$, where $\sigma^a$ ($a = 1, 2, 3$) are the Pauli matrices. Note that $\pi^3$ must be the neutral pion, since it couples diagonally to the two components of the doublet (of the three Pauli matrices, only $\sigma^3$ is diagonal). This interaction term is invariant under the transformation (6.95) provided that the fields $\sigma$ and $\pi$ transform as

$$\sigma \to \sigma - \alpha\, \pi^3, \quad \pi^{1,2} \to \pi^{1,2}, \quad \pi^3 \to \pi^3 + \alpha\sigma. \tag{6.101}$$

Moreover, the masses of nucleons are due to a spontaneous breaking of this symmetry, which also gives the $\sigma$ field a non-zero vacuum expectation value: $\langle\sigma\rangle \equiv f_\pi$. Thus the variation of the field $\pi^3$ is $\delta\pi^3 = f_\pi \alpha$.

When photons are added to this model, there is a priori no direct coupling between the neutral pion and the photon. But consider now the theory that would result from integrating out the quark fields. The anomaly (6.100) would produce a term

$$\mathcal{L}_{\text{anom}}(x) = -\frac{e^2 N_c}{48\pi^2} \epsilon_{\mu\nu\rho\sigma} F^{\mu\nu}(x) F^{\rho\sigma}(x) \alpha(x) \tag{6.102}$$

in the Lagrangian. But this term should now have a different origin, because we are talking about an effective theory involving only pions and photons, which no longer has any fermions. The resolution of this issue is that this effective theory must contain a coupling between the neutral pion and two photons, of the form

$$\mathcal{L}_{\pi^0\gamma\gamma} = -\frac{e^2 N_c}{48\pi^2 f_\pi} \epsilon_{\mu\nu\rho\sigma} F^{\mu\nu}(x) F^{\rho\sigma}(x) \pi^3(x). \tag{6.103}$$

The decay rate of a neutral pion into two photons can be easily determined from the effective coupling of eq. (6.103),

$$\Gamma(\pi^0 \to 2\gamma) = \frac{N_c^2 \alpha_{\text{em}}^2 m_\pi^3}{144\pi^3 f_\pi^2}. \tag{6.104}$$

This result could also be obtained by computing the one-loop transition amplitude from a neutral pion to two photons in the linear sigma model (this involves the calculation of a triangular quark loop – which is in fact another way of deriving the form of the anomaly, as we shall see in Chapter 12). The present considerations show that this decay is in fact controlled to a large extent by a quantum anomaly.

### 6.5.7 Atiyah–Singer Index Theorem

Covariant derivatives $D_\mu = \partial_\mu - ieA_\mu$ are anti-Hermitian, because the gauge potential $A_\mu$ is Hermitian (recall that an ordinary derivative is anti-Hermitian). However, $\gamma^0$ is Hermitian, while $\gamma^{1,2,3}$ are anti-Hermitian. Therefore, the Dirac operator $D_\mu \gamma^\mu$ in Minkowski space is neither Hermitian nor anti-Hermitian.

Let us introduce a Euclidean time $x_4 \equiv ix^0$. Likewise, we also define

$$\partial_4 = i\partial_0, \quad A_4 = iA_0, \quad \gamma_4 = i\gamma_0, \tag{6.105}$$

and the measure over space-time becomes $d^4x = i\, d^4x_E$, where $d^4x_E$ is the measure over four-dimensional Euclidean space ($d^4x_E = dx_1 dx_2 dx_3 dx_4$). The Dirac operator becomes:

$$\slashed{D} = \sum_{i=1}^{4}(\partial_i - ieA_i)\gamma^i. \tag{6.106}$$

## 6.5 QUANTUM ANOMALIES

Now, the Dirac matrices $\gamma^i$ are all anti-Hermitian, which implies that the Euclidean Dirac operator is Hermitian. It can therefore be diagonalized on an orthonormal basis of eigenfunctions $\phi_k$,

$$\slashed{D}_x \phi_k(x) = \lambda_k \phi_k(x), \quad \int d^4x_E \, \phi_k^\dagger(x)\phi_{k'}(x) = \delta_{kk'}, \tag{6.107}$$

with real eigenvalues $\lambda_k$. Note also that these eigenfunctions must obey the following completeness relation:

$$\sum_k \phi_k(x)\phi_k^\dagger(y) = \delta(x-y). \tag{6.108}$$

Consider now the case where t is the identity and use the completeness relation in order to express the delta function in the anomaly function $\mathcal{A}(x)$ in eq. (6.82):

$$\mathcal{A}(x) = -2 \lim_{y \to x, M \to +\infty} \mathrm{tr}\left\{\gamma^5 \mathcal{F}\left(-\frac{\slashed{D}_x^2}{M^2}\right) \sum_k \phi_k(x)\phi_k^\dagger(y)\right\}$$

$$= -2 \lim_{M \to +\infty} \sum_k \mathcal{F}\left(-\frac{\lambda_k^2}{M^2}\right) \phi_k^\dagger(x)\gamma^5\phi_k(x). \tag{6.109}$$

Thus, we obtain the following relationship:

$$\frac{e^2}{32\pi^2}\int d^4x_E \, \epsilon_{ijkl}\,\mathrm{tr}\left(F^{ij}(x)F^{kl}(x)\right)$$

$$= -\frac{1}{2}\int d^4x_E \, \mathcal{A}(x) = \lim_{M \to +\infty} \sum_k \mathcal{F}\left(-\frac{\lambda_k^2}{M^2}\right)\int d^4x_E \, \phi_k^\dagger(x)\gamma^5\phi_k(x), \tag{6.110}$$

between an integral that involves the field strength of a gauge field configuration and a sum over the spectrum of the Euclidean Dirac operator (in the same gauge field). Since $\gamma^5$ anticommutes with the Dirac operator, $\{\gamma^5, \slashed{D}\} = 0$, the state $\phi_{k'} \equiv \gamma^5\phi_k(x)$ is also an eigenfunction of $\slashed{D}_x$ with the eigenvalue $-\lambda_k$:

$$\slashed{D}_x(\gamma^5\phi_k(x)) = -\lambda_k(\gamma^5\phi_k(x)). \tag{6.111}$$

When $\lambda_k \neq 0$, the state $\phi_{k'}$ is therefore distinct from the state $\phi_k(x)$. Since $\slashed{D}_x$ is Hermitian, these two states are in fact orthogonal:

$$\int d^4x_E \, \phi_k^\dagger(x)\gamma^5\phi_k(x) = \int d^4x_E \, \phi_k^\dagger(x)\phi_{k'}(x) = 0. \tag{6.112}$$

This implies that none of the eigenfunctions $\phi_k$ with a non-zero eigenvalue can contribute to the right-hand side of eq. (6.110). The only contributions to eq. (6.110) come from the eigenfunctions for which $\lambda_k = 0$, i.e., the *zero modes* of the Euclidean Dirac operator. Since we have assumed that $\mathcal{F}(0) = 1$, we have

$$\frac{e^2}{32\pi^2}\int d^4x_E \, \epsilon_{ijkl}\,\mathrm{tr}\left(F^{ij}(x)F^{kl}(x)\right) = \sum_{k|\lambda_k=0}\int d^4x_E \, \phi_k^\dagger(x)\gamma^5\phi_k(x). \tag{6.113}$$

Thanks to $\{\gamma^5, \slashed{D}_x\} = 0$, we can choose these zero modes in such a way that they are also eigenmodes of $\gamma^5$, with eigenvalues $+1$ or $-1$. We can thus divide the zero modes into two families, the right-handed and the left-handed ones:

$$\slashed{D}_x \phi_{R,L}(x) = 0, \quad \gamma^5 \phi_{R,L}(x) = \pm \phi_{R,L}(x). \tag{6.114}$$

Using also the fact that the eigenfunctions are normalized as follows:

$$\int d^4 x_E \; \phi_R^\dagger(x) \phi_R(x) = 1, \quad \int d^4 x_E \; \phi_L^\dagger(x) \phi_L(x) = 1, \tag{6.115}$$

we obtain the following identity:

$$\frac{e^2}{32\pi^2} \int d^4 x_E \; \epsilon_{ijkl} \, \text{tr}\left(F^{ij}(x) F^{kl}(x)\right) = n_R - n_L, \tag{6.116}$$

where $n_R$ and $n_L$ are the numbers of right-handed and left-handed zero modes, respectively. This formula is the Atiyah–Singer index theorem. It tells us that the integral in the left-hand side is an integer, despite being the integral of a quantity that changes continuously when one deforms the gauge field. Different considerations, based on the study of Euclidean gauge field configurations known as *instantons* (see Chapter 13), will provide us another insight on this integral by relating it to the third homotopy group of the gauge group, i.e., the set of classes of mappings from the three-dimensional sphere $S_3$ to the gauge group. In the case of Abelian gauge theories, where the gauge group is $U(1)$, the third homotopy group is trivial since it contains only the null element and eq. (6.116) is just zero on both sides. However, in the case of non-Abelian gauge theories with an $SU(N)$ gauge group, the third homotopy group is $\mathbb{Z}$, and eq. (6.116) is a non-trivial identity.

## Exercises

**6.1** Show that a function $f(\psi_1, \psi_2, \cdots, \psi_n)$ of several Grassmann variables has an inverse if and only if the term of zeroth order in its Taylor expansion is non-zero. With a single variable, what is the inverse of $1 + \vartheta \psi$ (with $\vartheta$ a Grassmann constant)?

**6.2** Consider two functions $f, g$ of Grassmann variables $\psi_{1,2,\ldots,n}$, and $\partial_\psi$ a derivative with respect to one of these variables. Is $\partial_\psi(fg) = \partial_\psi(f) g + f \partial_\psi(g)$ true? Show that instead one has $\partial_\psi(fg) = \partial_\psi(f) g + \mathcal{P}(f) \partial_\psi(g)$, where $\mathcal{P}$ is a linear operator such that $\mathcal{P}(\psi_i) = -\psi_i$, $\mathcal{P}(\psi_1 \cdots \psi_p) = (-1)^p \psi_1 \cdots \psi_p$.

**\*6.3** The goal of this exercise is to extend Wick's theorem to the fermionic case. Consider the following integrals:

$$\langle \chi_{j_1} \cdots \chi_{j_p} \overline{\chi}_{i_1} \cdots \overline{\chi}_{i_q} \rangle \equiv \det{}^{-1}(M) \int \prod_{i=1}^n [d\chi_i d\overline{\chi}_i] \; \chi_{j_1} \cdots \chi_{j_p} \overline{\chi}_{i_1} \cdots \overline{\chi}_{i_q} \; \exp(\overline{\chi}^T M \chi).$$

- Why is the result zero if $p \neq q$?
- Compute these expectation values for $p = q = 1, 2$.
- Generalize the result to any $p$.

**\*6.4** Consider a function $f(\chi)$ of a Grassmann variable, and introduce a second Grassmann variable $\eta$ (that also anti-commutes with the first one). Calculate explicitly the following integral, $\tilde{f}(\eta) \equiv \int d\chi \, e^{\chi\eta} \, f(\chi)$. Let us denote $\delta(\eta)$ the result of the same transform applied to the constant function 1. Show that $\delta(\eta)$ behaves like the Grassmann generalization of a delta function.

**\*6.5** Consider two Grassmann variables $\theta_\pm$, and define the following three operators:

$$\tau_1 \equiv \frac{1}{2}\left(\theta_+ \frac{\partial}{\partial\theta_-} + \theta_- \frac{\partial}{\partial\theta_+}\right), \quad \tau_2 \equiv \frac{i}{2}\left(\theta_- \frac{\partial}{\partial\theta_+} - \theta_+ \frac{\partial}{\partial\theta_-}\right),$$

$$\tau_3 \equiv \frac{1}{2}\left(\theta_+ \frac{\partial}{\partial\theta_+} - \theta_- \frac{\partial}{\partial\theta_-}\right).$$

Calculate their commutators. Find their eigenvectors and eigenvalues. How can we interpret the variables $\theta_\pm$?

**6.6** Consider a two-dimensional square lattice. A dimer is an object that occupies two neighboring sites of this lattice. Two dimers cannot overlap. This setup is a model for diatomic molecules that can be adsorbed on the surface of a crystalline substrate, and are thus not allowed to overlap. The partition function of this model reads

$$Z[t] \equiv \sum_{\substack{\text{non overlapping} \\ \text{dimer configurations}}} \prod_{\text{bonds } b} t_b^{N_b},$$

where $N_b$ is the number of dimers on the bond $b$ (0 or 1), and $t_b$ is a parameter that allows each bond to have different affinities with the dimers. The fully packed dimer model is such that the total number of dimers is exactly the number of lattice sites divided by two. Show that its partition function may be obtained by assigning a Grassmann variable $\chi_i$ to each lattice site, via the following integral:

$$Z[t] = \int \prod_i d\chi_i \prod_{\{\langle ij \rangle\}} \left(1 + t_{\langle ij \rangle} \chi_i \chi_j\right),$$

where $\{\langle ij \rangle\}$ denotes the set of nearest-neighbor pairs of lattice sites.

**6.7** We consider QED with only photons (thus, it is a free theory). Given a closed contour $\gamma$ in space-time, we define the following quantity:

$$W_\gamma \equiv \langle 0 | T \exp\left(ie \int_\gamma dx^\mu \, A_\mu(x)\right) | 0 \rangle.$$

Express it in terms of the photon propagator. Calculate the coordinate space expression of the propagator to obtain a more explicit form of $W_\gamma$.

**6.8** Given N Grassmann variables $\chi \equiv (\chi_1, \cdots, \chi_N)$, any function $f(\chi)$ can be expanded as $f(\chi) = \sum_\alpha f_\alpha M_\alpha(\chi)$, where the $M_\alpha(\chi)$ are the $2^N$ terms in the following expansion: $(1+\chi_1)\cdots(1+\chi_N) \equiv \sum_\alpha M_\alpha(\chi)$. Endow this basis with an inner product $\langle M_\alpha | M_\beta \rangle = \delta_{\alpha\beta}$.

- Show that this induces the following inner product on functions:

$$\langle f|g\rangle = \sum_\alpha f^*_\alpha g_\alpha = \int \prod_{i=1}^N d\chi_i d\bar\chi_i \, e^{\sum_j \bar\chi_j \chi_j} \, \overline{f(\chi)} g(\chi), \quad \text{and} \quad f_\alpha = \langle M_\alpha|f\rangle.$$

- Show that $I(\chi,\bar\eta) \equiv \sum_\alpha M_\alpha(\chi)\overline{M_\alpha(\eta)} = \prod_{i=1}^N (1+\chi_i\bar\eta_i) = \exp(\sum_i \chi_i\bar\eta_i)$, and that

$$f(\chi) = \int \prod_i d\eta_i d\bar\eta_i \, e^{\sum_j \bar\eta_j \eta_j} \, I(\chi,\bar\eta) \, f(\eta).$$

# CHAPTER 7

# Non-Abelian Gauge Symmetry

Gauge theories are quantum field theories with matter fields (usually spin-1/2 fermions, but also possibly scalars) and gauge fields arranged in such a way that the Lagrangian is invariant under the action of a local continuous transformation. Quantum electrodynamics (QED) is the simplest such theory, with a local $U(1)$ invariance. Given $\Omega(x) \in U(1)$, the various objects that enter in the theory transform as follows:

$$\psi \;\to\; \Omega^{-1}\psi, \quad A^\mu \;\to\; A^\mu + \frac{i}{e}\Omega^{-1}\partial^\mu\Omega. \tag{7.1}$$

In view of applications to the electroweak and -strong interactions, our goal is to extend the concept of gauge theory to more general groups of transformations. In these two cases, the internal group of transformations is $SU(2)$ and $SU(3)$, respectively, but we shall consider in most of this chapter a general *Lie group*. Compared to the case of $U(1)$, the main distinctive feature will be that this group is in general not commutative (one also says *non-Abelian*).

## 7.1 Non-Abelian Lie Groups and Algebras

### 7.1.1 Lie Groups

Let us start by recalling that a Lie group is a group and also a smooth manifold. The group operation will be denoted multiplicatively, as in $\Omega_2\Omega_1$, and we will denote the identical element by $\mathbf{1}$ and the inverse of a group element $\Omega$ by $\Omega^{-1}$. The fact that a Lie group $\mathcal{G}$ is also a manifold allows us to import concepts of differential geometry in their study.

*Matrix Lie groups*, which will be our main concern in view of applications to quantum field theory, are closed subsets of $GL(n, \mathbb{C})$, the general linear group of $n \times n$ matrices on the field of complex numbers. Here is a list of some classical examples of matrix Lie groups, along with their definition:

| | | |
|---|---|---|
| Special linear groups: | $SL(n, \mathbb{C}|\mathbb{R})$ | $\det(\Omega) = 1$, |
| Special orthogonal group: | $SO(n)$ | $\Omega^T \Omega = 1$, $\det \Omega = 1$, |
| Unitary group: | $U(n)$ | $\Omega^\dagger \Omega = 1$, |
| Special unitary group: | $SU(n)$ | $\Omega^\dagger \Omega = 1$, $\det \Omega = 1$. (7.2) |

### 7.1.2 Lie Algebras

Geometrically, the *Lie algebra* $\mathfrak{g}$ is a vector space that may be viewed as tangent to the group at the identity $\Omega = 1$. Therefore, its dimension is the same as that of the group manifold. As we shall see, the group multiplication induces on the tangent space a non-associative multiplication, called the *Lie bracket*, thereby turning it into an algebra. The Lie algebra completely encapsulates the local properties of the underlying Lie group, and if the group is simply connected its Lie algebra defines it globally. Because they are linear spaces, Lie algebras are usually easier to study than their group counterpart, although they provide most of the information.

In the specific case of matrix Lie groups, the corresponding Lie algebra can be defined as the following set of matrices:[1]

$$\mathfrak{g} \equiv \{X | e^{itX} \in \mathcal{G}, \text{ for all real } t\}. \tag{7.3}$$

The matrix exponential $e^X$ is defined from the Taylor series of the exponential by

$$e^X \equiv \sum_{n=0}^{\infty} \frac{X^n}{n!}, \tag{7.4}$$

which converges for all finite-size matrices X since this series has an infinite radius of convergence. A crucial property of the matrix exponential is that

$$e^{X+Y} \neq e^X e^Y \quad \text{if } [X, Y] \neq 0. \tag{7.5}$$

Instead, one may use *Trotter's formula* (also known as the Lie product formula):[2]

$$e^{X+Y} = \lim_{n \to \infty} \left( e^{X/n} e^{Y/n} \right)^n. \tag{7.6}$$

(See Figure 7.2 for a geometrical interpretation of this formula.)

---

[1] The prefactor i inside the exponential is common in the quantum physics literature, but seldom used in mathematics. Its main benefit is to make X a Hermitian matrix when the group elements are unitary.

[2] A sketch of the proof is the following:
$$e^{X/n} e^{Y/n} = 1 + \frac{X}{n} + \frac{Y}{n} + \mathcal{O}(n^{-2}) = \exp\left(\frac{X+Y}{n} + \mathcal{O}(n^{-2})\right),$$
$$\left(e^{X/n} e^{Y/n}\right)^n = \exp\left(X + Y + \mathcal{O}(n^{-1})\right).$$

## 7.1 NON-ABELIAN LIE GROUPS AND ALGEBRAS

Note that for X to be in the algebra, it is sufficient that $\exp(tX) \in \mathcal{G}$ for t in a neighborhood of $t = 0$. Then, since the group $\mathcal{G}$ is closed under multiplication, this property extends to all t on the real axis. In fact, the mapping $t \to \exp(tX)$ is a group homomorphism (from the additive group $\mathbb{R}$ to $\mathcal{G}$) that spans a one-dimensional subgroup of $\mathcal{G}$. From the definition (7.3) of the Lie algebra, and using Trotter's formula, one can check that any real linear combination of elements of $\mathfrak{g}$ is in $\mathfrak{g}$, i.e., that $\mathfrak{g}$ *is a real vector space*. Therefore, every element of $\mathfrak{g}$ can be written as a linear combination of some basis elements $t^a$:

$$X = X_a t^a \qquad (X_a \in \mathbb{R}), \tag{7.7}$$

with an implicit sum on the index a. The $t^a$ are called the *generators* of the algebra.

Thanks to the exponential mapping (7.3), the properties of the Lie groups listed in eq. (7.2) translate into specific properties of the matrices X in the corresponding algebras:

| | | |
|---|---|---|
| Special linear groups: | $\mathfrak{sl}(n, \mathbb{C}\|\mathbb{R})$ | $\mathrm{tr}\,(X) = 0,$ |
| Special orthogonal group: | $\mathfrak{so}(n)$ | $X^T = -X,$ |
| Unitary group: | $\mathfrak{u}(n)$ | $X^\dagger = X,$ |
| Special unitary group: | $\mathfrak{su}(n)$ | $X^\dagger = X,\ \mathrm{tr}\,(X) = 0.$ (7.8) |

Note that the conditions imposed on $\Omega$ in eq. (7.2) are nonlinear, in contrast to the linear conditions obeyed by the matrices X in eq. (7.8). This is why a Lie group is a curved manifold, while a Lie algebra is a linear space.

### 7.1.3 Geometrical Interpretation

First note that we have

$$iX = \frac{d}{dt} e^{itX} \bigg|_{t=0}. \tag{7.9}$$

The group elements $\exp(itX)$ form a smooth curve on the group manifold ($t = 0$ corresponds to the identity), and $iX$ may be viewed as the vector tangent to this curve at the identity, as illustrated in Figure 7.1. The non-commutativity of the group is related to the curvature of the corresponding manifold.[3] Because of this curvature, a displacement $e^{iX}$ followed by a displacement $e^{iY}$ does not lead to the same point as the two displacements performed in reverse order. This geometrical representation also provides an interpretation of Trotter's formula for the exponential of a sum, as shown in Figure 7.2.

The dimension of the Lie algebra equals the number of independent directions on the group manifold. From the conditions listed in (7.8) on the matrices $X \in \mathfrak{g}$, it is easy to determine the dimension of these algebras (viewed as algebras over the field $\mathbb{R}$). The dimensions are listed in Table 7.1 for some common cases (see Exercise 7.1).

---

[3] This assertion could be made more precise as follows: It is possible to define a metric tensor on the group manifold, and the corresponding Ricci curvature tensor. This curvature may then be expressed in terms of the constants that define the commutators between the generators of the algebra (see eq. (7.13)).

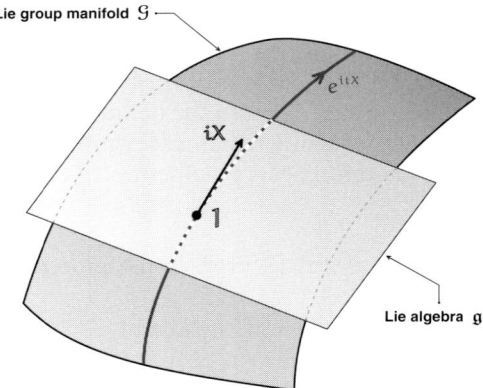

FIGURE 7.1: Lie group and Lie algebra.

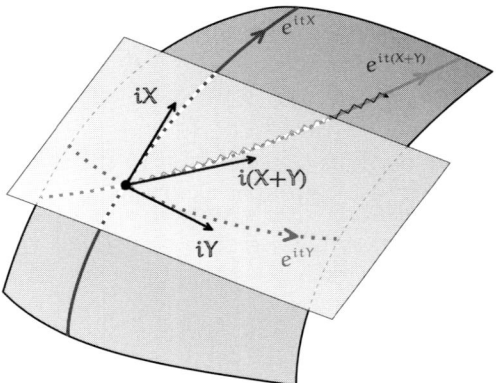

FIGURE 7.2: Geometrical interpretation of Trotter's formula: the broken path, made of a succession of elementary steps $e^{itX/n}$ and $e^{itY/n}$, approximates better and better the curve $e^{it(X+Y)}$ on the group manifold as $n \to \infty$.

Moreover, as we shall see in Section 7.1.5, the group multiplication can be inferred from that on the Lie algebra via the Baker–Campbell–Hausdorff formula. Despite these correspondences, the Lie algebra may not reflect the global properties of the group (e.g., whether it is connected), and distinct Lie groups may have isomorphic Lie algebras. This is, for instance, the case of $SO(3)$ and $SU(2)$, or $SU(2) \times SU(2)$ and $SO(4)$.

## 7.1.4 Lie Bracket and Structure Constants

Consider an element $\Omega$ of the Lie group and an element $X$ of the Lie algebra. For any real number $t$, we have

$$\exp\left(i t\, \Omega^{-1} X \Omega\right) = \sum_{n=0}^{\infty} \frac{(it)^n}{n!} \underbrace{\left(\Omega^{-1} X \Omega\right)^n}_{\Omega^{-1} X^n \Omega} = \underbrace{\Omega^{-1} \underbrace{e^{itX}}_{\in \mathcal{G}} \Omega}_{\in \mathcal{G}}, \qquad (7.10)$$

## 7.1 NON-ABELIAN LIE GROUPS AND ALGEBRAS

| Lie algebra | Dimension |
|---|---|
| $\mathfrak{sl}(n, \mathbb{R})$ | $n^2 - 1$ |
| $\mathfrak{so}(n)$ | $n(n-1)/2$ |
| $\mathfrak{u}(n)$ | $n^2$ |
| $\mathfrak{su}(n)$ | $n^2 - 1$ |

TABLE 7.1: Dimensions of a few common Lie algebras.

where the equality follows from the Taylor series of the exponential. From the definition of the Lie algebra, this implies that $\Omega^{-1} X \Omega \in \mathfrak{g}$. Therefore, if $X, Y \in \mathfrak{g}$ we also have

$$e^{-itX} Y e^{itX} \in \mathfrak{g}, \quad (7.11)$$

and the derivative with respect to t at $t = 0$ is also an element of the algebra,

$$-i[X, Y] \in \mathfrak{g}. \quad (7.12)$$

In other words, $-i$ times the *commutator* of two elements of a Lie algebra is another element of the algebra. Thus $-i[\cdot, \cdot]$ is the multiplication law[4] in $\mathfrak{g}$ (it is also called the *Lie bracket*). Therefore, the commutators between the generators can be written as

$$[t^a, t^b] = i f^{abc} t^c, \quad (7.13)$$

where the $f^{abc}$ are real numbers called the *structure constants*. The antisymmetry of the commutator implies the antisymmetry of the structure constants on their first two indices, $f^{abc} = -f^{bac}$. Given three elements $X, Y, Z \in \mathfrak{g}$ of the algebra, their commutator satisfies the *Jacobi identity*,

$$[X, [Y, Z]] + [Y, [Z, X]] + [Z, [X, Y]] = 0, \quad (7.14)$$

which implies the following relationship among the structure constants,

$$f^{ade} f^{bcd} + f^{bde} f^{cad} + f^{cde} f^{abd} = 0. \quad (7.15)$$

### 7.1.5 Baker–Campbell–Hausdorff Formula

Given an element $X \in \mathfrak{g}$, we may define a function from $\mathfrak{g}$ to $\mathfrak{g}$ as follows:

$$\text{ad}_X(Y) \equiv -i[X, Y]. \quad (7.16)$$

The function $\text{ad}_X$ is called the *adjoint mapping* at the point X. The exponential of the adjoint mapping plays an important role, thanks to the following formula:[5]

$$e^{\text{ad}_X} Y = e^{-iX} Y e^{iX}. \quad (7.17)$$

---

[4] In contrast, the ordinary product of two elements of the algebra is in general not in the algebra.

[5] This may be proven by considering a one-parameter family of such equalities,

$$e^{t \, \text{ad}_X} Y = e^{-itX} Y e^{itX},$$

and by noting that the left- and right-hand sides coincide at $t = 0$, and obey identical differential equations with respect to the parameter t.

This allows us to write the derivative of the exponential of a (matrix-valued) function as follows:

$$\frac{d}{dt} e^{iX(t)} = i\, e^{iX(t)}\, \frac{e^{ad_{X(t)}} - 1}{ad_{X(t)}}\, \frac{dX(t)}{dt}. \tag{7.18}$$

(This is known as Duhamel's formula.[6]) The non-trivial aspect of this formula is that it is true even when $X(t)$ does not commute with its derivative. Then, given $X, Y \in \mathfrak{g}$, let us define a matrix $Z(t)$ by

$$e^{iZ(t)} \equiv e^{iX}\, e^{itY}. \tag{7.19}$$

Differentiating both sides with respect to $t$ (using eq. (7.18) for the left-hand side), we obtain

$$\frac{dZ(t)}{dt} = \left[\frac{e^{ad_{Z(t)}} - 1}{ad_{Z(t)}}\right]^{-1} Y. \tag{7.20}$$

From eq. (7.17), we can also see that

$$e^{ad_{Z(t)}} = e^{t\, ad_Y}\, e^{ad_X}. \tag{7.21}$$

Integrating eq. (7.20) from $t=0$ to $t=1$, we obtain the following identity:

$$\ln\left(e^{iX} e^{iY}\right) = iX + i\int_0^1 dt\, F\left(e^{t\, ad_Y} e^{ad_X}\right) Y, \tag{7.22}$$

where the function $F(\cdot)$ is defined by

$$F(z) \equiv \frac{\ln(z)}{z - 1}. \tag{7.23}$$

Eq. (7.22) is the integral form of the Baker–Campbell–Hausdorff formula. In order to recover the more familiar expansion in nested commutators, note that

$$e^{t\, ad_Y} e^{ad_X} = 1 + t\, ad_Y + ad_X + \tfrac{1}{2}(t^2 ad_Y^2 + ad_X^2) + t\, ad_Y\, ad_X + \cdots,$$

$$F(z) = 1 - \frac{1}{2}(z-1) + \tfrac{1}{3}(z-1)^2 + \cdots \tag{7.24}$$

---

[6] Note that this formula is equivalent to

$$\frac{d}{dt} e^{iX(t)} = i \int_0^1 ds\, e^{i(1-s)X(t)}\, \frac{dX(t)}{dt}\, e^{isX(t)}.$$

This latter form can be proven by writing

$$e^{iX(t+\varepsilon)} = e^{i\left(X(t) + \varepsilon X'(t) + \mathcal{O}(\varepsilon^2)\right)} = \lim_{n\to\infty} \left(e^{i\frac{X(t)}{n}}\, e^{i\frac{\varepsilon}{n} X'(t) + \mathcal{O}(\varepsilon^2)}\right)^n,$$

(we use Trotter's formula to obtain the second equality) and by expanding the right-hand side to first order in $\varepsilon$.

This leads to

$$\ln\left(e^{iX}e^{iY}\right) = i(X+Y) - \frac{1}{2}[X,Y] - \frac{i}{12}\Big([X,[X,Y]] - [Y,[X,Y]]\Big) + \cdots \quad (7.25)$$

(Explicit expressions for all the coefficients of this series are given by *Dynkin's formula*.) In applications to quantum field theory, we usually need only the first two terms of this expansion because the commutators we encounter are commuting numbers and all the subsequent terms are zero. Besides being an intermediate step in the derivation of eq. (7.25), the integral form (7.22) also shows that the group product can be reconstructed from Lie algebra manipulations (since the right-hand side of this equation contains only objects that belong to the algebra). It also gives exact results for the product in some special cases (see Exercise 7.2).

### 7.1.6 Representations

A real *representation of a Lie group* $\mathcal{G}$ is a group homomorphism from elements of $\mathcal{G}$ to elements of $GL(n,\mathbb{R})$, i.e., a mapping $\pi$ from $\mathcal{G}$ to $GL(n,\mathbb{R})$ that preserves the group structure,

$$\pi(1) = 1, \qquad \pi(\Omega_2\Omega_1) = \pi(\Omega_2)\pi(\Omega_1). \quad (7.26)$$

A representation is said to be faithful if it is a one-to-one mapping.

Likewise, one may define representations of a Lie algebra as homomorphisms from $\mathfrak{g}$ to $\mathfrak{gl}(n,\mathbb{R})$, i.e., mappings $\pi$ that preserve the Lie algebra structure,

$$\pi(\alpha X + \beta Y) = \alpha\,\pi(X) + \beta\,\pi(Y), \qquad \pi([X,Y]) = \big[\pi(X),\pi(Y)\big]. \quad (7.27)$$

Note that if we define $t_\pi^a \equiv \pi(t^a)$ the images of the generators, then they obey $[t_\pi^a, t_\pi^b] = if^{abc} t_\pi^c$ with the same structure constants as in the original Lie algebra. Among the representations of a Lie algebra, the *irreducible representations* play a special role, since they are those for which the matrices that represent the generators cannot be written in block diagonal form with blocks that are themselves (smaller) representations. The irreducible representations are thus the "bricks" from which any other representation can be built.

**Singlet representation:** The *singlet representation*, or *trivial representation*, is the representation for which the mapping is $\pi(\Omega) = 1$ for all $\Omega$. The objects that belong to this representation space are invariant under the transformations of the group $\mathcal{G}$. In quantum field theory, one says that these objects are "neutral" (under the group $\mathcal{G}$).

**Fundamental representation:** The *fundamental representation*, or *standard representation*, is the smallest faithful representation. It is also the representation obtained when $\pi$ is the identical map. In other words, in the fundamental representation, the elements of a matrix Lie group are simply represented by themselves. In the case of compact simple Lie algebras (that give consistent non-Abelian gauge theories, as we shall see in the next section), it is possible to choose the generators $t_f^a$ (the subscript f denotes the fundamental representation) in such a way that $\mathrm{tr}\,(t_f^a t_f^b)$ is proportional to the identity. A customary choice of their normalization is to impose

$$\mathrm{tr}\,(t_f^a t_f^b) = \frac{\delta^{ab}}{2}. \quad (7.28)$$

This sets the normalization of the structure constants, through eq. (7.13). Then, one usually normalizes the generators of other representations in such a way that they fulfill the commutation relation (7.13) with the same structure constants (but the trace formula (7.28) with a prefactor 1/2 is in general only valid in the fundamental representation).

**Adjoint representation:** The *adjoint representation* of a Lie group $\mathcal{G}$ is a representation by linear operators that act on the Lie algebra $\mathfrak{g}$, defined by the following mapping:

$$\Omega \in \mathcal{G} \;\rightarrow\; \mathrm{Ad}_\Omega \in \mathrm{GL}(\mathfrak{g}) \quad \text{such that} \quad \mathrm{Ad}_\Omega(X) = \Omega^{-1} X \Omega. \tag{7.29}$$

If the dimension of the Lie algebra is d, then $\mathrm{Ad}_\Omega$ may be viewed as a d × d matrix. We may also define the adjoint representation of the algebra $\mathfrak{g}$ as follows:

$$X \in \mathfrak{g} \;\rightarrow\; \mathrm{ad}_X \in \mathrm{GL}(\mathfrak{g}) \quad \text{such that} \quad \mathrm{ad}_X(Y) = -i[X, Y]. \tag{7.30}$$

It is sufficient to know the adjoint representation of the generators $t^a$, for which one often uses the following notation

$$i\,\mathrm{ad}_{t^a} \equiv T^a_{\mathrm{adj}}. \tag{7.31}$$

Note that $T^a_{\mathrm{adj}}$ can be represented by a d × d matrix. Using Jacobi's identity, one may check that $[\mathrm{ad}_{t^a}, \mathrm{ad}_{t^b}] = -\mathrm{ad}_{i[t^a, t^b]} = f^{abc}\,\mathrm{ad}_{t^c}$. Therefore, the $T^a_{\mathrm{adj}}$ fulfill the same commutation relations as the $t^a$ themselves,

$$[T^a_{\mathrm{adj}}, T^b_{\mathrm{adj}}] = i f^{abc} T^c_{\mathrm{adj}}. \tag{7.32}$$

Using eq. (7.13), we find that the components of these matrices are given by

$$(T^a_{\mathrm{adj}})_{bc} = -i f^{cab}. \tag{7.33}$$

In other words, the adjoint representation is a representation by matrices whose size is the dimension of the algebra, and in which the components of the generators are the structure constants. That eqs. (7.32) and (7.33) are consistent is a consequence of the Jacobi identity (7.15) satisfied by the structure constants.

A common use of the adjoint representation is to rearrange expressions such as

$$e^{-iX} Y e^{iX} = e^{\mathrm{ad}_X} Y, \tag{7.34}$$

where X and Y are in some representation r of the Lie algebra. Using $X = X_a t^a_r$ and $Y = Y_a t^a_r$, we can rewrite this as follows:

$$e^{\mathrm{ad}_X} Y = e^{X_a \mathrm{ad}_{t^a}} Y_b t^b_r = t^c_r \left[ e^{-i X_a T^a_{\mathrm{adj}}} \right]_{cb} Y_b. \tag{7.35}$$

Thus, we have

$$\left[ e^{-iX} Y e^{iX} \right]_c = \left[ e^{-i X_{\mathrm{adj}}} \right]_{cb} Y_b, \tag{7.36}$$

where the left-hand side may be in any representation r. In other words, the right and left multiplication by a group element and its inverse can be rewritten as a left multiplication by the adjoint of this group element.

## 7.2 Yang-Mills Lagrangian

### 7.2.1 Covariant Derivative and Gauge Transformations

When trying to extend the concept of gauge theory to a non-Abelian symmetry, it is useful to first construct a *covariant derivative*. This object, denoted $D_\mu$, is a deformation of the ordinary derivative that transforms as follows:

$$D_\mu \quad \to \quad \Omega^{-1}(x)\, D_\mu\, \Omega(x), \tag{7.37}$$

where $\Omega(x)$ is a space-time dependent element of a Lie group $\mathcal{G}$. Let us look for a covariant derivative of the form

$$D_\mu \equiv \partial_\mu - ig\, A_\mu(x), \tag{7.38}$$

where g is a coupling constant similar to the constant e in QED and $A_\mu(x)$ a 4-vector (in quantum field theory, this field is called a gauge field). The transformation law (7.37) is satisfied provided that $A_\mu(x)$ transforms in a very specific way. Note first that the ordinary derivative $\partial_\mu$ is invariant (i.e., it belongs to the singlet representation). If we denote $A_\mu^\Omega(x)$ the transformed $A_\mu(x)$, then we must have

$$\begin{aligned}\partial_\mu - ig\, A_\mu^\Omega(x) &= \Omega^{-1}(x)\left[\partial_\mu - ig\, A_\mu(x)\right]\Omega(x) \\ &= \partial_\mu + \Omega^{-1}(x)(\partial_\mu \Omega(x)) - ig\, \Omega^{-1}(x)\, A_\mu(x)\, \Omega(x),\end{aligned} \tag{7.39}$$

from which we obtain the transformation law[7] of $A_\mu(x)$,

$$A_\mu(x) \quad \to \quad A_\mu^\Omega(x) \equiv \Omega^{-1}(x)\, A_\mu(x)\, \Omega(x) + \frac{i}{g}\Omega^{-1}(x)(\partial_\mu \Omega(x)). \tag{7.40}$$

From eqs. (7.16)–(7.18), we see that if $\Omega$ is an element of a Lie group $\mathcal{G}$, then $\Omega^{-1}\partial_\mu\Omega$ belongs to the Lie algebra $\mathfrak{g}$. Thus, if the second term on the right-hand side of eq. (7.40) belongs to the representation r of the Lie algebra, the first term should also be in this representation for consistency. The same applies to $A_\mu$, which we can decompose as $A_\mu(x) \equiv A_\mu^a(x)\, t_r^a$, where the $t_r^a$ are the generators of the algebra in the representation r.

**Infinitesimal transformations:** Equation (7.40) specifies how the field $A_\mu$ changes under any transformation in $\mathcal{G}$. However, it is sometimes useful to consider infinitesimal transformations, i.e., $\Omega$ close to 1. This is done by writing $\Omega = \exp(ig\,\theta_a t_r^a)$, with $|\theta_a| \ll 1$, and by expanding eq. (7.40) to order 1 in $\theta_a$. The variation of $A_\mu$ is given by

$$\delta A_\mu = -\partial_\mu \theta_r(x) + ig\,[A_\mu(x), \theta_r(x)] = -[D_\mu, \theta_r(x)], \tag{7.41}$$

where we have defined $\theta_r \equiv \theta_a t_r^a$. This can also be written more explicitly as

$$\delta A_\mu^a = -\partial_\mu \theta_a(x) + g\, f^{abc}\, \theta_b(x)\, A_\mu^c(x) = -\left(D_\mu^{\text{adj}}\right)_{ab}\theta_b(x), \tag{7.42}$$

where $D_\mu^{\text{adj}}$ is the covariant derivative in the adjoint representation.

---

[7] From this transformation law, we see that field configurations of the form $ig^{-1}\Omega^{-1}\partial_\mu\Omega$ may be transformed into the null field $A_\mu \equiv 0$. Such configurations are called *pure gauge fields*.

## 7.2.2 Non-Abelian Field Strength

In the previous section we introduced a vector field $A_\mu$ in order to define a covariant derivative. To interpret $A_\mu$ as describing a spin-1 particle, we should construct a kinetic term for this field, with the constraint that it is invariant under the transformations (7.40). In the case of QED, this Lagrangian was $-\frac{1}{4} F_{\mu\nu} F^{\mu\nu}$, where the field strength was defined as $F_{\mu\nu} \equiv \partial_\mu A_\nu - \partial_\nu A_\mu$. However, a direct verification indicates that this expression of the field strength cannot lead to an invariant Lagrangian in the case of a non-Abelian symmetry.

In order to mimic QED, we aim at constructing a Lagrangian with second-order derivatives. Indeed, since the field $A_\mu(x)$ has the dimension of a mass, two derivatives and two powers of the field would provide the required dimension four for a Lagrangian in four space-time dimensions. A useful intermediate step is the construction of a field that depends only on $A_\mu(x)$ and has a simple transformation law. From the transformation law of the covariant derivative, we see that the commutator $[D_\mu, D_\nu]$ transforms as

$$[D_\mu, D_\nu] \;\to\; \Omega^{-1}(x) [D_\mu, D_\nu] \Omega(x). \tag{7.43}$$

More explicitly, this commutator reads

$$[D_\mu, D_\nu] = -ig \Big( \underbrace{\partial_\mu A_\nu - \partial_\nu A_\mu - ig [A_\mu, A_\nu]}_{F_{\mu\nu}} \Big). \tag{7.44}$$

This generalizes the *field strength* $F_{\mu\nu}$ to an arbitrary Lie algebra $\mathfrak{g}$. Note the commutator between gauge fields, that did not exist in QED. By construction, the field strength is an element of the algebra, in the same representation as $A_\mu$, $F_{\mu\nu}(x) \equiv F_{\mu\nu}^a(x) t_r^a$, and its transformation law[8] is

$$F_{\mu\nu}(x) \;\to\; \Omega^{-1}(x) F_{\mu\nu}(x) \Omega(x). \tag{7.45}$$

As in QED, one may define (non-Abelian) electrical and magnetic fields by

$$E_a^i = F_a^{0i}, \quad B_a^i = \tfrac{1}{2} \epsilon_{ijk} F_a^{jk}, \tag{7.46}$$

but with an important difference: These **E** and **B** fields are not gauge invariant (they belong to the representation r of the Lie algebra). Instead, they transform covariantly as follows:

$$\mathbf{E}^i(x) \to \Omega^{-1}(x) \mathbf{E}^i(x) \Omega(x), \quad \mathbf{B}^i(x) \to \Omega^{-1}(x) \mathbf{B}^i(x) \Omega(x). \tag{7.47}$$

## 7.2.3 Lagrangian

In order to build a kinetic term from $F_{\mu\nu}$, we must contract all the Lorentz indices to have a Lorentz invariant Lagrangian. This forces us to have at least two Fs, since $g^{\mu\nu} F_{\mu\nu} = 0$. Therefore, if we restrict to objects of mass dimension four, this kinetic term should be quadratic in

---

[8] The field strength associated to a pure gauge field is zero, since there exists a transformation $\Omega$ for which $A_\mu$ becomes the null field.

## 7.2 YANG–MILLS LAGRANGIAN

F, with a dimensionless prefactor. The most general[9] term of this kind is

$$\mathcal{L}_A \equiv -\tfrac{1}{4} h_{ab} F^{a\,\mu\nu} F^b_{\mu\nu}, \qquad (7.48)$$

where $h_{ab}$ is a real constant matrix, symmetric in the group indices. In addition, for this Lagrangian to define a consistent field theory, the matrix $h_{ab}$ should be positive definite (otherwise some parts of the kinetic term would have the wrong sign and the energy of the system would not be bounded from below). Under an infinitesimal gauge transformation, the variation of this Lagrangian is

$$\delta \mathcal{L}_A = -\tfrac{1}{2} h_{ab} F^{a\,\mu\nu} \theta_c f^{dcb} F^d_{\mu\nu}, \qquad (7.49)$$

and for the kinetic term to be gauge invariant we must have

$$h_{ab} f^{dcb} \underbrace{F^{a\,\mu\nu} F^d_{\mu\nu}}_{\text{sym. in } a, d} = 0. \qquad (7.50)$$

This condition is satisfied for any gauge field configuration, provided that

$$f^{dcb} h_{ba} + f^{acb} h_{bd} = 0. \qquad (7.51)$$

Note that $h_{ab} \equiv \mathrm{tr}\,(t^a t^b)$ is a solution of this constraint, since $\mathrm{tr}\,(F_{\mu\nu} F^{\mu\nu})$ is obviously gauge invariant, given the transformation law (7.45) for the field strength, but the positivity condition imposes further restrictions on the kind of Lie algebra we are allowed to use.

### 7.2.4 Constraints on the Lie Algebra

Eq. (7.51), combined with the fact that $h_{ab}$ is a real symmetric positive-definite matrix, strongly constrains the Lie algebras that lead to consistent (gauge invariant, with a positive definite kinetic energy) non-Abelian gauge theories.

**Complete antisymmetry of the structure constants:** Let us start from a diagonalization of the matrix $h_{ab}$,

$$h_{ab} \equiv O^t_{ac} \lambda_c O_{cb} = O_{ca} O_{cb} \lambda_c, \qquad (7.52)$$

where $O_{ac}$ is a real orthogonal matrix. Since the matrix $h_{ab}$ is positive definite, all the eigenvalues $\lambda_c$ are positive, and we can define a square root of the matrix by

$$\Omega_{ab} \equiv O_{ca} O_{cb} \lambda_c^{1/2}, \quad h_{ab} = \Omega_{ac} \Omega_{cb}. \qquad (7.53)$$

Note that $\Omega_{ab}$ is itself a real symmetric matrix. Now, let us introduce a new basis for the algebra, defined by

$$\bar{t}^a \equiv \Omega^{-1}_{ab} t^b, \quad t^a = \Omega_{ab} \bar{t}^b. \qquad (7.54)$$

---

[9] We ignore for now the operator $\epsilon_{\mu\nu\rho\sigma} h_{ab} F^{\mu\nu}_a F^{\rho\sigma}_b$, which will be discussed in Section 7.5.

This is a legitimate change of basis for a real algebra since the matrix $\Omega$ is real and has no vanishing eigenvalue (all the eigenvalues $\lambda_c$ are strictly positive since $h_{ab}$ is positive definite). The commutator of two of these new generators is

$$[\bar{t}^a, \bar{t}^b] = i \underbrace{\Omega^{-1}_{aa'} \Omega^{-1}_{bb'} f^{a'b'c'} \Omega_{c'c}}_{\bar{f}^{abc}} \bar{t}^c. \tag{7.55}$$

By rewriting eq. (7.51) in terms of the new structure constants $\bar{f}^{abc}$ and by using the fact that $\Omega$ is invertible, we get

$$\bar{f}^{dca} + \bar{f}^{acd} = 0. \tag{7.56}$$

In other words, eq. (7.51) implies that there exists a basis in which the structure constants are also antisymmetric under the exchange of the first and third indices. From this, we conclude that they are in fact completely antisymmetric[10] (and not just in the first two indices, as implied by their definition in terms of the Lie bracket).

**Allowed subalgebras:** Consider now the generators in the adjoint representation, whose components are given by

$$(T^a_{\text{adj}})_{bc} = -i \bar{f}^{cab} = -i \bar{f}^{abc}. \tag{7.57}$$

Since the structure constants are real and antisymmetric, these generators are Hermitian matrices. Thanks to this property, there exists a basis in which all the adjoint generators have a block diagonal structure:

$$T^a_{\text{adj}} = \begin{pmatrix} D^a_{(1)} & 0 & 0 & \cdots \\ 0 & D^a_{(2)} & 0 & \cdots \\ 0 & 0 & D^a_{(3)} & \cdots \\ \vdots & \vdots & \vdots & \ddots \end{pmatrix}, \tag{7.58}$$

where the sizes of the blocks are the same for all the $T^a_{\text{adj}}$. This block decomposition can be obtained recursively until one gets blocks that are not further reducible. If d is the dimension of the adjoint representation (i.e., also the dimension of the Lie group), it corresponds to a decomposition of $\mathbb{R}^d$ into orthogonal subspaces that are invariant under the action of all the generators. Regarding the Lie algebra, this indicates that it is a direct sum of *simple subalgebras*,[11] and $u(1)$ *subalgebras* (if some diagonal blocks $D^a_{(n)}$ are zero for all $a$). In

---

[10]Assuming they are non-zero, this requires an algebra with at least three generators, since it is not possible to construct an antisymmetric rank-3 tensor with indices that take fewer than three values.

[11]A Lie algebra is *not simple* if there exists a set of generators $\mathcal{T}^\alpha$ (the number of which is strictly smaller than the dimension of the algebra) which is closed under commutation with the algebra, i.e., $[T^a_{\text{adj}}, \mathcal{T}^\alpha] = g^{a\alpha\beta} \mathcal{T}^\beta$ for all $a, \alpha$. If we write these new generators as linear combinations $\mathcal{T}^\alpha \equiv V^\alpha_a T^a_{\text{adj}}$, the closure of the subalgebra under commutation implies that the set of vectors $\{V^\alpha\}$ is the basis of a subspace invariant under the action of all the $T^a_{\text{adj}}$. Conversely, if we have an invariant subspace that cannot be reduced to a smaller one, then it corresponds to a simple subalgebra.

addition, these simple subalgebras are *compact* because $K_{ab} \equiv \text{tr}(T^a_{\text{adj}} T^b_{\text{adj}})$, restricted to the corresponding subspace, is positive definite.[12] Indeed, when the structure constants are totally antisymmetric, we have $K_{ab} X^a X^b = \sum_{c,d} (X^a \bar{f}^{acd})^2 \geq 0$ for any vector X. Moreover, there is no non-zero vector X for which this quadratic form is zero, because otherwise we would have $T^c_{\text{adj}} X = 0$ for all c, which means that this vector X would define a $\mathfrak{u}(1)$ subalgebra and cannot be part of the subspace associated with a simple subalgebra.

**Standard form of the Lagrangian:** Note now that the constraint (7.51) can also be written as $[T^c_{\text{adj}}, h] = 0$ (for all c). This implies that $h_{ab}$ has the same block decomposition as the adjoint generators (see eq. (7.58)), with diagonal blocks that are proportional to the identity (with positive prefactors)

$$h = \begin{pmatrix} \alpha^2_{(1)} \mathbb{1} & 0 & 0 & \cdots \\ 0 & \alpha^2_{(2)} \mathbb{1} & 0 & \cdots \\ 0 & 0 & \alpha^2_{(3)} \mathbb{1} & \cdots \\ \vdots & \vdots & \vdots & \ddots \end{pmatrix}. \tag{7.59}$$

The prefactors $\alpha^2_{(i)}$ can be absorbed into the normalization of the gauge field and the coupling constant of the corresponding subalgebra, by writing

$$\alpha_{(i)} F_{\mu\nu} = \partial_\mu A'_\nu - \partial_\nu A'_\mu - ig' [A'_\mu, A'_\nu] \tag{7.60}$$

with $A'_\mu \equiv \alpha_{(i)} A_\mu$ and $g' \equiv \alpha_{(i)}^{-1} g$. Therefore, we can always write the Lagrangian as a sum of terms (one for each simple and $\mathfrak{u}(1)$ subalgebra) having the following standard form:

$$\mathcal{L}_A = -\frac{1}{4} F^a_{\mu\nu}(x) F^{a\,\mu\nu}(x). \tag{7.61}$$

Despite its resemblance with the photon kinetic term in QED, this Lagrangian has a quite remarkable feature in the case of simple subalgebras: Due to the commutator term in $F_{\mu\nu}$, $\mathcal{L}_A$ contains terms that are cubic in $A_\mu$ and terms which are quartic in $A_\mu$. These terms are interactions between three and four of the spin-1 particles described by $A_\mu$, respectively. Thus, unlike in QED, the Lagrangian (7.61) has a very rich structure, and defines in itself a very interesting quantum field theory called *Yang–Mills theory*.

## 7.3 Non-Abelian Gauge Theories

A *non-Abelian gauge theory* is a quantum field theory that has at least a gauge field whose symmetry group is a non-Abelian group $\mathcal{G}$. Thus, the Lagrangian of all non-Abelian gauge theories contains a Yang–Mills term,

---

[12]The coefficients $K_{ab}$ enter in the *Killing form*, $K(X, Y) \equiv \text{tr}(\text{ad}_X \text{ad}_Y) = -K_{ab} X^a Y^b$, that defines an inner product on the Lie algebra which is invariant under the adjoint action of the group $X \to \Omega X \Omega^{-1}$. Since $K_{ab}$ is positive definite in compact Lie algebras, it naturally provides a distance $d^2(X, Y) \equiv K_{ab} (X - Y)^a (X - Y)^b$ on $\mathfrak{g}$. In turn, this inner product in $\mathfrak{g}$ gives rise to a metric for the underlying Lie group $\mathcal{G}$, which is invariant under the group action. Given two vectors $U, V$ in the tangent space to $\mathcal{G}$ at the point $\Omega$, the metric at $\Omega$ is defined by $g_\Omega(U, V) \equiv -K(\Omega^{-1} U, \Omega^{-1} V)$, where $\Omega^{-1} U, \Omega^{-1} V$ are elements of $\mathfrak{g}$ obtained by "pulling back" $U, V$ according to the left group multiplication map.

$$\mathcal{L}_A \equiv -\frac{1}{4} F^a_{\mu\nu}(x) F^{a\,\mu\nu}(x). \tag{7.62}$$

If $A_\mu$ is the only field of the theory, then it is a plain Yang–Mills theory.

### 7.3.1 Fermions

However, useful gauge theories in particle physics usually also have matter fields, i.e., fermions. Under the action of a Lie group $\mathcal{G}$, a fermion field transforms as

$$\psi(x) \;\to\; \Omega^{-1}(x)\,\psi(x), \quad \overline{\psi}(x) \;\to\; \overline{\psi}(x)\,\Omega^{-1\dagger}(x), \tag{7.63}$$

where $\Omega$ is an element of some representation r of $\mathcal{G}$. By an abuse of language, it is often said that the spinor $\psi$ "lives" in the representation r (although strictly speaking, the spinor belongs to the space on which the elements of the group representation are acting).

Consider now the Dirac Lagrangian constructed with a covariant derivative,

$$\mathcal{L}_D = \overline{\psi}(x)\left(i\slashed{D}_x - m\right)\psi(x). \tag{7.64}$$

Under a gauge transformation, it becomes

$$\mathcal{L}_D \;\to\; \overline{\psi}(x)\,\underbrace{\Omega^{-1\dagger}(x)\Omega^{-1}(x)}_{(\Omega\Omega^\dagger)^{-1}}\left(i\slashed{D}_x - m\right)\psi(x). \tag{7.65}$$

For this Lagrangian to be gauge invariant, $\Omega$ must be a unitary matrix, which restricts to unitary representations of the gauge group (all finite dimensional representations of compact Lie groups are equivalent to a unitary representation).

As in electrodynamics, the necessity of using a covariant derivative in order to have a Dirac Lagrangian invariant under local gauge transformations completely specifies the coupling between the fermions and the gauge field $A_\mu$,

$$\mathcal{L}_I = -ig\,\overline{\psi}_i \gamma^\mu A^a_\mu \left(t^a_r\right)_{ij} \psi_j, \tag{7.66}$$

where we have written explicitly the Lie algebra indices $i,j$ of all the objects. These indices, that run from 1 to the size of the representation r, label the "charge" carried by the fermions, while the index $a$ may be viewed as labeling the charge carried by the spin-1 particle associated to the vector field $A^a_\mu$ (this index runs from 1 to the dimension of the group).

### 7.3.2 Standard Model

Two important interactions in nature are described by non-Abelian gauge theories:

- *Quantum chromodynamics*, the quantum field theory of strong interactions, is of this type: The gauge fields are the gluons, and the matter fields are the quarks, of which exist six families, or flavors (up, down, strange, charmed, bottom, top). The charge associated with this gauge interaction is called *color*. The gauge group of QCD is $SU(3)$, and the quarks live in the fundamental representation (therefore, they can have three different colors). In QCD, the gluons interact equally with the right-handed and left-handed projections of the spinors: It is said to be a *non-chiral* interaction.

- Likewise, the *electroweak theory* is a non-Abelian gauge theory with the gauge group $SU(2)\times U(1)$, but with the peculiarity that the $SU(2)$ acts only on the left-handed projection of the fermions. In other words, the right-handed fermions belong to the singlet representation of $SU(2)$ (while the left-handed fermions are arranged in doublets, corresponding to the fundamental representation of $SU(2)$).

It is also possible to couple a (charged) scalar field $\phi(x)$ to a gauge potential $A_\mu$. Under a local gauge transformation, $\phi$ transforms as follows:

$$\phi(x) \quad\to\quad \Omega^\dagger(x)\,\phi(x), \tag{7.67}$$

and therefore the following Lagrangian density is invariant under local gauge transformations:

$$\mathcal{L}_{\text{scalar}} = \big(D_\mu\phi(x)\big)^\dagger\big(D^\mu\phi(x)\big) - m^2\,\phi^\dagger(x)\phi(x) - V\big(\phi^\dagger(x)\phi(x)\big). \tag{7.68}$$

(The potential should depend on the scalar field via the combination $\phi^\dagger\phi$ in order to be gauge invariant.) The most important example of such a scalar field in particle physics is the *Higgs boson*. In the Standard Model, the potential of the Higgs field is invariant under the gauge transformations, but has minima at non-zero values of the field $\phi$, leading to spontaneous symmetry breaking. Because of its coupling to the gauge potentials and to the fermions, the Higgs field vacuum expectation value turns them into massive particles (see the next section for a discussion of this phenomenon).

### 7.3.3 Classical Equations of Motion

From the Lagrangians (7.62), (7.64) and (7.68), it is straightforward to obtain the classical Euler–Lagrange equations of motion. For the fermions, we simply obtain the Dirac equation $(i\slashed{D}-m)\,\psi(x)=0$. For scalar fields, the classical equation of motion is a deformation of the Klein–Gordon equation, in which the ordinary derivatives are replaced by covariant derivatives,

$$\left[D_\mu D^\mu + m^2 + V'\big(\phi^\dagger(x)\phi(x)\big)\right]\phi(x) = 0. \tag{7.69}$$

For the gauge field $A_\mu$, the derivatives of the various pieces of the Lagrangian read

$$\partial_\mu\frac{\partial\mathcal{L}_A}{\partial(\partial_\mu A^a_\nu)} = -\partial_\mu F^{a\,\mu\nu},\quad \frac{\partial\mathcal{L}_A}{\partial A^a_\nu} = g\,f^{abc}\,A^b_\mu\,F^{c\,\mu\nu},$$

$$\frac{\partial\mathcal{L}_D}{\partial A^a_\nu} = g\,\overline{\psi}\,\gamma^\nu\,t^a\,\psi,\quad \frac{\partial\mathcal{L}_{\text{scalar}}}{\partial A^a_\nu} = ig\Big(\phi^\dagger\,t^a\,(D_\nu\phi) - (D_\nu\phi)^\dagger\,t^a\,\phi\Big). \tag{7.70}$$

This leads to the following equation of motion:

$$\big[D_\mu,F^{\mu\nu}\big]_a = -J^\nu_a,$$
$$J^\nu_a = g\,\overline{\psi}\,\gamma^\nu\,t^a\,\psi + ig\Big(\phi^\dagger\,t^a\,(D_\nu\phi) - (D_\nu\phi)^\dagger\,t^a\,\phi\Big), \tag{7.71}$$

known as the *Yang–Mills equation*. From the Dirac and Klein–Gordon equations, one may check that the color current $J_a^\nu$ is *covariantly conserved*: $[D_\nu, J^\nu] = 0$. The field strength also obeys another equation, known as the *Bianchi identity*,

$$[D_\mu, F^{\nu\rho}] + [D_\nu, F^{\rho\mu}] + [D_\rho, F^{\mu\nu}] = 0, \tag{7.72}$$

which follows from the Jacobi identity between covariant derivatives.

### 7.3.4 Useful $\mathfrak{su}(N)$ Identities

Feynman graphs relevant for the Standard Model involve manipulations of the $\mathfrak{su}(N)$ generators (for $N = 2, 3$), mostly in the fundamental representation (since all matter fields are in this representation). In this section, we derive some useful formulas that help in these calculations.

**Fierz identity:** In the case of $\mathfrak{su}(N)$, there are $N^2 - 1$ generators $t_f^a$, while the linear space of all $N \times N$ Hermitian matrices has a dimension $N^2$. A basis of the latter can be obtained by adding the identity matrix to the $t_f^a$. Thus, any $N \times N$ Hermitian matrix M can be written as

$$M = m_0\,\mathbf{1} + m_a\,t_f^a. \tag{7.73}$$

Since the $t_f^a$s are traceless, we have

$$m_0 = \frac{1}{N}\,\mathrm{tr}\,(M), \quad m_a = 2\,\mathrm{tr}\,(M\,t^a). \tag{7.74}$$

Considering the entry ij of the matrix M, we can write

$$M_{ij} = \frac{1}{N} M_{kk}\,\delta_{ij} + 2\,M_{lk}\,(t_f^a)_{kl}(t_f^a)_{ij} = M_{lk}\left[\frac{1}{N}\delta_{kl}\delta_{ij} + 2\,(t_f^a)_{kl}(t_f^a)_{ij}\right]. \tag{7.75}$$

Since this is true for any Hermitian matrix M, we must have

$$\frac{1}{N}\delta_{kl}\delta_{ij} + 2\,(t_f^a)_{kl}(t_f^a)_{ij} = \delta_{il}\delta_{jk}, \tag{7.76}$$

which is usually written as

$$(t_f^a)_{ij}(t_f^a)_{kl} = \frac{1}{2}\left[\delta_{il}\delta_{jk} - \frac{1}{N}\delta_{ij}\delta_{kl}\right]. \tag{7.77}$$

This formula is called a *Fierz identity* (see Exercise 7.4 for similar identities in other cases). It has a convenient diagrammatic representation,

$$(t_f^a)_{ij}(t_f^a)_{kl} = \begin{array}{c}j \longleftarrow \bullet \longleftarrow i \\ \S \\ k \longrightarrow \bullet \longrightarrow l\end{array} = \frac{1}{2}\;\Big)\Big(\; - \frac{1}{2N}\;\underline{\quad\quad}\;, \tag{7.78}$$

in which the solid blobs represent the $t_f^a$ matrices and the wavy line indicates that the indices $a$ are contracted. On the right-hand side, the solid lines indicate how the indices ijkl are

## 7.3 NON-ABELIAN GAUGE THEORIES

connected by the delta symbols. By contracting the indices jk in the Fierz identity (7.77), we obtain

$$(t_f^a t_f^a)_{il} = \frac{N^2 - 1}{2N} \delta_{il}. \tag{7.79}$$

The quadratic combination $t_f^a t_f^a$, called the *fundamental Casimir operator*, is proportional to the identity (and therefore commutes with everything) – see also Exercises 7.9 and 7.10. The prefactor is sometimes denoted $C_f \equiv (N^2 - 1)/(2N)$.

The diagrammatic representation (7.78) provides a very convenient way of obtaining certain identities involving the generators of the fundamental representation. As an illustration, let us consider the following example:

$$t_f^a t_f^b t_f^a = \frac{1}{2} - \frac{1}{2N}$$
$$= \frac{1}{2} \operatorname{tr}(t_f^b) \mathbf{1} - \frac{1}{2N} t_f^b = -\frac{1}{2N} t_f^b. \tag{7.80}$$

For the first term, we have used the fact that a closed loop in this diagrammatic representation corresponds to a trace over the color indices, and the tracelessness the generators. Likewise, one would obtain

$$t_f^a t_f^b t_f^c t_f^a t_f^b = \frac{1}{4}\left(1 + \frac{1}{N^2}\right) t_f^c. \tag{7.81}$$

**More $\mathfrak{su}(N)$ formulas:** Contrary to the commutator, the anti-commutator of two matrices of the algebra does not belong to the algebra. However, in the case of $\mathfrak{su}(N)$, it can be decomposed as a linear combination of the identity and the generators of the algebra,

$$\{t_f^a, t_f^b\} = \frac{\delta^{ab}}{N} \mathbf{1} + d^{abc} t_f^c. \tag{7.82}$$

The first term is obtained by taking the trace of the equation, using eq. (7.28) and the fact that the generators are traceless. The constants $d^{abc}$ are sometimes called the *symmetric structure constants*. Therefore, the product of two generators of the fundamental representation can be written as

$$t_f^a t_f^b = \frac{1}{2}\left(\frac{\delta^{ab}}{N} \mathbf{1} + (d^{abc} + i f^{abc}) t_f^c\right). \tag{7.83}$$

From this, we deduce the following identities (see Exercise 7.3):

$$f^{acd} d^{bcd} = 0, \quad \operatorname{tr}\left(t_f^a t_f^b t_f^c\right) = \frac{1}{4}\left(d^{abc} + i f^{abc}\right),$$
$$\operatorname{tr}\left(t_f^a t_f^b t_f^a t_f^c\right) = -\frac{1}{4N} \delta_{bc},$$

$$f^{acd}f^{bcd} = N\,\delta_{ab}, \quad d^{acd}d^{bcd} = \left(N - \frac{4}{N}\right)\delta_{ab},$$

$$f^{ade}f^{bef}f^{cfd} = \frac{N}{2}\,f^{abc}. \tag{7.84}$$

Note that the third of these equations provides the trace of the product of two generators in the adjoint representation:

$$\mathrm{tr}\left(T^a_{\mathrm{adj}}T^b_{\mathrm{adj}}\right) = N\,\delta_{ab}. \tag{7.85}$$

## 7.4 Spontaneous Gauge Symmetry Breaking

### 7.4.1 Dirac Fermion Masses and Chiral Symmetry

Chiral gauge theories, i.e., gauge theories in which the left- and right-handed fermions belong to distinct representations, are rather special regarding the masses of these fermions. Let us recall that the left and right spinors are defined by

$$\psi_R \equiv \frac{1+\gamma^5}{2}\,\psi, \quad \psi_L \equiv \frac{1-\gamma^5}{2}\,\psi,$$

$$\overline{\psi}_R = \psi^\dagger\,\frac{1+\gamma^5}{2}\,\gamma^0, \quad \overline{\psi}_L = \psi^\dagger\,\frac{1-\gamma^5}{2}\,\gamma^0. \tag{7.86}$$

Consider first the term of the Dirac Lagrangian that does not depend on the mass. It can be decomposed as follows in terms of the left- and right-handed spinors:

$$\overline{\psi}\slashed{D}\psi = \sum_{\epsilon,\epsilon'=\pm}\psi^\dagger\,\frac{1+\epsilon\gamma^5}{2}\,\gamma^0\slashed{D}\,\frac{1+\epsilon'\gamma^5}{2}\,\psi$$

$$= \sum_{\epsilon=\epsilon'=\pm}\psi^\dagger\,\frac{1+\epsilon\gamma^5}{2}\,\gamma^0\slashed{D}\,\psi = \overline{\psi}_R\slashed{D}\psi_R + \overline{\psi}_L\slashed{D}\psi_L. \tag{7.87}$$

Therefore, this term does not mix the left and right spinors, and is invariant under independent gauge transformations of the two spinor helicities. In particular, it is perfectly possible that they belong to different representations of the Lie algebra.

In contrast, a Dirac mass term $m\,\overline{\psi}\psi$ has the following decomposition in terms of the left- and right-handed spinors,

$$m\,\overline{\psi}\psi = m\sum_{\epsilon,\epsilon'=\pm}\psi^\dagger\,\frac{1+\epsilon\gamma^5}{2}\,\gamma^0\,\frac{1+\epsilon'\gamma^5}{2}\,\psi$$

$$= \sum_{\epsilon=-\epsilon'=\pm}\psi^\dagger\,\frac{1+\epsilon\gamma^5}{2}\,\gamma^0\,\psi = m\,\overline{\psi}_R\psi_L + m\,\overline{\psi}_L\psi_R. \tag{7.88}$$

If $\psi_R$ and $\psi_L$ belong to different representations and transform independently under the gauge transformations, such a term is not gauge invariant and is therefore not allowed. Therefore, generically, fermions must be massless in a chiral gauge theory.

## 7.4 SPONTANEOUS GAUGE SYMMETRY BREAKING

The most prominent example of this situation is the Standard Model, in which the gauge group is $SU(3) \times SU(2) \times U(1)$, and the left- and right-handed fermions transform differently under the $SU(2) \times U(1)$ part. More precisely, the two chiral components have different charges (called hypercharge in this context) under $U(1)$, and the left-handed fermions form $SU(2)$ doublets while the right-handed ones are singlet under $SU(2)$. Thus, in such a gauge theory, fermions should naively be massless, while experimental evidence shows that they are massive.

### 7.4.2 Coupling to a Scalar Field, Yukawa Terms

Let us focus on the case in which the right-handed fermions are singlet under the gauge group, while the left-handed ones belong to a non-trivial representation. This means that they transform as follows:

$$\psi_R \to \psi_R, \quad \overline{\psi}_R \to \overline{\psi}_R, \quad \psi_L \to \Omega^{-1}\psi_L, \quad \overline{\psi}_L \to \overline{\psi}_L \Omega. \tag{7.89}$$

Thus, a way out to construct an operator which is bilinear in the fermions, mixes the left and right components and does not contain derivatives is to introduce a scalar field $\Phi$ that transforms in the same way as $\psi_L$:

$$\Phi \to \Omega^{-1}\Phi. \tag{7.90}$$

This operator, called a Yukawa term, reads

$$\lambda \left( \overline{\psi}_{L\,ri} \Phi_i \right) \psi_{R\,r}, \tag{7.91}$$

where $\lambda$ is a coupling constant, $r$ is a Dirac index and $i$ is the index that labels the components of the Lie algebra representation to which $\psi_L$ and $\Phi$ both belong. From the way the indices are contracted, this term is both gauge and Lorentz invariant. Note that since the contraction of the Dirac indices between the two spinors already produces a Lorentz invariant object, the field $\Phi$ must be Lorentz invariant on its own, and thus must be a scalar.

At this point, the term of eq. (7.91) is not yet a mass term, but simply a trilinear interaction term between fermions and the newly introduced scalar field. However, a mass term is generated if the vacuum expectation value $\langle \Phi_i \rangle$ is non-zero (as we shall see, this is related to the spontaneous breaking of the gauge symmetry). Therefore, we may redefine the scalar field by writing (for the sake of this example, we choose this expectation value to point in the direction $i = 1$)

$$\Phi \equiv \Phi_v + \varphi, \qquad \Phi_v \equiv \begin{pmatrix} v \\ 0 \\ \vdots \\ 0 \end{pmatrix}, \tag{7.92}$$

and the term (7.91) becomes

$$\lambda \left( \overline{\psi}_{L\,ri} \Phi_i \right) \psi_{R\,r} = \underbrace{\lambda v}_{m} \, \overline{\psi}_{L\,r1} \psi_{R\,r} + \lambda \left( \overline{\psi}_{L\,ri} \varphi_i \right) \psi_{R\,r}. \tag{7.93}$$

If the fermion in the right-handed singlet matches the first component of the left-handed multiplet, then the first term in the right-hand side is a Dirac mass term for this fermion (the fermions corresponding to the other components of the multiplet remain massless[13]). The second term in the right-hand side is a genuine interaction term between the fermions and the fluctuating part of the scalar. Interestingly, with this mechanism, the strength of this interaction is proportional to the mass of the fermion. In a theory in which several fermions acquire their masses by coupling to the expectation value of the same scalar field, this leads to a definite prediction: The ratios of the couplings must equal the ratios of the masses (but the masses themselves are not predicted, since the Yukawa couplings $\lambda$ are free parameters).

**Family mixing:** When there are several families of fermions (which we label by an extra index f in this paragraph), the Yukawa term of eq. (7.91) can be generalized to

$$\lambda_{ff'}\left(\overline{\psi}_{L\,fri}\Phi_i\right)\psi_{R\,f'r}, \tag{7.94}$$

without spoiling Lorentz or gauge invariance. Thus, with a non-zero vacuum expectation value of the scalar field, we get a fermion mass matrix that is in general not diagonal in the fermion families. Note that here, we are implicitly choosing a basis of fermion fields in which the couplings to the gauge bosons are diagonal, i.e., for which the vertex with one gauge boson and two fermions does not mix the fermion families. Conversely, we could choose a basis of fermion fields in which the mass matrix is diagonal. In this alternate basis, the interactions with the gauge bosons are no longer diagonal, i.e., the coupling to a gauge boson may change the type of fermion. These non-diagonal interactions are described by a matrix known as the *Cabbibo–Kobayashi–Maskawa matrix* in the sector of quarks.

### 7.4.3 Higgs Mechanism

Until now, we have not made explicit the mechanism by which the scalar field $\Phi$ may have a non-zero vacuum expectation value. The simplest gauge invariant Lagrangian that exhibits this phenomenon is

$$\mathcal{L}_{\text{scalar}} = \left(D_\mu\Phi(x)\right)^\dagger\left(D^\mu\Phi(x)\right) + m^2\,\Phi^\dagger(x)\Phi(x) - \frac{\lambda}{4}\left(\Phi^\dagger(x)\Phi(x)\right)^2. \tag{7.95}$$

Note the unusual sign of the mass term. Because of this feature, the value $\Phi = 0$ is a local maximum of the potential, and cannot be a stable field configuration. Instead, this potential has minima for $\Phi^\dagger\Phi = 2m^2/\lambda$, which corresponds to a gauge invariant "shell" of non-trivial minima. The general arguments developed in Chapter 4 also apply here: In an infinite volume, the system chooses as its ground state one of these minima (as opposed to a symmetric linear combination of all the minima).

Let us denote $\Phi_v$ the ground state on which the system settles. The gauge group $\mathcal{G}$ contains a subgroup $\mathcal{H}$ that leaves $\Phi_v$ invariant, called the *stabilizer* of $\Phi_v$, and the set of the minima

---

[13]In the Standard Model, where the left-handed fermions belong to the fundamental representation of $SU(2)$, it is possible to give a mass to the second component of the doublet. Indeed, by noting that $\Omega t^2 \Omega^T = t^2$ for any matrix $\Omega$ in the fundamental representation of $SU(2)$, we see that the term $i\lambda(\overline{\psi}_L t^2 \Phi^*)\psi_R$ is gauge invariant and gives a mass $\lambda v$ to the second component of the left-handed doublet when the vacuum expectation value of the scalar field has a non-zero first component.

## 7.4 SPONTANEOUS GAUGE SYMMETRY BREAKING

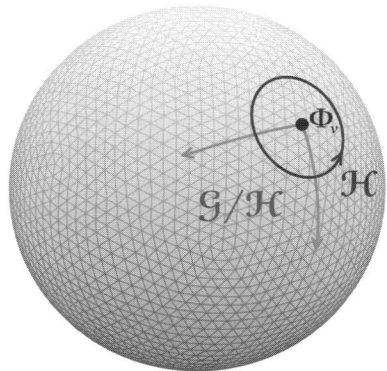

FIGURE 7.3: Illustration of the symmetry-breaking pattern in the case of a potential with $\mathcal{G} = O(3)$ symmetry. The set of minima of the potential is a two-dimensional sphere. The stabilizer of the minimum $\Phi_v$ is $\mathcal{H} = O(2)$. The dark circular arrow shows the action of the generator of $\mathfrak{h}$, while the lighter arrows show the action of the generators of the complementary set.

of the potential can be identified with the coset space[14] $\mathcal{G}/\mathcal{H}$ (see Figure 7.3). Then, the generators $t^a$ of the Lie algebra $\mathfrak{g}$ can be divided into two sets: a basis of $\mathfrak{h}$ (for $a > n$), and a complementary set (for $1 \leq a \leq n$):

$$\begin{aligned} 1 \leq a \leq n: &\quad t^a_{ij} \Phi_{vj} \neq 0, \\ a > n: &\quad t^a_{ij} \Phi_{vj} = 0. \end{aligned} \quad (7.96)$$

In the cases of interest in quantum field theory, the Lie algebra $\mathfrak{g}$ is a direct sum,

$$\mathfrak{g} = \mathfrak{h} \oplus \mathfrak{m}, \quad \text{with} \quad [\mathfrak{h}, \mathfrak{m}] \subset \mathfrak{m}, \qquad (7.97)$$

and the complementary set of generators is a basis of $\mathfrak{m}$. (This is called a reductive decomposition of $\mathfrak{g}$.) In this case, the tangent space to $\mathcal{G}/\mathcal{H}$ at $\Phi_v$ can be identified with $\mathfrak{m}$, via the following mapping

$$X \in \mathfrak{m} \to e^{itX} \mathcal{H} \in \mathcal{G}/\mathcal{H}. \qquad (7.98)$$

Thus, $\mathcal{G}/\mathcal{H}$ is obtained by exponentiation of the elements of $\mathfrak{m}$, and any configuration of the scalar field may be parametrized as follows:

$$\Phi(x) \equiv \underbrace{\exp\left(i \sum_{a=1}^{n} \vartheta_a(x) t^a\right)}_{\Omega^{-1}(x)} (\Phi_v + \mathbf{r}(x)). \qquad (7.99)$$

---

[14]Given a group $\mathcal{G}$ and $\mathcal{H}$ one of its subgroups, two elements $\Omega$ and $\Omega'$ are said to be $\mathcal{H}$-equivalent if $\Omega^{-1}\Omega' \in \mathcal{H}$. The quotient of the group $\mathcal{G}$ by this equivalence relationship, also called coset space and denoted $\mathcal{G}/\mathcal{H}$, is the set of the resulting equivalence classes. When $\mathcal{H}$ is a normal subgroup (i.e., $\Omega\mathcal{H}\Omega^{-1} = \mathcal{H}$ for all $\Omega \in \mathcal{G}$), the coset space is itself a group.

In this representation, $\mathbf{r}(x)$ denotes the "radial" field variables, while the $\vartheta_a(x)$ are the "angular" ones. From eq. (7.96), the latter correspond to the generators broken by the spontaneous symmetry breaking. Therefore, if it were not for the coupling of $\Phi$ to the gauge fields through the covariant derivatives, we would conclude from Goldstone's theorem that the modes $\mathbf{r}(x)$ are massive while the modes $\vartheta_a(x)$ are massless Nambu–Goldstone bosons. However, this conclusion is altered by the minimal coupling to a gauge field because it is possible to absorb the matrix $\Omega^{-1}(x)$, which contains the would-be Nambu–Goldstone modes, into a gauge transformation of that field. Indeed, we may write

$$D_\mu \Phi = D_\mu \Omega^{-1}(\Phi_v + \mathbf{r}) = \Omega^{-1} \underbrace{\Omega D_\mu \Omega^{-1}}_{D'_\mu}(\Phi_v + \mathbf{r}), \tag{7.100}$$

with

$$D'_\mu \equiv \partial_\mu - igA'_\mu, \quad A'_\mu \equiv \Omega A_\mu \Omega^{-1} + \frac{i}{g}\Omega \partial_\mu \Omega^{-1}. \tag{7.101}$$

We see that after this gauge transformation of $A_\mu$ (this choice of gauge is known as the unitary gauge), only the modes $\mathbf{r}(x)$ can still be considered as physical dynamical modes of the scalar field, and the kinetic term of the scalar Lagrangian can thus be rewritten as

$$\begin{aligned}
(D^\mu \Phi)^\dagger (D_\mu \Phi) &= \left(D'^\mu(\Phi_v + \mathbf{r})\right)^\dagger \left(D'_\mu(\Phi_v + \mathbf{r})\right) \\
&= (D'^\mu \mathbf{r})^\dagger (D'_\mu \mathbf{r}) \\
&\quad + (-igA'^\mu \Phi_v)^\dagger (D'_\mu \mathbf{r}) + (D'^\mu \mathbf{r})^\dagger (-igA'_\mu \Phi_v) \\
&\quad + \underbrace{(-igA'^\mu \Phi_v)^\dagger (-igA'_\mu \Phi_v)}_{\frac{1}{2}M_{ab} A'^a_\mu A'^{b\mu}}.
\end{aligned} \tag{7.102}$$

In this expression, the last term is particularly interesting, since it provides a mass for some of the gauge bosons. More explicitly, the mass matrix is given by

$$M_{ab} \equiv 2g^2 \, \Phi^\dagger_{vi} t^a_{ik} t^b_{kj} \Phi_{vj}. \tag{7.103}$$

Note that since

$$\frac{1}{2} M_{ab} X^a X^b = g^2 \sum_k \left| X^b t^b_{kj} \Phi_{vj} \right|^2 \geq 0, \tag{7.104}$$

this mass matrix is positive and has a number of flat directions equal to the number of generators $t^a$ that annihilate $\Phi_v$. From this, we conclude that the gauge bosons that become massive via this mechanism are those that couple to the generators of the broken symmetries.

## 7.5 θ-term and Strong-CP Problem

### 7.5.1 CP-Odd Gauge Invariant Operator

In the construction of the Lagrangian of Yang–Mills theory, we have argued that the only dimension-four gauge invariant local operator is an operator quadratic in the field strength

$F_a^{\mu\nu}$. All the Lorentz indices should be contracted in order to obtain a Lorentz invariant Lagrangian density. An obvious possibility is $F_a^{\mu\nu} F_{\mu\nu}^a$, which is the combination that appears in the Yang–Mills action. However, there exists another Lorentz invariant contraction, obtained by introducing the Levi-Civita symbol:

$$\mathcal{L}_\theta \equiv \frac{g^2 \theta}{32\pi^2} \epsilon_{\mu\nu\rho\sigma} \, \mathrm{tr}\,(F^{\mu\nu} F^{\rho\sigma}). \tag{7.105}$$

The prefactor $1/32\pi^2$ will appear convenient later, and the coupling constant in front of this term is usually denoted $\theta$. Consequently, this term is called the $\theta$-*term*.

### 7.5.2 Expression as a Total Derivative

First, we should clarify why we have not considered this term right away when we listed the possible gauge invariant operators that may enter in a non-Abelian gauge theory. As we shall prove now, the $\theta$-term is a total derivative. Therefore, it does not enter in the field equations of motion, and has also no influence on perturbation theory. Since our discussion has been so far centered on the perturbative expansion, this term was irrelevant. However, the $\theta$-term – that we cannot exclude on the grounds of symmetries – may lead to non-perturbative effects that we shall discuss in this section.

Let us consider the following vector:[15]

$$K^\mu \equiv \epsilon^{\mu\nu\rho\sigma} \left[ A_\nu^a F_{\rho\sigma}^a - \frac{g}{3} f^{abc} A_\nu^a A_\rho^b A_\sigma^c \right]. \tag{7.106}$$

The divergence of this vector is given by

$$\begin{aligned}\partial_\mu K^\mu &= \epsilon^{\mu\nu\rho\sigma} \Big[ (\partial_\mu A_\nu^a)(\partial_\rho A_\sigma^a - \partial_\sigma A_\rho^a + g f^{abc} A_\rho^b A_\sigma^c) \\
&\quad + A_\nu^a (\partial_\mu \partial_\rho A_\sigma^a - \partial_\mu \partial_\sigma A_\rho^a + g f^{abc} (\partial_\mu A_\rho^b) A_\sigma^c + g f^{abc} A_\rho^b (\partial_\mu A_\sigma^c)) \\
&\quad - \tfrac{g}{3} f^{abc}(\partial_\mu A_\nu^a) A_\rho^b A_\sigma^c - \tfrac{g}{3} f^{abc} A_\nu^a (\partial_\mu A_\rho^b) A_\sigma^c - \tfrac{g}{3} f^{abc} A_\nu^a A_\rho^b (\partial_\mu A_\sigma^c) \Big] \\
&= \tfrac{1}{2} \epsilon^{\mu\nu\rho\sigma} \Big[ F_{\mu\nu}^a F_{\rho\sigma}^a - g^2 f^{abc} f^{ade} A_\mu^b A_\nu^c A_\rho^d A_\sigma^e \\
&\quad + \tfrac{g}{3} f^{abc} \big( A_\mu^b A_\nu^c (\partial_\rho A_\sigma^a - \partial_\sigma A_\rho^a) - A_\rho^b A_\sigma^c (\partial_\mu A_\nu^a - \partial_\nu A_\mu^a) \big) \Big]. \end{aligned} \tag{7.107}$$

The two terms of the last line are antisymmetric under the exchange $(\mu\nu) \leftrightarrow (\rho\sigma)$, while the prefactor $\epsilon^{\mu\nu\rho\sigma}$ is symmetric under this exchange. These terms are therefore zero after summing over the indices $\nu\rho\sigma\mu$. Then, the second term on the first line can be written as follows:

$$\begin{aligned}&g^2 \epsilon^{\mu\nu\rho\sigma} \, \mathrm{tr}\,([A_\mu, A_\nu][A_\rho, A_\sigma]) \\
&= g^2 \epsilon^{\mu\nu\rho\sigma} \, \mathrm{tr}\,\big( A_\mu A_\nu A_\rho A_\sigma + A_\nu A_\mu A_\sigma A_\rho - A_\nu A_\mu A_\rho A_\sigma - A_\mu A_\nu A_\sigma A_\rho \big). \end{aligned} \tag{7.108}$$

---

[15] Note that this vector can also be expressed as a trace:
$$K^\mu \equiv 2\epsilon^{\mu\nu\rho\sigma} \, \mathrm{tr}\,\left[ A_\nu F_{\rho\sigma} + \tfrac{2ig}{3} A_\nu A_\rho A_\sigma \right].$$

Each term is a trace of four factors, and is invariant under cyclic permutations of the indices. Since cyclic permutations are odd in four dimensions, $\epsilon^{\mu\nu\rho\sigma}$ changes sign under such a permutation, and the contraction with the trace is zero. Therefore, we obtain

$$\partial_\mu K^\mu = \tfrac{1}{2}\epsilon^{\mu\nu\rho\sigma} F^a_{\mu\nu} F^a_{\rho\sigma}, \tag{7.109}$$

which is proportional to the θ-term. More precisely, we have

$$\mathcal{L}_\theta = \frac{g^2 \theta}{32\pi^2} \partial_\mu K^\mu. \tag{7.110}$$

### 7.5.3 Proof in Terms of Differential Forms

The elementary proof of this result that we have presented in the previous subsection is arguably a bit cumbersome. This could have been made much more compact by using the language of *differential forms*. The simplest differential forms are 1-forms, which one may think of as the contraction of a space-time dependent vector $a_\mu(x)$ and of the differential element $dx^\mu$, as in

$$\mathbf{A} \equiv a_\mu(x)\, dx^\mu. \tag{7.111}$$

1-forms measure the variation of a function along an infinitesimal one-dimensional path (thus, 1-forms may be integrated along a path γ, yielding a number).

Higher-degree forms may be constructed thanks to the *exterior product*, denoted ∧. A basis of 2-forms is provided by the products $dx^\mu \wedge dx^\nu$, which are the areas of the infinitesimal quadrangles of edges $(dx^\mu, dx^\nu, -dx^\mu, -dx^\nu)$. The exterior product is defined to be antisymmetric,

$$dx^\mu \wedge dx^\nu = -dx^\nu \wedge dx^\mu, \tag{7.112}$$

which corresponds to a definition of oriented areas, depending on the order in which the edges of the quadrangle are traveled:

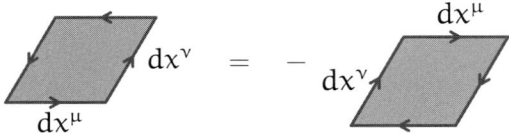

This also naturally implies that $dx^\mu \wedge dx^\mu = 0$, in accordance with the fact that a quadrangle of edges $(dx^\mu, dx^\mu, -dx^\mu, -dx^\mu)$ is reduced to a line segment and thus has zero area. The most general 2-form can be written as

$$\mathbf{F} \equiv f_{\mu\nu}(x)\, dx^\mu \wedge dx^\nu, \tag{7.113}$$

where $f_{\mu\nu}$ is antisymmetric under the exchange of the μ, ν indices. 2-forms can be integrated over a two-dimensional manifold to give a number. In d-dimensional space, one may iteratively construct p-forms for any $p \le d$ (higher degree forms are zero by antisymmetry of the

## 7.5 θ-TERM AND STRONG-CP PROBLEM

exterior product). In particular, in four dimensions, the volume element weighted by the fully antisymmetric tensor $\epsilon^{\mu\nu\rho\sigma}$ can be written as

$$d^4x\, \epsilon^{\mu\nu\rho\sigma} = dx^\mu \wedge dx^\nu \wedge dx^\rho \wedge dx^\sigma, \tag{7.114}$$

which allows the following compact notation:

$$\int d^4x\, \epsilon^{\mu\nu\rho\sigma} A_\mu A_\nu A_\rho A_\sigma = \int \mathbf{A} \wedge \mathbf{A} \wedge \mathbf{A} \wedge \mathbf{A}. \tag{7.115}$$

Here, it is important to note that $\mathbf{A} \wedge \mathbf{A} \neq 0$ when $A_\mu$ belongs to a non-Abelian Lie algebra, despite the antisymmetry of the exterior product.

Another important operation on differential forms is the *exterior derivative* d, defined as

$$d\omega \equiv dx^\mu \wedge \partial_\mu \omega. \tag{7.116}$$

Thus, the exterior derivative of a p-form is a $(p+1)$-form. For instance, given a 1-form $\mathbf{A} = a_\mu\, dx^\mu$, we have

$$d\mathbf{A} = dx^\mu \wedge \partial_\mu (a_\nu\, dx^\nu) = \tfrac{1}{2}(\partial_\mu a_\nu - \partial_\nu a_\mu)\, dx^\mu \wedge dx^\nu. \tag{7.117}$$

Note that, since ordinary derivatives commute, we have[16]

$$d^2 \omega = 0. \tag{7.118}$$

When applying the exterior derivative to the exterior product of two forms, one should distribute the partial derivative on the two factors, and account for the fact that the exterior derivative contains an anti-commuting $dx^\mu$. Thus, if $\mathbf{A}$ is a p-form, we have

$$d(\mathbf{A} \wedge \mathbf{B}) = d\mathbf{A} \wedge \mathbf{B} + (-1)^p\, \mathbf{A} \wedge d\mathbf{B}. \tag{7.119}$$

Differential forms also provide a unified version of various formulas of vector calculus (e.g., Kelvin–Stokes and Ostrogradsky–Gauss theorems), known as the Stokes theorem. Given a form $\omega$ and a manifold $\mathcal{M}$, the Stokes theorem states that

$$\int_{\partial \mathcal{M}} \omega = \int_{\mathcal{M}} d\omega, \tag{7.120}$$

where $\partial \mathcal{M}$ is the boundary of $\mathcal{M}$.

In order to cast the θ-term in the language of differential forms, let us first introduce the gauge potential 1-form,

$$\mathbf{A} \equiv ig\, A_\mu\, dx^\mu. \tag{7.121}$$

Then, we have

$$d\mathbf{A} = \frac{ig}{2}(\partial_\mu A_\nu - \partial_\nu A_\mu)\, dx^\mu \wedge dx^\nu, \quad \mathbf{A} \wedge \mathbf{A} = -\frac{g^2}{2}[A_\mu, A_\nu]\, dx^\mu \wedge dx^\nu, \tag{7.122}$$

---

[16] A differential form $\omega$ whose exterior derivative is zero ($d\omega = 0$) is said to be *closed*. A differential form $\chi$ which is the exterior derivative of another form ($\chi = d\omega$) is said to be *exact*.

and we see that the field strength $F_{\mu\nu}$ appears in the coefficients of the following 2-form:

$$\mathbf{F} \equiv d\mathbf{A} - \mathbf{A} \wedge \mathbf{A} = \frac{ig}{2} \big( \underbrace{\partial_\mu A_\nu - \partial_\nu A_\mu - ig\,[A_\mu, A_\nu]}_{F_{\mu\nu}} \big) dx^\mu \wedge dx^\nu. \tag{7.123}$$

Therefore, the integrand of the $\theta$-term may be written compactly as

$$d^4x\, \epsilon^{\mu\nu\rho\sigma}\, \mathrm{tr}\left(F_{\mu\nu} F_{\rho\sigma}\right) = -\frac{4}{g^2}\, \mathrm{tr}\left(\mathbf{F} \wedge \mathbf{F}\right). \tag{7.124}$$

Likewise, we have

$$d^4x\, K^\mu = -\frac{4}{g^2}\, dx^\mu \wedge \mathrm{tr}\left(\mathbf{A} \wedge \mathbf{F} + \frac{1}{3}\mathbf{A} \wedge \mathbf{A} \wedge \mathbf{A}\right). \tag{7.125}$$

Then, note that

$$d\left\{\mathrm{tr}\left(\mathbf{A} \wedge \mathbf{F} + \frac{1}{3}\mathbf{A} \wedge \mathbf{A} \wedge \mathbf{A}\right)\right\} = \mathrm{tr}\left\{d\mathbf{A} \wedge \mathbf{F} - \mathbf{A} \wedge d\mathbf{F} + (d\mathbf{A}) \wedge \mathbf{A} \wedge \mathbf{A}\right\}$$
$$= \mathrm{tr}\left\{d\mathbf{A} \wedge d\mathbf{A} - 2(d\mathbf{A}) \wedge \mathbf{A} \wedge \mathbf{A}\right\} = \mathrm{tr}\left(\mathbf{F} \wedge \mathbf{F}\right), \tag{7.126}$$

where we have used the cyclicity of the trace and the fact that commuting $d\mathbf{A}$ with other forms does not bring any sign since it is a 2-form. In order to obtain the last line, we have used

$$\mathrm{tr}\left(\mathbf{A} \wedge \mathbf{A} \wedge \mathbf{A} \wedge \mathbf{A}\right) = 0, \tag{7.127}$$

which is a consequence of the fact that a cyclic permutation of four objects is odd. Eq. (7.126) is the translation in terms of forms of the fact that the $\theta$-term is the derivative of the vector $K^\mu$. Thanks to the Stokes theorem, the integral of the $\theta$-term over a four-dimensional manifold $\mathcal{M}$ can be rewritten as an integral over its boundary (located at infinity if $\mathcal{M}$ is the entire $\mathbb{R}^4$),

$$\int_{\mathcal{M}} \mathrm{tr}\left(\mathbf{F} \wedge \mathbf{F}\right) = \int_{\partial\mathcal{M}} \mathrm{tr}\left(\mathbf{A} \wedge \mathbf{F} + \frac{1}{3}\mathbf{A} \wedge \mathbf{A} \wedge \mathbf{A}\right). \tag{7.128}$$

### 7.5.4 Effect of the $\theta$-Term on the Euclidean Path Integral

We have already encountered the integral of the $\theta$-term over Euclidean space-time in the context of anomalies and the Atiyah–Singer index theorem (see eq. (6.116)),

$$\int d^4x_{_E}\, \mathcal{L}_\theta = n\theta, \qquad n \in \mathbb{Z}, \tag{7.129}$$

where the integer $n$ is related to the chirality of the zero modes of the Dirac operator in the gauge field configuration. When added to the Yang–Mills action, the integral of the $\theta$-term modifies the Euclidean path integral as follows:

$$\int [DA_\mu \cdots]\, e^{-S[A,\cdots]} \to \int [DA_\mu \cdots]\, e^{-S[A,\cdots] - \int d^4x_{_E}\, \mathcal{L}_\theta}$$
$$= \sum_{n\in\mathbb{Z}} e^{-n\theta} \int [DA_\mu \cdots]_n\, e^{-S[A,\cdots]}, \tag{7.130}$$

## 7.5 θ-TERM AND STRONG-CP PROBLEM

where the measure $[DA_\mu]_n$ is restricted to the gauge fields of index $n$. Thus, the effect of the θ-term is to reweight the gauge field configurations by a factor $(e^{-\theta})^n$ that depends only on θ and on the index $n$. Note that since $n$ is an integer, the Euclidean path integral is periodic in θ with a period $2i\pi$.

### 7.5.5 Strong CP-Problem

As we saw in Section 6.5.6, an effective description of the interactions of nucleons with pions is provided by the linear σ model, whose interaction term is

$$\mathcal{L}_I \equiv \lambda \bar{\psi}(\sigma + i\boldsymbol{\pi} \cdot \boldsymbol{\sigma}\gamma^5)\psi. \tag{7.131}$$

However, this does not include any CP-violating interactions, such as those that may result from the θ-term. Its effects may be included in the effective theory by generalizing the interaction term into

$$\mathcal{L}_I \equiv \bar{\psi}(\lambda\sigma + \boldsymbol{\pi} \cdot \boldsymbol{\sigma}(i\lambda\gamma^5 + \bar{\lambda}))\psi. \tag{7.132}$$

By matching with the underlying theory, the new coupling $\bar{\lambda}$ can be related to the parameter θ by the following estimate (see Crewther, DiVecchia, Veneziano, Witten, Phys. Lett. B88 (1979) 123):

$$|\bar{\lambda}| \approx 0.038\,|\theta|. \tag{7.133}$$

Then, the effective theory (7.132) can be used to estimate the neutron electric dipole moment $D_N$ in the chiral limit where the pion mass $m_\pi$ is much smaller than the nucleon mass $m_N$. This leads to

$$D_N \approx \lambda\bar{\lambda}\,e\,\frac{\ln\left(\frac{m_N}{m_\pi}\right)}{4\pi^2 m_N} \approx 5 \times 10^{-16}\,\theta\,\,e\cdot\text{cm}. \tag{7.134}$$

Current experimental limits on the neutron electric dipole moment indicate that $|D_N| \leq 3 \times 10^{-26}$ e·cm, implying that $|\theta| \lesssim 10^{-10}$. We thus face a rather paradoxical situation. The gauge symmetry of quantum chromodynamics allows the addition of the θ-term to the Yang–Mills action, and without any prior knowledge of the coupling θ, one may expect that natural values are of order unity. This constitutes the *strong-CP problem*: Lacking a symmetry principle that would force θ to be zero, why is it nevertheless extremely small?

### 7.5.6 θ-Term and Quark Masses

There is an interesting interplay between the θ-term and chiral transformations of quark fields,

$$\psi_f \longrightarrow e^{i\gamma_5 \alpha_f}\psi_f, \tag{7.135}$$

where f is an index labeling the quark flavors and the $\alpha_f$ are real phases. Under this transformation, the functional measure for the quarks is not invariant, but transforms as follows (see eq. (6.93) in Section 6.5):

$$[D\psi D\bar{\psi}] \longrightarrow \exp\left(-\frac{i}{32\pi^2}\int d^4x\, \epsilon^{\mu\nu\rho\sigma} F^a_{\mu\nu} F^a_{\rho\sigma} \sum_f \alpha_f\right) [D\psi D\bar{\psi}]. \tag{7.136}$$

The same effect would have been obtained by a change of the angle $\theta$,

$$\theta \to \theta - 2\sum_f \alpha_f. \tag{7.137}$$

For the quarks, we can write generically the following mass term:[17]

$$\sum_f M_f\, \bar{\psi}_f \frac{1+\gamma_5}{2}\psi_f + \sum_f M_f^*\, \bar{\psi}_f \frac{1-\gamma_5}{2}\psi_f, \tag{7.138}$$

which transforms into the following under the above chiral transformation:

$$\sum_f e^{2i\alpha_f} M_f\, \bar{\psi}_f \frac{1+\gamma_5}{2}\psi_f + \sum_f e^{-2i\alpha_f} M_f^*\, \bar{\psi}_f \frac{1-\gamma_5}{2}\psi_f. \tag{7.139}$$

This is equivalent to transforming the quark masses according to $M_f \to e^{2i\alpha_f} M_f$. Since any change of $\theta$ can be absorbed by a chiral transformation of the quarks, whose effect is to multiply the quark masses by phases, physical quantities cannot depend separately on $\theta$ and on the quark masses. Instead, they can depend only on the combination $e^{i\theta}\prod_f M_f$, which is invariant. This discussion indicates that the $\theta$-term has no effect if at least one of the quarks is massless. Unfortunately, a massless up quark (the lightest quark) does not seem consistent with existing experimental evidence and lattice chromodynamics computations.

### 7.5.7 Link with the Topology of Gauge Fields

Using the Stokes theorem, the integral of the $\theta$-term over Euclidean space-time may be rewritten as an integral over a surface localized at infinity,

$$\int d^4x_E\, \mathcal{L}_\theta = \frac{g^2\theta}{32\pi^2}\int d^4x_E\, \partial_\mu K^\mu = \frac{g^2\theta}{32\pi^2} \lim_{R\to\infty} \int_{\mathcal{S}_{3,R}} dS_\mu\, K^\mu, \tag{7.140}$$

where $\mathcal{S}_{3,R}$ is a three-dimensional sphere of radius R and $dS_\mu$ the measure on this surface.

Let us now assume that the colored objects of the problem are comprised in a finite region of space-time, so that the gauge field configuration goes to a pure gauge at infinity. Such a field can be written as

$$A_\mu(x) = a_\mu(x) + \frac{i}{g}\Omega^\dagger(\hat{x})\, \partial_\mu\, \Omega(\hat{x}), \tag{7.141}$$

---

[17]If the masses are complex, then the symmetries P and CP are explicitly broken.

where $\Omega(\hat{x})$ is an element of the gauge group that depends only on the direction of the vector $x^\mu$, and $a_\mu(x)$ is the deviation from the asymptotic pure gauge. For the total field to be a pure gauge at infinity, this deviation must decrease faster than $|x|^{-1}$. When $|x| \to +\infty$, $A_\nu(x)$ goes to 0 as $|x|^{-1}$, while $F_{\rho\sigma}(x)$ goes to 0 *faster* than $|x|^{-2}$ (since $A_\nu(x)$ goes to a pure gauge), and we have

$$K^\mu \xrightarrow[|x|\to+\infty]{} \frac{4ig}{3} \epsilon^{\mu\nu\rho\sigma} \operatorname{tr}(A_\nu A_\rho A_\sigma) \sim |x|^{-3} \tag{7.142}$$

and

$$\int d^4x_E \, \mathcal{L}_\theta = \frac{\theta}{24\pi^2} \lim_{R\to\infty} \int_{S_{3,R}} dS \, \hat{x}_\mu \, \epsilon^{\mu\nu\rho\sigma}$$
$$\times \operatorname{tr}\left(\Omega^\dagger(\partial_\nu\Omega)\Omega^\dagger(\partial_\rho\Omega)\Omega^\dagger(\partial_\sigma\Omega)\right), \tag{7.143}$$

where we have used $dS_\mu = \hat{x}_\mu \, dS$, with $dS$ the element of area on the 3-sphere. Note that the integrand decreases as $R^{-3}$ because of the three derivatives, while $dS \sim R^3$. Therefore, the integral is in fact independent of the radius $R$ and we can drop the limit,

$$\int d^4x_E \, \mathcal{L}_\theta = \frac{\theta}{24\pi^2} \int_{S_3} dS \, \hat{x}_\mu \, \epsilon^{\mu\nu\rho\sigma} \operatorname{tr}\left(\Omega^\dagger(\partial_\nu\Omega)\Omega^\dagger(\partial_\rho\Omega)\Omega^\dagger(\partial_\sigma\Omega)\right). \tag{7.144}$$

Thus, the integral of the $\theta$-term depends only on the function $\Omega(\hat{x})$ that maps the three-dimensional sphere $S_3$ onto the gauge group,

$$\Omega : \quad S_3 \quad \longmapsto \quad \mathcal{G}. \tag{7.145}$$

It turns out that these mappings can be grouped in equivalence classes of the functions $\Omega$ that can be deformed continuously into one another. On the contrary, the functions $\Omega$ that belong to distinct classes cannot be related by a continuous deformation. The set of these classes possesses a group structure, and is called the *third homotopy group* of $\mathcal{G}$, denoted $\pi_3(\mathcal{G})$. For all $SU(N)$ groups with $N \geq 2$, the third homotopy group is isomorphic to $(\mathbb{Z}, +)$. The interpretation of eq. (7.144) is that the integral of the $\theta$-term depends only on the class to which $\Omega$ belongs, and is therefore a topological quantity that can change only in discrete amounts. This discussion provides another point of view on the Atiyah–Singer index theorem, where the same integral was related to the chirality imbalance between the zero modes of the Euclidean Dirac operator in a background gauge field.

## 7.6 Non-Local Gauge Invariant Operators

### 7.6.1 Two-Fermion Non-Local Operator

The discussion in the previous sections exhausts the *local* gauge invariant objects of dimension less than or equal to four. However, it is sometimes useful to construct gauge invariant non-local operators, for instance in the definition of *parton distributions*. The simplest operator of this type is an operator with two spinor fields at different space-time positions, $\overline{\psi}(y)\,W(y,x)\,\psi(x)$. Since the transformation laws of the two spinors involve different functions $\Omega$, such an operator is gauge invariant only if the object $W(y,x)$ between the spinors transforms as follows:

$$W(y,x) \quad\to\quad \Omega^\dagger(y)\,W(y,x)\,\Omega(x). \tag{7.146}$$

### 7.6.2 Wilson Lines

In order to construct such an object, let us define a path $\gamma^\mu(s)$ that goes from x to y,

$$\gamma^\mu(0) = x^\mu, \quad \gamma^\mu(1) = y^\mu, \tag{7.147}$$

and consider the following differential equation:

$$\frac{dW}{ds} \equiv \frac{d\gamma^\mu}{ds}\Big(D_\mu(\gamma(s))\,W\Big) = 0, \quad \text{with initial condition } W(0) = 1, \tag{7.148}$$

where the notation $D_\mu(\gamma(s))$ indicates that the gauge field in the covariant derivative must be evaluated at the point $\gamma^\mu(s)$. In other words, the covariant derivative of $W$, projected along the tangent vector to the path $\gamma^\mu(s)$, is zero. From this definition, it follows that $W(s)$ is an element of the representation r of the gauge group if $A_\mu$ is in the representation r of the algebra.

Note that when the gauge field $A_\mu$ is zero everywhere, then the solution is trivially $W(s) = 1$. For a generic gauge field, the value of the solution[18] at $s = 1$ is a property of the path $\gamma^\mu$ and of the gauge potential $A^\mu$. This object, that we will denote as

$$\mathcal{W}_{yx}[A;\gamma] \equiv W(1), \tag{7.149}$$

is called a *Wilson line*. Let us now study how it changes under a gauge transformation $\Omega$. From the transformation law of the covariant derivative, the differential equation that defines the transformed $W_\Omega(s)$ is

$$\frac{d\gamma^\mu}{ds}\Omega^\dagger(\gamma(s))D_\mu(\gamma(s))\Omega(\gamma(s))W_\Omega(s) = 0, \quad \text{with initial condition } W_\Omega(0) = 1. \tag{7.150}$$

If we define $Z(s) \equiv \Omega(\gamma(s))W_\Omega(s)$, this equation is equivalent to

$$\frac{d\gamma^\mu}{ds}D_\mu(\gamma(s))\,Z(s) = 0, \quad \text{with initial condition } Z(0) = \Omega(x). \tag{7.151}$$

Comparing this equation with the original equation (7.148), we obtain

$$Z(s) = W(s)\,\Omega(x), \quad \text{i.e.,} \quad W_\Omega(s) = \Omega^\dagger(\gamma(s))\,W(s)\,\Omega(x). \tag{7.152}$$

---

[18] Note that if the initial condition is $W(0) = \Omega_0$ instead of 1, then the solution would be changed as follows: $W(s) \to W(s)\Omega_0$.

## 7.6 NON-LOCAL GAUGE INVARIANT OPERATORS

Looking now at the point $s = 1$, we see that the Wilson line transforms as

$$\mathcal{W}_{yx}[A;\gamma] \quad \rightarrow \quad \Omega^\dagger(y)\, \mathcal{W}_{yx}[A;\gamma]\, \Omega(x). \tag{7.153}$$

Thus, the Wilson line transforms precisely as we wanted in eq. (7.146), and we conclude that the operator $\overline{\psi}(y)\mathcal{W}_{yx}[A;\gamma]\psi(x)$ is gauge invariant. Note that the Wilson line $\mathcal{W}_{yx}[A;\gamma]$, solution of eq. (7.148) at $s = 1$, can also be written as a *path-ordered exponential*,

$$\mathcal{W}_{yx}[A;\gamma] = \mathrm{P} \exp\left(ig \int_\gamma dx^\mu\, A_\mu(x)\right). \tag{7.154}$$

Although this compact notation is suggestive, it is often useful to return to the defining differential equation (7.148).

### 7.6.3 Path Dependence

By inserting a Wilson line between points $x$ and $y$, we can construct a gauge invariant non-local operator $\overline{\psi}(y)\cdots\psi(x)$. However, in doing so, we have introduced a path $\gamma$, for which there are infinitely many possible choices since only the endpoints are fixed. It turns out that, in general, the Wilson line depends on the path $\gamma$, i.e.,

$$\mathcal{W}_{yx}[A;\gamma] \neq \mathcal{W}_{yx}[A;\gamma']. \tag{7.155}$$

This implies that, although we may define gauge invariant non-local bilinear operators, their definition is not unique and each choice of the path connecting the two points leads to a different operator.

### 7.6.4 Case of Pure Gauge Fields

When the gauge potential is a pure gauge field, there exists a function $\Omega(x)$ such that

$$A_\mu^\Omega(x) = \frac{i}{g}\, \Omega^\dagger(x)\, \partial_\mu \Omega(x). \tag{7.156}$$

Since this field is a gauge transformation of the null field $A_\mu \equiv 0$, Wilson lines in this pure gauge field are given by

$$\mathcal{W}_{yx}[A^\Omega;\gamma] = \Omega^\dagger(y)\, \Omega(x). \tag{7.157}$$

In other words, in a pure gauge field, the Wilson lines depend only on their endpoints, but not on the path chosen to connect them. This is the only exception to the remark in the previous paragraph.

Conversely, a gauge potential $A^\mu(x)$ in which the Wilson lines depend only on the endpoints is a pure gauge. Indeed, the function $\Omega(x)$ that gives this gauge potential via eq. (7.156) can be constructed as a Wilson line from $x$ to some arbitrary base point $x_0$:

$$\Omega(x) = \mathcal{W}_{x_0 x}[A;\gamma]. \tag{7.158}$$

(The path $\gamma$ can be chosen arbitrarily.)

## 7.6.5 Wilson Loops

A *Wilson loop* is a special kind of Wilson line in which the initial point and endpoint are identical, $x = y$, and therefore the path $\gamma$ is a closed loop,

$$W[A;\gamma] = P \exp\left(ig \oint_\gamma dx^\mu A_\mu(x)\right). \tag{7.159}$$

Because the path has identical endpoints, the trace of a Wilson loop is gauge invariant. From the result of the previous paragraph, Wilson loops are equal to the identity in a pure gauge field, but they depend non-trivially on the path in a generic gauge field.[19]

In Abelian gauge theories, the Wilson loop can be rewritten in terms of the integral of the field strength $F_{\mu\nu}$ over a surface $\Sigma$ of boundary $\gamma$, by using the Stokes theorem,

$$\exp\left(ig \oint_\gamma dx^\mu A_\mu(x)\right) \underset{\text{Abelian}}{=} \exp\left(i\frac{g}{2}\int_\Sigma dx^\mu \wedge dx^\nu F_{\mu\nu}(x)\right). \tag{7.160}$$

Generalizations of this formula to the non-Abelian case exist, and involve a path-ordering on the left-hand side (thus giving a Wilson loop) and a *surface-ordering* on the right-hand side. For infinitesimally small closed loops, a more direct connection to the field strength may be established. Consider for instance a small square closed path in the (12) plane,

$$\gamma = \quad \boxed{\phantom{xxx} x^\bullet \phantom{xxx}} \quad a$$
$$\qquad\qquad a$$

In the following, we wish to calculate the trace of the Wilson loop along this path (also called a *plaquette*). Up to order two in the gauge potential, we may ignore the path ordering, thanks to the trace, and we have

$$\text{tr}(W[A;\gamma]) = \text{tr}(1) + ig\,\text{tr}\left(\oint_\gamma dr^\mu A_\mu(x+r)\right) - \frac{g^2}{2}\text{tr}\left(\oint_\gamma dr^\mu A_\mu(x+r)\right)^2 + \mathcal{O}(A^3), \tag{7.161}$$

where $x$ denotes the point at the center of the plaquette and $r$ is a coordinate relative to this center. The contour integral may be rewritten as a surface integral thanks to Stokes' theorem (since this integral is linear in the gauge field, the Abelian Stokes theorem holds). Expanding the gauge potential about the point $x$, we get

$$\oint_\gamma dr^\mu A_\mu(x+r) = a^2\left\{1 + \tfrac{a^2}{24}(\partial_1^2 + \partial_2^2)\right\}\underbrace{(\partial_1 A_2(x) - \partial_2 A_1(x))}_{F^{12}_{\text{Abel.}}(x)} + \mathcal{O}(a^6), \tag{7.162}$$

---

[19] Wilson loops are extensively used in lattice gauge theories. Moreover, Giles' theorem states that all the gauge invariant information contained in a gauge potential $A_\mu$ can be reconstructed from the trace of Wilson loops (assuming we know Wilson loops for arbitrary loops).

and

$$\text{tr}\left(\oint_\gamma dr^\mu A_\mu(x+r)\right)^2 = a^4\,\text{tr}\left(F^{12}_{\text{Abel.}}(x)F^{12}_{\text{Abel.}}(x)\right)$$
$$+ \frac{a^6}{12}\,\text{tr}\left(F^{12}_{\text{Abel.}}(x)(\partial_1^2+\partial_2^2)F^{12}_{\text{Abel.}}(x)\right) + \mathcal{O}(a^8). \tag{7.163}$$

(The absence of odd terms in $a$ follows from the fact that $x$ is at the center of the plaquette.) Among the higher order terms in the field in eq. (7.161), some merely restore the gauge invariance of $\text{tr}\,(W[A;\gamma])$ by promoting $F^{12}_{\text{Abel.}}$ into the non-Abelian field strength, and the ordinary derivatives into covariant ones. Using the fact that elements of the $\mathfrak{su}(N)$ algebra are traceless, we obtain the following expression for the trace of the plaquette:

$$\text{tr}\,(W[A;\gamma]) = \text{tr}\,(\mathbf{1}) - \frac{g^2 a^4}{2}\,\text{tr}\left(F^{12}(x)F^{12}(x)\right)$$
$$- \frac{g^2 a^6}{24}\,\text{tr}\left(F^{12}(x)(D_1^2+D_2^2)F^{12}(x)\right) + \mathcal{O}(a^8). \tag{7.164}$$

This equation, valid in any representation, is the basis of the discretization of the Yang–Mills action, the first step in formulating *lattice gauge theories* (see Chapter 16).

### 7.6.6 Wilson Lines and Eikonal Scattering

Wilson lines also appear in the high-energy limit of scattering by an external potential, known as the *eikonal limit*. Consider the S-matrix element

$$S_{\beta\alpha} \equiv \langle\beta_{\text{out}}|\alpha_{\text{in}}\rangle = \langle\beta_{\text{in}}|U(+\infty,-\infty)|\alpha_{\text{in}}\rangle \tag{7.165}$$

for the transition between two arbitrary states made of quarks, antiquarks and gluons, $\alpha$ and $\beta$. In the second equality, $U(+\infty,-\infty)$ is the evolution operator from the initial to the final state. It can be expressed as the time-ordered exponential of the interaction part of the Lagrangian,

$$U(+\infty,-\infty) = T\,\exp\left[i\int d^4x\,\mathcal{L}_I(\phi_{\text{in}}(x))\right], \tag{7.166}$$

where $\phi_{\text{in}}$ denotes generically the fields in the interaction picture. In this discussion, $\mathcal{L}_I$ contains both the self-interactions of the fields, and their interaction with the external field. Consider now the high-energy limit of this scattering amplitude,

$$S^{(\infty)}_{\beta\alpha} \equiv \lim_{\omega\to+\infty} \langle\beta_{\text{in}}|e^{-i\omega K^3} U(+\infty,-\infty)\,\underbrace{e^{+i\omega K^3}|\alpha_{\text{in}}\rangle}_{\text{boosted state}} \tag{7.167}$$

where $K^3$ is the generator of Lorentz boosts in the $+z$ direction.

Before doing any calculations, a simple argument can help to understand what happens in this limit. Quite generally, scattering amplitudes are proportional to the overlap in space-time between the wavefunctions of the two colliding objects. In the present case, it should scale as the time spent by the incoming state in the region occupied by the external field. This duration is inversely proportional to the energy of the incoming state, and goes to zero in the limit $\omega \to +\infty$. If the interaction between the projectile and the external field was via a scalar exchange, then the conclusion would be that the scattering amplitude vanishes in the high-energy limit (in other words, the S-matrix would become equal to 1). However, interactions with a color field involve a vector exchange, i.e., the external field couples to a 4-vector $J^\mu$ that represents the color current carried by the projectile, via a term of the form $\mathcal{A}_\mu J^\mu$. At high energy, the longitudinal component of this 4-vector increases proportionally to the energy, and compensates the small time spent in the interaction zone. Thus, for states that interact via a vector exchange,[20] we expect that scattering amplitudes have a finite high-energy limit (neither zero, nor infinite).

This calculation is best done using *light-cone coordinates*. For any 4-vector $a^\mu$, one defines

$$a^+ \equiv \frac{a^0 + a^3}{\sqrt{2}}, \quad a^- \equiv \frac{a^0 - a^3}{\sqrt{2}}. \tag{7.168}$$

These coordinates satisfy the following identities:

$$x \cdot y = x^+ y^- + x^- y^+ - \mathbf{x}_\perp \cdot \mathbf{y}_\perp, \quad d^4x = dx^+ dx^- d^2 \mathbf{x}_\perp$$
$$\Box = 2\partial^+ \partial^- - \boldsymbol{\nabla}_\perp^2 \quad \text{with} \quad \partial^+ \equiv \frac{\partial}{\partial x^-}, \; \partial^- \equiv \frac{\partial}{\partial x^+}. \tag{7.169}$$

Note also that the non-zero components of the metric tensor are

$$g^{+-} = g^{-+} = 1, \quad g^{11} = g^{22} = -1. \tag{7.170}$$

For a highly boosted projectile in the $+z$ direction, $x^+$ plays the role of the time, and the Hamiltonian is the $P^-$ component of the momentum. The generator of longitudinal boosts in light-cone coordinates is $K^3 = M^{+-}$. Using the commutation relations of the Poincaré algebra, this leads to the following identities:

$$e^{-i\omega K^3} P^- e^{i\omega K^3} = e^{-\omega} P^-, \quad e^{-i\omega K^3} P^+ e^{i\omega K^3} = e^{+\omega} P^+,$$
$$e^{-i\omega K^3} P^j e^{i\omega K^3} = P^j. \tag{7.171}$$

They express the fact that, under longitudinal boosts, the components $P^\pm$ of a 4-vector are simply rescaled, while the transverse components are left unchanged. Likewise, states, creation operators and field operators are transformed as follows:

$$e^{i\omega K^3} |\mathbf{p} \cdots \text{in}\rangle = |(e^\omega p^+, \mathbf{p}_\perp) \cdots \text{in}\rangle,$$
$$e^{i\omega K^3} a^\dagger_{\text{in}}(q) e^{-i\omega K^3} = a^\dagger_{\text{in}}(e^\omega q^+, e^{-\omega} q^-, \mathbf{q}_\perp),$$
$$e^{-i\omega K^3} \phi_{\text{in}}(x) e^{i\omega K^3} = \phi_{\text{in}}(e^{-\omega} x^+, e^\omega x^-, \mathbf{x}_\perp). \tag{7.172}$$

---

[20] By the same reasoning, gravitational interactions, which involve a spin-2 exchange, would lead to scattering amplitudes that grow linearly with energy.

## 7.6 NON-LOCAL GAUGE INVARIANT OPERATORS

Note that the last equation is valid only for a scalar field, or for the transverse components of a vector field. In addition, the $\pm$ components of a vector field receive an overall rescaling by a factor $e^{\pm\omega}$. Moreover, since a longitudinal boost does not alter the time ordering, we can also write

$$e^{-i\omega K^3} U(+\infty,-\infty) e^{i\omega K^3} = T \exp i \int d^4x \, \mathcal{L}_I(e^{-i\omega K^3} \phi_{in}(x) e^{i\omega K^3}). \tag{7.173}$$

The components of the vector current that couples to the target field transform as

$$\begin{aligned} e^{-i\omega K^3} J^i(x) e^{i\omega K^3} &= J^i(e^{-\omega}x^+, e^{\omega}x^-, \mathbf{x}_\perp), \\ e^{-i\omega K^3} J^-(x) e^{i\omega K^3} &= e^{-\omega} J^-(e^{-\omega}x^+, e^{\omega}x^-, \mathbf{x}_\perp), \\ e^{-i\omega K^3} J^+(x) e^{i\omega K^3} &= e^{\omega} J^+(e^{-\omega}x^+, e^{\omega}x^-, \mathbf{x}_\perp). \end{aligned} \tag{7.174}$$

Naturally, the target field $\mathcal{A}_\mu$ does not change when we boost the projectile. For simplicity, let us assume that $\mathcal{A}_\mu$ is confined in the region $-L \le x^+ \le +L$. We can thus split the evolution operator into three factors,

$$U(+\infty,-\infty) = U(+\infty,+L) \, U(+L,-L) \, U(-L,-\infty). \tag{7.175}$$

The factors $U(+\infty,+L)$ and $U(-L,-\infty)$ do not contain the external potential. For these two factors, the change of variables $e^{-\omega}x^+ \to x^+$, $e^{\omega}x^- \to x^-$ leads to

$$\begin{aligned} \lim_{\omega \to +\infty} e^{-i\omega K^3} U(+\infty,+L) e^{i\omega K^3} &= U_0(+\infty,0), \\ \lim_{\omega \to +\infty} e^{-i\omega K^3} U(-L,-\infty) e^{i\omega K^3} &= U_0(0,-\infty), \end{aligned} \tag{7.176}$$

where $U_0$ is the same as $U$, but defined with the self-interactions only (since these two factors correspond to the evolution of the projectile while outside of the target field). For the factor $U(+L,-L)$, the change $e^{\omega}x^- \to x^-$ gives

$$\lim_{\omega \to +\infty} e^{-i\omega K^3} U(+L,-L) e^{i\omega K^3} = T \exp\left[i \int d^2\mathbf{x}_\perp \, \chi(\mathbf{x}_\perp) \rho(\mathbf{x}_\perp)\right], \tag{7.177}$$

with
$$\begin{cases} \chi(\mathbf{x}_\perp) \equiv \int dx^+ \, \mathcal{A}^-(x^+, 0, \mathbf{x}_\perp), \\ \rho(\mathbf{x}_\perp) \equiv \int dx^- \, J^+(0, x^-, \mathbf{x}_\perp). \end{cases} \tag{7.178}$$

Thus, the high-energy limit of the scattering amplitude is

$$S^{(\infty)}_{\beta\alpha} = \langle \beta_{in} | U_0(+\infty,0) \, T \exp\left[i \int d^2\mathbf{x}_\perp \, \chi(\mathbf{x}_\perp) \rho(\mathbf{x}_\perp)\right] U_0(0,-\infty) | \alpha_{in} \rangle. \tag{7.179}$$

This formula is an exact result in the limit $\omega \to +\infty$. One may also note the following important properties:

- Only the $\mathcal{A}^-$ component of the external vector potential, integrated along the trajectory of the projectile, matters.

FIGURE 7.4: Illustration of the role of kinematics in the factorization of eq. (7.179). Left: before the boost is applied, quantum fluctuations of the incoming projectile may occur in the region of the external field. Right: after the boost, the region of the external field shrinks due to Lorentz contraction (in the frame of the projectile), and the effect of quantum fluctuations inside this region goes to zero.

- The self-interactions and the interactions with the external potential are factorized into three separate factors – this is a generic property of high-energy scattering. The role of the longitudinal boost in this factorization is illustrated in Figure 7.4.

Eq. (7.179) still contains the self-interactions of the fields to all orders. In order to evaluate it, one must insert the identity operator written as a sum over a complete set of states on each side of the exponential,

$$S^{(\infty)}_{\beta\alpha} = \sum_{\gamma,\delta} \langle \beta_{\text{in}} | U_0(+\infty, 0) | \gamma_{\text{in}} \rangle \langle \gamma_{\text{in}} | T \exp\left[ i \int d^2\mathbf{x}_\perp \, \chi(\mathbf{x}_\perp) \rho(\mathbf{x}_\perp) \right] | \delta_{\text{in}} \rangle \\ \times \langle \delta_{\text{in}} | U_0(0, -\infty) | \alpha_{\text{in}} \rangle. \qquad (7.180)$$

The factor

$$\sum_\delta | \delta_{\text{in}} \rangle \langle \delta_{\text{in}} | U_0(0, -\infty) | \alpha_{\text{in}} \rangle \qquad (7.181)$$

is the Fock expansion of the initial state: It accounts for the fact that the state $\alpha$ prepared at $x^+ = -\infty$ may have fluctuated into another state $\delta$ before it interacts with the external potential. The matrix elements of $U_0$ that appear in this expansion can be calculated perturbatively to any desired order. There is a similar factor for the final state evolution.

The interactions with the external field are in the central factor, $\langle \gamma_{\text{in}} | T \exp \ldots | \delta_{\text{in}} \rangle$. In order to rewrite it into a more intuitive form, let us first rewrite the operator $\rho$ in terms of creation and annihilation operators. For instance, the fermionic part of the current gives

$$\rho^a(\mathbf{x}_\perp)\Big|_{\text{fermions}} = g \int \frac{dp^+}{4\pi p^+} \frac{d^2\mathbf{p}_\perp}{(2\pi)^2} \frac{d^2\mathbf{q}_\perp}{(2\pi)^2} \, (t^a_f)_{ij} \left( a^\dagger_{si\,p+\mathbf{p}_\perp} a_{sj\,p+\mathbf{q}_\perp} \, e^{i(\mathbf{p}_\perp - \mathbf{q}_\perp)\cdot\mathbf{x}_\perp} \right. \\ \left. - b^\dagger_{si\,p+\mathbf{p}_\perp} b_{sj\,p+\mathbf{q}_\perp} \, e^{-i(\mathbf{p}_\perp - \mathbf{q}_\perp)\cdot\mathbf{x}_\perp} \right), \qquad (7.182)$$

where the $t^a_f$ are the generators of the fundamental representation of the $\mathfrak{su}(N)$ algebra and $a, b, a^\dagger, b^\dagger$ are the annihilation and creation operators for quarks and antiquarks. The operator $\rho^a$ also receives a contribution from gluons, not written here, obtained with the generators in the adjoint representation and the annihilation and creation operators of gluons instead. This formula captures the essence of eikonal scattering:

- Each annihilation operator has a matching creation operator – therefore, the number of quarks and gluons in the state does not change during the scattering, nor their flavor.
- The $p^+$ components of the momenta are not affected by the scattering.

- The spins are unchanged during the scattering.
- The colors and transverse momenta of the constituents of the state may change during the scattering.

Scattering amplitudes in the eikonal limit take a very simple form if one trades transverse momentum for a transverse position by a Fourier transform. For each intermediate state $\langle \delta_{in}| \equiv \langle \{k_i^+, k_{i\perp}\}|$, we first define the corresponding *light-cone wave function* by

$$\Psi_{\delta\alpha}(\{k_i^+, x_{i\perp}\}) \equiv \prod_{i\in\delta} \int \frac{d^2 k_{i\perp}}{(2\pi)^2} \, e^{-i k_{i\perp}\cdot x_{i\perp}} \, \langle \delta_{in}|U_0(0,-\infty)|\alpha_{in}\rangle, \qquad (7.183)$$

where the index $i$ runs over all the constituents of the state $\delta$. Then, each charged particle going through the external field acquires an $SU(N)$ factor that depends on the representation in which it lives,

$$\Psi_{\delta\alpha}(\{k_i^+, x_{i\perp}\}) \;\to\; \Psi_{\delta\alpha}(\{k_i^+, x_{i\perp}\}) \prod_{i\in\delta} U_i(x_{i\perp}),$$

$$U_i(x_{i\perp}) \equiv T \exp\left[ig \int dx^+ \, \mathcal{A}_a^-(x^+, 0, x_{i\perp}) \, t_{r_i}^a\right], \qquad (7.184)$$

where $r_i$ is the representation corresponding to the constituent $i$. We recognize in this formula Wilson lines defined on the light-cone direction that corresponds to the boosted projectile. The simplicity of this result is entirely due to kinematics: Thanks to the longitudinal boost, the external field is crossed in an infinitesimally short time, during which self-interactions have not enough time to happen and the transverse positions of the incoming quanta cannot vary.

## Exercises

*7.1 Check the dimensions of classical Lie algebras listed in Table 7.1.

7.2 Assume that $[X, Y] = icY$ where $c$ is a numerical constant. Use the all-orders Baker–Campbell–Hausdorff formula to calculate $\ln(e^{iX} e^{iY})$.

*7.3 Prove the identities listed in eq. (7.84).

7.4 Derive the Fierz identity for the Lie algebra $u(n)$, and the corresponding diagrammatical representation. Same question for $so(n)$.

7.5 Express the structure constants of the Lie algebra $su(n)$ in terms of traces of the generators in the fundamental representation. Then, use this in combination with the Fierz identity in order to calculate the trace $tr(T_{adj}^a T_{adj}^b)$ of the product of two adjoint generators. Same questions for $u(n)$ and $so(n)$.

7.6 Consider a non-Abelian gauge field $A^\mu(x)$, and define the following Wilson lines in the temporal direction:

$$U(x|+\infty,-\infty) \equiv T \exp\left(ig \int_{-\infty}^{+\infty} dx^0 \, A^0(x^0, x)\right),$$

$$V(x|+\infty,-\infty) \equiv T \exp\left(i\tfrac{g}{2} \int_{-\infty}^{+\infty} dx^0 \, A^0(x^0, x)\right).$$

- Prove that

$$\partial_i U(x|+\infty, -\infty) = \underbrace{\int_{-\infty}^{+\infty} dx^0\, U(x|+\infty, x^0)(ig\partial_i A^0(x^0, x))U(x|x^0, -\infty)}_{\equiv U\otimes(ig\partial_i A^0)\otimes U}.$$

- Prove that $2(U - V) = U \otimes (iA^0) \otimes V$.

**7.7** Consider temporal Wilson lines in the fundamental representation of $\mathfrak{su}(N)$,

$$U(x) \equiv T\exp\left(\int_{-\infty}^{+\infty} dx^0\, A_a^0(x^0, x)t_f^a\right),$$

where $A_a^0$ is a non-Abelian gauge field subject to Gaussian fluctuations such that

$$\langle A_a^0(x) \rangle = 0, \quad \langle A_a^0(x) A_b^0(y) \rangle = \alpha(x^0, x-y)\delta_{ab}\delta(x^0 - y^0).$$

(Note that fields at different times are not correlated.)

- Show that the average over the field fluctuations of one Wilson line is

$$\langle U(x) \rangle = \exp\left(-\tfrac{g^2}{2} C_f \alpha(0)\right) \quad \text{with} \quad \alpha(x-y) \equiv \int dx^0\, \alpha(x^0, x-y),$$

and $C_f = (N^2 - 1)/(2N)$.
- Show that the expectation value of the trace of two Wilson lines reads

$$\operatorname{tr}\langle U(x)U^\dagger(y)\rangle = N\exp(g^2 C_f(\alpha(x-y) - \alpha(0))).$$

- Same exercise for the trace of four Wilson lines, $\operatorname{tr}\langle U(x)U^\dagger(y)U(u)U^\dagger(v)\rangle$, and for $\langle \operatorname{tr}(U(x)U^\dagger(v))\operatorname{tr}(U(u)U^\dagger(y))\rangle$.

**\*7.8** Consider the classical Yang–Mills equation with an external current $J^\nu$.

- Show that in the Lorenz gauge ($\partial_\mu A^\mu = 0$), it takes the following form:

$$\Box A^\nu = J^\nu + ig[A_\mu, F^{\mu\nu} + \partial^\mu A^\nu].$$

- Consider now an external current of the form $J_a^\nu = \delta^{\nu-}\delta(x^+)\rho_a(x_\perp)$. (We use light-cone coordinates, $x^\pm \equiv (x^0 \pm x^3)/\sqrt{2}$, $x_\perp \equiv (x^1, x^2)$. The $\delta^{\nu-}$ means that only the $-$ component of the current is non-zero.) Find the solution of the above Yang–Mills equation that vanishes when $x^+ \to -\infty$. (Note: Be careful to ensure that the current is covariantly conserved, $[D_\nu, J^\nu] = 0$.)
- Convert the previous solution to the light-cone gauge $A^- = 0$.

**\*7.9** Show that $C_{2r} \equiv t_r^a t_r^a$ (summed on the index $a$) commutes with all the generators. (Schur's lemma then implies that this object must be proportional to the identity.) In the fundamental representation of $\mathfrak{su}(n)$, show that $C_{3r} \equiv d^{abc} t_f^a t_f^b t_f^c$ (called the *cubic Casimir*) is also proportional to the identity, and calculate the constant of proportionality.

# EXERCISES

**7.10** For the fundamental representation of $\mathfrak{su}(n)$, show that $f_{abc} t_f^a t_f^b = \frac{i}{2} n\, t_f^c$ and that $f_{abc} t_f^a t_f^b t_f^c = \frac{i}{4}(n^2 - 1) \mathbf{1}_r$. More generally, for a representation $r$ of $\mathfrak{su}(n)$, show that $f_{abc} t_r^a t_r^b t_r^c = \frac{i}{2} n\, C_{2r}$.

**7.11** Show that the $\mathfrak{su}(n)$ antisymmetric structure constants, $f^{abc}$, and the symmetric ones, $d^{abc}$, obey the following relations:

$$d^{abe} f^{ecd} + d^{bce} f^{ead} + d^{cae} f^{ebd} = 0,$$

$$\frac{2}{n}(\delta_{ab}\delta_{cd} - \delta_{ac}\delta_{bd}) + d^{abe} d^{ced} - d^{ace} d^{bed} + f^{bce} f^{aed} = 0.$$

# CHAPTER 8

# Quantization of Yang–Mills Theory

As we have seen in the previous chapter, the Lagrangian density of a generic non-Abelian gauge theory reads

$$\mathcal{L} \equiv \left(D_\mu \phi(x)\right)^\dagger \left(D^\mu \phi(x)\right) - m^2 \, \phi^\dagger(x)\phi(x) - V\!\left(\phi^\dagger(x)\phi(x)\right) \\ - \tfrac{1}{4} F^a_{\mu\nu}(x) F^{a\,\mu\nu}(x) + \overline{\psi}(x)\left(i\slashed{D}_x - m\right)\psi(x). \tag{8.1}$$

The local non-Abelian invariance of this Lagrangian does not change anything relating to the quantization of the scalar field $\phi$ and the spinor $\psi$, for which we may use the standard canonical or path integral approaches, with the result that the usual Feynman rules apply to them and to their couplings to the gauge field. The main complication resides in the pure Yang–Mills part of this Lagrangian, i.e., with the quantization of the gauge potential $A^\mu$. The identification of the degrees of freedom that are made redundant by the gauge symmetry is more complicated than in quantum electrodynamics (QED), and a lot more care is necessary in order to isolate the genuine dynamical variables of the theory.

## 8.1 Naive Quantization of the Gauge Bosons

In order to get a sense of the difficulty, let us try to mimic the QED case in order to guess the Feynman rules for non-Abelian gauge fields. Using the explicit form of the field strength,

$$F^a_{\mu\nu} = \partial_\mu A^a_\nu - \partial_\nu A^a_\mu + g\, f^{abc}\, A^b_\mu A^c_\nu, \tag{8.2}$$

## 8.1 NAIVE QUANTIZATION OF THE GAUGE BOSONS

we can rewrite the Yang–Mills Lagrangian as follows:

$$\mathcal{L}_A = \tfrac{1}{2} A_\mu^a \left( g^{\mu\nu} \Box - \partial^\mu \partial^\nu \right) A_\nu^a - g f^{abc} \left( \partial_\mu A_\nu^a \right) A^{b\,\mu} A^{c\,\nu} \\ - \tfrac{1}{4} g^2 f^{abc} f^{ade} A_\mu^b A_\nu^c A^{d\,\mu} A^{e\,\nu}, \tag{8.3}$$

where we have anticipated an integration by parts in the first (kinetic) term. Note that the kinetic term is formally identical to the kinetic term of photons, except for the color index $a$ carried by the gauge potential. Therefore, one may be tempted to generalize the QED Feynman rules to a non-Abelian gauge boson. As in the QED case, the quadratic part of the Lagrangian (8.3) poses a difficulty when trying to determine the free propagator, because the operator between $A_\mu^a \cdots A_\nu^a$ is not invertible. If we take for granted that a similar gauge fixing procedure (more on this later, as this is in fact the heart of the problem) can be applied here, we may assume that the free gauge boson propagator[1] in Feynman gauge is

$$G_{F\,ab}^{0\,\mu\nu}(p) = \underset{p}{\underbrace{\text{\small{ooooo}}}} = \frac{-i g^{\mu\nu} \delta_{ab}}{p^2 + i0^+}, \tag{8.4}$$

and one may read off directly from the Lagrangian (8.3) the following three-gluon and four-gluon vertices:

$$= -g f^{abc} \{ g^{\mu\nu} (k-p)^\rho \\ + g^{\nu\rho} (p-q)^\mu + g^{\rho\mu} (q-k)^\nu \}, \tag{8.5}$$

$$= -i g^2 \{ f^{abe} f^{cde} (g^{\mu\rho} g^{\nu\sigma} - g^{\mu\sigma} g^{\nu\rho}) \\ + f^{ace} f^{bde} (g^{\mu\nu} g^{\rho\sigma} - g^{\mu\sigma} g^{\nu\rho}) \\ + f^{ade} f^{bce} (g^{\mu\nu} g^{\rho\sigma} - g^{\mu\rho} g^{\nu\sigma}) \}. \tag{8.6}$$

All this seems fine, except for a rather subtle problem that would appear when using this perturbation theory (see Section 8.6.1): These Feynman rules lead to amplitudes that do not fulfill the usual Ward–Takahashi identities, even when all the external colored particles are on their mass-shell. From the discussion of perturbative unitarity for amplitudes with external gauge bosons in Section 3.6.2, the lack of Ward–Takahashi identities seems to imply a violation of unitarity in perturbation theory. Since unitarity is one of the cornerstones of any quantum theory, this is not a conclusion we are ready to accept, and we must conclude that something is missing in the above Feynman rules.

---

[1] In this chapter, we use the diagrammatic convention of QCD, where the gauge bosons (gluons) are represented as springs in Feynman diagrams. In the electroweak theory, it is more common to represent them as wavy lines, like the photon in QED.

## 8.2 Gauge Fixing

In our naive attempt to guess the Feynman rules appropriate for non-Abelian gauge bosons, we have implicitly assumed that the gauge fixing works in the same way as in QED, namely that the gauge fixing trivially leads to the factorization of an infinite factor in the path integral, with no other change to the degrees of freedom that are not constrained by the gauge condition. It turns out that this assumption is incorrect. Let us start from the path integral representation of the expectation value of some gauge invariant operator $\mathcal{O}(A_\mu)$,

$$\langle \mathcal{O} \rangle \equiv \int [DA^a_\mu(x)]\; \mathcal{O}(A_\mu)\; \exp\left\{i \underbrace{\int d^4x \left(-\frac{1}{4} F^a_{\mu\nu} F^{a\,\mu\nu}\right)}_{S_{YM}[A_\mu]}\right\}. \tag{8.7}$$

Local gauge transformations of the field $A_\mu$,

$$A_\mu(x) \quad\to\quad A^\Omega_\mu(x) \equiv \Omega^\dagger(x)\, A_\mu(x)\, \Omega(x) + \frac{i}{g}\, \Omega^\dagger(x)\, \partial_\mu \Omega(x), \tag{8.8}$$

leave the action and the observable unchanged. Moreover, the functional measure is also invariant. Indeed, it transforms as

$$[DA^\Omega_{a\mu}(x)] = [DA_{a\mu}(x)]\; \det\left(\frac{\delta A^\Omega_{a\mu}(x)}{\delta A_{b\nu}(y)}\right), \tag{8.9}$$

where the determinant is the Jacobian of the change of coordinates. Using eq. (7.36), this determinant can be rewritten as

$$\det\left(\frac{\delta A^\Omega_{a\mu}(x)}{\delta A_{b\nu}(y)}\right) = \det\left(\delta_\mu{}^\nu\, \delta(x-y)\, [\Omega^\dagger_{\mathrm{adj}}(x)]_{ab}\right) = 1, \tag{8.10}$$

since the group element $\Omega_{\mathrm{adj}}$ is a unitary matrix. Therefore, there is a large amount of redundancy in the above path integral, and it is in fact infinite. By applying a gauge transformation, each field configuration $A_\mu$ develops into a *gauge orbit* (see Figure 8.1), along which the physics is invariant. In order to eliminate this redundancy, we would like to impose a condition at every space-time point $x$ on the gauge fields,

$$G^a(A_\mu(x)) = 0, \tag{8.11}$$

in order to select a unique[2] field configuration along each orbit. Geometrically, the *gauge condition* (8.11) defines a manifold that intersects each orbit, as shown in Figure 8.1, and we choose this intersection as the representative of this field configuration.

---

[2]It turns out that this is not possible, due to the *Gribov ambiguity*: All gauge conditions of the form (8.11) have several solutions, called *Gribov copies*. However, only one of these solutions is a "small field," while the others are proportional to the inverse coupling $g^{-1}$. Since perturbation theory is an expansion around the vacuum (i.e., in the small field regime), these non-perturbatively large copies do not play any role in perturbation theory.

## 8.3 FADEEV–POPOV QUANTIZATION AND GHOST FIELDS

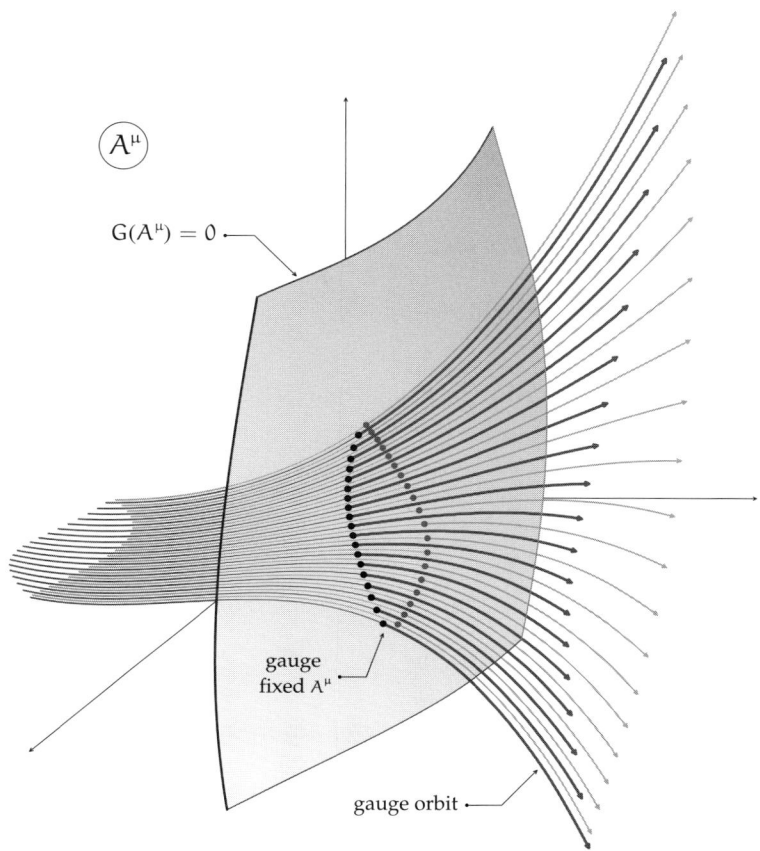

FIGURE 8.1: Illustration of the gauge fixing procedure. The lines represent the gauge field configurations spanned when varying $\Omega$. The shaded surface is the manifold where the gauge condition is satisfied, and the black dots are the gauge-fixed field configurations.

## 8.3 Fadeev–Popov Quantization and Ghost Fields

Thus, we would like to split the integration measure in eq. (8.7) into a physical component in the manifold $G(A) = 0$, and a redundant component along the gauge orbits that we should factor out. Unfortunately, achieving this in a non-Abelian gauge theory is far more complicated than in QED, because the modification of the gauge potential under a gauge transformation is nonlinear. In order to see the difficulty, let us define

$$\Delta^{-1}[A_\mu] \equiv \int \left[ D\Omega(x) \right] \delta[G^a(A_\mu^\Omega)]. \tag{8.12}$$

$\Delta[A_\mu]$ is the determinant of the derivative of the constraint $G(A_\mu)$ with respect to the gauge transformation $\Omega$, at the point where $G(A_\mu) = 0$,

$$\Delta(A_\mu) = \det \left( \frac{\delta G^a}{\delta \Omega} \right)_{G^a(A_\mu^\Omega)=0}. \tag{8.13}$$

In QED, for linear gauge fixing conditions, this derivative (and therefore the determinant) is independent of the gauge field, and can be trivially factored out of the path integral. This is not the case in non-Abelian gauge theories, and this determinant is the source of significant complications. One can first prove that the determinant $\Delta[A_\mu]$ is gauge invariant. Indeed, changing $A_\mu \to A_\mu^\Theta$, we have

$$\Delta^{-1}[A_\mu^\Theta] = \int [D\Omega(x)]\, \delta[G^a(A_\mu^{\overbrace{\Theta\Omega}^{\Omega'}})] = \int [D(\Theta^\dagger(x)\Omega'(x))]\, \delta[G^a(A_\mu^{\Omega'})]$$
$$= \int [D\Omega'(x)]\, \delta[G^a(A_\mu^{\Omega'})] = \Delta^{-1}[A_\mu]. \qquad (8.14)$$

Here, we have used the fact that there exists a group invariant integration measure on a Lie group. By inserting

$$1 = \Delta[A_\mu] \int [D\Omega(x)]\, \delta[G^a(A_\mu^\Omega)] \qquad (8.15)$$

inside the path integral (8.7), we obtain

$$\langle \mathcal{O} \rangle = \int [D\Omega(x)] \int [DA_\mu^a(x)]\, \Delta[A_\mu]\, \delta[G^a(A_\mu^\Omega)]\, \mathcal{O}(A_\mu)\, e^{i\mathcal{S}_{\rm YM}[A_\mu]}. \qquad (8.16)$$

Now, we change the integration variable of the second integral according to $A_\mu \to A_\mu^{\Omega^\dagger}$. In this transformation, the measure $[DA_\mu]$, the Yang–Mills action $\mathcal{S}_{\rm YM}[A_\mu]$, the observable $\mathcal{O}(A_\mu)$ and the determinant $\Delta[A_\mu]$ are all unchanged (because they are gauge invariant),

$$[DA_\mu^{\Omega^\dagger}] = [DA_\mu], \quad \mathcal{S}_{\rm YM}[A_\mu^{\Omega^\dagger}] = \mathcal{S}_{\rm YM}[A_\mu],$$
$$\mathcal{O}[A_\mu^{\Omega^\dagger}] = \mathcal{O}[A_\mu], \quad \Delta[A_\mu^{\Omega^\dagger}] = \Delta[A_\mu], \qquad (8.17)$$

while the field $A_\mu^\Omega$ becomes $A_\mu$. Therefore, we have

$$\langle \mathcal{O} \rangle = \int [D\Omega(x)] \int [DA_\mu^a(x)]\, \Delta[A_\mu]\, \delta[G^a(A_\mu)]\, \mathcal{O}(A_\mu)\, e^{i\mathcal{S}_{\rm YM}[A_\mu]}. \qquad (8.18)$$

At this point, the second integral does not contain the gauge transformation $\Omega$ anymore, and therefore we have managed to factorize the "integral along the orbits" in the form of the first integral over $[D\Omega]$. Dropping this constant factor, we can therefore write an integral free of any redundancy,

$$\langle \mathcal{O} \rangle = \int [DA_\mu^a(x)]\, \Delta[A_\mu]\, \delta[G^a(A_\mu)]\, \mathcal{O}(A_\mu)\, e^{i\mathcal{S}_{\rm YM}[A_\mu]}. \qquad (8.19)$$

In the above formula, the determinant $\Delta[A_\mu]$ depends on the gauge field and must therefore have an effect on the Feynman rules. The *Fadeev–Popov method* consists in rewriting this determinant as a path integral. Note that since $\Delta[A_\mu]$ appears in the numerator, we need

Grassmann variables in order to represent it as a path integral,[3] according to eq. (6.36):

$$\det(i\mathcal{M}) = \int [D\chi_a(x) D\overline{\chi}_a(x)] \exp\left\{i \int d^4x d^4y \, \overline{\chi}_a(x) \mathcal{M}_{ab}(x,y) \chi_b(y)\right\}. \tag{8.20}$$

An extra generalization, that we have already used in the path integral quantization of the photon (see eq. (6.50)), is to shift the gauge condition from $G^a(A) = 0$ to $G^a(A) = \omega^a$ and to perform a Gaussian integration over $\omega^a$. The final result takes the following form:

$$\langle \mathcal{O} \rangle = \int [DA_\mu^a(x)] [D\chi_a(x) D\overline{\chi}_a(x)] \, \mathcal{O}(A_\mu)$$
$$\times \exp i \int d^4x \left( \underbrace{-\frac{1}{4} F_{\mu\nu}^a F^{a\,\mu\nu}}_{\mathcal{L}_{YM}} \underbrace{-\frac{\xi}{2} (G^a(A_\mu))^2}_{\mathcal{L}_{GF}} + \underbrace{\overline{\chi}_a \mathcal{M}_{ab} \chi_b}_{\mathcal{L}_{FPG}} \right), \tag{8.21}$$

where $\mathcal{M}_{ab}$ is the derivative of $G^a(A^\Omega)$ with respect to the gauge transformation $\Omega$, at the point $\Omega = 1$ (here, we use the fact that the determinant is gauge invariant to choose freely the $\Omega$ at which we compute the derivative). The unphysical Grassmann fields $\chi$ and $\overline{\chi}$ introduced as a trick to express the determinant are called *Fadeev–Popov ghosts*, or simply ghosts. Note that they are scalar (they do not carry any Lorentz index) but anti-commuting, and therefore they do not obey the spin-statistics theorem. Although physical observables do not depend on these fictitious fields, there is in general a coupling between the ghosts and the gauge fields, because the matrix $\mathcal{M}_{ab}$ may contain the gauge field. This implies that the ghosts may appear in the form of loop corrections in the perturbative expansion. As we shall see shortly, they are in fact crucial for the consistency of perturbation theory in non-Abelian gauge theories. In particular, the ghosts ensure that the theory is unitary.

## 8.4 Feynman Rules for Non-Abelian Gauge Theories

Equation (8.21) contains all the necessary ingredients to complete the Feynman rules that we have started to derive heuristically at the beginning of this chapter. To turn this formula into explicit Feynman rules, we should first choose the gauge fixing function $G^a(A)$, since it enters directly in the term $\frac{\xi}{2}(G^a(A))^2$, and implicitly in the matrix $\mathcal{M}_{ab}$ that defines the ghost term. In the common situation where this gauge fixing function is linear in $A_\mu$ (all our examples will be of this type), then the terms that are quadratic in the gauge field are the same as in QED, and therefore the gauge boson propagator is also the same (except for an extra factor $\delta_{ab}$ that expresses the fact that the free propagation of a gluon does not change its color). Thus, our guess (8.4) for the Feynman gauge propagator was in fact correct. In addition, the gauge fixing term and the ghost term cannot contain terms of degree 3 or 4 in the gauge field, which implies that the vertices given in eqs. (8.5) and (8.6) are also correct.

---

[3]The factor $i$ in $\det(i\mathcal{M})$ has been included for aesthetic reasons, but does not change anything. In fact, any rescaling $\mathcal{M} \to \kappa \mathcal{M}$ would leave the results unchanged. Indeed, such a change would alter the ghost propagator according to $G \to \kappa^{-1} G$, and the ghost-gauge boson vertex by $V \to \kappa V$. Since the ghosts appear only in closed loops that contain an equal number of propagators and vertices, these factors $\kappa$ would cancel out.

## 8.4.1 Covariant Gauge

Let us now consider the general covariant gauge, also known as the $R_\xi$ gauge, already introduced in eq. (6.49) for QED. This amounts to choosing the gauge fixing function as

$$G^a(A) \equiv \partial^\mu A_\mu^a - \omega^a(x). \tag{8.22}$$

With this gauge fixing, the free gauge boson propagator is

$$G^{0\,\mu\nu}_{F\,ab}(p) = \begin{array}{c}p\\ \longrightarrow\\ \text{\footnotesize{\textit{\char"223E}}}\end{array} = \frac{-i g^{\mu\nu} \delta_{ab}}{p^2 + i0^+} + \frac{i \delta_{ab}}{p^2 + i0^+}\left(1 - \frac{1}{\xi}\right)\frac{p^\mu p^\nu}{p^2}. \tag{8.23}$$

(The simplest form is obtained in the limit $\xi \to 1$, giving the Feynman gauge.[4]) The matrix $\mathcal{M}_{ab}$ can be calculated by applying an infinitesimal gauge transformation $\Omega = \exp(i\theta_a t^a)$ to $A_\mu$. The variation of the gauge field is

$$\delta A_{a\,\mu}(x) = g f^{abc} \theta_b(x) A_{c\,\mu}(x) - \partial_\mu \theta_a(x), \tag{8.24}$$

and the variation of $G^a(A)$ at the point $x$ is

$$\delta G^a = g f^{abc} \left(\partial^\mu \theta_b(x)\right) A_{c\,\mu}(x) + g f^{abc} \theta_b(x) \left(\partial^\mu A_{c\,\mu}(x)\right) - \Box \theta_a(x). \tag{8.25}$$

Therefore, we have

$$\mathcal{M}_{ab} = \frac{\delta G^a(A)}{\delta \theta^b} = g f^{abc} \left(\partial^\mu A_{c\,\mu}(x)\right) + g f^{abc} A_{c\,\mu}(x) \partial^\mu - \delta_{ab} \Box, \tag{8.26}$$

and the terms that depend on the Fadeev–Popov ghosts can be encapsulated in the following effective Lagrangian:

$$\mathcal{L}_{FPG} = \overline{\chi}_a \left(-\delta_{ab} \Box + g f^{abc} \left(\partial^\mu A_{c\,\mu}(x)\right) + g f^{abc} A_{c\,\mu}(x) \partial^\mu\right) \chi_b. \tag{8.27}$$

The first term leads to the following propagator for the ghosts:

$$G_F^0(p) = \begin{array}{c}p\\ \dashrightarrow\end{array} = \frac{i \delta_{ab}}{p^2 + i0^+}. \tag{8.28}$$

Note that it has the form of a scalar propagator, although the ghosts are anti-commuting Grassmann variables. The vertex between ghosts and gauge bosons reads

$$= g f^{abc} (p_\mu + q_\mu) = g f^{abc} r_\mu. \tag{8.29}$$

---

[4]Another popular choice is the *Landau gauge*, obtained in the limit $\xi \to +\infty$, that corresponds to a strict enforcement of the condition $\partial_\mu A^\mu = 0$. Indeed, in this limit the exponential of $i\frac{\xi}{2}(\partial_\mu A^\mu)^2$ in the gauge fixed Lagrangian oscillates wildly – and produces cancellations – unless $\partial_\mu A^\mu = 0$. Equivalently, the Gaussian distribution for the function $\omega^a(x)$ has a vanishing width when $\xi$ becomes infinite, which forces the strict equality $\partial^\mu A_\mu^a = 0$.

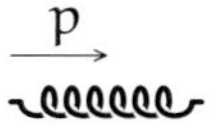

## 8.4 FEYNMAN RULES FOR NON-ABELIAN GAUGE THEORIES

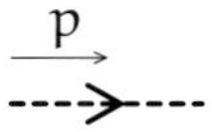

$$\begin{aligned}
\text{(gluon propagator)} &= \frac{-i g^{\mu\nu} \delta_{ab}}{p^2 + i0^+} + \frac{i \delta_{ab}}{p^2 + i0^+}\left(1 - \frac{1}{\xi}\right)\frac{p^\mu p^\nu}{p^2} \\
\text{(fermion propagator)} &= \frac{i \delta_{ij}}{\slashed{p} - m + i0^+} \\
\text{(ghost propagator)} &= \frac{i \delta_{ab}}{p^2 + i0^+}
\end{aligned}$$

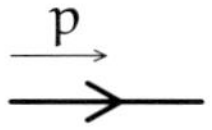

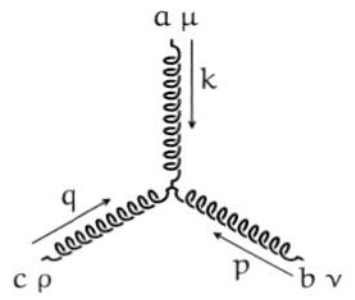

$$\text{(3-gluon vertex)} = -g f^{abc}\{g^{\mu\nu}(k-p)^\rho + g^{\nu\rho}(p-q)^\mu + g^{\rho\mu}(q-k)^\nu\}$$

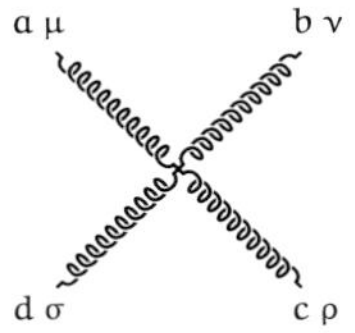

$$\text{(4-gluon vertex)} = -ig^2\{f^{abe}f^{cde}(g^{\mu\rho}g^{\nu\sigma} - g^{\mu\sigma}g^{\nu\rho}) + f^{ace}f^{bde}(g^{\mu\nu}g^{\rho\sigma} - g^{\mu\sigma}g^{\nu\rho}) + f^{ade}f^{bce}(g^{\mu\nu}g^{\rho\sigma} - g^{\mu\rho}g^{\nu\sigma})\}$$

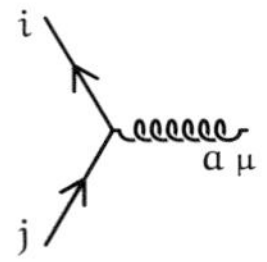

$$\text{(fermion-gluon vertex)} = -ig\gamma^\mu (t_r^a)_{ij}$$

$$\text{(ghost-gluon vertex)} = g f^{abc}(p_\mu + q_\mu) = g f^{abc} r_\mu$$

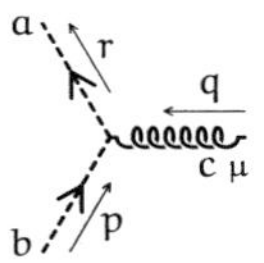

FIGURE 8.2: Feynman rules of non-Abelian gauge theories in covariant gauge. We also list the rules involving fermions for completeness. Latin characters $a, b, c$ refer to the adjoint representation, while the letters $i, j$ refer to the representation $r$ in which the fermions live (the fundamental representation for the quarks in QCD).

The Feynman rules for non-Abelian gauge theories in covariant gauge are summarized in Figure 8.2, where we have added for completeness the rules relative to fermions.

## 8.4.2 Axial Gauge

The *axial gauge* fixing consists in constraining the value of $n^\mu A_\mu^a$, where $n^\mu$ is a fixed 4-vector (when this vector is time-like, this gauge is called the *temporal gauge*, and when it is light-like, it is called the *light-cone gauge*). Therefore, the gauge fixing function is

$$G^a(A) \equiv n^\mu A_\mu^a - \omega^a(x). \tag{8.30}$$

After gauge fixing, the quadratic part of the effective Lagrangian reads

$$\frac{1}{2} A_\mu^a \left( g^{\mu\nu} \Box - \partial^\mu \partial^\nu - \xi n^\mu n^\nu \right) A_\nu^a, \tag{8.31}$$

and the free gauge boson propagator is obtained in momentum space by inverting

$$g^{\mu\nu} p^2 - p^\mu p^\nu + \xi n^\mu n^\nu. \tag{8.32}$$

The inverse of this matrix must be of the form

$$A g^{\mu\nu} + B p^\mu p^\nu + C n^\mu n^\nu + D (n^\mu p^\nu + n^\nu p^\mu). \tag{8.33}$$

(This is the most general symmetric tensor that one may construct with $g^{\mu\nu}$, $p^\mu$, and $n^\mu$.) This leads to the following propagator:

$$G^{0\,\mu\nu}_{F\,ab}(p) = \frac{-i\delta_{ab}}{p^2 + i0^+} \left[ g^{\mu\nu} - \frac{p^\mu n^\nu + p^\nu n^\mu}{p \cdot n} + \frac{p^\mu p^\nu}{(p \cdot n)^2} \left( n^2 + \xi^{-1} p^2 \right) \right]. \tag{8.34}$$

Note that this propagator does not vanish as $p^{-2}$ at large momentum, because of the term proportional to $\xi^{-1}$. With this gauge fixing, the variation of the gauge fixing function under an infinitesimal gauge transformation is given by

$$\delta G^a = g\, f^{abc}\, \theta_b(x)\, n^\mu A_{c\,\mu}(x) - n^\mu \partial_\mu\, \theta_a(x), \tag{8.35}$$

and the matrix $\mathcal{M}$ reads

$$\mathcal{M}_{ab} = g\, f^{abc}\, n^\mu A_{c\,\mu}(x) - \delta_{ab}\, n^\mu \partial_\mu. \tag{8.36}$$

Therefore, the Fadeev–Popov term in the effective Lagrangian is

$$\mathcal{L}_{FPG} = \overline{\chi}_a \left( -\delta_{ab}\, n^\mu \partial_\mu + g\, f^{abc}\, n^\mu A_{c\,\mu}(x) \right) \chi_b, \tag{8.37}$$

which leads to the following expressions for the ghost propagator and its coupling to the gauge boson:

$$G_F^0(p) = \quad \cdots\!\!\!\!\!\xrightarrow{p}\!\!\!\!\!\cdots \quad = -\frac{\delta_{ab}}{p \cdot n + i0^+},$$

$$\begin{array}{c} a \diagdown r \\ \phantom{a}\!\!\!\!\!\diagdown \\ \phantom{aa}\!\!\!\!\!\!\!\!\!\!\text{\scriptsize\textbackslash} \\ \phantom{aaa}\text{\tiny ooooo}\!\!\!\xleftarrow{q} \\ \phantom{aaaa}\text{\scriptsize c}\,\mu \\ \phantom{a}\diagup \\ b\diagup\,p \end{array} = i g\, f^{abc}\, n_\mu. \tag{8.38}$$

A significant simplification of these Feynman rules occurs in the limit $\xi \to \infty$ (that one may call the *strict axial gauge*, since the condition $n^\mu A^a_\mu = 0$ holds exactly in this limit). In this limit, the gauge boson propagator becomes

$$G^{0\,\mu\nu}_{F\,ab}(p) = \frac{-i\delta_{ab}}{p^2 + i0^+}\left[g^{\mu\nu} - \frac{p^\mu n^\nu + p^\nu n^\mu}{p \cdot n} + \frac{p^\mu p^\nu\, n^2}{(p \cdot n)^2}\right], \qquad (8.39)$$

and satisfies

$$n_\mu\, G^{0\,\mu\nu}_{F\,ab}(p) = n_\nu\, G^{0\,\mu\nu}_{F\,ab}(p) = 0. \qquad (8.40)$$

Therefore, the gauge boson propagator gives zero when contracted into the ghost-gauge boson vertex, which effectively decouples the ghosts from the gauge bosons. Thus, the limit $\xi \to \infty$ of the axial gauge is ghost-free (but its propagator is arguably much more complicated than the Feynman gauge propagator).

## 8.5 On-Shell Non-Abelian Ward–Takahashi Identities

In QED, the interpretation of Cutkosky's cutting rules as a perturbative realization of unitarity depends crucially on the Ward–Takahashi identities satisfied by amplitudes with external photons, namely

$$q_1^{\mu_1}\Gamma_{\mu_1\mu_2\cdots}(q_1, q_2, \cdots) = 0, \qquad (8.41)$$

valid when all the charged external lines are on-shell. Note that this identity does not require contracting the remaining photons with a polarization vector.

In Yang–Mills theory, it turns out that the identity (8.41) is in general not satisfied. Instead, it is replaced by a different on-shell identity, discovered by 't Hooft. In order to derive it, let us consider a generalized covariant gauge condition of the form

$$\partial_\mu A^\mu_a(x) - \zeta_a(x) = \omega_a(x). \qquad (8.42)$$

As in the Fadeev–Popov quantization, we integrate over the function $\omega_a$ with a Gaussian weight, which leads to the following gauge fixed Lagrangian:

$$\mathcal{L} \equiv -\frac{1}{4}F^a_{\mu\nu}F^{a\,\mu\nu} - \frac{\xi}{2}(\partial_\mu A^\mu_a - \zeta_a)^2 + \overline{\chi}_a \mathcal{M}_{ab}\chi_b. \qquad (8.43)$$

This Lagrangian is the same as the one encountered earlier (with $\zeta_a \equiv 0$), with the addition of two terms that depend on the arbitrary function $\zeta_a(x)$ introduced in the gauge fixing:

$$-\frac{\xi}{2}\zeta_a\zeta_a + \xi\zeta_a\partial_\mu A^\mu_a. \qquad (8.44)$$

Since $\zeta_a$ is not a dynamical field but merely a parameter, the first term can be factored out of the generating functional for correlation functions, and cancels once it is properly normalized.

The second term acts as a source coupled to the divergence of the gauge field. In momentum space, the Feynman rule for the insertion of such a source is

$$\underset{a\ \mu}{\otimes\underrightarrow{\overset{k}{\phantom{mmmm}}}} \quad = \quad i\xi k_\mu \widetilde{\zeta}_a(k), \tag{8.45}$$

where $\widetilde{\zeta}_a$ is the Fourier transform of $\zeta_a$. This source is always contracted into an external gluon propagator, leading to the combination[5]

$$i\xi k_\mu \widetilde{\zeta}_a(k)\, G^{0\,\mu\nu}_{F\,ab}(k) = \frac{\widetilde{\zeta}_b(k)\, k^\nu}{k^2}. \tag{8.46}$$

(The propagator is given in eq. (8.23).) In this contraction, the external gluon propagator is replaced by a factor $k^\nu$ directly contracted into the amputated correlation function, independently of the gauge parameter $\xi$.

Since the function $\zeta_a$ has been introduced as part of our choice of gauge fixing condition, gauge invariant quantities should not depend on it. Consequently, the sum of graphs contributing to gauge invariant quantities with a given non-zero number of insertions of the source $\zeta_a$ must be zero. Consider S-matrix elements, i.e., transition amplitudes between *physical* states. The graphs contributing to such a matrix element have a number of external gluons corresponding to the in- and out-states of the amplitude, plus possibly some insertions of the source $\zeta_a$,

$$\text{out}\left[\begin{array}{c}\Gamma\end{array}\right]\text{in} = \Gamma_{\{\mu a\}_{i\in[1,n]}\atop\{\nu b\}_{j\in[1,p]}}(\underbrace{k_1\cdots k_n}_{\text{on-shell}};\underbrace{q_1\cdots q_p}_{\text{off-shell}})$$

$$\times \left[\prod_{i=1}^n \underbrace{\epsilon^{\mu_i}(k_i)}_{\text{physical}}\right] \times \left[\prod_{j=1}^p \frac{q_j^{\nu_j}\widetilde{\zeta}_{b_j}(q_j)}{q_j^2}\right] = 0. \tag{8.47}$$

On the right-hand side, $\Gamma$ is an $(n + p)$-gluon amplitude, with amputated external lines. $n$ of these gluons correspond to the in- and out-states, with colors $\{a_i\}$. The corresponding momenta $\{k_i\}$ are on-shell and the Lorentz indices $\{\mu_i\}$ are contracted with physical polarization vectors. In contrast, the lines to which the sources $\zeta_{b_j}$ are attached are off-shell and contracted with their own momenta $\{q_j\}$. When including all the graphs contributing to a given order in the coupling and in $\zeta$, this expression must vanish if $p \geq 1$ because it is a contribution to a gauge invariant quantity. Thus, the Ward–Takahashi identity of eq. (8.41) can be adapted to non-Abelian gauge theories, with the restriction that all the gluon lines not contracted with $q_j^{\nu_j}$ must be on-shell and contracted with a physical polarization vector. For instance, the gluon self-energy obeys the following two relations:

$$k_\mu \epsilon_\nu(k)\, \Pi^{\mu\nu}(k)\Big|_{k^2=0} = 0, \quad k_\mu k_\nu\, \Pi^{\mu\nu}(k) = 0, \tag{8.48}$$

---

[5] In axial gauge, the insertion of a source $\zeta_a$ leads to a factor $\widetilde{\zeta}_b(k)\,k^\nu/(k\cdot n)$. Thanks to the factor $k^\nu$, the identity (8.47) is also valid in axial gauges.

while in QED it is sufficient to contract the self-energy with a single k (even off-shell) to obtain zero. Note that these identities are insufficient in order to obtain a unitary S-matrix with only internal gluons, because the tensor structure of the internal cut gluon propagators also involves polarizations which are neither physical nor proportional to $q^\nu$. The Zinn–Justin equation, which we derive in the next chapter, may be viewed as a generalization of these Ward–Takahashi identities to off-shell momenta and arbitrary polarizations.

## 8.6 Ghosts and Unitarity

### 8.6.1 Explicit Example

In Abelian gauge theories, we were able to show that cutting rules provide a perturbative realization of the optical theorem, by using the Ward–Takahashi identities obeyed by amplitudes when all the external charged particles are on-shell. These identities were sufficient to conclude that the unphysical polarizations carried by the internal photon lines of a graph cancel when these lines are cut. But in non-Abelian gauge theories, this reasoning faces two difficulties:

1. The Ward–Takahashi identities are weaker than the QED ones, and insufficient to prove unitarity.
2. Higher-order graphs in general have ghost loops, whose interpretation is at the moment unclear when such loops are cut.

As we shall see, these two issues are in fact related: The cut ghost lines precisely cancel the unphysical polarizations of the cut gluons. Let us first work out an explicit example that illustrates this assertion: the tree-level annihilation of a quark and an antiquark into two gluons in QCD. The corresponding diagrams are the following:

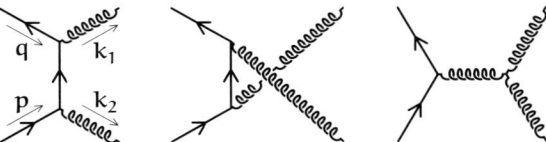

We denote p and q the momenta of the incoming quark and antiquark, respectively, and $k_{1,2}$ the momenta of the outgoing gluons (with Lorentz indices $\mu$, $\nu$ and colors a, b, respectively).

The contribution of the first two graphs is very similar to that of the analogous graphs in QED for the emission of two photons, except for the extra color matrices at the quark–gluon vertices:

$$i\mathcal{M}^{\mu\nu}_{ab}\big|_{1+2}(\mathbf{p},\mathbf{q}|k_1,k_2) = (ig)^2\, \bar{v}(\mathbf{q}) \Big\{ \gamma^\mu t^a \frac{i}{\slashed{k}_1 - \slashed{q} - m} \gamma^\nu t^b$$
$$+ \gamma^\nu t^b \frac{i}{\slashed{p} - \slashed{k}_1 - m} \gamma^\mu t^a \Big\} u(\mathbf{p}). \qquad (8.49)$$

By contracting this amplitude with the photon momentum $k_{1\mu}$, we get

$$k_{1\mu}\, i\mathcal{M}^{\mu\nu}_{ab}\big|_{1+2}(\mathbf{p},\mathbf{q}|k_1,k_2) = (ig)^2\, \bar{v}(\mathbf{q}) \Big\{ \slashed{k}_1 t^a \frac{i}{\slashed{k}_1 - \slashed{q} - m} \gamma^\nu t^b$$
$$+ \gamma^\nu t^b \frac{i}{\slashed{p} - \slashed{k}_1 - m} \slashed{k}_1 t^a \Big\} u(\mathbf{p}). \qquad (8.50)$$

In the numerator of the first term, we may write

$$\not{k}_1 = (\not{k}_1 - \not{q} - m) + (\not{q} + m), \tag{8.51}$$

and use the Dirac equation $\bar{v}(\mathbf{q})(\not{q} + m) = 0$. Likewise, we may simplify the second term by using

$$\not{k}_1 = (\not{p} - m) - (\not{p} - \not{k}_1 - m), \quad (\not{p} - m)u(\mathbf{p}) = 0, \tag{8.52}$$

which leads to

$$k_{1\mu} \, iM_{ab}^{\mu\nu}|_{1+2}(\mathbf{p}, \mathbf{q}|\mathbf{k}_1, \mathbf{k}_2) = i\,(ig)^2 \, \bar{v}(\mathbf{q}) \gamma^\nu [t^a, t^b] u(\mathbf{p}). \tag{8.53}$$

This is non-zero, because of the non-commutativity of the Lie generators in a non-Abelian gauge theory. However, by using $[t^a, t^b] = if^{abc}t^c$, this result may be related to the third graph, which contains a three-gluon vertex. If we use the Feynman gauge for the internal gluon propagator, its contribution can be written as

$$iM_{ab}^{\mu\nu}|_3(\mathbf{p},\mathbf{q}|\mathbf{k}_1,\mathbf{k}_2) = ig\,\bar{v}(\mathbf{q})\gamma_\rho t^c u(\mathbf{p}) \frac{-i}{k_3^2}$$
$$\times g\,f^{abc} [g^{\mu\nu}(k_2 - k_1)^\rho + g^{\nu\rho}(k_3 - k_2)^\mu + g^{\rho\mu}(k_1 - k_3)^\nu], \tag{8.54}$$

where we denote $k_3 \equiv -k_1 - k_2$. Contracting this amplitude with $k_{1\mu}$ gives

$$k_{1\mu} \, iM_{ab}^{\mu\nu}|_3(\mathbf{p},\mathbf{q}|\mathbf{k}_1,\mathbf{k}_2) = ig\,\bar{v}(\mathbf{q})\gamma_\rho t^c u(\mathbf{p}) \frac{-i}{k_3^2}$$
$$\times g\,f^{abc} [g^{\nu\rho}k_2^2 - k_2^\nu k_2^\rho - g^{\nu\rho}k_3^2 + k_3^\nu k_3^\rho]. \tag{8.55}$$

In this equation, the term in $k_3^\nu k_3^\rho$ vanishes once contracted with $\gamma_\rho$, since we can write

$$\bar{v}(\mathbf{q})\gamma_\rho t^c u(\mathbf{p}) k_3^\rho = -\bar{v}(\mathbf{q})[(\not{p} - m) + (\not{q} + m)]t^c u(\mathbf{p}) = 0. \tag{8.56}$$

However, this is not sufficient for eq. (8.55) to fully cancel eq. (8.53).

Setting $k_2^2 = 0$ kills another term in eq. (8.55). The term in $k_2^\nu k_2^\rho$ would be canceled if in addition we contract the amplitudes with a *transverse* polarization vector $\epsilon_{\lambda\nu}(\mathbf{k}_2)$ with $\lambda = 1, 2$, since $k_2^\nu \epsilon_{\lambda\nu}(\mathbf{k}_2) = 0$. We indeed have

$$k_{1\mu}\,\epsilon_{\lambda\nu}(\mathbf{k}_2) \left[iM_{ab}^{\mu\nu}|_{1+2}(\mathbf{p},\mathbf{q}|\mathbf{k}_1,\mathbf{k}_2) + iM_{ab}^{\mu\nu}|_3(\mathbf{p},\mathbf{q}|\mathbf{k}_1,\mathbf{k}_2)\right]_{k_2^2=0} = 0. \tag{8.57}$$

The same cancellation happens if we contract the amplitudes simultaneously with $k_{1\mu}$ and $k_{2\nu}$,

$$k_{1\mu}k_{2\nu}\left[iM_{ab}^{\mu\nu}|_{1+2}(\mathbf{p},\mathbf{q}|\mathbf{k}_1,\mathbf{k}_2) + iM_{ab}^{\mu\nu}|_3(\mathbf{p},\mathbf{q}|\mathbf{k}_1,\mathbf{k}_2)\right] = 0, \tag{8.58}$$

even if the momentum $k_2$ is not on-shell. Thus, we obtain for this process a Ward–Takahashi identity similar to the QED one, provided certain extra conditions are satisfied by the second gluon (both eqs. (8.57) and (8.58) are special cases of the non-Abelian Ward–Takahashi identity (8.47)). These restrictions weaken the resulting identity, and it is not sufficient to eliminate the longitudinal gluon polarizations when we try to recover the amplitude from the

## 8.6 GHOSTS AND UNITARITY

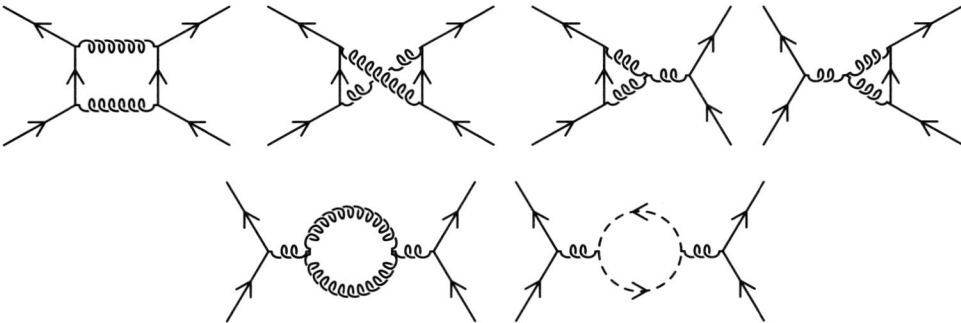

FIGURE 8.3: One-loop diagrams contributing to $q\bar{q} \to q\bar{q}$.

imaginary part of the $q\bar{q} \to q\bar{q}$ forward amplitude at one loop. In particular, some unphysical polarizations do not cancel in the following cut:

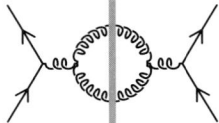

Except for a graph with a quark loop that does not play any role in the present discussion (since it does not give any two-gluon final state when cut), the complete list of graphs contributing to the $q\bar{q} \to q\bar{q}$ forward amplitude at one loop is shown in Figure 8.3. The contribution of the first five graphs (i.e., those with internal gluon lines) to the optical theorem can be calculated easily by noting that it can be expressed in terms of the amplitude we have just calculated,

$$i M_{ab}^{\mu\nu}(\mathbf{p}, \mathbf{q}|\mathbf{k}_1, \mathbf{k}_2) \equiv i M_{ab}^{\mu\nu}|_{1+2}(\mathbf{p}, \mathbf{q}|\mathbf{k}_1, \mathbf{k}_2) + i M_{ab}^{\mu\nu}|_3(\mathbf{p}, \mathbf{q}|\mathbf{k}_1, \mathbf{k}_2), \tag{8.59}$$

as follows:[6]

$$\begin{aligned}
\frac{1}{2} \int \frac{d^4 k_1}{(2\pi)^4} & \int \frac{d^4 k_2}{(2\pi)^4} (2\pi)^4 \delta^{(4)}(p + q - k_1 - k_2) \\
& \times 2\pi (-g_{\mu\rho}) \theta(k_1^0) \delta(k_1^2 - m^2) \, 2\pi (-g_{\nu\sigma}) \theta(k_2^0) \delta(k_2^2 - m^2) \\
& \times i M_{ab}^{\mu\nu}(\mathbf{p}, \mathbf{q}|\mathbf{k}_1, \mathbf{k}_2) \, (i M_{ab}^{\rho\sigma}(\mathbf{p}, \mathbf{q}|\mathbf{k}_1, \mathbf{k}_2))^*.
\end{aligned} \tag{8.60}$$

For a successful interpretation of this formula as a physical contribution in the optical theorem, only physical polarizations should survive after we have replaced the tensors $-g_{\mu\rho}$ and $-g_{\nu\sigma}$ by using (see eq. (3.108))

$$g^{\mu\nu} = \epsilon_+^\mu(k)\epsilon_-^\nu(k)^* + \epsilon_-^\mu(k)\epsilon_+^\nu(k)^* - \sum_{\lambda=1,2} \epsilon_\lambda^\mu(k)\epsilon_\lambda^\nu(k)^*, \tag{8.61}$$

---

[6] The factor $1/2$ is a symmetry factor due to the presence of two identical gluons in the final state.

where $\epsilon_\pm^\mu(\mathbf{k})$ are unphysical polarizations (with $\epsilon_+^\mu(\mathbf{k})$ proportional to $k^\mu$). After this substitution, several terms are not problematic:

- The terms that contain only the transverse polarizations $\epsilon_{1,2}^\mu$ are fully physical.
- The terms containing $\epsilon_{1,2}^\mu \epsilon_+^\nu$ or $\epsilon_+^\mu \epsilon_+^\nu$ vanish by virtue of eqs. (8.57) and (8.58).

Thus, we need only to focus on the following term:

$$\frac{1}{2}\left[(i M_{ab}^{\mu\nu}\epsilon_{-\mu}\epsilon_{+\nu})(i M_{ab}^{\rho\sigma}\epsilon_{+\rho}\epsilon_{-\sigma})^* + (i M_{ab}^{\mu\nu}\epsilon_{+\mu}\epsilon_{-\nu})(i M_{ab}^{\rho\sigma}\epsilon_{-\rho}\epsilon_{+\sigma})^*\right], \quad (8.62)$$

integrated over the on-shell momenta $k_1$ and $k_2$. Using $\epsilon_+^\mu(\mathbf{k}) = k^\mu/\sqrt{2}|\mathbf{k}|$ and eqs. (8.53) and (8.55), we obtain

$$\epsilon_{+\mu}(\mathbf{k}_1) i M_{ab}^{\mu\nu} = -\frac{g^2}{\sqrt{2}|\mathbf{k}_1|}\frac{1}{k_3^2}\bar{v}(\mathbf{q})\slashed{k}_2 k_2^\nu f^{abc} t^c u(\mathbf{p}). \quad (8.63)$$

Likewise with the other gluon, we have

$$\epsilon_{+\nu}(\mathbf{k}_2) i M_{ab}^{\mu\nu} = \frac{g^2}{\sqrt{2}|\mathbf{k}_2|}\frac{1}{k_3^2}\bar{v}(\mathbf{q})\slashed{k}_1 k_1^\nu f^{abc} t^c u(\mathbf{p}). \quad (8.64)$$

Using then $\epsilon_-^\mu(\mathbf{k}) = (k_0, -\mathbf{k})/\sqrt{2}|\mathbf{k}|$, we get

$$\epsilon_{-\nu}(\mathbf{k}_2)\epsilon_{+\mu}(\mathbf{k}_1) i M_{ab}^{\mu\nu} = -g^2 \frac{|\mathbf{k}_2|}{|\mathbf{k}_1|}\frac{1}{k_3^2}\bar{v}(\mathbf{q})\slashed{k}_2 f^{abc} t^c u(\mathbf{p}),$$

$$\epsilon_{+\nu}(\mathbf{k}_2)\epsilon_{-\mu}(\mathbf{k}_1) i M_{ab}^{\mu\nu} = +g^2 \frac{|\mathbf{k}_1|}{|\mathbf{k}_2|}\frac{1}{k_3^2}\bar{v}(\mathbf{q})\slashed{k}_1 f^{abc} t^c u(\mathbf{p}). \quad (8.65)$$

Furthermore, notice that

$$\bar{v}(\mathbf{q})(\slashed{k}_1 + \slashed{k}_2)u(\mathbf{p}) = \bar{v}(\mathbf{q})(\slashed{q} + m + \slashed{p} - m)u(\mathbf{p}) = 0. \quad (8.66)$$

Combining these equations, the non-physical contribution to the optical theorem of the diagrams with a gluon loop, eq. (8.62), can be written as follows:

$$g^4 \frac{1}{(k_3^2)^2}\left[\bar{v}(\mathbf{q})\slashed{k}_1 f^{abc} t^c u(\mathbf{p})\right]\left[\bar{v}(\mathbf{q})\slashed{k}_1 f^{abd} t^d u(\mathbf{p})\right]. \quad (8.67)$$

If this was all there is, as the naive Feynman rules we tried to guess at the beginning of this chapter would suggest, then we would have to conclude that Yang–Mills theories are inconsistent because they violate unitarity. Fortunately, there is one more graph in Figure 8.3, with a ghost loop. Let us first evaluate the annihilation amplitude of the quark–antiquark pair into a ghost–antighost pair:

$$i M_{q\bar{q}\to\chi\bar{\chi}} = i g \bar{v}(\mathbf{q})\gamma_\rho t^c u(\mathbf{p})\frac{i}{k_3^2}(g f^{abc} k_1^\rho). \quad (8.68)$$

Squaring this amplitude, and including the minus sign[7] associated to a ghost loop,[8] the contribution of the last graph of Figure 8.3 to the optical theorem becomes

$$-g^4 \frac{1}{(k_3^2)^2} \left[\overline{v}(q) k_1 f^{abc} t^c u(p)\right] \left[\overline{v}(q) k_1 f^{abd} t^d u(p)\right], \tag{8.69}$$

which exactly cancels the unphysical gluon contribution of eq. (8.67). In other words, the optical theorem is satisfied with only physical modes in the final state sum, thanks to a crucial cancellation that involves ghosts.

### 8.6.2 Becchi–Rouet–Stora–Tyutin Symmetry

The cancellation that occurred in the previous example is in fact general: For every gluon loop, there is a graph of identical topology where this loop is replaced by a ghost loop that cancels the contribution from the unphysical gluon polarizations in the optical theorem. However, it is difficult to transform the calculation of the previous subsection into a general proof. It turns out that this cancellation originates from a residual symmetry of the gauge fixed Lagrangian: Although the gauge fixing term explicitly breaks the gauge symmetry, the effective Lagrangian in eq. (8.21) has a remnant of the original gauge symmetry, known as *Becchi–Rouet–Stora–Tyutin symmetry* (BRST).

Recall that under an infinitesimal gauge transformation parametrized by $\theta_a(x)$, the gauge field and fermion field vary by

$$\delta A_\mu^a(x) = -\left(D_\mu^{adj}\right)_{ab} \theta_b(x), \quad \delta \psi(x) = -i g \theta_a(x) t_r^a \psi(x), \tag{8.70}$$

where $r$ is the representation in which the fermions live. A BRST transformation is similar to the above transformation, but with the substitution $\theta_a(x) \to -\vartheta \chi_a(x)$, where $\vartheta$ is a Grassmann constant,[9]

$$\delta_{\text{BRST}} A_\mu^a(x) = \left(D_\mu^{adj}\right)_{ab} \left[\vartheta \chi_b(x)\right], \quad \delta_{\text{BRST}} \psi(x) = i g \left[\vartheta \chi_a(x)\right] t_r^a \psi(x). \tag{8.71}$$

Since the BRST transformation is structurally identical to a local gauge transformation, any gauge invariant combination of gauge fields and fermions is also BRST invariant. This is therefore the case of the Yang–Mills Lagrangian and the Dirac Lagrangian with a minimal coupling of the fermions to the gauge fields. It is customary to introduce a generator $Q_{\text{BRST}}$ for this transformation, by denoting $\delta_{\text{BRST}} \equiv \vartheta \, Q_{\text{BRST}}$. Thus, we have

$$Q_{\text{BRST}} A_\mu^a(x) = \left(D_\mu^{adj}\right)_{ab} \chi_b(x), \quad Q_{\text{BRST}} \psi(x) = i g \chi_a(x) t_r^a \psi(x). \tag{8.72}$$

Eqs. (8.71) do not tell how ghost and antighost fields transform under BRST. For reasons that will become clear later, we shall impose that the BRST transformation is nilpotent, i.e., that $Q_{\text{BRST}}^2 = 0$ when applied to any of the fields of the theory. This requirement constrains

---

[7] There is no 1/2 symmetry factors for a ghost–antighost final state, because they are not identical.

[8] We see here how essential it is that ghosts are anti-commuting fields – otherwise, their contribution would not have the proper sign to cancel the unphysical gluon polarizations in the optical theorem.

[9] This Grassmann constant makes $\vartheta \chi_a(x)$ a commuting object like $\theta_a$.

the BRST transformation of the ghosts. Indeed, a double BRST transformation applied to fermions reads

$$\begin{aligned}Q_{\text{BRST}}^2 \psi(x) &= i g \left\{ (Q_{\text{BRST}} \chi_a(x)) t_r^a \psi(x) - \chi_a(x) t_r^a Q_{\text{BRST}} \psi(x) \right\} \\ &= i g (Q_{\text{BRST}} \chi_a(x)) t_r^a \psi(x) + g^2 \chi_a(x) \chi_b(x) t_r^a t_r^b \psi(x). \end{aligned} \quad (8.73)$$

(The BRST generator is an anti-commuting object, which leads to a minus sign in the second term of the first line when we push it through the Grassmann field $\chi_a$.) Since $\chi_a$ and $\chi_b$ anti-commute, we can replace $t_r^a t_r^b$ by $\frac{1}{2}[t_r^a, t_r^b] = \frac{i}{2} f^{abc} t_r^c$. We see that eq. (8.73) will identically vanish provided that

$$Q_{\text{BRST}} \chi_a(x) = -\frac{1}{2} g f^{abc} \chi_b(x) \chi_c(x). \quad (8.74)$$

Then we can calculate the action of a double BRST transformation on the gauge field,

$$\begin{aligned}Q_{\text{BRST}}^2 A_\mu^a &= (D_\mu^{\text{adj}})_{ab} (Q_{\text{BRST}} \chi_b) - g f^{abc} (Q_{\text{BRST}} A_\mu^c) \chi_b \\ &= (D_\mu^{\text{adj}})_{ab} \left[ -\tfrac{g}{2} f^{bcd} \chi_c(x) \chi_d(x) \right] - g f^{abc} \left[ \partial_\mu \chi_c - g f^{cde} A_\mu^e \chi_d \right] \chi_b. \end{aligned} \quad (8.75)$$

The terms linear in the gauge field cancel by using the anti-commuting nature of the $\chi$s and the Jacobi identity satisfied by the structure constants,

$$\tfrac{1}{2} g^2 f^{abe} f^{bcd} A_\mu^e \chi_c(x) \chi_d(x) + g^2 f^{abc} f^{cde} A_\mu^e \chi_d \chi_b \\ = \tfrac{1}{2} g^2 \underbrace{\left[ f^{ace} f^{cbd} - f^{abc} f^{cde} + f^{adc} f^{cbe} \right]}_{0} A_\mu^e \chi_b \chi_d, \quad (8.76)$$

and the terms with the derivative $\partial_\mu$ also vanish:

$$-\tfrac{1}{2} g f^{acd} \partial_\mu (\chi_c \chi_d) - g f^{abc} (\partial_\mu \chi_c) \chi_b \\ = \tfrac{1}{2} g f^{abc} \big[ \partial_\mu (\chi_c \chi_b) \underbrace{-(\partial_\mu \chi_c) \chi_b + (\partial_\mu \chi_b) \chi_c}_{\substack{-(\partial_\mu \chi_c) \chi_b - \chi_c (\partial_\mu \chi_b) \\ = -\partial_\mu (\chi_c \chi_b)}} \big] = 0. \quad (8.77)$$

The double BRST transformation of the ghost field also vanishes:

$$Q_{\text{BRST}}^2 \chi_a = \tfrac{g^2}{4} \underbrace{\left( f^{abc} f^{bde} + f^{acb} f^{bde} \right)}_{0} \chi_c \chi_d \chi_e. \quad (8.78)$$

Therefore, the prescription (8.74) for the BRST transformation of a ghost field leads to

$$Q_{\text{BRST}}^2 \psi = 0, \quad Q_{\text{BRST}}^2 A_\mu^a = 0, \quad Q_{\text{BRST}}^2 \chi_a = 0. \quad (8.79)$$

We need now to specify the BRST transformation of the antighost field. Note that in the path integral that gives the Fadeev–Popov determinant, the ghost and antighost fields are independent (this is how we get a determinant instead of the square root of a determinant); therefore, the BRST transformation of the antighost does not have to be related to that of the ghost. Let us denote

$$Q_{\text{BRST}} \bar{\chi}_a(x) \equiv B_a(x), \quad (8.80)$$

## 8.6 GHOSTS AND UNITARITY

where $B_a(x)$ is a commuting field. For $Q_{BRST}$ to be nilpotent, we must have in addition:

$$Q_{BRST} B_a(x) = 0. \tag{8.81}$$

(And of course $Q_{BRST}^2 B_a(x) = 0$.)

Consider now a local function $\Xi$ of all the fields (including $B_a$), and add its BRST variation to the Yang–Mills and Dirac Lagrangians:

$$\mathcal{L} \equiv \underbrace{\mathcal{L}_{YM} + \mathcal{L}_D}_{\text{BRST invariant}} + Q_{BRST} \Xi. \tag{8.82}$$

Since $Q_{BRST}$ is nilpotent, this Lagrangian is BRST invariant. Let us choose

$$\Xi \equiv \overline{\chi}_a(x) \left[ \frac{1}{2\xi} B^a(x) + G^a(A(x)) \right], \tag{8.83}$$

where $\xi$ is a parameter and $G^a(A)$ is the gauge fixing function. We can write[10]

$$Q_{BRST} \Xi = (Q_{BRST} \overline{\chi}_a) \left[ \frac{1}{2\xi} B^a + G^a \right] - \overline{\chi}_a \left[ \frac{1}{2\xi} (Q_{BRST} B^a) + \frac{\partial G^a}{\partial A_\mu^b} (Q_{BRST} A_\mu^b) \right]$$

$$= \frac{1}{2\xi} B^a B^a + B^a G^a + \underbrace{\overline{\chi}_a \frac{\partial G^a}{\partial A_\mu^b} (-D_\mu^{adj})_{bc} \chi_c}_{\mathcal{L}_{FPG}}. \tag{8.84}$$

Note that the last term is nothing but the Fadeev–Popov part of the Lagrangian we derived earlier in this chapter. Moreover, the field $B^a$ enters only quadratically in this Lagrangian, and the path integral on $B^a$ can be performed trivially,[11]

$$\int [DB^a(x)] \, e^{i \int d^4 x \left( \frac{1}{2\xi} B^a B^a + B^a G^a \right)} = e^{-i \frac{\xi}{2} \int d^4 x \, G^a G^a}. \tag{8.85}$$

Therefore, after integrating out the auxiliary field $B^a$, the resulting theory has exactly the same effective Lagrangian as the one resulting from the Fadeev–Popov procedure,

$$\mathcal{L}_{eff} = \mathcal{L}_{YM} + \mathcal{L}_D - \frac{\xi}{2} G^a G^a + \overline{\chi}_a \frac{\partial G^a}{\partial A_\mu^b} (-D_\mu^{adj})_{bc} \chi_c. \tag{8.86}$$

The formal construction we have followed in this section proves that $\mathcal{L}_{eff}$ is BRST invariant, but in a somewhat obfuscated manner after the auxiliary field $B^a$ has been integrated out. The BRST invariance of eq. (8.86) is realized if we define the BRST variation of the antighost field as follows:

$$Q_{BRST} \overline{\chi}_a = -\xi G^a, \tag{8.87}$$

which is reminiscent of the relationship between $B^a$ and $G^a$ when we do the Gaussian integration on $B^a$.

---

[10] Note that a minus sign arises when moving $Q_{BRST}$ through the anti-commuting field $\overline{\chi}_b$.

[11] Note that this is equivalent to evaluating the argument of the exponential at the stationary point $B^a = -\xi G^a$, since the stationary phase approximation is exact for Gaussian integrals.

### 8.6.3 BRST Current and Charge

The Lagrangian (8.82), with the choice (8.83) for the function $\Xi$, possesses the following symmetries:

- global gauge invariance (because all the color indices are contracted);
- BRST invariance; and
- ghost number conservation, if we assign a ghost number $+1$ to $\chi$ and $-1$ to $\bar{\chi}$.

The BRST invariance implies the existence of a conserved current,

$$J^\mu_{\text{BRST}} \equiv \sum_{\Phi \in \{A_\mu, \psi, \chi, \bar{\chi}, B\}} \frac{\partial \mathcal{L}}{\partial(\partial_\mu \Phi)} \left( \mathbf{Q}_{\text{BRST}} \, \Phi \right). \tag{8.88}$$

From the zeroth component of this current, we may obtain the BRST charge

$$\mathcal{Q}_{\text{BRST}} \equiv \int d^3x \, J^0_{\text{BRST}}(x^0, \mathbf{x}). \tag{8.89}$$

In fact, this charge generates the BRST transformation in the following sense:

$$i \left[ \mathcal{Q}_{\text{BRST}}, \Phi \right]_\pm = \mathbf{Q}_{\text{BRST}} \, \Phi \qquad (\Phi \in \{A_\mu, \psi, \chi, \bar{\chi}, B\}), \tag{8.90}$$

where $[\cdot, \cdot]_\pm$ is a commutator if $\Phi$ is a commuting field and an anti-commutator if $\Phi$ is anti-commuting. If we consider free fields (i.e., we set $g = 0$), and we Fourier decompose all the fields that appear in the (anti-)commutation relations (8.90),

$$A^\mu_a(x) \equiv \sum_{\lambda=1,2,+,-} \int \frac{d^3\mathbf{p}}{(2\pi)^3 2|\mathbf{p}|} \left\{ \epsilon^\mu_\lambda(\mathbf{p}) \, a^\dagger_{a\lambda\mathbf{p}} e^{+ip \cdot x} + \epsilon^{\mu*}_\lambda(\mathbf{p}) \, a_{a\lambda\mathbf{p}} e^{-ip \cdot x} \right\}$$

$$\psi(x) \equiv \sum_{s=\pm} \int \frac{d^3\mathbf{p}}{(2\pi)^3 2E_\mathbf{p}} \left\{ d^\dagger_{s\mathbf{p}} v_s(\mathbf{p}) e^{+ip \cdot x} + b_{s\mathbf{p}} u_s(\mathbf{p}) e^{-ip \cdot x} \right\}$$

$$\chi_a(x) \equiv \int \frac{d^3\mathbf{p}}{(2\pi)^3 2|\mathbf{p}|} \left\{ \alpha^\dagger_{a\mathbf{p}} e^{+ip \cdot x} + \alpha_{a\mathbf{p}} e^{-ip \cdot x} \right\}$$

$$\bar{\chi}_a(x) \equiv \int \frac{d^3\mathbf{p}}{(2\pi)^3 2|\mathbf{p}|} \left\{ \beta^\dagger_{a\mathbf{p}} e^{+ip \cdot x} + \beta_{a\mathbf{p}} e^{-ip \cdot x} \right\}, \tag{8.91}$$

we obtain (see Exercise 8.4)

$$\left[ \mathcal{Q}_{\text{BRST}}, a^\dagger_{a\lambda\mathbf{p}} \right] \propto \delta_{\lambda+} \alpha^\dagger_{a\mathbf{p}}, \quad \left\{ \mathcal{Q}_{\text{BRST}}, \alpha^\dagger_{a\mathbf{p}} \right\} = 0,$$
$$\left\{ \mathcal{Q}_{\text{BRST}}, \beta^\dagger_{a\mathbf{p}} \right\} \propto a^\dagger_{a-\mathbf{p}}, \quad \left\{ \mathcal{Q}_{\text{BRST}}, b^\dagger_{s\mathbf{p}} \right\} = \left\{ \mathcal{Q}_{\text{BRST}}, d^\dagger_{s\mathbf{p}} \right\} = 0. \tag{8.92}$$

### 8.6.4 BRST Cohomology, Physical States and Unitarity

The fact that the BRST charge is nilpotent, $\mathcal{Q}^2_{\text{BRST}} = 0$, has profound implications for the states of the system. The *kernel* of $\mathcal{Q}_{\text{BRST}}$ is the set of states annihilated by $\mathcal{Q}_{\text{BRST}}$,

$$\text{Ker}\left( \mathcal{Q}_{\text{BRST}} \right) \equiv \left\{ \psi \, \middle| \, \mathcal{Q}_{\text{BRST}} |\psi\rangle = 0 \right\}. \tag{8.93}$$

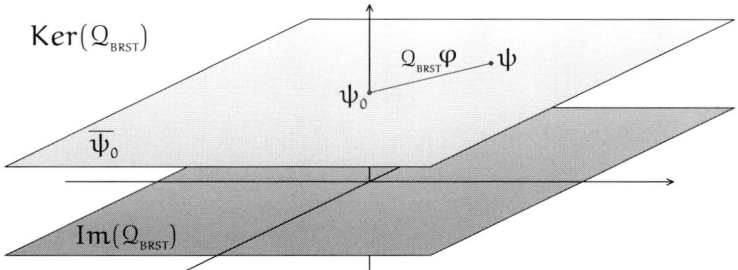

FIGURE 8.4: An equivalence class $\overline{\psi}_0$ may be viewed as a hyperplane parallel to $\text{Im}(\mathcal{Q}_{\text{BRST}})$. The BRST cohomology is the set of all such planes.

The set of states that can be obtained by the action of $\mathcal{Q}_{\text{BRST}}$ on another state is called the *image* of $\mathcal{Q}_{\text{BRST}}$,

$$\text{Im}(\mathcal{Q}_{\text{BRST}}) \equiv \{\mathcal{Q}_{\text{BRST}}|\psi\rangle\}. \tag{8.94}$$

Because $\mathcal{Q}_{\text{BRST}}$ is nilpotent, the image is a subset of the kernel,

$$\text{Im}(\mathcal{Q}_{\text{BRST}}) \subset \text{Ker}(\mathcal{Q}_{\text{BRST}}). \tag{8.95}$$

Note also that states in the image cannot be physical states, because they have a null norm,

$$\langle \psi | \psi \rangle = \langle \phi | \underbrace{\mathcal{Q}_{\text{BRST}} \mathcal{Q}_{\text{BRST}}}_{0} | \phi \rangle = 0. \tag{8.96}$$

Consider now the following equivalence relationship between states in the kernel: Two states are considered equivalent if their difference is in the image (see Figure 8.4):

$$|\psi\rangle \sim |\psi_0\rangle \quad \text{if} \quad |\psi\rangle = |\psi_0\rangle + \mathcal{Q}_{\text{BRST}}|\varphi\rangle. \tag{8.97}$$

The *cohomology* of $\mathcal{Q}_{\text{BRST}}$ is the set of classes of equivalent states,

$$\mathsf{H}(\mathcal{Q}_{\text{BRST}}) \equiv \text{Ker}(\mathcal{Q}_{\text{BRST}}) / \text{Im}(\mathcal{Q}_{\text{BRST}}). \tag{8.98}$$

It turns out that the physical states are the elements of the cohomology with non-zero norm.[12] Indeed, using eqs. (8.92), it is easy to prove (see Exercise 8.4) that if $|\psi\rangle$ is a state in the cohomology, then

$$a^\dagger_{a\{1,2\}\mathbf{p}}|\psi\rangle \in \mathsf{H}(\mathcal{Q}_{\text{BRST}}), \quad b^\dagger_{s\mathbf{p}}|\psi\rangle \in \mathsf{H}(\mathcal{Q}_{\text{BRST}}), \quad d^\dagger_{s\mathbf{p}}|\psi\rangle \in \mathsf{H}(\mathcal{Q}_{\text{BRST}}), \tag{8.99}$$

while

$$a^\dagger_{a\pm\mathbf{p}}|\psi\rangle \notin \mathsf{H}(\mathcal{Q}_{\text{BRST}}), \quad \alpha^\dagger_{\mathbf{p}}|\psi\rangle \notin \mathsf{H}(\mathcal{Q}_{\text{BRST}}), \quad \beta^\dagger_{\mathbf{p}}|\psi\rangle \notin \mathsf{H}(\mathcal{Q}_{\text{BRST}}). \tag{8.100}$$

---

[12]This restriction is necessary, because one of the classes in $\mathsf{H}(\mathcal{Q}_{\text{BRST}})$ is $\text{Im}(\mathcal{Q}_{\text{BRST}})$ itself, which we know has only zero-norm states.

In other words, adding to the state a physical particle (gluon with a physical polarization, or quark or antiquark) gives another state in the cohomology, while adding to the state a non-physical quantum (gluon with a non-physical polarization, ghost or antighost) takes the state out of the cohomology.

Furthermore, since the effective Lagrangian is BRST invariant, it corresponds to a Hamiltonian $\mathcal{H}$ that commutes with $\mathcal{Q}_{\text{BRST}}$. Therefore, a state in the kernel (i.e., for which $\mathcal{Q}_{\text{BRST}}|\psi\rangle = 0$) stays in the kernel under the time evolution generated by this Hamiltonian. Moreover, since the time evolution preserves the norm, states in the cohomology stay in the cohomology at all times. Therefore, starting from a physical state, the time evolution cannot produce unphysical objects in the final state. This explains why unphysical modes cancel in the sum over the final states in the optical theorem, despite the fact that the internal lines of Feynman graphs may propagate all sorts of non-physical excitations (see Exercise 8.5 for a more explicit discussion of this point).

## Exercises

**8.1** Consider the quantization of quantum electrodynamics with the following nonlinear gauge condition:

$$\partial_\mu A^\mu + \tfrac{\kappa}{2} A_\mu A^\mu = \omega,$$

where $\kappa$ is a constant.

- Assuming for the time being that the naive quantization procedure works, write the Lagrangian obtained after performing the Gaussian integration over $\omega(x)$. What new couplings does it contain?
- Draw the diagrams that contribute to the one-loop self-energy of the photon, and write the corresponding integrals. Does it satisfy the Ward–Takahashi identity?
- Reconsider the gauge fixing in the light of what has been discussed in this chapter. Why is it necessary to introduce ghosts in order to handle the above gauge condition? Determine the ghost term in the gauge fixed Lagrangian.
- Draw and calculate the additional contribution to the photon self-energy due to ghosts. Check that it restores the Ward–Takahashi identity.

**\*8.2** In this exercise, we study a simple example of QCD cross-section at tree level.

- Assuming massless quarks and electron, show that the total tree-level cross-section for the process $e^+e^- \to q\bar{q}$ is

$$\sigma_{\text{tot,LO}} = \frac{4\pi}{3s} \alpha^2 e_q^2 N,$$

where $e_q$ is the electrical charge of the quark (in units of the electron charge) and $N$ is the number of colors.
- Using the optical theorem, show that this total cross-section is related in the center-of-mass frame to the imaginary part $H^{\mu\nu} \equiv 2\,\text{Im}\,\Pi_q^{\mu\nu}$ of the quark loop contribution to the imaginary part of the photon self-energy by

$$\sigma_{\text{tot}} = \frac{e^2}{2s^3}(p_\mu q_\nu + p_\nu q_\mu - g_{\mu\nu} p\cdot q)\,H^{\mu\nu}(p+q),$$

# EXERCISES

where p, q are the incoming momenta of the electron and positron.
- Using the Ward–Takahashi identity, show that

$$\sigma_{\text{tot}} = -\frac{e^2}{4(p+q)^4} \frac{D-2}{D-1} H^\mu{}_\mu(p+q),$$

where we have kept the number of dimensions arbitrary when contracting the Lorentz indices. Using Cutkosky's rules, use this formula to recover the first expression of the total cross-section.

**\*8.3** This exercise is the continuation of 8.2.
- Draw all the contributions to the photon self-energy at order $g^2$, including the appropriate ultraviolet counterterms.
- We call "real" the contributions where a gluon is produced in the final state. Show that these terms are given by

$$-H^\mu{}_\mu(l)\big|_{\text{real}} = 2NC_f \frac{\alpha \alpha_s e_q^2 l^2}{\pi} \left[\frac{4\pi\mu^2}{l^2}\right]^{2\epsilon} \frac{1-\epsilon}{\Gamma(2-2\epsilon)} \left[\frac{1}{\epsilon^2} + \frac{3}{2\epsilon} - \frac{\pi^2}{2} + \frac{19}{4} + \mathcal{O}(\epsilon)\right],$$

$(C_f \equiv (N^2-1)/2N)$ where the poles in $\epsilon \equiv (4-D)/2$ come from infrared and collinear divergences.
- Consider now the terms with a $q\bar{q}$ final state and a virtual loop. Show that the self-energy corrections on the quark lines do not contribute. Show that the contributions with counterterms cancel thanks to $Z_1 = Z_2$. Calculate the remaining terms and show that its poles in $\epsilon$ precisely cancel those of the real terms.
- Show that the total cross-section at order $g^2$ is given by

$$\sigma_{\text{tot,NLO}} = \sigma_{\text{tot,LO}} \left[1 + \frac{3\alpha_s C_f}{4\pi}\right].$$

- These considerations tell us that the naive differential cross-section is plagued by infrared and collinear divergences. Given the pattern of cancellations observed above, what type of differential final state would lead to a finite cross-section?

**8.4** Check eqs. (8.92), (8.99) and (8.100).

**\*8.5** Consider S-matrix elements $S_{\beta\alpha} \equiv \langle \beta_{\text{in}} | U(+\infty,-\infty) | \alpha_{\text{in}} \rangle$, where $U(+\infty,-\infty)$ is the time evolution operator.
- Show that $S_{\beta\alpha}$ does not depend on the gauge fixing if $Q_{\text{BRST}} |\alpha, \beta_{\text{in}}\rangle = 0$.
- Show that S is unitary over the physical space, i.e., $\sum_\gamma S^\dagger_{\beta\gamma} S_{\gamma\alpha} = \delta_{\beta\alpha}$, where the intermediate sum runs only on the physical states.

# CHAPTER 9

# Renormalization of Gauge Theories

Due to the internal symmetry of non-Abelian gauge theories, the Yang–Mills Lagrangian has a very constrained form proportional to $F_{\mu\nu}F^{\mu\nu}$, which in turn completely controls the interaction vertices since they stem from cubic and quartic terms in this Lagrangian. Via the Fadeev–Popov quantization procedure, one has broken the gauge symmetry by the necessity of eliminating the redundant gauge field configurations, but the gauge-fixed Lagrangian has a remnant of this symmetry – the Becchi–Rouet–Stora–Tyutin symmetry (BRST) symmetry – which is strong enough to prove that the resulting theory is unitary. However, until now, all this discussion has taken place at the level of the classical action. For such a theory to make sense as a quantum theory, we must now show that these symmetries survive the renormalization procedure, i.e., that the counterterms that we must introduce in order to cancel the ultraviolet divergences do not spoil the delicate structure of the gauge fixed action.

## 9.1 Ultraviolet Power Counting

Before studying in more detail the renormalizability of gauge theories, one may assess the plausibility of this renormalizability by calculating the superficial degree of ultraviolet divergence of graphs in such a theory. Furthermore, this will guide us regarding which classes of graphs may contain divergences. For simplicity, we will consider here a pure Yang–Mills theory, without matter fields (keeping fermions would force us to distinguish the fermion propagators from the gluon and ghost propagators in the counting, because they have different behaviors at large momentum, but would not change the final conclusion).

## 9.1 ULTRAVIOLET POWER COUNTING

Note that the gluon propagator decreases as $(\text{momentum})^{-2}$ in the ultraviolet, in covariant gauge. This is also the behavior of the ghost propagator.[1] Moreover, the three-gluon vertex and the gluon–ghost–antighost vertex have the same scaling with momentum. Therefore, we need not distinguish in the ultraviolet power counting the ghosts and the gluons. Thus, let us consider a generic connected graph $\mathcal{G}$ with the following list of propagators and vertices:

- $n_E$ external lines (gluons or ghosts);
- $n_I$ internal lines (gluons or ghosts);
- $n_3$ trivalent vertices (three-gluon or gluon–ghost–antighost);
- $n_4$ four-gluon vertices; and
- $n_L$ loops.

These quantities are related by the following identities:

$$n_E + 2n_I = 3n_3 + 4n_4, \tag{9.1}$$

$$n_L = n_I - (n_3 + n_4) + 1. \tag{9.2}$$

The first equation states that each vertex must have all its "handles" attached to the endpoint of a propagator, and the second equation counts the number of internal momenta that are not determined by energy momentum conservation. In terms of these parameters, the ultraviolet degree of divergence of this graph (in four space-time dimensions) is

$$\omega(\mathcal{G}) = 4n_L - 2n_I + n_3. \tag{9.3}$$

Note that each trivalent vertex contains one power of momentum and therefore contributes $+1$ to this counting. Adding eq. (9.1) and four times eq. (9.2), we obtain

$$\omega(\mathcal{G}) = 4 - n_E, \tag{9.4}$$

which does not depend on any of the internal details of the graph. Moreover, the only functions that have intrinsic ultraviolet divergences are the two-point, three-point and four-point functions, which suggests that Yang–Mills theories may indeed be renormalizable. However, a Yang–Mills theory is not simply the addition of gluon and ghost kinetic terms, three- and four-gluon vertices, and a ghost–antighost–gluon vertex: All these terms of the Lagrangian are tightly constrained by gauge symmetry. For instance (but this is not the only constraint), all the vertices depend on a unique coupling constant g. Therefore, in order to establish the renormalizability of Yang–Mills theories, one needs to prove that the structure of the divergences in the above-listed functions is such that they can be absorbed into a redefinition of the classical Lagrangian that does not upset these tight constraints (up to a renormalization of the fields).

---

[1] In strict axial gauge, the ghost propagator behaves differently, but the ghosts decouple completely from the gluons.

## 9.2 Symmetries of the Quantum Effective Action

### 9.2.1 Linearly Realized Symmetries

After fixing the gauge with the Fadeev–Popov procedure, we obtain the following effective Lagrangian:

$$\mathcal{L}_{\text{eff}} = \mathcal{L}_{\text{YM}} + \mathcal{L}_{\text{D}} - \frac{\xi}{2} G^a G^a + \overline{\chi}_a \frac{\partial G^a}{\partial A^b_\mu} \left(-D^{\text{adj}}_\mu\right)_{bc} \chi_c. \tag{9.5}$$

Although the local gauge invariance of the Yang–Mills Lagrangian is now broken (this was precisely the goal of the gauge fixing procedure), this effective Lagrangian has a number of symmetries. One of them is the BRST symmetry, which we showed in the previous chapter. In addition, $\mathcal{L}_{\text{eff}}$ has the following symmetries:

- *Ghost number conservation:* The effective Lagrangian is invariant under global phase transformations of the ghost and antighost,

$$\chi \to e^{i\alpha}\chi, \qquad \overline{\chi} \to e^{-i\alpha}\overline{\chi}. \tag{9.6}$$

  Therefore, if we assign a *ghost number* $+1$ to the field $\chi$ and $-1$ to the field $\overline{\chi}$, this quantity is conserved by the Feynman rules of the gauge fixed theory.
- *Global gauge invariance:* Since all color indices are contracted in the effective Lagrangian, it is invariant under gauge transformations that do not depend on space-time.
- *Lorentz invariance* is of course also present in the effective Lagrangian.

For these three symmetries, the infinitesimal variation of the fields is linear in the fields (which is not the case of the BRST symmetry), and these linearly realized symmetries of the classical action are inherited directly by the quantum effective action.

In order to prove this assertion, let us consider a generic infinitesimal linear transformation of the fields,

$$\phi_n(x) \to \phi_n(x) + \varepsilon\, F_n[x;\phi], \tag{9.7}$$

where $\phi_1, \phi_2, \cdots$ denote the various fields of the theory (gauge fields, ghosts, etc.) and $F_n[x;\phi]$ is a local function of the fields (for now, we do not assume that it is linear in the fields). We assume that both the classical action and the functional measure are invariant under this symmetry. Consider now the generating functional $Z[j]$,

$$Z[j] \equiv \int [D\phi_n(x)]\, e^{i\left(S[\phi_n] + \int d^4x\, j_n(x)\phi_n(x)\right)}, \tag{9.8}$$

where there is one external source $j_n$ for each field $\phi_n$. Since $\phi_n(x)$ is a dummy integration variable in this path integral, we should obtain the same result after performing the change of variable (9.7). Using the fact that this transformation preserves the measure and the classical action, this implies that

$$Z[j] = \int [D\phi_n(x)]\, e^{i\left(S[\phi_n] + \int d^4x\, j_n(x)\phi_n(x) + \varepsilon \int d^4x\, j_n(x) F_n[x;\phi]\right)}$$

$$\approx Z[j] + i\varepsilon \int [D\phi_n(x)]\, e^{i\left(S[\phi_n] + \int d^4x\, j_n(x)\phi_n(x)\right)} \int d^4x\, j_n(x) F_n[x;\phi]. \tag{9.9}$$

## 9.2 SYMMETRIES OF THE QUANTUM EFFECTIVE ACTION

Therefore, for any sources $j_n$, we must have

$$\int d^4x \, j_n(x) \langle F_n[x; \phi(x)] \rangle_j = 0, \qquad (9.10)$$

where $\langle \cdots \rangle_j$ denotes the quantum average in the presence of an external source $j$,

$$\langle \mathcal{O}[\phi] \rangle_j \equiv \frac{1}{Z[j]} \int [D\phi_n(x)] \, e^{i\left(S[\phi_n] + \int d^4x \, j_n(x)\phi_n(x)\right)} \mathcal{O}[\phi]. \qquad (9.11)$$

(We have normalized it so that $\langle 1 \rangle_j = 1$.) Recall now that the sources and field can be related implicitly by using the quantum effective action,

$$j_{n;\phi}(x) = -\frac{\delta \Gamma[\phi]}{\delta \phi_n(x)}. \qquad (9.12)$$

Therefore, condition (9.10) is equivalent to

$$\int d^4x \, \langle F_n[x; \phi(x)] \rangle_{j_\phi} \frac{\delta \Gamma[\phi]}{\delta \phi_n(x)} = 0, \qquad (9.13)$$

now satisfied for any fields $\phi_n$. This is known as the *Slavnov–Taylor identity*. In other words, the functional $\Gamma[\phi]$ is invariant under the transformation

$$\phi_n(x) \quad \to \quad \phi_n(x) + \varepsilon \, \langle F_n[x; \phi] \rangle_{j_\phi}. \qquad (9.14)$$

It is crucial to note that, because the quantum average on the right-hand side is performed with the external source $j_{n;\phi}$ that depends implicitly on the fields $\phi_n$, this is a priori not the same transformation as in eq. (9.7).

Let us now consider the special case of a transformation of type (9.7) which is linear in the fields. In this case, we may write

$$F_n[x; \phi] = \int d^4y \, f_{nm}(x, y) \, \phi_m(y). \qquad (9.15)$$

(In most practical cases, the transformation will be local and the coefficients proportional to $\delta(x - y)$, but this restriction is not necessary for the following argument.) For such a linear transformation, we have

$$\langle F_n[x; \phi] \rangle_{j_\phi} = \int d^4y \, f_{nm}(x, y) \, \langle \phi_m(y) \rangle_{j_\phi}. \qquad (9.16)$$

Recalling that $j_\phi$ is the configuration of the source $j$ such that the quantum average $\langle \phi(x) \rangle_j$ precisely equals $\phi(x)$, this in fact reads

$$\langle F_n[x; \phi] \rangle_{j_\phi} = F_n[x; \phi]. \qquad (9.17)$$

It is this last step that fails when $F_n$ is nonlinear in fields. From eq. (9.17), we see that transformations (9.14) and (9.7) are identical. We have thus proven that all linearly realized symmetries of the classical action are also symmetries of the quantum effective action.

### 9.2.2 BRST Symmetry and Zinn–Justin Equation

Since an infinitesimal BRST variation is not linear in the fields, the BRST symmetry of the classical action is not inherited so simply by the quantum effective action. Instead, it leads to identities that may be viewed as the analogue of Ward–Takahashi identities for the BRST invariance, whose derivation follows the method of Section 6.4.2. Since we need to apply a BRST transformation to the Yang–Mills path integral, we should first study how this transformation affects the measure $[DA_\mu D\chi D\overline{\chi}]$. Under such a transformation, the fields transform into

$$A_\mu^a \to A_\mu^{a\prime} \equiv A_\mu^a + \vartheta\left(D_\mu^{\text{adj}}\right)_{ab}\chi_b = A_\mu^a + \vartheta\left(\partial_\mu \delta_{ab} - gf^{abc}A_\mu^c\right)\chi_b,$$
$$\chi_a \to \chi_a' \equiv \chi_a - g\frac{\vartheta}{2}f^{abc}\chi_b\chi_c,$$
$$\overline{\chi}_a \to \overline{\chi}_a' \equiv \overline{\chi}_a + \vartheta B^a = \overline{\chi}_a - \xi\vartheta G^a, \qquad (9.18)$$

where $\vartheta$ is a Grassmann constant. The Jacobian is a *supermatrix*,

$$\frac{\partial(A_\mu^{a\prime}, \chi_a', \overline{\chi}_a')}{\partial(A_\nu^b, \chi_b, \overline{\chi}_b)} = \delta(x-y)\begin{pmatrix} \overbrace{\delta_\mu^\nu(\delta_{ab} + g\vartheta f^{abc}\chi_c)}^{A} & \overbrace{* \qquad 0}^{B} \\ 0 & \delta_{ab} + g\vartheta f^{abc}\chi_c \quad 0 \\ \underbrace{*}_{C} & \underbrace{0 \qquad\qquad \delta_{ab}}_{D} \end{pmatrix}, \qquad (9.19)$$

which mixes ordinary and Grassmann objects (a $*$ denotes a non-zero entry whose precise value we do not need). Because of the difference in the changes of Grassmann and ordinary variables, the measure $[DA_\mu D\chi D\overline{\chi}]$ is multiplied by a *superdeterminant*, defined as $\det(\mathbf{A} - \mathbf{B}\mathbf{D}^{-1}\mathbf{C})/\det(\mathbf{D}) = 1$. (Note that $\mathbf{B}\mathbf{D}^{-1}\mathbf{C} = 0$.)

In the derivation, it is convenient to introduce sources $j_\mu^a, \overline{\eta}_a, \eta_a$ that couple respectively to $A_\mu^a, \chi_a, \overline{\chi}_a$, and also two extra sources that couple directly to $\mathbf{Q}_{\text{BRST}}A_\mu^a$ and $\mathbf{Q}_{\text{BRST}}\chi_a$,

$$Z[j,\eta,\overline{\eta};\zeta,\kappa] \equiv \int [DA_\mu D\chi D\overline{\chi}]\, \exp\left\{i\int d^4x\,\left(\mathcal{L}_{\text{eff}} + j_\mu^a A_a^\mu + \overline{\eta}_a\chi_a + \overline{\chi}_a\eta_a \right.\right.$$
$$\left.\left. + \zeta_a^\mu(\mathbf{Q}_{\text{BRST}}A_\mu^a) - \kappa_a(\mathbf{Q}_{\text{BRST}}\chi_a)\right)\right\}$$
$$= \int [DA_\mu D\chi D\overline{\chi}]\, \exp\left\{i\int d^4x\, \mathcal{L}_{\text{tot}}\right\}, \qquad (9.20)$$

where we use the shorthand $\mathcal{L}_{\text{tot}}$ for the sum of terms inside the exponential. Note that the coefficients of the new sources $\zeta_a^\mu$ and $\kappa_a$ are BRST invariant since the BRST transformation is nilpotent. Let us now perform a BRST transformation of the integration variables inside the path integral. This is just a change of variables, which does not change the value of the

## 9.2 SYMMETRIES OF THE QUANTUM EFFECTIVE ACTION

path integral. Using the fact that the measure and the Lagrangian $\mathcal{L}_{\text{eff}}$ are BRST invariant, we obtain

$$Z[j,\eta,\overline{\eta};\zeta,\kappa] = \int [DA_\mu D\chi D\overline{\chi}]\; \exp\left\{i\int d^4x\, \mathcal{L}_{\text{tot}}\right\}$$
$$\times \left[1+i\int d^4x\, \left(j_\mu^a \vartheta(Q_{\text{BRST}} A_a^\mu) + \overline{\eta}_a \vartheta(Q_{\text{BRST}} \chi_a) + \vartheta(Q_{\text{BRST}} \overline{\chi}_a)\eta_a\right)\right]$$
$$= Z[j,\eta,\overline{\eta};\zeta,\kappa] + i\vartheta \int d^4x\, \left(j_\mu^a(x)\frac{\delta Z}{i\delta\zeta_\mu^a(x)} + \overline{\eta}_a(x)\frac{\delta Z}{i\delta\kappa_a(x)}\right.$$
$$\left. - \xi\, G^a\left(\frac{\delta Z}{i\delta j(x)}\right) \eta_a(x)\right). \tag{9.21}$$

(Note that $\vartheta$ anti-commutes with $\overline{\eta}_a$.) Therefore, we conclude that

$$\int d^4x\, \left(j_\mu^a(x)\frac{\delta Z}{i\delta\zeta_\mu^a(x)} + \overline{\eta}_a(x)\frac{\delta Z}{i\delta\kappa_a(x)} - \xi\, G^a\left(\frac{\delta Z}{i\delta j(x)}\right)\eta_a(x)\right) = 0. \tag{9.22}$$

This is one of the forms of the announced identities. In this derivation, we see that introducing sources specifically coupled to the BRST variation of the gauge field $A_\mu^a$ and of the ghost $\chi_a$ avoided the need for terms with higher-order derivatives (indeed, these variations are non-linear in the fields and would have required more derivatives to be expressed as functional derivatives with respect to sources coupled to elementary fields). By writing $Z = \exp(W)$, we see that the same identity applies also to $W$:

$$\int d^4x\, \left(j_\mu^a(x)\frac{\delta W}{i\delta\zeta_\mu^a(x)} + \overline{\eta}_a(x)\frac{\delta W}{i\delta\kappa_a(x)} - \xi\, G^a\left(\frac{\delta W}{i\delta j(x)}\right)\eta_a(x)\right) = 0. \tag{9.23}$$

(Here, we have assumed that the gauge fixing function is linear in the gauge field.)

The next step is to convert this into an identity for the quantum effective action $\Gamma$ that generates the 1PI graphs. In the Legendre transformation, we keep the auxiliary sources $\zeta_\mu^a$ and $\kappa_a$ unmodified, as parameters. Thus, $\Gamma$ and $W$ are related by

$$-iW[j,\eta,\overline{\eta};\zeta,\kappa] = \Gamma[A,\chi,\overline{\chi};\zeta,\kappa]$$
$$+ \int d^4x\, \left(j_a^\mu(x) A_\mu^a(x) + \overline{\chi}_a(x)\eta^a(x) + \overline{\eta}^a(x)\chi_a(x)\right). \tag{9.24}$$

Fields and sources are related by the following quantum equations of motion:

$$\frac{\delta\Gamma}{\delta A_\mu^a(x)} + j_a^\mu(x) = 0,\quad \frac{\delta\Gamma}{\delta\overline{\chi}_a(x)} + \eta^a(x) = 0,\quad \frac{\delta\Gamma}{\delta\chi_a(x)} + \overline{\eta}^a(x) = 0, \tag{9.25}$$

and we also have

$$\frac{\delta W}{\delta j_a^\mu(x)} = i A_\mu^a(x),\quad \frac{\delta W}{\delta\zeta_\mu^a(x)} = i\frac{\delta\Gamma}{\delta\zeta_\mu^a(x)},\quad \frac{\delta W}{\delta\kappa_a(x)} = i\frac{\delta\Gamma}{\delta\kappa^a(x)}. \tag{9.26}$$

Therefore, the identity expressed in terms of the functional $\Gamma$ reads

$$\int d^4x \left( \frac{\delta \Gamma}{\delta A^a_\mu(x)} \frac{\delta \Gamma}{\delta \zeta^\mu_a(x)} + \frac{\delta \Gamma}{\delta \chi_a(x)} \frac{\delta \Gamma}{\delta \kappa^a(x)} - \xi \, G^a(A) \frac{\delta \Gamma}{\delta \overline{\chi}_a(x)} \right) = 0. \tag{9.27}$$

This equation can be simplified as follows. By inserting a derivative $\delta/\delta \overline{\chi}_a(x)$ under the integral in definition (9.21) of $Z$, we obtain zero since we now have the integral of a total derivative. Recalling that the Fadeev–Popov term in the effective Lagrangian is

$$\mathcal{L}_{FPG} = \overline{\chi}_a \frac{\partial G^a}{\partial A^b_\mu} (-D^{adj}_\mu)_{bc} \chi_c, \tag{9.28}$$

we can perform explicitly this derivative to obtain

$$0 = \int [DA_\mu D\chi D\overline{\chi}] \left[ \frac{\partial G^a}{\partial A^b_\mu} \underbrace{(-D^{adj}_\mu)_{bc} \chi_c(x)}_{\substack{-Q_{BRST} A^b_\mu(x) \\ = i \frac{\delta}{\delta \zeta^\mu_b(x)}}} + \eta_a(x) \right] e^{i \int d^4x \, \mathcal{L}_{tot}}. \tag{9.29}$$

This implies the following functional identity:

$$\left[ \eta_a(x) + i \frac{\partial G^a}{\partial A^b_\mu} \frac{\delta}{\delta \zeta^\mu_b(x)} \right] Z = 0, \tag{9.30}$$

and equivalent identities for $W$ and $\Gamma$,

$$\eta_a(x) + i \frac{\partial G^a}{\partial A^b_\mu} \frac{\delta W}{\delta \zeta^\mu_b(x)} = 0, \qquad \frac{\delta \Gamma}{\delta \overline{\chi}_a(x)} + \frac{\partial G^a}{\partial A^b_\mu} \frac{\delta \Gamma}{\delta \zeta^\mu_b(x)} = 0. \tag{9.31}$$

Furthermore, define a slightly modified effective action:

$$\overline{\Gamma} \equiv \Gamma + \frac{\xi}{2} \int d^4x \, G^a(A) G^a(A). \tag{9.32}$$

Now the BRST conservation identity takes the following compact form, known as the *Zinn–Justin equation*:

$$\int d^4x \left( \frac{\delta \overline{\Gamma}}{\delta A^a_\mu(x)} \frac{\delta \overline{\Gamma}}{\delta \zeta^\mu_a(x)} + \frac{\delta \overline{\Gamma}}{\delta \chi_a(x)} \frac{\delta \overline{\Gamma}}{\delta \kappa^a(x)} \right) = 0, \tag{9.33}$$

from which any explicit reference to the gauge fixing function $G^a(A)$ has disappeared, as well as the coupling constant $g$.

Eq. (9.33) applies to the full quantum effective action that encapsulates the results from all-order perturbation theory. In the next section, we will show that this identity (combined with the other symmetries of the effective action) completely constrains the structure of its local terms of dimension less than or equal to four, forcing them to be identical to those in the classical action (up to a rescaling of the fields and of the coupling constant).

## 9.3 Renormalizability

### 9.3.1 Constraints on the Counterterms

By taking the $\hbar \to 0$ limit in eq. (9.33), one immediately concludes that it is also satisfied by the classical action $\overline{S}$, supplemented with ghosts as well as the sources $\zeta_{\mu a}$ and $\kappa_a$:

$$\overline{S}[A, \chi, \overline{\chi}; \zeta, \kappa] = \int d^4x \left[ -\tfrac{1}{4} F_a^{\mu\nu} F_{\mu\nu}^a + (\zeta_a^\mu + \partial^\mu \overline{\chi}_a)(D_\mu^{\text{adj}})_{ab} \chi_b + \tfrac{g}{2} f^{abc} \kappa_a \chi_b \chi_c \right]. \tag{9.34}$$

(Here, we have written it for the case of covariant gauges.) By introducing the compact notation

$$(\mathcal{A}, \mathcal{B}) \equiv \int d^4x \left( \frac{\delta \mathcal{A}}{\delta A_\mu^a(x)} \frac{\delta \mathcal{B}}{\delta \zeta_a^\mu(x)} + \frac{\delta \mathcal{A}}{\delta \chi_a(x)} \frac{\delta \mathcal{B}}{\delta \kappa^a(x)} \right), \tag{9.35}$$

we therefore have

$$(\overline{S}, \overline{S}) = 0, \quad (\overline{\Gamma}, \overline{\Gamma}) = 0. \tag{9.36}$$

The first equation may be viewed as a constraint on the terms that can appear in the classical action, while the second equation constrains which divergences may appear in higher orders.

Let us now write the effective action as a loop expansion:

$$\overline{\Gamma} \equiv \overline{S} + \sum_{l=1}^\infty \overline{\Gamma}_l, \tag{9.37}$$

where $\overline{S}$ is given in eq. (9.34) and the subsequent terms $\overline{\Gamma}_l$ are of order $l$ in $\hbar$. The Zinn–Justin equation at order L thus reads:

$$\sum_{p+q=L} (\overline{\Gamma}_p, \overline{\Gamma}_q) = 0. \tag{9.38}$$

The renormalization procedure amounts to correcting order by order with counterterms the classical action $\overline{S}$,

$$\overline{S} \to \overline{S}_{(L)}, \tag{9.39}$$

such that $\overline{S}_{(L)}$ contains counterterms up to order L, and gives finite $\overline{\Gamma}_l$ for $l \leq L$ (but in general not beyond the order L).

The first step is to prove that it is possible to find counterterms such that the equation $(\overline{S}, \overline{S}) = 0$ is preserved at every order. Let us assume that we have achieved this up to the order $L-1$. All $\overline{\Gamma}_l$ for $l \leq L-1$ are now finite, while $\overline{\Gamma}_L$ still contains a divergent part, that we denote $\overline{\Gamma}_{L,\text{div}}$. We can rewrite the Zinn–Justin equation at order L as follows:

$$(\overline{S}, \overline{\Gamma}_L) + (\overline{\Gamma}_L, \overline{S}) = -\sum_{l=1}^{L-1} (\overline{\Gamma}_l, \overline{\Gamma}_{L-l}). \tag{9.40}$$

Only the left-hand side contains divergences, and we therefore have

$$(\overline{S}, \overline{\Gamma}_{L,\text{div}}) + (\overline{\Gamma}_{L,\text{div}}, \overline{S}) = 0, \qquad (9.41)$$

which constrains the structure of the divergences at order L. A natural candidate for the counterterm at order L is to simply add $-\overline{\Gamma}_{L,\text{div}}$ to the classical action,

$$\overline{S} \quad \to \quad \overline{S} - \overline{\Gamma}_{L,\text{div}}, \qquad (9.42)$$

since this automatically cancels the superficial divergence of $\overline{\Gamma}_L$ without affecting anything in the lower orders. However, this modified classical action does not obey exactly the Zinn–Justin equation, since

$$(\overline{S} - \overline{\Gamma}_{L,\text{div}}, \overline{S} - \overline{\Gamma}_{L,\text{div}}) = \underbrace{(\overline{S}, \overline{S})}_{\text{0 from lower orders}} - \underbrace{\left[(\overline{S}, \overline{\Gamma}_{L,\text{div}}) + (\overline{\Gamma}_{L,\text{div}}, \overline{S})\right]}_{\text{0 from eq. (9.41)}} + \underbrace{(\overline{\Gamma}_{L,\text{div}}, \overline{\Gamma}_{L,\text{div}})}_{\neq 0}. \qquad (9.43)$$

Note that the non-zero term on the right-hand side is of order strictly greater than L. It is possible to make it vanish by adding to the shift of eq. (9.42) some terms of order higher than L that do not change anything for any order $\leq$ L. The conclusion of this inductive argument is that one can shift the classical action at each order in such a way that the divergences in $\overline{\Gamma}$ are canceled, while always preserving $(\overline{S}, \overline{S}) = 0$.

### 9.3.2 Allowed Terms in the Classical Action

The second step in the discussion of the renormalization of Yang–Mills theory is to determine the terms that are allowed in the classical action. This action must satisfy the constraint $(\overline{S}, \overline{S}) = 0$, as well as Lorentz invariance, global gauge symmetry and ghost number conservation. In addition, from the power counting of Section 9.1 and Weinberg's theorem, we know that all the ultraviolet divergences in Yang–Mills theory will occur in *local* operators of dimension four at most.

In order to discuss the form of the allowed terms, let us first list the mass dimension and ghost number of the various fields that enter in $\overline{S}$:

| Field | $A_a^\mu$ | $\chi_a$ | $\overline{\chi}_a$ | $\zeta_a^\mu$ | $\kappa_a$ |
|---|---|---|---|---|---|
| Mass dimension | 1 | 1 | 1 | 2 | 2 |
| Ghost number | 0 | +1 | −1 | −1 | −2 |

All the allowed terms in $\overline{S}$ must obey the following conditions;
- mass dimension four or less;
- ghost number 0;
- Lorentz invariance; and
- global gauge invariance.

## 9.3 RENORMALIZABILITY

In addition, eq. (9.31) implies that the $\bar{\chi}$ and $\zeta$ dependences come in the form of a dependence on the combination

$$\zeta_\mu - \bar{\chi}\frac{\partial G}{\partial A^\mu} = \zeta_\mu + \partial_\mu \bar{\chi}, \qquad (9.44)$$

where on the right-hand side we have assumed the covariant gauge condition $G(A) = \partial_\mu A^\mu$ and anticipated an integration by parts. Finally, the Zinn–Justin equation $(\bar{S}, \bar{S}) = 0$ must be satisfied.

Since the sources $\zeta_a^\mu$ and $\kappa_a$ have mass dimension two, at most two of them may appear. However, terms with two such sources cannot contain any other field since mass dimension four is already reached, and they cannot have ghost number 0. Therefore, $\bar{S}$ can only contain terms that have degree 0 or 1 in $\zeta_a^\mu$ and $\kappa_a$.

The source $\zeta_a^\mu$ must be combined with another combination of fields that have one Lorentz index, one color index, mass dimension at most two and ghost number +1. The only operators that fulfill these conditions are (see Exercise 9.1)

$$f^{abc}\,\zeta_a^\mu A_\mu^b \chi_c \quad \text{and} \quad \zeta_a^\mu \partial_\mu \chi_a. \qquad (9.45)$$

Once the dependence on $\zeta_a^\mu$ is fixed, the dependence on the antighosts will be completely known from eq. (9.44). Likewise, $\kappa_a$ must be combined with an object that has one color index, mass dimension at most two and ghost number +2. The only possibility is

$$f^{abc}\,\kappa_a \chi_b \chi_c. \qquad (9.46)$$

From the information gathered so far, the classical action must have the following general form:

$$\bar{S}[A,\chi,\bar{\chi};\zeta,\kappa] = \bar{\Sigma}[A] + \int d^4x\,\Big[g\alpha\,f^{abc}(\zeta_a^\mu + \partial^\mu\bar{\chi}_a)A_\mu^b\chi_c$$
$$+ \beta\,(\zeta_a^\mu + \partial^\mu\bar{\chi}_a)\partial_\mu\chi_a + \tfrac{\gamma}{2}\,f^{abc}\kappa_a\chi_b\chi_c\Big], \qquad (9.47)$$

where $\alpha, \beta, \gamma$ are three arbitrary constants. The term $\bar{\Sigma}$ cannot depend on the sources $\zeta_a^\mu$ and $\kappa_a$ because we have already constructed explicitly all the allowed terms that contain these sources, and cannot depend on $\bar{\chi}$ because the antighost dependence is already encapsulated in the combination $\zeta_a^\mu + \partial^\mu\bar{\chi}_a$. A dependence on $\chi$ in $\bar{\Sigma}$ is also forbidden because $\chi$ would be the only field in $\bar{\Sigma}$ with a non-zero ghost number. The next step is to constrain the coefficients $g\alpha, \beta, \gamma$ and the functional $\bar{\Sigma}[A]$ in order to satisfy the Zinn–Justin equation (9.33). The functional derivatives that enter in (9.33) are given by

$$\frac{\delta \bar{S}}{\delta A_a^\mu} = \frac{\delta \bar{\Sigma}}{\delta A_a^\mu} - g\alpha\,f^{abc}(\zeta_{\mu b} + \partial_\mu \bar{\chi}_b)\chi_c,$$

$$\frac{\delta \bar{S}}{\delta \zeta_{\mu a}} = g\alpha\,f^{ade}A_d^\mu \chi_e + \beta\,\partial^\mu \chi_a,$$

$$\frac{\delta \bar{S}}{\delta \chi_a} = g\alpha\,f^{abc}(\zeta_{\mu b} + \partial_\mu\bar{\chi}_b)A_c^\mu + \beta\,(\zeta_{\mu a} + \partial_\mu\bar{\chi}_a)\partial^\mu + \gamma\,f^{abc}\kappa_b\chi_c,$$

$$\frac{\delta \bar{S}}{\delta \kappa_a} = \frac{\gamma}{2}\,f^{ade}\chi_d\chi_e. \qquad (9.48)$$

Thus, the Zinn–Justin equation reads

$$
\begin{aligned}
0 = \int d^4x \Big[ & \frac{\delta \overline{\Sigma}}{\delta A_a^\mu} \big[ g\alpha\, f^{ade} A_d^\mu \chi_e + \beta\, \partial^\mu \chi_a \big] \\
& + (\zeta_{\mu b} + \partial_\mu \overline{\chi}_b) \big[ - g\alpha\, f^{abc} \chi_c \big( g\alpha\, f^{ade} A_d^\mu \chi_e + \beta\, \partial^\mu \chi_a \big) \\
& \qquad\qquad + \big( g\alpha\, f^{abc} A_c^\mu + \beta\, \delta_{ab} \partial^\mu \big) \tfrac{\gamma}{2} f^{ade} \chi_d \chi_e \big] \\
& + \frac{\gamma^2}{2} f^{abc} f^{ade} \kappa_b \chi_c \chi_d \chi_e \Big].
\end{aligned}
\tag{9.49}
$$

Using the Jacobi identity satisfied by the structure constants, one may first check that the last term, in $\kappa\chi\chi\chi$, is identically zero, and therefore does not provide any constraint. Consider now the terms in $\zeta A \chi\chi$,

$$
g\alpha\, \zeta_{\mu b} \Big( - g\alpha\, f^{abc} f^{ade} A_d^\mu \chi_c \chi_e + \tfrac{\gamma}{2} \underbrace{f^{abc} f^{ade}}_{-f^{abd}f^{aec} - f^{abe}f^{acd}} A_c^\mu \chi_d \chi_e \Big)
$$

$$
= g\alpha\, (\gamma - g\alpha)\, f^{abc} f^{ade} \zeta_{\mu b} A_d^\mu \chi_c \chi_e.
\tag{9.50}
$$

Since this is the only term containing this combination of fields, it cannot be canceled by other terms, and therefore we must have

$$
g\alpha = \gamma.
\tag{9.51}
$$

Let us now study the terms in $\zeta \chi (\partial \chi)$,

$$
\beta\, \zeta_{\mu b} \Big( - g\alpha\, f^{abc} \chi_c \partial^\mu \chi_a + \tfrac{\gamma}{2} f^{bde} \partial^\mu (\chi_d \chi_e) \Big) = \beta\, (\gamma - g\alpha)\, f^{bac} \zeta_{\mu b} (\partial^\mu \chi_a) \chi_c.
\tag{9.52}
$$

Thus, the cancellation of this term does not bring any additional constraint beyond eq. (9.51). At this point, all the terms containing $\zeta_a^\mu$ have been canceled (and by extension also the terms with $\partial^\mu \overline{\chi}_a$), and the Zinn–Justin equation reduces to

$$
0 = \int d^4x\, \frac{\delta \overline{\Sigma}}{\delta A_a^\mu} \big[ g\alpha\, f^{acb} A_c^\mu \chi_b + \beta\, \partial^\mu \chi_a \big].
\tag{9.53}
$$

Let us first rewrite the second factor as

$$
\beta\, \big( \partial_\mu \delta_{ab} - ig\alpha\beta^{-1} \underbrace{(-if^{cab}) A_c^\mu}_{(A_{\mathrm{adj}}^\mu)_{ab}} \big) \chi_b,
\tag{9.54}
$$

and note that it has the structure of an adjoint covariant derivative acting on $\chi_b$,

$$
(\overline{D}_{\mathrm{adj}}^\mu)_{ab} \equiv \partial_\mu \delta_{ab} - ig\alpha\beta^{-1} (A_{\mathrm{adj}}^\mu)_{ab}.
\tag{9.55}
$$

Thus, eq. (9.53) is equivalent to

$$
0 = \int d^4x\, \frac{\delta \overline{\Sigma}}{\delta A_a^\mu}\, (\overline{D}_{\mathrm{adj}}^\mu)_{ab}\, \chi_b.
\tag{9.56}
$$

The second factor may be viewed as the variation of the gauge field under an infinitesimal gauge transformation,

$$A_a^\mu \to A_a^\mu + \vartheta \left(\overline{D}_{\text{adj}}^\mu\right)_{ab} \chi_b, \qquad (9.57)$$

where we have introduced a constant Grassmann variable $\vartheta$ to make the second term a commuting object. Therefore, for the integral to be zero for an arbitrary $\chi_b(x)$, the functional $\overline{\Sigma}[A]$ must be invariant under this transformation. Recalling our discussion of the local gauge invariant operators of mass dimension four or less, we conclude that the only possible form for $\overline{\Sigma}$ is

$$\overline{\Sigma}[A] = -\frac{\delta}{4} \int d^4x \, \overline{F}_a^{\mu\nu} \overline{F}_{\mu\nu}^a, \qquad (9.58)$$

where $\overline{F}^{\mu\nu}$ is the field strength constructed with the covariant derivative $\overline{D}^\mu$ and $\delta$ is another constant. Given all the above constraints, we must have

$$\overline{S}[A, \chi, \overline{\chi}; \zeta, \kappa] = \int d^4x \left[ -\frac{\delta}{4} \overline{F}_a^{\mu\nu} \overline{F}_{\mu\nu}^a + \beta \left(\zeta_a^\mu + \partial^\mu \overline{\chi}_a\right) \left(\overline{D}_\mu^{\text{adj}}\right)_{ab} \chi_b \right.$$
$$\left. + \frac{g\alpha}{2} f^{abc} \kappa_a \chi_b \chi_c \right]. \qquad (9.59)$$

Up to rescalings of the various fields and of the coupling constant $g$, this is structurally identical to the bare classical action of eq. (9.34). Note that this equation implies that the field renormalization factors for the gauge field $A_a^\mu$ and for the source $\kappa_a$ are equal, $Z_A = Z_\kappa$.

## 9.4 Background Field Method

### 9.4.1 Rescaled Fields

In this section we describe the calculation of the one-loop quantum corrections to the coupling constant by a method based on the quantum effective action combined with the so-called background field method. The first step of this method is to rescale the gauge field by the inverse of the coupling constant,

$$g A^\mu \to A^\mu. \qquad (9.60)$$

(See Exercise 9.4 for a justification of the fact that this rescaling does not alter the $g$ dependence of correlation functions.) By doing this, the various objects that appear in the Yang–Mills action are transformed as follows:

$$F^{\mu\nu} \to \frac{1}{g} \left(\partial^\mu A^\nu - \partial^\nu A^\mu - i[A^\mu, A^\nu]\right),$$
$$D^\mu \to \partial^\mu - i A^\mu. \qquad (9.61)$$

In other words, up to a rescaling in the case of the field strength $F^{\mu\nu}$, these objects are transformed into their counterparts for a coupling equal to unity. In the rest of this section, the notation $A^\mu, D^\mu, F^{\mu\nu}$ will refer to the rescaled quantities. In terms of the rescaled fields, the Yang–Mills action simply reads

$$S_{YM} = -\frac{1}{4g^2}\int d^4x\, \underbrace{F_a^{\mu\nu} F_{\mu\nu}^a}_{\text{no } g}, \tag{9.62}$$

where all the dependence on the coupling constant appears now in the prefactor $g^{-2}$. This action has a local non-Abelian gauge invariance analogous to the original one, but with $g=1$:

$$A_\mu \;\to\; A_\mu^\Omega \equiv \Omega^\dagger A_\mu \Omega + i\Omega^\dagger \partial_\mu \Omega. \tag{9.63}$$

### 9.4.2 Background Field Gauge

The *background field method* is defined by choosing a classical background field $\mathcal{A}_\mu^a(x)$, and writing the gauge field $A_\mu^a(x)$ as a deviation about this background:

$$A_\mu^a \equiv \mathcal{A}_\mu^a + a_\mu^a. \tag{9.64}$$

In this decomposition, the background field $\mathcal{A}_\mu$ is not a dynamical field: It will just act as a parameter that we shall not quantize, and the path integration is thus only on the deviation $a_\mu^a$ (one may thus view this as a shift of the integration variable). In terms of $\mathcal{A}_\mu^a$ and $a_\mu^a$, the field strength that enters in the Yang–Mills action can be written as

$$F^{\mu\nu} = \mathcal{F}^{\mu\nu} + \left(\partial^\mu a^\nu - i[\mathcal{A}^\mu, a^\nu]\right) - \left(\partial^\nu a^\mu - i[\mathcal{A}^\nu, a^\mu]\right) - i[a^\mu, a^\nu], \tag{9.65}$$

where $\mathcal{F}^{\mu\nu}$ is the field strength constructed with the background field. With explicit color indices, this reads

$$F_a^{\mu\nu} = \mathcal{F}_a^{\mu\nu} + \left(\mathcal{D}_{\text{adj}}^\mu\right)_{ab} a_b^\nu - \left(\mathcal{D}_{\text{adj}}^\nu\right)_{ab} a_b^\mu + f^{abc} a_b^\mu a_c^\nu, \tag{9.66}$$

where $\mathcal{D}_{\text{adj}}^\mu = \partial^\mu - i[\mathcal{A}^\mu, \cdot]$ is the adjoint covariant derivative associated to the background field $\mathcal{A}^\mu$. The original gauge transformation of $A^\mu$ corresponds to the following transformation of $a^\mu$:

$$a^\mu \;\to\; \Omega^\dagger a^\mu \Omega + \Omega^\dagger \mathcal{A}^\mu \Omega - \mathcal{A}^\mu + i\Omega^\dagger \partial^\mu \Omega. \tag{9.67}$$

If we parametrize $\Omega = \exp(i\theta_a t^a)$ and expand to first order in $\theta_a$, an infinitesimal gauge transformation of $a_a^\mu$ reads

$$a_a^\mu \;\to\; a_a^\mu - \left(\mathcal{D}_{\text{adj}}^\mu\right)_{ab}\theta_b + f^{abc}\theta_b a_c^\mu. \tag{9.68}$$

This invariance leads to the same pathologies as in the original theory, and we must fix the gauge in order to have a well-defined path integral. The *background field gauge* corresponds to the following condition on $a_\mu^a$:

$$G^a(A) \equiv \left(\mathcal{D}_{\text{adj}}^\mu\right)_{ab} a_\mu^b = \omega_a. \tag{9.69}$$

## 9.4 BACKGROUND FIELD METHOD

Let us recall that a gauge fixing function $G^a(A)$ leads to the following terms in the effective Lagrangian:

$$\mathcal{L}_{GF} = -\frac{\xi}{2g^2} G^a(A) G^a(A) \qquad \text{(gauge fixing term)},$$

$$\mathcal{L}_{FPG} = -\overline{\chi}_a \frac{\partial G^a}{\partial A^b_\mu} \left(D^{\text{adj}}_\mu\right)_{bc} \chi_c \qquad \text{(Fadeev–Popov ghosts)}. \qquad (9.70)$$

With the choice of eq. (9.69), the Fadeev–Popov term becomes

$$\mathcal{L}_{FPG} = -\overline{\chi}_a \left(\mathcal{D}^\mu_{\text{adj}}\right)_{ab} \left(D^{\text{adj}}_\mu\right)_{bc} \chi_c = \left(\mathcal{D}^\mu_{\text{adj}} \overline{\chi}\right)_a \left(D^{\text{adj}}_\mu \chi\right)_a, \qquad (9.71)$$

where in the second equality we have anticipated an integration by parts and used the notation $(D^{\text{adj}}_\mu \chi)_a \equiv (D^{\text{adj}}_\mu)_{ab} \chi_b$ (and a similar notation for $(\mathcal{D}^\mu_{\text{adj}} \overline{\chi})_a$).

### 9.4.3 Residual Symmetry of the Gauge Fixed Lagrangian

The effective Lagrangian $\mathcal{L}_{YM} + \mathcal{L}_{GF} + \mathcal{L}_{FPG}$ possesses a residual gauge symmetry that corresponds to gauge transforming in the same way the background field $\mathcal{A}^\mu$ and the total field $A^\mu$,

$$A_\mu \to \Omega^\dagger A_\mu \Omega + i \Omega^\dagger \partial_\mu \Omega,$$
$$\mathcal{A}_\mu \to \Omega^\dagger \mathcal{A}_\mu \Omega + i \Omega^\dagger \partial_\mu \Omega. \qquad (9.72)$$

Indeed, under this joint transformation we have

$$a_\mu \to \Omega^\dagger a_\mu \Omega,$$
$$D_\mu \to \Omega^\dagger D_\mu \Omega,$$
$$\mathcal{D}_\mu \to \Omega^\dagger \mathcal{D}_\mu \Omega,$$
$$\chi \to \Omega^\dagger \chi,$$
$$\overline{\chi} \to \overline{\chi} \Omega,$$
$$G(A) \to \Omega^\dagger G(A) \Omega. \qquad (9.73)$$

From this, we conclude that the gauge fixing Lagrangian $\mathcal{L}_{GF}$ and the Fadeev–Popov Lagrangian $\mathcal{L}_{FPG}$ are both invariant in this transformation, as well as the Yang–Mills Lagrangian. Since the path integration measure over $a_\mu, \chi, \overline{\chi}$ is also invariant under this transformation, the result of the path integral must be invariant under local gauge transformations of the background field $\mathcal{A}^\mu$.

### 9.4.4 One-Loop Running Coupling

Let us now turn to the calculation of the quantum effective action at one loop. For this, we use the results of Section 5.5.5, where we have shown that these one-loop corrections are obtained by expanding the classical action to quadratic order in deviations with respect to a background field, and by performing a Gaussian path integration over the deviations (which gives a functional determinant).

The first step is to expand the three terms of the gauge fixed Lagrangian to second order in the deviation $a_\mu$. In this calculation, we choose the gauge fixing parameter $\xi = 1$. The quadratic terms in the combined Yang–Mills and gauge fixing terms read

$$\begin{aligned}
\mathcal{L}_{YM} + \mathcal{L}_{GF} &= -\frac{1}{2g^2}\Big\{\tfrac{1}{2}\big((\mathcal{D}^\mu_{adj}a^\nu)_a - (\mathcal{D}^\nu_{adj}a^\mu)_a\big)^2 \\
&\quad + f^{abc}\mathcal{F}^{\mu\nu}_a a^b_\mu a^c_\nu + ((\mathcal{D}^\mu_{adj}a_\mu)^a)^2\Big\} \\
&= -\frac{1}{2g^2}\Big\{a^a_\mu\Big[-(\mathcal{D}_{adj})^2_{ac}g^{\mu\nu} - 2f^{abc}\mathcal{F}^{\mu\nu}_b\Big]a^c_\nu\Big\} \\
&= -\frac{1}{2g^2}\Big\{a^a_\mu\Big[-(\mathcal{D}_{adj})^2_{ac}g^{\mu\nu} + (\mathcal{F}^{\rho\sigma}_{adj})_{ac}(M^{(1)}_{\rho\sigma})^{\mu\nu}\Big]a^c_\nu\Big\},
\end{aligned} \quad (9.74)$$

where we have introduced $(M^{(1)}_{\rho\sigma})^{\mu\nu} \equiv i(\delta_\rho{}^\mu\delta_\sigma{}^\nu - \delta_\rho{}^\nu\delta_\sigma{}^\mu)$, the generators of the Lorentz transformations for 4-vectors (the Lorentz transformation corresponding to the transformation parameters $\omega^{\rho\sigma}$ reads $\Lambda^{\mu\nu} = \exp(\tfrac{i}{2}\omega^{\rho\sigma}(M^{(1)}_{\rho\sigma})^{\mu\nu})$). For the ghost term, the quadratic part is

$$\mathcal{L}_{FPG} = \overline{\chi}_a\Big[-(\mathcal{D}_{adj})^2_{ab}\Big]\chi_b. \quad (9.75)$$

Note that the operator that appears between the two ghost fields is the spin-0 analogue of the one that appears in eq. (9.74), since the generators of Lorentz transformations for spin-0 objects are identically zero ($M^{(0)}_{\rho\sigma} \equiv 0$). Although we have not considered fermions so far in this chapter, the Dirac Lagrangian would give a contribution equal to the determinant of $i\slashed{D}$, or equivalently the square root of the determinant of $(i\slashed{D})^2$. Note that

$$(i\slashed{D})^2 = -\mathcal{D}^2 + i(\tfrac{i}{2}[\gamma^\mu,\gamma^\nu])\mathcal{D}_\mu\mathcal{D}_\nu = -\mathcal{D}^2 + (\mathcal{F}^{\rho\sigma})M^{(1/2)}_{\rho\sigma}, \quad (9.76)$$

where the $M^{(1/2)}_{\rho\sigma} \equiv \tfrac{i}{4}[\gamma_\rho,\gamma_\sigma]$ are the generators of Lorentz transformations for spin-1/2 fields. Note that the covariant derivatives and the field strength are in the fundamental representation (assuming fermions that transform according to the fundamental representation, like quarks). Therefore, for each of the fields that appear in the quantum effective action (gauge fields, ghosts, fermions), we get a determinant $\Delta_{r,s}$ of an operator containing $-\mathcal{D}^2$ (in the representation r corresponding to the field under consideration) plus a "spin connection"[2] made of the contraction of the field strength with the Lorentz generators corresponding to the spin s of the field:

$$\begin{aligned}
\text{gauge fields:} \quad & \Delta_{adj,s=1} \equiv \det\Big(-\mathcal{D}^2_{adj} + \mathcal{F}^{\rho\sigma}_{adj}M^{(1)}_{\rho\sigma}\Big), \\
\text{ghosts:} \quad & \Delta_{adj,s=0} \equiv \det\Big(-\mathcal{D}^2_{adj} + \underbrace{\mathcal{F}^{\rho\sigma}_{adj}M^{(0)}_{\rho\sigma}}_{=0}\Big), \\
\text{fermions:} \quad & \Delta_{f,s=1/2} \equiv \det\Big(-\mathcal{D}^2_f + \mathcal{F}^{\rho\sigma}_f M^{(1/2)}_{\rho\sigma}\Big).
\end{aligned} \quad (9.77)$$

---

[2]This term describes the coupling between the magnetic moment of the particle and the background field. Its detailed form depends on the spin of the particle.

## 9.4 BACKGROUND FIELD METHOD

In terms of these determinants, the one-loop quantum effective action is given by

$$\Gamma[A,\chi,\psi] = S_r + \Delta S + \frac{i}{2}\ln\Delta_{adj,s=1} - \frac{in_f}{2}\ln\Delta_{f,s=1/2} - i\ln\Delta_{adj,s=0}, \qquad (9.78)$$

where $\Delta S$ denotes the one-loop counterterms, and $n_f$ is the number of fermion flavors. Using the invariance with respect to local gauge transformations of the background field, we must have

$$\ln\Delta_{r,s} = \frac{i}{4}C_{r,s}\int d^4x\, \mathcal{F}_a^{\mu\nu}\mathcal{F}_{\mu\nu}^a + \cdots, \qquad (9.79)$$

where the dots represent higher dimensional gauge invariant operators. Being of dimension higher than four, these operators do not contribute to the renormalization of the coupling. The constant $C_{r,s}$ depends on the representation r and spin s of the field. These coefficients are ultraviolet divergent,

$$C_{r,s} = c_{r,s}\ln\frac{\Lambda^2}{\kappa^2}, \qquad (9.80)$$

where $\Lambda$ is an ultraviolet scale and $\kappa$ the typical scale of inhomogeneities of the background field. After combining them with the counterterms from $\Delta S$, the ultraviolet scale is replaced by a renormalization scale $\mu$,

$$C_{r,s} \quad\to\quad C_{r,s} = c_{r,s}\ln\frac{\mu^2}{\kappa^2}. \qquad (9.81)$$

From eq. (9.78), we see that the one-loop renormalized coupling at the scale $\mu$ and the bare coupling must be related by

$$\begin{aligned}\frac{1}{g_b^2} &= \frac{1}{g_r^2(\mu)} + \frac{1}{2}C_{adj,1} - \frac{n_f}{2}C_{f,1/2} - C_{adj,0}\\ &= \frac{1}{g_r^2(\mu)} + \left(\frac{1}{2}c_{adj,1} - \frac{n_f}{2}c_{f,1/2} - c_{adj,0}\right)\ln\frac{\mu^2}{\kappa^2}.\end{aligned} \qquad (9.82)$$

The explicit calculation of the constants $c_{r,s}$ requires expanding the logarithm of the functional determinants to second order in the background field strength $\mathcal{F}^{\mu\nu}$ (Exercise 9.2). Thanks to the organization of eq. (9.77), this calculation needs to be performed only once, for generic representations of the gauge and Lorentz algebras. This leads to

$$c_{r,s} = \frac{1}{(4\pi)^2}\left[\frac{1}{3}d(s) - 4C(s)\right]N(r), \qquad (9.83)$$

where $d(s)$ is the number of field components (respectively $1, 4$ and $4$ for scalars, fermions, and vector particles) $C(s)$ is the normalization of the trace of two Lorentz generators,[3]

$$\mathrm{tr}\left(M_{\rho\sigma}^{(s)}M_{\alpha\beta}^{(s)}\right) = C(s)\left(g_{\rho\alpha}g_{\sigma\beta} - g_{\rho\beta}g_{\sigma\alpha}\right), \qquad (9.84)$$

---

[3] For spin-0, $\frac{1}{2}$ and $-1$, this constant is respectively 0, 1 and 2.

and $N(r)$ is the normalization of the trace of two generators of the Lie algebra in the representation $r$,

$$\text{tr}\left(t_r^a t_r^b\right) = N(r)\delta^{ab}. \tag{9.85}$$

For the fundamental and adjoint representations of $\mathfrak{su}(N)$, we have $N(f) = \tfrac{1}{2}$ and $N(\text{adj}) = N$. Therefore, the constants involved in the one-loop running coupling are

$$c_{\text{adj},0} = \frac{N}{3(4\pi)^2}, \quad c_{\text{adj},1} = -\frac{20\,N}{3(4\pi)^2}, \quad c_{f,1/2} = -\frac{4}{3(4\pi)^2}, \tag{9.86}$$

and the coupling evolves according to

$$\frac{1}{g_r^2(\mu)} = \frac{1}{g_b^2} + \frac{1}{(4\pi)^2}\underbrace{\left(\tfrac{11}{3}N - \tfrac{2}{3}n_f\right)}_{>0 \text{ for } n_f \leq \tfrac{11N}{2}} \ln\frac{\mu^2}{\kappa^2}. \tag{9.87}$$

Given two scales $\mu$ and $\mu_0$, the renormalized couplings at these scales are related by

$$\frac{1}{g_r^2(\mu)} - \frac{1}{g_r^2(\mu_0)} = \frac{1}{(4\pi)^2}\left(\tfrac{11}{3}N - \tfrac{2}{3}n_f\right)\ln\frac{\mu^2}{\mu_0^2}, \tag{9.88}$$

which may be rewritten as

$$g^2(\mu) = \frac{g^2(\mu_0)}{1 + \frac{g_r^2(\mu_0)}{(4\pi)^2}\left(\tfrac{11}{3}N - \tfrac{2}{3}n_f\right)\ln\frac{\mu^2}{\mu_0^2}}. \tag{9.89}$$

In quantum chromodynamics, where the gauge group is $SU(3)$ (i.e., $N = 3$) and where there are six flavors of quarks in the fundamental representation, the coefficient in front of the logarithm is positive, which indicates that the coupling constant decreases as the scale $\mu$ increases. The coupling constant in fact goes to zero when $\mu \to \infty$, a property known as *asymptotic freedom*. By parametrizing the coupling at the reference scale $\mu_0$ by

$$\frac{(4\pi)^2}{g^2(\mu_0)} \equiv \underbrace{\left(\tfrac{11}{3}N - \tfrac{2}{3}n_f\right)}_{\equiv\,\beta_0}\ln\left(\frac{\mu_0^2}{\Lambda_{\text{QCD}}^2}\right) \tag{9.90}$$

(this equation can be viewed as a definition of the scale $\Lambda_{\text{QCD}}$), we can rewrite eq. (9.89) as follows:

$$\alpha_s(\mu^2) \equiv \frac{g^2(\mu^2)}{4\pi} = \frac{4\pi}{\beta_0 \ln\left(\frac{\mu^2}{\Lambda_{\text{QCD}}^2}\right)}. \tag{9.91}$$

In this form, it becomes clear that the running coupling at one loop becomes infinite when $\mu$ approaches $\Lambda_{\text{QCD}}$ from above. Therefore, perturbation theory stops being applicable somewhat before one reaches these low momentum scales. Note also that thanks to eq. (9.83), it would have been easy to determine the one-loop running of the coupling in the presence of matter fields in arbitrary representations.

## Exercises

**\*9.1** Prove explicitly that the operators $f^{abc}\zeta_a^\mu A_\mu^b \chi_c$ and $f^{abc}\kappa_a \chi_b \chi_c$ in eqs. (9.45) and (9.46) are invariant under global gauge transformations.

**\*9.2** The goal of this exercise is to derive eq. (9.83).

- Expand $-\mathcal{D}^2 + \mathcal{F}_{\rho\sigma} M^{\rho\sigma}$ in powers of the background field as follows:

$$-\mathcal{D}^2 + \mathcal{F}M = -\Box + \underbrace{i(\mathcal{A}_\mu \partial^\mu + \partial_\mu \mathcal{A}^\mu)}_{O_1} + \underbrace{\mathcal{A}_\mu \mathcal{A}^\mu}_{O_2} + \underbrace{\mathcal{F}M}_{O_3}.$$

Then, expand the logarithm of its determinant. Why is it sufficient to consider only terms of second order in the background field in order to determine the coefficient of $\mathcal{F}\mathcal{F}$?

- Why can we drop terms that contain a single power of $O_3$? Show that at the desired order, we have

$$\log \det \left( -\mathcal{D}^2 + \mathcal{F}M \right) = -\tfrac{1}{2}\mathrm{tr}\left( \Box^{-1} O_1 \Box^{-1} O_1 \right)$$
$$- \mathrm{tr}\left( \Box^{-1} O_2 \right) - \tfrac{1}{2}\mathrm{tr}\left( \Box^{-1} O_3 \Box^{-1} O_3 \right) + \cdots$$

- Rewrite the right-hand side in momentum space, using dimensional regularization. Show that the second term is zero. Calculate the other terms.
- Relate the pole at $D = 4$ to the logarithm of an ultraviolet cutoff in order to obtain the coefficients $c_{r,s}$.

**\*9.3** Consider a non-Abelian gauge theory with the usual $su(N)$ gauge fields, $n_s$ complex scalar fields in the adjoint representation and $n_f$ Dirac fermions in the adjoint representation. Calculate the analogue of eq. (9.88) for this theory. For which values of $N$, $n_s$ and $n_f$ is the gauge coupling not running at one loop?

**9.4** With the rescaling $gA_a^\mu \to A_a^\mu$, explain how the powers of $g$ now appear in the Feynman rules. Show that the order in $g$ of correlation functions still has the same dependence on the number of loops and external lines, although the factors of $g$ have a different origin.

# CHAPTER 10

# Renormalization Group

Perturbation theory provides a powerful way to systematically compute observables order by order in powers of the coupling constants. However, through renormalization, powers of the renormalized coupling constant are usually accompanied by logarithms of the ratio between the renormalization scale M and a physical scale P of the problem, of the form $\left(\lambda_r^2 \ln(M/P)\right)^n$ (typically, $n = 1$ corresponds to one loop, and higher values of $n$ arise in graphs that contain $n$ such one-loop subgraphs). When the renormalization scale is far from the relevant physical scales, these logarithms may be large and offset the smallness of the coupling constant, thereby casting doubt on the relevance of fixed-order perturbative calculations. A way out is to deviate from fixed-order calculations by resumming these terms to all orders in $n$. Although it is by no means a complete all-orders calculation, this procedure would be an improvement because the terms not included have more powers of $\lambda_r^2$ than logarithms, and are in this sense subleading compared to the resummed terms. Instead of performing this resummation directly, the *renormalization group*, that we shall discuss in this chapter, is a way of obtaining differential equations whose solution is the resummed result.

## 10.1 Scale Dependence of Correlation Functions

### 10.1.1 Callan–Symanzik Equations

Let us consider a renormalizable quantum field theory, for instance a scalar theory with a $\phi^4$ interaction (renormalizable in $d \leq 4$ space-time dimensions). For simplicity, we assume a massless theory. The renormalized parameters of the Lagrangian are defined by imposing renormalization conditions on the two-point and four-point functions, for instance at a scale

## 10.1 SCALE DEPENDENCE OF CORRELATION FUNCTIONS

M in the space-like region,

$$\Gamma_4(p_1, p_2, p_3, p_4) = -i\lambda \quad \text{for} \quad (p_1+p_2)^2 = (p_1+p_3)^2 = (p_2+p_4)^2 = -M^2,$$

$$\Pi(p)|_{p^2=-M^2} = 0, \quad \left.\frac{d\Pi(p)}{dp^2}\right|_{p^2=-M^2} = 0, \tag{10.1}$$

where $\Pi(p)$ is the self-energy and $\Gamma_4$ the one-particle irreducible four-point function.

Two choices of M correspond to the same physical theory provided that the bare correlation functions, expressed in terms of the bare parameters of the Lagrangian, are identical. Indeed, the renormalization scale M appears only through the renormalized field $\phi_r \equiv \phi_b/\sqrt{Z}$ and the renormalized coupling constant $\lambda_r$. Under such a variation of the renormalization scale, the renormalized correlation function varies according to

$$\frac{dG_r^{(n)}}{dM} = \frac{\partial G_r^{(n)}}{\partial M} + \frac{\partial G_r^{(n)}}{\partial \lambda_r}\frac{\partial \lambda_r}{\partial M}. \tag{10.2}$$

On the other hand, the bare and renormalized correlation functions are related by

$$G_r^{(n)}(x_1, \cdots, x_n) = Z^{-n/2} G_b^{(n)}(x_1, \cdots, x_n), \tag{10.3}$$

which leads to another expression for the scale dependence of $G_r^{(n)}$,

$$\frac{dG_r^{(n)}}{dM} = -\frac{n}{2Z}\frac{\partial Z}{\partial M} G_r^{(n)}. \tag{10.4}$$

Combining the previous two results, we obtain

$$\left[M\frac{\partial}{\partial M} + \beta\frac{\partial}{\partial \lambda_r} + n\gamma\right] G_r^{(n)} = 0, \tag{10.5}$$

where we have defined

$$\beta \equiv M\frac{\partial \lambda_r}{\partial M}, \quad \gamma \equiv \frac{M}{2Z}\frac{\partial Z}{\partial M}. \tag{10.6}$$

Eq. (10.5) is known as the *Callan–Symanzik equation*, or *renormalization group* (RG) equation. (The reason why it is more convenient to write this equation with $M\partial_M$ instead of $\partial_M$ is that powers $M^\alpha$ are eigenfunctions of $M\partial_M$.) The quantity $\beta$, called the *beta function* of the theory, controls how the coupling constant varies with the scale M. $\gamma$, known as the *anomalous dimension* of the field $\phi$, controls the rescaling of the field when the scale M is changed. These two parameters play a crucial role when determining the scale dependence of renormalized correlation functions, since they encode the "trivial" scale dependence that depends solely on the number of vertices and the number of fields contained in the correlation function. The evolution in M is called a group for the following reason. Formally, the solutions of eq. (10.5) can be written as follows:

$$G_r^{(n)}(\cdots; M) = \mathcal{U}(M, M_0) G_r^{(n)}(\cdots; M_0), \tag{10.7}$$

where the evolution operator $\mathcal{U}(M, M_0)$ is a Green's function of the operator between the square brackets on the left-hand side of eq. (10.5). A one-dimensional group structure can be attached to this evolution by writing $\mathcal{U}(M_2, M_0) = \mathcal{U}(M_2, M_1)\mathcal{U}(M_1, M_0)$. In other words, a finite rescaling can be broken down into several smaller rescalings without affecting the final result.

## 10.1.2 One-Loop Calculation of β and γ

In practice, one can determine the anomalous dimension $\gamma$ and the beta function at one-loop from the wavefunction and vertex counterterms $\delta_z$ and $\delta_\lambda$. Since $Z = 1 + \delta_z$, we can directly write

$$\gamma = \frac{M}{2} \frac{\partial \delta_z}{\partial M}. \tag{10.8}$$

In order to determine the beta function for a four-leg vertex, one should start from the renormalized four-point function $G_r^{(4)}(p_1, \cdots, p_4)$. Diagrammatically, this function reads

$$\bigotimes = \bigotimes + \bigotimes + \sum_i \left[ \bigotimes + \bigotimes \right], \tag{10.9}$$

where the first term on the right-hand side is the 1PI four-point function surrounded by four free propagators and the second term is the corresponding vertex counterterm. The third and fourth terms are the self-energy corrections on the external lines and the corresponding counterterms. Up to one loop, this equation can be written more explicitly as follows:

$$G_r^{(4)}(p_1, \cdots, p_n) = \left( \prod_i \frac{i}{p_i^2} \right) \left( \Gamma_4 - i\delta_\lambda - i\lambda \sum_{i=1}^{4} \frac{1}{p_i^2} (\Pi(p_i) - p_i^2 \delta_z^i) \right). \tag{10.10}$$

In this equation, the dependence of the renormalized correlation function on the renormalization scale M arises from the counterterms $\delta_\lambda$ and $\delta_z^i$. By applying the Callan–Symanzik equation to this correlation function, we obtain at leading order

$$M \frac{\partial}{\partial M} \left( \delta_\lambda - \lambda \sum_i \delta_z^i \right) + \beta + \frac{\lambda}{2} \sum_i M \frac{\partial \delta_z^i}{\partial M} = 0, \tag{10.11}$$

where we have replaced the anomalous dimensions $\gamma^i$ attached to the external lines by their expression given by eq. (10.8) in terms of the corresponding counterterms $\delta_z^i$. Therefore, we obtain the following formula for the beta function:

$$\beta = M \frac{\partial}{\partial M} \left( -\delta_\lambda + \frac{\lambda}{2} \sum_i \delta_z^i \right). \tag{10.12}$$

(See Exercises 10.1 and 10.2 for examples of calculations of β in a scalar theory and in quantum electrodynamics (QED), and 10.3 for a discussion of typical behaviors of the coupling based on the beta function.)

## 10.1.3 Solution for the two-point function $G_r^{(2)}$

In a massless theory, we may always parametrize the two-point function by writing

$$G_r^{(2)}(p) = \frac{i}{p^2} g(-p^2/M^2), \tag{10.13}$$

## 10.1 SCALE DEPENDENCE OF CORRELATION FUNCTIONS

where $g(-p^2/M^2)$ is a function so far unknown. Since the M dependence arises solely from the ratio $-p^2/M^2$, we can rewrite the derivative with respect to M in the Callan–Symanzik equation in the form of a derivative with respect to p,

$$\left[p\frac{\partial}{\partial p} - \beta\frac{\partial}{\partial \lambda} + 2 - 2\gamma\right] G_r^{(2)}(p) = 0. \tag{10.14}$$

In order to solve this equation, let us introduce a function $\bar{\lambda}(p, \lambda)$ defined by

$$\frac{d\bar{\lambda}(p, \lambda)}{d\ln(p/M)} = \beta(\bar{\lambda}), \quad \bar{\lambda}(M, \lambda) = \lambda. \tag{10.15}$$

In other words, $\bar{\lambda}$ is the running coupling constant that takes the value $\lambda$ at the momentum scale M. We can then write a formal solution of the Callan–Symanzik equation in the following way (see Exercise 10.4):

$$G_r^{(2)}(p) = \frac{i}{p^2} \mathcal{G}(\bar{\lambda}(p, \lambda)) \exp\left[2\int_M^p \frac{dp'}{p'} \gamma(\bar{\lambda}(p', \lambda))\right], \tag{10.16}$$

where $\mathcal{G}(\bar{\lambda}(p, \lambda))$ is a function of the running coupling that cannot be determined from the renormalization group equations.[1] This function must be determined order by order from perturbative calculations. In the case of the two-point function, we have $\mathcal{G}(\bar{\lambda}(p, \lambda)) = 1 + \mathcal{O}(\bar{\lambda})$. In order to gain more intuition on this solution, it is useful to assume that the anomalous dimension is a constant. In this case, the exponential in eq. (10.16) becomes

$$(p/M)^{2\gamma} = \sum_{n=0}^{\infty} \frac{(2\gamma)^n}{n!} \ln^n\left(\frac{p}{M}\right). \tag{10.17}$$

First, since $\gamma$ is proportional to some power of the coupling constant ($\gamma \sim \lambda^2$ in $\phi^4$ theory), the expansion on the right-hand side shows that the solution of the renormalization group equation is indeed a summation to all orders of terms in $(\lambda^2 \ln(p/M))^n$. Moreover, this factor alters the power law dependence of the propagator with respect to momentum, changing a power $-2$ into $-2+2\gamma$. We can also see in eq. (10.16) the importance of having introduced the beta function and the anomalous dimensions in the renormalization group equation, since large chunks of the scale dependence are hidden in the scale dependence of the running coupling and in a simple integral of the anomalous dimension (the only remaining explicit M dependence is the lower bound of the integral in the exponential).

---

[1] An arbitrary function of the running coupling is allowed as a prefactor, since we have

$$\left[p\frac{\partial}{\partial p} - \beta(\lambda)\frac{\partial}{\partial \lambda}\right] \mathcal{G}(\bar{\lambda}(p, \lambda)) = \left[p\frac{\partial}{\partial p} - \beta(\bar{\lambda})\frac{\partial}{\partial \bar{\lambda}}\right] \mathcal{G}(\bar{\lambda}(p, \lambda)) = 0.$$

## 10.2 Correlators Containing Composite Operators

### 10.2.1 Callan–Symanzik Equations

An important extension of the previous formalism concerns the case of correlators that contain one or more composite operators, i.e., made of several fields evaluated at the same spacetime point. Similarly to the case of elementary operators, we must introduce a renormalization factor $Z_{\mathcal{O}}$, determined order by order in perturbation theory in order to fulfill a certain renormalization condition at the scale $M$. The renormalized operator $\mathcal{O}_r$ is related to the bare operator $\mathcal{O}_b$ by the relationship $Z_{\mathcal{O}} \mathcal{O}_r = \mathcal{O}_b$. Let us consider now a renormalized correlation function involving a composite operator $\mathcal{O}$ and $n$ elementary fields,

$$G_r^{(n;1)}(x_1, \cdots, x_n; y) \equiv \langle \phi(x_1) \cdots \phi(x_n) \mathcal{O}(y) \rangle. \tag{10.18}$$

The corresponding bare correlation function is given by

$$G_r^{(n;1)}(x_1, \cdots, x_n; y) = Z^{-n/2} Z_{\mathcal{O}}^{-1} G_b^{(n;1)}(x_1, \cdots, x_n; y). \tag{10.19}$$

By requesting that the bare correlation function remains unchanged upon changes of the renormalization scale $M$, we obtain the following equation satisfied by the renormalized correlation function:

$$\left[ M \frac{\partial}{\partial M} + \beta \frac{\partial}{\partial \lambda} + n\gamma + \gamma_{\mathcal{O}} \right] G_r^{(n;1)} = 0, \tag{10.20}$$

where we have defined the anomalous dimension of the composite operator $\mathcal{O}$ as follows:

$$\gamma_{\mathcal{O}} \equiv \frac{M}{Z_{\mathcal{O}}} \frac{\partial Z_{\mathcal{O}}}{\partial M}. \tag{10.21}$$

### 10.2.2 Anomalous Dimension of the Operator $\mathcal{O}$

The practical determination of the anomalous dimension $\gamma_{\mathcal{O}}$ of a composite operator $\mathcal{O}$ made of $m$ elementary fields $\phi$ can be done by studying the correlation function $G_r^{(m;1)}$ and by applying to it the Callan–Symanzik equation. This method is identical to the one used in order to obtain eq. (10.12) for the beta function, and leads to

$$\gamma_{\mathcal{O}} = M \frac{\partial}{\partial M} \left( -\delta_{\mathcal{O}} + \frac{1}{2} \sum_i \delta_z^i \right), \tag{10.22}$$

where $\delta_{\mathcal{O}}$ is the counterterm that one must adjust in order to satisfy the renormalization condition of the operator $\mathcal{O}$ at the scale $M$ (see Exercise 10.6 for a simple example).

### 10.2.3 Anomalous Dimension of a Conserved Current

A very useful example in practice is that of a conserved current such as

$$J^\mu \equiv \overline{\psi} \gamma^\mu \psi. \tag{10.23}$$

The anomalous dimension of such an operator is given by

$$\gamma_J = M \frac{\partial}{\partial M}(-\delta_J + \delta_\psi) = 0. \tag{10.24}$$

The equality of the counterterms $\delta_J$ and $\delta_\psi$ is a consequence of the Ward–Takahashi identities (in QED, they translate into the equality $Z_1 = Z_2$, that immediately implies the cancellation of the counterterms in the previous equation), i.e., of the gauge symmetry associated to charge conservation and to the conservation of the current $J^\mu$.

### 10.2.4 Renormalization of Operators of Arbitrary Dimensions

Let us denote by $\mathcal{L}_M$ the renormalized Lagrangian at the scale M. Consider now adding to this Lagrangian the following sum of interaction terms:

$$\mathcal{L}_M \to \mathcal{L}_M + \sum_i c_i \mathcal{O}_i(x), \tag{10.25}$$

where $\mathcal{O}_i$ are arbitrary local operators, not necessarily renormalizable in four dimensions. The Callan–Symanzik equation for a correlator containing $n$ elementary fields $\phi$ and an arbitrary number of these new interaction terms reads:

$$\left[ M \frac{\partial}{\partial M} + \beta \frac{\partial}{\partial \lambda} + n\gamma + \sum_i \gamma_i c_i \frac{\partial}{\partial c_i} \right] G_r^{(n)} = 0. \tag{10.26}$$

In this equation, $\gamma_i$ is the anomalous dimension of the operator $\mathcal{O}_i$ and the operator $c_i \partial/\partial c_i$ counts the number of occurrences of $\mathcal{O}_i$ inside the function $G_r^{(n)}$. If $d_i$ is the dimension of the operator $\mathcal{O}_i$ (in mass units), it is convenient to define a dimensionless coupling constant $\rho_i$ by the following relation:

$$c_i \equiv \rho_i M^{4-d_i}. \tag{10.27}$$

Thanks to this definition, the previous Callan–Symanzik equation becomes

$$\left[ M \frac{\partial}{\partial M} + \beta \frac{\partial}{\partial \lambda} + n\gamma + \sum_i \beta_i \frac{\partial}{\partial \rho_i} \right] G_r^{(n)} = 0, \tag{10.28}$$

where we denote $\beta_i \equiv \rho_i(\gamma_i + d_i - 4)$. With these notations, we see that the additional couplings $\rho_i$ play exactly the same role as the original coupling $\lambda$. We can therefore mimic the explicit solution found in the case of the two-point function in Section 10.1.3. Let us first introduce running couplings $\overline{\lambda}, \overline{\rho}_i$ as solutions of the following differential equations:

$$\frac{d\overline{\lambda}(p,\lambda)}{d\ln(p/M)} = \beta(\overline{\lambda}, \overline{\rho}_i), \quad \overline{\lambda}(M, \lambda) = \lambda,$$

$$\frac{d\overline{\rho}_i(p, \rho_i)}{d\ln(p/M)} = \beta_i(\overline{\lambda}, \overline{\rho}_i), \quad \overline{\rho}_i(M, \rho_i) = \rho_i. \tag{10.29}$$

In the weak coupling limit, the functions $\beta_i$ are given at lowest order by

$$\beta_i \approx (d_i - 4)\rho_i, \tag{10.30}$$

and the solution of the previous equations for $\bar{\rho}_i$ reads

$$\bar{\rho}_i(p) = \bar{\rho}_i(M) \left(\frac{p}{M}\right)^{d_i - 4}. \tag{10.31}$$

This result sheds some light on the fact that all fundamental interactions (except gravity, for which the underlying quantum theory is not known) appear to be described by renormalizable quantum field theories at the energy scales relevant for the Standard Model (i.e., $p \lesssim 1$ TeV). Indeed, let us assume that there exists at a much higher scale (typically $M \sim 10^{16}$ GeV, the conjectured scale for the unification of all gauge couplings) a more fundamental quantum field theory, comprising all sorts of interactions and whose couplings are of order 1 (at this unification scale, couplings that are allowed by symmetries have no reason to be much smaller than unity). After evolving the scale down to the sub-TeV scale of the Standard Model, all the couplings for which $d_i - 4 > 0$, i.e., all the operators that are not renormalizable in four space-time dimensions, have become much smaller than the others and have effectively disappeared from the Lagrangian.

## 10.3 Operator Product Expansion

### 10.3.1 Introduction

The *operator product expansion* (OPE) is a tool that allows study of the renormalization flow at the level of the operators themselves, instead of encapsulating them inside a correlator (although the derivation still requires that we consider a correlator). The intuitive idea is that a non-local product of operators may be approximated by a local composite operator when the separations between the original operators go to zero, possibly with a numerical prefactor that depends on the separation between the operators in the original product. However, since limits of operators are difficult to handle, it is convenient to first consider a weaker form of limit, in which the product of operators under consideration is encapsulated into a correlation function of the form

$$G_{12}^{(n)}(x; y_1, \cdots, y_n) \equiv \langle A_1(x) A_2(0) \phi(y_1) \cdots \phi(y_n) \rangle, \tag{10.32}$$

where $A_1$ and $A_2$ are local operators, and $\phi$ an elementary field. Let us consider a limit where the coordinates $y_i$ are fixed, while $x \to 0$. We can already note that, since the product of operators at the same point is ill-defined in general, we may expect divergences in this limit.

It turns out that the behavior of $G_{12}^{(n)}$ when $x \to 0$ is entirely determined by the operators $A_1$ and $A_2$ themselves, in a way that does not depend on the other fields $\phi(y_i)$ (provided they are kept at a finite distance from points 0 and x). In order to determine this behavior, Wilson proposed expanding the product $A_1(x) A_2(0)$ as a sum of composite local operators, with $x$ dependent coefficients,

$$A_1(x) A_2(0) = \sum_i C_{12}^i(x)\, \mathcal{O}_i(0), \tag{10.33}$$

## 10.3 OPERATOR PRODUCT EXPANSION

where the $\mathcal{O}_i$ are a basis of composite local operators that have the same quantum numbers as the product $A_1 A_2$. All the x dependence is carried by the *Wilson coefficients* $C_{12}^i(x)$. This decomposition can then be used in any correlation function where the product $A_1(x)A_2(0)$ appears. For instance, the correlation $G_{12}^{(n)}$ introduced at the beginning of this section would read

$$G_{12}^{(n)}(x;y_1,\cdots,y_n) = \sum_i C_{12}^i(x)\, G_i^{(n)}(y_1,\cdots,y_n), \tag{10.34}$$

where we denote

$$G_i^{(n)}(y_1,\cdots,y_n) \equiv \langle \mathcal{O}_i(0)\phi(y_1)\cdots\phi(y_n)\rangle. \tag{10.35}$$

### 10.3.2 Callan–Symanzik Equation for $C_{12}^i(x)$

Let us assume that we have defined the normalization of the operators $A_1, A_2, \mathcal{O}_i$ at the scale M. The coefficients $C_{12}^i(x)$ in eq. (10.33) should a priori also depend on M. In order to determine this dependence, let us first write the Callan–Symanzik equation for the renormalized correlator[2] $G_{12}^{(n)}$,

$$\left[ M\frac{\partial}{\partial M} + \beta\frac{\partial}{\partial \lambda} + n\gamma + \gamma_{A_1} + \gamma_{A_2} \right] G_{12}^{(n)} = 0, \tag{10.36}$$

where $\gamma, \gamma_{A_1}$ and $\gamma_{A_2}$ are the anomalous dimensions of the operators $\phi$, $A_1$ and $A_2$, respectively. Concerning the correlation functions $G_i^{(n)}$ that enter the right-hand side of eq. (10.34), we have the following equation:

$$\left[ M\frac{\partial}{\partial M} + \beta\frac{\partial}{\partial \lambda} + n\gamma + \gamma_i \right] G_i^{(n)} = 0, \tag{10.37}$$

where $\gamma_i$ is the anomalous dimension of $\mathcal{O}_i$. The left-hand side and right-hand side of eq. (10.34) are consistent provided that the coefficients $C_{12}^i$ obey

$$\left[ M\frac{\partial}{\partial M} + \beta\frac{\partial}{\partial \lambda} + \gamma_{A_1} + \gamma_{A_2} - \gamma_i \right] C_{12}^i = 0. \tag{10.38}$$

This equation confirms a posteriori the fact that the coefficients $C_{12}^i$ must depend on the renormalization scale M. Moreover, we see that this dependence only depends on the anomalous dimensions of the operators $A_1, A_2$ and $\mathcal{O}_i$, but not on the specific correlation function $G_{12}^{(n)}$ that was used in the derivation (in particular, eq. (10.38) does not depend on the number $n$ of fields $\phi$, nor on their anomalous dimension). It is this property that renders the operator product expansion universal.

---

[2] In the rest of this chapter, we do not write explicitly the subscript r to indicate the renormalized quantities, in order to simplify the notations. From the context, it is always clear when a quantity is renormalized.

## 10.3.3 Separation Dependence of $C^i_{12}(x)$

If the dimensions of $A_1$, $A_2$ and $\mathcal{O}_i$ are respectively $D_1$, $D_2$ and $d_i$, then the dimension of $C^i_{12}$ is $D_1 + D_2 - d_i$. Therefore, we may write

$$C^i_{12}(x;M) \equiv \frac{1}{|x|^{D_1+D_2-d_i}} \widetilde{C}^i_{12}(M|x|), \tag{10.39}$$

where $\widetilde{C}^i_{12}(Mx)$ is a dimensionless function of the sole variable $M|x|$. One can determine this function similarly to the case of the two-point function considered in Section 10.1.3, by introducing the running coupling $\bar{\lambda}(1/|x|)$. We obtain the following structure for the coefficient $C^i_{12}$:

$$C^i_{12}(x;M) = \frac{\mathcal{C}^i_{12}(\bar{\lambda}(1/|x|))}{|x|^{D_1+D_2-d_i}}$$
$$\times \exp\left[\int_{1/|x|}^{M} \frac{dp'}{p'} \left(\gamma_i(\bar{\lambda}(p')) - \gamma_{A_1}(\bar{\lambda}(p')) - \gamma_{A_2}(\bar{\lambda}(p'))\right)\right], \tag{10.40}$$

where $\mathcal{C}^i_{12}$ is a function of the running coupling that can be obtained by matching to perturbative calculations. We see that the leading short-distance behavior is controlled by the prefactor $|x|^{d_i - D_1 - D_2}$, which becomes singular if $d_i < D_1 + D_2$. Moreover, the contribution of the operators $\mathcal{O}_i$ whose dimension obeys $d_i > D_1 + D_2$ goes to zero when $x \to 0$. One does not need to consider such operators in the OPE when studying the short-distance limit.

In asymptotically free theories where the coupling goes to zero at short distance, such as QCD, we may carry a bit further the determination of the Wilson coefficients. Indeed, at the first order of perturbation theory, the anomalous dimensions are proportional to $g^2$ and we may write the anomalous dimension of any operator $\mathcal{O}$ as follows:

$$\gamma_{\mathcal{O}} \equiv -a_{\mathcal{O}} \frac{g^2}{(4\pi)^2}, \tag{10.41}$$

where $a_{\mathcal{O}}$ is a numerical constant (the minus sign is conventional). Therefore, we have

$$\gamma_i - \gamma_{A_1} - \gamma_{A_2} = (a_{A_1} + a_{A_2} - a_i)\frac{\alpha_s}{4\pi}, \tag{10.42}$$

with $\alpha_s \equiv g^2/4\pi$. At one loop, the running coupling $\alpha_s$ is given by eq. (9.91),

$$\alpha_s(Q^2) = \frac{4\pi}{\beta_0 \ln\left(\frac{Q^2}{\Lambda^2_{QCD}}\right)}, \tag{10.43}$$

where $\beta_0$ is the first Taylor coefficient of the QCD $\beta$ function. From this, we get

$$C^i_{12}(x;M) = \frac{\mathcal{C}^i_{12}(\bar{g}(1/|x|))}{|x|^{D_1+D_2-d_i}} \left[\frac{\ln(1/|x|^2\Lambda^2_{QCD})}{\ln(M^2/\Lambda^2_{QCD})}\right]^{\frac{a_i - a_{A_1} - a_{A_2}}{2\beta_0}}. \tag{10.44}$$

We see that, besides the trivial power law prefactor in $|x|^{d_i-D_1-D_2}$, there are corrections in the form of powers of logarithms that may be large when $x \to 0$. When $d_i = D_1 + D_2$, these logarithms are in fact the main source of $|x|$ dependence.

### 10.3.4 Operator Mixing

It may happen that several of the operators $\mathcal{O}_i$ that enter the OPE basis for the product $A_1(x)A_2(0)$ mix under the evolution of the scale $M$. This means that the anomalous dimensions $\gamma_i$ are in fact a matrix $\gamma_{ij}$ (when there is no mixing, this matrix is diagonal and the $\gamma_i$ that we have used so far are its diagonal elements) and the Callan–Symanzik equations for the correlators $G_i^{(n)}$ are coupled:

$$\sum_i \left[ \delta_{ij} \left( M \frac{\partial}{\partial M} + \beta \frac{\partial}{\partial \lambda} + n\gamma \right) + \gamma_{ij} \right] G_j^{(n)} = 0. \tag{10.45}$$

The equation for $G_{12}^{(n)}$ is unchanged, and we obtain the following equation for the Wilson coefficients:

$$\left[ M \frac{\partial}{\partial M} + \beta \frac{\partial}{\partial \lambda} + \gamma_{A_1} + \gamma_{A_2} \right] C_{12}^j - \sum_i \gamma_{ij} C_{12}^i = 0. \tag{10.46}$$

Note that when the operators $A_1$ and $A_2$ are conserved currents, their anomalous dimensions are zero, and this equation simplifies into:

$$\left[ M \frac{\partial}{\partial M} + \beta \frac{\partial}{\partial \lambda} \right] C_{12}^j - \sum_i \gamma_{ij} C_{12}^i = 0. \tag{10.47}$$

This situation turns out to be quite frequent in applications of the OPE.

## 10.4  Example: QCD Corrections to Weak Decays

### 10.4.1  Fermi Theory

In order to illustrate the use of the operator product expansion on a concrete case, let us consider the weak interactions between quarks and leptons (see Exercise 10.5 for another, less direct, application). In the Standard Model, the interactions between charged currents take the following form:

$$\mathcal{L}_I = \frac{g^2}{2} J_L^\mu(0) D_{\mu\nu}(0,x) J_L^{\nu\dagger}(x) + \text{h.c.}, \tag{10.48}$$

where $J_L^\mu$ is the left-handed charged current (containing a leptonic term and a term due to quarks) and $D_{\mu\nu}(0,x)$ is the propagator of the $W^\pm$ boson between points 0 and $x$.

At low energy, we may neglect the momentum carried by the $W^\pm$ boson propagator in front of the $W^\pm$ mass. In this approximation, the propagator becomes momentum independent, and its Fourier transform is proportional to $\delta(x)$. We may then replace the non-local interaction term of eq. (10.48) by a four-fermion (local) contact interaction, which is nothing but the interaction term of Fermi's theory. The prefactor of this interaction term, $g^2/2M_W^2$, is usually denoted $4G_F/\sqrt{2}$ where $G_F$ is Fermi's constant:

$$\mathcal{L}_{\text{int}} \approx \frac{4G_F}{\sqrt{2}} J_L^\mu(0) J_L^{\nu\dagger}(0) + \text{h.c.} \tag{10.49}$$

Thanks to the operator product expansion, one may study in greater detail the limit from the electroweak theory to Fermi's theory, i.e., the process by which one replaces the non-local product of two currents by one or more local interaction terms. This example will also illustrate how this decomposition in local operators depends on the energy scale of the processes under consideration, by including the strong interaction corrections at one loop.

Let us discuss first two trivial cases regarding the effect of QCD corrections at one loop. First, purely leptonic weak interactions are not affected by strong interactions at this order since leptons do not couple directly to gluons (but QCD corrections do exist at two loops and beyond). The other simple case is that of semi-leptonic weak interactions, involving a leptonic current and a current made of quarks. Indeed, the leptonic current is not renormalized by strong interactions. The quark current, conserved at leading order, is also not affected by strong interactions since its anomalous dimension is zero. Finally, a gluon cannot connect the lepton and the quark currents. Thus, semi-leptonic weak interactions are not affected by QCD corrections at one loop. The only relevant case at one loop, to which we will devote the rest of this section, is that of weak interactions between quark currents, i.e., the non-leptonic weak interactions. As an example, let us consider the QCD corrections to the weak decay of the strange quark, which in Fermi's theory comes from the following coupling: $(\bar{d}_L \gamma^\mu u_L)(\bar{u}_L \gamma_\mu s_L)$.

### 10.4.2 Operator Product Expansion

Let us consider the OPE of the product of currents $A_1^\mu(x) A_{2\mu}(0)$, with

$$A_1^\mu \equiv \bar{d}_L \gamma^\mu u_L, \quad A_2^\mu \equiv \bar{u}_L \gamma^\mu s_L. \tag{10.50}$$

When going from the standard model to Fermi's theory, the non-local dependence of the $W^\pm$ propagator is captured by the Wilson coefficients $C_{12}^i(x)$. Therefore, the typical separation $x$ is $x \sim M_W^{-1}$ (since the mass $M_W$ is the only dimensionful parameter in the propagator). On the other hand, the scale $M$ characteristic of kaon decays is of the order of the mass of a kaon, around 500 MeV. The simplest operators on which we may expand the product $A_1(x)A_2(0)$ are the following:

$$\mathcal{O}_1 \equiv (\bar{d}_L \gamma^\mu u_L)(\bar{u}_L \gamma_\mu s_L), \quad \mathcal{O}_2 \equiv (\bar{d}_L \gamma^\mu s_L)(\bar{u}_L \gamma_\mu u_L), \tag{10.51}$$

where in the second one two quark operators of different flavors have been exchanged. Note that the mass dimension of the operators $A_1$ and $A_2$ is 3, while that of $\mathcal{O}_1$ and $\mathcal{O}_2$ is 6. Therefore, we have $d_{A_1} + d_{A_2} - d_i = 0$, which means that the $x$ dependence of the Wilson coefficients comes entirely from the logarithms in eq. (10.44). The more complicated operators that could enter in this expansion all have a larger mass dimension, so that $d_{A_1} + d_{A_2} - d_i < 0$. Thanks to the prefactor in eq. (10.44), the corresponding Wilson coefficients are very small since $M|x| \sim M/M_W \ll 1$. Thus, one can restrict the OPE of $A_1(x)A_2(0)$ to the sole operators $\mathcal{O}_1$ and $\mathcal{O}_2$ in applications to the physics of kaon decays.

### 10.4.3 Evolution of the Wilson Coefficients

In order to determine the Wilson coefficients $C_{12}^i$ for the operators $\mathcal{O}_i$ with eq. (10.44), we first need to calculate the anomalous dimensions $\gamma_{A_1}, \gamma_{A_2}$, as well as $\gamma_1, \gamma_2$, for the operators $A_1, A_2, \mathcal{O}_1$ and $\mathcal{O}_2$. Since $A_1$ and $A_2$ are conserved at the first order, their anomalous dimension is zero,

$$\gamma_{A_1} = \gamma_{A_2} = 0. \tag{10.52}$$

In order to obtain the anomalous dimensions of the operators $\mathcal{O}_1$ and $\mathcal{O}_2$, let us introduce the following graphical representation for these operators:

$$\mathcal{O}_1 = \begin{array}{c}\includegraphics\end{array} \quad , \quad \mathcal{O}_2 = \begin{array}{c}\includegraphics\end{array} . \tag{10.53}$$

This representation renders explicit the fact that these operators are products of two currents. Thanks to eq. (10.22), the anomalous dimension of these operators is obtained by calculating the vertex counterterm and the counterterms associated to the external lines. All the order-$g^2$ strong interaction corrections are listed in Figure 10.1 in the case of $\mathcal{O}_1$. The contributions to $\gamma_1$ of the first three diagrams on the first line cancel, because their sum gives the anomalous dimension of a conserved (at first order) current. The same conclusion holds for the remaining three graphs of the first line. Thus, we need only to consider the diagrams of the second line. In Feynman gauge, the expression of the first diagram of the second line is given by

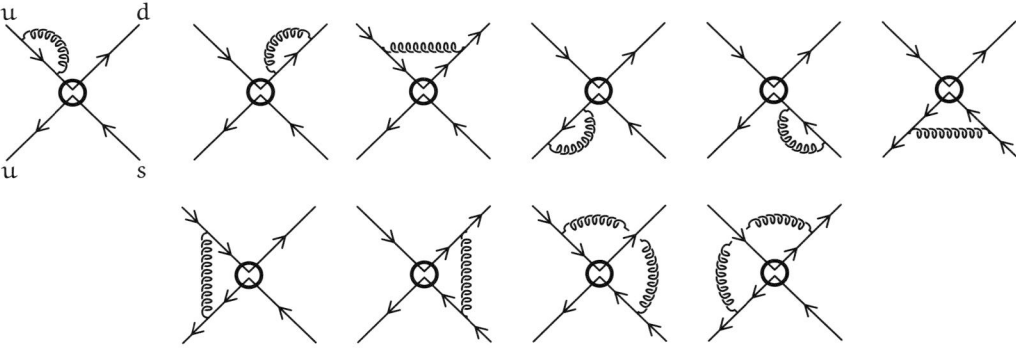

FIGURE 10.1: Order-$g^2$ QCD corrections to the operator $\mathcal{O}_1$.

$$= (-ig)^2 \int \frac{d^Dk}{(2\pi)^D} \frac{-i}{k^2} \left[ \overline{d}_L \gamma^\mu \frac{i\slashed{k}}{(k+p)^2} t_f^a \gamma^\lambda u_L \right] \left[ \overline{u}_L \gamma_\lambda t_f^a \frac{i\slashed{k}}{(k-q)^2} \gamma_\mu s_L \right],$$

(10.54)

where p and q are the (incoming) momenta carried by the quark lines to which the gluon is attached. The $t_f^a$ are the generators of the fundamental representation of the $\mathfrak{su}(3)$ algebra, which holds the quarks. In the numerator, some terms in $\slashed{p}$ and $\slashed{q}$ have been dropped because they do not contribute to the ultraviolet divergence of the graph. The integral over k can be rewritten as follows:[3]

$$\int \frac{d^Dk}{(2\pi)^D} \frac{k^\nu k^{\nu'}}{k^2(k+p)^2(k-q)^2} = \frac{g^{\nu\nu'}}{D} \int \frac{d^Dk}{(2\pi)^D} \frac{1}{(k+p)^2(k-q)^2}$$

$$= \frac{g^{\nu\nu'}}{D} \int_0^1 dx \int \frac{d^D\overline{k}}{(2\pi)^D} \frac{1}{(\overline{k}^2 + \Delta)^2}$$

$$= i\frac{g^{\nu\nu'}}{D} \int_0^1 dx \frac{\Gamma(2-\frac{D}{2})}{(4\pi)^{D/2}} \frac{1}{\Delta^{2-D/2}}, \quad (10.55)$$

where we denote $\overline{k} \equiv k + xp - (1-x)q$ and $\Delta \equiv x(1-x)(p+q)^2$. Since the renormalization scale is M, we may impose that the Lorentz invariant quantity $(p+q)^2$ is equal to $-M^2$, so that $\Delta$ is proportional to $M^2$. Since the power $2-D/2$ to which the denominator $\Delta$ is raised goes to zero in four dimensions, we may neglect the prefactor $x(1-x)$ inside $\Delta$ and the integral over the Feynman parameter x simply gives a factor equal to unity. If we take the limit $D \to 4$ in all the factors that do not diverge and do not depend on M, we obtain:

$$= \frac{g^2}{4} \frac{\Gamma(2-\frac{D}{2})}{(4\pi)^2} \frac{1}{M^{4-D}} \left[ \overline{d}_L \gamma^\mu \gamma^\nu t_f^a \gamma^\lambda u_L \right] \left[ \overline{u}_L \gamma_\lambda t_f^a \gamma_\nu \gamma_\mu s_L \right]. \quad (10.56)$$

The contribution of this graph to the counterterm for the normalization of $\mathcal{O}_1$ is given by the opposite of this result.

In order to simplify the combination of spinors, Dirac and color matrices that appear in the result of eq. (10.56), it is useful to use the chiral representation (also known as Weyl's representation) since only the left-handed component of the spinors enters this expression. In this representation, the Dirac matrices are given by

$$\gamma^\mu = \begin{pmatrix} 0 & \sigma^\mu \\ \overline{\sigma}^\mu & 0 \end{pmatrix}, \quad \gamma^5 = \begin{pmatrix} -1 & 0 \\ 0 & 1 \end{pmatrix}, \quad (10.57)$$

---

[3] The first equality disregards some terms that are ultraviolet finite.

## 10.4 EXAMPLE: QCD CORRECTIONS TO WEAK DECAYS

with $\sigma^\mu \equiv (1, \boldsymbol{\sigma})$ and $\overline{\sigma}^\mu \equiv (1, -\boldsymbol{\sigma})$ where $\boldsymbol{\sigma}$ is a vector made of the three Pauli matrices. In this representation, the left-handed projector $P_L \equiv (1 - \gamma^5)/2$ and the right-handed one $P_R \equiv (1 + \gamma^5)/2$ simplify into

$$P_L = \begin{pmatrix} 1 & 0 \\ 0 & 0 \end{pmatrix}, \quad P_R = \begin{pmatrix} 0 & 0 \\ 0 & 1 \end{pmatrix}, \qquad (10.58)$$

so that any four-component spinor can be viewed as two two-component spinors, one of which is right-handed and the other left-handed:

$$\psi = \begin{pmatrix} \psi_L \\ \psi_R \end{pmatrix}. \qquad (10.59)$$

Using this representation, we can for instance easily obtain

$$\overline{d}_L \gamma^\mu \gamma^\nu \gamma^\lambda t_f^a u_L = \overline{d}_L \overline{\sigma}^\mu \sigma^\nu \overline{\sigma}^\lambda t_f^a u_L. \qquad (10.60)$$

(This equation involves a small abuse of notations, since it contains the four-component spinors $(\psi_L, 0)$ on the left-hand side, while the right-hand side contains only the two-component left-handed spinors $\psi_L$.)

In order to reduce the combination of spinors that appear in eq. (10.56), we need to simplify the products $(\sigma^\mu)_{\alpha\beta}(\sigma_\mu)_{\gamma\delta}$ and $(\overline{\sigma}^\mu)_{\alpha\beta}(\overline{\sigma}_\mu)_{\gamma\delta}$ as well as $(t_f^a)_{ij}(t_f^a)_{kl}$. In both cases, this can be done by using the Fierz identity for the generators of the fundamental representation of the $\mathfrak{su}(n)$ algebra, introduced in Section 7.1.6. Let us recall this identity here:

$$(t_f^a)_{ij}(t_f^a)_{kl} = \frac{1}{2}\left[\delta_{il}\delta_{jk} - \frac{1}{n}\delta_{ij}\delta_{kl}\right]. \qquad (10.61)$$

For the contraction of color matrices $t_f^a$, we can apply it directly with $n = 3$:

$$(t_f^a)_{ij}(t_f^a)_{kl} = \frac{1}{2}\left[\delta_{il}\delta_{jk} - \frac{1}{3}\delta_{ij}\delta_{kl}\right]. \qquad (10.62)$$

For the contraction of the $\sigma^\mu$ or the $\overline{\sigma}^\mu$, let us recall that the Pauli matrices $\sigma^i$ are related to the generators $\tau^i$ (we use a different symbol in order to distinguish them from those of $\mathfrak{su}(3)$) of the fundamental representation of $\mathfrak{su}(2)$ by

$$\sigma^i = 2\tau^i. \qquad (10.63)$$

Using this relation and the Fierz identity for the fundamental representation of $\mathfrak{su}(2)$, we obtain

$$\begin{aligned}(\sigma^\mu)_{\alpha\beta}(\sigma_\mu)_{\gamma\delta} = (\overline{\sigma}^\mu)_{\alpha\beta}(\overline{\sigma}_\mu)_{\gamma\delta} &= \delta_{\alpha\beta}\delta_{\gamma\delta} - 4(\tau^i)_{\alpha\beta}(\tau^i)_{\gamma\delta} \\ &= \delta_{\alpha\beta}\delta_{\gamma\delta} - 2\left[\delta_{\alpha\delta}\delta_{\beta\gamma} - \frac{1}{2}\delta_{\alpha\beta}\delta_{\gamma\delta}\right] \\ &= 2[\delta_{\alpha\beta}\delta_{\gamma\delta} - \delta_{\alpha\delta}\delta_{\beta\gamma}]. \end{aligned} \qquad (10.64)$$

Thanks to eqs. (10.62) and (10.64), we have[4]

$$[\bar{d}_L \gamma^\mu \gamma^\nu t_f^a \gamma^\lambda u_L] [\bar{u}_L \gamma_\lambda t_f^a \gamma_\nu \gamma_\mu s_L]$$
$$= 2(\bar{u}_L \gamma^\mu u_L)(\bar{d}_L \gamma_\mu s_L) - \frac{2}{3}(\bar{d}_L \gamma^\mu u_L)(\bar{u}_L \gamma_\mu s_L). \tag{10.65}$$

We recognize the operators $O_1$ and $O_2$ in this expression. We are therefore in a situation in which renormalization introduces a mixing among operators. The second diagram is identical to the one we have just calculated.

The third diagram of the second line reads

$$= (-ig)^2 \int \frac{d^D k}{(2\pi)^D} \frac{-i}{k^2} \left[ \bar{d}_L \gamma^\mu \frac{i\slashed{k}}{(k+p)^2} t_f^a \gamma^\lambda u_L \right] \left[ \bar{u}_L \gamma_\mu \frac{-i\slashed{k}}{(k-r)^2} t_f^a \gamma_\lambda s_L \right], \tag{10.66}$$

where r is the momentum that flows into the diagram by the line carrying the s quark. The integration over k is similar to the previous case, and leads to

$$= -\frac{g^2}{4} \frac{\Gamma(2-\frac{D}{2})}{(4\pi)^2} \frac{1}{M^{4-D}} [\bar{d}_L \gamma^\mu \gamma^\nu t_f^a \gamma^\lambda u_L] [\bar{u}_L \gamma_\mu \gamma_\nu t_f^a \gamma_\lambda s_L]. \tag{10.67}$$

Likewise, we can simplify the Dirac and color matrices by using Fierz identities,

$$[\bar{d}_L \gamma^\mu \gamma^\nu t_f^a \gamma^\lambda u_L][\bar{u}_L \gamma_\mu \gamma_\nu t_f^a \gamma_\lambda s_L]$$
$$= 8(\bar{u}_L \gamma^\mu u_L)(\bar{d}_L \gamma_\mu s_L) - \frac{8}{3}(\bar{d}_L \gamma^\mu u_L)(\bar{u}_L \gamma_\mu s_L), \tag{10.68}$$

which is again a linear combination of $O_1$ and $O_2$. The last diagram gives the same result.

By combining the four contributions, we obtain the following form for the operator $O_1$, renormalized at the scale M, in terms of the bare operators,

$$O_{1r} = O_{1b} - \delta_{11} O_{1b} - \delta_{12} O_{2b}, \tag{10.69}$$

where the counterterms $\delta_{ij}$ are given by

$$\delta_{11} \equiv \frac{g^2}{(4\pi)^2} \frac{\Gamma(2-\frac{D}{2})}{M^{4-D}}, \quad \delta_{12} \equiv -3 \frac{g^2}{(4\pi)^2} \frac{\Gamma(2-\frac{D}{2})}{M^{4-D}}. \tag{10.70}$$

By calculating in the same way the one-loop corrections to the operator $O_2$, we obtain the counterterms $\delta_{22}$ and $\delta_{21}$ that are equal to

$$\delta_{21} = \delta_{12}, \quad \delta_{22} = \delta_{11}. \tag{10.71}$$

---

[4]The derivation can be made easier by using the graphical form (eq. (7.78)) of the Fierz identity.

Because of the mixing, the anomalous dimensions for the operators $\mathcal{O}_{1,2}$ form a non-diagonal matrix:

$$\gamma_{ij} = M \frac{\partial \delta_{ij}}{\partial M} = \frac{g^2}{(4\pi)^2} \begin{pmatrix} -2 & 6 \\ 6 & -2 \end{pmatrix}. \tag{10.72}$$

In order to solve the coupled Callan–Symanzik equations (10.47), we must find a basis of operators in which the matrix of anomalous dimensions becomes diagonal. This is achieved by choosing[5]

$$\mathcal{O}_{1/2} \equiv \frac{1}{2}[\mathcal{O}_1 - \mathcal{O}_2], \quad \mathcal{O}_{3/2} \equiv \frac{1}{2}[\mathcal{O}_1 + \mathcal{O}_2]. \tag{10.73}$$

The corresponding eigenvalues of the matrix $\gamma_{ij}$ are

$$\gamma_{1/2} = -8 \frac{g^2}{(4\pi)^2}, \quad \gamma_{3/2} = 4 \frac{g^2}{(4\pi)^2}. \tag{10.74}$$

Using eq. (10.44) (the functions $\mathcal{C}_{12}^i$ are equal to 1 at the first order of perturbation theory) at a distance scale $x \approx M_W^{-1}$, we obtain the following values for the Wilson coefficients:

$$C_{12}^{(1/2)}(M_W^{-1}; M) = \left[ \frac{\ln(M_W^2/\Lambda_{QCD}^2)}{\ln(M^2/\Lambda_{QCD}^2)} \right]^{\frac{4}{\beta_0}},$$

$$C_{12}^{(3/2)}(M_W^{-1}; M) = \left[ \frac{\ln(M_W^2/\Lambda_{QCD}^2)}{\ln(M^2/\Lambda_{QCD}^2)} \right]^{-\frac{2}{\beta_0}}. \tag{10.75}$$

Since $M_W \gg M$ and $\beta_0 = 11 - 2n_f/3$ is positive,[6] the operator $A_1(x)A_2(0)$ responsible for the weak decay of the quark s receives a larger contribution from the operator $\mathcal{O}_{1/2}$ than from $\mathcal{O}_{3/2}$ (roughly by a factor of 3.6 if we use $M \approx 500$ MeV, $\Lambda_{QCD} \approx 150$ MeV and 5 quark flavors). This calculation qualitatively[7] corroborates the empirical observation that weak decays of kaons correspond predominantly to an isospin variation of 1/2.

## 10.5 Non-Perturbative Renormalization Group

Until now, our discussion of renormalization has been strictly rooted in perturbation theory and limited to the context of renormalizable theories, with the exception of Section 10.2.4 where we discussed the running of the couplings in front of operators of any dimension. In

---

[5]The subscripts 1/2 and 3/2 are related to the isospin variation in the s quark decay mediated by these operators.

[6]In this problem, $n_f = 5$ flavors of quarks should be taken into account in the running of the strong coupling constant, in order to include all the quarks up to the mass of the $W^\pm$ bosons.

[7]The measured imbalance between the isospin variations 1/2 and 3/2 is larger, but a quantitative explanation would involve non-perturbative aspects of QCD.

this framework, the renormalization flow is formalized by the Callan–Symanzik equations that describe the scale dependence of correlation functions. However, the ideas behind renormalization have a much wider range of applications: They are also relevant non-perturbatively, and they may be applied directly at the level of actions rather than correlation functions. In this section, we first develop heuristically some general concepts related to the renormalization flow in an abstract space of theories. These ideas are then made more tangible in the form of a functional flow equation for the quantum effective action, whose solution interpolates between the classical action and the full quantum effective action.

## 10.5.1 Kadanoff's Blocking for Lattice Spin Systems

The general concepts of renormalization that we aim to introduce in this section can be first exposed by considering the simple example of a system of spins on a lattice, the simplest of which is the Ising model in two dimensions, which is exactly solvable for interactions limited to nearest neighbors. This model is known to have a disordered phase at high temperature, a ferromagnetic order at low temperature (where spins align with an external magnetic field, no matter how small), and a second-order phase transition at a critical temperature $T_*$. At the second-order transition, the correlation length of the system becomes infinite, despite the fact that the interactions are short-ranged. Roughly speaking, a measure of the complexity of the study of a discrete physical system (at least if one attempts to do it in terms of the interactions among the microscopic degrees of freedom) is the number of elementary degrees of freedom per correlation length. By this account, second-order phase transitions are among the hardest problems to analyze.

Kadanoff devised a method, called *block-spin renormalization*, to facilitate the study of such a situation. The basic ideas of this method are illustrated in Figure 10.2. First, one groups the spins into connected sets, for instance in $3 \times 3$ blocks as shown in the figure. Then, the spins inside each of these blocks are replaced by some sort of average spin. One possibility is to use the "rule of majority": The new spin is chosen to be up if five or more of the original spins were up, and down otherwise. The physical motivation for this replacement is that the calculation of *macroscopic observables* (e.g., the total magnetization in a large sample of the material under consideration) does not require knowing in detail the value of each of the elementary spins, and should be doable from these coarse-grained variables. Of course, one should adjust carefully the interactions among the newly introduced averaged spins, so that the macroscopic properties of the system are unchanged. One may, for instance, require that the partition function of the system is unmodified. In general, even if the original Hamiltonian had only short-range nearest-neighbors interactions, the Hamiltonian that describes the coarse-grained spins can have long-range interactions. The block-spin renormalization comprises a third step, that consists of a rescaling of distances so that each of the coarse-grained spins occupy the same area as one of the original elementary spins (this step is necessary for the transformation to have fixed points).

The combination of these three steps, called a (discrete) *renormalization group* step $\mathcal{R}$, may be viewed as transforming a bare action $\mathcal{S}_0$ into a renormalized action,

$$\mathcal{S}_r \equiv \mathcal{R}\,\mathcal{S}_0. \tag{10.76}$$

However, the real power of this idea comes by iterating the renormalization group steps $\mathcal{R}$ until there are only a few of the coarse-grained spins in a macroscopic area of the system.

## 10.5 NON-PERTURBATIVE RENORMALIZATION GROUP

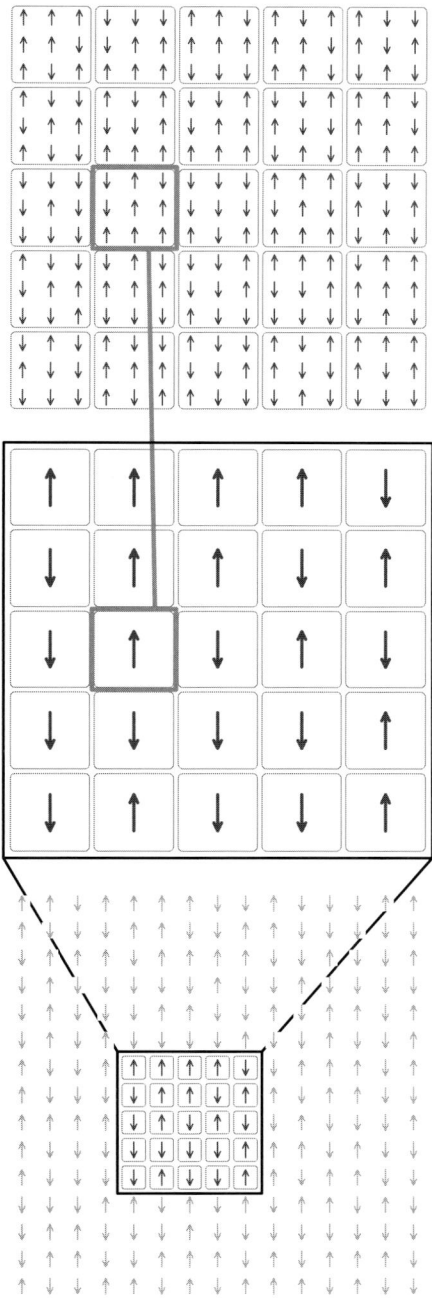

FIGURE 10.2: Kadanoff's block-spin renormalization. Top: The spins are grouped into $3 \times 3$ blocks. Middle: Each block of nine spins is replaced by a single spin determined by the rule of majority. Bottom: The lattice is scaled down (new spins come into the picture that were previously outside of the represented area).

Under such a sequence of renormalization steps, the actions are sequentially transformed as follows:

$$S_0 \underset{\mathcal{R}}{\longmapsto} S_1 \underset{\mathcal{R}}{\longmapsto} S_2 \underset{\mathcal{R}}{\longmapsto} S_3 \underset{\mathcal{R}}{\longmapsto} \cdots \qquad (10.77)$$

The behavior of the mapping $\mathcal{R}^n$ for large $n$ contains all the information we need about the macroscopic properties of the system. In particular, a critical point, where the system has an infinite correlation length and is self-similar, corresponds to a fixed point of this transformation, i.e., to an action $S_*$ that satisfies

$$S_* = \mathcal{R} S_*. \qquad (10.78)$$

The concept of renormalization group introduced so far in the case of a discrete system, consisting of a coarse-graining followed by a rescaling, can be generalized to a continuous system such as a quantum field theory. In this case, one introduces a length scale $\ell$, and the renormalization group transformation consists in integrating out the smaller length scales. One may denote $\tau \equiv \ln(\ell/\ell_0)$, where $\ell_0$ is the initial short-distance scale. Thus, $\tau = 0$ corresponds to the bare action at short distance, and $\tau = +\infty$ corresponds to macroscopic distances, and the discrete steps of eq. (10.77) are replaced by an equation of the form

$$\partial_\tau S_\tau = \mathcal{H} S_\tau, \qquad (10.79)$$

where the RG flow for an infinitesimal step $\Delta\tau$ is $\mathcal{R} = 1 + \Delta\tau\,\mathcal{H}$.

## 10.5.2 Wilsonian RG Flow in Theory Space

One may view a given action $S$ as a point in an abstract space, where each axis corresponds to the coupling constant in front of a given operator. For instance, in the case of a lattice spin system, there would an axis for the strength of the interactions among nearest neighbors, an axis for the strength of the interactions among sites whose distance is $\sqrt{2}$ lattice units, and so on. In a scalar quantum field theory, these could be the couplings for the operators $\phi\square\phi$, $\phi^2$, $\phi^4$, $\phi^6$, etc. A renormalization group transformation such as eq. (10.76) defines a mapping of the points in this *theory space*, either discrete or continuous, depending on the system. We have illustrated this in the continuous case in Figure 10.3, where the thick gray line shows how a bare action $S_0$ at short distance flows as the distance scale $\ell$ increases, leading to a theory that may have very different couplings at macroscopic scales. Note that only three out of many (possibly infinitely many for a continuous system) dimensions are shown in the figure.

As we mentioned in the previous section, a critical point must be a fixed point of this mapping, e.g., the point $S_*$ in Figure 10.3. Important properties of the renormalization group flow may be learned by linearizing the flow in the vicinity of such a fixed point, by writing

$$\begin{aligned} S &\equiv S_* + \Delta S, \quad \mathcal{H} S_* = 0, \\ \mathcal{H} S &= \mathbf{L}\,\Delta S + \cdots, \end{aligned} \qquad (10.80)$$

where $\mathbf{L}$ is a linear mapping. Then, one may define the eigenoperators of $\mathbf{L}$,

$$\mathbf{L}\,\mathcal{O}_n = \lambda_n\, \mathcal{O}_n, \qquad (10.81)$$

## 10.5 NON-PERTURBATIVE RENORMALIZATION GROUP

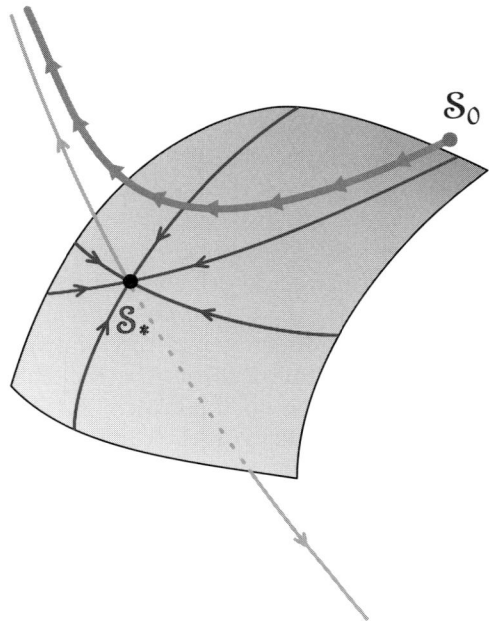

FIGURE 10.3: Renormalization group flow in theory space (the arrows go from UV to IR scales). The black dot is a critical fixed point $S_*$. The gray surface is the critical surface, i.e., the universality class made of all the theories that flow into the critical point. The light gray line, flowing away from the critical point, corresponds to the direction of a relevant operator. The thick gray line illustrates the flow from a generic initial action $S_0$.

where $\lambda_n$ is the corresponding eigenvalue. In the vicinity of the fixed point, we thus have

$$S \approx S_* + \sum_n c_n \, e^{\lambda_n \tau} \, \mathcal{O}_n, \tag{10.82}$$

where $c_n$ are coefficients determined by initial conditions. This expression leads to the following classification of operators:[8]

- $\lambda_n < 0$: Such an operator corresponds to an attractive direction in the vicinity of the fixed point. Even if the action contains this operator at some short-distance scale, its coupling vanishes as one gets close to the critical point. This operator is said to be *irrelevant*, because it plays no role in the long-distance critical phenomena.
- $\lambda_n > 0$: This operator corresponds to a repulsive direction in the vicinity of $S_*$. Any admixture of this operator will grow as one goes to larger distance scales. An operator with a positive eigenvalue is called *relevant*.

---

[8]This discussion does not exhaust all the possibilities. First, in a theory space with two or more dimensions, eigenvalues can be complex valued, corresponding to RG trajectories that spiral around the fixed point (spiraling inwards if the real part is negative and outwards if it is positive). Another possibility is *limit cycles* (i.e., closed RG trajectories), which play a role for instance in the Efimov effect (a scaling law in the binding energies of three-boson bound states when the two-body interaction is too weak to have a two-body bound state).

- $\lambda_n = 0$: Such an operator is called *marginal*. Usually, it means that the operator may either grow or shrink, but slower than exponentially (and a more refined calculation that goes beyond this linear analysis is necessary in order to decide between the two behaviors).

The previous discussion, based on a linear analysis near the critical point, may be extended globally as follows. One defines the *critical surface* as the domain of theory space which is attracted into the critical point as the length scale goes to infinity. All the bare actions that lie in this domain (the shaded surface in Figure 10.3) describe systems that have the same long-distance behavior. Despite the fact that these systems may correspond to completely different microscopic degrees of freedom and interactions, they are described by the same action $S_*$ at large distances. For this reason, this domain is also called the *universality class* of the critical point. The relevant operators correspond to the directions of theory space that are "orthogonal" to the critical surface. The term *relevant* follows from the fact that the coupling of these operators must be fine-tuned in order to be on the critical surface: In other words, the relevant couplings matter for making the system critical. A remarkable aspect of phase transitions is that the number of these relevant operators is small,[9] despite the fact that the microscopic interactions may require a very large number of distinct couplings. Heuristically, this follows from a dimensional argument: Since the action is dimensionless, the coupling constants of higher dimensional operators must have a negative mass dimension, and therefore they scale as inverse powers of the ultraviolet cutoff. Thus, these operators are irrelevant. Only operators of low dimensionality can be relevant, and there is usually a (small) finite number of them.[10]

Let us now consider the domain that originates from the fixed point (the light-colored line in Figure 10.3), sometimes called the *ultraviolet critical surface*. This is the domain spanned by the renormalization group flow if one starts from an infinitesimal region around the fixed point. Any theory that lies on the UV critical surface is renormalizable, since it evolves into the fixed point at short distance: This indeed means that one may safely send the ultraviolet cutoff to infinity in such a theory (this corresponds to moving in the direction opposite to the arrows in Figure 10.3). Note also that theories on the UV critical surface transform into one another under the renormalization flow, but the couplings of the various relevant operators depend on the scale. The following situations may occur:

- For such a theory to be renormalizable in the perturbative sense, the couplings should remain small all the way to the ultraviolet scales. This happens when the fixed point is a *Gaussian fixed point*, whose action $S_*$ contains only a kinetic term (i.e., is Gaussian in the fields). This is the case for quantum chromodynamics, thanks to asymptotic freedom.
- It may also happen that around a Gaussian fixed point, the only relevant operators are quadratic in the fields, like mass and kinetic terms. In this case, there is no interacting renormalizable action, and the theory is said to suffer from *triviality*. There is nowadays strong evidence that, in a pure real scalar field theory, the operator $\phi^4$ is not relevant in four space-time dimensions (it is relevant in three dimensions or fewer) and therefore such a field theory is trivial because only the non-interacting theory makes sense.

---

[9]In the case of the two-dimensional Ising model, the only parameters that need to be adjusted in order to reach the critical point are the temperature ($T_*^{-1} \approx 0.44$) and the external field (equal to zero).

[10]An exception to this assertion is the renormalization group on the light-cone used in the study of deep inelastic scattering. There, peculiarities of the kinematics lead to an infinite number of relevant operators.

## 10.5 NON-PERTURBATIVE RENORMALIZATION GROUP

- When the fixed point is a non-trivial interacting fixed point instead of a Gaussian one, the theories on the UV critical surface are also renormalizable, but their high-energy behavior cannot be studied by perturbative means. This situation is called *asymptotic safety*.[11]

To conclude this discussion, let us say a word about generic RG trajectories, i.e., neither located on the critical surface nor on the UV critical surface, such as the line originating from the short-distance action $S_0$ in Figure 10.3. Generically, when evolving toward larger length scales, the irrelevant couplings decrease and the relevant ones increase, and the action approaches that of a renormalizable theory. This sets in a more general perspective our observation of Section 10.2.4 (there, it was largely based on dimensional analysis). Moreover, if the microscopic action $S_0$ starts close to but not exactly on the critical surface, the theory first approaches the critical point upon increasing the length scale, but instead of reaching it, it departs from it on even larger scales to follow one of the repulsive directions. In such a system, the correlation length may be large but not infinite as it would be at the critical point (the turning point between the approach of the critical point and the subsequent departure from it happens roughly when the RG scale equals the correlation length).

### 10.5.3 Functional RG Equation for Scalar Theories

The block-spin renormalization procedure that we discussed in Section 10.5.1 can be extended to the case of a continuous system such as a quantum field theory. Moreover, while our discussion has been so far qualitative, we shall now derive an explicit RG flow equation for the quantum effective action, the solution of which would provide the full quantum content of the theory from tree-level contributions only.

**Reminders about the quantum effective action:** Let us first recall some basic results about the quantum effective action $\Gamma[\phi]$, taken from Section 5.5. It is related to the generating functional of connected Feynman graphs, $W[j]$, by

$$i\Gamma[\phi] = W[j_\phi] - i\int d^4x \, j_\phi(x)\phi(x), \tag{10.83}$$

where the current $j_\phi$ is defined implicitly by

$$\frac{\delta\Gamma[\phi]}{\delta\phi(x)} + j_\phi(x) = 0, \tag{10.84}$$

or equivalently in terms of $W$ by $\phi(x) = \delta W[j]/i\,\delta j(x)$ at $j = j_\phi$. In other words, $j_\phi(x)$ is the external source such that the expectation value of the field is $\phi(x)$. By combining the path integral representation of $W$,

$$e^{W[j]} = \int [D\phi(x)] \, \exp\left[iS[\phi(x)] + i\int d^4x \, j(x)\phi(x)\right], \tag{10.85}$$

---

[11] The concept of asymptotic safety was introduced by Weinberg as a logical possibility for a renormalizable quantum field theory of gravity.

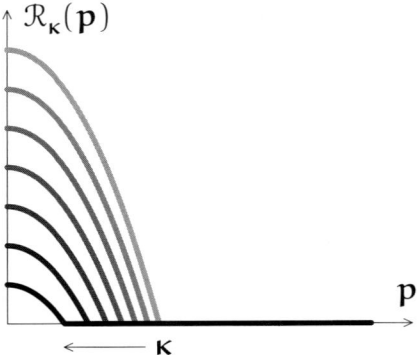

FIGURE 10.4: Sliding cutoff $\mathcal{R}_\kappa(p) \equiv \theta(\kappa^2 - p^2)(\kappa^2 - p^2)$.

with eqs. (10.83) and (10.84) we obtain the following functional equation satisfied by the effective action $\Gamma$:

$$e^{i\Gamma[\varphi]} = \int [D\phi(x)] \, \exp\left[iS[\phi + \varphi] - i\int d^4x \, \frac{\delta\Gamma[\varphi]}{\delta\varphi(x)} \phi(x)\right]. \tag{10.86}$$

(We have performed a shift $\phi \to \phi + \varphi$ in the dummy functional integration variable.) Although this equation formally defines the quantum effective action, its use is not convenient because it still contains a path integration. Physically, this difficulty is related to the fact that the equation integrates out all the length scales at once. The functional RG equation that we derive now circumvents this problem by integrating out quantum fluctuations only in a small range of scales at a time.

**Regularized generating functional:** Let us introduce a momentum scale $\kappa$ and define

$$e^{W_\kappa[j]} \equiv \exp\left\{i\Delta S_\kappa\left[\frac{\delta}{i\delta j}\right]\right\} Z[j]$$
$$= \int [D\phi(x)] \, \exp\left\{i\left(S[\phi] + \Delta S_\kappa[\phi]\right) + i\int j\phi\right\}, \tag{10.87}$$

where $Z[j]$ is the usual generating functional for time-ordered correlation functions and $\Delta S_\kappa$ is defined in terms of the Fourier transform of the fields as follows:

$$\Delta S_\kappa[\phi] \equiv \frac{1}{2}\int \frac{d^4p}{(2\pi)^4} \, \widetilde{\phi}(-p) \, \mathcal{R}_\kappa(p) \, \widetilde{\phi}(p). \tag{10.88}$$

$\mathcal{R}_\kappa$ is an ordinary function that plays the role of a cutoff in momentum. At low momentum $p/\kappa \ll 1$, it should be positive in order to give a mass for the soft modes, and thus provide an infrared regulator,

$$\lim_{p/\kappa \to 0} \mathcal{R}_\kappa(p) = \mu^2 > 0. \tag{10.89}$$

Moreover, this function is assumed go to zero when the scale $\kappa \to 0$,

$$\lim_{\kappa \to 0} \mathcal{R}_\kappa(p) = 0, \tag{10.90}$$

which means that the cutoff plays no role in this limit and we recover the full quantum theory. This is the limit we aim to reach at the end of the RG flow. In contrast, it should become large when $\kappa \to \infty$,

$$\lim_{\kappa \to \infty} \mathcal{R}_\kappa(p) = \infty. \tag{10.91}$$

This property ensures that, when $\kappa$ is large, quantum corrections due to loops are suppressed, so that the corresponding effective action equals the classical action. An example of function $\mathcal{R}_\kappa$ that fulfills these properties is shown in Figure 10.4.

**Scale dependence of $W_\kappa$:** By denoting $\tau \equiv \ln(\kappa/\Lambda)$ (where $\Lambda$ is the ultraviolet scale at which the classical action is defined), we have

$$\partial_\tau W_\kappa[j] = i \, \partial_\tau \Delta S_\kappa[\langle\phi\rangle_\kappa] + \frac{i}{2} \int \frac{d^4 p}{(2\pi)^4} \, \partial_\tau \mathcal{R}_\kappa(p) \, \widetilde{G}_\kappa(p), \tag{10.92}$$

where $G_\kappa(p)$ is the connected two-point function obtained from $W_\kappa[j]$,

$$G_\kappa(x, y) \equiv \frac{\delta^2 W_\kappa[j]}{i\delta j(x) i\delta j(y)}, \tag{10.93}$$

and $\langle\phi\rangle_\kappa$ is the corresponding one-point function, $\langle\phi(x)\rangle_\kappa \equiv \frac{\delta W_\kappa[j]}{i\delta j(x)}$.

**Scale-dependent effective action:** Let us now alter eq. (10.83) in order to make it depend on the scale $\kappa$, by writing

$$\Gamma_\kappa[\phi] + \Delta S_\kappa[\phi] = -i W_\kappa[j_\phi] - \int d^4 x \, j_\phi(x) \, \phi(x). \tag{10.94}$$

The left-hand side is written as $\Gamma_\kappa + \Delta S_\kappa$ in order not to include in the definition of the effective action the unphysical regulator $\Delta S_\kappa$. As in the original definition, the field $\phi$ and the current $j_\phi$ are related by

$$\phi(x) = \frac{\delta W_\kappa[j]}{i\delta j(x)}\bigg|_{j=j_\phi}. \tag{10.95}$$

In terms of $\Gamma_\kappa$ this relationship reads

$$j_\phi(x) + \frac{\delta \Gamma_\kappa[\phi]}{\delta \phi(x)} + \left[\mathcal{R}_\kappa \widetilde{\phi}\right](x) = 0. \tag{10.96}$$

Differentiating eq. (10.95) with respect to $j(y)$ and eq. (10.96) with respect to $\phi(y)$, and multiplying the results, we obtain the following identity:

$$i\delta(x - y) = \int d^4 z \, \underbrace{\frac{\delta^2 W_\kappa[j]}{i\delta j(y) i\delta j(z)}}_{G_\kappa(y,z)} \left[ \underbrace{\frac{\delta^2 \Gamma_\kappa[\phi_j]}{\delta \phi_j(z) \delta \phi_j(x)}}_{\Gamma_{\kappa,2}(z,x)} + \mathcal{R}_\kappa(x, y) \right], \tag{10.97}$$

which generalizes eq. (5.88).

**Flow equation for $\Gamma_\kappa$:** Now, we can differentiate eq. (10.94) with respect to the scale:

$$\partial_\tau \Gamma_\kappa[\phi] = -\partial_\tau \Delta S_\kappa[\phi] - i\,\partial_\tau W_\kappa[j_\phi] - \int d^4x\,\phi(x)\,\partial_\tau j_\phi(x)$$

$$= -\partial_\tau \Delta S_\kappa[\phi] - i\left[\partial_\tau W_\kappa[j]\right]_{j=j_\phi}$$

$$\quad - i \int d^4x\,\frac{\delta W_\kappa[j_\phi]}{\delta j_\phi(x)}\,\partial_\tau j_\phi(x) - \int d^4x\,\phi(x)\partial_\tau j_\phi(x)$$

$$= \frac{1}{2}\int \frac{d^4p}{(2\pi)^4}\,\partial_\tau \mathcal{R}_\kappa(p)\,\widetilde{G}_\kappa(p). \tag{10.98}$$

In the second line we have made explicit the fact that $W_\kappa[j_\phi]$ contains both an intrinsic scale dependence and an implicit one from the $\kappa$ dependence of its argument $j_\phi$. Using eq. (10.97), this can be put into the following compact form:

$$\partial_\tau \Gamma_\kappa = \frac{i}{2}\,\mathrm{tr}\left\{(\partial_\tau \mathcal{R}_\kappa)\left[\frac{\delta^2 \Gamma_\kappa[\phi]}{\delta\phi\delta\phi} + \mathcal{R}_\kappa\right]^{-1}\right\}, \tag{10.99}$$

which depends only on $\Gamma_\kappa$ (the integral over the momentum p has been written in the form of a trace). Let us make a few remarks concerning this equation:

- It describes the renormalization group trajectory of the effective action in theory space, starting from the bare classical action at $\kappa = \infty$ and going to the full quantum effective action when $\kappa \to 0$.
- This equation is a functional differential equation that does not involve any functional integral, unlike eq. (10.86). Nevertheless, it cannot be solved exactly in general, and various truncation schemes have been devised in order to obtain physical results (see Exercise 10.7 for the simplest truncation).
- The term $\mathcal{R}_\kappa$ in the denominator provides an infrared regularization (by adding a kind of mass term to the inverse propagator).
- The factor $\partial_\tau \mathcal{R}_\kappa$ is peaked around momentum modes of order $\kappa$. Thus, the right-hand side is rather localized in momentum space, in contrast with eq. (10.86) that includes all the momentum scales at once.
- The choice of the regularizing function $\mathcal{R}_\kappa$ is not unique, provided that it fulfills conditions (10.89–10.91). Consequently, the renormalization group trajectories depend somehow on this choice (this may be viewed as a dependence on the renormalization scheme). However, the fixed points of the renormalization group flow do not depend on this choice.

# Exercises

**\*10.1** Calculate the one-loop beta function of a scalar field theory with cubic interactions in six space-time dimensions.

**\*10.2** Calculate the one-loop beta function in QED. How does the electromagnetic coupling strength vary with distance? What is the physical interpretation of this behavior?

**\*10.3** Consider the beta function in an unspecified quantum field theory with a coupling g. What is the limit of $\beta(g)$ when $g \to 0$? Show that, up to two loops, the beta function

# EXERCISES

has at most one extremum as g varies. From the function $\beta(g)$, discuss the most common possibilities for the short-distance behaviors of this theory.

**\*10.4** Prove that eq. (10.16) solves the Callan–Symanzik equation.

**10.5** Consider the annihilation of an electron–positron pair into hadrons, $e^+e^- \to$ hadrons. In Exercise 8.2, we saw that this cross-section can be written as

$$\sigma_{tot} = \frac{e^2}{s^3}(p_\mu q_\nu + p_\nu q_\mu - g_{\mu\nu} p \cdot q) \, \text{Im} \, \Pi_q^{\mu\nu}(p+q),$$

where $\Pi_q^{\mu\nu}$ is the hadronic part of the photon self-energy (p and q are the electron and positron momenta, whose masses are neglected).

- Use the Ward–Takahashi identity to write $\Pi_q^{\mu\nu}(k) \equiv (k^2 g^{\mu\nu} - k^\mu k^\nu)\Pi_q(k^2)$ and then show that $\sigma_{tot} = -(e^2/s) \, \text{Im} \, \Pi_q(s)$.
- Explain why

$$-i\Pi_q^{\mu\nu}(k) = \int d^4x \, e^{ik \cdot x} \, \langle 0_{in} | T \left( J_q^\mu(x) J_q^\nu(0) \right) | 0_{in} \rangle,$$

where $J_q^\mu$ is the quark contribution to the electromagnetic current.

- We consider the high-energy limit, i.e., $s \equiv (p+q)^2 = k^2 \to +\infty$. Does this limit correspond to a short-distance limit in the above current–current correlator? In contrast, show that if $k^2 \to -\infty$, the two currents have a small space-time separation that allows us to use the operator product expansion. List the leading operators in this OPE, and give the scaling behavior of their coefficients.
- Consider now a fixed large $K^2 > 0$, and define the integral

$$\oint_\gamma \frac{ds}{2\pi i} \frac{\Pi_q(s)}{(s+K^2)^{n+1}},$$

where $\gamma$ is a closed contour circling around the point $s = -K^2$. Recalling that the imaginary part of $\Pi_q(s)$ corresponds to a branch cut on the positive real axis and evaluate this integral in two different ways to show that

$$\int_0^{+\infty} \frac{ds}{\pi} \frac{s\,\sigma_{tot}(s)}{(s+K^2)^{n+1}} = \frac{e^2}{n!} \frac{d^n \Pi_q(s)}{ds^n}\bigg|_{s=-K^2}.$$

**10.6** Calculate the one-loop anomalous dimension of the composite operator $\phi^2$ in a scalar theory with a $\phi^4$ interaction.

**\*10.7** By taking derivatives of the functional renormalization group equation (10.99), derive the flow equations for the first and second derivatives of the quantum effective action. Are these equations closed?

The *local potential approximation* consists in making the following approximate ansatz for the scale-dependent quantum effective action:

$$\Gamma_\kappa^{LPA}[\phi] \equiv \int d^4x \left( \frac{1}{2}(\partial_\mu \phi)(\partial^\mu \phi) - V_\kappa(\phi(x)) \right),$$

where $V_\kappa(\phi)$ is a local function of $\phi$, whose coefficients depend on $\kappa$. In this approximation, show that the functional renormalization group equation becomes a much simpler equation for the potential $V_\kappa(\phi)$.

# CHAPTER 11

# Effective Field Theories

Until now, we have discussed various quantum field theories (the electroweak theory and quantum chromodynamics) that are believed to provide a unified description of all particle physics up to the scale of electroweak symmetry breaking, i.e., roughly $\Lambda_{EW} \sim 200$ GeV. However, it is hard to imagine that there isn't some kind of new physical phenomena (new particles, new interactions) at higher energy scales (so far out of reach of experimental searches). An interesting question is therefore to understand why the Standard Model is such a good description of physics below the electroweak scale, despite the fact that it does not contain any of the physics from these higher scales. In other words, despite the fact that there is distinct physics on scales that span many orders of magnitude, why can "low-energy" phenomena be described by ignoring most of the higher scales? The same question could be asked in other areas: For instance, why can chemistry (i.e., phenomena of atomic bonding in molecules) get away without any of the complications of quantum electrodynamics (QED)? The general question asked here is that of the separation between various physical scales.

In the context of quantum field theory, such a low-energy description is called an *effective theory*. The basic idea is that most of the details of an underlying more fundamental (i.e., valid at higher energy) description are not important at lower energies, except for a small number of parameters. As we shall see in this chapter, effective field theories may occur in several situations:

- *Top-down*: The quantum field theory which is valid at higher energy is known, but it is unnecessarily complicated to describe phenomena at lower-energy scales. A typical example is that of a theory that contains particles that are much heavier than the energy scale of interest (e.g., the top quark in quantum chromodynamics, while one is interested in interactions at the GeV scale). In this case, the effective theory "integrates out" the higher-mass particles in order to obtain a simpler theory.

- *Bottom-up*: We have a theory believed to be valid at a given energy scale, but have no clear idea of what may exist at higher scales. In this case, one may view the existing theory as an effective description of some (so far unknown) more fundamental theory at higher energy, and try to complete it by adding new (higher dimensional, and therefore usually non-renormalizable in four dimensions) local interactions to it.
- *Symmetry driven*: Even when the underlying theory is known, its direct application may be rendered very impractical because the physics of interest involves some non-perturbative phenomena, such as the formation of bound states (for instance, in QCD at low energy, the quarks and gluons cease to be the relevant degrees of freedom and the physical excitations are the light hadrons). An effective theory for these bound states may be constructed from the requirement that it should be consistent with the symmetries of the underlying theory. This case differs from the top-down approach in the sense that the low energy description is not constructed by integrating out the high scales, but solely from symmetry considerations.

In the top-down approach, where the fundamental underlying theory is known, the goal of obtaining an effective description for low-energy phenomena could in principle be achieved by the renormalization group. In particular, the functional renormalization group introduced in Section 10.5.3 allows evolving from an ultraviolet classical action toward a low-energy quantum effective action by progressively integrating out layers of lower and lower momenta. There is nothing wrong with this approach, but one has to keep in mind that the effective action obtained in this way is usually extremely complicated and cumbersome to use in practical applications (in particular, it could have infinitely many effective interactions, all of which are in general non-local). In a sense, the quantum effective action that results from the RG evolution is much more complex that the original ultraviolet action, and the gain in terms of simplicity is rather dubious. In contrast, the concept of effective theory that we are aiming at in this chapter is a field theory in which the ultraviolet physics is encapsulated into a finite number of local operators, with coupling constants that may depend on the energy scale and on the properties of the degrees of freedom that have been integrated out.

## 11.1 General Principles of Effective Theories

### 11.1.1 Low-Energy Effective Action

For the purpose of this general discussion, let us consider a quantum field theory in which the fields are collectively denoted $\phi$ (this may be a single field, or a collection of several fields) and a classical action $S[\phi]$. We view this theory as the high-energy theory, and we wish to construct an alternative description applicable to low-energy phenomena, below some energy scale $\Lambda$. To this effect, let us assume that we can split the field into a low-frequency part (soft) and a high-frequency part (hard):

$$\phi \equiv \phi_{S} + \phi_{H}. \tag{11.1}$$

This separation may be achieved by a cutoff in Fourier space, but the details of how this is done are not important at this level of discussion. The classical action of the original theory is thus a function of $\phi_{S}$ and $\phi_{H}$ and the path integration is over the soft and the hard components of the field. Now, assume that we are interested in calculating the expectation value of an observable that depends only on the soft component of the field, $\mathcal{O}(\phi_s)$. Then, we may write

$$\langle \mathcal{O} \rangle = \int [D\phi_s D\phi_H]\, e^{iS[\phi_s,\phi_H]}\, \mathcal{O}(\phi_s) = \int [D\phi_s]\, e^{iS_\Lambda[\phi_s]}\, \mathcal{O}(\phi_s), \quad (11.2)$$

where in the second equality we have defined

$$e^{iS_\Lambda[\phi_s]} \equiv \int [D\phi_H]\, e^{iS[\phi_s,\phi_H]}. \quad (11.3)$$

$S_\Lambda[\phi_s]$ is the action of the low-energy effective theory. Using the operator product expansion, it may be written as a sum of local operators, possibly infinitely many of them:

$$S_\Lambda[\phi_s] \equiv \int d^d x \sum_n \lambda_n\, \mathcal{O}_n. \quad (11.4)$$

## 11.1.2 Power Counting

The behavior of the couplings $\lambda_n$ can be inferred from dimensional analysis. For the sake of this discussion, let us consider the case where $\phi$ is a scalar field whose mass dimension is $\phi \sim (\text{mass})^{(d-2)/2}$ in d space-time dimensions. If the operator $\mathcal{O}_n$ contains $N_n$ powers of the field $\phi$ and $D_n$ derivatives, its dimension is

$$\mathcal{O}_n \sim (\text{mass})^{d_n} \quad \text{with} \quad d_n = D_n + N_n \tfrac{d-2}{2}, \quad (11.5)$$

and it must be accompanied with a coupling $\lambda_n$ whose dimension is $(\text{mass})^{d-d_n}$. Assuming that the cutoff $\Lambda$ is the only dimensionful parameter that enters in the construction of the effective theory (except for the field operator and derivatives, that enter in the operators $\mathcal{O}_n$), we must have $\lambda_n = \Lambda^{d-d_n} g_n$, where $g_n$ is a dimensionless constant whose numerical value is typically of order 1.

Consider now the application of this effective theory to the study of a phenomenon characterized by a single energy scale E. On dimensional grounds, we have

$$\int d^d x\, \mathcal{O}_n \sim E^{d_n - d}. \quad (11.6)$$

Combined with the corresponding coupling constant, the contribution of this operator would be of order

$$\lambda_n \int d^d x\, \mathcal{O}_n \sim g_n \left(\frac{\Lambda}{E}\right)^{d-d_n}. \quad (11.7)$$

This estimate is the basis of the following classification of the operators that may enter in the action of the effective theory:

- $d_n > d$: The contribution of these operators is suppressed at low energy, i.e., when $E \ll \Lambda$. For this reason, these operators are called *irrelevant*. This does not mean that their contribution is not important and interesting, since there may be observables for which they are the sole contribution. Note also that these operators are non-renormalizable by the standard power counting rules.

## 11.1 GENERAL PRINCIPLES OF EFFECTIVE THEORIES

- $d_n = d$: The contribution of these operators does not depend on the ratio of scales $E/\Lambda$, except perhaps via logarithms. These operators are called *marginal*, and correspond to renormalizable operators.
- $d_n < d$: The contribution of these operators becomes more and more important as the energy scale decreases. These operators, called *relevant*, are super-renormalizable.

Recall also that a higher dimension $d_n$ corresponds to operators of greater complexity (since in $d > 2$ the dimension increases with more powers of the field or more derivatives). Therefore, there is in general only a finite number of operators whose dimension is below a given value. For a given cutoff $\Lambda$ and an energy scale $E$, one must therefore only consider a finite number of operators in order to reach a given accuracy.

In a conventional quantum field theory, one usually insists on including only renormalizable operators in order to avoid the proliferation of new couplings at each order of perturbation theory, and the usual statement of renormalizability amounts to saying that all infinities may be absorbed into the redefinition of a *finite* number of parameters of the theory, at every order of perturbation theory. In contrast, since a low-energy effective theory may contain operators of dimension $d_n > d$, it is usually not renormalizable in this usual sense, but the cutoff $\Lambda$ provides a natural way of keeping all the contributions finite. In this case, the power counting is organized by the fact that the cutoff $\Lambda$ is also the dimensionful scale that enters in the couplings that come with operators of mass dimension greater than four. For instance, an operator of dimension six has a coupling constant that scales as $\Lambda^{-2}$, and physical observables may be expanded in powers of $E/\Lambda$, where $E$ is some low-energy scale. In the presence of such higher dimensional operators, the usual statement of renormalizability must now be replaced by a weaker assertion: Namely, that all the ultraviolet divergences that occur at a given order in $E/\Lambda$ can be absorbed into the redefinition of a finite number of parameters. More precisely, in order to calculate consistently effects of order $\Lambda^{-r}$, we must include all operators up to a mass dimension of $4 + r$. Thus, the number of constants that must be adjusted in the renormalization process grows as we go to higher order.

In the case of top-down effective theories, the renormalizability of the underlying field theory implies that the low-energy physics depends on the ultraviolet only through the values of the relevant and marginal couplings. In addition, a small number of irrelevant couplings may matter in certain specific observables (e.g., if an irrelevant operator is the only one that contributes). In fact, if the cutoff of the effective theory is high enough compared to the physical energy scale of interest, the effective theory can have a very strong predictive power, despite the fact that it a priori contains an infinity of operators. But conversely, in a bottom-up approach in which we try to extend a renormalizable theory by adding to it higher dimensional operators, the fact that the low energy theory is renormalizable implies that it is not sensitive to the scale of new physics (in other words, a renormalizable low-energy theory cannot predict at which high-energy scale it breaks down and is superseded by another theory).

### 11.1.3 Relevant Operators

In fact, in an effective theory, the relevant operators (super-renormalizable) may be more troublesome than the irrelevant ones (non-renormalizable). Consider, for instance, the operator $\phi^2$, that corresponds to the mass term in the effective Lagrangian and has dimension $\phi^2 \sim (\text{mass})^{d-2}$. The corresponding coupling has dimension $(\text{mass})^2$, i.e., $\lambda = g\, \Lambda^2$. Thus, small masses are not natural in a low-energy effective theory: The natural scale of a mass is that of the cutoff $\Lambda$ (the dimensionless coupling $g$ is generically of order 1). In order to obtain small masses in a low-energy effective field theory, there must be some symmetry that prevents the corresponding mass term, e.g.,

- a gauge symmetry for spin-1 particles;
- a chiral symmetry for fermions;
- a spontaneous breaking of symmetry, so that some scalars are the corresponding massless Nambu–Goldstone bosons;
- supersymmetry may also reduce the sensitivity of a mass on the ultraviolet scale $\Lambda$.

By that account, the Standard Model (without any supersymmetric extension) is not natural, since it does not contain any mechanism to prevent the mass of the Higgs scalar boson to be at a cutoff scale (possibly much higher than the electroweak scale) where the Standard Model is superseded by a more fundamental theory.

More generally, the relevant interaction terms have a large contribution to low-energy observables that scales like

$$\left(\frac{\Lambda}{E}\right)^{d-d_n} \gg 1 \quad \text{with} \quad d > d_n. \tag{11.8}$$

Therefore, the existence of relevant interaction terms implies that the dynamics is strongly coupled at low energy. This may lead to the formation of bound states or condensates, which calls for a low-energy effective theory that contains different degrees of freedom. An example is that of the identity operator, which is not forbidden by any symmetry and has mass dimension 0 (therefore, it is a relevant operator). Although this operator has no effect if added to the Lagrangian of a field theory (since it amounts to adding a constant to the potential energy), its coefficient becomes a cosmological constant if this field theory is minimally coupled to gravity.[1] From the power counting of the previous section, the natural value of the coupling constant in front of this operator is $\Lambda^d$. Thus, if we view the Standard Model as an effective theory, the cosmological constant should be at least as large as the fourth ($d = 4$) power of the cutoff at which the Standard Model is replaced by some other theory. This is in sharp contrast with observations. Indeed, if the dark energy inferred from the measured acceleration of the expansion of the Universe is attributed to a cosmological constant, its value is many orders of magnitude below its natural value in quantum field theory (its corresponds to an energy density of the vacuum of the order of $10^{-47}$ GeV$^4$).

## 11.2 Example: Fermi Theory of Weak Decays

As a first illustration of the concept of effective field theory, let us consider the case of Fermi's theory of weak interactions. Historically, this model was constructed long before the advent of the electroweak gauge theory, and therefore it may be viewed as a bottom-up construction. Nowadays, since the electroweak theory provides us with a more fundamental description of weak interactions, we may derive Fermi's theory in a top-down fashion, as a low-energy approximation of a known high-energy theory.

---

[1]This example illustrates an ambiguity one faces when coupling a field theory to gravity: Only energy differences matter for the dynamics of the field theory, but the absolute value of the energy enters in the energy-momentum tensor that acts as a source in Einstein's equations.

## 11.2.1 Fermi Theory as a Phenomenological Description

If we consider the Standard Model at a scale of the order of the nucleon mass, i.e., around one GeV, it contains only the leptons, the light quarks, and the massless gauge bosons (photon and gluons). Thus, this low-energy truncation has no mechanism for weak decays. Nevertheless, one may write an effective coupling involving a proton, a neutron (here, we prefer to use hadrons, which are the states encountered in actual experimental situations), an electron and the corresponding neutrino. The most general local operator combining these four fields may be written as

$$\frac{g_{12}}{\Lambda^2} \left(\overline{\psi}_p \Gamma_1 \psi_n\right)\left(\overline{\psi}_e \Gamma_2 \psi_\nu\right), \tag{11.9}$$

where $g_{12}$ is a dimensionless constant, $\Lambda$ is a dimensionful scale and $\Gamma_{1,2}$ are matrices chosen in the following set:

$$\Gamma_{1,2} \in \{1, \gamma_5, \gamma^\mu, \gamma^\mu \gamma_5, \underbrace{\tfrac{i}{4}[\gamma^\mu, \gamma^\nu]}_{\sigma^{\mu\nu}}\}. \tag{11.10}$$

Note that $\sigma^{\mu\nu}\gamma_5$ is not linearly independent from these matrices, since $\sigma^{\mu\nu}\gamma_5 \propto \epsilon^{\mu\nu\rho\sigma}\sigma_{\rho\sigma}$, and therefore need not be included in this list. Thus, the most general Lorentz invariant Lagrangian involving these four fields reads:

$$\mathcal{L}_{\text{eff}} = \underbrace{\left(\overline{\psi}_p \gamma_\mu \psi_n\right)\left(\overline{\psi}_e \gamma^\mu (C_V + C'_V \gamma_5)\psi_\nu\right) + \left(\overline{\psi}_p \gamma_\mu \gamma_5 \psi_n\right)\left(\overline{\psi}_e \gamma^\mu \gamma_5 (C_A + C'_A \gamma_5)\psi_\nu\right)}_{\text{vector, axial}}$$
$$+ \underbrace{\left(\overline{\psi}_p \psi_n\right)\left(\overline{\psi}_e (C_S + C'_S \gamma_5)\psi_\nu\right) + \left(\overline{\psi}_p \gamma_5 \psi_n\right)\left(\overline{\psi}_e \gamma_5 (C_P + C'_P \gamma_5)\psi_\nu\right)}_{\text{scalar, pseudo-scalar}}$$
$$+ \underbrace{\left(\overline{\psi}_p \sigma_{\mu\nu} \psi_n\right)\left(\overline{\psi}_e \sigma^{\mu\nu}(C_T + C'_T \gamma_5)\psi_\nu\right)}_{\text{tensor}}. \tag{11.11}$$

Note that certain terms violate some discrete symmetries. For instance, the primed terms $C'_{V,S,P,T}$ all violate parity, and T-invariance requires that the ratio $C_i/C'_i$ be real for all $i \in \{V, A, S, P, T\}$. On the other hand, by confronting this effective Lagrangian with the existing data on weak decays, we learn that

$$C_V = \Lambda^{-2} \quad \text{with } \Lambda \sim 350 \text{ GeV}, \quad C_A \approx 1.25 \times C_V,$$
$$C_V \sim C'_V, \quad C_A \sim C'_A, \quad \frac{C_{S,P,T}}{C_V}, \frac{C'_{S,P,T}}{C_V} \lesssim 1 \text{ percent.} \tag{11.12}$$

The first of these results is an indication of the energy scale at which the Fermi theory breaks down and should be replaced by a more accurate microscopic description of weak decays, and the second one implies that this underlying theory is chiral. The fact that $C_{V,A} \sim C'_{V,A}$ is a sign of parity violation in weak interactions. Finally, the last property tells us that this microscopic interaction is not mediated by a scalar or a tensor with a mass less than $\sim 2$ TeV. All this information may be used in constraining the possible form of the theory that describes weak interactions at higher energies.

## 11.2.2 Fermi Theory from the Electroweak Theory

Let us now consider the opposite exercise: Namely, start from the Lagrangian of the Standard Model and obtain the low-energy effective theory of weak interactions by a matching procedure. We know that the $W^\pm$ bosons responsible for weak decays couple to left-handed fermions arranged in $SU(2)$ doublets,

$$\begin{pmatrix} \nu_e \\ e \end{pmatrix}_L, \quad \begin{pmatrix} d \\ u \end{pmatrix}_L, \tag{11.13}$$

where we have written only the relevant doublets for the decay $n \to p e \bar{\nu}_e$. In addition, we have to keep in mind that the mass eigenstates are misaligned with the weak interaction eigenstates in the quark sector. Thus, the vertex $W$-$u$-$d$ contains a factor $V_{ud}$ from the Cabbibo–Kobayashi–Maskawa matrix. With these ingredients, the tree-level decay amplitude $d \to u e \bar{\nu}_e$ reads

$$\mathcal{A} = \frac{g^2}{8} V_{ud} \frac{i}{k^2 - M_W^2} \left( \bar{u} \gamma^\mu (1 - \gamma_5) d \right) \left( \bar{e} \gamma_\mu (1 - \gamma_5) \nu_e \right), \tag{11.14}$$

where $k^\mu$ is the 4-momentum carried by the intermediate $W$ boson. In the low-momentum limit, $k^2 \ll M_W^2$, this amplitude becomes independent of the momentum transfer and could have been generated by the following contact interaction:

$$\mathcal{L}_{\text{eff}} = \frac{G_F}{\sqrt{2}} V_{ud} \left( \bar{\psi}_u \gamma^\mu (1 - \gamma_5) \psi_d \right) \left( \bar{\psi}_e \gamma_\mu (1 - \gamma_5) \psi_\nu \right) \quad \text{with} \quad \frac{G_F}{\sqrt{2}} \equiv \frac{g^2}{8 M_W^2}. \tag{11.15}$$

In order to obtain from this the physical decay amplitude $n \to p e \bar{\nu}_e$, we need the matrix element

$$\langle p | \bar{\psi}_u \gamma^\mu (1 - \gamma_5) \psi_d | n \rangle, \tag{11.16}$$

with initial and final nucleons instead of quarks. In the low-momentum limit, it may be related to a similar matrix element with the spinors of the proton and neutron by

$$\langle p | \bar{\psi}_u \gamma^\mu (1 - \gamma_5) \psi_d | n \rangle = \langle p | \bar{\psi}_p \gamma^\mu (g_V - g_A \gamma_5) \psi_n | n \rangle + \mathcal{O}(k^\mu), \tag{11.17}$$

where $g_{V,A}$ are two constants that may be viewed as the zero-momentum limit of some *form factors*.[2] Then, by comparing the decay amplitudes obtained from the low-energy effective theory guessed on the basis of phenomenological considerations, and the one obtained by starting from the electroweak theory, we obtain

$$C_V = -C_V' = g_V \frac{g^2}{8 M_W^2} V_{ud} = \frac{1}{\Lambda^2},$$

$$C_A = -C_A' = -g_A \frac{g^2}{8 M_W^2} V_{ud},$$

$$C_{S,P,T} = C_{S,P,T}' = 0. \tag{11.18}$$

---

[2] A form factor is a function that describes the coupling between an elementary gauge boson and a bound state, as a function of the virtuality of the gauge boson. Varying this virtuality amounts to changing the resolution scale at which the interior of the bound state is probed.

In this top-down approach, we see that the parity violation inferred from experimental evidence is in fact maximal in the electroweak theory, and that the scalar and tensor contributions are exactly zero. Note also that the scale $\Lambda$ that we introduced by hand in the low-energy effective theory does not coincide exactly with the mass of the heavy particle which is integrated out (in the present case, the $W$ boson), but has the same order of magnitude. Finally, even though we performed here the matching at tree level, it is in principle possible to correct the coefficients of the low-energy effective theory by electroweak and QCD loop corrections.

## 11.3 The Standard Model as an Effective Field Theory

### 11.3.1 Standard Model

The Standard Model unifies the strong and electroweak interactions into a unique renormalizable field theory. Although it agrees with most observed phenomena,[3] it is unreasonable to expect that the Standard Model remains an accurate description of particle physics to arbitrarily high-energy scales. A more modest point of view is to consider the Standard Model as a low-energy approximation of some more fundamental theory that we do not yet know. In this perspective, it would just be the zeroth order of some expansion,

$$\mathcal{L} = \underbrace{\mathcal{L}_{\text{SM}}}_{\Lambda^0} + \underbrace{\mathcal{L}^{(1)}}_{\Lambda^{-1}} + \underbrace{\mathcal{L}^{(2)}}_{\Lambda^{-2}} + \cdots , \qquad (11.19)$$

and a natural endeavor is to construct the terms $\mathcal{L}^{(1,2,\cdots)}$, made of operators with mass dimension greater than four. By power counting, these operators must be suppressed by coupling constants that are inversely proportional to powers of some high-energy scale $\Lambda$ at which corrections to the Standard Model become important. In the construction of these corrections, one usually abides by the following constraints:

- Lorentz invariance is preserved to all orders in $\Lambda^{-1}$.
- The $SU(3) \times SU(2) \times U(1)$ gauge symmetry of the Standard Model remains a symmetry of the higher-order corrections (the idea being that whatever is the more fundamental theory that underlies the Standard Model, it is more symmetric, not less).
- The corrections are built with the degrees of freedom of the Standard Model.
- The vacuum expectation value of the Higgs is not modified by the corrections.

As we mentioned earlier, since the Standard Model is renormalizable there is no way to determine the scale $\Lambda$ within the Standard Model itself. Instead, one should enumerate the higher dimensional operators up to a certain mass dimension (which corresponds to a certain order in $\Lambda^{-1}$) and investigate their possible observable consequences. Experiments can then search for these effects, and either provide the values of some of the parameters introduced in $\mathcal{L}^{(1,2,\cdots)}$, or give lower bounds on the scale of new physics in case of a null observation. Note that there are two main classes of higher dimensional operators, illustrated in Figure 11.1:

- Operators that lead to corrections to processes already allowed in the Standard Model. These corrections may become potentially visible in more precise experiments, provided

---

[3] One exception is the fact that neutrinos have masses, which does not have a very compelling explanation in the Standard Model – we shall return to this issue in the next subsection.

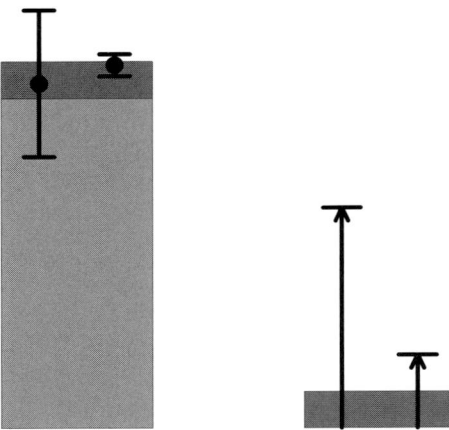

FIGURE 11.1: Left: A higher dimensional operator provides a correction (dark) to an observable which is non-zero in the Standard Model (light). Right: The higher dimensional operator allows a process that was impossible in the Standard Model. In the latter case, experiments usually provide an upper value for the yield of these very rare processes, which decreases as the sensitivity improves, thereby pushing higher up the energy scale of this new physics.

one also has a calculation of the Standard Model prediction for this process with a matching accuracy. An example of such a quantity is the anomalous magnetic moment of the muon.
- Operators that allow processes that were forbidden in the Standard Model. In this case, what is needed are more sensitive experiments, able to detect extremely rare events. For instance, experiments designed to observe proton decay fall within this category.

### 11.3.2 Dimension Five Operators and Neutrino Masses

The right-handed neutrinos are singlet under $SU(3)$ and $SU(2)$ and have a null electrical charge, which means they do not feel any of the interactions of the Standard Model. As a consequence, all the neutrinos detected in experiments (via their weak interactions with the matter of the detector) are left-handed neutrinos, implying that there is no direct evidence for the existence of right-handed neutrinos. For this reason, right-handed neutrinos are usually not considered as part of the Standard Model.

The observation of neutrino oscillations, i.e., the fact that the flavor of a neutrino can change as it propagates, implies that there are non-zero mass differences between neutrinos.[4]

---

[4]Consider for instance a β decay: It produces an electron anti-neutrino (i.e., a weak interaction eigenstate) of definite momentum. If mass eigenstates are misaligned with the weak interaction eigenstates, then this neutrino may project on several mass eigenstates. Since the time evolution of the phase of a wavefunction depends on the mass of the particle, these mass eigenstates evolve slightly differently in time (unless all the neutrino masses are identical). At the detection time, this leads to a flavor decomposition which is different from the one at the time of production. Thus, the original electron anti-neutrino will be a mixture of electron, muon and tau anti-neutrinos. Conversely, the observation of this change of flavor implies mass differences in the neutrino sector.

Therefore, at most one of the neutrinos can be massless, and at least two of them must be massive.

**Neutrino masses from the Higgs mechanism:** Since the electroweak theory is chiral (right-handed leptons are SU(2) singlet, while the left-handed ones belong to SU(2) doublets), a naive Dirac mass term of the form $m_D \overline{\psi}_L \psi_R$ is not invariant under SU(2). However, we may construct such a Dirac mass in the same way as for the other leptons, by starting from a Yukawa coupling involving the Higgs boson:

$$\lambda \left( \overline{\psi}_{L\,ri} \, \epsilon_{ij} \, \Phi_j^* \right) \psi_{R\,r}, \tag{11.20}$$

where $i, j$ are indices in the fundamental representation of SU(2) and $r$ is a Dirac index. The matrix $\epsilon \equiv i\sigma^2$ is proportional to the second generator of the fundamental representation of $\mathfrak{su}(2)$. Thanks to the contraction of the left-handed spinor doublet with the Higgs field, this combination is an SU(2) invariant combination (see footnote 13 in Chapter 7). Then, spontaneous symmetry breaking gives a non-zero expectation value $v$ to the Higgs field, and this interaction term becomes a Dirac mass term for the neutrino, with a mass $m_D = \lambda v$. Generating the neutrino mass by this mechanism would place the neutrinos almost on the same footing as the other leptons, provided we add right-handed neutrinos to the degrees of freedom of the Standard Model.[5] The only distinctive feature of the right-handed neutrinos would be that they do not feel any of the gauge interactions of the Standard Model. For this reason, they are sometimes called *sterile neutrinos*. The main drawback of this solution is that it requires an even larger range of values of the Yukawa couplings, with no natural explanation.

**Majorana neutrino masses:** An alternative would be to have a Majorana mass (see Exercise 11.2) for the left-handed neutrinos of the Standard Model. Instead of introducing this mass term by hand, it can be generated via spontaneous symmetry breaking from a *Weinberg operator*:

$$\frac{c}{\Lambda} \left( \psi_{L\,ri}^t \epsilon_{ij} \Phi_j \right) \mathbf{C}_{rs} \left( \Phi_k^t \epsilon_{kl} \psi_{L\,sl} \right), \tag{11.21}$$

where $\mathbf{C} \equiv \gamma^0 \gamma^2$ is the charge conjugation operator. First, note that this operator has mass dimension five, hence the coupling constant proportional to $\Lambda^{-1}$. In fact, this operator is the only lepton number violating five-dimensional operator that obeys the constraints listed in the previous section.[6] After spontaneous symmetry breaking, the Higgs field

---

[5]Whether this type of term is "beyond the Standard Model" is to a large extent a matter of definition. Before the observation of neutrino oscillations, the Standard Model was most often defined without right-handed neutrinos, and therefore had only massless neutrinos. But it would have been equally acceptable to include right-handed neutrinos from the start, with Yukawa couplings so small that their masses were too small to detect.

[6]$\psi_{L\,ri}^t \epsilon_{ij} \Phi_j$ and $\Phi_k^t \epsilon_{kl} \psi_{L\,sl}$ are both SU(2) invariant (but not Lorentz invariant because they have a dangling Dirac index), and the combination $\psi_{L\,ri}^t \mathbf{C}_{rs} \psi_{L\,sl}$ is Lorentz invariant. This combination is SU(3) (color) invariant only for the leptons (but not for the quarks) – see Exercise 11.1.

acquires a vacuum expectation value, leading to a Majorana mass term for the left-handed neutrinos,

$$\frac{cv^2}{\Lambda} \nu_L^t \, \mathbf{C} \, \nu_L, \tag{11.22}$$

which corresponds to a Majorana mass $m_M = cv^2 \Lambda^{-1}$. The appeal of this mechanism is that a small mass of the neutrinos is naturally explained by a high scale $\Lambda$ for the new physics. For instance, a neutrino mass of the order of 1 eV or below corresponds to $\Lambda \gtrsim 10^{13}$ GeV. As we have already mentioned, the operator in eq. (11.21) does not conserve lepton number, since it is not invariant under the following global transformation:

$$\psi \to e^{i\alpha}\psi, \quad \psi^\dagger \to e^{-i\alpha}\psi^\dagger, \quad \psi^t \to e^{i\alpha}\psi^t. \tag{11.23}$$

For this reason, this alternative mechanism is clearly beyond the Standard Model. However, as long as gauge symmetries are preserved, the violation of lepton number is not particularly dramatic. In a sense, one may view the lepton conservation that exists in the Standard Model as accidental, being a consequence of the fact that only dimension-four operators are included.

**Weinberg operator from the low-energy limit of another quantum field theory:** In the spirit of the bottom-up construction of an effective theory, the operator of eq. (11.21) can be obtained by exploring all the possibilities for dimension-five operators built with the degrees of freedom of the Standard Model and some symmetry requirements. However, this operator can also be obtained in the low-energy limit of a renormalizable quantum field theory. Consider an extension of the field content of the Standard Model, where we add a right-handed neutrino $\nu_R$ with a very large Majorana mass $M_R$ (much heavier than the electroweak scale), that also couples to the $SU(2)$ doublet containing the left-handed neutrino and to the Higgs field via a Yukawa coupling:

$$\begin{aligned}
\mathcal{L} &= \mathcal{L}_{SM} + \mathcal{L}_{\nu_R}, \\
\mathcal{L}_{\nu_R} &\equiv i\overline{\nu_R}\,\slashed{\partial}\,\nu_R + \tfrac{1}{2}\left(M_R\,\nu_R^t\,\mathbf{C}\,\nu_R + M_R^*\,\nu_R^{t*}\,\mathbf{C}\,\nu_R^*\right) \\
&\quad - y\left(\overline{\psi}_L\,\epsilon\Phi^*\right)\nu_R - y^*\,\overline{\nu_R}\left(\Phi^t\epsilon^\dagger\psi_L\right).
\end{aligned} \tag{11.24}$$

With two instances of the Yukawa coupling and a propagator of the heavy Majorana neutrino, it is possible to build a (non-local) four-point function involving two Higgs fields and two left-handed leptons (see the Figure 11.2). At energies much lower than the mass $M_R$ of the right-handed neutrino, the intermediate propagator may be approximated by a constant[7]

$$\frac{i(\slashed{p} + M_R)\mathbf{C}}{p^2 - M_R^2} \xrightarrow[p \ll M_R]{} -i\frac{\mathbf{C}}{M_R}, \tag{11.25}$$

which leads to the (local) Weinberg operator. The latter gives a Majorana mass for the left-handed neutrino after the Higgs field has acquired a non-zero vacuum expectation value

---

[7] The propagator of a Majorana fermion is that of a Dirac fermion multiplied by the charge conjugation matrix $\mathbf{C}$.

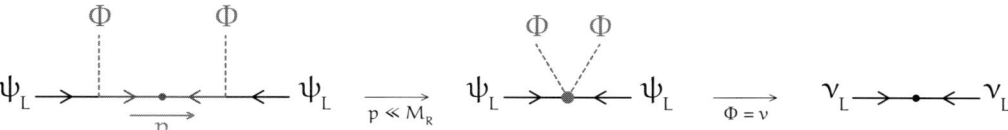

FIGURE 11.2: Diagrammatic illustration of the see-saw mechanism. Left: $\Phi\Phi\psi_L^t\psi_L$ 4-point function made with two Yukawa vertices and one insertion of the $\nu_R$ propagator. Middle: $\Phi\Phi\psi_L^t\psi_L$ local vertex obtained after integrating out the right-handed neutrino. Right: Majorana mass term for $\nu_L$ obtained after spontaneous symmetry breaking.

through spontaneous symmetry breaking. This mechanism is known as the *see-saw mechanism*.[8]

### 11.3.3 Higher Dimensional Operators

The number of operators of mass dimension six is much larger, and they lead to a broader array of possible phenomena. Even if we restrict to operators that conserve lepton and baryon number, there are 80 independent operators. Instead of listing them, let us discuss an important result used to reduce the list of possible operators down to a smaller set of independent operators, thanks to the field equations of motion that result from the zeroth-order Lagrangian (i.e., the Standard Model Lagrangian).

**Operator removal from the Lagrangian:** As an illustration of the principles at work in this reduction, consider the Lagrangian of a real scalar field with a quartic interaction, extended by two operators of dimension six:

$$\mathcal{L} = \underbrace{\frac{1}{2}(\partial_\mu\phi)(\partial^\mu\phi) - \frac{1}{2}m^2\phi^2 - \frac{\lambda}{4!}\phi^4}_{\mathcal{L}^{(0)}} + \frac{1}{\Lambda^2}\left(\lambda_1\phi^6 + \lambda_2\phi^3\Box\phi\right). \tag{11.26}$$

The equation of motion that follows from the zeroth-order Lagrangian is

$$(\Box + m^2)\phi + \frac{\lambda}{6}\phi^3 = 0. \tag{11.27}$$

Naively, it is tempting to replace the last term of the effective Lagrangian, $\phi^3\Box\phi$, by a sum of terms in $\phi^4$ and $\phi^6$. However, it is not totally clear that this is legitimate when this interaction term is inserted in a more complicated graph. A more robust justification goes as follows. Consider a new scalar field $\psi$ related to $\phi$ by

$$\phi = \psi + \lambda_2\Lambda^{-2}\psi^3. \tag{11.28}$$

---

[8] More precisely, it corresponds to the *Type-I* see-saw mechanism. Type-II and Type-III see-saw mechanisms exist, which differ in the nature of the heavy particle that connects the $\Phi\Phi\psi_L^t\psi_L$ fields in the original four-point function.

Note that both terms on the right-hand side have mass dimension one and transform as Lorentz scalars. Rewriting the terms of the above Lagrangian in terms of $\psi$, we obtain

$$\frac{1}{2}(\partial_\mu \phi)(\partial^\mu \phi) = \frac{1}{2}(\partial_\mu \psi)(\partial^\mu \psi) - \lambda_2 \Lambda^{-2} \psi^3 \Box \psi + \mathcal{O}(\Lambda^{-4}),$$
$$m^2 \phi^2 = m^2 \psi^2 + 2\lambda_2 \Lambda^{-2} \psi^4 + \mathcal{O}(\Lambda^{-4}),$$
$$\frac{\lambda}{4!} \phi^4 = \frac{\lambda}{4!} \psi^4 + 4\lambda_2 \Lambda^{-2} \psi^6 + \mathcal{O}(\Lambda^{-4}), \tag{11.29}$$

and finally

$$\mathcal{L} = \frac{1}{2}(\partial_\mu \psi)(\partial^\mu \psi) - \frac{1}{2}m^2 \psi^2 - \frac{\lambda'}{4!} \psi^4 + \frac{1}{\Lambda^2} \lambda'_1 \psi^6 + \mathcal{O}(\Lambda^{-4}), \tag{11.30}$$

where $\lambda', \lambda'_1$ are new coupling constants for the quartic and sextic terms. In the spirit of an effective field theory, we do not care about the terms of order $\Lambda^{-4}$ since they come with operators of dimension eight, which we are not considering here. Thus, by the change of variable of eq. (11.28), we can eliminate the term that seemed redundant in the Lagrangian. More generally, any term of the form

$$\Lambda^{-2} f(\phi) \underbrace{\left( \Box \phi + m^2 \phi + \frac{\lambda}{6} \phi^3 \right)}_{\text{l.h.s. of the EOM}}, \tag{11.31}$$

where $f(\phi)$ is any local function of the fields of mass dimension three (e.g., $\phi^3$, $m^2 \phi$, $\Box \phi$), can be removed from the effective Lagrangian by the following field redefinition:

$$\phi = \psi + \Lambda^{-2} f(\phi). \tag{11.32}$$

**Functional Jacobian:** Having removed the operator $\phi^3 \Box \phi$ from the Lagrangian is not enough, because the change of variable (11.28) also has implications elsewhere. First, in the path integral representation of the generating functional, this change of variable introduces a Jacobian since the functional integration measure is modified as follows:

$$[D\phi(x)] = [D\psi(x)] \det \left( \frac{\delta \phi(x)}{\delta \psi(y)} \right). \tag{11.33}$$

For a transformation of the type in eq. (11.28), the determinant depends on the field since we have

$$\frac{\delta \phi(x)}{\delta \psi(y)} = \Lambda^{-2} \delta(x - y) \left[ \Lambda^2 + 3\lambda_2 \psi^2(x) \right], \tag{11.34}$$

and therefore this determinant should not be disregarded. As in the Fadeev–Popov quantization procedure, we may express it as a path integral over fictitious Grassmann fields $\chi, \overline{\chi}$ by writing

$$\det \left( \frac{\delta \phi(x)}{\delta \psi(y)} \right) = \int [D\chi(x) D\overline{\chi}(x)] \exp \left\{ i \int d^4x \, \overline{\chi}(x) \left( \Lambda^2 + 3\lambda_2 \psi^2(x) \right) \chi(x) \right\}. \tag{11.35}$$

In the case of our simple example, the kinetic term of this ghost field is a bit peculiar since it does not contain any derivatives. However, it exhibits a feature which is completely generic, namely the fact that its mass is of order $\Lambda$. Since the ghosts can only appear in closed loops, their contribution is suppressed by inverse powers of $\Lambda$. In other words, the determinant depends on the field $\psi$, but this dependence is of higher order in $\Lambda^{-2}$ and will not affect our effective theory.

**Modifications at the external points:** Second, the change of variables in eq. (11.28) modifies the coupling to the fictitious source in the generating functional:

$$\int d^4x\, J(x)\phi(x) = \int d^4x\, J(x) \left(\psi(x) + \lambda_2 \Lambda^{-2} \psi^3(x)\right). \tag{11.36}$$

Thus, every functional derivative with respect to J brings a factor $\psi + \lambda_2 \Lambda^{-2} \psi^3$ in the time-ordered product of interest:

$$\langle 0_{\text{out}}|T\, \phi(x_1) \cdots |0_{\text{in}}\rangle = \langle 0_{\text{out}}|T\, \left(\psi(x_1) + \lambda_2 \Lambda^{-2} \psi^3(x_1)\right) \cdots |0_{\text{in}}\rangle. \tag{11.37}$$

At this point, we have shown that the only possible effect of the term we have removed from the effective Lagrangian is to modify the operators inside a time-ordered product of fields (in the form of extra terms that will appear on the external legs of the corresponding Feynman graphs). However, the physical quantities are not the above correlation functions themselves, but the on-shell transition amplitudes obtained with the Lehmann–Symanzik–Zimmermann (LSZ) reduction formulas, i.e., the residue of the one-particle poles in the Fourier transform of eq. (11.37). For instance, in a four-point function contributing to a $2 \to 2$ scattering amplitude, one would have a graph such as the following:

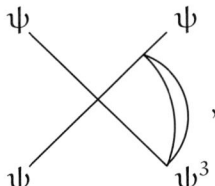

where one of the operators in the T-product is a $\psi^3$ (in this diagrammatic representation, we have not yet amputated the external propagators). We can readily see that one point of this function is not terminated by a propagator, and therefore does not exhibit a one-particle pole. Thus, such a graph does not contribute to the on-shell transition amplitude when inserted in the LSZ reduction formula. Although we have used a very simple example to illustrate the chain of arguments leading to this result, it is in fact completely general: If a term of the effective Lagrangian can be rewritten as a linear combination of other operators thanks to the leading order equation of motion, then this term can be ignored in the effective theory without changing anything in the S-matrix.

## 11.4 Effective Theories in QCD

Quantum chromodynamics (QCD) is also an area in which effective field theories are quite useful. Indeed, since QCD contains many dimensionful scales (the scale $\Lambda_{\text{QCD}}$ at which the coupling constant diverges, and the masses of the six families of quarks, which span a wide range of momentum scales), we may expect that some simplifications are possible if one is interested in processes in which some of these scales are irrelevant. Several effective theories have been developed in order to simplify the treatment of strong interactions in some special kinematical situations, and we shall discuss two of them in this section.

### 11.4.1 Heavy Quark Effective Theory

**Main ideas:** There are six families of quarks in nature: u, d, s, c, b and t. The u, d and s quarks are light in comparison to other QCD scales (in particular the confinement scale $\Lambda_{QCD}$), while the c, b and t are considered heavy. Besides the well-known nucleons (proton, neutron) and light mesons (pions, rho) that are made of u and d valence quarks, some hadrons contain heavy quarks (c and b only, since the t quark decays before a bound state can form). An obvious source of simplification in the presence of heavy quarks is asymptotic freedom, thanks to which the strong coupling constant at the scale $m_Q$ is not very large and thus the strong interactions are more like electromagnetic interactions. In particular, hadrons made of a pair of heavy quark and heavy antiquark $Q\overline{Q}$ have a size of order $(\alpha_s m_Q)^{-1}$. When this size is much smaller than $\Lambda_{QCD}^{-1}$, these bound states are quite similar to a hydrogen atom.

However, hadrons mixing heavy and light quarks are not as simple, because their size is of order $\Lambda_{QCD}^{-1}$ and the typical momentum transfer between the light and heavy quarks is of order $\Lambda_{QCD}$. Thus, in these heavy–light hadrons, one may view the heavy quark as surrounded by a non-perturbative cloud of light quarks and gluons. Such systems are characterized by two different scales:

- the heavy quark mass $m_Q$ and the corresponding Compton wavelength $\lambda_Q = m_Q^{-1}$;
- the confinement scale $\Lambda_{QCD}$, which controls the typical size of bound states, $R_h \sim \Lambda_{QCD}^{-1}$.

For a heavy quark, one has $\lambda_Q \ll R_h$. Thus, in a certain sense, the heavy quark may be viewed as a point-like object inside a much larger hadron. Loosely speaking, the quantum numbers of the heavy quark (flavor, spin) are confined in a volume of order of its Compton wavelength $\lambda_Q$, but the accompanying cloud of light quarks and gluons can only resolve distances as small as $\Lambda_{QCD}^{-1}$. Therefore, the light degrees of freedom are totally insensitive to the heavy quark quantum numbers, and they only feel its color field. Moreover, for a heavy–light hadron, the rest frame of the hadron is almost equivalent to the rest frame of the heavy quark. In this frame, the color field of the heavy quark is the Coulomb electrical field produced by a static color charge that does not depend on the heavy quark mass. Thus, we expect that the configuration of the light constituents is independent of $m_Q$ when $m_Q \to \infty$. These observations constitute what is called *heavy quark symmetry*, which we shall derive more formally later in this section. Note that, unlike chiral symmetry for massless quarks, it is not a symmetry of the QCD Lagrangian, but rather an approximate symmetry that arises in special kinematical conditions (namely, when a heavy quark interacts only with light degrees of freedom via soft exchanges). Heavy quark symmetry provides relationships between bound states that differ only in the flavor and/or the spin[9] of the heavy quark, for instance the B, D, B*, D* mesons, or the $\Lambda_b, \Lambda_c$ baryons. Heavy quark effective theory exploits

---

[9]This is analogous to the fact that isotopes have almost identical chemistry, since the cloud of electrons surrounding the nucleus is almost independent of its mass (in a first approximation, it depends only on its

## 11.4 EFFECTIVE THEORIES IN QCD

this separation of scales in a systematic way in order to calculate the dependence of various physical quantities on the mass of the heavy quark, by an expansion in powers of $m_Q^{-1}$.

**Spinor decomposition:** Let us assume that there is a large gap between the confinement scale $\Lambda_{QCD}$ and the heavy quark mass $m_Q$, and introduce an intermediate scale $\Lambda$ such that $\Lambda_{QCD} \ll \Lambda \ll m_Q$. Our goal is to construct an effective theory that is equivalent to QCD at long distance, i.e., for momenta below $\Lambda$ (but may differ from QCD above $\Lambda$). Heavy quark effective theory is somewhat special in that we do not completely integrate out the heavy quarks (since one of its applications is to describe bound states that contain heavy quarks), but we rather integrate out only a part of the heavy quark degrees of freedom. This is done by writing the momentum of a heavy quark as follows:

$$p^\mu \equiv m_Q v^\mu + q^\mu, \tag{11.38}$$

where $v^\mu$ is the *hadron* velocity (satisfying $v_\mu v^\mu = 1$) and $q^\mu$ is a residual momentum whose components are much smaller than $m_Q$. This decomposition just highlights the fact that the heavy quark moves almost with the hadron velocity. By definition, the term $m_Q v^\mu$ does not change, while $q^\mu$ fluctuates due to the interactions of the heavy quark with the light degrees of freedom. However, the typical changes of $q^\mu$ are of order $\Lambda_{QCD}$. Thus, the physical picture that emerges from this separation is that of a heavy quark that moves almost along a straight line, just undergoing little kicks from the surrounding cloud of light constituents.

By combining eq. (11.38) and the Dirac equation, we can see that the dominant space-time dependence of spinors is a phase $\exp(\pm i\, m_Q v \cdot x)$. Moreover, the velocity $v_\mu$ can be used to construct two spin projectors,

$$P_\pm \equiv \frac{1 \pm \slashed{v}}{2}, \tag{11.39}$$

thanks to which we may decompose the spinor $\psi$ of a heavy quark into

$$q_v(x) \equiv e^{i\, m_Q v \cdot x} P_+ \psi(x), \quad Q_v(x) \equiv e^{i\, m_Q v \cdot x} P_- \psi(x), \tag{11.40}$$

or conversely

$$\psi(x) = e^{-i\, m_Q v \cdot x} \left[ q_v(x) + Q_v(x) \right]. \tag{11.41}$$

By introducing this decomposition in the Dirac Lagrangian, we obtain

$$\begin{aligned}\mathcal{L} &= \overline{\psi}\, (i\slashed{D} - m_Q)\, \psi \\ &= (\overline{q}_v + \overline{Q}_v)(i\slashed{D} - m_Q + m_Q \slashed{v})(q_v + Q_v) \\ &= \overline{q}_v (iv\cdot D) q_v - \overline{Q}_v (iv\cdot D + 2m_Q) Q_v + \overline{q}_v (i\slashed{D}_\perp) Q_v + \overline{Q}_v (i\slashed{D}_\perp) q_v, \end{aligned} \tag{11.42}$$

where we have decomposed the covariant derivative as $D^\mu \equiv v^\mu (v \cdot D) + D_\perp^\mu$. From this form of the Lagrangian, we see that $q_v$ is a massless spinor while $Q_v$ has a mass $2m_Q$. Thus, the heavy quark effective theory will be obtained by integrating out $Q_v$.

---

electrical charge). Likewise, the independence with respect to the spin of the heavy quark is analogous to the near degeneracy of the hyperfine levels in atomic physics.

**Effective Lagrangian:** Let us consider the generating function of the correlation functions of the "light" field $q_v$, defined as

$$Z[\eta, \bar{\eta}] \equiv \int [Dq_v D\bar{q}_v DQ_v D\bar{Q}_v] \, e^{i \int d^4x \, (\mathcal{L} + \bar{\eta} q_v + \bar{q}_v \eta)}. \tag{11.43}$$

The path integration over the heavy field $Q_v$ is Gaussian and can be performed analytically, giving

$$Z[\eta, \bar{\eta}] = \int [Dq_v D\bar{q}_v] \, \Delta_v[A] \, e^{i \int d^4x \, (\mathcal{L}_{\text{eff}} + \bar{\eta} q_v + \bar{q}_v \eta)}, \tag{11.44}$$

with the effective Lagrangian

$$\mathcal{L}_{\text{eff}} \equiv \bar{q}_v (i v \cdot D) q_v + \bar{q}_v (i\slashed{D}_\perp) \frac{1}{2m_Q + i v \cdot D} (i\slashed{D}_\perp) q_v, \tag{11.45}$$

and where $\Delta_v[A]$ is the functional determinant produced by the Gaussian integral,

$$\Delta_v[A] \equiv \left( \det \left( 2m_Q + i v \cdot D \right) \right)^{1/2}. \tag{11.46}$$

Note that if one chooses the strict axial gauge $v \cdot A = 0$, then this determinant is constant and may be disregarded.

**Derivative expansion:** Because of the presence of derivatives in the denominator of eq. (11.45), the corresponding effective action is non-local. In order to obtain a local effective theory, we should expand this expression into a series of local operators. Such an expansion is legitimate because we have pulled out the fast phase $\exp(i \, m_Q \, v \cdot x)$ from the spinor. The resulting light field $q_v(x)$ has only a slow space-time dependence associated with the residual momentum $q^\mu \sim \Lambda_{\text{QCD}}$. Moreover, interactions with soft gluons involve a gauge field $A^\mu \sim \Lambda_{\text{QCD}}$, and we thus have

$$v \cdot D \sim \Lambda_{\text{QCD}} \ll m_Q, \tag{11.47}$$

which allows the following expansion:

$$\frac{1}{2m_Q + i v \cdot D} = \frac{1}{2m_Q} \sum_{n=0}^{\infty} \left( -i \frac{v \cdot D}{2m_Q} \right)^n. \tag{11.48}$$

Up to the terms of order $m_Q^{-1}$, the effective Lagrangian therefore reads

$$\mathcal{L}_{\text{eff}} = \underbrace{\bar{q}_v (i v \cdot D) q_v}_{\mathcal{L}_\infty} + \bar{q}_v \frac{(i D_\perp)^2}{2m_Q} q_v + \frac{g}{2m_Q} \bar{q}_v M^{ij} F_{ij} q_v + \mathcal{O}(m_Q^{-2}), \tag{11.49}$$

where $F_{ij}$ is the QCD field strength tensor, and $M^{ij} \equiv \frac{i}{4}[\gamma^i, \gamma^j]$ are the generators of the Poincaré algebra in the spin-1/2 representation (latin indices $i, j$ run only over the spatial components transverse to the velocity). The first term $\mathcal{L}_\infty$ is the only one that survives in the limit of infinite quark mass. The terms of order $m_Q^{-1}$ can be interpreted respectively as the contribution of the transverse motion to the kinetic energy and the interaction between the spin of the quark and the chromomagnetic field.

## 11.4 EFFECTIVE THEORIES IN QCD

**Heavy quark symmetry:** The leading term in the effective Lagrangian, $\mathcal{L}_\infty$, corresponds to the following Feynman rules:

$$\xrightarrow{p} \;=\; \frac{i\delta_{ij}\, P_+}{v\cdot p + i0^+}, \qquad \text{(vertex)} \;=\; -ig\, v^\mu \left(t^a_r\right)_{ij}.$$

Since there are no Dirac matrices in the expression of the vertex, the interactions with gluons do not alter the spin of the heavy quark at order $m_Q^0$. More formally, since $\mathcal{L}_\infty$ does not contain Dirac matrices, it is invariant under

$$q_v \to e^{i\theta^i S^i}\, q_v, \qquad \text{with } S^i \equiv \frac{1}{2}\begin{pmatrix} \sigma^i & 0 \\ 0 & \sigma^i \end{pmatrix}. \tag{11.50}$$

(The $\sigma^i$ are the Pauli matrices.) Since we have $[S^i, S^j] = i\epsilon_{ijk} S^k$, this corresponds to an $SU(2)$ invariance of $\mathcal{L}_\infty$. Moreover, since $\mathcal{L}_\infty$ is independent of $m_Q$, all heavy quarks play the same role. With $n_f$ flavors of heavy quarks, the leading effective Lagrangian,

$$\mathcal{L}_\infty = \sum_{f=0}^{n_f} \overline{q}_{vf}(iv\cdot D)q_{vf}, \tag{11.51}$$

has an $SU(2n_f)$ symmetry that constitutes the spin-flavor heavy quark symmetry. These symmetries are broken by the corrections in $m_Q^{-1}$, since they depend explicitly on the mass and contain Dirac matrices.

### 11.4.2 Color Glass Condensate

**Kinematics of high-energy collisions:** Another area of strong interactions which is hardly tractable in QCD itself, but where progress can be made with the help of an effective description, is that of collisions between hadrons (or nuclei) at very high energy. Consider, for instance, a proton. A naive picture is that it is made of three valence quarks, bound by gluon exchanges. However, in a relativistic quantum description, these constituents can all fluctuate into virtual quarks, antiquarks and gluons. The important point is that these fluctuations are short-lived: Roughly speaking, their lifetime spans all scales from zero to the proton size.

When a proton is probed in an experiment (for instance, by colliding it with another proton) characterized by a certain time resolution, the fluctuations of the proton whose lifetime is smaller than this resolution do not play an active role (see Figure 11.3). Through renormalization, their only effect is to set the values of the parameters of the Lagrangian (in particular, the coupling constant) at the scale relevant for this experiment. On the other hand, the fluctuations whose lifetime is large compared to the characteristic timescale of the probe are seen as actual constituents of the proton. For instance, if a quark has fluctuated into a long-lived (compared to the timescale probed in the experiment) quark–gluon state, then the experiment will see a quark plus a gluon, both on-shell. Note that on-shellness, i.e., the

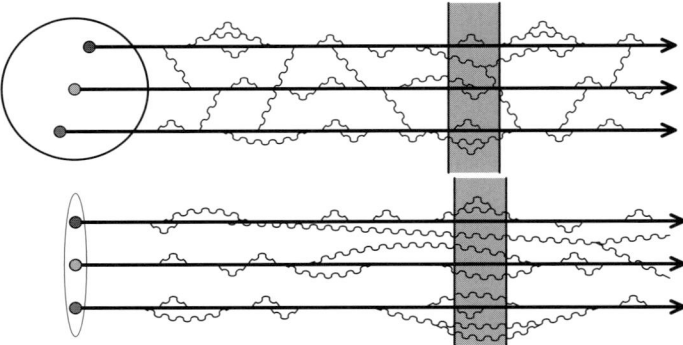

FIGURE 11.3: Cartoon of the fluctuations inside a nucleon. The shaded strip indicates the time resolution of some external probe. Top: slow nucleon. Bottom: boosted nucleon. All the internal time scales are dilated by a Lorentz factor, and new virtual fluctuations become accessible to the probe.

fact that a momentum satisfies $P^2 = m^2$ for a particle of mass $m$, should not be viewed in a strict sense in this context. On-shellness in principle requires that the energy be *exactly* $p_0 = \sqrt{\mathbf{p}^2 + m^2}$. But by the uncertainty principle, one would have to observe this particle for an infinitely long time to check an exact equality. Thus, in a measurement of duration $\Delta t$, any particle whose energy $p_0$ satisfies $\Delta t \left| p_0 - \sqrt{\mathbf{p}^2 + m^2} \right| \lesssim 1$ cannot be distinguished from an exactly on-shell particle.

This discussion provides the physical justification of the parton model, in which bound states such as protons are described by means of distributions of quarks, antiquarks and gluons (generically called "partons"). Except for the valence quarks, these constituents are in fact quantum fluctuations, but their long lifetime (compared to the interaction time) allows treating them as being on-shell. Moreover, these parton distributions must vary with the resolution scale (in space and time) with which the proton is probed, since a smaller resolution scale will resolve more partons in the measurement.

In particular, in a collision between two hadrons, the duration of the collision scales as the inverse of the collision energy (roughly speaking, this is the time necessary for the two Lorentz contracted hadrons to go through each other), and therefore the parton distributions grow with the collision energy. Let us now assume that in such a collision we are interested in processes characterized by a transverse momentum Q. This means that this measurement resolves a fixed transverse distance of the order of $Q^{-1}$, which may also be viewed as the minimal spatial extension of the wavefunction of the partons probed in this collision. Combining this fact with the increase of the number of partons with energy, we see that higher collision energy leads eventually to a situation in which the partons fill all the available volume in the hadron. This regime of strong interactions is known as *(gluon) saturation*. Note that the gluon occupation number cannot grow above $\alpha_s^{-1}$, because this is the value at which gluon splittings and gluon recombinations roughly balance each other.

**Degrees of freedom:** In order to discuss the relevant degrees of freedom, let us consider the point of view of an observer at rest, while the hadron moves with a very large momentum in

## 11.4 EFFECTIVE THEORIES IN QCD

the z direction. Due to the special kinematics of this problem, it is convenient to introduce *light-cone coordinates*, defined as

$$x^+ \equiv \frac{x^0 + x^3}{\sqrt{2}}, \quad x^- \equiv \frac{x^0 - x^3}{\sqrt{2}}. \tag{11.52}$$

(The remaining two coordinates are the transverse coordinates $x_\perp$.) Similar definitions can be introduced for 4-momenta. These coordinates transform very simply under boosts in the z direction, since they just undergo a rescaling:

$$x^+ \to e^\omega x^+, \quad x^- \to e^{-\omega} x^-, \quad x_\perp \to x_\perp. \tag{11.53}$$

To order the constituents by their longitudinal momentum, the most convenient variable is *rapidity*, defined as $y \equiv \frac{1}{2} \ln(p^+/p^-)$, since it is shifted by an additive constant under a boost in the z direction. By definition, $y = 0$ (i.e., $p_z = 0$) corresponds to objects with no longitudinal momentum in the observer's frame. Quantum fluctuations with a large positive rapidity appear to the observer as nearly on-shell constituents. At the largest rapidities (corresponding to the total $p_z$ of the hadron), there are few constituents, mostly the valence quarks. Because of their large longitudinal momentum, the dynamics of these constituents is considerably slowed down by time dilation, and therefore they appear static to the observer. The only relevant information about these fast partons is the color current they carry. This current is longitudinal, and because these constituents are static, it does not depend on the light-cone variable[10] $x^+$ and takes the following form:

$$J^\mu_a(x) \equiv \delta^{\mu+} \rho_a(x^-, x_\perp), \tag{11.54}$$

where the function $\rho_a$ is the spatial distribution of color charge. For a high-energy hadron, Lorentz contraction implies that the $x^-$ dependence of this function is very peaked around $x^- \approx 0$. On the other hand, the $x_\perp$ dependence reflects the distribution of the constituents of the hadron in the plane transverse to the collision axis. Since this depends on the peculiar spatial arrangement of the constituents at the time of the collision, the function $\rho_a(x^-, x_\perp)$ is not known and may be considered as a random variable with a probability distribution $W[\rho]$. When one repeats many similar collisions, the expectation value of an observable is obtained by a statistical average,

$$\langle \mathcal{O} \rangle = \int [D\rho] \, W[\rho] \, \mathcal{O}[\rho], \tag{11.55}$$

where $\mathcal{O}[\rho]$ is the value of this observable calculated with an arbitrary instance of the distribution $\rho_a$. This description of a fast hadronic projectile is known as the McLerran–Venugopalan model.

In contrast, the constituents that lie at small rapidity in the observer's frame have a time evolution that cannot be neglected. These modes are thus described according to the original Yang–Mills action, as illustrated in Figure 11.4. Moreover, due to the hierarchy between the longitudinal momenta of the modes described as a color current and those described as

---

[10] The evolution in $x^+$ is generated by the component $P^-$ of the momentum. However, for massless on-shell modes, we have $P^- = P_\perp^2/(2P^+) \to 0$ for the fast-moving modes in the z direction.

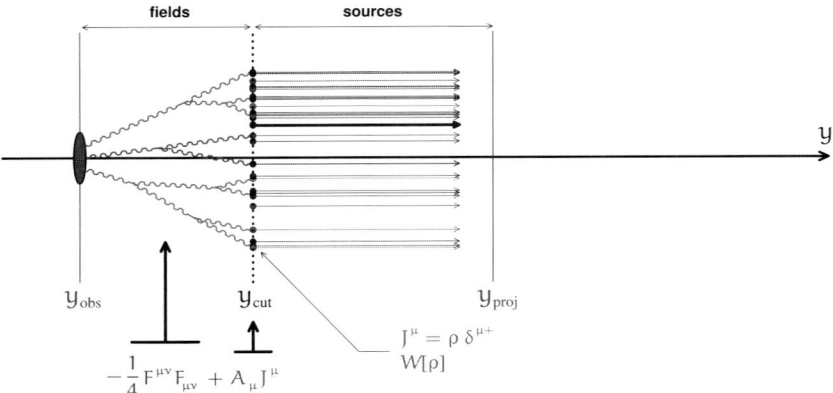

FIGURE 11.4: Degrees of freedom in the color glass condensate effective description of a high-energy hadron.

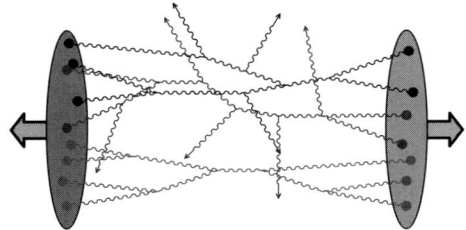

FIGURE 11.5: Typical graph in a hadron–hadron collision, in which both hadrons are described with the CGC effective theory. The solid dots represent insertions of the color current $J_\mu$.

regular gauge fields, the coupling between them may be approximated as eikonal, i.e., by a term of the form $J_\mu A^\mu$, and therefore the action of the effective theory reads

$$S = \int d^4x \left( -\frac{1}{4} F^a_{\mu\nu} F^{\mu\nu\,a} + J_{\mu\,a} A^\mu_a \right). \tag{11.56}$$

This effective theory is called the *color glass condensate (CGC)*.

**Power counting in the saturation regime:** The power counting for the graphs that appear in this effective theory (see Figure 11.5 for an example) is a bit peculiar in the saturation regime. Indeed, this situation corresponds to a gluon occupation number of order $g^{-2}$, which is achieved with a color current of order $g^{-1}$. The order of a connected graph $\mathcal{G}$ with $n_E$ external gluons, $n_L$ loops and $n_J$ insertions of the color current is given by

$$\mathcal{G} \sim g^{-2}\, g^{n_E}\, g^{2n_L}\, (gJ)^{n_J}, \tag{11.57}$$

where $J$ denotes the typical magnitude of the current. Thus, in the saturated regime where $J \sim g^{-1}$, the magnitude of connected graphs does not depend on $n_J$, which means that all observables depend non-perturbatively on the color current. In contrast, the loop expansion

## 11.4 EFFECTIVE THEORIES IN QCD

still corresponds to an expansion in powers of $g^2$. Observables at tree level are given by an infinite sum of tree diagrams (corresponding to an arbitrary number of insertions of $J^\mu$), which can be expressed in terms of classical solutions of the Yang–Mills equations of motion in the presence of an external source:

$$\left[D_\mu, F^{\mu\nu}\right]_a = -J_a^\nu. \tag{11.58}$$

In order to have a unique solution, these equations must be supplemented with boundary conditions. One may show that in the case of inclusive observables,[11] the appropriate boundary condition is a retarded one, in which the initial fields (and their time derivative) are zero in the remote past (i.e. long before the collision). The classical field obtained by solving the above equation of motion is a strong field,

$$\mathcal{A}^\mu \sim g^{-1}, \tag{11.59}$$

which leads to interesting technical complications that will be discussed in Chapter 18. Higher-order contributions correspond to loops evaluated in the presence of this classical field that plays the role of a background.

**Cutoff dependence:** In addition, this effective description must be endowed with a cutoff (denoted $y_{\rm cut}$ in Figure 11.4) in rapidity in order to separate the two types of degrees of freedom. This cutoff does not appear explicitly in the above classical action of eq. (11.56), and therefore observables do not depend on it at tree level. But it enters in loop corrections as an upper limit in the integral over the longitudinal momentum circulating in the loop. Indeed, including in the loop modes that have a rapidity larger than $y_{\rm cut}$ would lead to double counting, since these modes are already included via the color current $J_\mu$. Generically, this cutoff introduces a linear dependence on $y_{\rm cut}$ in the one-loop correction of observables. In fact, one may show that for all inclusive observables, the cutoff dependence at one loop can be written as

$$\delta\mathcal{O}_{\rm NLO}[\rho] = y_{\rm cut}\, \mathbf{H}\, \mathcal{O}_{\rm LO}[\rho] + \text{terms that do not depend on } y_{\rm cut}, \tag{11.60}$$

where $\mathbf{H}$ is a universal (i.e., the same for all inclusive observables) operator containing second-order derivatives with respect to $\rho_a$. An important property of this operator is that it is self-adjoint:

$$\int [D\rho]\, A[\rho]\, \left(\mathbf{H}\, B[\rho]\right) = \int [D\rho]\, \left(\mathbf{H}\, A[\rho]\right)\, B[\rho]. \tag{11.61}$$

However, the cutoff is not a physical parameter since it was introduced by hand in order to separate the two types of degrees of freedom, and therefore it should not appear in physical quantities. The way out of this situation is to realize that by changing the value of the cutoff, one is also modifying which modes are described by the color current $J_\mu$. Consequently, the distribution $W[\rho]$ should in fact depend on $y_{\rm cut}$. Using eqs. (11.60) and (11.61), we see

---

[11] Inclusive observables are measurements for which one sums over all the possible final states without excluding any of them. For instance, the average particle multiplicity in the final state is an inclusive observable, while the probability of producing exactly three particles is not.

immediately that the cutoff dependence coming from the loop correction to observables can be canceled if we also change

$$W[\rho] \to W[\rho] - y_{\text{cut}}\, \mathbf{H}\, W[\rho]. \tag{11.62}$$

More precisely, this substitution cancels the linear dependence on $y_{\text{cut}}$. A more rigorous procedure is to apply it to an infinitesimal variation $\delta y_{\text{cut}}$ of the cutoff, for which the quadratic terms are truly negligible. By doing so, the change of eq. (11.62) becomes a differential equation,

$$\frac{\partial W[\rho]}{\partial y_{\text{cut}}} = -\mathbf{H}\, W[\rho], \tag{11.63}$$

which controls how the probability distribution $W[\rho]$ changes as one varies the cutoff (this equation is called the *JIMWLK equation* – see Exercise 11.4 for the derivation of a closely related equation).

## 11.5 EFT of Spontaneous Symmetry-Breaking

### 11.5.1 Nambu–Goldstone Bosons

Spontaneous breaking of a global continuous symmetry leads to the emergence of massless spin 0 bosons – one for each broken generator of the original symmetry – the Nambu–Goldstone bosons. The other fields of the theory remain massive. Therefore, at low energy we expect that the physics is dominated by the fluctuations of the Nambu–Goldstone bosons, and that we may neglect all the other excitations. Nonlinear sigma models[12] provide an effective description that contains only the massless particles.

Let our starting point be an action of the form

$$S \equiv \int d^d x \left\{ \tfrac{1}{2}(\partial_\mu \phi(x))(\partial^\mu \phi(x)) - V(\phi(x)) \right\}, \tag{11.64}$$

assumed to be invariant under the global action of a Lie group $\mathcal{G}$. The metric of d-dimensional space-time is chosen to be Minkowskian (but this discussion is equally applicable to Euclidean space). In addition, the potential $V(\phi)$ has non-trivial minima at some $\phi \neq 0$. Due to the $\mathcal{G}$-invariance of the action, the non-trivial minima cannot be unique. Given a certain minimum $\phi_c$, all the field configurations that may be reached from $\phi_c$ by the action of $\mathcal{G}$ are also minima. If we assume that there are no accidental (i.e., not caused by the symmetry of the action) degeneracies of the minima, the set of all minima can therefore be written as

$$\mathcal{M}_0 \equiv \{g\phi_c | g \in \mathcal{G}\}. \tag{11.65}$$

If $\mathcal{H}$ is the subgroup of $\mathcal{G}$ that leaves $\phi_c$ invariant (this subgroup is called the *stabilizer* of $\phi_c$), then $\mathcal{M}_0$ is also the coset $\mathcal{G}/\mathcal{H}$ (see Figure 11.6 for an example).

---

[12] Their name comes from applications to the physics of pions, which may be viewed as Goldstone bosons of the (approximate for small but non-zero quark masses) chiral symmetry $SU(2) \times SU(2)$ that exists in the light quark (u and d) sector of quantum chromodynamics. This symmetry is spontaneously broken to a residual $SU(2)$ symmetry in the vacuum of QCD, leading to the appearance of three nearly massless scalar particles.

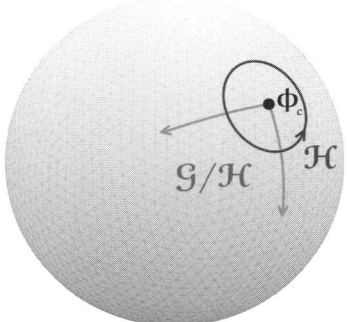

FIGURE 11.6: Illustration of the symmetry-breaking pattern in the case of a theory with $\mathcal{G} = O(3)$ symmetry. The set $\mathcal{M}_0$ of the minima of the potential is a two-dimensional sphere. The stabilizer of the minimum $\phi_c$ is $\mathcal{H} = O(2)$. The light-colored arrows show the field configurations obtained by applying $\mathcal{G}/\mathcal{H}$ to $\phi_c$, i.e., the allowed values of the Nambu–Goldstone fields.

### 11.5.2 Nonlinear Sigma Model

**Definition:** At low energy, the quantum fluctuations of the fields that have remained massive can be neglected, and the massless components of $\phi$ may be obtained by acting on $\phi_c$ by a matrix representation of $\mathcal{G}$,

$$\phi_i = R_{ij}(g)\,\phi_{cj}. \tag{11.66}$$

Note that, given an element $h \in H$, we have

$$R_{ij}(gh)\,\phi_{cj} = R_{ij}(g)\,\underbrace{R_{jk}(h)\,\phi_{ck}}_{\phi_{cj}} = R_{ij}(g)\,\phi_{cj}. \tag{11.67}$$

Thus, the field $\phi$ given by eq. (11.66) is not really a function of the full group $\mathcal{G}$, but depends only on elements of the coset $\mathcal{G}/\mathcal{H}$. Let us now split the generators $t^a$ of the Lie algebra $\mathfrak{g}$ into those (for $n < a$) that correspond to $\mathfrak{h}$, and the complement (for $1 \leq a \leq n$). From the definition of $\mathcal{H}$ as the stabilizer of $\phi_c$, we have

$$\begin{aligned} 1 \leq a \leq n: \quad & t^a_{ij}\,\phi_{cj} \neq 0, \\ a > n: \quad & t^a_{ij}\,\phi_{cj} = 0. \end{aligned} \tag{11.68}$$

Thus, the matrix $R(g)$ can be written as

$$R(\theta) = \exp\left(i \sum_{a=1}^{n} \theta^a t^a\right). \tag{11.69}$$

The value of the potential does not change under the action of $\mathcal{G}$ on $\phi_c$, and we are free to choose the value of its minimum to be $V(\phi_c) = 0$. Thus, the action becomes

$$S = \frac{1}{2}\int d^dx\,\phi_{ci}\left(\partial_\mu R^{-1}_{ik}(\theta)\right)\left(\partial^\mu R_{kj}(\theta)\right)\phi_{cj}$$

$$= -\frac{1}{2}\int d^d x \, \phi_{ci}\left(A_\mu(\theta)A^\mu(\theta)\right)_{ij}\phi_{cj}, \tag{11.70}$$

where in the second expression we have introduced $A_\mu \equiv R^{-1}\partial_\mu R$ (an element of the algebra).

Eq. (11.70) gives the action in terms of the "coordinates" $\theta^a$ on the coset $\mathcal{G}/\mathcal{H}$, corresponding to a certain choice of the generators $t^a$. However, it is interesting to express the action in terms of a completely arbitrary system of coordinates on $\mathcal{G}/\mathcal{H}$, that we may denote $\vartheta^m$. Since eq. (11.70) has only two derivatives $\partial_\mu \cdots \partial^\mu$, the same must be true of its expression in any system of coordinates. On the other hand, it may contain terms of arbitrarily high degree in $\vartheta$. Thus, the most general action is of the form

$$\mathcal{S} = \frac{1}{2}\int d^d x \, g_{mn}(\vartheta)\left(\partial_\mu \vartheta^m\right)\left(\partial^\mu \vartheta^n\right), \tag{11.71}$$

where the coefficients $g_{mn}(\vartheta)$ can be related to $R(\theta)$ as follows,

$$g_{mn}(\vartheta) \equiv -4\left[\phi_{ci}t^a_{ik}t^b_{kj}\phi_{cj}\right] \, \text{tr}\left[t^a R^{-1}\frac{\partial R}{\partial \vartheta^m}\right] \, \text{tr}\left[t^b R^{-1}\frac{\partial R}{\partial \vartheta^n}\right]. \tag{11.72}$$

They form a metric tensor on $\mathcal{G}/\mathcal{H}$, if the coset is viewed as a Riemannian manifold. Indeed, if we use a different system of coordinates $\varpi_p$ on $\mathcal{G}/\mathcal{H}$, $g_{mn}(\vartheta)$ would be replaced by

$$g_{pq}(\varpi) \equiv -4\left[\phi_{ci}t^a_{ik}t^b_{kj}\phi_{cj}\right] \, \text{tr}\left[t^a R^{-1}\frac{\partial R}{\partial \varpi^p}\right] \, \text{tr}\left[t^b R^{-1}\frac{\partial R}{\partial \varpi^q}\right]$$

$$= \left(\frac{\partial \vartheta^m}{\partial \varpi^p}\right)\left(\frac{\partial \vartheta^n}{\partial \varpi^q}\right) g_{mn}(\vartheta), \tag{11.73}$$

which is indeed the expected transformation law of a metric tensor under a change of coordinates. The field theory described by the action (11.71) is called a *nonlinear sigma model*. Note that the derivative $\partial_\mu \vartheta^m$ of the coordinate $\vartheta^m$ is a vector that lives on the tangent space to the manifold $\mathcal{G}/\mathcal{H}$ at the point $\vartheta$. Therefore, the action (11.71), in which the tensor $g_{mn}$ is contracted with two vectors, is a scalar – invariant under changes of coordinates on the manifold.

The Taylor expansion of the metric in powers of the field $\vartheta$ determines which couplings exist in the classical action. Interestingly, even though the kinetic term of the original action was quadratic in the fields, we now have a term with possibly arbitrarily high orders in the field. Loosely speaking, this is due to the fact that spontaneous symmetry breaking has restricted the fields from a space $\mathbb{R}^n$ in which the symmetry $\mathcal{G}$ was linearly realized, down to a curved manifold in which it is realized nonlinearly. In addition, it is worth stressing that the final action is uniquely determined from eq. (11.69), but may take various explicit forms depending on the choice of coordinates $\vartheta^m$ on $\mathcal{G}/\mathcal{H}$. In other words, the nonlinear sigma model has an intrinsic geometrical meaning, that does not depend on the system of coordinates one uses.

**Path integral quantization:** The quantization of the nonlinear sigma model can be achieved via path integration. The action is quadratic in derivatives of the field, but with the unusual feature that these derivatives are multiplied by a function of the field. In order to ascertain the consequence of this property, it is necessary to start from the Hamilton formulation of the path integral, and to perform explicitly the integral over the conjugate momenta. For a Lagrangian density

$$\mathcal{L} = \frac{1}{2} g_{mn}(\vartheta) \left(\partial_\mu \vartheta^m\right) \left(\partial^\mu \vartheta^n\right), \tag{11.74}$$

the conjugate momenta read

$$\pi_m \equiv \frac{\partial \mathcal{L}}{\partial \partial^0 \vartheta^m} = g_{mn}(\vartheta) \partial^0 \vartheta^n, \tag{11.75}$$

and the Hamiltonian density is given by

$$\mathcal{H} = \pi_m \left(\partial^0 \vartheta^m\right) - \mathcal{L} = \frac{1}{2} g^{mn}(\vartheta) \pi_m \pi_n + \frac{1}{2} g_{mn}(\vartheta) \left(\boldsymbol{\nabla}\vartheta^m\right) \cdot \left(\boldsymbol{\nabla}\vartheta^n\right), \tag{11.76}$$

where $g^{mn}$ is the inverse of the metric tensor, $g^{mn} g_{np} = \delta^m{}_p$. The Hamiltonian is quadratic in the momenta, but since the coefficient in front of $\pi_m \pi_n$ depends on the field, the determinant produced in the Gaussian integration over the momenta cannot be disregarded. After this integral has been performed, the generating functional is given by the following formula:

$$Z[j_m] = \int \left[\sqrt{g(x)} \prod_m D\vartheta^m(x)\right] \exp\left\{\frac{i}{\hbar} \int d^d x \left(\mathcal{L}(\vartheta) + j_m \vartheta^m\right)\right\}, \tag{11.77}$$

where we denote $g(x) \equiv \det\left(g_{mn}(\vartheta(x))\right)$. Interestingly, the field dependence of $g_{mn}(\vartheta)$ alters the path integral in a rather natural way: $\left[\prod_m D\vartheta^m\right]$ is replaced by $\left[\sqrt{g} \prod_m D\vartheta^m\right]$, which is invariant under changes of coordinates on the manifold $\mathcal{G}/\mathcal{H}$ (see Exercise 11.5).

Note that in eq. (11.77) we have introduced an explicit $\hbar$, which will be useful later to keep track of the number of loops. The perturbative expansion in the nonlinear sigma model corresponds to an expansion in powers of $\hbar$. From the path integral, we can infer that the typical field amplitudes scale as

$$\vartheta \sim \sqrt{\hbar}, \tag{11.78}$$

which means that the perturbative expansion is also an expansion around $\vartheta = 0$ (i.e., around $\phi = \phi_c$). For such small fields, the effects of the curvature of the manifold are perturbative (see Figure 11.7), and we can expand the metric tensor in powers of the field (an explicit choice of coordinates must be made for this). The bare propagator of the $\vartheta$ fields is given by

$$G_{mn}(p) = \frac{i \delta_{mn}}{p^2 + i0^+}. \tag{11.79}$$

**Renormalization:** Dimensional analysis tells us that the field $\vartheta$ has the dimension

$$\vartheta \sim (\text{mass})^{(d-2)/2} \tag{11.80}$$

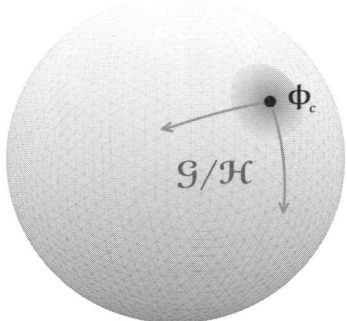

FIGURE 11.7: Perturbative expansion in the nonlinear sigma model: only field configurations near $\phi_c$ are explored.

(in a system of units where $\hbar = 1$). From this, we see that there are three cases regarding the ultraviolet power counting in the nonlinear sigma model:

- $d < 2$: The Taylor coefficients of the metric tensor all have a positive mass dimension, and are therefore super-renormalizable.
- $d = 2$: The Taylor coefficients are dimensionless, and the theory is renormalizable.
- $d > 2$: The Taylor coefficients have a negative mass dimension and are all non-renormalizable by power counting.

The most interesting situation is therefore the two-dimensional case. It differs somewhat from the renormalization of the quantum field theories we have encountered until now, since the action contains an infinite series of terms (of increasing degree in $\vartheta$), and an important question is whether the action (11.71) conserves its structure under renormalization.

Recall that the fields $\vartheta^m$ transform under a nonlinear representation of the group $\mathcal{G}$. Thus, their variation under an infinitesimal transformation of parameters $\epsilon_a$ may be written as

$$\delta\vartheta^m \equiv \epsilon^a\, T_a^m(\vartheta), \tag{11.81}$$

where the $T_a^m(\vartheta)$ are smooth functions of the fields. Under the same transformation, the variation of the action reads

$$\delta S = \epsilon^a \int d^2x\, T_a^m(\vartheta)\, \frac{\delta S}{\delta\vartheta^m(x)}, \tag{11.82}$$

and the invariance of the action under $\mathcal{G}$ thus requires that (see Exercise 11.6)

$$T_a^p\, \frac{\partial g_{mn}}{\partial\vartheta^p} + g_{pn}\, \frac{\partial T_a^p}{\partial\vartheta^m} + g_{mp}\, \frac{\partial T_a^p}{\partial\vartheta^n} = 0. \tag{11.83}$$

In other words, the possible forms of the metric tensor are constrained by the symmetry $\mathcal{G}$. Indeed, the coset $\mathcal{G}/\mathcal{H}$ is a *homogeneous space*,[13] i.e., a manifold that possesses additional

---

[13]Thanks to their connections to Lie algebras, a systematic classification of homogeneous spaces is possible.

symmetries that reduce the dimension of the space of allowed metrics. More precisely, a homogeneous space is such that given any pair of points $\vartheta$ and $\vartheta'$ on the manifold, there is an isometry (i.e., a distance-preserving transformation) that maps $\vartheta$ to $\vartheta'$. If in addition the space is isotropic, then it is said to be *maximally symmetric*.[14] In an N-dimensional maximally symmetric space, there is a particularly simple relationship between the metric and curvature tensors:

$$R_{mn} = \frac{R}{N} g_{mn} \qquad (R \equiv R^m{}_m),$$
$$R_{mnpq} = \frac{R}{N(N-1)} (g_{mp}g_{nq} - g_{mq}g_{np}). \tag{11.84}$$

(These two identities imply that the scalar curvature R is constant over the entire manifold for a dimension $N > 2$.)

A possible strategy for studying the renormalization of the sigma model is to introduce an analogue of the BRST transformation of non-Abelian gauge theories, and the associated Slavnov–Taylor identities obeyed by the quantum effective action. These identities, combined with dimensional and symmetry arguments that restrict the terms that may arise in the renormalized action, are sufficient to show that the renormalized action is structurally identical to eq. (11.71), with a group-invariant metric tensor that obeys a renormalized version of eqs. (11.83).

**Example of $\mathcal{G} = O(n)$:** A scalar field $\phi^i$ with $n$ components has an $O(n)$ symmetry if the action depends only on the combination $\phi^i\phi^i$. Potentials with non-trivial minima (i.e., at $\phi \neq 0$) in fact have infinitely degenerate minima that form an $(n-1)$-dimensional sphere $\mathcal{S}_{n-1}$ (see Figure 11.6 for an illustration in the case $n = 3$). Each minimum has a stabilizer subgroup $\mathcal{H} = O(n-1)$ (the smaller group of rotations around the direction fixed by this minimum), and we indeed have $\mathcal{S}_{n-1} = O(n)/O(n-1)$. A possible explicit parametrization of the field $\phi$ consists of writing $\phi \equiv \{\sigma, \pi\}$, where $\sigma$ has one component and $\pi$ has $n-1$ components. Assuming the parameters of the potential are adjusted so that the sphere $\mathcal{S}_{n-1}$ of minima has radius $|\phi| = 1$, we must impose the constraint $\sigma^2 + \pi^2 = 1$, which means that $\sigma$ may be viewed as a dependent field that depends nonlinearly on $\pi$. Usually, these coordinates are chosen in such a way that the symmetry-breaking vacuum is $\phi_c = (\sigma = 1, \pi = 0)$. In the vicinity of $\phi_c$, $\sigma$ is the "radial" massive field, while the $\pi^i$ are the "angular" variables corresponding to the massless Nambu–Goldstone bosons.

Then, we may split the generators of the $o(n)$ algebra into those of the stabilizer $o(n-1)$ and the complementary set of generators (see Figure 11.8):

- The generators of $o(n-1)$ act linearly on $\pi$. More precisely, they leave $\sigma^2 + \pi^2$ invariant by leaving both $\sigma$ and $\pi^2$ unchanged (thus simply rotating the $n-1$ components of $\pi$).
- In contrast, the generators of the complementary set also preserve $\sigma^2 + \pi^2$, but mix $\sigma$ and $\pi$ as follows:

$$\sigma \to \sigma - \epsilon^i \pi^i, \quad \pi^i \to \pi^i + \epsilon^i \sqrt{1 - \pi^2}, \tag{11.85}$$

and therefore they act nonlinearly on $\pi$.

---

[14] A maximally symmetric manifold of dimension N has $N(N+1)/2$ distinct isometries. In Euclidean space, this corresponds to N translations and $N(N-1)/2$ rotations, but this maximal number of isometries is the same in curved N-dimensional manifolds.

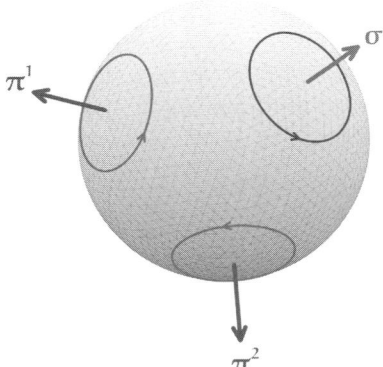

FIGURE 11.8: Illustration of the $(\sigma, \pi)$ coordinates for an O(3) model. The dark circle corresponds to the transformations that preserve $\sigma$ and act linearly on $\pi$ (as an O(2) rotation). The light-colored circles are the transformations that mix $\sigma$ and $\pi$ (and transform the latter nonlinearly).

The most general $O(n)$-invariant action with only second derivatives reads

$$S = \tfrac{1}{2} \int d^d x \left\{ (\partial_\mu \sigma)(\partial^\mu \sigma) + (\partial_\mu \boldsymbol{\pi})(\partial^\mu \boldsymbol{\pi}) \right\} = \tfrac{1}{2} \int d^d x \, g_{ij}(\boldsymbol{\pi}) \, (\partial_\mu \pi^i)(\partial^\mu \pi^j), \quad (11.86)$$

where in the second equality we have eliminated $\sigma$ by using $\sigma^2 + \pi^2 = 1$ and we have defined

$$g_{ij}(\boldsymbol{\pi}) \equiv \delta_{ij} + \frac{\pi^i \pi^j}{1 - \pi^2}. \quad (11.87)$$

The tensor $g_{ij}$ is the metric on the $S_{n-1}$ sphere, in the system of coordinates provided by the $\pi^i$. The couplings of this theory are determined by the Taylor expansion of the metric tensor, which in this case is completely specified by the choice of the coordinates and by the symmetries of the problem. In $d = 2$ dimensions, this theory is renormalizable by power counting. Although it contains an infinite number of couplings, it is not necessary to renormalize each of them individually. Instead, the renormalization preserves the structure of the action (11.86) with a metric tensor that remains dictated by the $O(n)$ symmetry.

**Chiral Lagrangians:** Let us now present a more concrete example of nonlinear sigma model, in the context of strong nuclear forces, that provides an effective theory for the interactions of low-energy pions. The starting point is the linear sigma model described in Section 4.5, without the coupling to the fermion doublet and without the explicit symmetry-breaking term. Its Lagrangian is

$$\mathcal{L} = \frac{1}{2}(\partial_\mu \sigma)(\partial^\mu \sigma) + \frac{1}{2}(\partial_\mu \pi^a)(\partial^\mu \pi^a) - \frac{\mu^2}{2}(\sigma^2 + \pi^2) - \frac{\lambda^2}{4}(\sigma^2 + \pi^2)^2. \quad (11.88)$$

If we denote $\Sigma \equiv \sigma + i\pi^a \sigma^a$, this Lagrangian is invariant under the following global $SU(2) \times SU(2)$ transformations:

$$\Sigma \to U^\dagger \Sigma V, \quad \text{with } U, V \in SU(2). \quad (11.89)$$

This can be seen by rewriting the Lagrangian as

$$\mathcal{L} = \frac{1}{4}\mathrm{tr}\left(\partial_\mu \Sigma \partial^\mu \Sigma^\dagger\right) - \frac{\mu^2}{4}\mathrm{tr}\left(\Sigma\Sigma^\dagger\right) - \frac{\lambda^2}{16}\left(\mathrm{tr}\left(\Sigma\Sigma^\dagger\right)\right)^2. \tag{11.90}$$

When $\mu^2 < 0$, the potential has minima at $\sigma^2 + \pi^2 = -\mu^2/\lambda^2$, and the system undergoes spontaneous symmetry breaking. Let us assume that the ground state resulting from symmetry breaking is $\langle(\sigma, \pi^a)\rangle = (f_\pi, 0)$, with $f_\pi \equiv |\mu|/\lambda$. The group of symmetries that leaves this field configuration invariant is the subgroup of diagonal transformations with $U = V$. Therefore, the resulting theory has $6 - 3$ massless Nambu–Goldstone bosons (the pions). If we neglect the radial (massive) excitations, the dynamics of the massless modes happens on the coset $SU(2) \times SU(2)/SU(2)$, that we can parametrize by

$$\Sigma = f_\pi \Omega, \quad \text{with } \Omega \equiv \exp\left(i\pi^a \sigma^a / f_\pi\right), \tag{11.91}$$

and the Lagrangian (called a *chiral Lagrangian* in this context) of the associated nonlinear sigma model is

$$\mathcal{L}_{\mathrm{NLSM}} = \frac{f_\pi^2}{4}\,\mathrm{tr}\left(\partial_\mu \Omega \, \partial^\mu \Omega^\dagger\right). \tag{11.92}$$

By expanding the $\Omega$ we see that this Lagrangian contains interaction terms of arbitrarily high dimensions (the field $\pi^a$ has the dimension of a mass in four space-time dimensions), and this theory is not renormalizable in four dimensions. But it is nevertheless applicable to phenomenology (see Exercise 11.7), as long as the typical momentum scale involved is small compared to the mass of the neglected radial excitations.

### 11.5.3 Nonlinear Sigma Model on a Generic Riemannian Manifold

We have derived the nonlinear sigma model as the effective action that describes the dynamics of the massless Nambu–Goldstone bosons after a spontaneous breaking of symmetry. In this case, the fields of the nonlinear sigma model live on a manifold which is also a homogeneous space thanks to the symmetries of the original problem. These symmetries severely constrain the possible forms of the metric, and play an important role in constraining the form of the loop corrections.

However, it is possible to consider an action of the form (11.71) for fields $\vartheta^m$ living on a generic smooth Riemannian manifold that does not possess any special symmetry. The power counting argument made earlier is unchanged, and we expect that this more general kind of sigma model is also renormalizable in two dimensions. For these generalized models, it has been shown that the dependence of the metric tensor (i.e., the function that defines all the couplings of the model) on the renormalization scale $\mu$ is governed by the following Callan–Symanzik equation:

$$\mu \frac{\partial}{\partial \mu} g^{mn} = -\frac{1}{2\pi} R^{mn} - \frac{1}{8\pi^2} R^{mpqr} R^n{}_{pqr} + \text{higher orders}. \tag{11.93}$$

FIGURE 11.9: Left to right: successive stages of the Ricci flow on a two-dimensional manifold.

Note that if we apply this equation in the case of a maximally symmetric space, for which the curvature tensors have simple expressions in terms of the metric tensor, it reduces to

$$\mu \frac{\partial}{\partial \mu} g^{mn} = -\frac{R}{2\pi N} g^{mn} \left[ 1 + \frac{R}{2\pi N(N-1)} + \cdots \right]. \tag{11.94}$$

Thus, in this special case, the metric is rescaled but retains its form under changes of scale (because it is constrained by the isometries of the manifold). On a generic manifold, the scale evolution governed by eq. (11.93) explores a much broader space of metrics. Generally speaking, the renormalization flow tends to expand the regions of negative curvature and to shrink those of positive curvature.

There is an interesting analogy between the renormalization group equation (11.93) and the *Ricci flow*,

$$\partial_\tau g^{mn} = -2 R^{mn}, \tag{11.95}$$

introduced independently in mathematics by Hamilton in 1981 as a tool for studying the geometrical classification of three-dimensional manifolds.[15] In a sketchy way, the idea is to start with a generic metric tensor on the manifold, and to smoothen this metric by evolution with the Ricci flow (the Ricci flow is somewhat analogous to a heat equation that tends to uniformize the temperature distribution). For instance, if the metric evolves into one that has a constant positive curvature, one would have proven that the original manifold is homeomorphic to a sphere. For two-dimensional manifolds, this is indeed what happens: The Ricci flow evolves the metric tensor into one that has a constant scalar curvature (see Figure 11.9), corresponding to one of the three possible geometries. Applications of the Ricci flow to three-dimensional manifolds turned out to be complicated by singularities that develop as the metric

---

[15] In two dimensions, connected manifolds are known to fall into three geometrical classes: flat, spherical or hyperbolic, depending on their curvature. More precisely, any such two-dimensional manifold can be endowed with a metric that has a constant scalar curvature, either null, positive or negative. *Thurston geometrization conjecture* proposed a similar – but much more complicated – classification of three-dimensional manifolds. In particular, this conjecture contains as a special case *Poincaré's conjecture*, stating that every closed simply connected three-dimensional manifold is homeomorphic to a 3-sphere. The geometrization conjecture was proven in 2003 by Perelman, with techniques in which the Ricci flow plays a central role.

evolves, and required additional steps known as "surgery" to excise the singularities. At the time of this writing, there is some speculation about whether the additional terms in eq. (11.93) compared to eq. (11.95) have a regularizing effect that may prevent the appearance of these singularities and thus make the surgical steps unnecessary.

## Exercises

*11.1  Prove the symmetries of the Weinberg operator stated in footnote 6.

*11.2  This exercise discusses various aspects of *Majorana fermions*.

- Show that if the Dirac matrices are purely imaginary, then the Dirac equation has solutions for which the spinor is real. Show that this describes anti-particles that are identical to the particles.
- Check that the following matrices are imaginary and satisfy the Dirac algebra:

$$\gamma^0 \equiv \begin{pmatrix} 0 & \sigma^2 \\ \sigma^2 & 0 \end{pmatrix}, \quad \gamma^1 \equiv \begin{pmatrix} i\sigma^3 & 0 \\ 0 & i\sigma^3 \end{pmatrix}, \quad \gamma^2 \equiv \begin{pmatrix} 0 & -\sigma^2 \\ \sigma^2 & 0 \end{pmatrix}, \quad \gamma^3 \equiv \begin{pmatrix} -i\sigma^1 & 0 \\ 0 & -i\sigma^1 \end{pmatrix}.$$

Is this solution unique?
- In this representation, show that the $\gamma^5$ matrix and the charge conjugation operator are given by

$$\gamma^5 = \begin{pmatrix} \sigma^2 & 0 \\ 0 & -\sigma^2 \end{pmatrix}, \quad C = \gamma^0.$$

From this, deduce that the charge conjugate spinor is $\psi^c = \psi^*$, and that if we restrict to real spinors (check that this reality condition is preserved by Lorentz transformations) then they must describe electrically neutral particles (called Majorana fermions).
- How are the left- and right-handed components of a Majorana fermion related? In particular, show that the charge conjugate of a left-handed spinor is right-handed. Check that $\overline{(\psi_L)^c} = \psi_L^t C^{-1}$ and $\overline{(\psi_R)^c} = \psi_R^t C^{-1}$.
- Construct a mass term that involves only the left-handed spinor.

*11.3  Consider a nucleon moving in the $-z$ direction, described in the color glass condensate effective theory by the color current $J^\mu_a \equiv \delta^{\mu-}\delta(x^+)\rho_a(x_\perp)$ (this form is valid in a frame where the nucleon is very fast). Recall from Exercise 7.8 the solution in Lorenz gauge of the classical Yang–Mills equations with this source. Calculate at tree level the scattering amplitude of a quark off this nucleon. Square it to obtain the cross-section, differential with respect to the transverse momentum of the deflected quark.

11.4  Consider the same nucleon as in the previous exercise, and the scattering of a virtual photon off this nucleon (the results of Section 7.6.6 may be useful).

- Show that at leading order, the color-dependent part of the total cross-section comes from a factor $\mathrm{tr}\,(U(x_\perp)U^\dagger(y_\perp))$, where $U$ is the fundamental representation Wilson line in the color field created by the nucleon color current.
- Consider the strong corrections of order $g^2$ to this result. Draw the contributing diagrams.
- Let us admit that the emission of a gluon in the photon wavefunction amounts to a correction by a factor $2gt^a_f\,(\epsilon_\lambda \cdot k_\perp)/k_\perp^2$, where $k_\perp$ is the transverse momentum of the photon and $\epsilon_\lambda$ its polarization vector. Rewrite this rule in coordinate space.

- Write down the expression of the next-to-leading order corrections, show that they suffer from a divergence in the integration over the longitudinal component of the gluon momentum, and introduce a cutoff to regularize them.
- Interpret this cutoff in the context of the CGC effective theory, and derive an evolution equation for the dependence of $\text{tr}\,(U(\mathbf{x}_\perp)U^\dagger(\mathbf{y}_\perp))$ with respect to this cutoff.

**11.5** Explain why the measure $\sqrt{g}\prod_m D\vartheta^m(x)$ that appears in eq. (11.77) is invariant under changes of coordinates.

**11.6** Derive eq. (11.83). Show that these equations are satisfied by the metric tensor (11.87) in the case of the spontaneously broken $O(n)$ model. Conversely, show that with an $O(n)$ symmetry (the symmetry transformations under consideration define the $T_a^m(\vartheta)$), eq. (11.83) implies the form (11.87) for the metric, up to a rescaling.

**\*11.7** Use the chiral Lagrangian of eq. (11.92) in order to calculate the four-point amplitude $\mathcal{A}_4(1_a 2_b 3_c 4_d)$ at lowest order.

**11.8** Consider the collision of two hadrons or nuclei described by the color glass condensate, with currents $J_1^-, J_2^+$, respectively. The goal of this exercise is to study the solution of the classical Yang–Mills equations (with a null retarded boundary condition at $x^0 = -\infty$). From these currents, define

$$\alpha_1^- \equiv -\frac{1}{\partial_\perp^2} J_1^-,\quad \mathcal{U}_1 \equiv T\exp ig\int dx^+\,\alpha_1^-,\quad \beta_1^i \equiv \tfrac{i}{g}\mathcal{U}_1^\dagger \partial^i \mathcal{U}_1,$$

$$\alpha_2^+ \equiv -\frac{1}{\partial_\perp^2} J_2^+,\quad \mathcal{U}_2 \equiv T\exp ig\int dx^-\,\alpha_2^+,\quad \beta_2^i \equiv \tfrac{i}{g}\mathcal{U}_2^\dagger \partial^i \mathcal{U}_2.$$

- Show by causality arguments that space-time may be divided into four quadrants, only one of which contains a gauge potential that depends on the two projectiles. We will denote these quadrants as follows: bottom ($x^\pm < 0$); left ($x^- > 0, x^+ < 0$); right ($x^+ > 0, x^- < 0$); top ($x^\pm > 0$).
- Consider the gauge $x^+ A^- + x^- A^+ = 0$, known as the *Fock–Schwinger gauge*. What is the main advantage of this gauge concerning the covariant conservation of the current?
- Show that in this gauge, the non-zero components of the gauge field in the left and right quadrants are given by $A_R^i = \beta_1^i$, $A_L^i = \beta_2^i$.
- To map the upper quadrant, introduce *Milne coordinates*, $\tau \equiv \sqrt{2x^+ x^-}$, $\eta \equiv \tfrac{1}{2}\ln(x^+/x^-)$. Show that the gauge condition is also $A^\tau = 0$ (thus, it may be viewed as a temporal gauge if one uses proper time instead of Cartesian time).
- Show that a possible parametrization of the gauge field components in the upper quadrant is $A^+ = x^+\alpha(\tau, \mathbf{x}_\perp)$, $A^- = -x^-\alpha(\tau, \mathbf{x}_\perp)$, $A^i = \beta^i(\tau, \mathbf{x}_\perp)$. What are the Yang–Mills equations in the upper quadrant, when written in terms of $\alpha$ and $\beta^i$?

**11.9** This is a continuation of the previous exercise. Let us admit that the fields $\alpha$ and $\beta^i$ on the surface $x^+ x^- = 0^+$ are given by $\beta^i = \beta_1^i + \beta_2^i$, $\partial_\tau \beta^i = 0$, $\alpha = \tfrac{ig}{2}[\beta_1^i, \beta_2^i]$.

- Show that the chromoelectric and chromomagnetic fields are parallel to the z direction (i.e., to the collision axis).
- Prove that this implies the following structure for the energy–momentum tensor: $T^{\mu\nu} = \text{diag}\,(\epsilon, \epsilon, \epsilon, -\epsilon)$ (one may use eq. (12.123) for the energy–momentum tensor in Yang–Mills theory).

# EXERCISES

- At $x^+ x^- = 0^+$, show that the energy density can be expressed as

$$T^{00}\Big|_{\tau=0^+} = \frac{1}{4} F_a^{ij} F_a^{ij} + 2\alpha_a \alpha_a.$$

- Assuming the following form for the average of the Fourier transform of the fields $\beta^i_{1,2}$ over the fluctuations of the color sources,

$$\langle \beta^i_{ma}(\mathbf{p}_\perp) \beta^j_{nb}(\mathbf{q}_\perp) \rangle \equiv (2\pi)^2 \delta_{mn} \delta_{ab} \delta(\mathbf{p}_\perp + \mathbf{q}_\perp) \frac{p^i_\perp p^j_\perp}{p^2_\perp} G_m(\mathbf{p}_\perp),$$

show that the average initial energy density reads

$$\langle T^{00} \rangle\Big|_{\tau=0^+} = \frac{g^2 N(N^2-1)}{2} \prod_{m=1,2} \int \frac{d^2 \mathbf{p}_\perp}{(2\pi)^2} G_m(\mathbf{p}_\perp).$$

# CHAPTER 12

# Quantum Anomalies

Noether's theorem states that for each continuous symmetry of a classical Lagrangian, there exists a corresponding conserved current. By construction, this conservation law holds at tree level, and a very important question is whether it is preserved by quantum corrections in higher orders of the theory. *Quantum anomalies* are situations in which a classical symmetry is violated by quantum effects. We have already encountered anomalies in Section 6.5, where we saw that the fermionic functional measure is not invariant under chiral transformations of massless fermions, which had interesting connections with the index of the Dirac operator (its zero modes in the presence of an external field).

When such an anomaly arises in a global symmetry like chiral symmetry, its effect is just to introduce a corrective term into the conservation equation of the corresponding current (which may have some physical consequences, however). But when it affects a local gauge symmetry, its effects are devastating, since the renormalizability and unitarity of gauge theories rely on the validity to all orders of the gauge symmetry. In general, gauge theories with an anomalous gauge symmetry do not make sense, and it is therefore of utmost importance to check that no such gauge anomaly is present in theories of phenomenological relevance.

## 12.1 Axial Anomalies in a Gauge Background

### 12.1.1 Two-Dimensional Example: Schwinger Model

The simplest example of theory that exhibits a quantum anomaly is quantum electrodynamics (QED) in two dimensions with massless fermions, also known as the Schwinger model.

## 12.1 AXIAL ANOMALIES IN A GAUGE BACKGROUND

The Lagrangian of this theory reads

$$\mathcal{L} \equiv i\overline{\Psi}\slashed{D}\Psi - \frac{1}{4}F_{\mu\nu}F^{\mu\nu}, \qquad (12.1)$$

where $D_\mu \equiv \partial_\mu - ieA_\mu$ and $F_{\mu\nu} \equiv \partial_\mu A_\nu - \partial_\nu A_\mu$. This Lagrangian is invariant under (local) $U(1)$ transformations,

$$\Psi(x) \to e^{-i\theta(x)}\Psi(x), \qquad (12.2)$$

which, by Noether's theorem, implies the existence of a conserved electromagnetic current,

$$J^\mu \equiv \overline{\Psi}\gamma^\mu\Psi, \quad \partial_\mu J^\mu = 0. \qquad (12.3)$$

(In the following, this current will be called a *vector current*.) Being a gauge symmetry, this invariance is crucial for the unitarity of the theory, since it ensures via Ward–Takahashi identities that the unphysical photons do not contribute as initial or final states of physical amplitudes.

Because the fermions are massless, this theory has another symmetry. In order to see it, let us introduce[1] a two-dimensional $\gamma^5$ matrix,

$$\gamma^5 \equiv \frac{1}{2}\epsilon_{\mu\nu}\gamma^\mu\gamma^\nu = \gamma^0\gamma^1, \qquad (12.4)$$

where $\epsilon_{\mu\nu}$ is the two-dimensional completely antisymmetric tensor, normalized by $\epsilon_{01} = +1$. Using $\gamma^5$, one may decompose $\Psi$ into its left- and right-handed components,

$$\Psi = \Psi_R + \Psi_L, \quad \Psi_R \equiv \frac{1+\gamma^5}{2}\Psi, \quad \Psi_L \equiv \frac{1-\gamma^5}{2}\Psi, \qquad (12.5)$$

and the fermionic part of the Lagrangian can be rewritten as

$$i\overline{\Psi}\slashed{D}\Psi = i\Psi_R^\dagger \gamma^0 \slashed{D}\Psi_R + i\Psi_L^\dagger \gamma^0 \slashed{D}\Psi_L. \qquad (12.6)$$

In other words, the kinetic term does not mix the left- and right-hand components (this would not be true with a mass term). As a consequence, the Lagrangian is invariant if we multiply the left- and right-hand components by independent phases:

$$\Psi_R \to e^{-i\alpha}\Psi_R, \quad \Psi_L \to e^{-i\beta}\Psi_L. \qquad (12.7)$$

Note that this is a global invariance, unlike the gauge symmetry discussed previously. Equivalently, the massless Dirac Lagrangian is invariant under the following global transformation:

$$\Psi \to e^{-i\theta\gamma^5}\Psi, \qquad (12.8)$$

---

[1] More generally, it is possible to define $\gamma^5$ in any even space-time dimension $D = 2r$ as follows:

$$\gamma^5 \equiv \frac{i^{r-1}}{(2r)!}\epsilon_{\mu_1\mu_2\cdots\mu_{2r}}\gamma^{\mu_1}\gamma^{\mu_2}\cdots\gamma^{\mu_{2r}}.$$

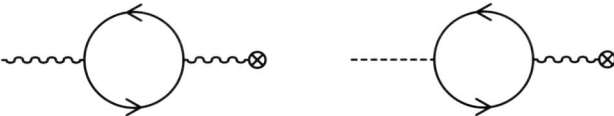

FIGURE 12.1: Left: One-loop contribution to the vector current in a background gauge potential (the wavy line terminated by a cross represents the background field). Right: One-loop contribution to the axial current.

which amounts to multiplying by conjugate phases the left- and right-hand components (because of the $\gamma^5$ in the exponential). Since this is a continuous symmetry, Noether's theorem also applies here and tells us that the *axial current* is conserved:

$$J_5^\mu \equiv \bar\Psi \gamma^\mu \gamma^5 \Psi, \quad \partial_\mu J_5^\mu = 0. \tag{12.9}$$

The conservation laws (eqs. (12.3) and (12.9)) have been obtained with Noether's theorem, from the fact that the *classical* Lagrangian possesses certain continuous symmetries. Let us now study how the vector and axial currents are modified at one loop. Here, we consider a fixed configuration of the gauge potential $A_\mu(x)$, which acts as an external background field (this also means that the photon kinetic term plays no role in this discussion). The lowest order one-loop graphs that contribute to these currents are shown in Figure 12.1. The one-loop expectation values of the vector and axial currents are given by

$$\langle J^\mu(x)\rangle = -\int d^4y\ \text{tr}\left(\gamma^\mu S_F^0(x,y)(-ie\gamma^\nu)S_F^0(y,x)\right) A_\nu(y),$$

$$\langle J_5^\mu(x)\rangle = -\int d^4y\ \text{tr}\left(\gamma^\mu \gamma^5 S_F^0(x,y)(-ie\gamma^\nu)S_F^0(y,x)\right) A_\nu(y), \tag{12.10}$$

where the overall minus sign comes from the fermion loop. The only difference between the two currents is the $\gamma^5$ inside the trace, which comes from the definition of the axial current. These expectation values are easier to calculate in momentum space. After a Fourier transform,

$$\widetilde{J}_{(5)}^\mu(q) \equiv \int d^4x\ e^{-iq\cdot x}\ J_{(5)}^\mu(x), \quad \widetilde{A}_\nu(q) \equiv \int d^4x\ e^{-iq\cdot x}\ A_\nu(x), \tag{12.11}$$

we obtain

$$\langle \widetilde{J}^\mu(q)\rangle = -ie\widetilde{A}_\nu(q) \int \frac{d^D k}{(2\pi)^D} \frac{\text{tr}\left(\gamma^\mu \slashed{k} \gamma^\nu (\slashed{k}+\slashed{q})\right)}{(k^2+i0^+)((k+q)^2+i0^+)},$$

$$\langle \widetilde{J}_5^\mu(q)\rangle = -ie\widetilde{A}_\nu(q) \int \frac{d^D k}{(2\pi)^D} \frac{\text{tr}\left(\gamma^\mu \gamma^5 \slashed{k} \gamma^\nu (\slashed{k}+\slashed{q})\right)}{(k^2+i0^+)((k+q)^2+i0^+)}. \tag{12.12}$$

In order to secure the subsequent manipulations, let us assume that some ultraviolet regularization has been applied to the momentum integrals (we have written them with dimensional regularization in mind, but a different regularization could be used – with some caveats to be discussed below). The denominators can be arranged into a single factor by using Feynman's parametrization,

$$\frac{1}{(k^2+i0^+)((k+q)^2+i0^+)} = \int_0^1 dx\ \frac{1}{(l^2+\Delta(x))^2}, \tag{12.13}$$

## 12.1 AXIAL ANOMALIES IN A GAUGE BACKGROUND

where we have introduced $l \equiv k + x q$ and $\Delta(x) \equiv x(1-x) q^2$. After calculating the trace and dropping terms that are odd in $l$, the integral that appears in the vector current can be written as follows:

$$-ie \int \frac{d^D k}{(2\pi)^D} \frac{\operatorname{tr}(\gamma^\mu \slashed{k} \gamma^\nu (\slashed{k}+\slashed{q}))}{(k^2 + i0^+)((k+q)^2 + i0^+)} = A(q^2) g^{\mu\nu} - B(q^2) \frac{q^\mu q^\nu}{q^2}, \qquad (12.14)$$

where the coefficients $A(q^2)$ and $B(q^2)$ are defined as

$$A(q^2) \equiv -iDe \int \frac{d^D l}{(2\pi)^D} \int_0^1 dx \, \frac{\Delta(x) + \left(\frac{2}{D} - 1\right) l^2}{(l^2 + \Delta(x))^2},$$

$$B(q^2) \equiv -iDe \int \frac{d^D l}{(2\pi)^D} \int_0^1 dx \, \frac{2 \Delta(x)}{(l^2 + \Delta(x))^2}. \qquad (12.15)$$

In $D = 2$ space-time dimensions, the second integral is finite and gives

$$B(q^2) \underset{D=2}{=} -\frac{e}{\pi}, \qquad (12.16)$$

while the first integral is ambiguous. Indeed, the term in $l^2$ in the numerator leads to an ultraviolet divergence, but it is multiplied by the factor $\frac{2}{D} - 1$ that vanishes precisely when $D = 2$. If we used a cutoff as the ultraviolet regulator, this term would vanish and we would have $A = B/2$, which would violate the conservation of the vector current at one loop. In dimensional regularization, in contrast, the factor $\frac{2}{D} - 1$ compensates a pole in $1/(D - 2)$ that comes from evaluating the integral in $D$ dimensions, leaving a finite but non-zero result. In fact, in dimensional regularization we obtain $A = B$ (see Exercise 12.1), and the conservation of the vector current holds at one loop. No matter which regularization procedure we adopt, it must give $A = B$ in order to preserve vector current conservation, i.e., for preserving gauge symmetry at one loop.

Consider now the axial current. From the definition of $\gamma^5$, we get (see Exercise 12.2)

$$\operatorname{tr}(\gamma^5 \gamma^\mu \gamma^\nu) = -D \, \epsilon^{\mu\nu} \qquad (12.17)$$

and

$$\operatorname{tr}(\gamma^5 \gamma^\mu \slashed{A} \gamma^\nu \slashed{B}) = A^\nu \operatorname{tr}(\gamma^5 \gamma^\mu \slashed{B}) + B^\nu \operatorname{tr}(\gamma^5 \gamma^\mu \slashed{A}) - A \cdot B \operatorname{tr}(\gamma^5 \gamma^\mu \gamma^\nu)$$
$$= -D \, \epsilon^{\mu\sigma} \Big[ B_\sigma A^\nu + A_\sigma B^\nu - (A \cdot B) g_\sigma{}^\nu \Big]. \qquad (12.18)$$

This identity leads to

$$-ie \int \frac{d^D k}{(2\pi)^D} \frac{\operatorname{tr}(\gamma^\mu \gamma^5 \slashed{k} \gamma^\nu (\slashed{k}+\slashed{q}))}{(k^2 + i0^+)((k+q)^2 + i0^+)} = \epsilon^{\mu\sigma} \Big[ A(q^2) g_\sigma{}^\nu - B(q^2) \frac{q_\sigma q^\nu}{q^2} \Big], \qquad (12.19)$$

where $A$ and $B$ are the same coefficients as in eq. (12.14). Therefore, the divergence of the axial current is given by

$$q_\mu \langle \tilde{J}_5^\mu(q) \rangle = A(q^2) \, \epsilon^{\mu\nu} \, q_\mu \, \tilde{A}_\nu(q). \qquad (12.20)$$

If we have adopted a regularization that preserves gauge symmetry, i.e., such that A = B, this divergence is non-zero and reads

$$q_\mu \langle \widetilde{J}_5^\mu(q) \rangle = -\frac{e}{\pi} \epsilon^{\mu\nu} q_\mu \widetilde{A}_\nu(q), \qquad (12.21)$$

or, going back to coordinate space,

$$\partial_\mu \langle J_5^\mu(x) \rangle = -\frac{e}{\pi} \epsilon^{\mu\nu} \partial_\mu A_\nu(x) = -\frac{e}{2\pi} \epsilon^{\mu\nu} F_{\mu\nu}(x). \qquad (12.22)$$

The non-conservation of the axial current at one loop is the unavoidable conclusion in any regularization scheme that preserves the conservation of the vector current. Moreover, since when this is the case A becomes equal to the ultraviolet finite coefficient B, it does not suffer from any scheme dependence, and the above result may thus be viewed as a scheme-free result. The result (12.22) is known as an *axial anomaly*. A somewhat milder conclusion of this two-dimensional exercise is that it is not possible to preserve both vector and axial current conservation at one loop. We could in principle adopt a regularization scheme that conserves the axial current, which requires A = 0. But the price to pay would be the loss of gauge invariance at one loop. Since gauge invariance is deemed more fundamental (in particular, it ensures the unitarity of the theory), this route is generally not considered further.

Note that ultraviolet divergences are necessary[2] for the existence of this anomaly. Indeed, at the classical level, the Lagrangian density is invariant under the global transformation:

$$\Psi \to e^{-i\theta \gamma^5}\Psi, \quad \Psi^\dagger \to \Psi^\dagger e^{i\theta \gamma^5}. \qquad (12.23)$$

The Feynman graphs that contribute to the expectation value of the axial current in a background electromagnetic field have an equal number of $\Psi$ and $\Psi^\dagger$ (this statement is true to all orders of perturbation theory). Since the axial symmetry is global, when we apply the above axial transformation to a graph, all the factors $\exp(\pm i\theta\gamma^5)$ should naively cancel, leaving a result that does not depend on $\theta$. This conclusion would indeed be correct if all the integrals were finite, but may be invalidated by the subtraction procedure necessary to obtain finite results in the presence of divergences. In the explicit example that we have studied, the ultraviolet regularizations that are consistent with gauge symmetry all spoil axial symmetry.

Beyond one loop, a graph contributing to the expectation value of the axial current may contain subgraphs that are ultraviolet divergent. However, since QED is renormalizable, these sub-divergences will all have been made finite thanks to counterterms calculated in the previous orders of the perturbative expansion. Thus, we need only to study the intrinsic ultraviolet divergence of the graph under consideration, an indicator of which is given by its superficial degree of divergence. For the sake of definiteness, let us assume that the graph $\mathcal{G}$ has $n_\psi$ fermion propagators, $n_\gamma$ photon propagators, $n_V$ photon–fermion–fermion vertices, $n_A$ insertions of the external electromagnetic field and $n_L$ loops (plus one extra vertex where

---

[2]In a certain sense, the axial anomaly is also an infrared effect since it exists only for massless fermions (for massive fermions, there is no axial symmetry to begin with). Moreover, as we have already seen when discussing the Atiyah–Singer index theorem, the axial anomaly is related to the zero modes of the Dirac operator in a background field.

the axial current is attached). These quantities are not independent, but obey the following identities:

$$2n_\gamma = n_V,$$
$$2n_\psi = 2 + 2(n_V + n_A),$$
$$n_L = n_\psi + n_\gamma - n_A - n_V. \tag{12.24}$$

Using these relations, the superficial degree of divergence of the graph reads

$$\omega(\mathcal{G}) \equiv 2n_L - n_\psi - 2n_\gamma = 2 - n_\psi. \tag{12.25}$$

The simplest graph that contributes to the axial current, shown in Figure 12.1, has $n_\psi = 2$ and therefore has a logarithmic ultraviolet divergence. More complicated graphs, either with more insertions of the external field or with more than one loop, all have $n_\psi > 2$ and are therefore convergent after all their sub-divergences have been subtracted. This argument, although it lacks some rigor, indicates that the axial anomaly does not receive any correction beyond the one-loop result, and that eq. (12.22) is therefore an exact result. An alternate justification of this property is based on the derivation of the axial anomaly from the fermionic path integral, which gives the determinant of the Dirac operator in the background field. Indeed, as we saw in Section 5.4, functional determinants correspond to one-loop diagrams.

## 12.1.2 Axial Anomaly in Four Dimensions

**$\gamma^5$ in four dimensions:** Let us now turn to a more realistic four-dimensional example that also has some relevance in understanding the decay of pseudo-scalar mesons like the $\pi^0$. The setup is exactly the same as in the previous section, except that we now consider four space-time dimensions. The main modification is the definition of the $\gamma^5$ matrix:

$$\gamma^5 \underset{D=4}{=} \frac{i}{4!} \epsilon_{\mu\nu\rho\sigma} \gamma^\mu \gamma^\nu \gamma^\rho \gamma^\sigma = i\gamma^0 \gamma^1 \gamma^2 \gamma^3. \tag{12.26}$$

The traces of a $\gamma^5$ with any odd number of ordinary Dirac matrices are all zero:

$$\mathrm{tr}\,(\gamma^5 \gamma^{\mu_1} \cdots \gamma^{\mu_{2n+1}}) = 0. \tag{12.27}$$

In order to evaluate the traces of $\gamma^5$ with an even number of Dirac matrices, let us first recall the general formula for a trace of an even number of Dirac matrices,

$$\mathrm{tr}\,(\gamma^{\mu_1} \cdots \gamma^{\mu_{2n}}) = D \sum_{\text{pairings } \mathcal{P}} \mathrm{sign}\,(\mathcal{P}) \prod_{s \in \mathcal{P}} g^{\mu_{s_1} \mu_{s_2}}, \tag{12.28}$$

where a pairing $\mathcal{P}$ is a set of pairs $\mathcal{P} = \{(s_1 s_2), (s'_1 s'_2), \cdots\}$ made of the integers in $[1, 2n]$. The signature of $\mathcal{P}$, denoted $\mathrm{sign}\,(\mathcal{P})$, is the signature of the permutation that reorders the sequence $s_1 s_2 s'_1 s'_2 \cdots$ into $1234 \cdots$. Since the Minkowski metric tensor $g^{\mu\nu}$ is diagonal, each Lorentz index carried by one of the Dirac matrices must coincide with the Lorentz index of another matrix in order to obtain a non-vanishing result. Hence, we have

$$\mathrm{tr}\,(\gamma^5) = i\,\mathrm{tr}\,(\gamma^0 \gamma^1 \gamma^2 \gamma^3) = 0. \tag{12.29}$$

The same is true if the $\gamma^5$ is accompanied by only two ordinary Dirac matrices,

$$\text{tr}\left(\gamma^5\gamma^\mu\gamma^\nu\right) = i\,\text{tr}\left(\gamma^0\gamma^1\gamma^2\gamma^3\gamma^\mu\gamma^\nu\right) = 0, \tag{12.30}$$

and the simplest non-zero trace is $\text{tr}\left(\gamma^5\gamma^\mu\gamma^\nu\gamma^\rho\gamma^\sigma\right)$. By the previous argument, each of the indices $\mu\nu\rho\sigma$ must match one of the indices 0123 hidden in $\gamma^5 = i\gamma^0\gamma^1\gamma^2\gamma^3$. Therefore, $\mu\nu\rho\sigma$ must be a permutation of 0123. Since the four Dirac matrices are all distinct, they all anti-commute, and the result is completely antisymmetric in $\mu\nu\rho\sigma$, so that we have

$$\text{tr}\left(\gamma^5\gamma^\mu\gamma^\nu\gamma^\rho\gamma^\sigma\right) = A\,\epsilon^{\mu\nu\rho\sigma}. \tag{12.31}$$

In order to calculate the prefactor, we just need to evaluate the trace for a particular assignment of the indices, for instance $\mu\nu\rho\sigma = 3210$,

$$A\,\underbrace{\epsilon^{3210}}_{-1} = \text{tr}\left(\gamma^5\gamma^3\gamma^2\gamma^1\gamma^0\right) = i\,\text{tr}\left(\gamma^0\gamma^1\gamma^2\underbrace{\gamma^3\gamma^3}_{+1}\gamma^2\gamma^1\gamma^0\right) = -4i. \tag{12.32}$$

This gives $A = 4i$, i.e.,

$$\text{tr}\left(\gamma^5\gamma^\mu\gamma^\nu\gamma^\rho\gamma^\sigma\right) = 4i\,\epsilon^{\mu\nu\rho\sigma}. \tag{12.33}$$

**Order 1 in the external field:** Let us now turn to the calculation of the expectation value of the axial current in four dimensions. The simplest graph to consider is again the graph on the right of Figure 12.1. Its contribution to the axial current is

$$\langle \widetilde{J}_5^\mu(q)\rangle = -ie\,\widetilde{A}_\nu(q)\underbrace{\int\frac{d^D k}{(2\pi)^D}\frac{\text{tr}\left(\gamma^\mu\gamma^5\slashed{k}\gamma^\nu(\slashed{k}+\slashed{q})\right)}{(k^2+i0^+)((k+q)^2+i0^+)}}_{\mathcal{J}^{\mu\nu}}. \tag{12.34}$$

By introducing a Feynman parametrization to handle the denominators, and the new integration variable $l \equiv k + xq$, the integral can be rewritten as

$$\mathcal{J}^{\mu\nu} = \int\frac{d^D l}{(2\pi)^D}\int_0^1 dx\,\frac{\text{tr}\left(\gamma^\mu\gamma^5(\slashed{l}-x\slashed{q})\gamma^\nu(\slashed{l}+(1-x)\slashed{q})\right)}{(l^2+\Delta(x))^2}. \tag{12.35}$$

The trace that appears in the numerator is proportional to

$$\epsilon^{\mu\alpha\nu\beta}(l-xq)_\alpha(l+(1-x)q)_\beta \propto \epsilon^{\mu\alpha\nu\beta}l_\alpha q_\beta, \tag{12.36}$$

and is thus odd in the momentum $l$. Therefore, the momentum integral vanishes, and this graph does not contribute to the axial current.

## 12.1 AXIAL ANOMALIES IN A GAUGE BACKGROUND

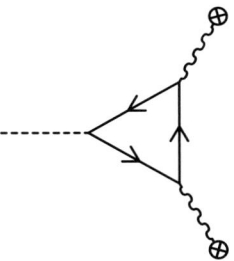

FIGURE 12.2: Graph contributing to the chiral anomaly in a gauge background in four space-time dimensions.

**Order 2 in the external field:** At second order in the external field, we encounter the graph shown in Figure 12.2. In coordinate space, its contribution to the expectation value of the axial current reads

$$\langle J_5^\mu(x) \rangle = -(-ie)^2 \int d^4y\, d^4z\, \text{tr}\left(\gamma^\mu \gamma^5 S_F^0(x,y) \gamma^\nu S_F^0(y,z) \gamma^\rho S_F^0(z,x)\right) A_\nu(y) A_\rho(z). \tag{12.37}$$

Applying a Fourier transform gives

$$\langle \tilde{J}_5^\mu(q) \rangle = -ie^2 \int \frac{d^4k_1 d^4k_2 d^4k}{(2\pi)^{12}} (2\pi)^4 \delta(q + k_1 + k_2)\, \tilde{A}_\nu(-k_1)\tilde{A}_\rho(-k_2)$$
$$\times \frac{\text{tr}\left(\gamma^\mu \gamma^5 (\slashed{k}+\slashed{k}_1) \gamma^\nu \slashed{k} \gamma^\rho (\slashed{k}-\slashed{k}_2)\right)}{((k+k_1)^2 + i0^+)(k^2 + i0^+)((k-k_2)^2 + i0^+)}. \tag{12.38}$$

For later convenience, it is useful to symmetrize the integrand in the variables $k_{1,2}$, and to apply separate shifts to the integration momentum k in the two terms resulting from the symmetrization. This leads to

$$\langle \tilde{J}_5^\mu(q) \rangle = \frac{1}{2!} \int \frac{d^4k_1 d^4k_2}{(2\pi)^8} (2\pi)^4 \delta(q + k_1 + k_2)$$
$$\times \Gamma_5^{\mu\nu\rho}(q, k_1, k_2)\, \tilde{A}_\nu(-k_1)\tilde{A}_\rho(-k_2), \tag{12.39}$$

with

$$\Gamma_5^{\mu\nu\rho}(q, k_1, k_2) \equiv$$
$$\equiv -ie^2 \int \frac{d^4k}{(2\pi)^4} \frac{\text{tr}\left(\gamma^\mu \gamma^5 (\slashed{k}+\slashed{a}+\slashed{k}_1)\gamma^\nu (\slashed{k}+\slashed{a})\gamma^\rho(\slashed{k}+\slashed{a}-\slashed{k}_2)\right)}{((k+a+k_1)^2+i0^+)((k+a)^2+i0^+)((k+a-k_2)^2+i0^+)}$$
$$- ie^2 \int \frac{d^4k}{(2\pi)^4} \frac{\text{tr}\left(\gamma^\mu \gamma^5 (\slashed{k}+\slashed{b}+\slashed{k}_2)\gamma^\rho (\slashed{k}+\slashed{b})\gamma^\nu(\slashed{k}+\slashed{b}-\slashed{k}_1)\right)}{((k+b+k_2)^2+i0^+)((k+b)^2+i0^+)((k+b-k_1)^2+i0^+)}. \tag{12.40}$$

The shifts $k \to k+a$, $k \to k+b$ would of course have no effect on convergent integrals, since they just correspond to a change of integration variable. However, we are here in the presence of linearly divergent integrals, and these shifts have a non-trivial interplay with the ultraviolet regularization. Note that since $\{\gamma^5, \gamma^\alpha\} = 0$, we may move the $\gamma^5$ just after the matrices $\gamma^\nu$ or $\gamma^\rho$ without changing the integrand, as if the axial current was attached at the other summits of the triangle (where the momenta $k_1$ or $k_2$ enter, respectively).

Next, in order to test the conservation of the axial current, we contract this amplitude with $q_\mu$, that we may rewrite as follows:

$$\begin{aligned} q_\mu &= -(k_1 + k_2)_\mu = (k + a - k_2)_\mu - (k + a + k_1)_\mu \\ &= (k + b - k_1)_\mu - (k + b + k_2)_\mu. \end{aligned} \tag{12.41}$$

This leads to

$$q_\mu \Gamma_5^{\mu\nu\rho}(q, k_1, k_2) = 4e^2 \int \frac{d^4k}{(2\pi)^4} \, \epsilon^{\alpha\nu\beta\rho}$$
$$\times \left\{ \frac{(k_1)_\alpha (k+a)_\beta}{((k+a)^2 + i0^+)((k+a+k_1)^2 + i0^+)} + \frac{(k_2)_\alpha (k+a)_\beta}{((k+a)^2 + i0^+)((k+a-k_2)^2 + i0^+)} \right.$$
$$\left. - \frac{(k_1)_\alpha (k+b)_\beta}{((k+b)^2 + i0^+)((k+b-k_1)^2 + i0^+)} - \frac{(k_2)_\alpha (k+b)_\beta}{((k+b)^2 + i0^+)((k+b+k_2)^2 + i0^+)} \right\}. \tag{12.42}$$

By taking $a = b = 0$, and assuming a regularization that preserves Lorentz invariance, each term leads to a vanishing integral. Consider, for instance, the first term. Since $k_1$ is the only 4-vector that enters in the integrand besides the integration variable $k$, the result of its integral is proportional to $\epsilon^{\alpha\nu\beta\rho}(k_1)_\alpha (k_1)_\beta = 0$. Since the same reasoning applies to the four terms, we would therefore naively conclude that the axial current is conserved. However, we should make sure that the vector currents are also conserved. For this, we also need to calculate $(k_1)_\nu \Gamma_5^{\mu\nu\rho}$ and $(k_2)_\rho \Gamma_5^{\mu\nu\rho}$. The same method as above gives

$$(k_1)_\nu \Gamma_5^{\mu\nu\rho}(q, k_1, k_2) = -4e^2 \int \frac{d^4k}{(2\pi)^4} \, \epsilon^{\alpha\mu\beta\rho}$$
$$\times \left\{ \frac{(k+a)_\alpha (k+a-k_2)_\beta}{((k+a)^2 + i0^+)((k+a-k_2)^2 + i0^+)} - \frac{(k+a+k_1)_\alpha (k+a-k_2)_\beta}{((k+a+k_1)^2 + i0^+)((k+a-k_2)^2 + i0^+)} \right.$$
$$\left. + \frac{(k+b+k_2)_\alpha (k+b-k_1)_\beta}{((k+b+k_2)^2 + i0^+)((k+b-k_1)^2 + i0^+)} - \frac{(k+b+k_2)_\alpha (k+b)_\beta}{((k+b)^2 + i0^+)((k+b+k_2)^2 + i0^+)} \right\} \tag{12.43}$$

and

$$(k_2)_\rho \Gamma_5^{\mu\nu\rho}(q, k_1, k_2) = -4e^2 \int \frac{d^4k}{(2\pi)^4} \, \epsilon^{\alpha\mu\beta\nu}$$
$$\times \left\{ \frac{(k+a+k_1)_\alpha (k+a-k_2)_\beta}{((k+a+k_1)^2 + i0^+)((k+a-k_2)^2 + i0^+)} - \frac{(k+a+k_1)_\alpha (k+a)_\beta}{((k+a+k_1)^2 + i0^+)((k+a)^2 + i0^+)} \right.$$
$$\left. + \frac{(k+b)_\alpha (k+b-k_1)_\beta}{((k+b)^2 + i0^+)((k+b-k_1)^2 + i0^+)} - \frac{(k+b+k_2)_\alpha (k+b-k_1)_\beta}{((k+b+k_2)^2 + i0^+)((k+b-k_1)^2 + i0^+)} \right\}. \tag{12.44}$$

## 12.1 AXIAL ANOMALIES IN A GAUGE BACKGROUND

It turns out that the choice $a = b = 0$ leads to non-vanishing results for the conservation of the vector currents. Consider, for instance, $(k_1)_\nu \Gamma_5^{\mu\nu\rho}$. With $a = b = 0$ and a regularization that preserves Lorentz invariance as well as reflection symmetry $k \to -k$, we have

$$(k_1)_\nu \Gamma_5^{\mu\nu\rho}(q, k_1, k_2) = -8e^2 \int \frac{d^4k}{(2\pi)^4} \, \epsilon^{\alpha\mu\beta\rho} \frac{(k+k_2)_\alpha (k-k_1)_\beta}{((k+k_2)^2 + i0^+)((k-k_1)^2 + i0^+)}$$
$$\propto \epsilon^{\alpha\mu\beta\rho}(k_2)_\alpha (k_1)_\beta \neq 0. \qquad (12.45)$$

A systematic search (see Exercise 12.3) indicates that the only choice of a and b that gives a null result for both eqs. (12.43) and (12.44) is

$$a - b = 2(k_2 - k_1). \qquad (12.46)$$

Note that shifting $k$ simultaneously in the four terms of eqs. (12.42), (12.43) and (12.44) amounts to shifting a, b with $a-b$ fixed. Therefore, we could in fact choose $a = -b = k_2 - k_1$. Since the conservation of the vector current is necessary in order to preserve gauge symmetry, and the latter is a requirement for unitarity, we must adopt this choice. Returning to eq. (12.42) for the axial current with these values of a and b, we obtain

$$q_\mu \Gamma_5^{\mu\nu\rho}(q, k_1, k_2) = 16e^2 \int \frac{d^4k}{(2\pi)^4} \, \epsilon^{\alpha\nu\beta\rho} \frac{(k_1)_\alpha}{(k+k_2)^2 + i0^+} \frac{(k+k_2-k_1)_\beta}{(k+k_2-k_1)^2 + i0^+}. \qquad (12.47)$$

Let us define

$$F^{\nu\rho}(k) \equiv \epsilon^{\alpha\nu\beta\rho} \frac{(k_1)_\alpha}{k^2 + i0^+} \frac{(k-k_1)_\beta}{(k-k_1)^2 + i0^+}, \qquad (12.48)$$

and note that

$$\int \frac{d^4k}{(2\pi)^4} F^{\nu\rho}(k) = 0 \qquad (12.49)$$

(because with a Lorentz invariant regularization the result can only depend on the vector $k_1$, which would unavoidably give zero when contracted with the two free slots of the Levi-Civita symbol $\epsilon^{\alpha\nu\beta\rho}$.) Therefore, we can write

$$q_\mu \Gamma_5^{\mu\nu\rho}(q, k_1, k_2) = 16e^2 \int \frac{d^4k}{(2\pi)^4} \left[ F^{\nu\rho}(k + k_2) - F^{\nu\rho}(k) \right]$$
$$= 16e^2 \int \frac{d^4k}{(2\pi)^4} \left[ k_2^\sigma \frac{\partial F^{\nu\rho}(k)}{\partial k^\sigma} + \frac{k_2^\sigma k_2^\tau}{2} \frac{\partial^2 F^{\nu\rho}(k)}{\partial k^\sigma \partial k^\tau} + \cdots \right]. \qquad (12.50)$$

Since the integrand now contains only derivatives, we can use the Stokes theorem in order to rewrite the divergence of the axial current as a surface integral on the boundary at infinity of momentum space. If we view this boundary as the limit $k_* \to \infty$ of a sphere of radius $k_*$, the "area" of this boundary grows like $k_*^3$ in four dimensions. On the other hand, the function $F^{\nu\rho}(k)$ behaves as $k^{-3}$ at large $k$, and each subsequent derivative decreases faster by one additional power of $k^{-1}$. Therefore, the result is given in full by the first term of the expansion,

$$\begin{aligned}q_\mu \Gamma_5^{\mu\nu\rho}(q,k_1,k_2) &= 16e^2 \int \frac{d^4k}{(2\pi)^4} k_2^\sigma \frac{\partial F^{\nu\rho}(k)}{\partial k^\sigma}\\ &= \frac{16ie^2}{(2\pi)^4} \epsilon^{\alpha\nu\beta\rho}(k_1)_\alpha (k_2)^\sigma \lim_{k_*\to\infty} \underbrace{\int_{S_3(k_*)} d^3S \frac{k_\sigma}{k}\frac{k_\beta}{k^4}}_{\frac{\pi^2 g_{\sigma\beta}}{2}}\\ &= i\frac{e^2}{2\pi^2} \epsilon^{\alpha\nu\beta\rho}(k_1)_\alpha (k_2)_\beta, \end{aligned} \quad (12.51)$$

In the second line, $S_3(k_*)$ is the 3-sphere of radius $k_*$ (i.e., the boundary of a 4-ball of radius $k_*$), $k_\sigma/k$ is the unit vector normal to the sphere and the factor $i$ arises when going to Euclidean momentum space. Note that we have anticipated the limit $k \to \infty$ in order to simplify the function $F^{\nu\rho}(k)$ by keeping only its asymptotic form. Therefore, the contribution of the triangle graph to the divergence of the axial current reads

$$\begin{aligned}q_\mu \langle \widetilde{J}_5^\mu(q)\rangle = i\frac{e^2}{4\pi^2} \epsilon^{\alpha\nu\beta\rho} \int \frac{d^4k_1 d^4k_2}{(2\pi)^4} \delta(q+k_1+k_2)\\ \times (k_1)_\alpha (k_2)_\beta \widetilde{A}_\nu(-k_1)\widetilde{A}_\rho(-k_2),\end{aligned} \quad (12.52)$$

or in coordinate space:[3]

$$\partial_\mu \langle J_5^\mu(x)\rangle = \frac{e^2}{16\pi^2} \epsilon^{\alpha\nu\beta\rho} F_{\alpha\nu}(x) F_{\beta\rho}(x). \quad (12.53)$$

This is the main result of this section, namely the existence of an anomalous divergence of the axial current in the presence of a background electromagnetic field. In the course of the calculation, we have seen that depending on the labeling of the integration momentum, we can make the anomaly appear in any of the three external currents. In the situation considered here, with one axial current corresponding to a global symmetry, and two vector currents stemming from a local gauge symmetry, we must enforce the conservation of the vector currents and therefore assign in full the anomaly to the axial one. But the same calculation would arise in the context of a chiral gauge theory (where the left- and right-handed fermions belong to different representations of the gauge group). In this case, the natural choice would be to regularize the triangle so that the symmetry among the three currents is preserved, and the anomaly would then be equally shared by the three currents.

**Corrections:** Let us now discuss potential corrections to the result (12.53). First, we should examine one-loop graphs with more than two photons in addition to the insertion of the axial

---

[3]This result is identical to eq. (6.94), obtained by studying the variation of the fermionic functional measure under chiral transformations. The relative minus sign between eqs. (6.94) and (12.53) stems from the fact that the axial current was defined with the opposite sign in Section 6.5.5.

current. A simple dimensional argument can exclude that such graphs contribute to the divergence of the axial current. Indeed, $\partial_\mu J_5^\mu$ has mass dimension four. In an Abelian gauge theory, each external photon must appear on the right-hand side in the form of the field strength $F^{\mu\nu}$, which has mass dimension two. A local operator with n photons would thus have mass dimension 2n, and require a prefactor of mass dimension $4 - 2n$ to be a valid contribution to the divergence of the axial current. But since the fermions we are considering are massless and the coupling constant is dimensionless in four dimensions, there is no dimensionful parameter in the theory for making up such a prefactor.

Let us now consider higher loop corrections. From the calculation that led to eq. (12.53), the anomaly results from the integration over the momentum that runs in the fermion loop, provided that the integrand has mass dimension four or higher. Note that some of the higher-order corrections just renormalize the objects that appear on the right-hand side of eq. (12.53), such as the photon field strength and the coupling constant, without changing the structure of the anomaly (including the numerical prefactor). Quite generally, however, adding an internal photon line requires adding more fermion propagators in the main loop, which reduces its degree of ultraviolet divergence. Of course, the integration over the momentum of this internal photon may itself be ultraviolet divergent, but it can be regularized in a way that does not interfere with axial symmetry and thus does not contribute to the anomaly.

## 12.2 Generalizations

### 12.2.1 Axial Anomaly in a Non-Abelian Background

In the previous section, we have discussed axial anomalies in an Abelian gauge theory. However, a similar anomaly arises in the presence of a non-Abelian background gauge field. Let us assume that the fermions are in a representation of the gauge algebra where the generators are $t^a$. The calculation of the triangle graph proceeds almost in the same way as in the Abelian case, except for the Lie algebra generators, and eq. (12.53) becomes

$$\partial_\mu \langle J_5^\mu(x) \rangle = \frac{e^2}{4\pi^2} \, \text{tr}\left(t^a t^b\right) \epsilon^{\alpha\nu\beta\rho} \left(\partial_\alpha A_\nu^a(x)\right) \left(\partial_\beta A_\rho^b(x)\right). \tag{12.54}$$

This is not gauge invariant, but it is easy to guess what should be the right-hand side to restore gauge invariance:

$$\partial_\mu \langle J_5^\mu(x) \rangle = \frac{e^2}{16\pi^2} \, \text{tr}\left(t^a t^b\right) \epsilon^{\alpha\nu\beta\rho} F_{\alpha\nu}^a(x) F_{\beta\rho}^b(x). \tag{12.55}$$

The same dimensional argument that we have used in the Abelian case also applies here: There cannot be contributions to the anomaly of degree higher than two in the field strength. Note that when expanded in terms of the gauge potential $A_\mu^a$, eq. (12.55) contains terms of degree three and four, which exist only in a non-Abelian background. Diagrammatically, they correspond to contributions coming from the following two diagrams:

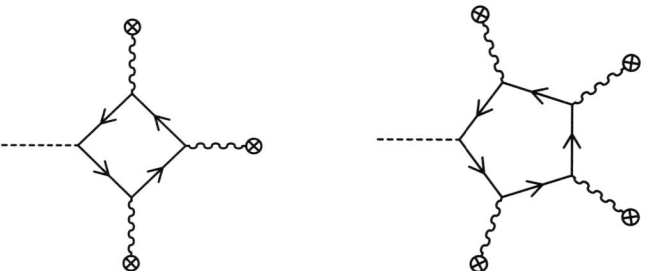

(But the direct extraction of the anomaly contained in these graphs would be very cumbersome due to the numerous terms arising from permutations of the external gauge fields.)

## 12.2.2  Axial Anomaly in a Gravitational Background

Another situation in which an axial anomaly is present is the case of a gravitational background. Of course, this is to a large extent an academic question since the resulting anomaly is extremely small due to the weakness of the gravitational coupling at the usual scales of particle physics. Nevertheless, since every field is in principle coupled to gravity, the anomalies caused by a gravitational background are unavoidable unless the matter fields of the theory are arranged in a specific way. Interestingly, the calculation of this gravitational anomaly can be performed even if we do not have a consistent quantum theory of gravity, since it does not involve quantum fluctuations of the gravitational field (the only loop is a fermion loop).

At tree level, the couplings between gravity and ordinary fields are determined from the principle of general covariance. Let us sketch here how such a calculation is done, without entering into too many technical details. The first step is to obtain a generally covariant generalization of the Dirac operator for an arbitrary metric tensor $g^{\mu\nu}$, from which we can read off the coupling of the fermion to the background gravitational field. In a curved space-time, we wish to generalize the Dirac matrices so that they satisfy

$$\{\gamma^\mu(x), \gamma^\nu(x)\} = 2\, g^{\mu\nu}(x). \tag{12.56}$$

(In this section, we use the Greek letters $\mu$, $\nu$, $\rho$, $\sigma$ for indices related to curved coordinates, and Greek letters from the beginning of the alphabet $\alpha$, $\beta$, $\gamma$, $\delta$ for indices related to flat Minkowski coordinates.) In a curved space-time, the covariant derivative of the metric tensor vanishes, and it is therefore natural to request the same for the Dirac matrices. However, this requires that we introduce a *spin connection*, which is a matrix $\Gamma_\mu$ defined so that

$$\nabla_\mu \gamma_\nu \equiv \partial_\mu \gamma_\nu - \Gamma^\lambda_{\mu\nu} \gamma_\lambda - \Gamma_\mu \gamma_\nu + \gamma_\nu \Gamma_\mu = 0, \tag{12.57}$$

where $\Gamma^\lambda_{\mu\nu}$ is the usual Christoffel's symbol. The covariant derivative acting on a spinor is $(\partial_\mu - \Gamma_\mu)\Psi$ and the generally covariant Dirac equation for a massless fermion reads

$$i\gamma^\mu (\partial_\mu - \Gamma_\mu)\Psi = 0. \tag{12.58}$$

## 12.2 GENERALIZATIONS

In order to construct a Lagrangian that transforms as a scalar, we need a matrix $\Gamma$ such that $\psi^\dagger \Gamma \psi$ is a real scalar. This is the case if the following conditions are satisfied:

$$\Gamma = \Gamma^\dagger,$$
$$\Gamma \gamma^\mu = \gamma^{\mu\dagger} \Gamma,$$
$$\nabla_\mu \Gamma = \partial_\mu \Gamma + \Gamma_\mu^\dagger \Gamma + \Gamma \Gamma_\mu. \tag{12.59}$$

We then define $\overline{\Psi} \equiv \Psi^\dagger \Gamma$, and the Lagrangian density is

$$\mathcal{L} \equiv i \sqrt{-g}\, \overline{\Psi} \gamma^\mu \nabla_\mu \Psi. \tag{12.60}$$

(g is the determinant of the metric tensor.) The vector current and its conservation law generalize into

$$J^\mu \equiv \overline{\Psi} \gamma^\mu \Psi, \quad \nabla_\mu J^\mu = 0. \tag{12.61}$$

In the massless case, we can in addition define a conserved axial current:

$$J_5^\mu \equiv \overline{\Psi} \gamma^\mu \gamma^5 \Psi, \quad \nabla_\mu J_5^\mu = 0. \tag{12.62}$$

However, as we shall see, this conservation law suffers from an anomaly in a curved space-time. First, let us introduce a representation of the Dirac matrices for a generic curved space-time, which makes an explicit connection with the metric tensor. This is achieved by introducing four vector fields $e^\alpha{}_\mu(x)$ (called a *vierbein*, or *tetrad*) such that[4]

$$g_{\mu\nu}(x) = \eta_{\alpha\beta}\, e^\alpha{}_\mu(x)\, e^\beta{}_\nu(x), \tag{12.63}$$

where in this section we use the notation $\eta_{\alpha\beta}$ for the Minkowski metric tensor (see Exercise 12.4 for an explicit example). This is equivalent to introducing at each point x a local Minkowski frame with coordinates $y^\alpha$. Note that $e^\alpha{}_\mu$ transforms as a vector under diffeomorphisms (a *coordinate vector*) with respect to the index $\mu$, and as an ordinary 4-vector under Lorentz transformations (called a *tetrad vector* in this context) with respect to the index $\alpha$. The indices $\alpha, \beta, \cdots$ are raised and lowered with the Minkowski metric tensor, while the indices $\mu, \nu, \cdots$ are raised and lowered with the curved space metric $g_{\mu\nu}(x)$. Since in the right-hand side of eq. (12.63) the indices $\alpha$ and $\beta$ are contracted with the Lorentz tensor $\eta_{\alpha\beta}$, the result is a scalar under Lorentz transformations, but a rank-2 tensor under diffeomorphisms. The Dirac matrices in curved space-time ($\gamma^\mu(x)$) can then be related to those in flat space-time ($\gamma^\alpha$) by

$$\gamma^\mu(x) = e_\alpha{}^\mu(x)\, \gamma^\alpha, \tag{12.64}$$

and a spin connection $\Gamma_\mu$ that satisfies eq. (12.57) (and reduces to zero in flat space-time) is given by

$$\Gamma_\mu(x) = -\frac{1}{8} [\gamma_\alpha, \gamma_\beta]\, e^{\alpha\rho}(x)\, \nabla_\mu e^\beta{}_\rho(x), \tag{12.65}$$

---

[4] In this section, we denote $\eta_{\alpha\beta} \equiv \text{diag}(1, -1, -1, -1)$ the flat space-time Minkowski metric, in order to distinguish it from $g_{\mu\nu}$.

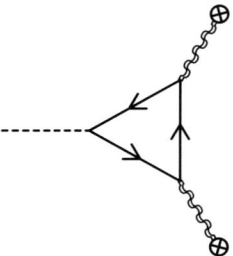

FIGURE 12.3: Graph contributing to the chiral anomaly in a gravitational background in four space-time dimensions.

with $\nabla_\mu e^\beta{}_\rho = \partial_\mu e^\beta{}_\rho - \Gamma^\gamma_{\mu\rho} e^\beta{}_\nu$ (since $e^\beta{}_\rho$ is a coordinate vector with respect to the index $\rho$). A matrix $\Gamma$ that fulfills eq. (12.59) is the flat space-time $\gamma^0$, and the matrix $\gamma^5$ is still given in terms of the flat space-time Dirac matrices by $\gamma^5 = i\gamma^0 \gamma^1 \gamma^2 \gamma^3$.

We have now a representation of the Dirac operator in an arbitrary curved space-time, expressed in terms of the vierbein $e^\alpha{}_\mu$ that encodes the curved metric, from which we may read off the coupling between a spin-1/2 field and the external gravitational field. We will not go into the technology required for calculating a fermion loop like the graph shown in Figure 12.3, and just quote the final result for the divergence of the axial current,

$$\nabla_\mu \langle J_5^\mu(x) \rangle = \frac{1}{384\pi^2} \epsilon^{\alpha\beta\gamma\delta} R_{\alpha\beta}{}^{\mu\nu}(x) R_{\gamma\delta\mu\nu}(x), \tag{12.66}$$

where $R_{\mu\nu\rho\sigma}$ is the curvature tensor (it plays in gravity the same role as the field strength $F_{\mu\nu}$ in a non-Abelian gauge theory). This formula indicates that a curved space-time, i.e., an external gravitational field, leads to an anomalous contribution to the divergence of the axial current. This effect is, of course, tiny in ordinary situations where gravity is weak. But it should in principle be kept in mind when attempting to construct an anomaly-free chiral gauge theory if one wishes this theory to remain consistent all the way up to the Planck scale.

### 12.2.3 Anomalies in Chiral Gauge Theories

In all the examples that we have considered until now in this chapter, the anomaly appeared in the conservation of a current associated to a global symmetry such as chiral symmetry. Although it indicates a violation of this symmetry by quantum corrections, the anomaly does not make the theory inconsistent in this case. However, in graphs mixing the axial current and insertions of external gauge fields, we made sure that the ultraviolet regularization does not spoil the Ward–Takahashi identity associated with the gauge symmetry.

But we may also consider chiral gauge theories, in which the left- and right-handed components of the fermions transform according to different representations of the gauge group. This is, for instance, the case in the Standard Model, where the electroweak interaction is chiral (the left-handed fermions form $SU(2)$ doublets, while the right-handed fermions are singlet under $SU(2)$). In such a theory, the gauge coupling between fermions and gauge fields involve the left or right projectors $P_{R,L} \equiv (1 \pm \gamma^5)/2$, and the generators of the Lie algebra that appear in these vertices are $t^a_{R,L}$, respectively (the left and right generators would be equal in a theory in which the two fermion chiralities belong to the same representation).

## 12.2 GENERALIZATIONS

The triangle diagram that gave the axial anomaly in four dimensions is replaced by a graph with three external gauge bosons, with chiral couplings to the fermion loop. When the fermion in the loop is massless, the left and right chiralities do not mix, and the multiple occurrences of the projectors simplify into a single one, thanks to

$$P_R P_L = 0, \quad P_L^2 = P_L, \quad P_R^2 = P_R,$$
$$[P_{R,L}, \gamma^\mu \gamma^\nu] = 0,$$
$$\mathrm{tr}\left(P_R \gamma^\mu \not{p}_1 P_R \gamma^\nu \not{p}_2 P_R \gamma^\rho \not{p}_3\right) = \mathrm{tr}\left(P_R \gamma^\mu \not{p}_1 \gamma^\nu \not{p}_2 \gamma^\rho \not{p}_3\right),$$
$$\mathrm{tr}\left(P_L \gamma^\mu \not{p}_1 P_L \gamma^\nu \not{p}_2 P_L \gamma^\rho \not{p}_3\right) = \mathrm{tr}\left(P_L \gamma^\mu \not{p}_1 \gamma^\nu \not{p}_2 \gamma^\rho \not{p}_3\right). \qquad (12.67)$$

The $\gamma^5$ contained in the projectors $P_{R,L}$ may lead to an anomaly, with a relative sign between the right and left chiralities. The calculation is almost identical to the case of a global axial symmetry, except that now we should choose the shifts $a$ and $b$ so that the resulting three-point function is symmetric in the external fields, since they play identical roles. But this choice does not eliminate the anomaly; it just distributes it evenly among the three external currents, leading to an anomaly proportional to $\mathrm{tr}\left(t^a\{t^b, t^c\}\right)$. When there are both right and left fermions in the loop, the anomaly is proportional to

$$D_{abc} \equiv \mathrm{tr}\left(t_R^a\{t_R^b, t_R^c\}\right) - \mathrm{tr}\left(t_L^a\{t_L^b, t_L^c\}\right). \qquad (12.68)$$

Obviously, this is zero in a vector theory, where the right and left fermions couple in the same way to the gauge bosons.

**Anomaly cancellation in the Standard Model:** Unlike anomalies of global symmetries, an anomaly of a gauge symmetry makes the theory immediately inconsistent because it would, for instance, spoil its unitarity and renormalizability. For this reason, most chiral gauge theories do not make sense. The only ones that actually do are those for which the fermion fields are arranged in representations of the gauge group such that $D_{abc} = 0$. This turns out to be the case for the Standard Model with its known matter fields: All the gauge anomalies cancel (within each generation of fermions) thanks in particular to the peculiar values of the weak hypercharges of the quarks and leptons. In order to proceed with this verification, we need the quantum numbers listed in Table 12.1 for the fermions of the Standard Model.

The weak isospin and hypercharge are the quantum numbers of the fermion under $SU(2) \times U(1)$. Both of these gauge interactions are chiral, since the charges $T_3$ and $Y$ of the left- and right-handed fermions are different. After spontaneous symmetry breaking via the Higgs mechanism, the fields $B_3^\mu$ (third component of $SU(2)$) and $A^\mu$ ($U(1)$) mix to give the Z boson and the photon fields. The electrical charges of the fermions are then given by $Q = T_3 + \frac{Y}{2}$ (since the electrical charges are the same for left and right fermions, the resulting $U(1)_{em}$ of electromagnetism is a non-chiral gauge interaction).

The simplest case of anomaly cancellation is the three-gluon triangle, which is not anomalous because the strong interaction vertex is a vector coupling,

R − L cancellation (see eq. (12.68)).

For the triangle involving three $SU(2)$ bosons, the anomaly cancels thanks to a peculiar identity obeyed by the $\mathfrak{su}(2)$ generators:

| | Weak isospin $T_3$ | Weak hypercharge $Y$ | Electrical charge |
|---|---|---|---|
| *Left-handed fermions* | | | |
| $\nu_e, \nu_\mu, \nu_\tau$ | $+\frac{1}{2}$ | $-1$ | $0$ |
| $e, \mu, \tau$ | $-\frac{1}{2}$ | $-1$ | $-1$ |
| $u, c, t$ | $+\frac{1}{2}$ | $+\frac{1}{3}$ | $+\frac{2}{3}$ |
| $d, s, b$ | $-\frac{1}{2}$ | $+\frac{1}{3}$ | $-\frac{1}{3}$ |
| *Right-handed fermions* | | | |
| $e_R, \mu_R, \tau_R$ | $0$ | $-2$ | $-1$ |
| $u_R, c_R, t_R$ | $0$ | $+\frac{4}{3}$ | $+\frac{2}{3}$ |
| $d_R, s_R, b_R$ | $0$ | $-\frac{2}{3}$ | $-\frac{1}{3}$ |

TABLE 12.1: Weak isospin, hypercharge and electrical charge of the fermions of the Standard Model.

$$\mathrm{tr}_{su(2)}\left(t^i\{t^j, t^k\}\right) = 0.$$

In triangles that have a single SU(3) or a single SU(2) boson, the anomaly cancels because the corresponding generators are traceless:

$$\mathrm{tr}_{su(3)}(t^a) = 0,$$

$$\mathrm{tr}_{su(2)}(t^i) = 0.$$

In triangles with a single U(1) boson and a pair of SU(2) or SU(3) bosons, the anomaly cancels thanks to the specific linear combination of weak hypercharges one gets by summing over all the allowed fermions in the loop:

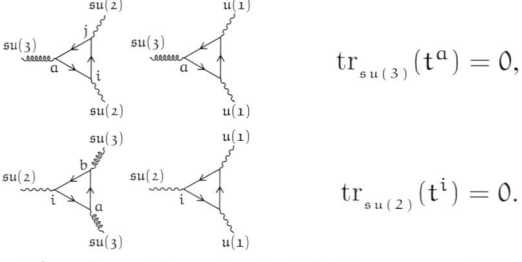

$$\sum_{\text{quarks}} y = 2\left[-\frac{1}{3}\right] + \frac{4}{3} - \frac{2}{3} = 0,$$

$$\sum_{\substack{\text{left-handed}\\\text{fermions}}} y = 3\left[-\frac{1}{3}\right] + 1 = 0.$$

(In the first of these cancellations, there is a factor of 2 in the first term to account for the fact that the left-handed quarks form SU(2) doublets, and in the second equality the first term has a factor of 3 because the quarks can have three colors.) Note also that loops with left-handed fermions should be counted with a minus sign, according to eq. (12.68). Finally, the triangle with three U(1) bosons has no anomaly, thanks to the fact that the sum of the cubes of the weak hypercharges over all fermions is zero:

$$\sum y^3 = 6\left[-\tfrac{1}{3}\right]^3 + 3\left[\tfrac{4}{3}\right]^3 + 3\left[-\tfrac{2}{3}\right]^3 + 2 + \left[-2\right]^3 = 0.$$

(Again, the numerical prefactors count the number of SU(3) and SU(2) states for each fermion.) Interestingly, gravitational anomalies also cancel in the standard model. Indeed, an anomaly may potentially exist in the triangle with a U(1) boson and two gravitons. But this anomaly would be proportional to the sum of the weak hypercharges of all fermions, which turns out to be zero,

$$\sum y = 6\left[-\tfrac{1}{3}\right] + 3\left[\tfrac{4}{3}\right] + 3\left[-\tfrac{2}{3}\right] + 2 + \left[-2\right] = 0.$$

One can see the crucial role played by the weak hypercharges assigned to the various fermions of the Standard Model in these cancellations. Conversely, one may determine these hypercharges so that all anomalies cancel within each fermion family (see Exercise 12.5). Up to a permutation of the right-handed quarks ($u_R$ and $d_R$ in the first family), there are only two solutions, say $U(1)_A$ and $U(1)_B$ (one of them corresponds to the Standard Model). Moreover, these two solutions cannot be mixed in the same theory, because one would have a non-canceling anomaly in a triangle that mixes the $U(1)_A$ and $U(1)_B$ gauge bosons.

## 12.3 Wess–Zumino Consistency Conditions

### 12.3.1 Consistency Conditions

In Section 12.2.1, where we have derived the axial anomaly in a non-Abelian background field, we first obtained a partial answer with only the terms quadratic in the external field, and then we used gauge symmetry in order to reconstruct the missing terms (of order 3 and 4 in the external field). However, how to promote such a partial result into the full expression of the anomaly is not always so obvious, for instance in the case of chiral gauge theories in which the gauge symmetry itself is anomalous (in this case, we cannot invoke gauge invariance to restore the full answer). The *Wess–Zumino consistency conditions* are a set of equations satisfied by the anomaly function, that are powerful enough to allow reconstructing the anomaly from the knowledge of its lowest order in the gauge fields.

Even in the case where the anomalous symmetry is global, it is convenient to couple a (fictitious in that case) gauge field $A^\mu$ to the corresponding current $J^\mu$ whose conservation is violated by the anomaly. By doing this, we promote the symmetry to a local gauge invariance (violated by the anomaly), and we may return to a global symmetry by letting the gauge coupling go to zero. Let us denote $\Gamma[A]$ the effective action for the gauge field (i.e., the effective action in which the fermions are included only in the form of loop corrections).

In the absence of anomaly, $\Gamma[A]$ would be invariant under gauge transformations of the field $A^\mu$,

$$\begin{aligned}
0\underset{\text{no anomaly}}{=}\delta_\theta\Gamma[A] &= -\int d^4x\,\left((D_\mu^{\text{adj}})_{ab}\theta_b(x)\right)\frac{\delta\Gamma[A]}{\delta A_\mu^a(x)}\\
&= \int d^4x\,\theta_b(x)\underbrace{(D_\mu^{\text{adj}})_{ba}\frac{\delta}{\delta A_\mu^a(x)}}_{i\,\mathcal{T}_b(x)}\Gamma[A].
\end{aligned} \qquad (12.69)$$

When this symmetry is spoiled by an anomaly, the effective action is no longer invariant, and we may write

$$\mathcal{T}_a(x)\,\Gamma[A] \equiv G_a[x;A], \qquad (12.70)$$

where the function $G_a[x;A]$ encodes the anomaly. This function is closely related to the non-zero right-hand side of the anomalous conservation law for the current associated to the symmetry, since the effective action and the current are related by

$$J^{\mu a}(x) + \frac{\delta\Gamma[A]}{\delta A_\mu^a(x)} = 0, \qquad (12.71)$$

which implies

$$(D_\mu^{\text{adj}})_{ba} J^{\mu a}(x) = -G_b[x;A]. \qquad (12.72)$$

Since the anomaly is local, $G_b[x;A]$ should be a local (at the point x) polynomial in the gauge field and its derivatives. One may then check that the operators $\mathcal{T}_a(x)$ obey the following commutation relation,

$$[\mathcal{T}_a(x),\mathcal{T}_b(y)] = i g\,f^{abc}\,\delta(x-y)\,\mathcal{T}_c(x), \qquad (12.73)$$

where $f^{abc}$ are the structure constants of the Lie algebra. From this, we deduce the following identity:

$$\mathcal{T}_a(x)\,G_b[y;A] - \mathcal{T}_b(y)\,G[x;A] = i g\,f^{abc}\,\delta(x-y)\,G_c[x;A], \qquad (12.74)$$

called the Wess–Zumino consistency conditions. Since this identity is linear in the anomaly function $G_a$, it cannot constrain its overall normalization (for this, it is usually necessary to compute the triangle diagram). However, this equation is strong enough to fully constrain its dependence on the gauge field from the term of lowest order in A.

## 12.3.2 BRST Form of the Wess–Zumino Condition

The consistency condition can be recast into a more convenient form that involves BRST symmetry. Let us introduce a ghost field $\chi_a$, and recall that the BRST transformation reads

$$Q_{\text{BRST}}A_\mu^a(x) = (D_\mu^{\text{adj}})_{ab}\chi_b(x),\qquad Q_{\text{BRST}}\chi_a(x) = -\frac{g}{2}f^{abc}\chi_b(x)\chi_c(x). \qquad (12.75)$$

## 12.3 WESS–ZUMINO CONSISTENCY CONDITIONS

Then, let us encapsulate the anomaly function into the following local functional of ghost number $+1$,

$$\mathcal{G}[A,\chi] \equiv \int d^4x\, \chi_a(x)\, \mathcal{G}_a[x;A]. \tag{12.76}$$

We obtain

$$\begin{aligned}\mathbf{Q}_{\text{BRST}}\mathcal{G}[A,\chi] &= i\int d^4x d^4y\, \chi_a(x)\chi_b(y)\, \mathcal{T}_b(y)\mathcal{G}_a[x;A] \\ &\quad - \frac{g}{2}\int d^4x\, f^{abc}\chi_a(x)\chi_b(x)\, \mathcal{G}_c[x;A] \\ &= \frac{i}{2}\int d^4x d^4y\, \chi_a(x)\chi_b(y)\, \underbrace{\bigl\{\mathcal{T}_b(y)\mathcal{G}_a[x;A] - \mathcal{T}_a(x)\mathcal{G}_b[y;A]}_{=0} \\ &\qquad\qquad\qquad\qquad \underbrace{+ i g\,\delta(x-y)\, f^{abc}\mathcal{G}_c[x;A]\bigr\}}_{=0}.\end{aligned} \tag{12.77}$$

Therefore, the Wess–Zumino consistency conditions are equivalent to the statement that the functional $\mathcal{G}[A,\chi]$ is BRST invariant:

$$\mathbf{Q}_{\text{BRST}}\mathcal{G}[A,\chi] = 0. \tag{12.78}$$

Since $\mathbf{Q}_{\text{BRST}}$ is nilpotent, a trivial solution of this equation is of course

$$\mathcal{G}[A,\chi] = \mathbf{Q}_{\text{BRST}}\, h[A], \tag{12.79}$$

where $h[A]$ does not depend on the ghost field (indeed, $\mathbf{Q}_{\text{BRST}}$ increases the ghost number by one unit, and $\mathcal{G}[A,\chi]$ must have ghost number unity). But since $h[A]$ is a local functional of the gauge field, it may be subtracted from the action to cancel the anomaly. Thus, genuine anomalies are given by local functionals $\mathcal{G}[A,\chi]$ of ghost number $+1$ that satisfy the consistency condition (12.78), modulo a term obtained by acting with $\mathbf{Q}_{\text{BRST}}$ on a functional of $A$ only. Note that if we write $\mathcal{G}[A,\chi]$ as the integral of a local density,

$$\mathcal{G}[A,\chi] \equiv \int d^4x\, \mathcal{G}(x), \tag{12.80}$$

then the BRST action on the density should be a total derivative

$$\mathbf{Q}_{\text{BRST}}\, \mathcal{G}(x) = \partial_\mu \zeta^\mu. \tag{12.81}$$

### 12.3.3 Solution of the Consistency Condition

In order to determine how the Wess–Zumino equation constrains $\mathcal{G}(x)$, the language of *differential forms* introduced in Section 7.5.3 is very handy, as a way to encapsulate both Lorentz and group indices in compact objects. The 1-forms $dx^\mu$ anti-commute among themselves under the exterior product $\wedge$. In addition, they also anti-commute with the ghost field and the BRST generator $Q_{BRST}$. The volume element weighted by the fully antisymmetric tensor $\epsilon^{\mu\nu\rho\sigma}$ can therefore be written as

$$d^4x\, \epsilon^{\mu\nu\rho\sigma} = dx^\mu \wedge dx^\nu \wedge dx^\rho \wedge dx^\sigma. \tag{12.82}$$

Then, given a vector $V_\mu$ and the corresponding 1-form

$$\mathbf{V} \equiv V_\mu dx^\mu, \tag{12.83}$$

we may write in a compact manner

$$\int d^4x\, \epsilon^{\mu\nu\rho\sigma} V_\mu V_\nu V_\rho V_\sigma = \int \mathbf{V} \wedge \mathbf{V} \wedge \mathbf{V} \wedge \mathbf{V}. \tag{12.84}$$

The exterior derivative $\mathbf{d} \equiv \partial_\mu\, dx^\mu \wedge$ satisfies

$$\mathbf{d}^2 = 0, \qquad Q_{BRST}\mathbf{d} + \mathbf{d}Q_{BRST} = 0. \tag{12.85}$$

If we also denote

$$\mathbf{A} \equiv ig\, A_\mu^a t^a dx^\mu, \qquad \mathbf{\chi} \equiv ig\, \chi_a t^a \tag{12.86}$$

(for convenience, we absorb a factor $i$ in the definitions of $\mathbf{A}$ and $\mathbf{\chi}$), the BRST transformations take the following form:

$$Q_{BRST}\mathbf{A} = -\mathbf{d\chi} - \underbrace{\mathbf{A} \wedge \mathbf{\chi} + \mathbf{\chi} \wedge \mathbf{A}}_{\{\mathbf{A},\mathbf{\chi}\}}, \qquad Q_{BRST}\mathbf{\chi} = -\mathbf{\chi} \wedge \mathbf{\chi}. \tag{12.87}$$

On dimensional grounds, the anomaly function $\mathcal{G}[A,\chi]$ may contain the following terms:

$$\begin{aligned}\mathcal{G}[A,\chi] = -iC \int d^4x\, \epsilon^{\mu\nu\rho\sigma}\, \chi_a\, \mathrm{tr}\Big\{t^a\Big(&(\partial_\mu A_\nu)(\partial_\rho A_\sigma) \\ + ia_1\, (\partial_\mu A_\nu)A_\rho A_\sigma &+ ia_2\, A_\mu(\partial_\nu A_\rho)A_\sigma + ia_3\, A_\mu A_\nu(\partial_\rho A_\sigma) \\ - b\, A_\mu A_\nu A_\rho A_\sigma\Big)\Big\}.&\end{aligned} \tag{12.88}$$

The term on the first line comes from the triangle diagram, whose explicit calculation gives the overall coefficient C. The terms of the second and third lines come from the square and pentagon diagrams, respectively. Alternatively, they can be obtained from the consistency conditions. First, the previous equation may be rewritten as a sum of forms:

$$\begin{aligned}\mathcal{G}[A,\chi] = \gamma \int \mathrm{tr}\Big\{\mathbf{\chi} \wedge \big((\mathbf{dA}) \wedge (\mathbf{dA})\big) \\ + \alpha_1\, (\mathbf{dA}) \wedge \mathbf{A} \wedge \mathbf{A} + \alpha_2\, \mathbf{A} \wedge (\mathbf{dA}) \wedge \mathbf{A} \\ + \alpha_3\, \mathbf{A} \wedge \mathbf{A} \wedge (\mathbf{dA}) + \beta\, \mathbf{A} \wedge \mathbf{A} \wedge \mathbf{A} \wedge \mathbf{A}\Big\},\end{aligned} \tag{12.89}$$

where $\gamma$, $\alpha_{1,2,3}$, $\beta$ are constants related to C, $a_{1,2,3}$, b. Consider first the BRST transform of the last term,

$$Q_{BRST}\,\mathrm{tr}\,\{\chi \wedge A \wedge A \wedge A \wedge A\} = -\mathrm{tr}\,\{\chi \wedge A \wedge A \wedge A \wedge A \wedge \chi\}$$
$$+ \text{ terms in } \chi \wedge (d\chi) \wedge A \wedge A \wedge A. \tag{12.90}$$

Since $Q_{BRST}$ cannot increase the degree in $A$, the term in $\chi \wedge A \wedge A \wedge A \wedge A \wedge \chi$ cannot be canceled by the terms in $\alpha_{1,2,3}$, and therefore we must have $\beta = 0$. By evaluating similarly the BRST transforms of the other terms, one can check that when $\alpha_1 = -\alpha_2 = \alpha_3 = 1/2$ the BRST transform of the anomaly functional is the integral of an exact form (see Exercise 12.6) and therefore vanishes:

$$Q_{BRST}\,\mathcal{G}[A,\chi] = \gamma \int_{\mathbb{R}^4} d\mathcal{F} = \gamma \int_{\partial \mathbb{R}^4} \mathcal{F} = 0. \tag{12.91}$$

This is in fact the only possibility. Introducing the field strength 2-form,

$$F \equiv dA - A \wedge A = \frac{ig}{2} t^a F^a_{\mu\nu}\, dx^\mu dx^\nu, \tag{12.92}$$

the anomaly functional for these values of the coefficients can then be rewritten as

$$\mathcal{G}[A,\chi] = \gamma \int \mathrm{tr}\,\Big\{\chi \wedge d\,\Big[A \wedge F + \frac{1}{2} A \wedge A \wedge A\Big]\Big\}. \tag{12.93}$$

Therefore, except for the prefactor $\gamma$ whose determination requires calculating the triangle diagram, the consistency relations completely determine the dependence of the anomaly function on the gauge field.

## 12.4 't Hooft Anomaly Matching

Some models of physics beyond the Standard Model conjecture that the quarks and leptons are bound states made of more fundamental degrees of freedom, confined by some strong gauge interaction at a scale $\Lambda \gg \Lambda_{\text{electroweak}}$. A difficulty with this picture is to explain the fact that quarks and leptons are light (in fact, massless, if it were not for electroweak symmetry breaking), while being bound states of some strong interaction at a much higher scale. Indeed, the naive mass of these confined states is naturally of order $\Lambda$ (the Goldstone mechanism cannot give light *fermions*, only scalar particles).

As shown by 't Hooft, one way this may happen is to have in the underlying fundamental theory a global chiral symmetry with generators $T^a$, such that the anomaly function $\mathrm{tr}\,(T^a\{T^b, T^c\})$ is non-zero. In the low-energy sector of the spectrum of this theory, there must be spin-1/2 massless bound states, on which this chiral symmetry acts with generators $\mathbb{T}^a$, and whose anomaly coefficients are identical to the high-energy ones,

$$\mathrm{tr}\,(\mathbb{T}^a\,\{\mathbb{T}^b, \mathbb{T}^c\}) = \mathrm{tr}\,(T^a\,\{T^b, T^c\}). \tag{12.94}$$

The proof of this assertion goes as follows. Let us first couple a fictitious weakly coupled gauge boson to the generators $T^a$. We also introduce additional fictitious massless fermions coupled only to the fictitious gauge boson, but not to the strongly interacting gauge bosons responsible for the confinement, tuned so that their contribution exactly cancels the anomaly,

$$\left[\text{tr}\left(T^a\{T^b,T^c\}\right)\right]_{\substack{\text{physical}\\\text{high energy}}} + \left[\text{tr}\left(T^a\{T^b,T^c\}\right)\right]_{\substack{\text{fictitious}\\\text{fermions}}} = 0. \tag{12.95}$$

Let us now examine the low-energy part of the spectrum of this theory, i.e., at energies much lower than the strong scale $\Lambda$. Since they are not coupled in any way to the strong sector, this low-energy spectrum contains the fictitious gauge bosons and massless fermions, unmodified compared to what we have introduced at high energy. In addition, this spectrum contains the bound states made of the trapped fermions and strongly interacting gauge bosons. For consistency, this low-energy description must also be anomaly-free, which means that the bound states must transform under the chiral symmetry with generators $\mathbb{T}^a$, such that

$$\left[\text{tr}\left(\mathbb{T}^a\{\mathbb{T}^b,\mathbb{T}^c\}\right)\right]_{\substack{\text{physical}\\\text{bound states}}} + \left[\text{tr}\left(T^a\{T^b,T^c\}\right)\right]_{\substack{\text{fictitious}\\\text{fermions}}} = 0. \tag{12.96}$$

The crucial point in this argument is that the contribution of the fictitious fermions is the same in eqs. (12.95) and (12.96) because these fermions are not coupled to the strongly interacting sector. Eqs. (12.95) and (12.96) immediately give (12.94). In other words, *the anomalies of the trapped elementary fermions must be mimicked by those of the massless spin-1/2 bound states they are confined to.*

## 12.5 Scale Anomalies

### 12.5.1 Classical Scale Invariance

Until now, the quantum anomalies we have encountered in this chapter are related to chiral couplings of fermions. But there exists another anomaly, ubiquitous in most quantum field theories, related to quantum violations of scale invariance. Consider a quantum field theory whose Lagrangian does not contain any dimensionful parameter. In four space-time dimensions, this means that it contains only operators of mass dimension exactly equal to four, which excludes all mass terms. This is the case for instance for a massless scalar field theory with a quartic coupling, whose action reads

$$S[\phi] \equiv \int d^4x \left\{\frac{g_{\mu\nu}}{2}(\partial_x^\mu\phi(x))(\partial_x^\nu\phi(x)) - \frac{\lambda}{4!}\phi^4(x)\right\}. \tag{12.97}$$

A scaling transformation amounts to multiplying all length scales by some factor

$$x^\mu \quad\to\quad y^\mu \equiv e^\vartheta x^\mu. \tag{12.98}$$

In this transformation, a field $\phi(x)$ of dimension $(\text{mass})^{d_\phi}$ and its derivative transform as follows:

$$\begin{aligned}\phi(x) &\quad\to\quad \phi'(y) \equiv e^{-\vartheta d_\phi}\phi(e^{-\vartheta}y),\\ \partial_x^\mu\phi(x) &\quad\to\quad \partial_y^\mu\phi'(y) = e^{-\vartheta(d_\phi+1)}\partial_x^\mu\phi(x)\Big|_{x=e^{-\vartheta}y},\end{aligned} \tag{12.99}$$

## 12.5 SCALE ANOMALIES

while the integration measure over space-time is rescaled by

$$d^4x \quad \to \quad d^4y = e^{4\vartheta}\, d^4x. \tag{12.100}$$

Consider now the transformed action,

$$\begin{aligned}
S[\phi'] &= \int d^4y \left\{ \frac{g_{\mu\nu}}{2}(\partial^\mu_y \phi'(y))(\partial^\nu_y \phi'(y)) - \frac{\lambda}{4!}\phi'^4(y) \right\} \\
&= e^{4\vartheta} \int d^4x \left\{ e^{-2\vartheta(1+d_\phi)}\frac{g_{\mu\nu}}{2}(\partial^\mu_x \phi(x))(\partial^\nu_x \phi(x)) - e^{-4\vartheta d_\phi}\frac{\lambda}{4!}\phi^4(x) \right\} \\
&= S[\phi],
\end{aligned} \tag{12.101}$$

where we have used the fact that the mass dimension of $\phi$ is $d_\phi = 1$ in four space-time dimensions. The action defined in eq. (12.97) is thus invariant under scale transformations. The same conclusion holds for any classical action that does not contain any dimensionful parameter, provided the appropriate dimension $d_\phi$ is used for each field. This is for instance the case of pure Yang–Mills theory in four dimensions, or quantum chromodynamics in which we neglect the quark masses.

### 12.5.2 Dilatation Current

Since the transformation of eq. (12.99) is continuous, Noether's theorem implies that there is a corresponding conserved current. On the one hand, the infinitesimal variation of the field is

$$\delta\phi(x) \equiv \phi'(x) - \phi(x) = -\vartheta\left(d_\phi + x^\mu \partial_\mu\right)\phi(x) + \mathcal{O}(\vartheta^2). \tag{12.102}$$

On the other hand, the scale transformation (12.98) directly applied to the integrand of the action gives a variation:

$$\begin{aligned}
\delta\!\left[d^4x\, \mathcal{L}(x)\right] &= -\vartheta\, d^4x \left(4 + x^\mu \partial_\mu\right) \mathcal{L}(x) + \mathcal{O}(\vartheta^2) \\
&= -\vartheta\, d^4x\, \partial_\mu\!\left(x^\mu\, \mathcal{L}(x)\right) + \mathcal{O}(\vartheta^2).
\end{aligned} \tag{12.103}$$

It is important to include the measure in this calculation, since it is not invariant under scale transformations. The variation of the measure gives the 4 in the first line, which is crucial for obtaining a total derivative in the second line. Then, from the derivation of Noether's theorem, we conclude that

$$\partial_\mu \Big( \underbrace{\frac{\partial \mathcal{L}}{\partial(\partial_\mu \phi)}(d_\phi + x_\nu \partial^\nu)\phi - x^\mu \mathcal{L}}_{D^\mu} \Big) = 0. \tag{12.104}$$

The vector $D^\mu$ is called the *dilatation current*. In the case of the scalar field theory used earlier as an example, the explicit form of $D^\mu$ is

$$D^\mu = x_\nu \underbrace{\left((\partial^\mu\phi)(\partial^\nu\phi) - g^{\mu\nu}\mathcal{L}\right)}_{\Theta^{\mu\nu}} + \underbrace{\phi(\partial^\mu\phi)}_{\frac{1}{2}\partial^\mu\phi^2}. \tag{12.105}$$

In this formula, we recognize that the factor multiplying $x_\nu$ is the energy–momentum tensor $\Theta^{\mu\nu}$, whose divergence is zero thanks to translation invariance, i.e., $\partial_\mu \Theta^{\mu\nu} = 0$. This observation facilitates the calculation of $\partial_\mu D^\mu$, since we have

$$\begin{aligned}\partial_\mu D^\mu &= \Theta^\mu{}_\mu + (\partial^\mu \phi)(\partial_\mu \phi) + \phi(\Box \phi) \\ &= 2(\partial^\mu \phi)(\partial_\mu \phi) + \phi(\Box \phi) - 4\mathcal{L} \\ &= \phi \underbrace{\left( \Box \phi + \frac{\lambda}{6} \phi^3 \right)}_{=0}.\end{aligned} \quad (12.106)$$

The final zero follows from the classical equation of motion of the field.

### 12.5.3 Link with the Energy–Momentum Tensor

In the previous section we saw that the energy–momentum tensor appears in the expression of the dilatation current. More precisely, this energy–momentum tensor is the canonical one (i.e., the one obtained as Noether's current associated to translation invariance). Then, the divergence of the dilatation tensor is the trace of this energy–momentum tensor ($\Theta^\mu{}_\mu = -\frac{1}{2} \Box \phi^2$), plus an additional term that turns out to cancel it exactly.

In fact, in such a scale-invariant theory, it is possible to introduce a traceless definition of the energy–momentum tensor, which we shall denote $T^{\mu\nu}$, such that

$$T^\mu{}_\mu = 0, \quad \partial_\mu T^{\mu\nu} = 0, \quad (12.107)$$

and a valid definition of the dilatation current is

$$\mathcal{D}^\mu \equiv x_\nu T^{\mu\nu}. \quad (12.108)$$

(As we shall see shortly, this new dilatation current gives the same conserved charge as the current $D^\mu$ introduced earlier.) The tracelessness of $T^{\mu\nu}$ is then equivalent to the conservation of $\mathcal{D}^\mu$.

In the case of a massless $\phi^4$ scalar field theory in four dimensions, this improved energy–momentum tensor reads

$$T^{\mu\nu} \equiv \Theta^{\mu\nu} + \frac{1}{6} \left( g^{\mu\nu} \Box - \partial^\mu \partial^\nu \right) \phi^2. \quad (12.109)$$

This tensor is traceless, because the trace of the additional term is $+\frac{1}{2} \Box \phi^2$. Moreover, this additional term has a null divergence, and therefore we have $\partial_\mu T^{\mu\nu} = 0$. Note also that the component $\mu = 0$ of the added term is a total spatial derivative,

$$\left( g^{0\nu} \Box - \partial^0 \partial^\nu \right) \phi^2 = \begin{cases} -\partial^i \partial^i \phi^2 & (\nu = 0) \\ -\partial^i \partial^0 \phi^2 & (\nu = i) \end{cases}, \quad (12.110)$$

which implies that the conserved charges (i.e., the momenta $P^\nu$) obtained from $\Theta^{\mu\nu}$ and $T^{\mu\nu}$ are the same.

## 12.5 SCALE ANOMALIES

Since $T^{\mu\nu}$ is traceless, the current $\mathcal{D}^\mu = x_\nu T^{\mu\nu}$ is conserved. But we should also check that we have not modified the corresponding conserved charge. We have

$$\begin{aligned}
\mathcal{D}^0 = x_\nu T^{0\nu} &= x_\nu \Theta^{0\nu} + \frac{1}{6} x_\nu \left(g^{0\nu}\Box - \partial^0\partial^\nu\right)\phi^2 \\
&= x_\nu \Theta^{0\nu} + \frac{1}{6}\left(x^i\partial^0\partial^i - x^0\partial^i\partial^i\right)\phi^2 \\
&= x_\nu \Theta^{0\nu} + \frac{1}{6}\left(-\underbrace{(\partial^i x^i)}_{=-3}\partial^0 + \partial^i(x^i\partial^0 - x^0\partial^i)\right)\phi^2 \\
&= \underbrace{x_\nu \Theta^{0\nu} + \frac{1}{2}\partial^0\phi^2}_{D^0} + \partial^i\left[\frac{1}{6}(x^i\partial^0 - x^0\partial^i)\phi^2\right].
\end{aligned} \qquad (12.111)$$

The first two terms are identical to the original $D^0$ of eq. (12.105), and the third term is a total spatial derivative. Therefore, when we integrate this charge density over all space, the new definition of the dilatation current gives the same conserved charge as eq. (12.105).

### 12.5.4 Energy–Momentum Tensor via Coupling to Gravity

The discussion of the previous section highlights the fact that, even in classical field theory, the energy–momentum tensor is not uniquely defined. It is possible to add a term that does not alter its conservation and does not change the conserved charges, but that modifies its trace. There are also cases (e.g., Yang–Mills theory), in which the canonical energy–momentum tensor $\Theta^{\mu\nu}$ is not even symmetric, but can be improved into a symmetric one.

An alternate method of deriving the energy–momentum tensor, that leads directly to a symmetric tensor, is to minimally couple the theory to gravity, and to vary the metric. To that effect, consider an infinitesimal space-time-dependent translation:

$$x^\mu \quad \to \quad x^\mu + \xi^\mu(x). \qquad (12.112)$$

Under such a transformation, the metric tensor varies by[5]

$$\delta g^{\mu\nu}(x) = \nabla^\mu \xi^\nu(x) + \nabla^\nu \xi^\mu(x), \qquad (12.113)$$

where $\nabla^\mu$ is the covariant derivative. Let us recall for later use an important identity:

$$\int d^4x \sqrt{-g}\, A\left(\nabla^\mu B\right) = -\int d^4x \sqrt{-g}\, \left(\nabla^\mu A\right) B, \qquad (12.114)$$

---

[5]Note that, although $x^\mu$ is not a vector, the infinitesimal variation $\xi^\mu(x)$ is a vector, tangent to the coordinate manifold at point x. Therefore, it makes sense to act on it with a covariant derivative.

where $g \equiv \det(g_{\mu\nu})$. Since eq. (12.113) is merely a change of a dummy integration variable, the action is not modified. Therefore, we may write

$$0 = \delta S = \int d^4x \left( \frac{\delta S}{\delta g^{\mu\nu}(x)} \underbrace{(\nabla^\mu \xi^\nu(x) + \nabla^\nu \xi^\mu(x))}_{\delta g^{\mu\nu}(x)} + \underbrace{\frac{\delta S}{\delta \phi(x)}}_{=0} \delta\phi(x) \right)$$

$$= -\int d^4x \sqrt{-g} \, \nabla^\mu \left( \frac{2}{\sqrt{-g}} \frac{\delta S}{\delta g^{\mu\nu}(x)} \right) \xi^\nu(x). \qquad (12.115)$$

In the first line, the second term vanishes when the field $\phi$ is a solution of the classical equation of motion. For this to be true for an arbitrary variation $\xi^\nu(x)$, we must have

$$\nabla^\mu T_{\mu\nu} = 0, \quad \text{with} \quad T_{\mu\nu} \equiv \frac{2}{\sqrt{-g}} \frac{\delta S}{\delta g^{\mu\nu}}. \qquad (12.116)$$

By construction, this tensor is symmetric and (covariantly) conserved, and the nature of the coordinate transformation (12.113) makes it clear that it is related to translation invariance.[6] In order to obtain the flat space energy–momentum tensor, one should set $g^{\mu\nu}$ to the Minkowski metric tensor after evaluating the derivative.

Moreover, if we apply a scale transformation to the coordinates,

$$x^\mu \quad \to \quad e^\vartheta x^\mu, \qquad (12.117)$$

the metric tensor is simply rescaled:

$$g^{\mu\nu} \quad \to \quad e^{2\vartheta} g^{\mu\nu}. \qquad (12.118)$$

Moreover, if the classical action does not contain any dimensionful parameter, it is invariant under this rescaling, and we can write

$$0 = \delta S = \int d^4x \left( 2\vartheta \frac{\delta S}{\delta g^{\mu\nu}(x)} g^{\mu\nu}(x) + \underbrace{\frac{\delta S}{\delta \phi(x)}}_{=0} \delta\phi(x) \right). \qquad (12.119)$$

This equation implies that the derivative of the action with respect to the metric, and therefore the energy–momentum tensor $T_{\mu\nu}$, is traceless.

In order to illustrate this method, let us consider Yang–Mills theory, whose action coupled to gravity reads

$$S = -\frac{1}{4} \int d^4x \sqrt{-g} \, g^{\mu\rho} g^{\nu\sigma} F^a_{\mu\nu} F^a_{\rho\sigma}. \qquad (12.120)$$

---

[6]It is important to note that the derivation implicitly assumes that the parameters in the action, such as the coupling constants, do not depend explicitly on the position.

## 12.5 SCALE ANOMALIES

In order to calculate the derivative of this action with respect to the metric, we need the variation of $\sqrt{-g}$, which can be obtained as follows:

$$\begin{aligned}
\det(g_{\mu\nu} + \delta g_{\mu\nu}) &= e^{\operatorname{tr}\ln(g_{\mu\nu}+\delta g_{\mu\nu})} \\
&= e^{\operatorname{tr}\ln(g_{\mu\rho}(\delta^\rho{}_\nu + g^{\rho\sigma}\delta g_{\sigma\nu}))} \\
&\approx \det(g_{\mu\nu})\left(1 + g^{\mu\nu}\delta g_{\mu\nu}\right).
\end{aligned} \quad (12.121)$$

Hence,

$$\frac{\partial\sqrt{-g}}{\partial g^{\mu\nu}} = -\sqrt{-g}\,\frac{g_{\mu\nu}}{2}, \quad (12.122)$$

and we obtain the following expression for the energy–momentum tensor:

$$T^{\mu\nu} = -F^{\mu\alpha\,a}\,F^\nu{}_\alpha{}^a + \frac{g^{\mu\nu}}{4}F^a_{\alpha\beta}F^{\alpha\beta\,a}, \quad (12.123)$$

whose trace is zero (see Exercise 12.7).

### 12.5.5 Scale Anomaly and Beta Function

Until now, our analysis of the dilatation current has been purely classical, since we have shown its conservation from the classical action. At this level, it follows from the absence of any dimensionful parameters in the theory. However, this may not remain true when loop corrections are taken into account, because of ultraviolet divergences. This is quite clear if we regularize these divergences by introducing an ultraviolet cutoff, but it is also true in dimensional regularization. In the latter case, the fact that $d \neq 4$ implies that the coupling constants become dimensionful, which also breaks scale invariance. Taking the trace of eq. (12.123) in d dimensions, we obtain

$$T^\mu_\mu = \frac{d-4}{4}F^a_{\alpha\beta}F^{\alpha\beta\,a}. \quad (12.124)$$

The expectation value of the right-hand side has ultraviolet divergences that, in dimensional regularization, become poles in $(d-4)^{-1}$. These terms cancel the prefactor, leaving a non-zero result for the trace of the energy–momentum tensor even in the limit $d \to 4$.

Another point of view is to introduce counterterms to subtract the ultraviolet divergences, and then remove the regulator that controlled the ultraviolet behavior. But after this procedure, the bare coupling constant in the action becomes scale dependent, which also breaks scaling symmetry. From this hand-waving discussion, we expect that the divergence of the dilatation current, i.e., the trace of the energy–momentum tensor, is related to the beta function that controls the running of the coupling,

$$\mu\frac{\partial g}{\partial \mu} = \beta(g). \quad (12.125)$$

Moreover, even if the classical scale invariance is broken by the renormalization group flow, it should be recovered at the fixed points of the RG flow. For instance, a quantum field theory is scale invariant at critical points.

In Yang–Mills theory, we can derive the form of this trace in the following (non-rigorous) manner. Let us start from the Yang–Mills action, written in terms of rescaled fields, so that the coupling appears in the form of a prefactor $g^{-2}$,

$$S = \int d^4x \sqrt{-g} \left[ -\frac{1}{4 g_b^2} F^a_{\mu\nu} F^{\mu\nu\,a} \right], \tag{12.126}$$

where $g_b$ is the bare coupling constant. When this theory is regularized by a gauge invariant cutoff $\mu$ (e.g., a lattice regularization), the bare coupling becomes cutoff dependent in order for the renormalized quantities to have a proper ultraviolet limit. Then, consider again the scaling transformation defined in eqs. (12.117) and (12.118). With a scale-dependent coupling, the physics is invariant provided we also change the scale at which the coupling is evaluated:

$$g_b(\mu) \quad \to \quad g_b(e^{-\vartheta}\mu). \tag{12.127}$$

The infinitesimal form of this transformation is

$$\delta g_{\mu\nu} = -2\vartheta\, g_{\mu\nu}, \quad \delta g_b = -\vartheta\, \underbrace{\mu \frac{\partial g_b}{\partial \mu}}_{\beta\ \text{function}}. \tag{12.128}$$

Then, by writing explicitly the two sources of $\vartheta$ dependence in the variation of the action, we get

$$0 = \delta S \bigg|_{g_{\mu\nu}=\eta_{\mu\nu}} = \vartheta \int d^4x \left[ T^\mu{}_\mu - \frac{\beta(g_b)}{2 g_b^3} F^a_{\mu\nu} F^{\mu\nu\,a} \right]. \tag{12.129}$$

Therefore, we obtain the following form of the anomalous divergence of the dilatation current:

$$\partial_\mu D^\mu = T^\mu{}_\mu = \frac{\beta(g)}{2 g^3} F^a_{\mu\nu} F^{\mu\nu\,a}. \tag{12.130}$$

(This is the expression in terms of the rescaled field strength; with the original field strength, the denominator is $2g$ instead of $2g^3$.)

This derivation is only heuristic, but a more rigorous treatment using properly renormalized operators would lead to the same result. This anomaly can also be derived in perturbation theory, from the loop corrections to the dilatation current:

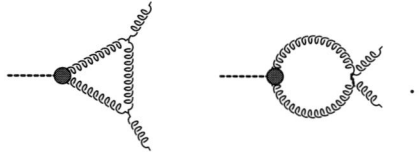

(The dotted line terminated by the dark blob denotes the vertex between two gluons and the dilatation current.) Note that, thanks to asymptotic freedom, Yang–Mills theory becomes more and more scale invariant as the energy scale increases. Finally, when one adds quarks in order to obtain QCD, the right-hand side of the previous equation contains also terms in $m\bar{\psi}\psi$, due to the explicit breaking of scale invariance (already in the classical theory, therefore this is not a quantum anomaly) by the masses of the quarks.

## Exercises

**\*12.1** Carry out explicitly the calculation of the functions A and B in eq. (12.15).

**12.2** Prove eqs. (12.17) and (12.18).

**\*12.3** After eq. (12.45), check that the two vector currents are conserved with shifts that obey $a - b = 2(k_2 - k_1)$, and that this is the only choice.

**12.4** Consider the system of coordinates $\tau \equiv \sqrt{x_0^2 - x_3^2}$ (proper time), $\eta \equiv \frac{1}{2}\ln((x^0 + x^3)/(x^0 - x^3))$ (rapidity) and $x_\perp \equiv (x^1, x^2)$ (transverse coordinates). What is the metric tensor with these coordinates? Calculate a tetrad that relates it to the Minkowski metric, the curved coordinates Dirac matrices $\gamma^\mu(x)$ and the spin connection $\Gamma_\mu$. How is the resulting Dirac equation related to the usual one in Cartesian coordinates? Is the choice of tetrad (and therefore the resulting Dirac equation) unique?

**\*12.5** Find the assignments of $U(1)$ hypercharges for the fermions of the Standard Model that would lead to a cancellation of all anomalies.

**12.6** Check that eq. (12.91) holds with $\alpha_1 = -\alpha_2 = \alpha_3 = 1/2$.

**\*12.7** Consider a classical solution $A^\mu$ of the Yang–Mills equations. Show explicitly that the energy–momentum tensor evaluated for this field configuration is traceless and conserved.

**12.8** Consider the two-particle irreducible generating functional (5.142). Given a certain approximation of $\Gamma_2[\phi, G]$, derive the corresponding energy–momentum tensor.

**\*12.9** *Conformal transformations* are space-time transformations that preserve the Minkowski metric up to a rescaling factor $s(x)$:

$$x^\mu \to y^\mu, \quad \text{such that} \quad \frac{\partial x^\rho}{\partial y^\mu}\frac{\partial x^\sigma}{\partial y^\nu} g_{\rho\sigma} = s(x) g_{\mu\nu}.$$

Obviously, Lorentz transformations and dilatations ($x^\mu \to y^\mu = \lambda x^\mu$) are conformal transformations. Given a 4-vector $b^\mu$, consider the *special conformal transformations*

$$x^\mu \to y^\mu = \frac{x^\mu + b^\mu x^2}{1 + 2b \cdot x + b^2 x^2}.$$

- Show that this transformation is equivalent to the composition $\mathcal{I}\mathcal{T}_b\mathcal{I}$, where $\mathcal{T}_b$ is the translation by $b^\mu$ and $\mathcal{I}$ is the inversion, $x^\mu \to x^\mu/x^2$.
- Show that the inversion is a conformal transformation.
- What is the generator of infinitesimal special conformal transformations?

# CHAPTER 13

# Localized Field Configurations

All the applications of quantum field theory we have encountered so far amount to studying situations that are small perturbations above the vacuum state – i.e., interactions involving states that contain only a few particles. Besides the fact that these situations are actually encountered in scattering experiments, their importance stems from the stability of the vacuum, which makes it a natural state to expand around.

In this chapter, we will study other field configurations, classically stable, that may also be sensible substrates for expansions that differ from the standard perturbative expansion that we have studied until now. However, under normal circumstances, a localized "blob" of fields is not stable: It will usually decay into a field that is zero everywhere. As we shall see, the stability of the field configurations considered in this chapter is due to topological obstructions that prevent a smooth transformation between the field configuration of interest and the null field that corresponds to the vacuum. These field configurations can be classified according to their space-time structure:

- *Event-like:* Localized both in time and space (e.g., instantons). These may be viewed as local extrema of the four-dimensional action, and therefore may give a (non-perturbative) contribution to path integrals.
- *Worldline-like:* Localized in space, independent of time (e.g., skyrmions, monopoles). These field configurations behave very much like stable particles (at least classically), and their non-trivial topology gives them conserved charges.
- *Strings, domain walls:* Extended in one or two spatial dimensions, independent of time.

## 13.1 Domain Walls

A domain wall is a two-dimensional[1] interface between two regions of space where a discrete symmetry is broken in different ways. Their simplest realization arises in a real scalar field theory, symmetric under $\phi \to -\phi$, but with a potential that leads to spontaneous symmetry breaking, such as

$$V(\phi) \equiv V_0 - \frac{\mu^2}{2}\phi^2 + \frac{\lambda}{4!}\phi^4, \tag{13.1}$$

where the constant shift $V_0$ is chosen so that the minima of this potential are zero (see Figure 13.1). There are two such minima, at field values

$$\phi = \pm\phi_*, \quad \phi_* \equiv \sqrt{\frac{6\mu^2}{\lambda}}. \tag{13.2}$$

In order to simplify the discussion, let us consider field configurations that depend only on $x$, and are independent of time, as well as of the transverse coordinates $y, z$. We seek field configurations that obey the classical field equation of motion,

$$-\partial_x^2 \phi + V'(\phi) = 0, \tag{13.3}$$

and have a finite energy (per unit of transverse area),

$$\frac{d\mathcal{E}}{dy\,dz} = \int_{-\infty}^{+\infty} dx \left\{ \tfrac{1}{2}(\partial_x \phi(x))^2 + V(\phi(x)) \right\} < \infty. \tag{13.4}$$

This energy density is the sum of two positive definite terms (since we have adjusted the potential so that its minima are $V(\pm\phi_*) = 0$). For the integral over $x$ to converge when

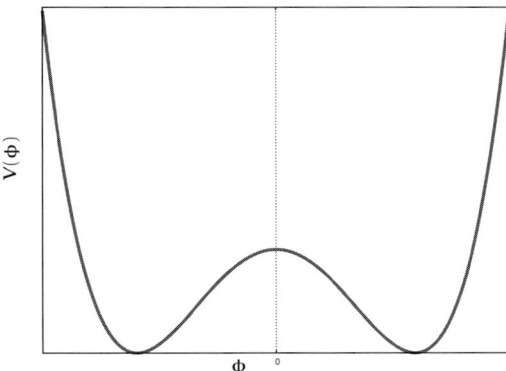

FIGURE 13.1: Quartic potential (13.1) exhibiting spontaneous symmetry breaking.

---

[1]This is for four-dimensional space-time. In D-dimensional space-time, domain walls have dimension $D - 2$.

$x \to \pm\infty$, it is necessary that $\phi(x)$ becomes constant when $|x| \to \infty$, and that this constant be $+\phi_*$ or $-\phi_*$. There are therefore four possibilities for the values of the field at $x = \pm\infty$:

1. $\phi(-\infty) = +\phi_*, \quad \phi(+\infty) = +\phi_*,$
2. $\phi(-\infty) = -\phi_*, \quad \phi(+\infty) = -\phi_*,$
3. $\phi(-\infty) = -\phi_*, \quad \phi(+\infty) = +\phi_*,$
4. $\phi(-\infty) = +\phi_*, \quad \phi(+\infty) = -\phi_*.$ (13.5)

The first two of these possibilities do not lead to stable field configurations of positive energy, because they can be continuously deformed (while holding the asymptotic values unchanged) into the constant fields $\phi(x) = +\phi_*$ or $\phi(x) = -\phi_*$, respectively, that have zero energy. Physically, this means that if one creates a field configuration with these boundary values, it will decay into a constant field (i.e., the regions where the field was excited to values different from $\pm\phi_*$ will dilute away to $|x| = \infty$).

The interesting cases are encountered when the field takes values corresponding to opposite minima at $x = -\infty$ and $x = +\infty$. If one holds the asymptotic values of the field fixed, then it is not possible to deform continuously such a field configuration into one that would have zero energy. Thus, there must be stable field configurations of positive energy with these boundary values. A very handy trick, due to Bogomol'nyi, is to rewrite the energy density as follows:

$$\frac{d\mathcal{E}}{dy\,dz} = \frac{1}{2} \int_{-\infty}^{+\infty} dx \left(\partial_x \phi(x) \pm \sqrt{2V(\phi(x))}\right)^2 \mp \int_{\phi(-\infty)}^{\phi(+\infty)} d\phi \sqrt{2V(\phi)}. \qquad (13.6)$$

In cases 1 and 2 above, the second term vanishes and the energy density can reach zero by having a constant field equal to $\pm\phi_*$. Let us consider now case 3. In this case, it is convenient to choose the minus sign in the first term, so that

$$\frac{d\mathcal{E}}{dy\,dz} = \frac{1}{2} \int_{-\infty}^{+\infty} dx \left(\partial_x \phi(x) - \sqrt{2V(\phi(x))}\right)^2 + \underbrace{\int_{-\phi_*}^{+\phi_*} d\phi \sqrt{2V(\phi)}}_{>0}. \qquad (13.7)$$

The second term is now strictly positive, and does not depend on the details of $\phi(x)$ (except its boundary values). Since the first term is the integral of a square, this implies that there is no field configuration of zero energy with this boundary condition. The minimal energy density possible with this boundary condition is

$$\left.\frac{d\mathcal{E}}{dy\,dz}\right|_{\min} = \int_{-\phi_*}^{+\phi_*} d\phi \sqrt{2V(\phi)}, \qquad (13.8)$$

reached for a field configuration that obeys

$$\partial_x \phi(x) = \sqrt{2V(\phi(x))}. \qquad (13.9)$$

Taking one more derivative implies that

$$\partial_x^2 \phi = \frac{(\partial_x \phi)\, V'(\phi)}{\sqrt{2V(\phi)}} = V'(\phi), \qquad (13.10)$$

## 13.1 DOMAIN WALLS

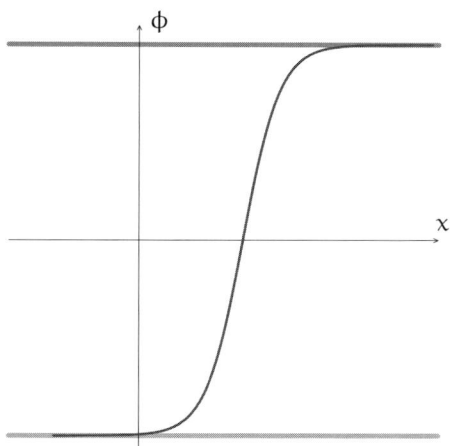

FIGURE 13.2: Domain wall profile corresponding to the potential of Figure 13.1.

which is nothing but the classical equation of motion (13.3). Solutions of this equation with prescribed boundary values $\pm\phi_*$ at $x = \pm\infty$ interpolate between the two ground states of the potential of Figure 13.1. The ground state $\phi = +\phi_*$ is realized at $x \to +\infty$, while the other ground state is realized at $x \to -\infty$. Since these two vacua correspond to two different ways of spontaneously breaking the $\phi \to -\phi$ symmetry, there must exist an interface between the two phases, called a *domain wall* (see Figure 13.2). From eq. (13.9), we may write

$$x(\phi) = x_0 + \int_0^\phi \frac{d\xi}{\sqrt{2\,V(\xi)}}, \tag{13.11}$$

where $x_0$ is an integration constant that can be interpreted as the coordinate where the field $\phi$ is zero. In other words, $x_0$ is the location of the center of the domain wall that separates the regions of different vacua. The domain wall is a local minimum of the energy density (and the absolute minimum for the mixed boundary conditions of case 3). Moreover, it is separated from the (lower-energy) configurations of cases 1 and 2 that have a constant field by an infinite energy barrier.[2] Indeed, going from case 3 to case 1 implies shifting the value of the field from $-\phi_*$ to $+\phi_*$ in the (infinite) vicinity of $x = -\infty$. In the middle of this process, the field in this region will be $\phi = 0$, at which $V(\phi) = V_0 > 0$, a configuration that has an infinite energy density. Thus, the domain wall solution is stable, except for shifts of $x_0$ (since the energy density is independent of $x_0$); the domain wall may move along the $x$ axis, but cannot disappear.

Let us finish by a note on the $y, z$ dependence that has been neglected so far. Reintroducing the transverse dependence adds the term $\frac{1}{2}\big((\partial_y\phi)^2 + (\partial_z\phi)^2\big)$ to the integrand of the energy density in eq. (13.4). This term is positive, or zero for fields that do not depend on $y$ and $z$. Therefore, the minimum of energy density is reached for domain walls that are invariant by translation in the transverse directions. Domain walls that are not translation invariant are not stable, but will relax to this $y, z$ invariant configuration. Physically, one may view the term

---

[2] From this fact, we may infer that domain walls are also stable quantum mechanically.

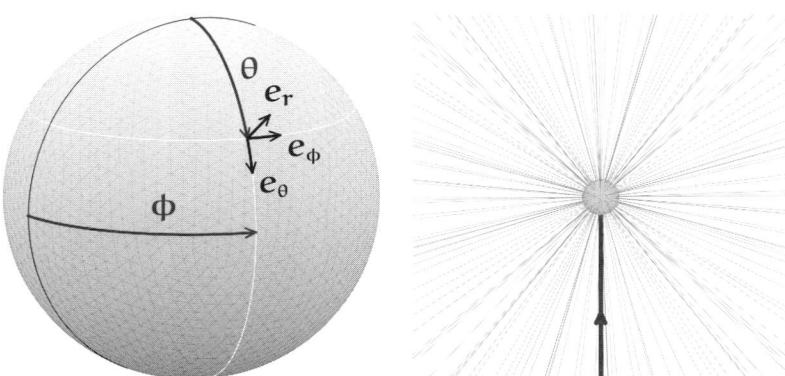

FIGURE 13.3: Left: Notations for the polar coordinates local frame used in eq. (13.13). Right: Magnetic field lines of the Dirac monopole, corresponding to the vector potential of eq. (13.13).

$\frac{1}{2}\big((\partial_y \phi)^2 + (\partial_z \phi)^2\big)$ as a surface tension energy, and the energetically favored configurations are those for which the interface has the lowest curvature.

## 13.2 Monopoles

### 13.2.1 Dirac Monopole

Magnetic monopoles are not forbidden in quantum electrodynamics (QED), but their existence would automatically lead to the quantization of electrical charge, as first noted by Dirac. Let us reproduce this argument here. Consider the radial magnetic field of a would-be monopole:

$$\mathbf{B} = g \frac{\widehat{x}}{|x|^2}. \tag{13.12}$$

Maxwell's equation $\boldsymbol{\nabla} \cdot \mathbf{B} = 0$ implies that we cannot find a vector potential $\mathbf{A}$ for this magnetic field in all space. But it is possible to find one that works almost everywhere; for instance:

$$\mathbf{A}(x) = g \frac{1 - \cos \theta}{|x| \sin \theta} \, \mathbf{e}_\phi, \tag{13.13}$$

where $\theta$ is the polar angle, $\phi$ is the azimuthal angle and $\mathbf{e}_\phi$ is the unit vector tangent to the circle of constant $|x|$ and $\theta$. This vector potential is not defined on the semi-axis $\theta = \pi$ (i.e., the semi-axis of negative $z$). One may argue that on this semi-axis, we have in addition to the monopole field a singular $B_z$ whose magnetic flux precisely cancels the magnetic flux of the monopole, so that the total flux on any closed surface containing the origin is zero, as illustrated in Figure 13.3. Thus, in this solution, the magnetic flux $\Phi_m \equiv 4\pi g$ of the monopole is "brought from infinity" by an infinitely thin "solenoid." Even if it is infinitely thin, such a solenoid may in principle be detected by looking for interferences between the wavefunctions of charged particles that have propagated left and right of the solenoid (this corresponds to the *Aharonov–Bohm effect*). For a particle of electrical charge e, the corresponding phase shift is $e\Phi_m = 4\pi eg$. Dirac pointed out that this interference is absent when the phase shift is a multiple of $2\pi$, i.e., when the electric and magnetic charges are related by

$$ge = \frac{n}{2}, \quad n \in \mathbb{Z}. \tag{13.14}$$

Thus, electrodynamics can perfectly accommodate genuine magnetic monopoles, provided this condition is satisfied, since the annoying solenoid that comes with the above vector potential is totally undetectable. In particular, this would imply that all electrical charges should be multiples of some elementary quantum of electrical charge if monopoles exist. Note that in QED, while the electric and magnetic charges must be related by eq. (13.14), there is no constraint a priori on the mass of monopoles and it should be regarded as a free parameter.

Let us mention briefly an alternative argument that does not involve discussing the detectability of Dirac's solenoid. Instead of the vector potential of eq. (13.13), one could instead have chosen

$$\mathbf{A}'(x) = -g \frac{1 + \cos\theta}{|x|\sin\theta} \mathbf{e}_\phi, \tag{13.15}$$

which has a singularity on the semi-axis $\theta = 0$. When Dirac's quantization condition is satisfied, one may patch eqs. (13.13) and (13.15) in order to obtain a vector potential which is regular in all space (except at the origin, where the monopole is located). To see this, consider a region $\Omega_1$ corresponding to $0 \leq \theta \leq 3\pi/4$ and a region $\Omega_2$ corresponding to $\pi/4 \leq \theta \leq \pi$. Then, we choose $\mathbf{A}$ in $\Omega_1$ and $\mathbf{A}'$ in $\Omega_2$. In the overlap of the two regions, $\pi/4 \leq \theta \leq 3\pi/4$, we have

$$(\mathbf{A} - \mathbf{A}') \cdot d\mathbf{x} = 2g\, d\phi, \tag{13.16}$$

and we can write

$$\mathbf{A} - \mathbf{A}' = \boldsymbol{\nabla}\chi, \quad \text{with } \chi(\phi) \equiv 2g\,\phi. \tag{13.17}$$

For this to be an acceptable gauge transformation, the phase by which it multiplies the wavefunction of a charged particle should be single-valued, i.e.,

$$e^{ie\chi(\phi+2\pi)} = e^{ie\chi(\phi)}, \tag{13.18}$$

which is precisely the case when condition (13.14) is satisfied.

This argument can even be made without any reference to the explicit solutions of eqs. (13.13) and (13.15). Let us consider a large sphere surrounding the origin. Divide it into upper and lower hemispheres (see Figure 13.5), and denote $\mathbf{A}$ and $\mathbf{A}'$ the vector potentials that represent the monopole magnetic field in these two hemispheres. On the equator, their difference should be a pure gauge:

$$\mathbf{A} - \mathbf{A}' = \frac{i}{e} \Omega^\dagger(x) \boldsymbol{\nabla}\, \Omega(x). \tag{13.19}$$

Along the equator, we thus have

$$\Omega(\phi) = \Omega(0) \, \exp\left\{ - ie \int_{\gamma[0,\phi]} (\mathbf{A} - \mathbf{A}') \cdot d\mathbf{x} \right\}, \tag{13.20}$$

where the integration path $\gamma[0, \phi]$ is the portion of the equator that extends between the azimuthal angles 0 and $\phi$. After a complete revolution, we have

$$\Omega(2\pi) = \Omega(0) \, \exp\left\{ - ie \underbrace{\oint_{\text{Equator}} (\mathbf{A} - \mathbf{A}') \cdot d\mathbf{x}}_{} \right\}$$

$$= \Omega(0) \, \exp\left\{ - ie \big( \underbrace{\Phi_{\text{U}} + \Phi_{\text{L}}}_{\text{flux} = 4\pi g} \big) \right\} = \Omega(0) \, e^{-4\pi i \, eg}. \tag{13.21}$$

To obtain the first equality on the second line, we use the Stokes theorem to rewrite the contour integrals of $\mathbf{A}$ and $\mathbf{A}'$ as surface integrals of the corresponding magnetic field. Therefore, we obtain the magnetic fluxes through the upper and lower hemispheres, respectively, whose sum is the total flux $4\pi g$ of the monopole. Requesting the single-valuedness of $\Omega$ leads to Dirac's condition on $eg$.

### 13.2.2 Monopoles in Non-Abelian Gauge Theories

There are also non-Abelian field theories that exhibit $U(1)$ magnetic monopoles, as classical solutions whose stability is ensured by topology. The simplest example is an $SU(2)$ gauge theory coupled to a scalar field in the adjoint representation,[3] whose Lagrangian density reads

$$\mathcal{L} \equiv -\frac{1}{4} F^a_{\mu\nu} F^{a,\mu\nu} + \frac{1}{2} (D_\mu \Phi)_a (D^\mu \Phi)_a - V(\Phi), \tag{13.22}$$

with

$$V(\Phi) \equiv \frac{\lambda}{8} \left( \Phi^a \Phi^a - v^2 \right)^2,$$
$$(D_\mu \Phi)_a = \partial_\mu \Phi_a - e \, \epsilon_{abc} A^b_\mu \Phi^c,$$
$$F^a_{\mu\nu} = \partial_\mu A^a_\nu - \partial_\nu A^a_\mu - e \, \epsilon_{abc} A^b_\mu A^c_\nu, \tag{13.23}$$

where we have written explicitly the structure constants of the $\mathfrak{su}(2)$ algebra. In order to study static classical solutions, it is simpler to consider the minima of the energy,

$$\mathcal{E} \equiv \int d^3x \left\{ \frac{1}{2} \left( E^a_i E^a_i + B^a_i B^a_i + (D_i \Phi)_a (D_i \Phi)_a \right) + V(\Phi) \right\}, \tag{13.24}$$

where $E^a_i \equiv F^a_{0i}$ is the (non-Abelian) electrical field and $B^a_i \equiv \frac{1}{2} \epsilon_{ijk} F^a_{jk}$ is the magnetic field.

---

[3] This model is known as the Georgi–Glashow model. It was considered at some point as a possible candidate for a field theory of electroweak interactions, until the neutral vector boson $Z^0$ was discovered. Here, we use it as a didactical example of a theory with classical solutions that are magnetic monopoles.

## 13.2 MONOPOLES

It is possible to choose a gauge (called the unitary gauge) in which the scalar field triplet takes the form

$$\Phi^a = (0, 0, v + \varphi). \tag{13.25}$$

In this equation, we have anticipated spontaneous symmetry breaking, which will give to the scalar field a vacuum expectation value $v$, and we have made a specific choice about the orientation of the vacuum in $SU(2)$. The field $\varphi$ is thus the quantum fluctuation of the scalar about its expectation value. In this process, the fields $A_\mu^{1,2}$ will become massive (with a mass $M_W = ev$), as well as the scalar field (with mass $M_\Phi = \sqrt{\lambda} v$), while the field $A_\mu^3$ remains massless (it corresponds to a residual unbroken $U(1)$ symmetry). The classical vacuum of this theory corresponds to

$$\varphi = 0, \quad A_\mu^a = 0. \tag{13.26}$$

Now, we seek stable classical field configurations that are local minima of the energy, but are not equivalent to the vacuum in the entire space. To prove the existence of such fields, it is sufficient to exhibit a field configuration of non-zero energy that cannot be continuously deformed into the null fields of eq. (13.26) (up to a gauge transformation). In order to have a finite energy, the scalar field $\Phi^a$ should reach a minimum of the potential $V(\Phi)$ at large distance $|x| \to \infty$ (we have shifted the potential so that its minimum is zero), but it may approach different minima depending on the direction $\hat{x}$ in space.

The allowed asymptotic behaviors of $\Phi^a$ define a mapping from the sphere $S_2$ (the orientations $\hat{x}$ for three spatial dimensions) to the sphere $\Phi^a \Phi^a = v^2$ of the minima of $V(\Phi)$. Since $\mathfrak{su}(2)$ is three-dimensional, the set of zeroes of the scalar potential is also a sphere $S_2$, and it is natural to consider the following configuration:[4]

$$\Phi^a(\hat{x}) \equiv v \hat{x}^a, \tag{13.27}$$

sometimes called a "hedgehog field" because the direction of internal space pointed to by the scalar field is locked to the spatial direction, as shown in Figure 13.4. Any smooth field $\Phi^a$ that obeys this boundary condition at infinite spatial distance must vanish at some point in the interior of the sphere. Therefore, it cannot simply be a gauge transform[5] of the constant field $\Phi^a = v \delta_{3a}$ (the expectation value of the scalar field in the vacuum). The classes of fields that can be continuously deformed into one another are given by a homotopy group, in this case the second homotopy group $\pi_2(\mathcal{M}_0)$ where $\mathcal{M}_0$ is the manifold of the minima of the scalar potential. For the $SU(2)$ group, $\mathcal{M}_0$ is topologically equivalent to the 2-sphere $S_2$, and the equivalence classes of the mappings $S_2 \mapsto S_2$ are indexed by the integers, since $\pi_2(S_2) = \mathbb{Z}$. The hedgehog field of eq. (13.27) has topological number $+1$, while the vacuum has topological number $0$.

---

[4]Here, we see that it is crucial that the scalar potential has non-trivial minima. If $\Phi^a \equiv 0$ was the only minimum, it would not be possible to construct solutions of finite energy that are not topologically equivalent to the vacuum.

[5]Recall that $\Phi_a \to \Omega_{ab}^\dagger \Phi_b$ under a gauge transformation, and therefore a non-zero scalar field cannot be made to vanish by changing the gauge.

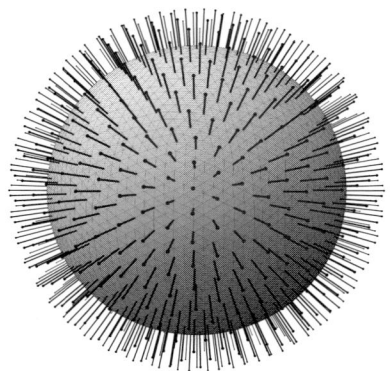

FIGURE 13.4: Cartoon of the hedgehog configuration of eq. (13.27). Each needle indicates the internal orientation of $\Phi^a$ at the corresponding point on the sphere.

At spatial infinity, the hedgehog configuration of eq. (13.27) is gauge equivalent to the standard scalar vacuum aligned with the third color direction, $\Phi^a = v\delta_{3a}$. In order to see this, let us introduce the following $SU(2)$ transformation that depends on the polar angle $\theta$ and azimuthal angle $\phi$:[6]

$$\Omega(\theta, \phi) \equiv -\cos\frac{\theta}{2}\sin\phi + 2i\left(\sin\frac{\theta}{2}\,t_f^1 + \cos\frac{\theta}{2}\cos\phi\,t_f^3\right), \tag{13.28}$$

where $t_f^a$ are the generators of the fundamental representation of $\mathfrak{su}(2)$. Then, one may check explicitly that (see Exercise 13.1)

$$\delta_{3a}\,\Omega^\dagger t^a \Omega = \sin\theta\left(\cos\phi\,t_f^1 + \sin\phi\,t_f^2\right) + \cos\theta\,t_f^3 = \widehat{x}_a t_f^a. \tag{13.29}$$

Thus, $\Omega$ transforms the usual scalar vacuum into the hedgehog configuration at infinity. Note that eq. (13.28) is not a valid gauge transformation over the entire space because it is not well defined at the origin.

The choice of eq. (13.27) for the asymptotic behavior of the scalar field was motivated by the requirement that the potential $V(\Phi)$ gives a finite contribution to the energy. The term in $(D_i\Phi)_a^2$ should also give a finite contribution. However, note that

$$\partial_i \Phi^a(\widehat{x}) = \frac{v}{|x|}\left(\delta^{ia} - \widehat{x}^i\widehat{x}^a\right) \tag{13.30}$$

is not square integrable. We must therefore adjust the asymptotic behavior of the gauge potential in order to cancel this term in the covariant derivative, by requesting that

$$\epsilon_{abc}\,A_i^b\,\widehat{x}^c \underset{|x|\to\infty}{=} \frac{\delta^{ia} - \widehat{x}^i\widehat{x}^a}{e\,|x|}, \tag{13.31}$$

---

[6]When an $SU(2)$ transformation in the fundamental representation is written as
$$\Omega \equiv u_0 + 2i\,u_a\,t_f^a,$$
its unitarity ($\Omega^\dagger\Omega = 1$) is equivalent to $u_0^2 + u_1^2 + u_2^2 + u_3^2 = 1$.

## 13.2 MONOPOLES

which is satisfied if

$$A_i^b \underset{|x|\to\infty}{=} \frac{\epsilon_{ibd}\,\widehat{x}^d}{e\,|x|} + \text{term in } \widehat{x}^b. \tag{13.32}$$

The corresponding field strength and magnetic field are given by

$$F_{ij}^a \underset{|x|\to\infty}{=} \frac{1}{e\,|x|^2}\left(2\,\epsilon_{ija} + 2\left(\epsilon_{iad}\widehat{x}^j - \epsilon_{jad}\widehat{x}^i\right)\widehat{x}^d - \epsilon_{ijd}\,\widehat{x}^a\widehat{x}^d\right),$$

$$B_i^a \underset{|x|\to\infty}{=} \frac{\widehat{x}^i\widehat{x}^a}{e\,|x|^2}. \tag{13.33}$$

Therefore, at large distance (these considerations do not give the precise form of the fields at finite distance) there is a purely radial magnetic field that vanishes like $|x|^{-2}$, i.e., according to Coulomb's law, thus suggesting that a magnetic monopole is present at the origin. For a more robust interpretation, we should apply a gauge transformation that maps the asymptotic hedgehog scalar field into the usual scalar vacuum, aligned with the third color direction. Thanks to eq. (13.29), we see that in this process the magnetic field of eq. (13.33), proportional to $\widehat{x}^a$, will become proportional to $\delta_{3a}$. But the third color direction precisely corresponds to the gauge potential that remains massless in the spontaneous symmetry breaking $SU(2) \to U(1)$. Therefore, eq. (13.33) is indeed the magnetic field of a $U(1)$ magnetic monopole. Its flux through a sphere surrounding the origin is

$$\Phi_m = \frac{4\pi}{e}, \tag{13.34}$$

equivalent to that of a magnetic charge $g \equiv e^{-1}$ at the origin.

Until now, we have only discussed the implications of requiring a finite energy on the asymptotic form of the scalar field and of the gauge potentials. In order to obtain their values at finite distance, one may make the following ansatz:

$$\Phi^a(x) = \frac{v\,\widehat{x}^a}{\xi}\,f(\xi), \quad A_i^a(x) = \frac{v\epsilon_{iab}\widehat{x}^b}{\xi}(1 - g(\xi)), \tag{13.35}$$

where $f, g$ are two functions of $\xi \equiv ev|x|$ that can be determined from the classical equations of motion (see exercise 13.2). From this solution over the entire space, one sees that the monopole is an extended object made of two parts:

- A compact core, of radius $R_m \sim M_W^{-1}$, in which the $SU(2)$ symmetry is unbroken and the vector bosons are all massless. One may view the core as a cloud of highly virtual gauge bosons and scalars.
- Beyond this radius, a halo in which the $SU(2)$ symmetry is spontaneously broken. In this halo, up to a gauge transformation, the scalar field is that of the ordinary broken vacuum, the vector bosons $A_{1,2}$ are massive and the $A_3$ field is massless, with a tail that corresponds to a radial $U(1)$ magnetic field.

Given these fields, the total energy of the field configuration can be identified with the mass (in contrast with Dirac's point-like monopole in QED, whose mass is not constrained) of the monopole (since it is static). It takes the form

$$M_m = \frac{4\pi}{e^2} M_w \, C(\lambda/e^2), \qquad (13.36)$$

where $C(\lambda/e^2)$ is a function of the ratio of coupling constants, of order unity. Note also that the core and the halo contribute comparable amounts to this mass. Interestingly, the size $M_w^{-1}$ of this monopole is much larger (by a factor $\alpha^{-1} = 4\pi/e^2$) than its Compton wavelength $M_m^{-1}$. Therefore, when $\alpha \ll 1$, the monopole receives very small quantum corrections and is a nearly classical object.

We have argued earlier that the topologically non-trivial configurations of the scalar field that lead to a finite energy can be classified according to the homotopy group $\pi_2(S_2)$. Since this group is the group $\mathbb{Z}$ of the integers, there are monopole solutions with any magnetic charge multiple of $e^{-1}$ (the solution we have constructed explicitly above has topological number 1), i.e.,

$$ge = n, \quad n \in \mathbb{Z}. \qquad (13.37)$$

Therefore, in this field theoretical monopole solution, the electrical charge would also be naturally quantized. At first sight, eqs. (13.37) and (13.14) appear to differ by a factor of $1/2$. Note, however, that in the $SU(2)$ model we are considering in this section, it is possible to introduce matter fields in the fundamental representation[7] that carry a $U(1)$ electrical charge $\pm e/2$ (this is the smallest possible electrical charge in this model). Thus, if rewritten in terms of this minimal electrical charge, the monopole quantization condition of eq. (13.37) is in fact identical to Dirac's condition. Although the Georgi–Glashow model studied in this section is no longer considered as phenomenologically relevant, theories that unify the strong and electroweak interactions into a unique compact Lie group (such as $SU(5)$, for instance) do have magnetic monopoles.

### 13.2.3 Topological Considerations

In the previous two subsections, we have encountered two seemingly different topological classifications of magnetic monopoles. The Dirac monopole appeared closely related to the mappings from a circle (the equator between the two hemispheres in Figure 13.5) to the group $U(1)$, whose classes are the elements of the homotopy group $\pi_1(U(1)) = \mathbb{Z}$. In contrast, the monopole discussed in the Georgi–Glashow model was related to the behavior of the scalar field at large distance, i.e., to mappings from the 2-sphere $S_2$ to the manifold $\mathcal{M}_0$ of the minima

---

[7]If $\Psi$ is a doublet that lives in this representation, the covariant derivative acting on it reads

$$D_\mu \Psi = \partial_\mu \Psi - i e A_\mu^a t_f^a \Psi = \partial_\mu \Psi - i \frac{e}{2} A_\mu^3 \begin{pmatrix} 1 & 0 \\ 0 & -1 \end{pmatrix} \Psi.$$

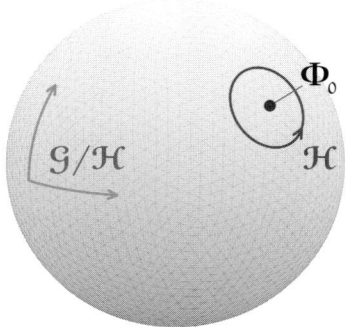

FIGURE 13.5: Illustration of the symmetry-breaking pattern. $\mathcal{H}$ is the residual invariance after choosing a minimum $\Phi_0$. The coset $\mathcal{G}/\mathcal{H}$ is the manifold that holds the minima of $V(\Phi)$ (a 2-sphere in the case of the Georgi–Glashow model).

of the scalar potential $V(\Phi)$, whose equivalence classes are the elements of the homotopy group $\pi_2(\mathcal{M}_0) = \mathbb{Z}$.

Let us now argue that these two ways of viewing monopoles are in fact equivalent. In order to make this discussion more general, consider a gauge theory with internal group $\mathcal{G}$, coupled to a scalar, spontaneously broken to a residual gauge symmetry of group $\mathcal{H}$. Let us denote $\mathcal{M}_0$ the manifold of the minima of the scalar potential. This manifold is invariant under transformations of $\mathcal{G}$. Given a minimum $\Phi_0$, the other minima can be obtained by multiplying $\Phi_0$ by the elements of $\mathcal{G}$:

$$\mathcal{M}_0 = \Big\{ \Phi \,\Big|\, \Phi = \Omega \Phi_0; \Omega \in \mathcal{G} \Big\}. \tag{13.38}$$

(Here, we are assuming that there are no accidental degeneracies among the minima, i.e., no minima $\Phi_0$ and $\Phi_0'$ that are not related by a gauge transformation.) The manifold defined in eq. (13.38) is in fact the coset $\mathcal{G}/\mathcal{H}$,

$$\mathcal{M}_0 = \mathcal{G}/\mathcal{H}. \tag{13.39}$$

This pattern of spontaneous symmetry breaking is illustrated in Figure 13.6.

The first way of classifying monopoles is to consider the gauge field on a sphere, as was done in Section 13.2.1. At large distance compared to the inverse mass of the bosons that became massive due to spontaneous symmetry breaking, only the massless gauge bosons contribute, and the corresponding gauge fields live in the algebra $\mathfrak{h}$ of the residual group $\mathcal{H}$. We can reproduce the argument made at the end of Section 13.2.1. The gauge potentials in the upper and lower hemispheres are related on the equator by a gauge transformation $\Omega(\phi) \in \mathcal{H}$, which must be single-valued as the azimuthal angle $\phi$ wraps around the equator. $\Omega(\phi)$ is therefore a mapping from the circle $S_1$ to the residual gauge group $\mathcal{H}$. These mappings can be grouped into classes that differ by their winding number (the set of these classes is the first homotopy group $\pi_1(\mathcal{H})$). In this general setting, we may adopt the winding number as the *definition* of the product $eg$ of the electric charge by the magnetic charge comprised within the sphere – see Exercise 13.3 for an example. Note that $\pi_1(\mathcal{H})$ is discrete, and therefore

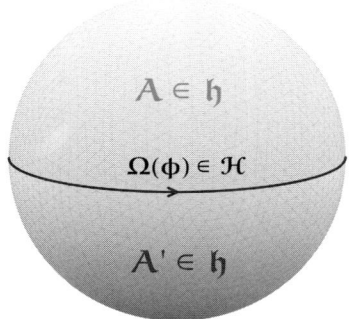

FIGURE 13.6: Decomposition of the sphere into two hemispheres with gauge potentials $\mathbf{A}$ and $\mathbf{A}'$.

the winding number can vary only by finite jumps.[8] Moreover, the mapping $\Omega(\phi)$ on the equator is a smooth function of the azimuthal angle $\phi$ and of the radius R of the sphere. Consequently, the winding number must be independent of the radius R. From this fact, two different situations may arise:

- The relevant gauge fields belong to $\mathfrak{h}$ all the way down to zero radius. In this case, the magnetic charge is independent of the radius of the sphere at all R, which means that the monopole is a point-like singularity at the origin, like the original Dirac monopole.
- There exists a short-distance core in which the gauge fields live in an algebra which is larger than $\mathfrak{h}$ (possibly the algebra $\mathfrak{g}$ before symmetry breaking). Inside this core, the above argument is no longer valid, and the magnetic charge inside the sphere may vary continuously with the radius. In this case, the monopole is an extended object whose size is the radius of the core (its magnetic charge is spread out in the core).

Alternatively, we may construct a monopole as a non-trivial classical field configuration that minimizes the energy, by starting from the behavior at infinity of the scalar field. In order to have a finite energy, the scalar field should go to a minimum of $V(\Phi)$ when $|x| \to \infty$. The asymptotic scalar field is therefore a mapping from the 2-sphere $S_2$ to $\mathcal{M}_0 = \mathcal{G}/\mathcal{H}$, and it leads to a classification of the classical field configurations based on the homotopy group $\pi_2(\mathcal{G}/\mathcal{H})$. The correspondence between the two points of view is based on the following relationship:

$$\pi_2(\mathcal{G}/\mathcal{H}) = \pi_1(\mathcal{H})/\pi_1(\mathcal{G}). \tag{13.40}$$

For a simply connected Lie group $\mathcal{G}$ (e.g., all the SU(N)), the first homotopy group is trivial, $\pi_1(\mathcal{G}) = \{0\}$, and we have

$$\pi_2(\mathcal{G}/\mathcal{H}) = \pi_1(\mathcal{H}), \tag{13.41}$$

hence the equivalence between the two ways of classifying monopoles.

---

[8]Therefore, it must be conserved by time evolution. Indeed, time evolution is continuous, and the only way for a discrete quantity to evolve continuously is to be constant.

## 13.3 Instantons

Until now, all the extended field configurations we have encountered were time independent. After integration over time, their action is infinite, and therefore they do not contribute to path integrals. In this section, we will discuss field configurations of finite action, called *instantons*, that are localized both in space and in time. Consider a Yang–Mills theory in D-dimensional Euclidean space, whose action reads

$$S[A] \equiv \frac{1}{4} \int d^D x \, F_a^{ij}(x) F_a^{ij}(x). \tag{13.42}$$

(We use Latin indices i, j, k, $\cdots$ for Lorentz indices in Euclidean space.) Instantons are non-trivial (i.e., not pure gauges in the entire space-time) gauge field configurations that realize local minima of this action.

### 13.3.1 Asymptotic Behavior

In order to have a finite action, these fields must go to a pure gauge when $|x| \to \infty$:

$$A_a^i t^a \underset{|x| \to \infty}{\to} \frac{i}{g} \Omega^\dagger(\widehat{x}) \partial^i \Omega(\widehat{x}), \tag{13.43}$$

where $\Omega(\widehat{x})$ is an element of the gauge group that depends only on the orientation $\widehat{x}$. Since multiplying $\Omega(\widehat{x})$ by a constant group element $\Omega_0$ does not change the asymptotic gauge potential, we can always arrange that $\Omega(\widehat{x}_0) = 1$ for some fixed orientation $\widehat{x}_0$. Note that a gauge potential such as in eq. (13.43) that becomes a pure gauge at large distance (see Figure 13.7), must decrease at least as fast as $|x|^{-1}$. More precisely, we may write

$$A_a^i(x) t^a = \underbrace{\frac{i}{g} \Omega^\dagger(\widehat{x}) \partial^i \Omega(\widehat{x})}_{|x|^{-1}} + \underbrace{a^i(x)}_{\ll |x|^{-1}}. \tag{13.44}$$

The field strength associated to such a field decreases faster than $|x|^{-2}$, and therefore the corresponding action is finite in D = 4 dimensions. There is in fact a scaling argument showing that instanton solutions can only exist in four dimensions. Given an instanton field configuration $A^i(x)$ and a scaling factor R, let us define

$$A_R^i(x) \equiv \frac{1}{R} A^i(x/R). \tag{13.45}$$

Since classical Yang–Mills theory is scale invariant, the field $A_R^i$ is also an extremum of the action (i.e., a solution of the classical Yang–Mills equations) if $A^i$ is. The action of this rescaled field is given by

$$S[A_R] = R^{D-4} S[A]. \tag{13.46}$$

Therefore, given an instanton $A^i(x)$, we may continuously deform it into another field configuration $A_R^i(x)$ whose action is multiplied by $R^{D-4}$. Unless D = 4, this action has a higher or

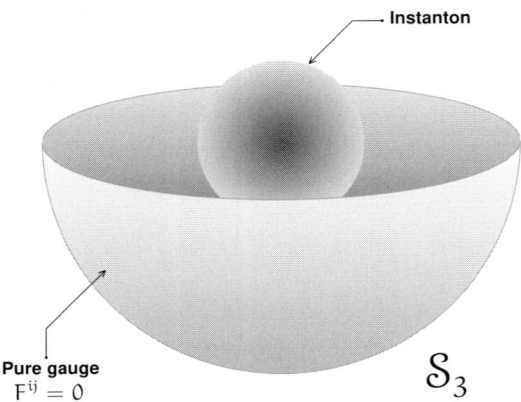

FIGURE 13.7: Cartoon of an instanton (the illustration is for D = 3, although instantons actually exist in D = 4). The sphere $S_3$ is in fact infinitely far away from the center of the instanton.

lower value, in contradiction with the fact that $A^i$ was a local extremum.[9] Thus, non-trivial local extrema of the classical Euclidean Yang–Mills action can only exist in D = 4. In four dimensions, if $A^i$ is an instanton, then $A^i_R$ is also an instanton (with the same value of the action). Thus, classical instantons can exist with any size. But this degeneracy is lifted by quantum corrections that introduce a scale into Yang–Mills theory via the running coupling.

### 13.3.2 Bogomol'nyi Inequality and Self-Duality Condition

In the study of instantons, a useful variant of the Bogomol'nyi trick is to start from the following obvious inequality:

$$0 \leq \int d^4x \, (F^a_{ij} \mp \frac{1}{2}\epsilon_{ijkl}F^a_{kl})^2, \tag{13.47}$$

which leads to

$$0 \leq \int d^4x \left( F^a_{ij}F^a_{ij} \mp \epsilon_{ijkl}F^a_{ij}F^a_{kl} + \frac{1}{4}\epsilon_{ijkl}\epsilon_{ijmn}F^a_{kl}F^a_{mn} \right)$$

$$= \int d^4x \left( F^a_{ij}F^a_{ij} \mp \epsilon_{ijkl}F^a_{ij}F^a_{kl} + \frac{1}{2}(\delta_{km}\delta_{ln} - \delta_{kn}\delta_{lm})F^a_{kl}F^a_{mn} \right)$$

$$= \int d^4x \left( 2F^a_{ij}F^a_{ij} \mp \epsilon_{ijkl}F^a_{ij}F^a_{kl} \right). \tag{13.48}$$

By choosing appropriately the sign, this can be rearranged into a lower bound for the action,

$$S[A] \geq \frac{1}{8} \left| \epsilon_{ijkl} \int d^4x \, F^a_{ij}F^a_{kl} \right|, \tag{13.49}$$

---

[9]The only exception to this reasoning occurs if $S[A] = 0$. But this happens only in the trivial situation where $A^i$ is a pure gauge in the entire space-time.

known as *Bogomol'nyi's inequality*. Interestingly, we recognize on the right-hand side an integral identical to the one that enters in the θ-term of Yang–Mills theories (see Section 7.5) or in the anomaly function (see Section 6.5 and Chapter 12). This inequality becomes an equality when

$$F^a_{ij} = \pm \frac{1}{2} \epsilon_{ijkl} F^a_{kl}. \tag{13.50}$$

A solution that obeys this condition is by construction a minimum of the Euclidean action $S[A]$, and therefore a solution of the classical Yang–Mills equations. But as in the case of domain walls, finding field configurations that fulfill this self-duality condition is somewhat simpler than solving directly the Yang–Mills equations. Thus, from now on, we will look for gauge fields that fulfill eq. (13.50) and go to a pure gauge as $|x| \to \infty$.

### 13.3.3 Topological Classification

In $D = 4$, the functions $\Omega(\hat{x})$ that define the asymptotic behavior of instantons map the 3-sphere $S_3$ into the gauge group $\mathcal{G}$,

$$\Omega : \quad S_3 \mapsto \mathcal{G}, \tag{13.51}$$

with a fixed value $\Omega(\hat{x}_0) = 1$. These functions can be grouped into topological classes, such that mappings belonging to the same class can be continuously deformed into one another. The set of these classes can be endowed with a group structure, called the third homotopy group of $\mathcal{G}$ and denoted $\pi_3(\mathcal{G})$ (for any SU(N) group with N ≥ 2, we have $\pi_3(\mathcal{G}) = \mathbb{Z}$). Note that the asymptotic forms of the fields $A^i$ and $A^i_R$ are identical, implying that these two instantons belong to the same topological class. Since their actions are identical in four dimensions, this scaling provides a continuous family of instantons that belong to the same topological class and have the same action. This is in fact more general: We will show later that the action of an instanton depends only on the topological class of the instanton, and therefore can only vary by discrete amounts.

### 13.3.4 Minimal Action

Let us assume that we have found a self-dual gauge field configuration that realizes the equality in eq. (13.49). In order to calculate its action, we can use the fact that $\epsilon_{ijkl} F^a_{ij} F^a_{kl}$ is a total derivative,

$$\frac{1}{2} \epsilon_{ijkl} F^a_{ij} F^a_{kl} = \partial_i \underbrace{\left[ \epsilon_{ijkl} \left( A^a_j F^a_{kl} - \frac{g}{3} f^{abc} A^a_j A^b_k A^c_l \right) \right]}_{K^i}. \tag{13.52}$$

(This property was derived in Section 7.5.) The vector $K^i$ can also be written as a trace of objects belonging to the fundamental representation,

$$K^i = 2 \epsilon_{ijkl} \, \text{tr} \left( A_j F_{kl} + \frac{2ig}{3} A_j A_k A_l \right). \tag{13.53}$$

Since the integrand on the right-hand side of eq. (13.49) is a total derivative, one may use the Stokes theorem to rewrite the integral as a three-dimensional integral extended to a spherical hypersurface $\mathcal{S}_R$ of radius $R \to \infty$,

$$\mathcal{S}_{\min}[A] \equiv \frac{1}{8}\epsilon_{ijkl}\int d^4x\, F_{ij}^a F_{kl}^a = \lim_{R\to\infty}\frac{1}{4}\int_{\mathcal{S}_R} d^3S_i\, K^i. \tag{13.54}$$

Thus, the minimum of the action depends only on the behavior of the gauge field at large distance (this does not mean that the action does not depend on details of the gauge field in the interior, but more simply that the gauge fields that realize the minima are fully determined in the bulk by their asymptotic behavior). From the earlier discussion of the asymptotic behavior of instanton solutions, we know that

$$A_i(x) \underset{|x|\to\infty}{\sim} |x|^{-1}, \quad F_{ij}(x) \underset{|x|\to\infty}{\ll} |x|^{-2}. \tag{13.55}$$

Therefore, in the current $K^i$, the term $A_j F_{kl}$ is negligible in front of the term $A_j A_k A_l$ at large distance, and we can also write

$$\begin{aligned}
\mathcal{S}_{\min}[A] &= \lim_{R\to\infty}\frac{ig}{3}\int_{\mathcal{S}_R} d^3S_i\, \epsilon_{ijkl}\,\text{tr}\left(A_j A_k A_l\right) \\
&= \lim_{R\to\infty}\frac{1}{3g^2}\int_{\mathcal{S}_R} d^3S_i\, \epsilon_{ijkl}\,\text{tr}\left(\Omega^\dagger(\partial_j\Omega)\Omega^\dagger(\partial_k\Omega)\Omega^\dagger(\partial_l\Omega)\right),
\end{aligned} \tag{13.56}$$

where $\Omega(\hat{x})$ is the group element that defines the asymptotic pure gauge behavior of the gauge potential in the direction $\hat{x}$. In this expression, each derivative brings a factor $R^{-1}$, while the domain of integration scales as $R^3$. The result is therefore independent of the radius of the sphere and we can ignore the limit $R \to \infty$.

On this sphere, let us choose a system of coordinates made of three variables $(\theta_1, \theta_2, \theta_3)$, such that the volume element in $\mathcal{S}_R$ is $d\theta_1 d\theta_2 d\theta_3$. To rewrite the previous integral more explicitly in terms of these variables, it is convenient to introduce a fourth – radial – coordinate $\theta_0 \equiv |x|$. The coordinates $(\theta_0, \theta_1, \theta_2, \theta_3)$ are thus coordinates in $\mathbb{R}^4$, and $d^4x = d\theta_0 d\theta_1 d\theta_2 d\theta_3$. The volume element on the sphere $\mathcal{S}_R$ is $d\theta_1 d\theta_2 d\theta_3 = d^4x\, \delta(\theta_0 - R)$. Noting that $\hat{x}_i = \partial\theta_0/\partial x_i$, we can write

$$d^3S_i = \hat{x}_i\, d\theta_1 d\theta_2 d\theta_3 = \frac{\partial\theta_0}{\partial x_i}\,d\theta_1 d\theta_2 d\theta_3 = d^4x\,\delta(\theta_0 - R)\frac{\partial\theta_0}{\partial x_i} \tag{13.57}$$

and the minimal action becomes

$$\begin{aligned}
\mathcal{S}_{\min}[A] = \frac{1}{3g^2}\int d^4x\,\delta(\theta_0 - R)\,\epsilon_{ijkl}\frac{\partial\theta_0}{\partial x_i}\frac{\partial\theta_a}{\partial x_j}\frac{\partial\theta_b}{\partial x_k}\frac{\partial\theta_c}{\partial x_l} \\
\times \text{tr}\left\{\Omega^\dagger(\theta)\frac{\partial\Omega(\theta)}{\partial\theta_a}\Omega^\dagger(\theta)\frac{\partial\Omega(\theta)}{\partial\theta_b}\Omega^\dagger(\theta)\frac{\partial\Omega(\theta)}{\partial\theta_c}\right\},
\end{aligned} \tag{13.58}$$

where we have rewritten the derivatives with respect to $x_i$ in terms of derivatives with respect to $\theta_a$ (the implicit sums on $a, b, c$ run over the indices $1, 2, 3$ only, because the group element $\Omega$ depends only on the orientation $\hat{x}$). Finally, we may use

$$\epsilon_{lijk}\frac{\partial\theta_0}{\partial x_i}\frac{\partial\theta_a}{\partial x_j}\frac{\partial\theta_b}{\partial x_k}\frac{\partial\theta_c}{\partial x_l} = \det\left(\frac{\partial(\theta_0\theta_1\theta_2\theta_3)}{\partial(x_1x_2x_3x_4)}\right)\underbrace{\epsilon_{0abc}}_{=\epsilon_{abc}}. \tag{13.59}$$

The determinant is nothing but the Jacobian of the coordinate transformation $\{x_i\} \to \{\theta_a\}$. Therefore, we obtain

$$S_{\min}[A] = \frac{1}{3g^2}\int d\theta_1 d\theta_2 d\theta_3\, \epsilon_{abc}\, \text{tr}\left\{\Omega^\dagger(\theta)\frac{\partial\Omega(\theta)}{\partial\theta_a}\Omega^\dagger(\theta)\frac{\partial\Omega(\theta)}{\partial\theta_b}\Omega^\dagger(\theta)\frac{\partial\Omega(\theta)}{\partial\theta_c}\right\}. \tag{13.60}$$

### 13.3.5 Cartan–Maurer Invariant

**Definition:** In order to calculate the integral that appears in eq. (13.60), let us make a mathematical digression. Consider a d-dimensional manifold $S$, of coordinates $(\theta_1, \theta_2, \cdots, \theta_d)$, a manifold $M$ that may be viewed as a matrix representation of a Lie group and a mapping $\Omega$ from $S$ to $M$:

$$(\theta_1, \theta_2, \cdots, \theta_d) \in S \longrightarrow \Omega(\theta_1, \theta_2, \cdots, \theta_d) \in M. \tag{13.61}$$

The *Cartan–Maurer form* $\mathcal{F}[\Omega]$ is an integral that generalizes the one encountered earlier,

$$\mathcal{F}[\Omega] \equiv \int d\theta_1 \cdots d\theta_d\, \epsilon_{i_1\cdots i_d}\, \text{tr}\left\{\Omega^\dagger(\theta)\frac{\partial\Omega(\theta)}{\partial\theta_{i_1}}\cdots\Omega^\dagger(\theta)\frac{\partial\Omega(\theta)}{\partial\theta_{i_d}}\right\}, \tag{13.62}$$

where $\epsilon_{i_1\cdots i_d}$ is the d-dimensional completely antisymmetric tensor, normalized according to $\epsilon_{12\cdots d} = +1$. In d dimensions, this tensor transforms as follows under circular permutations:

$$\epsilon_{i_1\cdots i_d} = (-1)^{d-1}\,\epsilon_{i_2\cdots i_d i_1}. \tag{13.63}$$

Using the cyclicity of the trace, we conclude that $\mathcal{F}[\Omega] = 0$ if the dimension d is even. In the following, we thus restrict the discussion to the case where d is odd.[10]

**Coordinate independence:** Consider now another system of coordinates on $S$, that we denote $\theta'_i$. We have

$$\epsilon_{i_1\cdots i_d}\frac{\partial\theta'_{j_1}}{\partial\theta_{i_1}}\cdots\frac{\partial\theta'_{j_d}}{\partial\theta_{i_d}} = \det\left(\frac{\partial(\theta'_i)}{\partial(\theta_j)}\right)\epsilon_{j_1\cdots j_d}, \tag{13.64}$$

and the determinant on the right-hand side is the Jacobian of the coordinate transformation. We thus obtain

$$\mathcal{F}[\Omega] = \int d\theta'_1 \cdots d\theta'_d\, \epsilon_{j_1\cdots j_d}\, \text{tr}\left\{\Omega^\dagger(\theta)\frac{\partial\Omega(\theta)}{\partial\theta'_{j_1}}\cdots\Omega^\dagger(\theta)\frac{\partial\Omega(\theta)}{\partial\theta'_{j_d}}\right\}, \tag{13.65}$$

which is identical to eq. (13.62), except for the fact that it is expressed in terms of the new coordinates $\theta'_i$. This proves that $\mathcal{F}[\Omega]$ is independent of the choice of the coordinate system on $S$, and is a property of the manifold $S$ itself.

---

[10]This is the case in the study of instantons, since in this case the manifold $S$ is the 3-sphere $S_3$.

**Change under a small variation of $\Omega$:** Let us now study the change of $\mathcal{F}[\Omega]$ when we vary the mapping $\Omega$ by $\delta\Omega$. Thanks to the cyclicity of the trace, the variation of each factor $\Omega^\dagger \partial\Omega/\partial\theta_i$ gives the same contribution to the variation of $\mathcal{F}[\Omega]$. Therefore, it is sufficient to consider one of these variations, and to multiply its contribution by the number of factors, i.e., d:

$$\delta\mathcal{F}[\Omega] = d \int d\theta_1 \cdots d\theta_d\; \epsilon_{i_1\cdots i_d}\; \mathrm{tr}\left\{\Omega^\dagger(\theta)\frac{\partial\Omega(\theta)}{\partial\theta_{i_1}} \cdots \delta\left(\Omega^\dagger(\theta)\frac{\partial\Omega(\theta)}{\partial\theta_{i_d}}\right)\right\}. \tag{13.66}$$

The variation of the last factor inside the trace can be written as

$$\delta\left(\Omega^\dagger(\theta)\frac{\partial\Omega(\theta)}{\partial\theta_{i_d}}\right) = \underbrace{-\Omega^\dagger(\theta)\delta\Omega(\theta)\,\Omega^\dagger(\theta)\frac{\partial\Omega(\theta)}{\partial\theta_{i_d}}}_{-\frac{\partial\Omega^\dagger(\theta)}{\partial\theta_{i_d}}\Omega(\theta)} + \Omega^\dagger(\theta)\frac{\partial\delta\Omega(\theta)}{\partial\theta_{i_d}}$$

$$= \Omega^\dagger(\theta)\frac{\partial\delta\Omega(\theta)\Omega^\dagger(\theta)}{\partial\theta_{i_d}}\Omega(\theta). \tag{13.67}$$

Then, integrating by parts with respect to $\theta_{i_d}$, we obtain

$$\delta\mathcal{F}[\Omega] = -d\int d\theta_1 \cdots d\theta_d\; \epsilon_{i_1\cdots i_d}$$
$$\times\;\mathrm{tr}\left\{\frac{\partial}{\partial\theta_{i_d}}\left(\frac{\partial\Omega(\theta)}{\partial\theta_{i_1}}\Omega^\dagger(\theta)\cdots\frac{\partial\Omega(\theta)}{\partial\theta_{i_{d-1}}}\Omega^\dagger(\theta)\right)\delta\Omega(\theta)\Omega^\dagger(\theta)\right\}. \tag{13.68}$$

All the terms containing a factor $\partial^2\Omega/\partial\theta_{i_d}\partial\theta_{i_a}$ vanish because the second derivative is symmetric under the exchange of of the indices $i_d$ and $i_a$, while the prefactor $\epsilon_{i_1\cdots i_d}$ is antisymmetric. The remaining terms are those where the derivative with respect to $\theta_d$ acts on one of the factors $\Omega^\dagger$. There are $d-1$ such terms, which after some reorganization can be written as

$$\delta\mathcal{F}[\Omega] = -d\int d\theta_1 \cdots d\theta_d \underbrace{\sum_{\substack{\sigma\text{ cyclic perm.}\\ \text{of }2\cdots d}}\epsilon_{i_1 i_{\sigma(2)}\cdots i_{\sigma(d)}}\;\mathrm{tr}\left\{\frac{\partial\Omega^\dagger}{\partial\theta_{i_1}}\Omega\frac{\partial\Omega^\dagger}{\partial\theta_{i_2}}\Omega\cdots\Omega\frac{\partial\Omega^\dagger}{\partial\theta_{i_d}}\delta\Omega\right\}}_{0}. \tag{13.69}$$

Note that $\epsilon_{i_1\cdots i_d}$ changes sign under a one-step cyclic permutation of its last $d-1$ indices. Therefore, the $d-1$ terms in the sum exactly cancel since $d-1$ is even, and we have $\delta\mathcal{F}[\Omega] = 0$. Thus, $\mathcal{F}[\Omega]$ is invariant under small changes of $\Omega$, which implies that $\mathcal{F}[\Omega]$ can only vary by discrete jumps. In particular, when $\mathcal{S}$ is the d-sphere $\mathcal{S}_d$, $\mathcal{F}[\Omega]$ depends only on the homotopy class of $\Omega$. These classes form a group $\pi_d(\mathcal{M})$. Moreover, $\mathcal{F}[\Omega]$ provides a representation of $\pi_d(\mathcal{M})$: If $\overline{\Omega}$ denotes the homotopy class to which $\Omega$ belongs, we have

$$\mathcal{F}[\overline{\Omega}_1 \times \overline{\Omega}_2] = \mathcal{F}[\overline{\Omega}_1] + \mathcal{F}[\overline{\Omega}_2]. \tag{13.70}$$

(We denote by $\times$ the group composition in $\pi_d(\mathcal{M})$.) As a consequence, if there exists an $\Omega$ for which the Cartan–Maurer invariant is non-zero, then all its integer multiples can also be obtained, thereby proving that the homotopy group $\pi_d(\mathcal{M})$ contains $\mathbb{Z}$.

## 13.3 INSTANTONS

**Case of a Lie group target manifold:** Let us now specialize to the case where the target manifold $\mathcal{M}$ is a d-dimensional Lie group $\mathcal{H}$, and exploit its group structure in order to obtain simpler expressions. In this case, the $\theta_a$ can also be used as coordinates on $\mathcal{H}$. Consider two elements $\Omega_1$ and $\Omega_2$ of $\mathcal{H}$, represented respectively by the coordinates $\theta_a$ and $\phi_a$. Their product $\Omega_2\Omega_1$ is an element of $\mathcal{H}$ of coordinates $\psi(\theta, \phi)$ (the group multiplication determines how $\psi$ depends on $\theta$ and $\phi$). Since we have shown that the choice of coordinates on $\mathcal{S}$ is irrelevant, we may choose them in such a way that the function $\Omega(\theta)$ is a representation of the group $\mathcal{H}$, i.e., $\Omega(\phi)\Omega(\theta) = \Omega(\psi(\theta,\phi))$. By differentiating this equality with respect to $\psi_j$ at fixed $\phi$, we obtain

$$\Omega(\phi)\frac{\partial\Omega(\theta)}{\partial\theta_i}\frac{\partial\theta_i}{\partial\psi_j} = \frac{\partial\Omega(\psi)}{\partial\psi_j}, \quad \text{and} \quad \Omega^\dagger(\theta)\frac{\partial\Omega(\theta)}{\partial\theta_i} = \frac{\partial\psi_j}{\partial\theta_i}\,\Omega^\dagger(\psi)\frac{\partial\Omega(\psi)}{\partial\psi_j}. \tag{13.71}$$

Using eq. (13.64), the integrand of $\mathcal{F}[\Omega]$ at point $\theta$ can be expressed as

$$\epsilon_{i_1\cdots i_d}\,\mathrm{tr}\left\{\Omega^\dagger(\theta)\frac{\partial\Omega(\theta)}{\partial\theta_{i_1}}\cdots\Omega^\dagger(\theta)\frac{\partial\Omega(\theta)}{\partial\theta_{i_d}}\right\}$$
$$= \det\left(\frac{\partial(\psi_i)}{\partial(\theta_j)}\right)\epsilon_{j_1\cdots j_d}\,\mathrm{tr}\left\{\Omega^\dagger(\psi)\frac{\partial\Omega(\psi)}{\partial\psi_{j_1}}\cdots\Omega^\dagger(\psi)\frac{\partial\Omega(\psi)}{\partial\psi_{j_d}}\right\}, \tag{13.72}$$

where $\psi$ can be any fixed reference point in the group. On the right side, the integration variable $\theta$ now appears only inside the determinant.

The Lie group $\mathcal{H}$ being a smooth manifold, it can be endowed with a metric tensor $\gamma_{ij}(\theta)$, which transforms as follows in a change of coordinates:

$$\gamma_{ij}(\psi) = \frac{\partial\theta_k}{\partial\psi_i}\frac{\partial\theta_l}{\partial\psi_j}\gamma_{kl}(\theta). \tag{13.73}$$

Given a mapping $\Omega(\theta)$ between coordinates and group elements, a possible choice for the metric is given by[11] (see Exercise 13.5)

$$\gamma_{ij}(\theta) = -\frac{1}{2}\mathrm{tr}\left\{\Omega^\dagger(\theta)\frac{\partial\Omega(\theta)}{\partial\theta_i}\Omega^\dagger(\theta)\frac{\partial\Omega(\theta)}{\partial\theta_j}\right\}. \tag{13.74}$$

Moreover, for any such metric $\gamma_{ij}(\theta)$, we have

$$\det\left(\frac{\partial(\psi_i)}{\partial(\theta_j)}\right) = \sqrt{\frac{\det\gamma(\theta)}{\det\gamma(\psi)}}. \tag{13.75}$$

Therefore, the Cartan–Maurer invariant $\mathcal{F}[\Omega]$ takes the following form:

$$\mathcal{F}[\Omega] = \epsilon_{j_1\cdots j_d}\,\mathrm{tr}\left\{\Omega^\dagger(\psi)\frac{\partial\Omega(\psi)}{\partial\psi_{j_1}}\cdots\Omega^\dagger(\psi)\frac{\partial\Omega(\psi)}{\partial\psi_{j_d}}\right\}\int d^d\theta\,\left(\frac{\det\gamma(\theta)}{\det\gamma(\psi)}\right)^{1/2}. \tag{13.76}$$

---

[11] In the algebra of a compact Lie group, the Killing form $\mathbf{K}(X,Y) \equiv \mathrm{tr}\,(\mathrm{ad}_X\,\mathrm{ad}_Y)$ is a negative definite inner product (see Section 7.2.4), from which one can define a metric on the group manifold, invariant under left and right group action. Eq. (13.74) provides an explicit representation of this metric tensor.

In fact, $d^d\theta \sqrt{\det \gamma(\theta)}$ is an invariant measure on the Lie group, and the integral is therefore the volume of the group. In other words, the previous formula exploits the group invariance in order to rewrite the Cartan–Maurer invariant as the product of the integrand evaluated at a fixed point by the volume of the group. Since $\psi$ is arbitrary in this expression, we may choose the value $\psi_0$ that corresponds to the group identity. Furthermore, group elements and their derivatives in the vicinity of the identity may be written as

$$\Omega(\psi) \underset{\psi \to \psi_0}{\approx} 1 + 2i\,(\psi - \psi_0)_i\,t^i, \qquad \left.\frac{\partial \Omega(\psi)}{\partial \psi_i}\right|_{\psi_0} = 2i\,t^i, \tag{13.77}$$

where the $t^i$ are the generators of the Lie algebra $\mathfrak{h}$. From this, we obtain the following compact expression for $\mathcal{F}[\Omega]$:

$$\mathcal{F}[\Omega] = (2i)^d\, \epsilon_{i_1 \cdots i_d}\, \mathrm{tr}\{t^{i_1} \cdots t^{i_d}\}\, \frac{1}{\sqrt{\det \gamma(\psi_0)}} \int d^d\theta\, \sqrt{\det \gamma(\theta)}. \tag{13.78}$$

**Cartan–Maurer invariant for $\mathcal{H} = SU(2)$:** Consider the following mapping from the 3-sphere $\mathcal{S}_3$ to the fundamental representation of $SU(2)$:

$$\Omega(\theta) = \begin{pmatrix} \theta_4 + i\theta_3 & \theta_2 + i\theta_1 \\ -\theta_2 + i\theta_1 & \theta_4 - i\theta_3 \end{pmatrix} = \theta_4 + 2i\,\theta_i t^i, \tag{13.79}$$

with $t^{1,2,3}$ the generators of the $\mathfrak{su}(2)$ algebra (for the fundamental representation, the Pauli matrices divided by 2) and $\theta_1^2 + \theta_2^2 + \theta_3^2 + \theta_4^2 = 1$. The following identities hold:

$$\det \Omega(\theta) = 1, \quad \Omega^\dagger(\theta) = \theta_4 - 2i\,\theta_i t^i, \quad \frac{\partial \Omega(\theta)}{\partial \theta_i} = 2i\,t^i - \frac{\theta_i}{\theta_4}. \tag{13.80}$$

(We denote $\theta^2 \equiv \theta_1^2 + \theta_2^2 + \theta_3^2$.) In the evaluation of eq. (13.74), we need traces of products of up to four $t^i$ matrices. In the fundamental representation, they read

$$\mathrm{tr}(t^i) = 0, \quad \mathrm{tr}(t^i t^j) = \frac{1}{2}\delta^{ij},$$

$$\mathrm{tr}(t^i t^j t^k) = \frac{i}{4}\epsilon_{ijk}, \quad \mathrm{tr}(t^i t^j t^k t^l) = \frac{1}{8}(\delta^{ij}\delta^{kl} + \delta^{il}\delta^{jk} - \delta^{ik}\delta^{jl}). \tag{13.81}$$

Then, the metric tensor of eq. (13.74) and its determinant are given by

$$\gamma_{ij}(\theta) = \delta_{ij} + \frac{\theta_i \theta_j}{1 - \theta^2}, \quad \det(\gamma(\theta)) = \frac{1}{1 - \theta^2}. \tag{13.82}$$

Combining the above results, we obtain the following expression for the Cartan–Maurer invariant of the homotopy class of $\Omega$ in $\pi_3(SU(2))$:

$$\mathcal{F}[\Omega] = (2i)^3\, \epsilon_{ijk}\, \mathrm{tr}(t^i t^j t^k) \int \frac{2\,d^3\theta}{\sqrt{1 - \theta^2}}. \tag{13.83}$$

The factor 2 comes from the fact that there are two allowed values of $\theta_4$ for each $\theta_{1,2,3}$. Finally, we have (see Exercise 13.6)

$$\mathcal{F}[\Omega] = 96\pi \int_0^1 \frac{d\theta\, \theta^2}{\sqrt{1-\theta^2}} = 24\pi^2. \tag{13.84}$$

In fact, the mapping of eq. (13.79) wraps only once in $SU(2)$, and the above result therefore corresponds to the topological index $+1$. Since $24\pi^2$ is non-zero, there are other classes of $\Omega$ whose Cartan–Maurer invariants are the integer multiples of this result, and the second homotopy group is $\pi_3(SU(2)) = \mathbb{Z}$. Note also that this result extends to any Lie group that contains an $SU(2)$ subgroup.

### 13.3.6  Explicit Instanton Solution

In a gauge theory whose gauge group contains an $SU(2)$ subgroup, the mapping of eq. (13.79) can be used to construct the asymptotic form of an instanton of topological index $+1$,

$$A_i(x) \underset{|x|\to\infty}{=} \frac{i}{g} \Omega^\dagger(\hat{x}) \partial_i \Omega(\hat{x}), \tag{13.85}$$

with $\Omega(\hat{x}) \equiv \hat{x}_4 + 2i\hat{x}_i t^i$. One may then prove that the self-dual field configuration in the bulk that has this large-distance behavior is given by (see Exercise 13.7)

$$A_i(x) = \frac{i}{g} \frac{r^2}{r^2 + R^2} \Omega^\dagger(\hat{x}) \partial_i \Omega(\hat{x}), \tag{13.86}$$

with an arbitrary radius R. From the result (13.84) of the previous subsection, we find that the minimum of the action that corresponds to this solution is $S_{\min}[A] = 8\pi^2/g^2$. Up to translations, dilatations or gauge transformations, this is the only field configuration that gives this action. The field strength corresponding to eq. (13.86) is localized in Euclidean space-time, with a size of order R. One may also superimpose several such solutions. Provided that their centers are separated by distances much larger than R, this sum is also a solution of the classical equations of motion, and its action is a multiple of $8\pi^2/g^2$. Note that gauge fields that produce an action of order $g^{-2}$ must themselves be of order $g^{-1}$, and therefore cannot be treated perturbatively. In particular, the calculation of observables in an instanton background in principle requires treating the instanton field to all orders (see Chapter 18 for a discussion of some aspects of this strong field regime).

### 13.3.7  Instantons and the θ-Term in Yang–Mills Theory

Since we have uncovered classical field configurations of non-zero topological index with finite action, a legitimate question is their role in an Euclidean path integral, since functional integration a priori sums over all classical field configurations. For more generality, we may assume that in the path integral the fields of topological index $n$ are weighted with a factor $P(n)$ that may vary with $n$ (this generalization would allow for instance to exclude fields of topological index different from zero). Thus, the expectation value of an observable $\mathcal{O}$ may be written as

$$\langle \mathcal{O} \rangle = Z^{-1} \sum_{n \in \mathbb{Z}} P(n) \int [DA]_n \, \mathcal{O}[A] \, e^{-S[A]}, \qquad (13.87)$$

where $[DA]_n$ is the functional measure restricted to gauge fields of topological index $n$. The normalization factor $Z$ is given by the same path integral without the observable.

The dependence of $P(n)$ on the topological index cannot be arbitrary. In order to see this, let us consider two space-time subvolumes $\Omega_1$ and $\Omega_2$, non-overlapping and such that $\Omega_1 \cup \Omega_2 = \mathbb{R}^4$. Assume further that the support of the observable $\mathcal{O}$ is entirely inside $\Omega_1$. The topological number, which may be obtained as the integral over space-time of $\epsilon_{ijkl} F^a_{ij} F^a_{kl}$, is additive[12] and we may define topological numbers $n_1$ and $n_2$ for $\Omega_1$ and $\Omega_2$, respectively. The total topological number $n$ is given by $n = n_1 + n_2$. In the expectation value of eq. (13.87), we can therefore split the integration into the domains $\Omega_1$ and $\Omega_2$ as

$$\langle \mathcal{O} \rangle = Z^{-1} \sum_{n_1, n_2 \in \mathbb{Z}} P(n_1 + n_2) \int [DA]_{n_1} \, \mathcal{O}[A] \, e^{-S_{\Omega_1}[A]} \int [DA]_{n_2} \, e^{-S_{\Omega_2}[A]}, \qquad (13.88)$$

where $[DA]_{n_i}$ is the functional measure for gauge fields with topological number $n_i$ in the domain $\Omega_i$. Since the observable is localized inside the domain $\Omega_1$, we should be able to remove any dependence on the domain $\Omega_2$ from its expectation value. This dependence cancels between the numerator and the factor $Z^{-1}$ in the previous expression provided that the weight $P(n_1 + n_2)$ factorizes as follows:

$$P(n_1 + n_2) = P(n_1) P(n_2), \qquad (13.89)$$

which implies that $P(n) = e^{-n\theta}$, where $\theta$ is an arbitrary constant. From the previous results, the topological number of a field configuration is given by the integral

$$n = \frac{g^2}{64\pi^2} \int d^4x \, \epsilon_{ijkl} F^a_{ij} F^a_{kl}. \qquad (13.90)$$

Therefore, we may capture the effect of the topological weight $P(n)$ by adding to the Lagrangian density the following term:

$$\mathcal{L}_\theta \equiv \frac{\theta g^2}{64\pi^2} \epsilon_{ijkl} F^a_{ij} F^a_{kl}. \qquad (13.91)$$

After this term has been added, it is no longer necessary to split the path integral into separate topological sectors. The previous Lagrangian is nothing but the $\theta$-term that we have already encountered in the discussion of non-Abelian gauge theories. There, it appeared as a term

---

[12] But note that this integral does not have to be an integer when the integration domain is not the entire space-time. However, it is approximately an integer when the size of the domain is much larger than the instanton size.

## 13.3 INSTANTONS

that cannot be excluded on the grounds of gauge symmetry. In the present discussion, we see that the θ-term results from a non-uniform weighting of the field configurations of different topological index (θ = 0 corresponds to a path integration in which all the fields are weighted equally, regardless of their topological index). In Exercises 13.8, 13.9 and 13.10, we also discuss how instantons are related to the non-trivial ground state structure of Yang–Mills theory.

### 13.3.8 Quantum Fluctuations Around an Instanton

Consider an instanton solution $A^\mu_{n,\alpha}(x)$, which provides a local minimum of the Euclidean action, where the subscript $n$ is the topological index of the instanton, and $\alpha$ collectively denotes all the other parameters that characterize the instanton (its center, its size, its orientation in color space). The expectation value of an observable reads

$$\langle \mathcal{O} \rangle = Z^{-1} \int [DA] \, e^{-S[A]} \, \mathcal{O}(A) = Z^{-1} \int [Da] \, e^{-S[A_{n,\alpha}+a]} \, \mathcal{O}(A_{n,\alpha}+a), \qquad (13.92)$$

where we denote $a^\mu$ the difference $A^\mu - A^\mu_{n,\alpha}$. Since the instanton is an extremum of the action, the dependence of the action on $a^\mu$ begins with quadratic terms,

$$S[A_{n,\alpha}+a] = \frac{8\pi^2 |n|}{g^2} + \frac{1}{2} \int d^4x \, d^4y \, \mathcal{G}^{-1}_{n\alpha,m\beta}(x,y) \, a(x) a(y) + \cdots \qquad (13.93)$$

It is important to note that the action has *flat directions* in the space of field configurations, which correspond to changing the parameters of the instanton inside its topological class. For instance, changing the center coordinates of the instanton does not modify the value of its action. Along these directions, the second derivative of the action vanishes. This means that the matrix of second-order coefficient $\mathcal{G}^{-1}_{n\alpha,m\beta}(x,y)$ has a number of vanishing eigenvalues, corresponding to these flat directions.

If we expand the action only to quadratic order in $a^\mu$, which amounts to a one-loop approximation in the background of the instanton, a typical contribution to the expectation value of eq. (13.92) is a product of dressed propagators $\mathcal{G}_{n\alpha,m\beta}(x,y)$ connecting pairwise the gauge fields contained in the observable $\mathcal{O}$ and a determinant,

$$\langle \mathcal{O} \rangle = Z^{-1} \, e^{-8\pi^2 |n|/g^2} \left\{ \left( \det \mathcal{G} \right)^{1/2} \prod \mathcal{G} + \cdots \right\}. \qquad (13.94)$$

Our goal here is simply to extract the dependence of such an expectation value on the topological index $n$. Besides the obvious exponential prefactor, a dependence on $n$ hides in the determinant. Le us rewrite it as a product on the spectrum of $\mathcal{G}^{-1}$,

$$\left( \det \mathcal{G} \right)^{1/2} = \prod_s \lambda_s^{-1/2}, \qquad (13.95)$$

where the $\lambda_s$ are the eigenvalues of $\mathcal{G}^{-1}$. If we rescale the gauge fields by a power of the coupling $g$, $gA \to A$, the only dependence on $g$ in the Yang–Mills action is a prefactor $g^{-2}$, and all the eigenvalues $\lambda_s$ are also proportional to $g^{-2}$. Moreover, as explained above, we should remove the zero modes from this product, since they do not give a quadratic term in eq. (13.93). If we are interested only in the powers of g, we may write

$$\left(\det \mathcal{G}\right)^{1/2} \sim \prod_{\text{all modes}} g \prod_{\text{zero modes}} g^{-1}. \tag{13.96}$$

The first factor, which involves a (continuous) infinity of modes, is not well defined but it does not depend on the details of the instanton background. In contrast, the second factor brings one factor of $g^{-1}$ for each collective coordinate of the instanton. For an instanton of topological number $n = 1$, these collective coordinates are:

- the four coordinates of the center of the instanton;
- the size R of the instanton;
- the three angles that determine the orientation of the instanton; and
- for SU(2), the three parameters defining a global gauge rotation.

Of the last six parameters, three correspond to simultaneous spatial and color rotations that produce the same instanton solution, and they should not be counted. There are therefore eight collective coordinates for the $n = 1$ SU(2) instanton,[13] and its contribution to expectation values scales as

$$\left\langle \mathcal{O} \right\rangle_{n=1} \sim e^{-8\pi^2/g^2} \, g^{-8}. \tag{13.97}$$

Because of the exponential factor that contains the inverse coupling, all the Taylor coefficients of this function vanish at $g = 0$. Thus, such a contribution never shows up in perturbation theory.

## 13.4 Skyrmions

### 13.4.1 Definition and Topological Classification

Skyrmions are field configurations that arise in models resulting from a spontaneous symmetry breaking, such as a nonlinear sigma model. Consider for instance the following action:

$$S[\pi] = \int d^D x \left\{ \frac{1}{2} \sum_{a,b} g_{ab}(\pi) (\partial_i \pi^a)(\partial_i \pi^b) + \cdots \right\}, \tag{13.98}$$

where the fields $\pi^a$ are the Nambu–Goldstone bosons of a broken symmetry from the symmetry group $\mathcal{G}$ down to $\mathcal{H}$. The matrix $g_{ab}(\pi)$ is positive definite, and in general field dependent. The dots represent terms with higher derivatives, that we have not written explicitly. In such a model, the Nambu–Goldstone fields $\pi^a$ may be viewed as elements of the coset $\mathcal{G}/\mathcal{H}$.

In order to have a finite action, the derivatives of the fields should decrease faster than $|x|^{-D/2}$ at large distance,

$$\left|\partial_i \pi^a(x)\right|_{|x| \to \infty} \ll |x|^{-D/2}, \tag{13.99}$$

---

[13] This counting is more involved for an SU(3) instanton. In this case, there are seven collective coordinates corresponding to rotations and gauge transformations, hence a total of 12 collective coordinates.

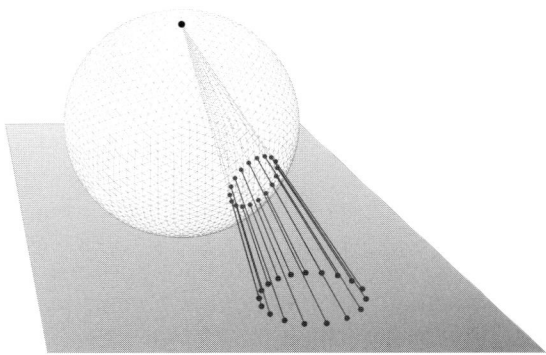

FIGURE 13.8: Stereographic projection that maps the plane $\mathbb{R}^2$ to the sphere $\mathcal{S}_2$. All the points at infinity in the plane are identified, and mapped to the north pole of the sphere.

which means that the field $\pi^a(x)$ should go to a constant, with a remainder that decreases faster than $|x|^{1-D/2}$.

The constant value of $\pi^a$ at infinity can be chosen to be some fixed predefined element of $\mathcal{G}/\mathcal{H}$. Thus, we may view the field $\pi^a(x)$ as a mapping

$$\pi^a: \quad \mathcal{S}_D \mapsto \mathcal{G}/\mathcal{H}, \tag{13.100}$$

where $\mathcal{S}_D$ is the D-dimensional sphere, topologically equivalent to the Euclidean space $\mathbb{R}^D$ with all the points $|x| = \infty$ identified as a single point. This equivalence may be made manifest by a stereographic projection, illustrated for $D = 2$ in Figure 13.8. These mappings, taking a fixed value at $|x| = \infty$, can be organized into topological classes containing functions that can be continuously deformed into one another. The set of these classes is a group, known as the Dth homotopy group of $\mathcal{G}/\mathcal{H}$, denoted $\pi_D(\mathcal{G}/\mathcal{H})$.

Note that the model defined by eq. (13.98), with only second-order derivatives, cannot have stable solutions, a result known as *Derrick's theorem*. In order to see this, consider a skyrmion solution $\pi^a(x)$, and construct another field by a rescaling:

$$\pi_R^a(x) \equiv \pi^a(x/R). \tag{13.101}$$

The action becomes $S[\pi_R] = R^{D-2} S[\pi]$. In $D > 2$ dimensions, we may make it decrease continuously to zero, despite the fact that $\pi^a$ and $\pi_R^a$ have the same topology. Such a solution may be stabilized by adding a term with higher derivatives, such as

$$V[\pi] \equiv \int d^D x \, h_{abcd}(\pi) \left( \partial_i \pi^a \partial_i \pi^b \right) \left( \partial_j \pi^c \partial_j \pi^d \right). \tag{13.102}$$

Under the same rescaling, we now have $V[\pi_R] = R^{D-4} V[\pi]$. In $D = 3$ spatial dimensions, the term with second derivatives decreases to zero when $R \to 0$, while the above quartic term increases to $+\infty$. Their sum therefore exhibits an extremum at some finite scale $R_*$. Although we obtain in this way non-trivial stable solutions, there is a priori no reason to limit ourselves to terms with four derivatives, and therefore the predictive power of such a model is limited by the many possible choices for these higher-order terms.

## 13.4.2 SU(2) Skyrmion

The original version of this model was intended to describe nucleons as a topologically stable configuration of the pion field. In this case, there are $D = 3$ spatial dimensions, and one considers the nonlinear sigma model that results from the spontaneous breakdown of chiral symmetry ($SU(2) \times SU(2) \to SU(2)$). The coset in which $\pi^a$ lives is $SU(2)$, and the relevant homotopy group is $\pi_3(SU(2)) = \mathbb{Z}$. The integer that enumerates the topological classes is then identified with the baryon number.

To make this exposition more explicit, let us start from the linear sigma model (see eq. (4.66)), focusing on the part that involves only the scalar fields. Combining the four scalar fields into a $2 \times 2$ matrix $\Sigma \equiv \sigma + i\pi^a \sigma^a$ (the $\sigma^{1,2,3}$ are the Pauli matrices, not to be confused with the field $\sigma$), the Lagrangian of this model takes a very compact form:

$$\mathcal{L} = \frac{1}{4} \operatorname{tr}\left(\partial_\mu \Sigma \, \partial^\mu \Sigma\right) - V(\det \Sigma). \tag{13.103}$$

This Lagrangian is invariant under independent left and right multiplication by $SU(2)$ matrices, $\Sigma \to U\Sigma V^\dagger$, as long as $U, V$ do not depend on space-time. If the potential $V$ has nontrivial minima at $\sigma^2 + \pi^2 = f_\pi^2 \neq 0$, there is a spontaneous breakdown of the $SU(2) \times SU(2)$ symmetry. Without loss of generality, we may assume that only the $\sigma$ field acquires a vacuum expectation value in the ground state, $\langle \sigma \rangle = f_\pi$. If we neglect the fluctuations of the radial field and denote $\Sigma \equiv f_\pi \Omega$, we obtain the corresponding nonlinear sigma model,

$$\mathcal{L}_{\text{NLSM}} = \frac{f_\pi^2}{4} \operatorname{tr}\left(\partial_\mu \Omega \, \partial^\mu \Omega^\dagger\right). \tag{13.104}$$

Since we have dropped the radial excitations, the fields are now confined to the 3-sphere $\sigma^2 + \pi^2 = f_\pi^2$, i.e., $\Omega\Omega^\dagger = 1$ (thus, $\Omega \in SU(2)$). Thanks to this fact, we can also rewrite this Lagrangian as

$$\mathcal{L}_{\text{NLSM}} = -\frac{f_\pi^2}{4} \operatorname{tr}\left(A_\mu A^\mu\right), \tag{13.105}$$

where $A_\mu \equiv \Omega^\dagger \partial_\mu \Omega$ ($A_\mu$ belongs to the $\mathfrak{su}(2)$ algebra). Note that $A_\mu$ is formally identical to a pure gauge field, and that the previous Lagrangian remains invariant under $\Omega(x) \to \Omega(x)U$ with $U$ a constant $SU(2)$ matrix ($A^\mu(x) \to U^\dagger A^\mu(x) U$ under this global gauge transformation). Since $A^\mu$ is a pure gauge, the associated field strength is identically zero, $F^{\mu\nu} = 0$, and the formula that expresses the fact that the $\theta$-term (see Section 7.5) is a total derivative becomes a conservation equation,

$$\partial_\mu \underbrace{\left[-\tfrac{i}{96\pi^2} \epsilon^{\mu\nu\rho\sigma} \epsilon^{abc} A_\nu^a A_\rho^b A_\sigma^c\right]}_{B^\mu} = 0. \tag{13.106}$$

(The $\epsilon^{abc}$ are the $\mathfrak{su}(2)$ structure constants.) This current leads to a conserved charge,

$$B \equiv \int d^3x \, B^0 = -\tfrac{i}{96\pi^2} \int d^3x \, \epsilon^{ijk} \epsilon^{abc} A_i^a A_j^b A_k^c$$
$$= \tfrac{1}{24\pi^2} \int d^3x \, \epsilon_{ijk} \operatorname{tr}\left(\Omega^\dagger(\partial_i \Omega) \Omega^\dagger(\partial_j \Omega) \Omega^\dagger(\partial_k \Omega)\right). \tag{13.107}$$

Recall now that in order to have a finite energy, the fields under consideration must go to the same constant in all directions at spatial infinity. Thus, $\Omega(x)$ is in fact a function from the sphere $S_3$ to $SU(2)$. Therefore, the integral in the previous equation is proportional to the Cartan–Maurer invariant (in three dimensions) studied in the previous section. This integral varies only in discrete steps, according to which equivalence class the function $\Omega$ belongs to in $\pi_3(SU(2))$. Moreover, given the normalization we have chosen for the current $B^\mu$, the charge B takes integer values, which are identified with the baryon number.

As argued earlier in the general discussion about skyrmions, the pure nonlinear sigma model with only two derivatives leads to unstable solutions, but it can be stabilized by adding a term with four derivatives. In order to preserve the conservation of the charge B, this additional term should also be expressed in terms of the pure gauge field $A_\mu$. As shown by Skyrme, stabilization may be achieved with the following Lagrangian:

$$\mathcal{L} = -\frac{f_\pi^2}{4}\,\mathrm{tr}\,(A_\mu A^\mu) + \frac{1}{g^2}\,\mathrm{tr}\,([A_\mu, A_\nu][A^\mu, A^\nu]), \tag{13.108}$$

where the value of the dimensionless prefactor $g^{-2}$ determines the radius of the stable solutions. In phenomenological applications, one may adjust its value so that the configuration of minimal energy within the topological class $B = +1$ has an energy equal to the proton mass (the value of the constant $f_\pi$, that sets all energy scales in the model, is obtained from measurements of pion decays).

### 13.4.3 Magnetic Skyrmions in the Heisenberg Model

One may also encounter skyrmion configurations in the Heisenberg model of ferromagnetism, already discussed in Section 4.6. Here, we consider a continuous version of this model, with a unit spin vector $\mathbf{s}(\mathbf{x})$ defined over two-dimensional space. At infinity, we assume that the system is in a ferromagnetic state, with all spins aligned in the same direction. Since $\mathbf{s}(\mathbf{x})$ takes the same value in all directions at spatial infinity, the plane $\mathbb{R}^2$ is topologically equivalent to the 2-sphere $S_2$. Since $\mathbf{s}(\mathbf{x})$ is a three-dimensional unit vector, it also lives on the same sphere, and the spin configurations are mappings $S_2 \mapsto S_2$. Since $\pi_2(S_2) = \mathbb{Z}$, there must be spin configurations that have a non-trivial topology and therefore cannot be deformed continuously into a uniform ferromagnetic state. In this case, the topological charge density is defined as

$$Q(\mathbf{x}) \equiv \tfrac{1}{8\pi}\,\epsilon_{ijk}\epsilon_{\mu\nu}\,s_i(\partial_\mu s_j)(\partial_\nu s_k), \tag{13.109}$$

where $i, j, k$ label the spin components and $\mu, \nu$ label the spatial directions. It is convenient to represent the spin vector in spherical coordinates, $\mathbf{s} \equiv (\sin\theta\cos\psi, \sin\theta\sin\psi, \cos\theta)$ and the position in polar coordinates, $\mathbf{x} \equiv (r\cos\phi, r\sin\phi)$. In the special case of an axially symmetric skyrmion, $\theta = \theta(r)$, $\psi = \psi(\phi)$, and the density has the following explicit form:

$$Q(\mathbf{x}) = \frac{\sin\theta}{4\pi r}\frac{d\theta}{dr}\frac{d\psi}{d\phi}. \tag{13.110}$$

After integration over two-dimensional space, the total charge is

$$Q = -\frac{1}{4\pi}\underbrace{\bigl[\cos\theta\bigr]_{r=0}^{r=\infty}}_{=\,2}\,\underbrace{\bigl[\psi\bigr]_{\phi=0}^{\phi=2\pi}}_{=\,2\pi n} = -n \in \mathbb{Z}. \tag{13.111}$$

Here, we have assumed that the spins are up ($\theta = 0$) at $r = \infty$ and down at $r = 0$. The factor $2\pi n$ comes from the fact that the angle $\psi$ must be single-valued modulo $2\pi$ as one performs a full rotation in two-dimensional space. Spin configurations with $Q \neq 0$ are stable, even though they have flipped spins at the origin whose natural tendency is to want to be aligned with the spins in the surrounding ferromagnet.

## Exercises

**\*13.1** Prove eq. (13.29).

**13.2** Given the parametrization in eq. (13.35), derive the equations for f and g such that energy of the field configuration is minimized.

**\*13.3** What would be the allowed values for the magnetic charge of a monopole if the residual gauge group is $\mathcal{H} = SO(3)$?

**13.4** Check that the result obtained for the $SU(2)$ Cartan–Maurer invariant is consistent with the Atiyah–Singer theorem. Use this correspondence to conclude that the invariant would take the same discrete values for any $SU(N)$ with $N \geq 2$, and that $\pi_3(SU(N)) = \mathbb{Z}$.

**13.5** Justify why eq. (13.74) is a valid definition for a metric tensor on a Lie group. *Hint: Use the Killing form to define a distance in the vicinity of the identity, then extend it globally to the entire group.*

**\*13.6** Check the details of the $SU(2)$ calculation from eq. (13.79) to eq. (13.84).

**13.7** Check that the radial dependence of the instanton given in eq. (13.86) is indeed the bulk dependence of the field configuration that minimizes the action. *Hint: Check that it satisfies the self-duality condition.*

**\*13.8** Consider Yang–Mills theory in the temporal gauge $A^0 = 0$, with the coupling set to $g = 1$ for simplicity.

- What is the conjugate momentum of $A^i$? Derive the Hamiltonian, and check that Hamilton's equations are the usual Yang–Mills equations. Why do we need in addition to impose *Gauss' law*, i.e., $G[A] \equiv [D_i, \partial^0 A^i] = 0$, on the classical solutions? Likewise, physical quantum states should be annihilated by $G[A]$.
- Yang–Mills theory in the temporal gauge has a residual invariance under purely spatial gauge transformations $\Omega(\mathbf{x})$. Explain why the gauge transformation $\Omega(\mathbf{x}) \equiv e^{i\vartheta_a(\mathbf{x})t^a}$ is generated by

$$Q_\vartheta = -\int d^3x \, (\partial_0 A^i_a)(D^i_{ab}\vartheta_b).$$

If $\Omega(\mathbf{x})$ goes to the identity at spatial infinity, show that this generator becomes proportional to Gauss' constraint, implying that physical states are invariant under the action of $e^{iQ_\vartheta}$.
- From the above considerations, show that a vacuum state is a superposition of pure gauge fields $i U^\dagger \partial^i U$, related to one another by spatial gauge transformations $\Omega(\mathbf{x})$ that go to the identity at infinity.

- Consider transition amplitudes between two vacua, the first one corresponding to $U_1 \equiv 1$ and the second one to $U_2(\mathbf{x})$. Show that $U_2(\mathbf{x})$ should go to a constant at infinity for this amplitude to be non-zero.
- Conclude that $SU(N)$ Yang–Mills theory has a countable infinity of inequivalent vacua, indexed by an integer given by the Cartan–Maurer invariant.

**\*13.9** Consider a double well potential $V(x) \equiv \lambda(x^2 - x_*^2)^2$ in non-relativistic quantum mechanics, that has two degenerate minima located at $x = \pm x_*$.

- Recall the path integral expression of a transition amplitude between positions $-x_*$ and $+x_*$, $\mathcal{A} \equiv \langle +x_* | e^{-iHt} | -x_* \rangle$. Explain why there is no path making the action in this path integral extremal.
- Perform a Wick rotation to the time variable by defining $\tau = it$, and write the resulting expression of the transition amplitude.
- Show that a saddle point approximation is now possible, and that the leading behavior of the amplitude is given by the quantum mechanical analogue of instantons. Solve the imaginary time equation of motion to obtain this dominant classical path, and calculate the corresponding action. Are there other minima of the action, that would give subleading contributions?
- The previous calculation does not give the prefactor in $\mathcal{A}(t)$. How can it be obtained? Without doing an explicit calculation, explain why this prefactor is proportional to the time t. What is the range of t where this semi-classical approximation is valid? What contributions should one include to extend it to larger times?
- Let us denote $|\psi_{0,1,2,\cdots}\rangle$ the eigenstates of the Hamiltonian, with energies $E_{0,1,2,\cdots}$. Express the transition amplitude $\mathcal{A}(t)$ in terms of these eigenstates. Use this alternate representation to show that the tunneling amplitude $\mathcal{A}(t)$ is approximately proportional to the splitting between the energies of the first two levels, $\delta E \equiv E_1 - E_0$.

**\*13.10** This exercise is a continuation of Exercise 13.8. Let us denote by $|0_n\rangle$ the vacuum state of Yang–Mills theory with topological index n.

- A representative field configuration of the vacuum $|0_0\rangle$ is $A_a^i(\mathbf{x}) \equiv 0$. How does this field (or any field gauge equivalent to it) evolve under the classical Yang–Mills equation? Conclude from this that the transitions $\langle 0_{n\,\text{out}} | 0_{0\,\text{in}} \rangle$ are classically forbidden and necessarily happen by quantum tunneling.
- By analogy with the previous exercise, write the path integral representation of this transition amplitude after going to imaginary time, and argue that the amplitude is dominated by instanton solutions.
- Recall that the topological index of the instanton is given by

$$N = \int \frac{d^4x}{64\pi^2} \, \epsilon_{ijkl} F_{ij}^a F_{kl}^a = \lim_{R \to \infty} \int_{S_R} \frac{d^3 S_i}{32\pi^2} K^i.$$

Argue that the sphere $S_\infty$ at infinity can be deformed into the surface of a hypercube, and show that in the $A^0 = 0$ gauge this integral is the difference of two integrals on surfaces at $x^0 = \pm\infty$, respectively. Check that these integrals are Cartan–Maurer invariants giving the topological indices of vacuum states, and conclude that the instanton of index N interpolates between vacua of index 0 and N.

# CHAPTER 14

# Modern Tools for Tree Amplitudes

Transition amplitudes play a central role in quantum field theory, since they are the building blocks of most observables. Their square gives transition probabilities that enter in measurable cross-sections. Until now, we have exposed the traditional workflow for calculating these amplitudes. Starting from a classical action that encapsulates the bare couplings of a given quantum field theory, one can derive Feynman rules for propagators and vertices (listed in Figure 8.2 for Yang–Mills theory in covariant gauges), whose application provides a straightforward algorithm for the evaluation of amplitudes. However, the use of these Feynman rules is very cumbersome for the following reasons:

- Even at tree level, the number of distinct graphs contributing to a given amplitude increases very rapidly with the number of external lines, as shown in Table 14.1 for amplitudes with external gluons only (see Exercise 14.1).

- The internal gluon propagators of these diagrams carry unphysical degrees of freedom, which contributes to the great complexity of each individual diagram.

- The Feynman rules are sufficiently general to compute amplitudes with arbitrary external momenta (not necessarily on-shell) and polarizations (not necessarily physical), although this is not useful for amplitudes that will be used in cross-sections. One would hope for a leaner formalism that only calculates what is strictly necessary for physical quantities.

The situation becomes even worse with loop diagrams. Another situation with an even higher degree of complexity, even at tree level, is that of gravity. It would be desirable to be able to calculate tree-level amplitudes with gravitons, since they enter for instance in the study of the scattering of gravitational waves by a distribution of masses. But because the graviton has spin-2, the corresponding Feynman rules are considerably more complicated (especially the self-couplings of the graviton) than those of Yang–Mills theory.

## 14.1 COLOR DECOMPOSITION OF GLUON AMPLITUDES

| Number of gluons | Number of diagrams | n! |
|---|---|---|
| 4 | 4 | 24 |
| 5 | 25 | 120 |
| 6 | 220 | 720 |
| 7 | 2,485 | 5,040 |
| 8 | 34,300 | 40,320 |
| 9 | 559,405 | 362,880 |
| 10 | 10,525,900 | 3,628,800 |

TABLE 14.1: Number of Feynman diagrams contributing to tree-level amplitudes with external gluons only. (The third column indicates the values of $n!$, for comparison. We see that the number of graphs grows faster than the factorial of the number of external gluons.)

It turns out that physical on-shell amplitudes in gauge theories are considerably simpler than one may expect from the Feynman rules and the intermediate steps of their calculation by the usual perturbation theory, and a legitimate query is whether there is a more direct route to reach these compact answers. The goal of this chapter is to give a glimpse (in particular, our discussion will be restricted to tree-level amplitudes, but a significant part of the many recent developments deal with loop corrections – see Chapter 19) of some of the recent developments that led to powerful new methods for calculating amplitudes. A recurring theme of these methods is to avoid as much as possible references to the Lagrangian, which may be viewed as the main source of the complications in standard perturbation theory (for instance, the gauge invariance of the Lagrangian is the reason why non-physical gluon polarizations appear in the Feynman rules). Instead, these methods try to gather as much information as possible on amplitudes based on symmetries and kinematics.

## 14.1 Color Decomposition of Gluon Amplitudes

### 14.1.1 Rules for the Removal of Color Factors

Let us first focus on the color structure of tree Feynman diagrams, in order to organize and simplify it. Although the techniques we expose here can be extended to quarks, we consider tree amplitudes that contain only gluons for simplicity, in the case of the $SU(N)$ gauge group. The structure constants $f^{abc}$ of the group appear in the three-gluon and four-gluon vertices. The first step is to rewrite the structure constants in terms of the generators $t_f^a$ of the fundamental representation of $\mathfrak{su}(N)$. Using the following relations among the generators,

$$[t_f^a, t_f^b] = i f^{abc} t_f^c, \quad \text{tr}(t_f^a t_f^b) = \frac{\delta^{ab}}{2}, \tag{14.1}$$

we can write

$$i f^{abc} = 2 \, \text{tr}(t_f^a t_f^b t_f^c) - 2 \, \text{tr}(t_f^b t_f^a t_f^c), \tag{14.2}$$

which has also the following diagrammatic representation:

$$i f^{abc} = 2 \left\{ \text{[diagram]} - \text{[diagram]} \right\}. \tag{14.3}$$

The black dots indicate the fundamental representation generators $t_f^a$. Note that the "loops" in this representation are not actual fermion loops, they are just a graphical cue indicating how the indices carried by the $t_f^a$ are contracted in the traces. We may also apply this trick to the four-gluon vertex, which from the point of view of its color structure (but not for what concerns its momentum dependence) is equivalent to a sum of three terms with two three-gluon vertices:

$$\text{[diagram]} = \text{[diagram]} + \text{[diagram]} + \text{[diagram]}. \tag{14.4}$$

Since the gluon propagators are diagonal in color (i.e., proportional to a $\delta_{ab}$), the $t_f^a$ that are attached to the endpoints of the internal gluon propagators have their color indices contracted and summed over. The result of this contraction is given by the following $\mathfrak{su}(N)$ Fierz identity:

$$(t_f^a)_{ij}(t_f^a)_{kl} = \text{[diagram]} = \frac{1}{2} \text{[diagram]} - \frac{1}{2N} \text{[diagram]}. \tag{14.5}$$

Thus, it seems that these contractions produce $2^n$ terms for $n$ internal gluon propagators, but this can in fact be simplified tremendously by noticing that the second term of the Fierz identity corresponds to the exchange of a colorless object,[1] that does not couple to gluons. All these terms in $1/N$ must therefore cancel in purely gluonic amplitudes (this is not true anymore if quarks are involved, either as external lines or via loop corrections).

---

[1] A more rigorous justification is to note that $SU(N) \times U(1) = U(N)$, where $U(N)$ is the group of the $N \times N$ unitary matrices. For the fundamental generators of the $\mathfrak{u}(N)$ algebra, the Fierz identity is

$$\text{[diagram]} \underset{\mathfrak{u}(N)}{=} \frac{1}{2} \text{[diagram]}.$$

The $U(N)$ gauge theory differs from the $SU(N)$ one by the extra $U(1)$, and the comparison of their Fierz identities indicates that the term in $1/2N$ in eq. (14.5) is due to this $U(1)$ factor. Being Abelian, this extra factor corresponds to a photon-like mode that does not couple to gluons.

## 14.1 COLOR DECOMPOSITION OF GLUON AMPLITUDES

We illustrate in the following equation a few of the color structures generated by this procedure in the case of a tree-level five-gluon diagram:

$$\text{(diagram)} = \text{(diagram)} + \text{(diagram)} + \cdots \quad (14.6)$$

Each of the terms contains a *single trace* of five $t_f^a$, one for each external gluon (the color matrices attached to the internal gluon lines have all disappeared when using the Fierz identity). The terms on the right-hand side correspond to the various ways of choosing the clockwise or counterclockwise loop for each $f^{abc}$ (see eq. (14.3)). "Twists" such as the one appearing in the second term of the previous equation arise when two such adjacent loops have opposite orientations.

### 14.1.2 Example of a Four-Point Diagram

In order to see more precisely how the color and kinematics can be disentangled by this method, let us consider the example of one of the graphs that contributes to the four-gluon amplitude, namely the graph that gives the s-channel of the process $12 \to 34$,

$$\mathcal{M}_{4s}(1234) \equiv \text{(diagram)} = \left(if^{a_1 a_2 c}\right)\left(if^{c a_3 a_4}\right) \mathcal{A}_{4s}(1234), \quad (14.7)$$

where in the last equality we have denoted by $\mathcal{A}_{4s}(1234)$ the part of this Feynman diagram (kinematical factors from the three-gluon vertices, intermediate propagator and polarization vectors of the external legs) that does not involve the structure constants. Let us now rearrange the color factors with the method described above:

$$\begin{aligned}\mathcal{M}_{4s}(1234) &= 4\,\mathcal{A}_{4s}(1234)\,\mathrm{tr}\left(t_f^{a_1} t_f^{a_2} t_f^{c} - t_f^{a_2} t_f^{a_1} t_f^{c}\right)\mathrm{tr}\left(t_f^{c} t_f^{a_3} t_f^{a_4} - t_f^{c} t_f^{a_4} t_f^{a_3}\right) \\ &= 2\,\mathcal{A}_{4s}(1234)\,\mathrm{tr}\left(t_f^{a_1} t_f^{a_2} t_f^{a_3} t_f^{a_4} - t_f^{a_1} t_f^{a_2} t_f^{a_4} t_f^{a_3}\right. \\ &\quad \left. - t_f^{a_2} t_f^{a_1} t_f^{a_3} t_f^{a_4} + t_f^{a_2} t_f^{a_1} t_f^{a_4} t_f^{a_3}\right). \end{aligned} \quad (14.8)$$

Note now that the kinematical part of the Feynman rule for the three-gluon vertex is antisymmetric under the exchange of two lines, so that the order in which we write the labels is significant:

$$\mathcal{A}_{4s}(1234) = -\mathcal{A}_{4s}(2134) = -\mathcal{A}_{4s}(1243) = \mathcal{A}_{4s}(2143). \quad (14.9)$$

Using these identities, we can rewrite this Feynman graph as

$$\begin{aligned}\mathcal{M}_{4s}(1234) = 2\,\Big\{&\mathcal{A}_{4s}(1234)\,\mathrm{tr}\left(t_f^{a_1} t_f^{a_2} t_f^{a_3} t_f^{a_4}\right) \\ +&\mathcal{A}_{4s}(1243)\,\mathrm{tr}\left(t_f^{a_1} t_f^{a_2} t_f^{a_4} t_f^{a_3}\right) \\ +&\mathcal{A}_{4s}(2134)\,\mathrm{tr}\left(t_f^{a_2} t_f^{a_1} t_f^{a_3} t_f^{a_4}\right) \\ +&\mathcal{A}_{4s}(2143)\,\mathrm{tr}\left(t_f^{a_2} t_f^{a_1} t_f^{a_4} t_f^{a_3}\right)\Big\}, \end{aligned} \quad (14.10)$$

with only positive signs and where, in each term, the order of the labels in $\mathcal{A}_{4s}$ matches the order of the labels carried by the matrices inside the trace. Note that orderings of the labels that would give the same trace due to its cyclic invariance appear only once, and that some of the orderings do not appear in this expression (e.g., 1324). Applying the same manipulations to the other three graphs that contribute to the four-point amplitude (t and u channels, plus one graph with a four-gluon vertex) also gives expressions that have the same structure as eq. (14.10), but possibly with other sets of cyclic orderings.

### 14.1.3 Color-Ordered Feynman Rules

Quite generally, any n-gluon tree amplitude $\mathcal{M}_n(1\cdots n)$ can be decomposed as a sum of terms corresponding to the allowed color structures. These color structures are single traces of fundamental representation color matrices carrying the color indices of the external gluons. A priori, these matrices could be reshuffled by an arbitrary permutation in $\mathfrak{S}_n$, but thanks to the cyclic invariance of the trace we can reduce the sum to the quotient set $\mathfrak{S}_n/\mathbb{Z}_n$ of permutations modulo a cyclic permutation:[2]

$$\mathcal{M}_n(1\cdots n) \equiv 2 \sum_{\sigma \in \mathfrak{S}_n/\mathbb{Z}_n} \mathrm{tr}\,(t_f^{a_{\sigma(1)}} \cdots t_f^{a_{\sigma(n)}})\, \mathcal{A}_n(\sigma(1)\cdots\sigma(n)), \tag{14.11}$$

where the prefactor 2 combines the factors 2 from eq. (14.2) and the factors $\tfrac{1}{2}$ from the first term of the Fierz identity (14.5). The object $\mathcal{A}_n(\sigma(1)\cdots\sigma(n))$ is called a *color-ordered partial amplitude*. By construction, it depends only on the momenta and polarizations of the external gluons, but not on their colors since they have already been factored out in the trace. Therefore, the partial amplitudes are gauge invariant. From eq. (14.11), the squared amplitude summed over all colors can be written as

$$\sum_{\text{colors}} \left|\mathcal{M}_n(1\cdots n)\right|^2 = 4 \sum_{\sigma,\rho \in \mathfrak{S}_n/\mathbb{Z}_n} \sum_{\text{colors}} \mathrm{tr}\,(t_f^{a_{\sigma(1)}} \cdots t_f^{a_{\sigma(n)}})\, \mathrm{tr}^*(t_f^{a_{\rho(1)}} \cdots t_f^{a_{\rho(n)}})$$
$$\times\, \mathcal{A}_n(\sigma(1)\cdots\sigma(n))\, \mathcal{A}_n^*(\rho(1)\cdots\rho(n)). \tag{14.12}$$

The sum over colors of the product of two traces that appears in the first line can be performed using the $\mathfrak{su}(N)$ Fierz identity (14.5). For instance,

$$\mathrm{tr}\,(t^a t^b t^c t^d t^e)\, \mathrm{tr}^*(t^b t^a t^c t^d t^e) = \left\{\vcenter{\hbox{\includegraphics{}}}\right\}, \tag{14.13}$$

which can then be expressed as a function of N by repeated use of the Fierz identity.

---

[2]This is equivalent to considering permutations that have the fixed point $\sigma(1) = 1$, i.e., permutations that only reshuffle the set $\{2\cdots n\}$. For n external gluons, there are $(n-1)!$ independent color structures. The basis provided by these traces is over-complete, and there exist linear relationships among the tree-level partial amplitudes, known as the *Kleiss–Kuijf relations* (see Exercise 14.12 for a simple example). These relations reduce the number of partial amplitudes from $(n-1)!$ to $(n-2)!$. Additional relationships, known as the *Bern–Carrasco–Johansson relations*, further reduce this number to $(n-3)!$.

## 14.1 COLOR DECOMPOSITION OF GLUON AMPLITUDES

| Number of gluons | Number of graphs | Number of cyclic-ordered graphs |
|---|---|---|
| 4 | 4 | 3 |
| 5 | 25 | 10 |
| 6 | 220 | 38 |
| 7 | 2,485 | 154 |
| 8 | 34,300 | 654 |
| 9 | 559,405 | 2,871 |
| 10 | 10,525,900 | 12,925 |

TABLE 14.2: Comparison between the number of Feynman graphs and the number of cyclic-ordered graphs for tree-level amplitudes with external gluons only.

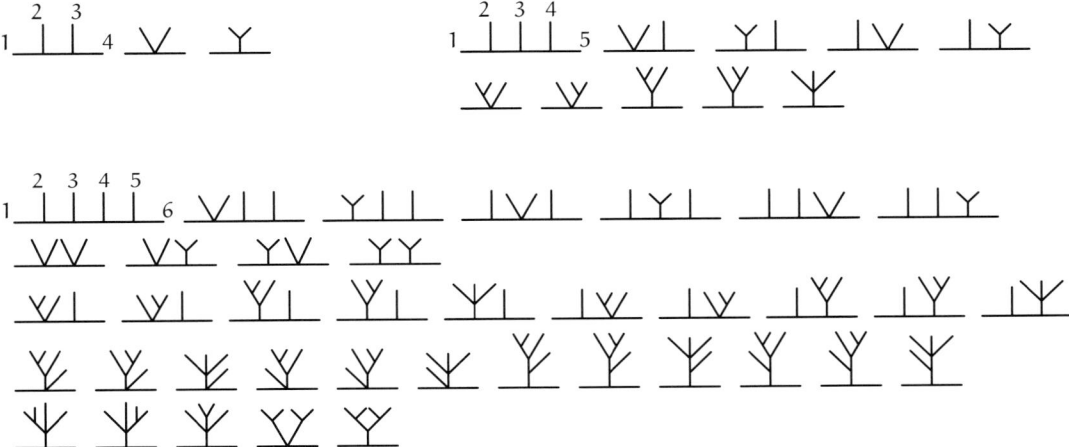

FIGURE 14.1: Diagrams contributing to the four-, five-, six-point color-ordered amplitudes in Yang–Mills theory. The external points are labeled 1 to $n = 4, 5, 6$ in the clockwise direction. The solid lines represent gluons.

At this point, we have isolated the color dependence of the amplitude from its momentum and polarization dependences that are factorized into the partial amplitudes. Of course, calculating the latter is still not easy, but the task is significantly reduced for two reasons:

- The color-ordered partial amplitude $\mathcal{A}_n(12\cdots n)$ receives contributions only from planar diagrams in which the gluons are ordered clockwise according to the sequence $12\cdots n$. The number of these graphs, called cyclic-ordered, grows much slower than the total number of graphs, as shown in Table 14.2. The graphs contributing to the four-, five-, six-point color-ordered amplitudes are listed in Figure 14.1.
- The Feynman rules for calculating the cyclic color-ordered amplitudes, listed in Figure 14.2, are significantly simpler than the original Yang–Mills Feynman rules because the vertices are stripped of all their color factors.

In the case of the four-gluon vertex, we have included only the terms that correspond to the cyclic ordering $\mu\nu\rho\sigma$ (note that it is invariant under cyclic permutations, i.e., the Feynman rule is the same for the vertices $\nu\rho\sigma\mu$, $\rho\sigma\mu\nu$ and $\sigma\mu\nu\rho$). We can already see a considerable simplification of the Feynman rules, since all the color factors have disappeared, and the Lorentz structure of the four-gluon vertex is also much simpler than in the original Feynman rules.

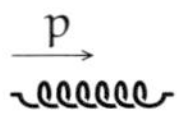


$$= \frac{-i\,g^{\mu\nu}}{p^2 + i0^+} + \frac{i}{p^2 + i0^+}\left(1 - \frac{1}{\xi}\right)\frac{p^\mu p^\nu}{p^2}$$

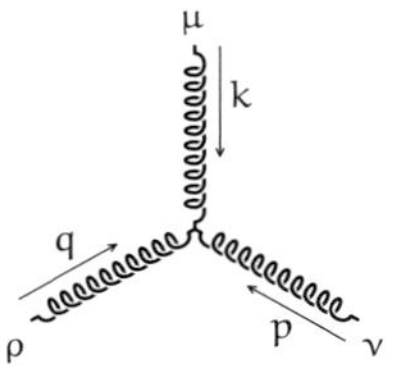


$$= g\,\{g^{\mu\nu}\,(k-p)^\rho + g^{\nu\rho}\,(p-q)^\mu + g^{\rho\mu}\,(q-k)^\nu\}$$

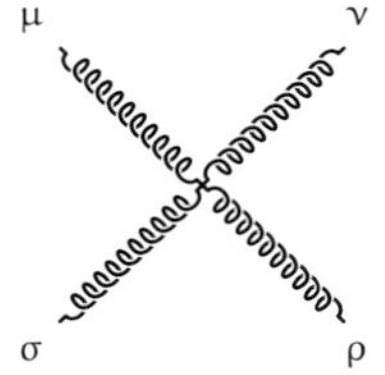


$$= -\,i\,g^2\,(2\,g^{\mu\rho}g^{\nu\sigma} - g^{\mu\sigma}g^{\nu\rho} - g^{\mu\nu}g^{\rho\sigma})$$

FIGURE 14.2: Rules for color-ordered graphs in Yang–Mills theory.

But even after isolating the color structure, the remaining color-ordered amplitudes are still complicated. As an illustration of the color-ordered Feynman rules, let us consider the partial amplitude $\mathcal{A}_4(1,2,3,4)$ that contributes to one of the color structures in the $gg \to gg$ amplitude. Because of color ordering, only three graphs contribute to this partial amplitude:

$$\mathcal{A}_4(1,2,3,4) = \quad (14.14)$$

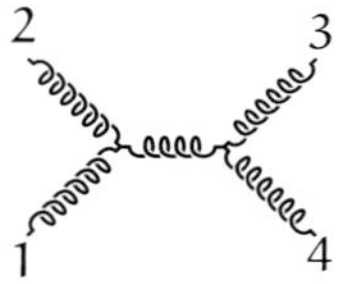


$+$

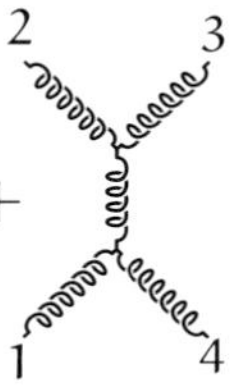


$+$

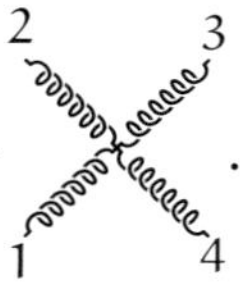


For definiteness, let us assume that the external momenta $p_1 \cdots p_4$ are defined as incoming, and denote $\epsilon_1 \cdots \epsilon_4$ the four polarization vectors. Using the rules listed in Figure 14.2, we obtain:

$$\begin{aligned}
\mathcal{A}_4(1,2,3,4) = &\\
&\frac{-i\,g^2}{(p_1+p_2)^2}\Big[(2p_2+p_1)\cdot\epsilon_1\,\epsilon_2^\lambda - (2p_1+p_2)\cdot\epsilon_2\,\epsilon_1^\lambda \\
&\qquad + \epsilon_1\cdot\epsilon_2\,(p_1-p_2)^\lambda\Big]\Big[(p_3+2p_4)\cdot\epsilon_3\,\epsilon_{4\lambda} \\
&\qquad - (2p_3+p_4)\cdot\epsilon_4\,\epsilon_{3\lambda} + \epsilon_3\cdot\epsilon_4\,(p_3-p_4)_\lambda\Big] \\
&+ \frac{-i\,g^2}{(p_2+p_2)^2}\Big[(p_2+2p_3)\cdot\epsilon_2\,\epsilon_3^\lambda - (2p_2+p_3)\cdot\epsilon_3\,\epsilon_2^\lambda \\
&\qquad + \epsilon_2\cdot\epsilon_3\,(p_2-p_3)^\lambda\Big]\Big[(2p_1+p_4)\cdot\epsilon_4\,\epsilon_{1\lambda} \\
&\qquad - (p_1+2p_4)\cdot\epsilon_1\,\epsilon_{4\lambda} + \epsilon_1\cdot\epsilon_4\,(p_4-p_1)_\lambda\Big] \\
&- i\,g^2\Big[2(\epsilon_1\cdot\epsilon_3)(\epsilon_2\cdot\epsilon_4) - (\epsilon_1\cdot\epsilon_4)(\epsilon_2\cdot\epsilon_3) - (\epsilon_1\cdot\epsilon_2)(\epsilon_3\cdot\epsilon_4)\Big].
\end{aligned} \tag{14.15}$$

Although this is simpler than the full four-gluon amplitude, it remains quite difficult to extract physical results from such an expression.

## 14.2 Spinor-Helicity Formalism

### 14.2.1 Motivation

Part of the complexity of eq. (14.15) lies in the fact that this formula still contains a large amount of unnecessary information, since each polarization may be shifted by a 4-vector proportional to the momentum of the corresponding external gluon, thanks to gauge invariance. For instance, the transformation

$$\epsilon_1^\mu \to \epsilon_1^\mu + \kappa p_1^\mu, \tag{14.16}$$

leaves the amplitude unchanged on-shell. However, it is not clear how to optimally choose the polarization vectors in order to simplify an expression such as eq. (14.15). In other words, the question is how to represent the spin degrees of freedom of the external particles in order to make the amplitude as simple as possible. In the traditional approach to the calculation of amplitudes, one usually refrains from introducing any explicit form for the polarization vectors. Instead, one first squares the amplitude written in terms of generic polarization vectors, such as eq. (14.15), and then the sum over the polarizations of the external gluons is performed by using

$$\sum_{\text{physical pol.}} \epsilon^{\mu*}(\mathbf{p})\epsilon^\nu(\mathbf{p}) = -g^{\mu\nu} + \frac{p^\mu n^\nu}{p \cdot n} + \frac{n^\mu p^\nu}{p \cdot n} \tag{14.17}$$

where $n^\mu$ is some arbitrary light-like vector. Note that this is the formula for summing over all physical polarizations, which is necessary when calculating unpolarized cross-sections. For cross-sections involving polarized particles, one would perform only a partial sum, which leads to a different projector on the right-hand side. If the amplitude is a sum of $N_t$ terms, then this process generates $3N_t^2$ terms in the squared amplitude summed over polarizations. In contrast, the spinor-helicity method that we shall expose below aims at obtaining the amplitude with explicit polarization vectors for a given assignment of the helicities $\{h_1 = \pm, \cdots, h_n = \pm\}$ of the external gluons, in the form of an expression made of $N_t$ terms that can be easily evaluated (numerically at least). The sum of these $N_t$ terms is done first, and then squared, which is an $\mathcal{O}(1)$ computational task (simply squaring a complex number). Thus, the total cost scales as $2^n N_t$ in this approach. Since $N_t$ grows very quickly with $n$, this is usually better.

### 14.2.2 Representation of 4-Vectors as Bi-Spinors

In the previous section, we saw how the adjoint color degrees of freedom may be represented in terms of the smaller fundamental representation. Likewise, we will now represent the Lorentz structure associated to spin-1 particles in terms of spin-1/2 variables. From a mathematical standpoint, this representation exploits the fact that elements of the Lorentz group $SO(3,1)$ can be mapped to $2 \times 2$ complex matrices of unit determinant, i.e., elements of the group SL(2, $\mathbb{C}$). Likewise, 4-momenta can be mapped to $2 \times 2$ complex matrices. In order to make this mapping explicit, let us introduce a set of four matrices $\sigma^\mu$ defined by

$$\sigma^\mu \equiv (\mathbf{1}, \sigma^i), \tag{14.18}$$

where $\sigma^{1,2,3}$ are the Pauli matrices. In terms of these matrices, a 4-vector $p^\mu$ can be mapped into

$$p^\mu \quad \to \quad \mathbf{P} \equiv p_\mu \sigma^\mu = \begin{pmatrix} p_0 + p_3 & p_1 - ip_2 \\ p_1 + ip_2 & p_0 - p_3 \end{pmatrix}. \tag{14.19}$$

(In the second equality, we have used the explicit representation of the Pauli matrices.) For amplitudes involving only external gluons, the momentum $p^\mu$ has a vanishing invariant norm, $p_\mu p^\mu = 0$, which translates into

$$\begin{aligned} 0 = p_\mu p^\mu &= p_0^2 - p_1^2 - p_2^2 - p_3^2 \\ &= (p_0 + p_3)(p_0 - p_3) - (p_1 + ip_2)(p_1 - ip_2) = \det(\mathbf{P}). \end{aligned} \tag{14.20}$$

Thus, the massless on-shell condition is equivalent to the determinant of the matrix $\mathbf{P}$ being zero. For a $2 \times 2$ matrix, a null determinant means that the matrix can be factorized as the direct product of two vectors,

$$\mathbf{P}_{ab} = \lambda_a \xi_b, \tag{14.21}$$

where $\lambda, \xi$ are complex vectors known as *Weyl spinors*. An explicit representation of these vectors is

$$\lambda_a \equiv \begin{pmatrix} \sqrt{p_0 + p_3} \\ \frac{p_1 + ip_2}{\sqrt{p_0 + p_3}} \end{pmatrix}, \quad \xi_b \equiv \begin{pmatrix} \sqrt{p_0 + p_3} \\ \frac{p_1 - ip_2}{\sqrt{p_0 + p_3}} \end{pmatrix}. \tag{14.22}$$

For a real-valued 4-vector, $\lambda_a$ and $\xi_a$ are mutual complex conjugates. However, when we later analytically continue the external momenta in the complex plane, this will no longer be the case. To make the notations more compact, it is customary to introduce the following notations:

$$|\mathbf{p}] = \lambda_a, \quad \langle \mathbf{p}| = \xi_a, \tag{14.23}$$

so that the matrix $\mathbf{P}$ may be written as

$$\mathbf{P} = |\mathbf{p}]\langle \mathbf{p}|. \tag{14.24}$$

It is also convenient to define spinors with raised indices, related to the previous ones as follows:

$$\lambda^a \equiv \lambda_b \epsilon^{ba} = [\mathbf{p}|, \quad \xi^a \equiv \epsilon^{ab} \xi_b = |\mathbf{p}\rangle, \tag{14.25}$$

where $\epsilon^{ab}$ is the completely antisymmetric tensor in two dimensions, normalized with $\epsilon^{12} = +1$. From these spinors with raised indices, we may define an alternate $2 \times 2$ matrix representation of the 4-vector $p^\mu$,

$$\overline{\mathbf{P}} \equiv |\mathbf{p}\rangle[\mathbf{p}|. \tag{14.26}$$

Note that this alternate representation corresponds to the definition[3] $\overline{P} = -p_\mu \overline{\sigma}^\mu$, with $\overline{\sigma}^\mu \equiv (1, -\sigma^i)$.

The fact that we are dealing with on-shell momenta is already built into the factorized representation of eq. (14.21). Amplitudes depend on kinematical invariants such as $(p+q)^2$, for which it is straightforward to check that[4]

$$(p+q)^2 = 2p \cdot q = \langle pq \rangle [pq], \qquad (14.27)$$

where the brackets are defined by contracting upper and lower spinor indices, as in

$$\langle pq \rangle \equiv \langle p|_a |q \rangle^a. \qquad (14.28)$$

These brackets are antisymmetric ($\langle pq \rangle = -\langle qp \rangle$), since they may also be written as

$$\langle pq \rangle = \epsilon^{ab} \xi_a(p) \xi_b(q), \quad [pq] = \epsilon^{ab} \lambda_a(p) \lambda_b(q). \qquad (14.29)$$

Note that the mixed brackets are zero, $\langle pq ] = 0$, as well as the angle and square brackets with twice the same momentum, $\langle pp \rangle = [pp] = 0$.

It is useful to work out the form of momentum conservation in the spinor formalism. For an amplitude with external momenta $\{p_i\}$, chosen to be all incoming, let us denote $|i\rangle, |i], \cdots$ the corresponding spinors. For any arbitrary on-shell momenta p and q, we may then write

$$0 = \langle p | \sum_i \overline{P}_i | q ] = \sum_i \langle p | i \rangle [i | q]. \qquad (14.30)$$

Another interesting identity follows from the fact that three two-component spinors cannot be linearly independent. Thus, given $|p\rangle, |q\rangle$ and $|r\rangle$, we must have a relationship of the form

$$|r\rangle = \alpha |p\rangle + \beta |q\rangle. \qquad (14.31)$$

Contracting this equation with $\langle p|$ and $\langle q|$ gives the explicit expression of the coefficients $\alpha$ and $\beta$,

$$\alpha = \frac{\langle qr \rangle}{\langle qp \rangle}, \qquad \beta = \frac{\langle pr \rangle}{\langle pq \rangle}. \qquad (14.32)$$

This leads to

$$|p\rangle \langle qr \rangle + |q\rangle \langle rp \rangle + |r\rangle \langle pq \rangle = 0, \qquad (14.33)$$

known as the *Schouten identity*. A similar identity holds with square brackets:

$$|p][qr] + |q][rp] + |r][pq] = 0. \qquad (14.34)$$

---

[3] We may use $\epsilon^{ac} \epsilon^{bd} \delta_{dc} = \delta_{ab}$ and $\epsilon^{ac} \epsilon^{bd} \sigma^i_{dc} = -\sigma^i_{ab}$ (see Exercise 14.2).
[4] A quick way to prove it is to use $\langle pq \rangle [pq] = -\text{tr}(\overline{P} Q) = p_\mu q_\nu \text{tr}(\sigma^\mu \overline{\sigma}^\nu) = 2 p_\mu q_\nu g^{\mu\nu}$. Note also that for real momenta, angle and square brackets are complex conjugates, and $(p+q)^2$ is a real quantity.

## 14.2.3 Polarization Vectors

At this point, we have a representation in terms of spinors for the on-shell momenta that appear on the external legs of amplitudes. We also need a similar representation for the polarization vectors. The polarization vectors for a gluon of momentum p with positive and negative helicities may be represented as follows:

$$\epsilon_+^\mu(\mathbf{p};\mathbf{q}) \equiv \frac{\langle \mathbf{q}|\bar{\sigma}^\mu|\mathbf{p}]}{\sqrt{2}\langle \mathbf{qp}\rangle}, \qquad \epsilon_-^\mu(\mathbf{p};\mathbf{q}) \equiv \frac{\langle \mathbf{p}|\bar{\sigma}^\mu|\mathbf{q}]}{\sqrt{2}\,[\mathbf{pq}]}, \tag{14.35}$$

where $\mathbf{q}$ is an arbitrary reference momentum, whose presence is allowed by the gauge invariance (eq. (14.16)). It does not have to correspond to any of the physical momenta upon which the amplitude depends, and can be chosen in such a way that it simplifies the amplitude. This auxiliary vector can be different for each external line, but it must be the same in each contribution to a given process (this is because a single graph usually does not give a gauge invariant contribution when considered alone). Let us mention a useful Fierz identity for contracting two of the numerators that appear in the above polarization vectors[5] (see Exercise 14.2),

$$\langle 1|\bar{\sigma}^\mu|2]\langle 3|\bar{\sigma}_\mu|4] = 2\langle 13\rangle\,[24], \tag{14.36}$$

from which we obtain the contractions between polarization vectors (see Exercise 14.3)

$$\epsilon_-(\mathbf{p};\mathbf{q}) \cdot \epsilon_-(\mathbf{p}';\mathbf{q}') = \frac{\langle \mathbf{pp}'\rangle[\mathbf{qq}']}{[\mathbf{qp}][\mathbf{q}'\mathbf{p}']}, \qquad \epsilon_+(\mathbf{p};\mathbf{q}) \cdot \epsilon_-(\mathbf{p}';\mathbf{q}') = \frac{\langle \mathbf{qp}'\rangle[\mathbf{pq}']}{\langle \mathbf{qp}\rangle[\mathbf{p}'\mathbf{q}']},$$

$$\epsilon_+(\mathbf{p};\mathbf{q}) \cdot \epsilon_+(\mathbf{p}';\mathbf{q}') = \frac{[\mathbf{pp}']\langle \mathbf{qq}'\rangle}{\langle \mathbf{qp}\rangle\langle \mathbf{q}'\mathbf{p}'\rangle}. \tag{14.37}$$

Using $p_\mu\bar{\sigma}^\mu = -|\mathbf{p}\rangle[\mathbf{p}|$, we also obtain the following identities:

$$p \cdot \epsilon_\pm(\mathbf{p};\mathbf{q}) = q \cdot \epsilon_\pm(\mathbf{p};\mathbf{q}) = 0,$$

$$k \cdot \epsilon_+(\mathbf{p};\mathbf{q}) = -\frac{\langle \mathbf{qk}\rangle[\mathbf{kp}]}{\sqrt{2}\langle \mathbf{qp}\rangle}, \qquad k \cdot \epsilon_-(\mathbf{p};\mathbf{q}) = -\frac{\langle \mathbf{pk}\rangle[\mathbf{kq}]}{\sqrt{2}\,[\mathbf{pq}]}. \tag{14.38}$$

## 14.2.4 Three-Point Amplitudes in Yang–Mills Theory

Let us now discuss the very important case of three-particle amplitudes in the massless case, since they will appear later as the building blocks of more complicated amplitudes. Such an amplitude depends on three on-shell momenta $p_{1,2,3}$ such that $p_1 + p_2 + p_3 = 0$. This implies that

$$\langle 12\rangle[12] = 2\,p_1 \cdot p_2 = (p_1 + p_2)^2 = p_3^2 = 0. \tag{14.39}$$

---

[5] We may use $(\bar{\sigma}^\mu)_{ab}(\bar{\sigma}_\mu)_{cd} = 2(\delta_{ab}\delta_{cd} - \delta_{ad}\delta_{bc}) = 2\,\epsilon^{ac}\epsilon^{bd}$.

## 14.2 SPINOR-HELICITY FORMALISM

Therefore, either $\langle 12 \rangle = 0$ or $[12] = 0$. Let us assume that $\langle 12 \rangle \neq 0$. We also have

$$\langle 12 \rangle [23] = \langle 1|\overline{P}_2|3] = -\langle 1|\overline{P}_1 + \overline{P}_3|3] = -\underbrace{\langle 11 \rangle}_{0}[13] - \langle 13 \rangle \underbrace{[33]}_{0} = 0, \quad (14.40)$$

which implies that $[23] = 0$. Likewise, $[13] = 0$. Therefore, all the square brackets are zero if $\langle 12 \rangle \neq 0$. Conversely, all the angle brackets would be zero if instead we had assumed that $[12] \neq 0$. From this discussion, we conclude that massless on-shell three-point amplitudes may depend either on square brackets or on angle brackets, but not on a mixture of both. Recall now that, for real momenta, angle and square brackets are related by complex conjugation. Thus, three-point amplitudes can only exist for complex momenta. This is of course a trivial consequence of kinematics: Momentum conservation $p_1 + p_2 + p_3 = 0$ is impossible for three real-valued light-like momenta, except on a measure-zero subset of exceptional configurations.

Let us now be more explicit and calculate the three-point amplitudes in Yang–Mills theory. For generic polarization vectors $\epsilon_{1,2,3}$, the second Feynman rule of Figure 14.2 leads to

$$\mathcal{A}_3(123) = 2g\left[(\epsilon_1 \cdot \epsilon_2)(p_1 \cdot \epsilon_3) + (\epsilon_2 \cdot \epsilon_3)(p_2 \cdot \epsilon_1) + (\epsilon_3 \cdot \epsilon_1)(p_3 \cdot \epsilon_2)\right], \quad (14.41)$$

where we have used $p_i \cdot \epsilon_i = 0$ to cancel several terms. Consider first the helicities $--+$. Using eqs. (14.36) and (14.38), we obtain

$$\mathcal{A}_3(1^- 2^- 3^+) = -\frac{\sqrt{2}\,g}{[q_1 1][q_2 2]\langle q_3 3 \rangle}\Big\{\langle 12 \rangle [q_1 q_2]\langle q_3 1 \rangle [13] \\
+ \langle 2 q_3 \rangle [q_2 3]\langle 12 \rangle [2 q_1] + \langle q_3 1 \rangle [3 q_1]\langle 23 \rangle [3 q_2]\Big\}. \quad (14.42)$$

Each of the three terms contains in the numerator an angle bracket between the external momenta (respectively $\langle 12 \rangle$, $\langle 12 \rangle$ and $\langle 23 \rangle$). Therefore, for this amplitude to be non-zero, we must adopt the choice of spinor representation where it is the square brackets that are zero. With this choice, the first term vanishes since it contains $[13]$:

$$\mathcal{A}_3(1^- 2^- 3^+) = -\sqrt{2}\,g\,\frac{\langle 2 q_3 \rangle [q_2 3]\langle 12 \rangle [2 q_1] + \langle q_3 1 \rangle [3 q_1]\langle 23 \rangle [3 q_2]}{[q_1 1][q_2 2]\langle q_3 3 \rangle}. \quad (14.43)$$

Using momentum conservation (14.30) in the form of

$$\underbrace{\langle 11 \rangle}_{0}[1 q_1] + \langle 12 \rangle [2 q_1] + \langle 13 \rangle [3 q_1] = 0, \quad (14.44)$$

and the Schouten identity (14.33), we arrive at

$$\mathcal{A}_3(1^- 2^- 3^+) = \sqrt{2}\,g\,\langle 12 \rangle \frac{[q_1 3][q_2 3]}{[q_1 1][q_2 2]}. \quad (14.45)$$

Momentum conservation also implies

$$\frac{[q_1 3]}{[q_1 1]} = \frac{\langle 12 \rangle}{\langle 23 \rangle}, \quad \frac{[q_2 3]}{[q_2 2]} = \frac{\langle 12 \rangle}{\langle 31 \rangle}, \tag{14.46}$$

which leads to a form of the amplitude that does not contain the auxiliary vectors $q_{1,2}$ anymore:

$$\mathcal{A}_3(1^- 2^- 3^+) = \sqrt{2}\, g\, \frac{\langle 12 \rangle^3}{\langle 23 \rangle \langle 31 \rangle}. \tag{14.47}$$

We have thus obtained a remarkably compact expression of the three-point amplitude in terms of spinor variables, which is explicitly independent of all the auxiliary vectors $q_i$. Likewise, a similar calculation would give the following answer for the $++-$ amplitude:

$$\mathcal{A}_3(1^+ 2^+ 3^-) = \sqrt{2}\, g\, \frac{[12]^3}{[23][31]}. \tag{14.48}$$

(The $+++$ and $---$ amplitudes are zero in Yang–Mills theory, as argued in the next subsection.) Eqs. (14.47) and (14.48) are both much simpler than the Feynman rule for the three-gluon vertex. This is the simplest illustration of an assertion we have made earlier, namely that on-shell amplitudes with physical polarizations are much simpler than one may expect from the traditional perturbative expansion. In the case of the three-gluon amplitude, we may think that the simplicity comes from the fact that it receives contributions from a single diagram. However, this is not true. As a teaser for the next section, let us give the answers for some four-gluon and five-gluon amplitudes in the spinor-helicity formalism,

$$\mathcal{A}_4(1^- 2^- 3^+ 4^+) = i(\sqrt{2}\, g)^2\, \frac{\langle 12 \rangle^3}{\langle 23 \rangle \langle 34 \rangle \langle 41 \rangle},$$

$$\mathcal{A}_4(1^- 2^- 3^+ 4^+ 5^+) = i^2(\sqrt{2}\, g)^3\, \frac{\langle 12 \rangle^3}{\langle 23 \rangle \langle 34 \rangle \langle 45 \rangle \langle 51 \rangle}, \tag{14.49}$$

that appear to generalize trivially eq. (14.47), although they result from the sum of three and ten Feynman graphs (see Figure 14.1), respectively. In this section, we have followed a pedestrian approach that consists of starting from the usual Feynman rules, and translating all their building blocks in the spinor-helicity language. However, the simplicity of the results suggests that there must be a better way to obtain them, that bypasses the traditional Feynman rules and provides the answer much more directly.

## 14.2.5 Little Group Scaling

It turns out that massless on-shell three-point amplitudes are almost completely constrained by a scaling argument, except for an overall prefactor. Thus, the Lagrangian is in a sense not necessary for specifying their form (it only plays a marginal role in setting their normalization).

From eqs. (14.24) and (14.26), it is clear that the representation of massless on-shell 4-momenta as bi-spinors is invariant under the following rescaling:

$$|\mathbf{p}\rangle \to \lambda |\mathbf{p}\rangle, \qquad |\mathbf{p}] \to \lambda^{-1} |\mathbf{p}], \qquad (14.50)$$

known as *little group scaling*. The terminology follows from the fact that there is a one-parameter SO(2) subgroup (the rotations in the plane transverse to **p**) of the Lorentz group that leaves invariant the vector $p^\mu$. Such a residual symmetry that leaves a vector invariant is called *little group*. In the spinor formulation, this residual symmetry precisely corresponds to the transformation of eq. (14.50).

Under little group scaling of $|\mathbf{p}\rangle$ and $|\mathbf{p}]$, the polarization vectors of eq. (14.35) scale as follows:

$$\epsilon_+^\mu(\mathbf{p};\mathbf{q}) \to \lambda^{-2} \epsilon_+^\mu(\mathbf{p};\mathbf{q}), \qquad \epsilon_-^\mu(\mathbf{p};\mathbf{q}) \to \lambda^2 \epsilon_-^\mu(\mathbf{p};\mathbf{q}), \qquad (14.51)$$

i.e., a scaling by a factor $\lambda^{-2h}$ for a helicity h. Note that the polarization vectors are invariant under little group scaling of the auxiliary vector **q**. In an amplitude, the internal ingredients (propagators and vertices) are not affected by little group scaling. Therefore, if we apply the little group scaling $\lambda_i$ to an external momentum i of an amplitude, its expression in terms of square and angle spinors must transform as

$$\mathcal{A}_n(1\cdots i^{h_i}\cdots n) \quad \to \quad \lambda_i^{-2h_i} \mathcal{A}_n(1\cdots i^{h_i}\cdots n), \qquad (14.52)$$

where $h_i$ is the helicity of the external line i (we do not need to specify the helicities of the other external lines).

It turns out that the structure of all three-point amplitudes[6] is completely fixed by this property. Let us start from the following generic expression:

$$\mathcal{A}_3(1^{h_1} 2^{h_2} 3^{h_3}) = C \langle 12 \rangle^\alpha \langle 23 \rangle^\beta \langle 31 \rangle^\gamma, \qquad (14.53)$$

with $\alpha, \beta, \gamma$ undetermined exponents and C a numerical prefactor. Little group scaling implies that

$$-2h_1 = \alpha + \gamma, \quad -2h_2 = \alpha + \beta, \quad -2h_3 = \beta + \gamma, \qquad (14.54)$$

whose solution is

$$\alpha = h_3 - h_1 - h_2, \quad \beta = h_1 - h_2 - h_3, \quad \gamma = h_2 - h_3 - h_1. \qquad (14.55)$$

Therefore the three-point amplitude must have the following structure:

$$\mathcal{A}_3(1^{h_1} 2^{h_2} 3^{h_3}) = C \langle 12 \rangle^{h_3-h_1-h_2} \langle 23 \rangle^{h_1-h_2-h_3} \langle 31 \rangle^{h_2-h_3-h_1}, \qquad (14.56)$$

in which only the numerical prefactor remains to be determined. The $--+$ three-gluon amplitude derived in the previous subsection indeed has this structure.

---

[6]This reasoning cannot be extended to higher n-point amplitudes because they can depend on both square and angle brackets, and because the number of constraints provided by the helicities of the external lines is not sufficient to fix the unknown exponents.

Note that instead of eq. (14.53), we could have chosen an ansatz that involves the square brackets,

$$\mathcal{A}_3(1^{h_1}2^{h_2}3^{h_3}) = C\,[12]^{\alpha'}[23]^{\beta'}[31]^{\gamma'}. \tag{14.57}$$

(This is the only alternative, since we are not allowed to mix square and angle brackets in a three-point amplitude for massless particles.) Little group scaling would now lead to

$$\alpha' = -h_3 + h_1 + h_2, \quad \beta' = -h_1 + h_2 + h_3, \quad \gamma' = -h_2 + h_3 + h_1, \tag{14.58}$$

and consequently

$$\mathcal{A}_3(1^{h_1}2^{h_2}3^{h_3}) = C\,[12]^{-h_3+h_1+h_2}[23]^{-h_1+h_2+h_3}[31]^{-h_2+h_3+h_1}. \tag{14.59}$$

The expected dimension of the amplitude is sufficient to choose between eqs. (14.56) and (14.59). Indeed, both angle and square brackets have mass dimension 1, while the three-gluon amplitude should have dimension 1 in four-dimensional Yang–Mills theory (for which the coupling constant is dimensionless). Since all the kinematical dependence is carried by the brackets, the prefactor C can only be made of coupling constants and numerical factors, and must therefore be dimensionless in Yang–Mills theory. Consider first the $--+$ amplitude: eq. (14.56) gives a mass dimension $+1$, while eq. (14.59) gives a mass dimension $-1$. Therefore, the $--+$ amplitude must be expressed by eq. (14.56) in terms of angle brackets. The same argument tells us that the $++-$ amplitude must be given by eq. (14.59) in terms of square brackets.

Let us consider now the $---$ amplitude. Little group scaling tells us that

$$\mathcal{A}_3(1^-2^-3^-) = C\,\langle 12\rangle\langle 23\rangle\langle 31\rangle. \tag{14.60}$$

Therefore, the prefactor C should have mass dimension $-2$, which cannot be constructed from the dimensionless coupling constant of Yang–Mills theory, unless $C=0$ (the same conclusion holds if we try to construct this amplitude with square brackets). Likewise, we conclude that the $+++$ amplitude is zero as well. See also Exercises 14.4–14.6 for other constraints derived from little group scaling.

### 14.2.6 Maximally Helicity Violating Amplitudes

Let us consider a tree Feynman diagram contributing to an n-point amplitude, with $n_3$ three-gluon vertices, $n_4$ four-gluon vertices and $n_I$ internal propagators. These quantities are related by

$$n + 2n_I = 3n_3 + 4n_4, \quad n_I = n_3 + n_4 - 1. \tag{14.61}$$

The second equation is the statement that this graph has no loops. From these equations we get the following identities:

$$n = n_3 + 2n_4 + 2, \quad n_3 - 2n_I = 4 - n. \tag{14.62}$$

## 14.2 SPINOR-HELICITY FORMALISM

The contribution of this Feynman graph to the amplitude is made of $n$ polarization vectors, $n_I$ denominators coming from the internal propagators and $n_3$ powers of momentum in the numerator, which come from the three-gluon vertices:[7]

$$\mathcal{A}_n(1\cdots n) \sim \frac{\left[\prod_{i=1}^n \epsilon_i^{\mu_i}\right]\left[\prod_{j=1}^{n_3} L_j^{\nu_j}\right]}{\prod_{k=1}^{n_I} K_k^2} \sim \frac{(\text{mass})^{n_3}}{(\text{mass})^{2n_I}} \sim (\text{mass})^{4-n}. \tag{14.63}$$

First, we see that the mass dimension of the $n$-point amplitude is $4 - n$. Moreover, the amplitude $\mathcal{A}_n$ does not carry any Lorentz index. Therefore, in the numerator all the Lorentz indices $\mu_i$ and $\nu_j$ must be contracted pairwise. These contractions lead to three types of factors:

$$\epsilon_i \cdot \epsilon_{i'}, \quad \epsilon_i \cdot L_j, \quad L_j \cdot L_{j'}. \tag{14.64}$$

**Only-+ amplitude:** Now, consider an amplitude with only positive helicities. From eq. (14.37), we see that all contractions between polarization vectors are proportional to

$$\epsilon_+(i; q_i) \cdot \epsilon_+(i'; q_{i'}) \propto \langle q_i q_{i'} \rangle. \tag{14.65}$$

By choosing the auxiliary momenta $q_i$ to be all equal to $q$, we make all these contractions vanish. Therefore, to obtain a non-zero contribution, it is necessary to contract all the polarization vectors with momenta from the three-gluon vertices, $\epsilon_i \cdot L_j$. But from the first of eqs. (14.62), we see that $n > n_3$, which means that it is impossible to contract all the $n$ polarization vectors with the $n_3$ momenta from the vertices. Thus, the all-plus amplitude is zero:

$$\mathcal{A}_n(1^+ 2^+ \cdots n^+) = 0. \tag{14.66}$$

By the same reasoning, we conclude that the all-minus amplitude is also zero. We can see here the power that stems from the freedom of choosing the auxiliary vectors $q_i$; for generic $q_i$, this amplitude would still be zero (since it does not depend on the $q_i$), but this zero would result from intricate cancellations among the many graphs that contribute to $\mathcal{A}_n$. Instead, with a smart choice of the auxiliary vectors, we can make this cancellation happen graph by graph.

**$-+\cdots+$ amplitude:** Consider now an amplitude with one negative helicity carried by the first external leg, and $n - 1$ positive helicities carried by the external legs 2 to $n$. Now, it is convenient to choose the auxiliary vectors as follows:

$$q_2 = q_3 = \cdots = q_n = p_1. \tag{14.67}$$

Again, all the contractions between pairs of polarization vectors cancel, since we have $\epsilon_+(i; q_i) \cdot \epsilon_+(i'; q_{i'}) = 0$ for $i, i' \geq 2$, and $\epsilon_-(1; q_1) \cdot \epsilon_+(i; q_i) = 0$ for $i \geq 2$. Since $n > n_3$, it is not possible to contract all the polarization vectors with momenta from the three-gluon vertices, and these amplitudes also vanish at tree level:

$$\mathcal{A}_n(1^- 2^+ \cdots n^+) = 0. \tag{14.68}$$

(We also have $\mathcal{A}_n(1^+ 2^- \cdots n^-) = 0$ at tree level.)

---

[7]We assume for simplicity Feynman gauge, in which the numerator of the gluon propagator does not depend on momentum.

**Maximally helicity violating amplitudes:** Let us flip one more helicity, e.g., with the assignment $1^-2^-3^+\cdots n^+$. This time, a useful choice of auxiliary vectors is

$$q_1 = q_2 = p_n, \qquad q_3 = q_4 = \cdots = q_n = p_1. \tag{14.69}$$

With this choice, all the contractions of polarization vectors are zero, except

$$\epsilon_-(2;q_2)\cdot \epsilon_+(i;q_i) \neq 0 \quad \text{for } i = 3,\cdots, n-1. \tag{14.70}$$

Thus, this time, we need to contract the remaining $n-2$ polarization vectors with the $n_3$ momenta from the three-gluon vertices, which is possible (provided that $n_4 = 0$, which means that diagrams containing four-gluon vertices do not contribute to the $--+\cdots+$ amplitude for our choice of auxiliary vectors). Therefore, this assignment of helicities gives a non-zero amplitude,

$$\mathcal{A}_n(1^-2^-3^+\cdots n^+) \neq 0. \tag{14.71}$$

These amplitudes, called the *maximally helicity violating (MHV) amplitudes*, are the simplest non-zero amplitudes at tree level.[8] As we shall see later, they are given at tree level by very compact formulas in terms of square and angle brackets (note that up to $n = 5$ external lines, all the non-zero amplitudes are MHV amplitudes). Generically, the complexity of amplitudes increases with the number of negative helicities, culminating with amplitudes that have comparable numbers of negative and positive helicities (increasing further the number of negative helicities then reduces the complexity).

## 14.3 Britto–Cachazo–Feng–Witten On-Shell Recursion

As we have seen, the main obstacle to the calculation of amplitudes by the usual Feynman rules is the proliferation of graphs as one increases the number of external legs. This problem remains true even after one has factorized the color factors, even if it is somewhat mitigated by the fact that the number of cyclic-ordered graphs grows at a slower pace.

This issue could be avoided if there was a way to break down a tree amplitude into smaller pieces (themselves tree amplitudes) that have a smaller number of external legs. It turns out that an amplitude naturally factorizes into two sub-amplitudes when one of its internal propagators goes on-shell. The physical reason of such a factorization is that on-shell momenta correspond to infinitely long-lived particles. Thus, the two sub-amplitudes on each side of this on-shell propagator do not talk to one another. The other advantage of this situation is that the two sub-amplitudes would themselves be on-shell, and therefore we may use for them spinor-helicity formulas that could have been previously obtained for amplitudes with fewer external legs. If this were possible, we would thus obtain a recursive relationship (in the number of external legs) for on-shell amplitudes.

---

[8] At higher orders, the all-plus and $-+\cdots+$ amplitudes are non-zero in Yang–Mills theory. It turns out that these amplitudes are zero (to all orders) in supersymmetric Yang–Mills theories. At tree level, the supersymmetric partners do not appear in amplitudes with only external gluons, which is the reason why these amplitudes also vanish in non-supersymmetric Yang–Mills theory. At the loop level, the cancellation requires a compensation between the various supersymmetric partners running in the loops, which is not happening in the plain Yang–Mills theory.

## 14.3.1 Analytical Properties of Amplitudes with Shifted Momenta

Unfortunately, with fixed generic external momenta, tree amplitudes do not have internal on-shell propagators. The trick is to consider a one-parameter *complex* deformation of the external momenta, adjusted in order to make an internal denominator vanish:

$$\mathcal{A}_n(12\cdots n) \quad \to \quad \mathcal{A}_n(12\cdots n; z), \tag{14.72}$$

where $z$ is a complex variable that controls the deformation. The singularities of tree Feynman graphs come from the zeroes of the denominators of its internal propagators, which give poles in $z$. Our goal will be to choose this deformation in such a way that the total momentum remains conserved, and the deformed external momenta are still on-shell. With such a choice, we will be able to reuse the on-shell formulas obtained for smaller amplitudes.

Let us consider the ratio $\mathcal{A}_n(\cdots; z)/z$. Besides the poles coming from the internal propagators, the ratio also has a simple pole at $z = 0$. Let us assume that $\mathcal{A}_n(\cdots; z)$ vanishes when $|z| \to \infty$, so that the integral of $\mathcal{A}_n(\cdots; z)/z$ on a contour at infinity in the complex plane vanishes. Then, if the contour $\gamma$ is a circle whose radius goes to infinity, we may write

$$0 = \oint_\gamma \frac{dz}{2\pi i} \frac{\mathcal{A}_n(\cdots; z)}{z} = \mathcal{A}_n(\cdots; 0) + \sum_{z_i \in \{\text{poles of } \mathcal{A}_n\}} \left. \text{Res}\, \frac{\mathcal{A}_n(\cdots; z)}{z} \right|_{z_i}. \tag{14.73}$$

The first term, $\mathcal{A}_n(\cdots; z = 0)$, is nothing but the amplitude we aim at calculating. This formula therefore expresses it in terms of the residues of $\mathcal{A}_n(\cdots; z)/z$ at the simple poles corresponding to the internal propagators of the amplitude. Moreover, these residues will be factorizable into smaller on-shell amplitudes, precisely because the poles $z_i$ correspond to the on-shellness of some internal propagator.

## 14.3.2 Minimal Momentum Shifts

There are many ways to implement a complex shift of the external momenta, but all of them must fulfill the following conditions:

- The sum of the shifted incoming momenta should remain zero. Therefore, we must shift at least two momenta (and the simplest is to shift only two).
- The shifted external momenta should stay on-shell at all $z$.
- The amplitude evaluated at the shifted momenta should go to zero as $|z| \to \infty$.

The condition of momentum conservation is trivially satisfied by choosing two momenta $i, j$ to be shifted, and by giving them opposite shifts,

$$p_i \to \widehat{p}_i = p_i(z) \equiv p_i + z k, \quad p_j \to \widehat{p}_j = p_j(z) \equiv p_j - z k, \tag{14.74}$$

where we denote with a hat the shifted momenta. All the momenta $p_k$ for $k \neq i, j$ are left unmodified. The on-shell conditions for $\widehat{p}_{i,j}$ are satisfied provided that

$$k^2 = 0, \quad p_i \cdot k = 0, \quad p_j \cdot k = 0. \tag{14.75}$$

It turns out that these equations have two solutions (up to an arbitrary prefactor), provided we allow complex momenta. In the spinor notation, the first condition is automatically satisfied if $\mathbf{K}$ can be factorized as in eq. (14.21), while the second and third conditions become

$$\langle ik\rangle[ik] = 0, \quad \langle jk\rangle[jk] = 0. \tag{14.76}$$

This explains why we need a complex momentum $k^\mu$. Indeed, for a real $k^\mu$, $|k]$ and $|k\rangle$ are related by complex conjugation, and the above conditions reduce to $\langle ik\rangle = \langle jk\rangle = 0$. With two-component spinors, this implies $|k\rangle \propto |i\rangle$ and $|k\rangle \propto |j\rangle$, which is in general impossible. By allowing a complex momentum $k^\mu$, we let $|k]$ and $|k\rangle$ be independent, which allows solving the above conditions by having, for instance,

$$|k\rangle = |i\rangle, \quad |k] = |j]. \tag{14.77}$$

(The other independent solution is obtained by exchanging the roles of $i$ and $j$.) The bi-spinors corresponding to the shifted momenta are

$$\widehat{P}_i = |i]\langle i| + z|j]\langle i| = (|i] + z|j])\langle i|, \quad \widehat{P}_j = |j]\langle j| - z|j]\langle i| = |j](\langle j| - z\langle i|), \tag{14.78}$$

from which we read the shifted spinors,

$$|\hat{\imath}\rangle = |i\rangle, \quad |\hat{\jmath}\rangle = |j\rangle - z|i\rangle, \quad |\hat{\imath}] = |i] + z|j], \quad |\hat{\jmath}] = |j]. \tag{14.79}$$

### 14.3.3 Behavior at $|z| \to \infty$

Until now, our description of this method has been completely generic and applicable to all sorts of quantum field theories, since no reference has been made to the details of its Lagrangian. These details become important when discussing the condition that $\mathcal{A}_n(\cdots;z)$ vanishes at infinity. Let us discuss the behavior at large $z$ in the case of Yang–Mills theory (see Exercise 14.8 for the case of a scalar theory). First, a $z$ dependence enters in the polarization vectors of the external lines $i$ and $j$. For generic auxiliary vectors, we have

$$\epsilon_+^\mu(\hat{\imath};q) = \frac{\langle q|\overline{\sigma}^\mu|\hat{\imath}]}{\sqrt{2}\langle q\hat{\imath}\rangle} \sim z, \quad \epsilon_+^\mu(\hat{\jmath};q) = \frac{\langle q|\overline{\sigma}^\mu|\hat{\jmath}]}{\sqrt{2}\langle q\hat{\jmath}\rangle} \sim z^{-1},$$

$$\epsilon_-^\mu(\hat{\imath};q) = \frac{\langle \hat{\imath}|\overline{\sigma}^\mu|q]}{\sqrt{2}[\hat{\imath}q]} \sim z^{-1}, \quad \epsilon_-^\mu(\hat{\jmath};q) = \frac{\langle \hat{\jmath}|\overline{\sigma}^\mu|q]}{\sqrt{2}[\hat{\jmath}q]} \sim z. \tag{14.80}$$

Inside a graph contributing to this n-point amplitude, we can follow a string of propagators that all carry shifted momenta, from the external line $i$ to the external line $j$, as illustrated in Figure 14.3. For all these propagators, since $k^2 = 0$, the denominators are linear in $z$. In addition, the three-gluon vertices along this string of propagators are linear in the momenta, and therefore scale as $z$. Along this string, there are $\nu$ vertices (three-gluon or four-gluon vertices), and $\nu - 1$ propagators, hence a global behavior at most $\sim z$ at large $z$ (obtained when

## 14.3 BRITTO–CACHAZO–FENG–WITTEN ON-SHELL RECURSION

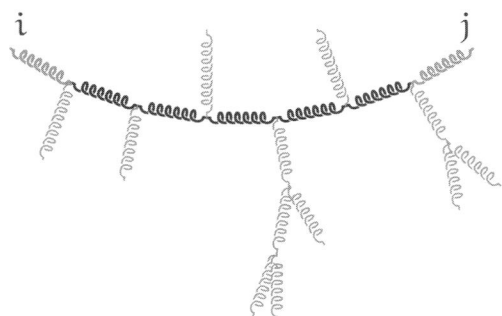

FIGURE 14.3: Propagators affected by the momentum shift (shown as dark color) in a tree amplitude. The lighter-colored lines do not depend on z. The propagators on the external lines are not actually part of the expression of the amplitude.

all these vertices are three-gluon vertices, that scale as z). For the assignment $\{h_i = -, h_j = +\}$ of polarizations, we thus find an overall behavior[9] in $z^{-1}$, valid graph by graph.

For other combinations of polarizations on the lines $i, j$, this diagrammatic argument suggests that they may not go to zero. However, the actual behavior for $\{h_i = -, h_j = -\}$ and $\{h_i = +, h_j = +\}$ is better than the one suggested by this graph-by-graph estimate. First, note that this problem is reminiscent of the eikonal approximation, in which a hard on-shell particle punches through a background of much softer particles that very mildly disturb its motion. This can be studied by splitting the gauge field $A^\mu$ into a hard component $a^\mu$ that describes the gluons along the string with shifted momenta and a soft background $\mathcal{A}^\mu$ (describing the unshifted gluons attached to the hard ones):

$$A^\mu \equiv \mathcal{A}^\mu + a^\mu. \tag{14.81}$$

When rewriting the Yang–Mills Lagrangian in terms of these fields, it is sufficient to keep terms that are quadratic in the hard field, since in our problem exactly two external lines are shifted,

$$\mathcal{L}_{\text{YM}} = \cdots - \frac{1}{4}\text{tr}\left((\mathcal{D}_\mu a_\nu - \mathcal{D}_\nu a_\mu)(\mathcal{D}^\mu a^\nu - \mathcal{D}^\nu a^\mu)\right) + \frac{ig}{2}\text{tr}\left([a_\mu, a_\nu]\mathcal{F}^{\mu\nu}\right) + \cdots, \tag{14.82}$$

where $\mathcal{D}_\mu$ and $\mathcal{F}^{\mu\nu}$ are constructed with the background field. When splitting the gauge potential as in eq. (14.81), one may fix independently the gauge for the background and for the fluctuation $a^\mu$. For the latter, a convenient choice is the background field gauge,

$$\mathcal{D}_\mu a^\mu(x) = \omega(x). \tag{14.83}$$

After adding the gauge fixing term, the quadratic part of the Lagrangian becomes

$$\mathcal{L}_{\text{YM+GF}} = \cdots - \frac{1}{4}\text{tr}\left((\mathcal{D}_\mu a_\alpha)(\mathcal{D}^\mu a^\alpha)\right) + \frac{ig}{2}\text{tr}\left([a_\mu, a_\nu]\mathcal{F}^{\mu\nu}\right) + \cdots \tag{14.84}$$

---

[9] When the shifted amplitude decreases faster than $z^{-1}$, one may obtain a more compact expression by integrating $\mathcal{A}_n(\cdots; z)(1 - z/z_*)/z$, where $z_*$ is one of the poles of $\mathcal{A}_n$. There is no boundary term thanks to the faster decrease of $\mathcal{A}_n$, and the additional subtraction removes the contribution from the pole $z_*$, leading to an expression with one fewer term.

In this equation, the first term possesses an extended Lorentz symmetry, since it is invariant under independent Lorentz transformations of the fluctuations and of the background, while the second term is only invariant under simultaneous Lorentz transformations of $\mathcal{A}^\mu$ and $a^\mu$.

Let us denote $\mathcal{M}_{\alpha\beta}[\mathcal{A}]$ the propagator of the fluctuation $a^\mu$, amputated of its final lines. This propagator contains three-gluon couplings to the background field, which come from the first term of eq. (14.84), and four-gluon couplings to the background field coming also from the second term. With only three-gluon couplings, we have $\mathcal{M}_{\alpha\beta} \sim z$ (because there is one more vertex than propagators), and each four-gluon vertex removes one power of z. Given the Lorentz structure of eq. (14.84), we may write

$$\mathcal{M}_{\alpha\beta} = \left(c_1 z + c_0 + c_{-1} z^{-1} + \cdots\right) g_{\alpha\beta} + A_{\alpha\beta} + z^{-1} B_{\alpha\beta} + \cdots \qquad (14.85)$$

In this formula, the term in $g_{\alpha\beta}$ comes entirely from the first term of eq. (14.84), whose extended Lorentz symmetry leads to the factor $g_{\alpha\beta}$. All the coefficients in this expansion are functionals of the soft field. The term $A_{\alpha\beta}$, which comes from a single insertion of $[a_\mu, a_\nu] \mathcal{F}^{\mu\nu}$, is antisymmetric. The subsequent terms correspond to two or more insertions of the four-gluon vertex. These terms have no definite symmetry, but they are not needed in the discussion. The amputated two-point function $\mathcal{M}_{\alpha\beta}$ also obeys the following on-shell Ward–Takahashi identity:

$$\widehat{p}_i^\alpha \, \mathcal{M}_{\alpha\beta} \, \epsilon_{h_j}^\beta(\widehat{j}; q') = 0, \quad \epsilon_{h_i}^\alpha(\widehat{i}; q) \, \mathcal{M}_{\alpha\beta} \, \widehat{p}_j^\beta = 0, \qquad (14.86)$$

with shifted on-shell momenta and polarization vectors. Note that, unlike in an Abelian gauge theory, it is necessary to contract one side of the function with a physical polarization vector for the identity to hold (see Section 8.5).

The shifted n-point amplitude $\mathcal{A}_n$ is obtained by keeping $n-2$ powers of the background field in $\mathcal{M}_{\alpha\beta}$, and by contracting with the appropriate polarization vectors,

$$\mathcal{A}_n \sim \epsilon_{h_i}^\alpha(\widehat{i}; q) \, \mathcal{M}_{\alpha\beta} \, \epsilon_{h_j}^\beta(\widehat{j}; q'). \qquad (14.87)$$

Choosing the auxiliary vectors to be $q \equiv p_i$ and $q' \equiv p_j$, the explicit form of the polarization vectors is[10]

$$\epsilon_+^\mu(\widehat{i}; q) = \frac{\sqrt{2}}{\langle ji \rangle} \left(k^{*\mu} + z \, p_j^\mu\right), \qquad \epsilon_-^\mu(\widehat{i}; q) = \frac{\sqrt{2}}{[ij]} k^\mu,$$

$$\epsilon_+^\mu(\widehat{j}; q') = \frac{\sqrt{2}}{\langle ij \rangle} k^\mu, \qquad \epsilon_-^\mu(\widehat{j}; q') = \frac{\sqrt{2}}{[ji]} \left(k^{*\mu} - z \, p_i^\mu\right). \qquad (14.88)$$

Note that with this choice of auxiliary vectors, we have lost a power of z in the denominators of $\epsilon_-^\mu(\widehat{i}; q)$ and $\epsilon_+^\mu(\widehat{j}; q')$. This will not change the final results, since on-shell amplitudes do not depend on the auxiliary vectors. As a check of the insight gained from Feynman diagrams, let us first consider the case $\{h_i = -, h_j = +\}$. The shifted amplitude behaves as

$$\mathcal{A}_{n;-+} \sim \underbrace{(k \cdot k)}_{0}(c_1 z + \cdots) + \underbrace{k^\alpha k^\beta}_{0} A_{\alpha\beta} + \mathcal{O}(z^{-1}) \sim \mathcal{O}(z^{-1}) \qquad (14.89)$$

---

[10]In order to obtain these expressions, one may contract the polarization vectors with $\sigma_\mu$ to first obtain the corresponding $2 \times 2$ matrix, which can be done by using $\left(\overline{\sigma}^\mu\right)^{ab} \left(\sigma_\mu\right)_{cd} = 2 \delta^a{}_d \, \delta^b{}_c$, $|j] \langle i| = k_\mu \sigma^\mu$, and $|i\rangle [j| = k_\mu^* \overline{\sigma}^\mu$.

The first term vanishes because k is on-shell, and the second one thanks to the antisymmetry of $A_{\alpha\beta}$. Next, consider the case $\{h_i = -, h_j = -\}$, for which we obtain

$$\begin{aligned}\mathcal{A}_{n;--} &\sim k^\alpha \, \mathcal{M}_{\alpha\beta} \, \epsilon_-^\beta(\hat{\jmath}; \mathbf{q}') = -z^{-1} \, p_i^\alpha \, \mathcal{M}_{\alpha\beta} \, \epsilon_-^\beta(\hat{\jmath}; \mathbf{q}') \\ &\sim z^{-1} p_i \cdot (k^* - z\, p_i)(c_1 z + \cdots) + z^{-1} p_i^\alpha (k^{*\beta} - z\, p_i^\beta) A_{\alpha\beta} + \mathcal{O}(z^{-1}) \\ &\sim \mathcal{O}(z^{-1}).\end{aligned} \qquad (14.90)$$

The first line is obtained by using the Ward–Takahashi identity, and in the second line all terms that could be larger than $z^{-1}$ vanish due to $p_i^2 = p_i \cdot k^* = 0$ and thanks to the antisymmetry of $A_{\alpha\beta}$. The case $\{h_i = +, h_j = +\}$ is very similar and leads to

$$\begin{aligned}\mathcal{A}_{n;++} &\sim \epsilon_+^\alpha(\hat{\imath}; \mathbf{q}) \, \mathcal{M}_{\alpha\beta} \, k^\beta = z^{-1} \epsilon_+^\alpha(\hat{\imath}; \mathbf{q}) \, \mathcal{M}_{\alpha\beta} \, p_j^\beta \\ &\sim z^{-1} (k^* + z\, p_j) \cdot p_j (c_1 z + \cdots) + z^{-1} (k^{*\alpha} + z\, p_j^\alpha) p_j^\beta A_{\alpha\beta} + \mathcal{O}(z^{-1}) \\ &\sim \mathcal{O}(z^{-1}).\end{aligned} \qquad (14.91)$$

Finally, in the last case, $\{h_i = +, h_j = -\}$, we obtain $\mathcal{A}_{n;+-} \sim \mathcal{O}(z^3)$, and therefore we cannot use such a shift in eq. (14.73).

### 14.3.4 Recursion Formula

Since all-plus amplitudes are zero, the assignment of helicities that we shall consider is generically of the form $1^- \cdots r^-(r+1)^+ \cdots n^+$, and the shift applied to the lines $i = 1, j = n$ leads to a vanishing amplitude when $|z| \to \infty$. We can therefore apply eq. (14.73) and write the amplitude in the following way:

$$\mathcal{A}_n(\cdots) = -\sum_{z_i \in \{\text{poles of } \mathcal{A}_n\}} \text{Res} \, \frac{\mathcal{A}_n(\cdots; z)}{z} \bigg|_{z_i}. \qquad (14.92)$$

As explained earlier, the poles $z_i$ come from the vanishing denominators of the internal propagators, i.e., one of the dark-colored propagators in Figure 14.4. Let us denote $K_I$ the momentum (before the shift) carried by the propagator producing the pole, with the convention that it is oriented in the same direction as $p_1$. The shift changes this momentum into

$$K_I \to \widehat{K}_I \equiv K_I + z\, k, \qquad (14.93)$$

and the condition that the denominator of the propagator vanishes after the shift is

$$0 = \widehat{K}_I^2 = K_I^2 + 2 z_I \, K_I \cdot k, \quad \text{i.e. } z_I = -\frac{K_I^2}{2 K_I \cdot k}. \qquad (14.94)$$

The singular propagator divides the amplitude into left and right sub-amplitudes, so that we may write

$$\mathcal{A}_n(\hat{1}\,2\cdots(n-1)\hat{n}; z) \equiv \sum_{h=\pm} \mathcal{A}_L(\hat{1}\,2\cdots -\widehat{K}_I^{+h}; z) \, \frac{i}{\widehat{K}_I^2} \, \mathcal{A}_R(\widehat{K}_I^{-h}\cdots(n-1)\hat{n}; z), \qquad (14.95)$$

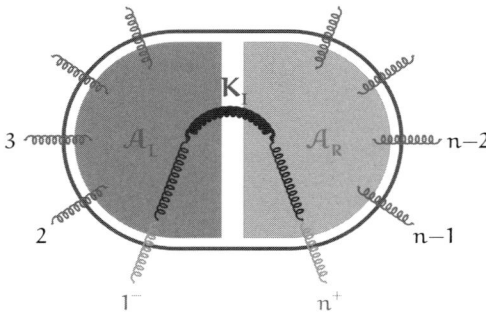

FIGURE 14.4: Setup for the BCFW recursion formula with shifts applied to the external lines 1 and n. The pole comes from the propagator carrying the momentum $K_I$ (dark color). This singular propagator divides the graph into left and right sub-amplitudes, $\mathcal{A}_L$ and $\mathcal{A}_R$.

with a sum over the helicity h of the intermediate gluon.[11] From this expression, the residue at the pole $z_I$ of $\mathcal{A}_n(\cdots;z)/z$ takes the form

$$\operatorname{Res} \frac{\mathcal{A}_n(\cdots;z)}{z}\bigg|_{z_I} = -\sum_{h=\pm} \mathcal{A}_L(\hat{1}\,2\cdots-\widehat{K}_I^{+h};z_I)\,\frac{i}{K_I^2}\,\mathcal{A}_R(\widehat{K}_I^{-h}\cdots(n-1)\hat{n};z_I). \quad (14.96)$$

Both $\mathcal{A}_L$ and $\mathcal{A}_R$ have strictly less than n external lines, which means that the formula is recursive: it expresses an amplitude in terms of smaller amplitudes, eventually breaking it down to three-point amplitudes. Moreover, the crucial point here is that, when evaluated at the value $z_I$ that gives $\widehat{K}_I^2 = 0$, the left and right sub-amplitudes have only *on-shell* (but complex) external momenta. Therefore, this recursion never requires off-shell amplitudes, which is of utmost importance for keeping out of the calculation unnecessarily complicated kinematics and unphysical degrees of freedom. Since each internal z-dependent propagator can be singular for some z, eq. (14.92) contains one term for each such propagator. There are at most $n-3$ terms in this sum, corresponding to the partitions of $[2,n-1] = [2,l]\cup[l+1,n-1]$ with $2 \leq l \leq n-2$.

### 14.3.5 Parke–Taylor Formula for MHV Amplitudes

**MHV recursion formula:** As an illustration of the BCFW recursion formula, let us determine the explicit expression of the MHV amplitudes[12] $\mathcal{A}_n(1^-2^-3^+\cdots n^+)$. We show all the helicity assignments, including those of the singular propagator, in Figure 14.5. In order to avoid having an all-plus sub-amplitude on the right, we must choose $h = +$. This choice makes

---

[11] Both $\mathcal{A}_L$ and $\mathcal{A}_R$ are defined with all gluons incoming. This is why one has argument $-\widehat{K}_I$ and the other one $+\widehat{K}_I$. For the same reason, the helicity is $+h$ on one side and $-h$ on the other side.

[12] This assignment of helicities, with the negative helicities carried by adjacent lines, is the simplest. MHV amplitudes with non-adjacent negative helicities are also given by the Parke–Taylor formula, but the proof is a bit more complicated in this case (see Exercise 14.10).

## 14.3 BRITTO-CACHAZO-FENG-WITTEN ON-SHELL RECURSION

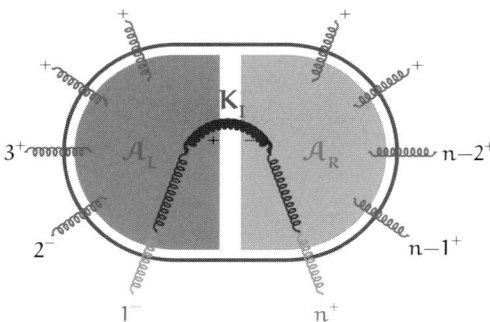

FIGURE 14.5: Setup for applying the BCFW recursion formula to the calculation of the $--+\cdots+$ MHV amplitude. We have indicated explicitly all the helicities. Note that only $h = +$ is allowed in the sum over the helicity of the singular propagator (otherwise the right-side sub-amplitude would be a vanishing all-plus amplitude).

$\mathcal{A}_R$ an $-+\cdots+$ amplitude, which is also zero unless it is a three-point amplitude. Thus, the BCFW formula reduces to a single term,

$$\mathcal{A}_n(1^-2^-3^+\cdots n^+) = \mathcal{A}_{n-1}(\hat{1}^-\, 2^- 3^+ \cdots (n-2)^+ -\widehat{K}_I^+;z_I)\, \frac{i}{K_I^2}$$
$$\times \mathcal{A}_3(\widehat{K}_I^- (n-1)^+ \hat{n}^+;z_I), \qquad (14.97)$$

where the momentum carried by the singular propagator is (before the shift)

$$K_I = -(p_{n-1} + p_n). \qquad (14.98)$$

On the right-hand side of eq. (14.97), the factor on the right is an already known three-point amplitude, and the factor on the left is an MHV amplitude with $n-1$ external legs.

**Four-point MHV amplitude:** Let us now calculate the first few iterations of this recursion, in order to guess a formula for the MHV amplitude that will be our hypothesis for an inductive proof. First, consider the $--++$ four-point MHV amplitude. In this case, the BCFW recursion formula gives

$$\mathcal{A}_4(1^-2^-3^+4^+) = \mathcal{A}_3(\hat{1}^-\, 2^- -\widehat{K}_I^+;z_I)\, \frac{i}{K_I^2}\, \mathcal{A}_3(\widehat{K}_I^- 3^+ \hat{4}^+;z_I), \qquad (14.99)$$

and both amplitudes on the right-hand side are known. This gives

$$\mathcal{A}_4(1^-2^-3^+4^+) = 2ig^2\, \frac{\langle \hat{1}2\rangle^3}{\langle 2\widehat{K}_I\rangle\langle \widehat{K}_I\hat{1}\rangle}\, \frac{1}{\langle 12\rangle[12]}\, \frac{[3\hat{4}]^3}{[\hat{4}\widehat{K}_I][\widehat{K}_I 3]}. \qquad (14.100)$$

Using the fact that

$$|\hat{1}\rangle = |1\rangle, \quad |\hat{1}] = |1] + z|4], \quad |\hat{4}\rangle = |4\rangle - z|1\rangle, \quad |\hat{4}] = |4], \qquad (14.101)$$

we obtain

$$|\widehat{K}_I\rangle[\widehat{K}_I| = |1\rangle[1| + |2\rangle[2| + z_I\,|1\rangle[4|,$$
$$\langle 2\widehat{K}_I\rangle[\widehat{K}_I 4] = \langle 21\rangle[14], \quad \langle 1\widehat{K}_I\rangle[\widehat{K}_I 3] = \langle 12\rangle[23], \qquad (14.102)$$

which leads to

$$\mathcal{A}_4(1^-2^-3^+4^+) = 2\,i\,g^2\,\frac{[34]^3}{[41][12][23]}. \qquad (14.103)$$

This formula, which depends only on square brackets, can also be expressed in terms of angle brackets. Let us multiply the numerator and denominator by $\langle 12\rangle^3$. Then, momentum conservation leads to

$$[41]\langle 12\rangle = -[43]\langle 32\rangle, \quad \langle 12\rangle[23] = -\langle 14\rangle[43],$$
$$[12]\langle 12\rangle = (p_1+p_2)^2 = (p_3+p_4)^2 = [34]\langle 34\rangle, \qquad (14.104)$$

and we finally obtain

$$\mathcal{A}_4(1^-2^-3^+4^+) = 2\,i\,g^2\,\frac{\langle 12\rangle^3}{\langle 23\rangle\langle 34\rangle\langle 41\rangle}. \qquad (14.105)$$

This formula could in principle have been obtained from eq. (14.15) by putting the external lines on-shell and by using $--++$ polarization vectors, at the cost of considerable effort. We see here the power of on-shell recursion: Since one only manipulates on-shell sub-amplitudes with physical polarizations, the complexity of all the intermediate expressions is comparable to that of the final result, unlike with the standard method.

**Five-point MHV amplitude:** Consider now the amplitude $\mathcal{A}_5(1^-2^-3^+4^+5^+)$. The BCFW recursion formula (14.97) now reads

$$\mathcal{A}_5(1^-2^-3^+4^+5^+) = \mathcal{A}_4(\widehat{1}^-\,2^-3^+\,-\widehat{K}_I^+;z_I)\,\frac{i}{K_I^2}\,\mathcal{A}_3(\widehat{K}_I^-4^+\widehat{5}^+;z_I)$$
$$= (\sqrt{2}\,g)^3\,i^2\,\frac{\langle\widehat{1}2\rangle^3[4\widehat{5}]^3}{\langle 23\rangle\langle 3\widehat{K}_I\rangle\langle\widehat{K}_I1\rangle[45]\langle 45\rangle[\widehat{5}\widehat{K}_I][\widehat{K}_I4]}, \qquad (14.106)$$

where we have chosen to express $K_I^2$ as $(p_4+p_5)^2 = [45]\langle 45\rangle$. This time, we use

$$|\widehat{1}\rangle = |1\rangle, \quad |\widehat{1}] = |1] + z|5], \quad |\widehat{5}\rangle = |5\rangle - z|1\rangle, \quad |\widehat{5}] = |5],$$
$$|\widehat{K}_I\rangle[\widehat{K}_I| = -|4\rangle[4| - |5\rangle[5| + z_I\,|1\rangle[5|,$$
$$\langle 3\widehat{K}_I\rangle[\widehat{K}_I\widehat{5}] = -\langle 34\rangle[45], \quad \langle\widehat{1}\widehat{K}_I\rangle[\widehat{K}_I 4] = -\langle 51\rangle[45], \qquad (14.107)$$

which gives

$$\mathcal{A}_5(1^-2^-3^+4^+5^+) = (\sqrt{2}\,g)^3\,i^2\,\frac{\langle 12\rangle^3}{\langle 23\rangle\langle 34\rangle\langle 45\rangle\langle 51\rangle}. \qquad (14.108)$$

## 14.3 BRITTO–CACHAZO–FENG–WITTEN ON-SHELL RECURSION

This remarkably simple formula, which encapsulates the sum of ten cyclic-ordered Feynman diagrams (in QCD, this corresponds to 25 diagrams before color ordering), in fact exhausts all the possibilities for five-point functions (the $++---$ amplitude is given by the same formula with square brackets instead of angle brackets).

**Parke–Taylor formula:** The previous results for three-, four- and five-point MHV amplitudes lead us to conjecture the following general formula:

$$\mathcal{A}_n(1^-2^-3^+\cdots n^+) = (\sqrt{2}\,g)^{n-2}\, i^{n-3}\, \frac{\langle 12\rangle^3}{\langle 23\rangle\langle 34\rangle\cdots\langle (n-1)n\rangle\langle n1\rangle}, \qquad (14.109)$$

known as the *Parke–Taylor formula*. Let us assume the formula to be true for all $p < n$, and consider now the case of the n-point MHV amplitude. The BCFW recursion formula reads

$$\begin{aligned}
\mathcal{A}_n(1^-2^-3^+\cdots n^+) &= \mathcal{A}_{n-1}(\hat{1}^-\,2^-3^+\cdots(n-2)^+ -\hat{K}_I^+; z_I) \\
&\quad \times \frac{i}{K_I^2}\, \mathcal{A}_3(\hat{K}_I^-(n-1)^+\hat{n}^+; z_I) \\
&= (\sqrt{2}\,g)^{n-2}\, i^{n-3}\, \frac{\langle \hat{1}2\rangle^3}{\langle 23\rangle\cdots\langle (n-2)\hat{K}_I\rangle\langle \hat{K}_I 1\rangle} \\
&\quad \times \frac{1}{[(n-1)n]\langle (n-1)n\rangle}\, \frac{[(n-1)\hat{n}]^3}{[\hat{n}\hat{K}_I][\hat{K}_I(n-1)]},
\end{aligned}$$
$$(14.110)$$

where we have used our induction hypothesis for the $(n-1)$-point MHV amplitude that appears in the left sub-amplitude. The spinor manipulations that are necessary to simplify this expression are the same as in the case of the five-point amplitude, and lead to

$$\begin{aligned}
\langle (n-2)\hat{K}_I\rangle [\hat{K}_I\hat{n}] &= -\langle (n-2)(n-1)\rangle [(n-1)n], \\
\langle \hat{1}\hat{K}_I\rangle [\hat{K}_I(n-1)] &= -\langle n1\rangle [(n-1)n],
\end{aligned} \qquad (14.111)$$

thanks to which we obtain eq. (14.109) for n points. In Exercise 14.10, we show that the Parke–Taylor formula is also valid for non-adjacent negative helicities, provided that we write it as

$$\mathcal{A}_n(\cdots i^-\cdots j^-\cdots) = (\sqrt{2}\,g)^{n-2}\, i^{n-3}\, \frac{\langle ij\rangle^4}{\langle 12\rangle\langle 23\rangle\langle 34\rangle\cdots\langle (n-1)n\rangle\langle n1\rangle}. \qquad (14.112)$$

Up to five points, all amplitudes are MHV (or anti-MHV, i.e., $++---$). Beyond five points, there exist non-MHV amplitudes that are not given by the Parke–Taylor formula. However, multiple MHV amplitudes can be sewed together in order to construct the non-MHV ones, with a set of rules known as the *Cachazo–Svrcek–Witten (CSW) rules*, derived in Section 14.5. Such an expansion is much more efficient that the usual perturbation theory, because it is in terms of on-shell gauge invariant building blocks (the MHV amplitudes) that already encapsulate a lot of the underlying complexity.

## 14.4 Tree-Level Gravitational Amplitudes

### 14.4.1 Textbook Approach for Amplitudes with Gravitons

In the previous section we derived the BCFW recursion formula and applied it to the calculation of the tree-level MHV amplitudes in Yang–Mills theory. However, the validity of this recursion is by no means limited to a gauge theory with spin-1 bosons such as gluons. It may in fact be applied to any quantum field theory provided that

- we have expressions for the on-shell three-point amplitudes;
- the shifted amplitudes vanish when $|z| \to \infty$.

In particular, it could be interesting to apply it to the calculation of scattering amplitudes that involve gravitons.[13] The Feynman rules for Einstein gravity can be obtained from the Hilbert–Einstein action:

$$S_{HE} \equiv \int d^4x \sqrt{-g} \left\{ \frac{2}{\kappa^2} R - \frac{g^{\mu\nu}g^{\rho\sigma}}{4} F_{\mu\rho}F_{\nu\sigma} + \frac{g^{\mu\nu}}{2}(\partial_\mu \phi)(\partial_\nu \phi) - \frac{m^2}{2}\phi^2 \right\}, \quad (14.113)$$

where $g^{\mu\nu}$ is the metric tensor, $g$ its determinant, R is the Ricci curvature and $\kappa$ is a coupling constant related to Newton's constant by $\kappa^2 = 32\pi G_N$. In this action, we have also added the minimal coupling to a gauge field and to a scalar field, in order to investigate gravitational interactions with light and matter. The rules for the propagators and vertices involving gravitons are obtained by expanding the metric around flat space,

$$g^{\mu\nu} = \eta^{\mu\nu} + \kappa h^{\mu\nu}. \quad (14.114)$$

($\eta^{\mu\nu}$ is the flat-space Minkowski metric.) Let us make a remark on dimensions: Newton's constant has mass dimension $-2$, $\kappa$ has mass dimension $-1$, the Ricci curvature has mass dimension 2 and therefore $h^{\mu\nu}$ has mass dimension $+1$ (like the scalar field $\phi$ and the gauge field $A_\mu$). The expansion in powers of $h^{\mu\nu}$ leads to an infinite series of terms (because the Ricci tensor contains the inverse $g_{\mu\nu}$ of the metric tensor, and also because of the expansion of the square root $\sqrt{-g}$). Schematically, the expansion of the Hilbert–Einstein action starts with the following terms:

$$S_{HE} \sim \int d^4x \left\{ h\partial^2 h + \kappa h^2 \partial^2 h + \kappa^2 h^3 \partial^2 h + \cdots + \kappa h \phi \partial^2 \phi + \kappa h F^2 + \cdots \right\}. \quad (14.115)$$

This sketch only indicates the number of powers of $h$ and the number of derivatives contained in each term, but of course the actual structure of these terms is much more complicated. For instance, the vertex describing the coupling $\phi\phi h$ between two scalars and a graviton reads

$$\Gamma^{\mu\nu}(p_1, p_2) = -\frac{i\kappa}{2} \left[ p_1^\mu p_2^\nu + p_1^\nu p_2^\mu - \eta^{\mu\nu}(p_1 \cdot p_2 - m^2) \right], \quad (14.116)$$

---

[13] At tree level, these amplitudes are completely prescribed by the equivalence principle and general relativity, and their calculation does not require having a consistent theory of gravitational quantum fluctuations.

## 14.4 TREE-LEVEL GRAVITATIONAL AMPLITUDES

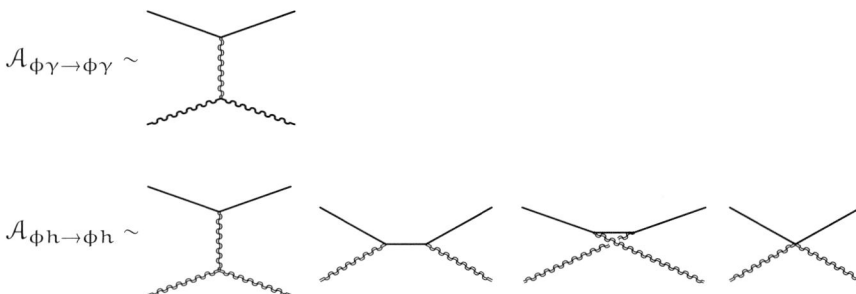

FIGURE 14.6: Feynman diagrams contributing to the gravitational photon-scalar and graviton-scalar scattering amplitudes. The wavy double lines represent gravitons.

where $p_{1,2}$ are the momenta carried by the two scalar lines (since the graviton has spin-2, this vertex has two Lorentz indices). But the $\gamma\gamma h$ coupling is more complicated and the $hhh$ tri-graviton vertex is even more intricate, leading to extremely cumbersome perturbative calculations if performed within the traditional approach.

It turns out that tree amplitudes in Einstein gravity have a simple form in the spinor-helicity formalism, very much like their Yang–Mills analogue. The goal of this section is to illustrate through two examples the use of the spinor-helicity formalism, combined to the BCFW recursion, in order to calculate some amplitudes that have relevance in gravitational physics: the gravitational bending of light by a mass, and the scattering of a gravitational wave off a mass. In both examples, the mass acting as a source of gravitational field is taken to be a scalar particle. In the approach based on conventional Feynman perturbation theory, these processes are given by the diagrams shown in Figure 14.6. In particular, the second example (bending of a gravitational wave by a mass) would be an extremely difficult calculation, because of the complexity of the three-graviton vertex.

### 14.4.2 Three-Point Amplitudes with Gravitons

In order to obtain these amplitudes with the formalism previously exposed in the case of Yang–Mills theory, the first step is to obtain the three-point amplitudes involving scalars, photons and gravitons. External scalar particles must have helicity $h = 0$, photons can have helicities $h = \pm 1$ and gravitons can have helicities $h = \pm 2$. One may choose a gauge in which their polarization vectors are "squares" of the gluon polarization vectors:

$$\epsilon_{2h}^{\mu\nu}(\mathbf{p};\mathbf{q}) = \epsilon_h^\mu(\mathbf{p};\mathbf{q})\,\epsilon_h^\nu(\mathbf{p};\mathbf{q}). \tag{14.117}$$

For three-point amplitudes that involve only massless particles (photons and gravitons), little group scaling is sufficient to constrain completely their form. We obtain

$$\mathcal{A}_{h\gamma\gamma}(1^{\pm 2}2^+3^+) = \mathcal{A}_{h\gamma\gamma}(1^{\pm 2}2^-3^-) = 0,$$
$$\mathcal{A}_{h\gamma\gamma}(1^{+2}2^+3^-) = -\tfrac{\kappa}{2}\,[12]^4\,[23]^{-2},$$
$$\mathcal{A}_{h\gamma\gamma}(1^{+2}2^-3^+) = -\tfrac{\kappa}{2}\,[23]^{-2}\,[31]^4,$$

$$\mathcal{A}_{h\gamma\gamma}(1^{-2}2^+3^-) = -\tfrac{\kappa}{2}\langle 23\rangle^{-2}\langle 31\rangle^4,$$
$$\mathcal{A}_{h\gamma\gamma}(1^{-2}2^-3^+) = -\tfrac{\kappa}{2}\langle 12\rangle^4\langle 23\rangle^{-2}. \qquad(14.118)$$

In order to obtain the zeroes of the first line, and to choose between square and angle brackets for the non-zero results, we use the fact that the three-point amplitude must have mass dimension $+1$, with a prefactor made up only of numerical constants and one power of $\kappa$ (that has mass dimension $-1$). The value of the prefactor is obtained by inspecting the term of order $\kappa$ in the expansion of $\sqrt{-g}\,F^2$. For the three-graviton amplitudes, little group scaling leads to

$$\mathcal{A}_{hhh}(1^{+2}2^{+2}3^{+2}) = \mathcal{A}_{hhh}(1^{-2}2^{-2}3^{-2}) = 0,$$
$$\mathcal{A}_{hhh}(1^{-2}2^{-2}3^{+2}) \propto \kappa\,\langle 12\rangle^6\langle 23\rangle^{-2}\langle 31\rangle^{-2},$$
$$\mathcal{A}_{hhh}(1^{+2}2^{+2}3^{-2}) \propto \kappa\,[12]^6[23]^{-2}[31]^{-2}. \qquad(14.119)$$

Interestingly, the kinematical part of the non-zero three-graviton amplitudes is simply the square[14] of that of the three-gluon amplitudes with like-sign helicities (see eqs. (14.47) and (14.48)), despite a considerably more complicated Feynman rule for the three-graviton vertex. This is yet another illustration of the fact that traditional Feynman rules carry a lot of unnecessary information that disappears in on-shell amplitudes with physical polarizations.

For the $\phi\phi h$ amplitude, we cannot rely on little group scaling because the scalar field is massive. Instead, we simply contract eq. (14.116) with the polarization vector of eq. (14.117) of the graviton, and take the external momenta on mass-shell. For a graviton of helicity $+2$, we have

$$\mathcal{A}_{\phi\phi h}(1^0 2^0 3^{+2}) = -i\kappa\,(p_1\cdot\epsilon_+(p_3;q))(p_2\cdot\epsilon_+(p_3;q))$$
$$= -\frac{i\kappa}{2}\frac{\langle q|\overline{P}_1|p_3]\langle q|\overline{P}_2|p_3]}{\langle qp_3\rangle^2}, \qquad(14.120)$$

where $p_1+p_2+p_3 = 0$. With a graviton of helicity $-2$, we have

$$\mathcal{A}_{\phi\phi h}(1^0 2^0 3^{-2}) = -\frac{i\kappa}{2}\frac{\langle p_3|\overline{P}_1|q]\langle p_3|\overline{P}_2|q]}{[qp_3]^2}. \qquad(14.121)$$

Note that

$$\langle p_3|\overline{P}_2|q] = -\langle p_3|\overline{P}_1|q] - \underbrace{\langle p_3|\overline{P}_3|q]}_{0} = -\langle p_3|\overline{P}_1|q],$$
$$\langle q|\overline{P}_2|p_3] = -\langle q|\overline{P}_1|p_3] - \underbrace{\langle q|\overline{P}_3|p_3]}_{0} = -\langle q|\overline{P}_1|p_3], \qquad(14.122)$$

---

[14] This property of three-point purely gravitational amplitudes has a generalization for n-point amplitudes, known as the *Kawai–Lewellen–Tye (KLT) relations*. These relations have also been interpreted as a form of *color-kinematics duality* by Bern, Carrasco and Johansson (see Exercises 14.13 and 14.14 for examples).

## 14.4 TREE-LEVEL GRAVITATIONAL AMPLITUDES

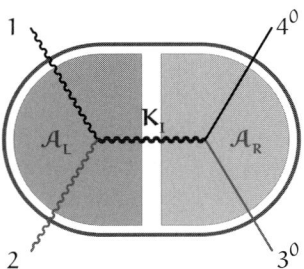

FIGURE 14.7: BCFW shift for the calculation of the $\gamma\gamma\phi\phi$ amplitude.

which allows the following simplification of the above three-point amplitudes:

$$\mathcal{A}_{\phi\phi h}(1^0 2^0 3^{+2}) = \frac{i\kappa}{2} \frac{\langle q|\overline{P}_1|p_3]^2}{\langle qp_3\rangle^2}, \quad \mathcal{A}_{\phi\phi h}(1^0 2^0 3^{-2}) = \frac{i\kappa}{2} \frac{\langle p_3|\overline{P}_1|q]^2}{[qp_3]^2}. \tag{14.123}$$

Note that since $p_1^2 = m^2 \neq 0$, the bi-spinor $\overline{P}_1$ does not admit a factorized form, and this cannot be simplified further.

### 14.4.3 Gravitational Bending of Light

**Shifted momenta:** Consider now the amplitude $\mathcal{A}_{\gamma\gamma\phi\phi}(1^+ 2^- 3^0 4^0)$, and apply the shift to lines 2 and 3, as illustrated in Figure 14.7:[15]

$$\widehat{p}_2 \equiv p_2 + zk, \quad \widehat{p}_3 \equiv p_3 - zk, \quad k^2 = 0, \quad k \cdot p_2 = k \cdot p_3 = 0. \tag{14.124}$$

Since $p_2$ is massless, the condition $p_2 \cdot k = 0$ can be satisfied by choosing, for instance,

$$|k\rangle = |2\rangle \tag{14.125}$$

However, since $p_3^2 = m^2$, the bi-spinor $\mathbf{P}_3$ that represents the momentum $p_3$ cannot be factorized. Instead, we may write

$$0 = 2k \cdot p_3 = -\langle k|\overline{P}_3|k], \tag{14.126}$$

which can be solved by[16]

$$|k] = \mathbf{P}_3|2\rangle. \tag{14.127}$$

---

[15] Note that the factorization with one scalar and one photon on each side of the singular propagator is not allowed: Indeed, the intermediate propagator would need to carry a scalar, and we would have two $\phi\phi\gamma$ sub-amplitudes, which are zero per our assumption that the scalar field is not electrically charged.

[16] We use $\langle k|\overline{P}_3|k] = \langle 2|\overline{P}_3 P_3|2\rangle = m^2 \langle 22\rangle = 0$. When $p_3$ is massless, $\mathbf{P}_3$ factorizes as $\mathbf{P}_3 = |3]\langle 3|$, and this solution becomes $|k] = |3]\langle 32\rangle$. Up to a rescaling, this is the solution we have previously used in the massless case.

The shifted bi-spinors read

$$\widehat{P}_2 = |2]\langle 2| + z\,|k]\langle k| = \underbrace{\bigl(\,|2] + z\,P_3|2\rangle\bigr)}_{|\widehat{2}]}\langle 2|,$$
$$\widehat{P}_3 = P_3 - z\,|k]\langle k| = P_3 - z\,P_3|2\rangle\langle 2|. \tag{14.128}$$

Note that the second one is not factorizable, because line 3 carries a massive particle.

**Scattering amplitude:** With this choice of shifts, the BCFW recursion formula can be written as follows:

$$\mathcal{A}_{\gamma\gamma\phi\phi}(1^+2^-3^04^0) = \frac{i}{K_I^2} \sum_{h=\pm 2} \mathcal{A}_{\gamma\gamma h}(1^+\widehat{2}^- - \widehat{K}_I^{+h}; z_I)\, \mathcal{A}_{h\phi\phi}(\widehat{K}_I^{-h}\widehat{3}^04^0; z_I), \tag{14.129}$$

where the shifted momenta in the three-point amplitudes are evaluated at $z_I$ for which the shifted momentum $\widehat{K}_I$ of the intermediate graviton is on-shell. The condition for the intermediate momentum to be on-shell reads

$$0 = \widehat{K}_I^2 = (p_1 + \widehat{p}_2)^2 = 2p_1 \cdot \widehat{p}_2 = \langle 12\rangle\Bigl(\underbrace{[12] + z_I\,[1|P_3|2\rangle}_{[1\widehat{2}]=0}\Bigr), \tag{14.130}$$

whose solution is $z_I = -[12]/[1|P_3|2\rangle$. Plugging in the results for the three-point amplitudes and summing explicitly over the two helicities of the intermediate graviton, the $\gamma\gamma\phi\phi$ amplitude can be written as

$$\mathcal{A}_{\gamma\gamma\phi\phi}(1^+2^-3^04^0) = \frac{\kappa^2}{4}\frac{1}{\langle 12\rangle[12]}\left\{\frac{[\widehat{K}_I 1]^4}{[1\widehat{2}]^2}\frac{\langle \widehat{K}_I|\overline{P}_4|q]^2}{[q\widehat{K}_I]^2} + \frac{\langle \widehat{K}_I \widehat{2}\rangle^4}{\langle 1\widehat{2}\rangle^2}\frac{\langle q|\overline{P}_4|\widehat{K}_I]^2}{\langle q\widehat{K}_I\rangle^2}\right\}. \tag{14.131}$$

For the first term, we may write

$$\frac{[\widehat{K}_I 1]^4}{[1\widehat{2}]^2} = \frac{[\widehat{K}_I 1]^4 \langle \widehat{K}_I 1\rangle^4}{[1\widehat{2}]^2 \langle \widehat{K}_I 1\rangle^4} = \frac{(2p_1\cdot\widehat{p}_2)^4}{[1\widehat{2}]^2\langle \widehat{K}_I 1\rangle^4} = \frac{\langle 1\widehat{2}\rangle^4[1\widehat{2}]^2}{\langle \widehat{K}_I 1\rangle^4} = 0. \tag{14.132}$$

The final zero occurs when we evaluate the expression at $z_I$, as a consequence of eq. (14.130). Therefore, the amplitude reduces to a single term. Furthermore, we are still free to choose the auxiliary vector $q$. A convenient choice turns out to be $q = p_2$, which leads to

$$\mathcal{A}_{\gamma\gamma\phi\phi}(1^+2^-3^04^0) = \frac{\kappa^2}{4}\frac{\langle \widehat{K}_I 2\rangle^2\langle 2|\overline{P}_4|\widehat{K}_I]^2}{\langle 12\rangle^3[12]}. \tag{14.133}$$

Then, notice that

$$\langle 2|\overline{P}_4|\widehat{K}_I]\langle \widehat{K}_I 2\rangle = \langle 2|\overline{P}_4|1]\langle 12\rangle, \tag{14.134}$$

## 14.4 TREE-LEVEL GRAVITATIONAL AMPLITUDES

which gives the following extremely compact form for the amplitude:

$$\mathcal{A}_{\gamma\gamma\phi\phi}(1^+2^-3^04^0) = \frac{\kappa^2}{4} \frac{\langle 2|\overline{P}_4|1]^2}{\langle 12 \rangle [12]}. \tag{14.135}$$

**Cross-section and deflection angle:** Using $\langle 2|\overline{P}_4|1]^\dagger = \langle 1|\overline{P}_4|2]$, the squared modulus of the amplitude is

$$\left|\mathcal{A}_{\gamma\gamma\phi\phi}(1^+2^-3^04^0)\right|^2 = \frac{\kappa^4}{16} \frac{\langle 2|\overline{P}_4|1]^2 \langle 1|\overline{P}_4|2]^2}{\langle 12 \rangle^2 [12]^2}. \tag{14.136}$$

Note that

$$\langle 2|\overline{P}_4|1]\langle 1|\overline{P}_4|2] = \langle 2|_a \overline{P}_4^{ab}|1]_b \langle 1|_c \overline{P}_4^{cd}|2]_d = \overline{P}_4^{ab}|1]_b \langle 1|_c \overline{P}_4^{cd}|2]_d \langle 2|_a$$
$$= \mathrm{tr}\,(\overline{P}_4 P_1 \overline{P}_4 P_2) = p_{4\mu}p_{1\nu}p_{4\rho}p_{2\sigma}\,\mathrm{tr}\,(\overline{\sigma}^\mu \sigma^\nu \overline{\sigma}^\rho \sigma^\sigma)$$
$$= 2\,p_{4\mu}p_{1\nu}p_{4\rho}p_{2\sigma}\left(\eta^{\mu\nu}\eta^{\rho\sigma} - \eta^{\mu\rho}\eta^{\nu\sigma} + \eta^{\mu\sigma}\eta^{\nu\rho} + i\epsilon^{\mu\nu\rho\sigma}\right)$$
$$= 2\left(2(p_1 \cdot p_4)(p_2 \cdot p_4) - p_4^2\,(p_1 \cdot p_2)\right)$$
$$= s_{13}s_{14} - m^4, \tag{14.137}$$

where we have introduced the Lorentz invariants $s_{ij} \equiv (p_i + p_j)^2$ and used $s_{24} = s_{13}$ and $s_{12} + s_{13} + s_{14} = 2m^2$ (both follow from momentum conservation). Therefore, the squared amplitude reads

$$\left|\mathcal{A}_{\gamma\gamma\phi\phi}(1^+2^-3^04^0)\right|^2 = \frac{\kappa^4}{16} \frac{(s_{13}s_{14} - m^4)^2}{s_{12}^2}. \tag{14.138}$$

The differential cross-section with respect to the solid angle of the outgoing photon is given by

$$\frac{d\sigma}{d\Omega} = \frac{1}{64\pi^2\,s_{14}}\left|\mathcal{A}_{\gamma\gamma\phi\phi}(1^+2^-3^04^0)\right|^2. \tag{14.139}$$

Let us now consider the limit of long wavelength photons, namely $\omega = |\mathbf{p}_{1,2}| \ll m$. In this limit, the Lorentz invariants that appear in the cross-section simplify into[17]

$$s_{12} \approx 4\omega^2 \sin^2\tfrac{\theta}{2},$$
$$s_{13} \approx m^2 - 2m\omega - 4\omega^2 \sin^2\tfrac{\theta}{2},$$
$$s_{14} \approx m^2 + 2m\omega, \tag{14.140}$$

---

[17] If we are only interested in small $\omega$ and small deflection angles $\theta$, then the somewhat complicated calculation of the numerator done in eq. (14.137) can be avoided. Indeed, in this limit, the massive scalar is at rest and $\overline{P}_4^{ab} \approx m\,\delta^{ab}$. Moreover, the 3-momenta of the incoming and outgoing photons are nearly parallel to the z axis. This implies that $\langle 1|_a \approx |1]_a \approx -\langle 2|_a \approx -|2]_a \approx \binom{\sqrt{2\omega}}{0}$, and $\langle 2|\overline{P}_4|1] \approx \langle 1|\overline{P}_4|2] \approx -2m\omega$.

where $\omega$ is the photon energy and $\theta$ its deflection angle in the center-of-mass frame (which is also the frame of the massive scalar particle in this limit). For large enough impact parameters, the deflection angle is small, $\theta \ll 1$. Thus, we obtain in this limit

$$\frac{d\sigma}{d\Omega} \approx \frac{16\, G_N^2\, m^2}{\theta^4}. \tag{14.141}$$

In order to determine the deflection angle as a function of the impact parameter b, consider a flux $\mathcal{F}$ of photons along the z direction, with the massive scalar at rest at the origin. Within this flux, consider specifically the incoming photons in a ring of radius b and width db. The number of photons flowing per unit time through this ring is

$$2\pi\, b\, \mathcal{F}\, db. \tag{14.142}$$

All these photons are scattered in the range of polar angles $[\theta(b) + d\theta, \theta(b)]$ (note that $d\theta$ is negative for $db > 0$, because the deflection angle decreases at larger b), which corresponds to a solid angle:

$$d\Omega = -2\pi \sin\bigl(\theta(b)\bigr)\, d\theta. \tag{14.143}$$

By definition, the number of scattering events is the flux times the cross-section, i.e.,

$$2\pi\, b\, \mathcal{F}\, db = \mathcal{F}\, \frac{d\sigma}{d\Omega}\, d\Omega, \tag{14.144}$$

which can be integrated for small angles into

$$\theta(b) = \frac{4\, G_N\, m}{b}. \tag{14.145}$$

(The integration constant is chosen so that the deflection vanishes when $b \to \infty$.) This is indeed the standard formula from general relativity, which can be derived by considering geodesics in the Schwarzchild metric.

## 14.4.4  Scattering of Gravitational Waves by a Mass

Let us now study the scattering amplitude between a scalar and a graviton, whose low energy limit will provide us information about the scattering of a long wavelength gravitational wave by a mass. A priori, each of the two gravitons may have a helicity $\pm 2$, but the cases $\{+2, +2\}$ and $\{-2, -2\}$ correspond to a helicity flip of the graviton, which is suppressed at low frequency. Therefore, let us consider the amplitude $\mathcal{A}_{hh\phi\phi}(1^{-2}2^{+2}3^0 4^0)$. When writing the BCFW recursion for this amplitude, the simplest shift is one that affects lines 1 and 2; more specifically:

$$|\hat{2}] = |2], \quad |\hat{2}\rangle = |2\rangle - z|1\rangle, \quad |\hat{1}] = |1] + z|2], \quad |\hat{1}\rangle = |1\rangle. \tag{14.146}$$

Because the polarization vectors of the gravitons are squares of the spin-1 ones, this shift can be proven to lead to a vanishing amplitude when $|z| \to \infty$ simply by power counting. With

## 14.4 TREE-LEVEL GRAVITATIONAL AMPLITUDES

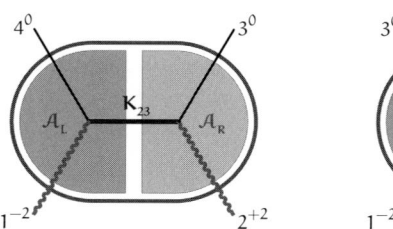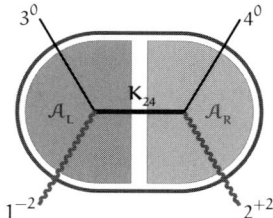

FIGURE 14.8: BCFW shift for the calculation of the hhφφ amplitude.

the shift (14.146), the intermediate propagator carries a scalar, and therefore it has only the $h = 0$ helicity. The BCFW recursion formula contains two terms,

$$\mathcal{A}_{hh\phi\phi}(1^{-2}2^{+2}3^04^0) = \mathcal{A}_{h\phi\phi}(\hat{1}^{-2}4^0\widehat{K}_{23}^0)\frac{i}{K_{23}^2 - m^2}\mathcal{A}_{h\phi\phi}(\hat{2}^{+2}3^0 - \widehat{K}_{23}^0)$$
$$+ \mathcal{A}_{h\phi\phi}(\hat{1}^{-2}3^0\widehat{K}_{24}^0)\frac{i}{K_{24}^2 - m^2}\mathcal{A}_{h\phi\phi}(\hat{2}^{+2}4^0 - \widehat{K}_{24}^0), \quad (14.147)$$

that differ by a permutation of the external scalars (see Figure 14.8). In the above equation, we have made explicit the intermediate momentum, $\widehat{K}_{23} \equiv \hat{p}_2 + p_3$ in the first term and $\widehat{K}_{24} \equiv \hat{p}_2 + p_4$ in the second one. The explicit forms of the first and second terms are

$$i\frac{\mathcal{A}_{h\phi\phi}(\hat{1}^{-2}4^0\widehat{K}_{23}^0)\mathcal{A}_{h\phi\phi}(\hat{2}^{+2}3^0 - \widehat{K}_{23}^0)}{K_{23}^2 - m^2} = \frac{-i\kappa^2}{4(K_{23}^2-m^2)}\frac{\langle\hat{1}|\mathbf{P}_4|\mathbf{q}]^2\langle\mathbf{q}'|\mathbf{P}_3|\hat{2}]^2}{[\hat{1}\mathbf{q}]^2\langle\mathbf{q}'\hat{2}\rangle^2},$$

$$i\frac{\mathcal{A}_{h\phi\phi}(\hat{1}^{-2}3^0\widehat{K}_{24}^0)\mathcal{A}_{h\phi\phi}(\hat{2}^{+2}4^0 - \widehat{K}_{24}^0)}{K_{24}^2 - m^2} = \frac{-i\kappa^2}{4(K_{24}^2-m^2)}\frac{\langle\hat{1}|\mathbf{P}_3|\mathbf{q}]^2\langle\mathbf{q}'|\mathbf{P}_4|\hat{2}]^2}{[\hat{1}\mathbf{q}]^2\langle\mathbf{q}'\hat{2}\rangle^2}. \quad (14.148)$$

A convenient choice of auxiliary vectors is $\mathbf{q} = \mathbf{p}_2$ and $\mathbf{q}' = \mathbf{p}_1$, which leads to

$$\mathcal{A}_{hh\phi\phi}(1^{-2}2^{+2}3^04^0) = -i\frac{\kappa^2}{4}\frac{\langle 1|\overline{\mathbf{P}}_3|2]^4}{\langle 12\rangle^2[12]^2}\left\{\frac{1}{K_{23}^2 - m^2} + \frac{1}{K_{24}^2 - m^2}\right\}$$
$$= i\frac{\kappa^2}{16}\frac{\langle 1|\overline{\mathbf{P}}_3|2]^4}{\langle 12\rangle[12]}\frac{1}{(p_2 \cdot p_3)(p_2 \cdot p_4)}. \quad (14.149)$$

The square of this amplitude can be related to that of photon-scalar gravitational scattering by

$$\left|\mathcal{A}_{hh\phi\phi}(1^{-2}2^{+2}3^04^0)\right|^2 = \left|\mathcal{A}_{\gamma\gamma\phi\phi}(1^-2^+3^04^0)\right|^2\left\{1 - \frac{m^2 s_{12}}{(s_{13}-m^2)(s_{14}-m^2)}\right\}. \quad (14.150)$$

In the limit of a graviton of small energy (i.e., a gravitational wave of long wavelength) and small deflection angle (i.e., at large impact parameter), the second factor on the right-hand side becomes equal to 1, and we have

$$\left|\mathcal{A}_{hh\phi\phi}(1^{-2}2^{+2}3^04^0)\right|^2 \underset{\substack{\omega \ll m \\ \theta \ll 1}}{\approx} \left|\mathcal{A}_{\gamma\gamma\phi\phi}(1^{-2}2^{+}3^04^0)\right|^2. \tag{14.151}$$

This implies that in this limit the bending of a gravitational wave by a mass is the same as the bending of a light ray (but there are some differences beyond this limit).

## 14.5 Cachazo–Svrcek–Witten Rules

### 14.5.1 Off-Shell Continuation of MHV Amplitudes

Our proof of the Parke–Taylor formula, based on BCFW recursion, is not faithful to the actual chronology, since the formula was conjectured in 1986 and a proof was found in 1988 using an off-shell recursion method derived by Berends and Giele, well before on-shell recursion. The *Cachazo–Svrcek–Witten rules*, also anterior to BCFW recursion, provide a way to construct the tree-level non-MHV amplitudes by an expansion in which the MHV ones play the role of vertices, as in the following diagram:

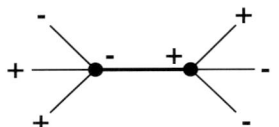

where the two "vertices" are $++--$ four-point MHV amplitudes, sewn together in order to make a contribution to a non-MHV six-point amplitude. In this section, we will use ideas inspired by the derivation of on-shell recursion in order to establish these rules.

First, for energy–momentum conservation to hold in the vertices of such a diagram, the intermediate propagator linking the two vertices must generically carry an off-shell momentum. This means that in such a construction, we need first to generalize the MHV amplitudes to external off-shell momenta. As we shall see shortly, the CSW rules only use as vertices the MHV amplitudes that have exactly two negative helicities, i.e., those that are expressible in terms of angle brackets only. Thus, we would like to have an off-shell extension of these brackets. The main issue here is that for an off-shell momentum $p^\mu$, the $2 \times 2$ matrix $\mathbf{P}$ by which it is represented in the spinor-helicity method has a non-vanishing determinant, and is therefore not factorizable as $\mathbf{P} = |\mathbf{p}]\langle\mathbf{p}|$. Assuming that $\eta^\mu$ is a light-like auxiliary vector, then we may always write

$$p^\mu \equiv \mathcal{P}^\mu + z\eta^\mu, \tag{14.152}$$

with $\mathcal{P}^2 = 0$. The on-shellness of $\mathcal{P}^\mu$ determines uniquely the value of $z$,

$$z = \frac{p^2}{2p \cdot \eta}. \tag{14.153}$$

## 14.5 CACHAZO-SVRCEK-WITTEN RULES

Since $\mathcal{P}^\mu$ is on-shell, the corresponding matrix $\mathcal{P}$ is factorizable into the direct product of a square and angle spinors, $\mathcal{P} = |\mathcal{P}]\langle\mathcal{P}|$. Then, we may write

$$[\eta|\mathbf{P} = [\eta|\mathcal{P} + z\,[\eta|\eta = [\eta\mathcal{P}]\langle\mathcal{P}| + z\,\underbrace{[\eta\eta]}_{0}\langle\eta|, \qquad (14.154)$$

which gives

$$\langle\mathcal{P}| = \frac{[\eta|\mathbf{P}}{[\eta\mathcal{P}]}. \qquad (14.155)$$

Note that this identity contains the angle spinor $\langle\mathcal{P}|$ in the numerator and the square spinor $|\mathcal{P}]$ in the denominator, consistent with the fact that rescaling one with $\lambda$ and the other with $\lambda^{-1}$ gives an equally valid representation. For this reason we may simply ignore the denominator and define

$$\langle\mathcal{P}| = [\eta|\mathbf{P}. \qquad (14.156)$$

In the following, we adopt this formula as the definition of the angle spinor associated with the off-shell momentum $p^\mu$. Note that when $p^\mu$ is on-shell, then $p = \mathcal{P}$, and the angle spinor $\langle\mathcal{P}|$ defined in eq. (14.156) is indeed proportional to the usual $\langle p|$.

### 14.5.2 CSW Rules for Next-to-MHV Amplitudes

Consider now the case of amplitudes with exactly three negative helicities, carried by the external lines $i, j, k$, and consider the following deformed square spinors:

$$|\hat{\imath}] \equiv |i] + z\langle jk\rangle|\eta], \quad |\hat{\jmath}] \equiv |j] + z\langle ki\rangle|\eta], \quad |\hat{k}] \equiv |k] + z\langle ij\rangle|\eta], \qquad (14.157)$$

while the corresponding angle spinors are left unchanged.[18] The shifted external momenta are defined as direct products of these shifted square spinors and the original angle spinors, and are therefore still on-shell. Note also that

$$|\hat{\imath}]\langle i| + |\hat{\jmath}]\langle j| + |\hat{k}]\langle k| = \underbrace{|i]\langle i| + |j]\langle j| + |k]\langle k|}_{0}$$

$$+ z|\eta]\underbrace{\left(\langle jk\rangle\langle i| + \langle ki\rangle\langle j| + \langle ij\rangle\langle k|\right)}_{0}. \qquad (14.158)$$

The first zero is due to momentum conservation in the unshifted amplitude, and the second one follows from Schouten identity. Thus, the above shift preserves momentum conservation.

---

[18] Recall that the usual BCFW shift acts only on a pair $i, j$ of external lines, by shifting the angle spinor of one and the square spinor of the other.

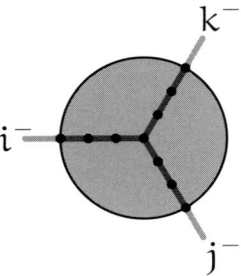

FIGURE 14.9: Propagators affected by the shift of eq. (14.157) in a generic amplitude.

The propagators affected by the shift of eq. (14.157) form three lines starting at the three external points of negative helicity, that meet somewhere inside the graph. Since the shift modifies only the square spinors, the polarization vectors of negative helicity scale as $z^{-1}$. Moreover, from Figure 14.9, we see that there are $v$ vertices and $v-1$ propagators along the affected lines. Even in the worst case where all these vertices are three-gluon vertices that scale like $z$, the overall scaling of the graph is bounded by $z^{-2}$, and therefore it goes to zero as $|z| \to \infty$. Then, we proceed as in the derivation of BCFW's recursion formula, by integrating $\mathcal{A}_n(\cdots;z)/z$ over $z$ on a circle of infinite radius. The behavior of the deformed amplitude at large $z$ ensures that there is no boundary term, and we obtain

$$\mathcal{A}_n(\cdots) = -\sum_{z_* \in \{\text{poles of } \mathcal{A}_n\}} \operatorname{Res} \left. \frac{\mathcal{A}_n(\cdots;z)}{z} \right|_{z_*}$$

$$= \sum_{\substack{I,h=\pm \\ i \in \mathcal{A}_L}} \mathcal{A}_L(\cdots - \widehat{K}_I^{-h}; z_I) \frac{i}{K_I^2} \mathcal{A}_R(\cdots \widehat{K}_I^h; z_I), \qquad (14.159)$$

where $K_I$ is the momentum of the intermediate propagator producing the pole. By construction, the singular propagator separating the left and right sub-amplitudes must be one of the dark propagators in Figure 14.9. Thus, when writing this factorized formula, we may decide that the external line $i$ belongs to the left sub-amplitude, and at least one of $j,k$ belongs to the right sub-amplitude. The propagator carrying the momentum $K_I$ can be on any of the three branches highlighted in Figure 14.9. In all three cases, there is only one choice of the intermediate helicity $h$ that gives a non-zero result, and this $h$ is such that both $\mathcal{A}_L$ and $\mathcal{A}_R$ are MHV amplitudes with two negative helicities. The next-to-MHV amplitude under consideration takes the following diagrammatic form:

$$\mathcal{A}_n = i \!-\!\!\!\begin{array}{c}k\\[-2pt]\text{\tiny diagram}\end{array}\!\!\!\begin{array}{c}i\\[-2pt]\\j\end{array} + \begin{array}{c}k\\[-2pt]\text{\tiny diagram}\\j\end{array}\!-\!k + \begin{array}{c}k\\[-2pt]\text{\tiny diagram}\\i\end{array}\!-\!j, \qquad (14.160)$$

where we have only indicated the relevant helicity assignments (all the thin lines carry positive helicities). Thus, as far as the helicities are concerned, the vertices in these graphs are MHV amplitudes with exactly two negative indices.

## 14.5 CACHAZO-SVRCEK-WITTEN RULES

Note that in eq. (14.160) the shift has no influence on the external lines because the MHV amplitudes with two negative helicities depend only on the angle spinors. The MHV vertices in this equation also depend on the angle spinor $\langle \widehat{K}_I |$ that corresponds to the on-shell (because we evaluate the amplitude at the value of $z$ for which the intermediate propagator is singular) shifted momentum. Consider for instance the first diagram, obtained when the singular propagator is on the line that stems from the external line $i$. In this case, we have

$$|\widehat{K}_I] \langle \widehat{K}_I| = \widehat{\mathbf{K}}_I = \mathbf{K}_I + z_* \langle jk \rangle |\eta] \langle i|, \qquad (14.161)$$

where $z_*$ is the value of $z$ at the pole. By contracting this relation with $[\eta|$, we obtain

$$[\eta \widehat{K}_I] \langle \widehat{K}_I| = [\eta | \mathbf{K}_I. \qquad (14.162)$$

Thus, the angle spinor $\langle \widehat{K}_I |$ in the MHV vertices resulting from the theorem of residues is proportional to the off-shell extension $[\eta | \mathbf{K}_I$ that we have proposed in the previous subsection. Finally, note that the factors $[\eta \widehat{K}_I]$ all cancel, because the line of momentum $K_I$ has opposite helicities on either side of the propagator that links the two MHV amplitudes.[19] Therefore, in the MHV diagrams of eq. (14.160), we do not need to find the poles $z_*$ and we may directly evaluate the MHV amplitudes that play the role of vertices with the off-shell angle spinor $[\eta | \mathbf{K}_I$.

Let us summarize here the CSW rules for calculating amplitudes with exactly three negative helicities:

- Start from the three skeleton diagrams of eq. (14.160), and interpret the vertices as MHV amplitudes with one off-shell external leg. Note that the actual number of MHV graphs depends on the number of positive helicity external lines, since they may be attached to either of the two MHV vertices (provided we do not change the cyclic ordering of the external lines, and that all the vertices obtained in this way have at least three lines).
- The intermediate propagator is simply a scalar propagator $i/K_I^2$, with the value of $K_I$ determined by momentum conservation at the vertices.
- Replace the MHV vertices by their Parke–Taylor expression, with the angle spinor of the intermediate off-shell line given by $\langle K_I | \equiv [\eta | \mathbf{K}_I$ (from now on, we omit the hat on $\langle K_I |$, since shifting the momenta was just an intermediate device for establishing the CSW rules).

Note that in this derivation of the CSW rules, we are guaranteed that the final result does not depend on the choice of the spinor $|\eta]$ introduced in the shift. Indeed, the pole at $z = 0$ if $\mathcal{A}_n(\cdots;z)/z$ is by construction the amplitude we are looking for. The main advantage of the CSW rules is that they express amplitudes in terms of high-level building blocks that already

---

[19] Recall that under a rescaling by a factor $\lambda$ of an angle spinor, the MHV amplitudes with exactly two negative helicities scale by $\lambda^2$ if the scaling affects an external line of negative helicity, and by $\lambda^{-2}$ if it affects an external line of positive helicity.

contain a large number of color-ordered graphs, as illustrated in the following example of a two-vertex MHV graph contributing to the $1^-2^-3^-4^+5^+6^+$ six-point amplitude:

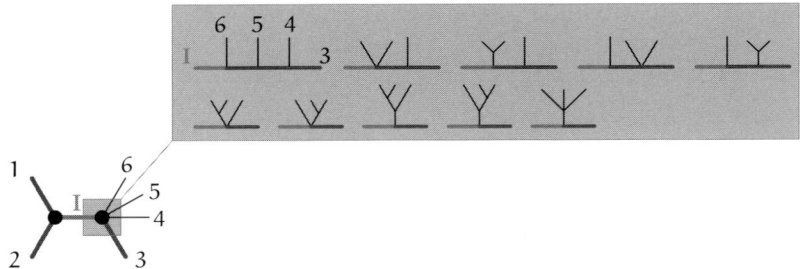

### 14.5.3 Examples

**Four-point $---+$ amplitude:** Let us first use the CSW rules to evaluate the $1^-2^-3^-4^+$ amplitude. Of course, we know beforehand that the result should be zero, so this is no more than a trivial illustration of the rules. In this case, the CSW rules give the following graphs:

$$\mathcal{A}_4(1^-2^-3^-4^+) = \quad\text{(graph 1)} \quad + \quad \text{(graph 2)} \quad + \quad \text{(graph 3)}. \tag{14.163}$$

But the last one is trivially zero because one of the vertices $(--)$ does not exist. In the first graph, the intermediate momentum (oriented from left to right) is $K_I = p_1 + p_4 = -(p_2 + p_3)$, and its angle spinor $\langle K_I| \equiv [\eta|K_I$ obeys

$$\langle K_I 1\rangle = -[\eta 4]\langle 14\rangle, \quad \langle K_I 2\rangle = [\eta 3]\langle 23\rangle,$$
$$\langle K_I 3\rangle = -[\eta 2]\langle 23\rangle, \quad \langle K_I 4\rangle = [\eta 1]\langle 14\rangle, \tag{14.164}$$

while the CSW rules give

$$\mathcal{A}_{4,1}(1^-2^-3^-4^+) = 2ig^2 \frac{\langle 1 K_I\rangle^3}{\langle K_I 4\rangle\langle 41\rangle} \frac{1}{K_I^2} \frac{\langle 23\rangle^3}{\langle 3 K_I\rangle\langle K_I 2\rangle}$$
$$= -2ig^2 \frac{[\eta 4]^3}{[\eta 1][\eta 2][\eta 3]} \frac{\langle 14\rangle}{[23]}. \tag{14.165}$$

In the second graph, the intermediate momentum is $L_I = p_1 + p_2 = -(p_3 + p_4)$, and the corresponding angle spinor satisfies

$$\langle L_I 1\rangle = -[\eta 2]\langle 12\rangle, \quad \langle L_I 2\rangle = [\eta 1]\langle 12\rangle,$$
$$\langle L_I 3\rangle = [\eta 4]\langle 34\rangle, \quad \langle L_I 4\rangle = -[\eta 3]\langle 34\rangle. \tag{14.166}$$

## 14.5 CACHAZO-SVRCEK-WITTEN RULES

This time, the CSW rules give

$$A_{4,2}(1^-2^-3^-4^+) = 2ig^2 \frac{\langle 12\rangle^3}{\langle 2L_I\rangle\langle L_I 1\rangle} \frac{1}{L_I^2} \frac{\langle L_I 3\rangle^3}{\langle 34\rangle\langle 4L_I\rangle}$$
$$= 2ig^2 \frac{[\eta 4]^3}{[\eta 1][\eta 2][\eta 3]} \frac{\langle 34\rangle}{[12]}. \tag{14.167}$$

Therefore, the sum of these two contributions is

$$A_4(1^-2^-3^-4^+) = 2ig^2 \frac{[\eta 4]^3}{[\eta 1][\eta 2][\eta 3]} \underbrace{\left\{\frac{\langle 34\rangle}{[12]} - \frac{\langle 14\rangle}{[23]}\right\}}_{0} = 0, \tag{14.168}$$

where the final cancellation is due to momentum conservation.

**Six-point** $---+++$ **amplitude:** Let us consider now a genuine example of next-to-MHV amplitude, the six-point $[123]^-[456]^+$ amplitude. The MHV graphs that contribute to this function are

$$A_6([123]^-[456]^+) = \quad + \quad , \tag{14.169}$$

where the shaded ellipses indicate that lines $(4,5)$ (in the left graph) or $(5,6)$ (in the right graph) can either be attached to the left or right MHV vertex, provided the cyclic order is not modified. Each term in eq. (14.169) therefore corresponds to three MHV graphs, i.e., a total of six.[20] For this amplitude, the CSW rules give

$$A_6([123]^-[456]^+) = -4ig^4 \Bigg\{ \sum_{i=3}^{5} \frac{\langle 1K_i\rangle^3}{\langle K_i i+1\rangle\langle i+1\,i+2\rangle\cdots\langle 61\rangle} \frac{1}{K_i^2} \frac{\langle 23\rangle^3}{\langle K_i 2\rangle\langle 34\rangle\cdots\langle iK_i\rangle}$$
$$+ \sum_{j=4}^{6} \frac{\langle 12\rangle^3}{\langle 2L_j\rangle\langle L_j j+1\rangle\cdots\langle 61\rangle} \frac{1}{L_j^2} \frac{\langle L_j 3\rangle^3}{\langle 34\rangle\cdots\langle j-1\,j\rangle\langle jL_j\rangle} \Bigg\}, \tag{14.170}$$

where the intermediate momenta are respectively $K_i \equiv -(p_2+p_3+\cdots+p_i)$ and $L_j \equiv -(p_3+p_4+\cdots+p_j)$, and the corresponding angle spinors are $\langle K_i| \equiv [\eta|K_i$ and $\langle L_j| \equiv [\eta|L_j$. For a numerical evaluation of this amplitude, one may take any auxiliary spinor $[\eta|$, since the amplitude does not depend on this choice. A somewhat simpler analytic expression may also be obtained by choosing $[\eta| \equiv [2|$. Indeed, since $K_i = L_i - p_2$, this choice leads to the following simplification:

$$\langle K_i j\rangle = \langle L_i j\rangle, \quad \text{for all } i,j, \tag{14.171}$$

---

[20] This number should be contrasted with the 38 six-point color-ordered graphs (see Figure 14.1). Moreover, each of these color-ordered graphs is considerably more complicated than the MHV diagrams.

thanks to which many factors become identical in the two terms of eq. (14.170). With this choice, the terms $i = 4, 5$ in the first line and $j = 4, 5$ in the second line combine in the following compact expression:

$$A_{6,1} = -\frac{4ig^4}{\langle 34 \rangle \langle 45 \rangle \langle 56 \rangle \langle 61 \rangle} \sum_{i=4,5} \frac{\langle ii+1 \rangle}{\langle K_i i \rangle \langle K_i i+1 \rangle \langle K_i 2 \rangle}$$
$$\times \left[ \frac{\langle 23 \rangle^3 \langle K_i 1 \rangle^3}{K_i^2} + \frac{\langle 12 \rangle^3 \langle K_i 3 \rangle^3}{(K_i + p_2)^2} \right]. \tag{14.172}$$

The terms $i = 3$ in the first line and $j = 6$ in the second line of eq. (14.170) must be handled more carefully. Indeed, they both contain a denominator that vanishes when $|\eta| \equiv |2|$, due to a factor $[\eta 2]$. In order to calculate these terms, we must leave $|\eta|$ unspecified, but such that

$$[\eta 2] \ll [\eta j] \quad \text{for } j \neq 2, \tag{14.173}$$

and expand in powers of $[\eta 2]$. After simplifications involving the Schouten identity, the sum of these two terms is found to be finite when $|\eta| \to |2|$ and equal to

$$A_{6,2} = \frac{4ig^4 \langle 13 \rangle^2}{\langle 34 \rangle \langle 45 \rangle \langle 56 \rangle \langle 61 \rangle} \left[ \frac{s_{13} + 2(s_{12} + s_{32})}{[12][32]} + \frac{\langle 12 \rangle \langle 36 \rangle}{[12] \langle 16 \rangle} + \frac{\langle 23 \rangle \langle 14 \rangle}{[23] \langle 34 \rangle} \right], \tag{14.174}$$

where we denote $s_{ij} \equiv \langle ij \rangle [ij] = (p_i + p_j)^2$. Combining all the contributions, the six-point amplitude reads

$$A_6([123]^-[456]^+) = \frac{4ig^4}{\langle 34 \rangle \langle 45 \rangle \langle 56 \rangle \langle 61 \rangle}$$
$$\times \left\{ -\sum_{i=4,5} \frac{\langle ii+1 \rangle}{\langle K_i i \rangle \langle K_i i+1 \rangle \langle K_i 2 \rangle} \left[ \frac{\langle 23 \rangle^3 \langle K_i 1 \rangle^3}{K_i^2} + \frac{\langle 12 \rangle^3 \langle K_i 3 \rangle^3}{(K_i + p_2)^2} \right] \right.$$
$$\left. + \langle 13 \rangle^2 \left[ \frac{s_{13} + 2(s_{12} + s_{32})}{[12][32]} + \frac{\langle 12 \rangle \langle 36 \rangle}{[12] \langle 16 \rangle} + \frac{\langle 23 \rangle \langle 14 \rangle}{[23] \langle 34 \rangle} \right] \right\}. \tag{14.175}$$

This is the simplest of the six-point amplitudes with three negative helicities, because the legs with negative helicities are adjacent (see Exercise 14.11 for another way of obtaining this amplitude). The non-adjacent cases lead to more complex expressions, but nevertheless considerably simpler than what one would get from traditional perturbation theory.

### 14.5.4 General CSW Rules

In order to prove the CSW rules in general, we now proceed by induction, i.e., we assume that the CSW rules are applicable to all *on-shell* amplitudes with up to $N - 1$ negative helicities and we consider an amplitude with $N$ negative helicities. Let us denote $\mathcal{N}$ the set of external

## 14.5 CACHAZO-SVRCEK-WITTEN RULES

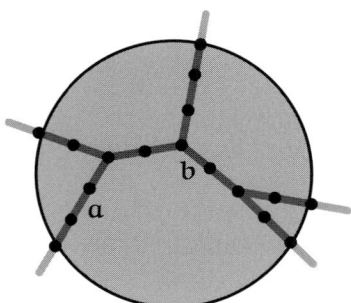

FIGURE 14.10: Propagators affected by the shift of eq. (14.176) in a generic amplitude. They correspond to the subgraph obtained by following the momentum that flows from the external lines of negative helicity, assuming that the momenta of the positive helicity ones are held fixed.

lines of negative helicity. We then introduce an auxiliary square spinor $[\eta|$, and we apply the following complex shift to these lines:

$$[\widehat{\imath}| \equiv [i| + r_i z [\eta|, \quad |\widehat{\imath}\rangle \equiv |i\rangle \quad \text{for } i \in \mathcal{N}, \tag{14.176}$$

where the $r_i$ are coefficients chosen in such a way that momentum conservation is preserved at any $z$. Namely, they must satisfy

$$\sum_{i \in \mathcal{N}} r_i |i\rangle = 0. \tag{14.177}$$

(We further assume that partial sums are not zero, so that the internal propagators connected to the external negative helicities all carry a $z$-dependent momentum.)

When $|z| \to \infty$, the shifted amplitude goes to zero like $|z|^{1-N}$ (graph by graph), which ensures that there is no boundary term when we integrate over $z$ on the circle at infinity. The propagators that contribute poles to this integral are among those represented in dark color in Figure 14.10 (we may assume that they become singular at distinct values of $z$, so that all the poles are simple). When we assign helicities to these singular propagators, two cases may arise:

- The singular propagator is directly connected to one of the external lines of negative helicity (e.g., the propagator labeled "a" in the figure). In this case, only one choice of helicity is allowed, and the amplitude factorizes into an amplitude with two negative helicities and one with $N - 1$ negative helicities.
- The singular propagator is an inner propagator, such as the one labeled "b" in the figure. It means that both the left and the right sub-amplitudes contain at least two of the $N$ external lines of negative helicity. In this case, both choices of helicity are valid for the singular propagator, and in both cases the left and right sub-amplitudes have at most $N - 1$ negative helicities.

In these two cases, the theorem of residues divides the amplitude into left and right on-shell sub-amplitudes that have at most $N - 1$ negative helicities, for which we may use the CSW rules assumed to be valid by the induction hypothesis.

After the left and right sub-amplitudes have been replaced by using the CSW rules proven for less than N negative helicities, all the poles produce terms that correspond to the same topology of MHV graph, but whose expressions differ because the value of $z_*$ is different for each pole. How this sum of contributions produces the product of denominators one would obtain by applying directly the CSW rules for amplitudes with N negative helicities requires some clarification. Let us consider a graph with $n_I$ internal shifted propagators. The application of the theorem of residues to the shifted amplitude divided by $z$ produces $n_I$ terms, whose sum corresponds to the following combination of denominators:

$$D_{\text{poles}} = \sum_{I=1}^{n_I} \frac{1}{K_I^2} \prod_{J \neq I} \frac{1}{\widehat{K}_{J,I}^2}. \tag{14.178}$$

Each term in this sum corresponds to the vanishing of one of the shifted internal propagators. The factor $K_I^2$ comes from the residue of the pole $z_I$ for which $\widehat{K}_I^2 = 0$, and $\widehat{K}_{J,I}^2$ denotes the value taken by $\widehat{K}_J^2$ at this $z_I$.

From eq. (14.176), we can write the $SL(2, \mathbb{C})$ matrices that represent the shifted internal momenta as follows:

$$\widehat{\mathbf{K}}_I = \mathbf{K}_I + z\, |\eta\rangle\langle I|, \tag{14.179}$$

where $\langle I|$ is a linear combination of the $r_i \langle i|$ that depends on the precise topology of the MHV graph and of the internal propagator under consideration. This implies that

$$\widehat{K}_I^2 = K_I^2 - z\, [\eta|\mathbf{K}_I|I\rangle, \tag{14.180}$$

and the corresponding denominator vanishes at $z_I = K_I^2/[\eta|\mathbf{K}_I|I\rangle$. Therefore, we have

$$D_{\text{poles}} = \sum_{I=1}^{n_I} \frac{1}{K_I^2} \prod_{J \neq I} \frac{1}{K_J^2 - K_I^2 \frac{[\eta|\mathbf{K}_I|J\rangle}{[\eta|\mathbf{K}_I|I\rangle}} = \prod_{I=1}^{n_I} \frac{1}{K_I^2}. \tag{14.181}$$

The last equality[21] shows that the $n_I$ terms produced by the theorem of residues combine into an expression which is nothing but the product of unshifted off-shell denominators that would appear in the CSW rules. This completes the proof of the general CSW diagrammatic rules:

- Draw all the diagrams with the required assignment of external helicities, and such that all vertices have exactly two negative helicities. With N external negative helicities, these graphs all have $N - 1$ vertices. For instance, the $[1234]^-[5\cdots n]^+$ amplitude receives

---

[21] This may be proven by integrating over a circle at infinity the following function:

$$f(z) \equiv z^{-1} \prod_{I=1}^{n_I} \frac{1}{K_I^2 - z\, [\eta|\mathbf{K}_I|I\rangle}.$$

contributions from the following three classes of MHV diagrams:

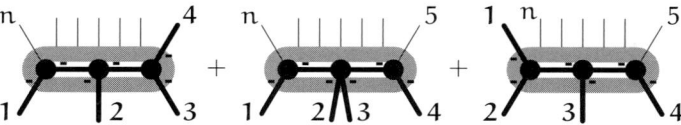

(Only the negative helicities are indicated explicitly.)
- The intermediate propagators are scalar propagators $i/K_I^2$, with the value of $K_I$ determined by momentum conservation at the vertices.
- Replace the vertices by MHV amplitudes, with some off-shell legs for which the angle spinor is defined as $\langle K_I| \equiv [\eta|K_I$, and use the Parke–Taylor formula to obtain their expression.

## Exercises

**\*14.1** Counting tree graphs amounts to replacing all the propagators and vertices by 1.

- Consider first a scalar $\phi^3$ theory. Explain why the number of tree graphs contributing to the n-point amplitude is the coefficient of $j^{n-1}/(n-1)!$ in the solution of $X = j + X^2/2$. Solve this equation and calculate the first few coefficients. Show that the number of n-point tree graphs in this theory is $(2n-5)!! = 1 \cdot 3 \cdots (2n-5)$.
- Consider now Yang–Mills theory, and derive the equation whose solution gives the number of graphs contributing to the tree-level n-gluon amplitude. Solve this equation, and do a Taylor expansion to recover the numbers listed in Table 14.1.

**14.2** Check the identities in footnotes 3 and 5, and the Fierz identity in eq. (14.36).

**14.3** Check eqs. (14.37) and (14.38).

**\*14.4** What does little group scaling tell us about the on-shell three-scalar amplitude? How can we reconcile this result with a Lagrangian that contains a derivative coupling such as $\phi(\partial\phi)^2$? How could the triviality of this interaction be seen directly from the action?

**\*14.5** Use little group scaling arguments to recover the fact that an on-shell three-photon amplitude is zero. Show that this obstruction can be evaded with spin-1 particles that carry a quantum number such that the three-point vertex is fully antisymmetric in this quantum number.

**14.6** How does the locality of cubic interactions constrain the dimension of a three-point amplitude? Combining this with little group scaling, what is the most general form of this amplitude? In the spin-1 case, is this sufficient to forbid the $+++$ and $---$ amplitudes? What type of local operator could produce them?

**\*14.7** Consider a massless local renormalizable theory of spin-1 bosons. With only local interactions, amplitudes cannot have multiple poles. Consider a four-point amplitude $\mathcal{A}_4$ (with s, t, u the Mandelstam invariants) in a spin-1 theory where the particles carry a quantum number $a$. Let $f^{abc}$ be the coupling constant between three such bosons.

- Explain why $\lim_{s\to 0} s\mathcal{A}_4$ factorizes as the product of two three-point amplitudes. What does this imply for the dimension of four-point amplitudes?
- Focus on the following helicity and quantum number assignment: $1^-_a 2^-_b 3^+_c 4^+_d$. Using little group scaling, show that the amplitude must have the form

$$\mathcal{A}_4(1^-_a 2^-_b 3^+_c 4^+_d) = \langle 12 \rangle^2 [34]^2 \left\{ \frac{C_{st}}{st} + \frac{C_{tu}}{tu} + \frac{C_{us}}{us} \right\},$$

where $C_{st}, C_{tu}, C_{us}$ are dimensionless constants.
- Using the above factorization, show that

$$f^{abe} f^{cde} = C_{st} - C_{us}.$$

- Using similar factorizations when $u, t \to 0$, prove that the $f^{abc}$ obey Jacobi identity (even though we have nowhere assumed that our model is a gauge theory).
- Extra question: Using similar arguments, show that the four-graviton amplitude with helicities $1^{--} 2^{--} 3^{++} 4^{++}$ has the following form:

$$\mathcal{A}_4(1^{--} 2^{--} 3^{++} 4^{++}) = \text{const} \times \frac{\langle 12 \rangle^4 [34]^4}{stu}.$$

**14.8** Is it possible to use BCFW recursion in $\phi^4$ scalar theory? Could we have guessed beforehand that this was bound to fail?

**\*14.9** Consider the $1^- 2^- 3^+ 4^+$ amplitude with a BCFW shift applied on lines $1, 2$. A priori there are two terms, corresponding to the two helicities of the intermediate line, but one of them turns out to be zero. Why?

**14.10** The goal of this exercise is to prove the Parke–Taylor formula for MHV amplitudes whose two minus helicities are not adjacent. Without loss of generality, we may assume that one of these helicities is carried by the first line.
- First, establish the formula in the case $1^- 2^+ 3^- 4^+ \cdots n^+$, by using a BCFW shift on lines $1, 2$.
- Then, consider the case of $1^- 2^+ \cdots j^+ \cdots n^+$, with $3 < j < n$. Use BCFW recursion to obtain an induction formula that relates the n-point and $(n-1)$-point cases.

**14.11** Using a BCFW shift on lines $3, 4$, derive an expression of the color-ordered amplitude $1^- 2^- 3^- 4^+ 5^+ 6^+$ that has only two terms.

**\*14.12** Consider the amplitude $\mathcal{A}_4(1^-_a 2^-_b 3^+_c 4^+_d)$ in Yang–Mills theory. Naively, its color decomposition is a sum of $(4-1)! = 6$ color-ordered terms.
- Use the Parke–Taylor formula to check the following symmetries of the four-point color-ordered amplitudes:

$$\mathcal{A}_4(1^- 2^- 3^+ 4^+) = \mathcal{A}_4(2^- 3^+ 4^+ 1^-) \quad \text{(cyclic invariance)},$$
$$\mathcal{A}_4(1^- 2^- 3^+ 4^+) = \mathcal{A}_4(4^+ 3^+ 2^- 1^-) \quad \text{(reflexion invariance)}.$$

Use these two properties to reduce the color decomposition to three terms.

- Prove now that

$$\mathcal{A}_4(1^-2^-3^+4^+) + \mathcal{A}_4(1^-3^+2^-4^+) = -\mathcal{A}_4(1^-2^-4^+3^+)$$

(this is an example of the *Kleiss–Kuijf relations*), and use this identity to reduce the color decomposition to two terms only:

$$\mathcal{A}_4(1_a^-2_b^-3_c^+4_d^+) = \tfrac{i^2}{2} f^{abe} f^{cde} \mathcal{A}_4(1^-2^-3^+4^+) - \tfrac{i^2}{2} f^{cae} f^{bde} \mathcal{A}_4(1^-3^+2^-4^+).$$

**\*14.13** This exercise builds on Exercises 14.7 and 14.12. Consider again the tree-level four-point amplitude $\mathcal{A}_4(1_a^-2_b^-3_c^+4_d^+)$. We want to write it as follows:

$$\mathcal{A}_4(1_a^-2_b^-3_c^+4_d^+) = \frac{C_s N_s}{s} + \frac{C_t N_t}{t} + \frac{C_u N_u}{u},$$

where the color indices are contained only in the coefficients $C_{s,t,u}$ defined by

$$C_s \equiv f^{abe} f^{cde}, \quad C_t \equiv f^{bce} f^{ade}, \quad C_u \equiv f^{cae} f^{bde} \quad (\text{hence, } C_s + C_t + C_u = 0).$$

- By identifying this form of the amplitude with the one obtained in Exercise 14.7, determine the coefficients $N_{s,t,u}$ (depending only on momenta) to make this expression valid. Check that the solution is not unique, but that all obey $N_s + N_t + N_u = 0$.
- Verify that this form of the amplitude is consistent with the color decomposition obtained in the previous exercise (use the Parke–Taylor formula to check that the kinematical factors indeed agree).
- Replace $C_{s,t,u} \to N_{s,t,u}$ in this amplitude and show that this substitution gives the corresponding four-graviton amplitude (also discussed in Exercise 14.7).

**\*14.14** This exercise builds on Exercises 11.7 and 14.13. Consider the $\pi^a \pi^b \to \pi^c \pi^d$ scattering amplitude obtained at lowest order in the nonlinear sigma model:

$$\mathcal{A}_4^{\text{NLSM}}(1_a 2_b 3_c 4_d) = \frac{1}{f_\pi^2} \left( s\, \delta_{ab}\delta_{cd} + t\, \delta_{ad}\delta_{bc} + u\, \delta_{ac}\delta_{bd} \right).$$

(This is the result of Exercise 11.7.) Show that this amplitude admits the following algebra-kinematics factorized form:

$$\mathcal{A}_4^{\text{NLSM}}(1_a 2_b 3_c 4_d) = \frac{C_s \tilde{N}_s}{s} + \frac{C_t \tilde{N}_t}{t} + \frac{C_u \tilde{N}_u}{u},$$

where $C_{s,t,u}$ are color factors defined as in the previous exercise (with the structure constants of $\mathfrak{su}(2)$) and the $\tilde{N}_{s,t,u}$ are purely kinematical factors obeying $\tilde{N}_s + \tilde{N}_t + \tilde{N}_u = 0$. Thus, the combined result of this exercise and the previous one is a pair of dualities relating tree-level four-point amplitudes in Yang–Mills theory, in the nonlinear sigma model and in classical gravity:

$$\mathcal{A}_4^{\text{NLSM}}(1_a 2_b 3_c 4_d) \xleftarrow{\tilde{N} \leftarrow N} \mathcal{A}_4^{\text{YM}}(1_a^- 2_b^- 3_c^+ 4_d^+) \xrightarrow{C \to N} \mathcal{A}_4^{\text{grav}}(1^{--} 2^{--} 3^{++} 4^{++})$$

# CHAPTER 15

# Worldline Formalism

In the previous chapter, we have exposed the spinor-helicity language in which the building blocks of scattering amplitudes are expressed in terms of two-component spinors. As we have seen, when combined with techniques such as on-shell recursion, this leads to great simplifications in the evaluation of on-shell tree amplitudes with physical polarizations. To a large extent, this simplification stems from the fact that the calculation of amplitudes based on these methods bypasses the usual representation in terms of Feynman diagrams.

In fact, the spinor-helicity method is not the only one that relegates Feynman diagrams to a minor secondary role. Another approach, that we discuss in this chapter, is the *worldline formalism*. The name comes from the fact that in this approach, Feynman graphs are replaced by a representation in terms of a path integral over a function $z^\mu(\tau)$ (plus additional auxiliary variables in the case of fields with internal degrees of freedom, such as spin or color), that defines a line embedded in space-time. This function can be viewed as a parametrization of the whole history of a point-like particle. Historically, this method was first derived by starting from string theory and by taking the limit of infinite string tension. Subsequently, it was re-derived in a more mundane manner in a first quantized framework. This is the point of view that we shall adopt in this chapter.

## 15.1 Worldline Representation

### 15.1.1 Heat Kernel

In order to illustrate the principles of the worldline representation, consider a scalar field theory of Lagrangian

$$\mathcal{L} \equiv \frac{1}{2}(\partial_\mu \phi)(\partial^\mu \phi) - \frac{1}{2}m^2\phi^2 - V(\phi). \tag{15.1}$$

## 15.1 WORLDLINE REPRESENTATION

Let us assume that we wish to obtain the tree-level propagator $G(x,y)$ in a background field $\varphi$. Up to a factor $i$, this propagator is the inverse of the operator $\Box + m^2 + V''(\varphi)$,

$$(\Box_x + m^2 + V''(\varphi(x)))\, G(x,y) = -i\,\delta(x-y). \tag{15.2}$$

This equation must be supplemented by boundary conditions that depend on the type of propagator one wishes to obtain (time-ordered, retarded, etc.), but since we will later restrict to Euclidean time, these differences will become irrelevant. Formally, we may write

$$(\Box + m^2 + V''(\varphi))^{-1} = \int_0^\infty dT\, e^{-T(\Box + m^2 + V''(\varphi))}. \tag{15.3}$$

(The integrand in this formula is sometimes called a *heat kernel*, by analogy with the propagator of a heat equation.) However, for this integral to make sense in the limit $T \to \infty$, it is necessary that all the eigenvalues of the operator $\Box + m^2 + V''(\varphi)$ be positive. The high-lying eigenvalues of this operator do not depend on the background field $\varphi(x)$ (assuming that it is smooth enough), and are of the form

$$-g_{\mu\nu} k^\mu k^\nu + m^2. \tag{15.4}$$

In order to be positive for any momentum $k^\mu$, it is therefore necessary that the metric be Euclidean, with only minus signs. For this reason, we restrict our discussion to a Euclidean field theory from now on, so that we may write $-g_{\mu\nu} k^\mu k^\nu = k^i k^i$, and $\Box = -\partial^2$.

### 15.1.2 Propagator in a Background Field

The propagator $G(x,y)$ is obtained by evaluating eq. (15.3) between states of definite position,

$$G(x,y) = -i \int_0^\infty dT\, \langle y | e^{-T(\Box + m^2 + V''(\varphi))} | x \rangle. \tag{15.5}$$

Such a matrix element is quite common in ordinary quantum mechanics, and its representation as a path integral is well known. For a non-relativistic Hamiltonian of the form

$$H \equiv \frac{P^2}{2M} + V(Q), \tag{15.6}$$

we have

$$\langle y | e^{-i(t_1 - t_0)H} | x \rangle = \int_{\substack{q(t_0)=x \\ q(t_1)=y}} [Dq(t)]\, e^{i \int_{t_0}^{t_1} dt\, (\frac{M}{2} \dot{q}^2(t) - V(q(t)))}. \tag{15.7}$$

Eq. (15.5) can be similarly expressed as a path integral if we use the following dictionary:

$$\begin{aligned}
i(t_1 - t_0) &\to T, \\
it &\to \tau, \\
M &\to \frac{1}{2}, \\
V(Q) &\to m^2 + V''(\varphi(x)).
\end{aligned} \tag{15.8}$$

This leads to

$$G(x,y) = -i \int_0^\infty dT \int_{\substack{z(0)=x \\ z(T)=y}} [Dz(\tau)] \, e^{-\int_0^T d\tau \, (\frac{1}{4}\dot{z}^2(\tau) + m^2 + V''(\varphi(z(\tau))))}, \qquad (15.9)$$

where the dot denotes a derivative with respect to $\tau$. For simplicity, we denote the integration variable $z(\tau)$ instead of $z^\mu(\tau)$, although it takes values in a d-dimensional Euclidean space-time. This expression is known as the worldline representation of the tree propagator in a background field. Very much as in ordinary quantum mechanics, the function $z^\mu(\tau)$ explores all the paths that start at $x$ and end at $y$. Note also that the formula contains an integral over the "duration" (we use quotes here because $\tau$ is not a physical time) of this evolution.

### 15.1.3 Alternate Derivation

Starting from eq. (15.3), it is possible to follow a slightly different route (that one may view as another derivation of the path integral formulation of quantum mechanics), that has the virtue of providing more control on all the prefactors. Consider first the case of the theory in the vacuum, i.e., with no background field. An important result is the following formula (see Exercise 15.1),

$$e^{T\partial^2} f(x) = \int_{-\infty}^{+\infty} \frac{dz}{\sqrt{4\pi T}} \, e^{-\frac{z^2}{4T}} f(x+z), \qquad (15.10)$$

known as the *Weierstrass transform* of f, which may be proven by Fourier transform. In words, this formula means that an operator Gaussian in a derivative is equivalent to a Gaussian smearing. Note here that it is crucial that the squared derivative has a positive prefactor inside the exponential, otherwise on the right-hand side we would have a Gaussian with a wrong sign and the result would be ill-defined. From this formula, we get

$$\langle y | e^{-T(\Box + m^2)} | x \rangle = e^{-Tm^2} \int_{-\infty}^{+\infty} \frac{d^d z}{(4\pi T)^{d/2}} \, e^{-\frac{z^2}{4T}} \underbrace{\langle y | x+z \rangle}_{\delta(x+z-y)}$$

$$= \frac{e^{-Tm^2}}{(4\pi T)^{d/2}} e^{-\frac{(y-x)^2}{4T}}, \qquad (15.11)$$

which, apart from the prefactor $\exp(-Tm^2)$, is a Gaussian probability distribution normalized to unity. This Gaussian distribution may be viewed as a Green's function for the diffusion equation

$$\partial_T f(T, x) = -\Box_x f(T, x), \qquad (15.12)$$

which highlights the connection that exists between the propagator associated to an elliptic differential operator and diffusion (i.e., Brownian motion). By comparing eqs. (15.9) (without

## 15.1 WORLDLINE REPRESENTATION

external field) and (15.11), one can obtain the following formula for the absolute normalization of the integral over closed loops in d dimensions

$$\int_{z(0)=z(T)} [Dz(\tau)] \exp\left(-\int_0^T d\tau\, \frac{\dot z^2}{4}\right) = \frac{1}{(4\pi T)^{d/2}} \underset{d=4}{=} \frac{1}{(4\pi T)^2}. \tag{15.13}$$

The next step is to note that such a Gaussian distribution may be written as the convolution of two similar distributions defined on a halved interval,

$$\frac{e^{-\frac{(y-x)^2}{4T}}}{(4\pi T)^{d/2}} = \int d^d z\, \frac{e^{-\frac{(y-z)^2}{2T}}}{(2\pi T)^{d/2}}\, \frac{e^{-\frac{(z-x)^2}{2T}}}{(2\pi T)^{d/2}}. \tag{15.14}$$

By taking $n-1$ of these intermediate points, we arrive at

$$\langle y | e^{-T(\Box + m^2)} | x \rangle = e^{-Tm^2} \int \frac{d^d z_1 d^d z_2 \cdots d^d z_{n-1}}{(4\pi\epsilon)^{nd/2}}\, e^{-\frac{\epsilon}{4}\sum_{i=1}^{n} \frac{(z_i - z_{i-1})^2}{\epsilon^2}}, \tag{15.15}$$

where we denote $\epsilon \equiv T/n$, $z_0 \equiv x$ and $z_n \equiv y$. In the limit where $n \to \infty$, the argument of the exponential becomes an integral, and we obtain the following path integral:

$$\langle y | e^{-T(\Box + m^2)} | x \rangle = e^{-Tm^2} \int_{\substack{z(0)=x \\ z(T)=y}} [Dz(\tau)]\, e^{-\frac{1}{4}\int_0^T d\tau\, \dot z^2(\tau)}. \tag{15.16}$$

Taking into account the term $V''(\varphi)$ due to a background field poses no difficulty if one breaks the interval $[0, T]$ into many small intervals. Indeed, even though $V''(\varphi(x))$ does not commute with $\Box_x$, the Baker–Campbell–Hausdorff formula indicates that the exponential of their sum is equal to the product of their respective exponentials, up to terms of higher order in $\epsilon = T/n$, that do not matter in the limit $n \to \infty$.

### 15.1.4 One-Loop Effective Action

A minor modification of this derivation also applies to the quantum effective action at one loop in the background field $\varphi$:

$$\Gamma[\varphi] = -\frac{1}{2} \mathrm{tr}\, \ln\left[\frac{\Box + m^2 + V''(\varphi)}{\Box + m^2}\right]. \tag{15.17}$$

The denominator inside the logarithm is not crucial since it is independent of the background field, but it produces an ultraviolet subtraction since the large eigenvalues of the numerator and denominator are almost equal. First, the logarithm may be represented as follows:

$$\ln\left[\frac{\Box + m^2 + V''(\varphi)}{\Box + m^2}\right] = -\int_0^\infty \frac{dT}{T} \left(e^{-T(\Box + m^2 + V''(\varphi))} - e^{-T(\Box + m^2)}\right), \tag{15.18}$$

which is very similar to eq. (15.3), except for the denominator $1/T$. The same restrictions on the signs of the eigenvalues apply here, forcing us to consider again a Euclidean theory. The proof of this formula goes along the following lines:

$$\ln\left(\frac{A}{B}\right) = \int_A^B \frac{dY}{Y} = \int_A^B dY \int_0^\infty dT\, e^{-TY} = \int_0^\infty \frac{dT}{T}\left(e^{-TB} - e^{-TA}\right). \tag{15.19}$$

Then, the trace is given in coordinate space by

$$\mathrm{tr}\left(\cdots\right) = \int d^d x\, \langle x|\cdots|x\rangle. \tag{15.20}$$

Therefore, we obtain a path integral representation similar to eq. (15.9), but with a path that starts and ends at the same point:

$$\Gamma[\varphi] = \mathrm{const} + \frac{1}{2}\int_0^\infty \frac{dT}{T} \int_{z(0)=z(T)} [Dz(\tau)]\, e^{-\int_0^T d\tau\,\left(\frac{1}{4}\dot{z}^2(\tau) + m^2 + V''(\varphi(z(\tau)))\right)}. \tag{15.21}$$

(We have not written explicitly the term coming from the denominator $\Box + m^2$ – it is contained in the unspecified additive constant.) In this case, the worldlines are closed, and therefore form loops in space-time.

### 15.1.5 Length Scales

Let us now discuss some qualitative aspects of eqs. (15.9) and (15.21). First, note that the parameter $T$ has the dimension of an inverse squared mass,

$$T \sim (\mathrm{mass})^{-2}. \tag{15.22}$$

On dimensional grounds, one sees that the typical diameter of the loops[1] $z(\tau)$ that appear in eq. (15.21) is

$$\Delta z \sim \sqrt{T}. \tag{15.23}$$

In contrast, the perimeter of these loops scales as $T$. These scaling laws are consistent with a Brownian motion of duration $T$ (see Figure 15.1).

The integration measure $dT/T$ corresponds to a uniform distribution of the values of $\ln(T)$. Thus, the loop sizes are uniformly distributed on a log scale. Large loops (i.e., large values of $T$) encode the infrared sector of the theory. In a massless theory, loops of arbitrarily high size are allowed. With a non-zero mass, the factor $\exp(-Tm^2)$ suppresses the values $Tm^2 \gg 1$, i.e., the loops of size larger than the inverse mass (the Compton wavelength of the particle). By suppressing the probability of occurrence of large loops, the mass thus regulates the infrared. Note also how the second derivative $V''(\varphi(z))$ acts as a position-dependent squared mass.

---

[1] For a uniform background field, the exponential depends only on the derivative $\dot{z}$. As a consequence, this exponential weight constrains the size of the loops, but not the location of their barycenter.

## 15.2 QUANTUM ELECTRODYNAMICS

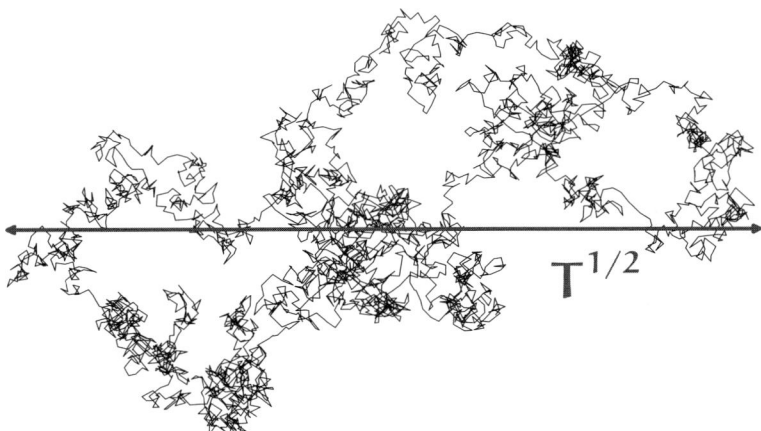

FIGURE 15.1: Typical worldloop that contributes in eq. (15.21). While its length scales as T, its extent in space-time only grows like $T^{1/2}$.

On the other hand, small loops encode the ultraviolet behavior of the theory. When the extent of the loop becomes smaller than the typical scale over which the background field $\varphi(x)$ varies, the loop sees only a constant background field, whose sole effect in eq. (15.21) is an overall rescaling. Therefore, these small loops behave as in the vacuum, and the ultraviolet sector of the theory does not depend on the background field.

## 15.2 Quantum Electrodynamics

### 15.2.1 Scalar QED

Let us now consider the case where the background field is an Abelian vector field $\mathcal{A}^\mu(x)$, while the particle in the loop is a (complex) scalar. The one-loop effective action is now given by

$$\Gamma[\mathcal{A}] \equiv -\operatorname{tr} \ln \left[ \frac{-(\partial - ie\mathcal{A})^2 + m^2}{\Box + m^2} \right]. \tag{15.24}$$

Note that there is no prefactor $1/2$ because the scalar field in this theory is a complex field. First, we obtain

$$\Gamma[\mathcal{A}] = \text{const} + \int_0^\infty \frac{dT}{T} \int d^d x \, \langle x | e^{-T(m^2 - (\partial - ie\mathcal{A})^2)} | x \rangle. \tag{15.25}$$

Then, note that the exponential contains an operator which is very similar to the Hamiltonian of a charged particle in an external electromagnetic field. We can use this analogy in order to obtain a path integral representation of the matrix element under the integral in the previous equation. This gives the following representation (see Exercise 15.2):

$$\Gamma[\mathcal{A}] = \mathrm{const} + \int_0^\infty \frac{\mathrm{d}T}{T} \int_{z(0)=z(T)} [Dz(\tau)]\, e^{-\int_0^T \mathrm{d}\tau\, (\frac{1}{4}\dot{z}^2(\tau)+ie\dot{z}(\tau)\cdot \mathcal{A}(z(\tau))+m^2)}. \tag{15.26}$$

Likewise, the tree-level scalar propagator in a background electromagnetic field is given by

$$G(x,y) = -i \int_0^\infty \mathrm{d}T \int_{\substack{z(0)=x \\ z(T)=y}} [Dz(\tau)]\, e^{-\int_0^T \mathrm{d}\tau\, (\frac{1}{4}\dot{z}^2(\tau)+ie\dot{z}(\tau)\cdot \mathcal{A}(z(\tau))+m^2)}. \tag{15.27}$$

Under an Abelian gauge transformation of the electromagnetic potential,

$$\mathcal{A}^\mu(z) \quad \to \quad \mathcal{A}^\mu(z) + \partial^\mu \chi(z), \tag{15.28}$$

the term in $\dot{z} \cdot \mathcal{A}$ transforms as follows:

$$\dot{z}\cdot \mathcal{A} \quad \to \quad \dot{z}\cdot(\mathcal{A}+\partial_z \chi) = \dot{z}\cdot \mathcal{A} + \partial_\tau \chi(z(\tau)). \tag{15.29}$$

Thus, the gauge transformation modifies this term by the addition of a total derivative with respect to $\tau$, whose integral is

$$\int_0^T \mathrm{d}\tau\, \partial_\tau \chi(z(\tau)) = \chi(z(T)) - \chi(z(0)). \tag{15.30}$$

In the calculation of the one-loop effective action, the trajectories $z(\tau)$ have equal initial and final points, and this shift is therefore zero. Thus, the expression (15.26) of the one-loop effective action is explicitly gauge invariant. If we were considering instead the scalar propagator $G(x, y)$, this term would be

$$\int_0^T \mathrm{d}\tau\, \partial_\tau \chi(z(\tau)) = \chi(y) - \chi(x), \tag{15.31}$$

and as a consequence the propagator transforms as

$$G(x,y) \quad \to \quad e^{ie\chi(x)}\, G(x,y)\, e^{-ie\chi(y)}, \tag{15.32}$$

which is indeed the correct gauge transformation law of the propagator in scalar electrodynamics.

### 15.2.2 Spinor QED

When the particle in the loop is a spin-1/2 fermion, the one-loop effective action involves the Dirac operator:

$$\Gamma[\mathcal{A}] \equiv \mathrm{tr}\, \ln \left[ \frac{iD_\mu \gamma^\mu + m}{i\partial_\mu \gamma^\mu + m} \right]. \tag{15.33}$$

## 15.2 QUANTUM ELECTRODYNAMICS

The first step is to note that (see Exercise 15.3)

$$\det(m + i\slashed{D}) = \det(m - i\slashed{D}) = \left[\det(m^2 + \slashed{D}^2)\right]^{1/2}, \qquad (15.34)$$

which leads to

$$\Gamma[A] \equiv \frac{1}{2}\,\mathrm{tr}\,\ln\left[\frac{\slashed{D}^2 + m^2}{\slashed{\partial}^2 + m^2}\right]. \qquad (15.35)$$

Then, we may use

$$\begin{aligned}\slashed{D}^2 &= D_\mu D_\nu \gamma^\mu \gamma^\nu = D_\mu D_\nu \left(\tfrac{1}{2}\{\gamma^\mu,\gamma^\nu\} + \tfrac{1}{2}[\gamma^\mu,\gamma^\nu]\right) \\ &= -D^2 + \tfrac{1}{4}[D_\mu, D_\nu][\gamma^\mu,\gamma^\nu] = -D^2 - e\,F_{\mu\nu} M^{\mu\nu},\end{aligned} \qquad (15.36)$$

where $M^{\mu\nu} \equiv \tfrac{i}{4}[\gamma^\mu,\gamma^\nu]$. (We have assumed a Euclidean metric tensor with only minus signs in the second line.) This gives the following representation of the one-loop effective action:

$$\Gamma[A] = \mathrm{const} - \frac{1}{2}\,\mathrm{tr}\int_0^\infty \frac{dT}{T}\,e^{-T(m^2 - D^2 - e F_{\mu\nu} M^{\mu\nu})}. \qquad (15.37)$$

The term $m^2 - D^2$, identical to the operator encountered in the case of scalar QED, is now supplemented by a potential

$$U(x) \equiv -e\,F_{\mu\nu}(x)\,M^{\mu\nu}. \qquad (15.38)$$

However, because $U(x)$ still contains non-commuting Dirac matrices, the world-line representation of the exponential is now more complicated, and the overall trace applies both to the space-time dependence and to the Dirac indices. A first possibility is to reproduce the method used in the previous sections, where we introduce a path integral over classical trajectories $z^\mu(\tau)$. When doing this, the matrix Dirac structure inside the exponential is not touched, and is handled by a path ordering:

$$\begin{aligned}\langle x|e^{-T(m^2 - D^2 - e F_{\mu\nu} M^{\mu\nu})}|x\rangle = \int_{\substack{z(0)=x \\ z(T)=x}} [Dz(\tau)] \\ \times\, P\left(e^{-\int_0^T d\tau\,(\frac{1}{4}\dot z^2(\tau) + i e \dot z(\tau)\cdot A(z(\tau)) + m^2 + U(z(\tau)))}\right).\end{aligned} \qquad (15.39)$$

But it is in fact possible to remove the path ordering by introducing some auxiliary variables. In the procedure that leads to eq. (15.39), one breaks the interval $[0, T]$ into infinitesimal subintervals and one inserts a complete sum of states between each factor. When the evolution operator to be evaluated contains extra internal degrees of freedom (in the present case, the spin degree of freedom encoded in the Dirac matrices), the intermediate states inserted in the expression must contain information about this internal structure for the matrix elements produced in the process to be ordinary commuting numbers.

Let us define the following operators from the Dirac matrices:

$$c_1^\pm \equiv \frac{i\gamma_1 \pm \gamma_2}{2}, \qquad c_2^\pm \equiv \frac{i\gamma_3 \pm \gamma_4}{2}. \tag{15.40}$$

From the anti-commutation relation obeyed by the Dirac matrices, we have

$$\{c_r^+, c_s^-\} = \delta_{rs}, \quad \{c_r^+, c_s^+\} = \{c_r^-, c_s^-\} = 0. \tag{15.41}$$

Therefore, the operators $c_r^+$ are fermionic creation operators (creating independent fermions), and the $c_i^-$ are the corresponding annihilation operators. By inverting eq. (15.40),

$$\begin{aligned}\gamma_1 &= -i(c_1^+ + c_1^-), & \gamma_2 &= c_1^+ - c_1^-, \\ \gamma_3 &= -i(c_2^+ + c_2^-), & \gamma_4 &= c_2^+ - c_2^-,\end{aligned} \tag{15.42}$$

the potential $U(x)$ may be viewed as a Hamiltonian quadratic in fermionic creation and annihilation operators, and a time evolution operator constructed with this Hamiltonian may be written as a Grassmann path integral. In order to see this, let us start from a state $|0\rangle$ which is annihilated by $c_{1,2}^-$,

$$c_{1,2}^- |0\rangle = 0, \tag{15.43}$$

and construct populated Fock states by applying $c_{1,2}^+$ to it,

$$|n_1 n_2\rangle \equiv (c_1^+)^{n_1} (c_2^+)^{n_2} |0\rangle, \quad \langle n_2 n_1| \equiv \langle 0| (c_2^-)^{n_2} (c_1^-)^{n_1}. \tag{15.44}$$

Here, one has to keep in mind the order of the operators in this definition, because this order does not appear explicitly in the notation used for the Fock state in the left-hand side. Consider now a pair of complex Grassmann variables $\xi_{1,2}$ such that

$$\begin{aligned}\{\xi_i, \xi_j\} &= \{\bar\xi_i, \xi_j\} = \{\xi_i, \bar\xi_j\} = \{\bar\xi_i, \bar\xi_j\} = 0, \\ \{\xi_i, c_j^\pm\} &= \{\bar\xi_i, c_j^\pm\} = 0.\end{aligned} \tag{15.45}$$

A fermionic coherent state $|\xi\rangle$ may be defined as

$$|\xi\rangle \equiv e^{-\xi c^+} |0\rangle, \qquad \langle\xi| = \langle 0| e^{\bar\xi c^-}. \tag{15.46}$$

Like bosonic coherent states, they are eigenstates of the annihilation operators,

$$c_i^- |\xi\rangle = \xi_i |\xi\rangle, \qquad \langle\xi| c_i^+ = \bar\xi_i \langle\xi|. \tag{15.47}$$

In addition, the overlap between two such coherent states is given by

$$\langle\xi|\zeta\rangle = e^{\bar\xi\zeta}, \tag{15.48}$$

and one may construct the identity operator as a superposition of projectors on these coherent states:

$$1 = \int \underbrace{d\bar\xi_1 d\xi_1 d\bar\xi_2 d\xi_2}_{\equiv d\bar\xi d\xi}\, e^{-\bar\xi\xi}\, |\xi\rangle\langle\xi|. \tag{15.49}$$

Moreover, if $A(\mathbf{c}^+, \mathbf{c}^-)$ is a normal ordered operator made of the creation and annihilation operators, then its matrix element between two coherent states is given by

$$\langle \xi | A(\mathbf{c}^+, \mathbf{c}^-) | \zeta \rangle = e^{\overline{\xi}\zeta} A(\overline{\xi}, \zeta), \tag{15.50}$$

and the trace over the Dirac indices of an operator A may be written as

$$\mathrm{tr}\,(A) = \int d\overline{\xi}d\xi\, e^{-\overline{\xi}\xi} \langle -\xi | A | \xi \rangle. \tag{15.51}$$

(One may easily check that this gives 4 when A is the identity. See Exercise 15.4 for a derivation.) Note that in the calculation of this trace, the coherent states that appear on the left and on the right are defined with opposite Grassmann variables $\xi$ and $-\xi$. This is a standard property of fermionic traces, whose path integral representation must obey an anti-periodic boundary condition in time.

This formalism can be used to transform the Dirac structure in eq. (15.39) into a Grassmann path integral. To achieve this, we follow the standard procedure of breaking the interval $[0, T]$ into N small sub-intervals, and we insert a unit operator given by eq. (15.49) at the boundaries of the sub-intervals. This produces matrix elements of the form

$$\langle \xi_{i+1} | e^{\epsilon e F_{\mu\nu}(z(\tau_i)) M^{\mu\nu}} | \xi_i \rangle, \tag{15.52}$$

($\epsilon \equiv T/N$) that may be evaluated by replacing the Dirac matrices in $M^{\mu\nu}$ by their expression in terms of the operators $\mathbf{c}^\pm_{1,2}$ and by using the properties of fermionic coherent states. This leads to the following worldline representation for the one-loop effective action in QED, with a spin-1/2 field in the loop (see Exercise 15.5):

$$\Gamma[\mathcal{A}] = \mathrm{const} - \frac{1}{2} \int_0^\infty \frac{dT}{T} \int_{\substack{z(0)=z(T) \\ \psi(0)=-\psi(T)}} [Dz(\tau)D\psi(\tau)]$$

$$\times e^{-\int_0^T d\tau\, (\frac{1}{4}\dot{z}^2 + ie\dot{z}\cdot\mathcal{A}(z) + m^2 + \frac{1}{2}\psi_\mu\dot{\psi}^\mu - ie\psi^\mu F_{\mu\nu}(z)\psi^\nu)}, \tag{15.53}$$

where $\psi_\mu$ is a collection of four Grassmann variables that combine the $\xi_{1,2}, \overline{\xi}_{1,2}$ at each intermediate time. In this formula, the ordering that was necessary to handle the non-commutative nature of the Dirac matrices has now been replaced by a path integral over fermionic internal degrees of freedom.

## 15.3 Schwinger Mechanism

Since it provides expressions for propagators and effective actions in a background field, the worldline formalism is well suited to studying phenomena that occur in the presence of such an external field, such as the splitting of a photon into two photons in an external magnetic field ($\gamma \to 2\gamma$ is forbidden in the vacuum, but becomes possible if an external electromagnetic field provides a fourth photon), or the bremsstrahlung radiation by a charged particle in a magnetic field.

Another interesting process that can be addressed by the worldline formalism is the *Schwinger mechanism*, which amounts to the spontaneous production of $e^+e^-$ pairs by a

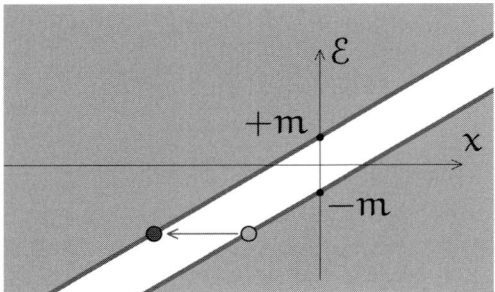

FIGURE 15.2: Schematic picture of the tunneling process involved in the Schwinger mechanism. The empty band is the gap between the anti-electron Dirac sea and the positive energy electron continuum, tilted by the potential energy $eV(x) = -eEx$ in the presence of an external electrical field E.

static and homogeneous electrical field. That this is possible may be understood intuitively as follows. Without any external field, QED has an empty band of states with energy larger than m corresponding to free electrons and anti-electrons, and a filled band of states with energy lower than $-m$ that corresponds to "trapped" particles (the so-called *Dirac sea*). A minimal energy of 2m (the minimal energy of an $e^+e^-$ pair) must be provided to move one of these particles from the Dirac sea to the positive band of free particles. Consider now an electrical field E in the x direction. If this field is static and homogeneous, we may find a gauge in which it is represented by a vector potential whose only non-zero component is $A_4 = -Ex$, which tilts the boundaries of the band of free states and of the Dirac sea, as shown in Figure 15.2. Now, a pair from the Dirac sea can move to the band of free particles by a tunneling process that does not require any energy. Standard results of quantum mechanics indicate that the tunneling probability scales like $\exp(-\text{const} \times m^2/(eE))$. Note that this expression is non-analytic in the coupling constant e, and therefore impossible to obtain at any finite order of the standard perturbative expansion.

Although the Schwinger mechanism was computed a long time ago by resummed perturbation theory, the worldline formalism provides a straightforward way to calculate it and offers very interesting new insights about the space-time development of the particle production process. Let us consider the case of scalar QED in order to illustrate this in a simpler setting. The probability of pair production may be inferred from the vacuum-to-vacuum transition amplitude, which can be written as an exponential,

$$\langle 0_{\text{out}} | 0_{\text{in}} \rangle = e^{i\mathcal{V}}, \tag{15.54}$$

$i\mathcal{V}$ being the sum of all the connected vacuum diagrams. The possibility of particle production is intimately related to the imaginary part of $\mathcal{V}$, since the total probability of producing particles reads

$$P_{\text{prod}} = 1 - |\langle 0_{\text{out}} | 0_{\text{in}} \rangle|^2 = 1 - e^{-2\,\text{Im}\,\mathcal{V}}. \tag{15.55}$$

In scalar QED, the graphs made of one scalar loop embedded in a background electromagnetic field lead to the following contribution to $\mathcal{V}$:

$$\mathcal{V}_{1\,\text{loop}} = \ln \det \left( g_{\mu\nu} D^\mu D^\nu + m^2 \right), \tag{15.56}$$

## 15.3 SCHWINGER MECHANISM

where $D^\mu$ is the covariant derivative in the background field. The metric should be Euclidean in order to apply the worldline formalism, i.e., $g_{\mu\nu} = -\delta_{\mu\nu}$. At one loop, the sum of the connected vacuum graphs in the presence of a background electromagnetic field is given by

$$\mathcal{V}_{1\,\text{loop}} = \int_0^\infty \frac{dT}{T} e^{-m^2 T} \int_{z(0)=z(T)} [Dz(\tau)] \, e^{-\int_0^T d\tau \, (\frac{1}{4}\dot{z}^2(\tau) + ie\dot{z}(\tau)\cdot A(z(\tau)))}. \tag{15.57}$$

This formula involves a double integration: a path integral over all the worldlines $z(\tau)$, i.e., closed paths in Euclidean space-time parametrized by the fictitious time $\tau \in [0,T]$, and an ordinary integral over the length T of these paths. The sum over all the worldlines can be viewed as a materialization of the quantum fluctuations in space-time, and the prefactor $\exp(-m^2 T)$ suppresses the very long worldlines that explore regions of space-time that are much larger than the Compton wavelength of the particles.

In eq. (15.57), the path integral can be factored into an integral over the barycenter Z of the worldline and the position $\zeta(\tau)$ about this barycenter,

$$z(\tau) \equiv Z + \zeta(\tau), \qquad \int_0^T d\tau \, \zeta(\tau) = 0. \tag{15.58}$$

After this separation, all the information about the background field contained in eq. (15.57) comes via a Wilson line,

$$W_Z[\zeta] \equiv \exp\left(-ie \int_0^T d\tau \, \dot{\zeta}(\tau) \cdot A(Z + \zeta(\tau))\right), \tag{15.59}$$

averaged over all closed loops of length T,

$$\langle W_Z \rangle_T \equiv \frac{\int_{\zeta(0)=\zeta(T)} [D\zeta(\tau)] \, W_Z[\zeta] \, \exp\left(-\int_0^T d\tau \, \frac{\dot{\zeta}^2}{4}\right)}{\int_{\zeta(0)=\zeta(T)} [D\zeta(\tau)] \, \exp\left(-\int_0^T d\tau \, \frac{\dot{\zeta}^2}{4}\right)}. \tag{15.60}$$

This path average is determined by an ensemble of loops localized around the barycenter Z, and $\langle W_Z \rangle_T$ encapsulates the local properties of the quantum field theory in the vicinity of Z (roughly up to a distance of order $T^{1/2}$). In terms of this averaged Wilson loop, the one-loop Euclidean connected vacuum amplitude reads

$$\mathcal{V}_{1\,\text{loop}} = \frac{1}{(4\pi)^2} \int d^4 Z \int_0^\infty \frac{dT}{T^3} e^{-m^2 T} \langle W_Z \rangle_T. \tag{15.61}$$

(In this formula, the prefactor and power of T in the measure assume four space-time dimensions.) The imaginary part of $\mathcal{V}_{1\,\text{loop}}$ comes from the existence of poles in $\langle W_Z \rangle_T$ at real values of the fictitious time T. In terms of these poles, the imaginary part reads

$$\text{Im}\,(\mathcal{V}_{1\,\text{loop}}) = \frac{\pi}{(4\pi)^2} \int d^4 Z \, \text{Re} \sum_{\text{poles } T_n} \frac{e^{-m^2 T_n}}{T_n^3} \, \text{Res}\left(\langle W_Z \rangle_{T_n}, T_n\right). \tag{15.62}$$

Let us now be more specific and consider a static and uniform electrical field E. Since one can choose a gauge potential which is linear in the coordinates Z, ζ, the path integral that gives the average Wilson loop is Gaussian and can therefore be performed in closed form, leading to

$$\langle W_z \rangle_T = \frac{eET}{\sin(eET)}. \tag{15.63}$$

(Note that it does not depend on the barycenter Z since the field is constant.) This quantity has an infinite series of simple poles along the positive real axis, located at $T_n = n\pi/(eE)$ ($n = 1, 2, 3, \cdots$), that give the following expression for the imaginary part:

$$\text{Im}\,(\mathcal{V}_{1\text{ loop}}) = \frac{V_4}{16\pi^3}(eE)^2 \sum_{n=1}^{\infty} \frac{(-1)^{n-1}}{n^2} e^{-n\pi m^2/(eE)}. \tag{15.64}$$

In this formula, $V_4$ is the volume in space-time over which the integration over the barycenter Z is carried out. After exponentiation, this formula gives the vacuum survival probability $P_0 = \exp(-2\,\text{Im}\,\mathcal{V})$. A more detailed study would reveal that the term of index $n$ comes from Bose–Einstein correlations among $n$ produced pairs, while the first pole $\tau_1$ only contains information about the uncorrelated part of the spectrum. Given the origin of these terms in the present derivation, as coming from poles $T_n$ that are more distant from $T = 0$, we see that increasingly intricate (the index $n$ is the number of correlated particles) quantum correlations come from worldlines that explore larger and larger portions of space-time. This supports the intuitive image that quantum fluctuations and correlations are encoded in the fact that the worldlines explore an extended region around the base point Z (see also Exercise 15.6).

## 15.4 Calculation of One-Loop Amplitudes

The worldline formalism can also be used in order to derive expressions for one-loop amplitudes. The main difference, compared to the calculation of the Schwinger mechanism, is that in the case of amplitudes the momenta carried by the lines attached to the loop are fixed instead of integrated over. The expected result is therefore a function of N momenta (or coordinates), rather than just a number.

### 15.4.1 $\phi^3$ Scalar Field Theory

As a first simple illustration of the method, let us consider a scalar field theory with a cubic coupling $V(\phi) = \frac{\lambda}{3!}\phi^3$, for which the worldline representation of the one-loop effective action is

$$\Gamma[\varphi] = \text{const} + \frac{1}{2}\int_0^\infty \frac{dT}{T} \int_{z(0)=z(T)} [Dz(\tau)]\, e^{-\int_0^T d\tau\,(\frac{1}{4}\dot{z}^2 + m^2 + \lambda\varphi(z))}. \tag{15.65}$$

Again, we first split $z(\tau)$ into the barycenter and a deviation about it, $z(\tau) \equiv Z + \zeta(\tau)$. In the case of amplitudes, the integration over Z will simply produce the delta function of

## 15.4 CALCULATION OF ONE-LOOP AMPLITUDES

overall energy–momentum conservation. Using the T-periodicity of the paths over which we integrate, the term in $\dot{\zeta}^2$ inside the exponential can be integrated by parts,

$$-\frac{1}{4}\int_0^T d\tau\, \dot{\zeta}^2 = \frac{1}{4}\int_0^T d\tau\, \zeta\ddot{\zeta}, \tag{15.66}$$

and the integral on $\zeta(\tau)$ involves the inverse $G(\tau,\tau')$ of the operator $\frac{1}{2}\partial_\tau^2$. This inverse exists thanks to the fact that we have subtracted the barycenter from $z(\tau)$, which amounts to removing the zero mode from $\zeta(\tau)$. Indeed, a T-periodic function can be written as

$$\zeta(\tau) = \sum_{n\in\mathbb{Z}} \zeta_n\, e^{2i\pi n \frac{\tau}{T}}, \tag{15.67}$$

and excluding the zero mode is done by setting $\zeta_0 = 0$. A very useful identity is

$$\sum_{n\in\mathbb{Z}} e^{2i\pi n \frac{\tau}{T}} = T \sum_{n\in\mathbb{Z}} \delta(\tau - nT). \tag{15.68}$$

Using this formula, we can check that the propagator $G(\tau,\tau')$ is given by

$$G(\tau,\tau') = 2T \sum_{n\in\mathbb{Z}_*} \frac{1}{(2i\pi n)^2}\, e^{2i\pi n \frac{(\tau-\tau')}{T}}. \tag{15.69}$$

Note that this function is even in $\tau - \tau'$ and T-periodic. Integrating[2] eq. (15.68) twice from 0 to $\tau - \tau'$, we obtain (see Exercise 15.7)

$$G(\tau,\tau') = |\tau - \tau'| - \frac{(\tau-\tau')^2}{T} - \frac{T}{6}. \tag{15.70}$$

From the quantum effective action, one-particle irreducible amplitudes are obtained by differentiating with respect to the field, as many times as there are external legs, and by setting $\varphi \equiv 0$ afterwards. Thus, the N-point function is given by

$$\Gamma_N(x_1,\cdots,x_N) = \frac{(-\lambda)^N}{2} \int_0^\infty \frac{dT}{T}\, e^{-m^2 T} \int_{z(0)=z(T)} [Dz(\tau)]$$

$$\times \prod_{i=1}^N \int_0^T d\tau_i\, \delta(z(\tau_i) - x_i)\, e^{-\int_0^T d\tau\, \frac{1}{4}\dot{z}^2}. \tag{15.71}$$

In this formula, the path integral is over all closed paths that pass at all the coordinates $x_1,\cdots,x_N$, in any order, which provides a rather intuitive picture of the worldline representation of the amplitude. Let us now Fourier transform this expression in order to obtain the

---

[2] We adopt a symmetric convention for handling the delta function $\delta(\tau)$, which amounts to $\int_0^\infty d\tau'\, \delta(\tau') = \frac{1}{2}$.

amplitude in momentum space. The Fourier integrals over the $x_i$ are trivial thanks to the N delta functions, and we obtain

$$\Gamma_N(p_1,\cdots,p_N) = \frac{(-\lambda)^N}{2} \int_0^\infty \frac{dT}{T} e^{-m^2 T} \int d^d Z \int_{\zeta(0)=\zeta(T)} [D\zeta(\tau)]$$
$$\times \prod_{i=1}^N \int_0^T d\tau_i \; e^{ip_i\cdot(Z+\zeta(\tau_i))} \; e^{\int_0^T d\tau \frac{1}{4}\dot\zeta\dot\zeta}, \tag{15.72}$$

where we have also separated the barycenter coordinate Z from the deviation $\zeta$ and integrated by parts the term in $\dot\zeta^2$. The integral over Z produces a delta function of the sum of the momenta, and the path integral over $\zeta$ is Gaussian, leading to

$$\Gamma_N(p_1,\cdots,p_N) = \frac{(-\lambda)^N}{2^{1+d/2}} (2\pi)^{d/2} \delta\left(\sum_i p_i\right) \int_0^\infty \frac{dT}{T^{1+d/2}} e^{-m^2 T}$$
$$\times \int_0^T \prod_{i=1}^N d\tau_i \; \exp\left(\tfrac{1}{2}\sum_{i,j} G(\tau_i,\tau_j)(p_i\cdot p_j)\right). \tag{15.73}$$

This is the worldline expression of a one-loop N-point scalar amplitude (see Exercise 15.8 for a comparison with one-loop results obtained by standard methods). One may make a number of remarks about this formula:

- In contrast with the formulas obtained from Feynman diagrams, there is no loop momentum. It is replaced by the variable T that measures the length of the worldloops. As we said earlier, there is a loose connection between T and a momentum, since small values of T correspond to the ultraviolet and large values of T to the infrared.
- The dependence on the external momenta is directly expressed in terms of the Lorentz invariants $p_i\cdot p_j$. In a formula that contains a loop momentum k, one would also have all the $k\cdot p_i$.
- The integral on T may be divergent at small T, because of the factor $1/T^{1+d/2}$. However, the integrals of the second line roughly behave as $T^N$ (since there are N integrals over an interval of size T, with an integrand of order 1). Thus, the overall behavior of the T integral is $dT\,T^{N-1-d/2}$. This integral is convergent at small T if $N-1-d/2 > -1$, i.e., $N > d/2$. In four space-time dimensions, this is $N > 2$, in agreement with conventional power counting that indicates that all one-loop functions with $N \geq 3$ are finite in the $\phi^3$ scalar theory.
- Each ordering of the fictitious times $\tau_i$ corresponds to a given cyclic ordering of the momenta $p_i$ around the loop. The term corresponding to one such ordering in the formula (15.73) can be mapped to the expression of the corresponding Feynman diagram. The $\tau_i$ correspond to the N Feynman parameters introduced to combine the N denominators into a single one,[3] and T is related to the squared momentum that appears in this unique denominator.

---

[3] Note that there are only $N-1$ independent Feynman parameters, since their sum is constrained to be 1, but because of the periodicity and translation invariance of the propagators $G(\tau_i,\tau_j)$, it is possible to choose one of the $\tau_i$ to be equal to zero, hence only $N-1$ of them are truly independent.

## 15.4 CALCULATION OF ONE-LOOP AMPLITUDES

- The constant term $-T/6$ in the propagator of eq. (15.70) does not contribute in eq. (15.73). Indeed, its contribution inside the exponential is

$$-\frac{T}{12}\sum_{i,j} p_i \cdot p_j = -\frac{T}{12}\left(\sum_i p_i\right)\cdot\left(\sum_j p_j\right) = 0, \qquad (15.74)$$

which is zero thanks to momentum conservation.

### 15.4.2 Scalar Quantum Electrodynamics

As a slightly more complicated example of application, let us now derive the expression of the one-loop N-photon amplitude in scalar QED. The starting point is the one-loop quantum effective action in an Abelian background gauge field:

$$\Gamma[\mathcal{A}] = \text{const} + \int_0^\infty \frac{dT}{T} e^{-m^2 T} \int_{z(0)=z(T)} [Dz(\tau)] \; e^{-\int_0^T d\tau\,(\frac{1}{4}\dot{z}^2 + ie\dot{z}\cdot\mathcal{A}(z))}. \qquad (15.75)$$

Differentiating N times with respect to $\mathcal{A}^\mu(x)$ and setting the background field to zero afterwards, we obtain

$$\Gamma_N^{\mu_1\cdots\mu_N}(x_1,\cdots,x_N) = (-ie)^N \int_0^\infty \frac{dT}{T} e^{-m^2 T} \int_{z(0)=z(T)} [Dz(\tau)] \; e^{-\int_0^T d\tau\,\frac{1}{4}\dot{z}^2}$$

$$\times \int_0^T \prod_{i=1}^N d\tau_i \; \delta(z(\tau_i) - x_i) \; \dot{z}^{\mu_i}(\tau_i). \qquad (15.76)$$

Next, we Fourier transform this expression and contract a polarization vector to each external Lorentz index, and we isolate the integral over the barycenter Z,

$$\Gamma_N(p_1\epsilon_1,\cdots,p_N\epsilon_N) = (-ie)^N (2\pi)^d \delta\Big(\sum_i p_i\Big) \int_0^\infty \frac{dT}{T} e^{-m^2 T} \int_{\zeta(0)=\zeta(T)} [D\zeta(\tau)]$$

$$\times e^{\int_0^T d\tau\,\frac{1}{4}\dot{\zeta}\dot{\zeta}} \int_0^T \prod_{i=1}^N d\tau_i \; e^{ip_i\cdot\zeta(\tau_i)} \; (\dot{\zeta}(\tau_i)\cdot\epsilon_i). \qquad (15.77)$$

The path integral is still Gaussian, but the factors $\dot{\zeta}(\tau_i)\cdot\epsilon_i$ complicate it significantly compared to the $\phi^3$ theory. In particular, the answer will now contain derivatives of the propagator:

$$\dot{G}(\tau,\tau') \equiv \partial_\tau G(\tau,\tau') = \text{sign}(\tau-\tau') - \frac{2(\tau-\tau')}{T},$$

$$\ddot{G}(\tau,\tau') \equiv \partial_\tau\partial_{\tau'} G(\tau,\tau') = 2\,\delta(\tau-\tau') - \frac{2}{T}. \qquad (15.78)$$

(The first derivative of the propagator with respect to the second time is the opposite of $\dot{G}$ defined above since the propagator G is even.) Note again that the term $-T/6$ in G does not contribute, thanks to momentum conservation. A convenient trick to perform this integral is to write

$$\prod_i \dot{\zeta}(\tau_i) \cdot \epsilon_i = \exp\left(\sum_i \dot{\zeta}(\tau_i) \cdot \epsilon_i\right)_{\text{multi-linear}}, \qquad (15.79)$$

where the subscript "multi-linear" means that we keep only the term in $\epsilon_1 \epsilon_2 \cdots \epsilon_N$ in the Taylor expansion of the exponential. This leads to

$$\Gamma_N(p_1\epsilon_1, \cdots, p_N\epsilon_N) = \frac{(-ie)^N}{2^{d/2}} (2\pi)^{d/2} \delta\Big(\sum_i p_i\Big) \int_0^\infty \frac{dT}{T^{1+d/2}} e^{-m^2 T}$$

$$\times \int_0^T \prod_{i=1}^N d\tau_i \, \exp\bigg\{\sum_{i,j}\bigg[\frac{1}{2} G(\tau_i, \tau_j)(p_i \cdot p_j)$$

$$+ i\, \dot{G}(\tau_i, \tau_j)(p_i \cdot \epsilon_j) + \frac{1}{2}\ddot{G}(\tau_i, \tau_j)(\epsilon_i \cdot \epsilon_j)\bigg]\bigg\}_{\text{multi-linear}}. \qquad (15.80)$$

The expansion of the exponential and extraction of the term that contains each polarization vector exactly once leads to an expression of the form

$$\exp\{\cdots\}_{\text{multi-linear}} = P_N(\dot{G}, \ddot{G}) \, \exp\bigg(\frac{1}{2}\sum_{i,j} G(\tau_i, \tau_j)(p_i \cdot p_j)\bigg), \qquad (15.81)$$

where $P_N$ is a polynomial in the derivatives of the propagator, with coefficients made of the Lorentz invariants $p_i \cdot \epsilon_j$ and $\epsilon_i \cdot \epsilon_j$. By integration by parts, it is possible to replace the second derivatives $\ddot{G}$ by first derivatives $\dot{G}$. In this operation, the polynomial $P_N$ is replaced by another polynomial $Q_N$ that depends only on the $\dot{G}$, hence

$$\exp\{\cdots\}_{\text{multi-linear}} = Q_N(\dot{G}) \, \exp\bigg(\frac{1}{2}\sum_{i,j} G(\tau_i, \tau_j)(p_i \cdot p_j)\bigg). \qquad (15.82)$$

The polynomial $Q_N(\dot{G})$ corresponds to the combination of numerators that would appear in the expression of this amplitude obtained from the usual Feynman rules.

### 15.4.3 Spinor QED

In QED with spin-1/2 matter fields, the one-loop effective action in a photon background is given by

$$\Gamma[\mathcal{A}] = \text{const} - \frac{1}{2}\int_0^\infty \frac{dT}{T} \int\limits_{\substack{z(0)=z(T)\\ \psi(0)=-\psi(T)}} \big[Dz(\tau)D\psi(\tau)\big]$$

$$\times e^{-\int_0^T d\tau \, (\frac{1}{4}\dot{z}^2 + ie\dot{z}\cdot\mathcal{A}(z) + m^2 + \frac{1}{2}\psi_\mu\dot{\psi}^\mu - ie\psi^\mu F_{\mu\nu}(z)\psi^\nu)}. \qquad (15.83)$$

## 15.4 CALCULATION OF ONE-LOOP AMPLITUDES

Now, we have a second path integral that involves the anti-periodic Grassmann variables $\psi^\mu$. This additional integral is also Gaussian, and its result can be expressed in terms of the inverse of the operator $\tfrac{1}{2}\partial_\tau$ over the space of anti-periodic functions,[4] whose expression reads (see Exercise 15.7)

$$S(\tau,\tau') = 2 \sum_{n\in\mathbb{Z}} \frac{1}{2i\pi(n+\tfrac{1}{2})} e^{2i\pi(n+\tfrac{1}{2})\frac{\tau-\tau'}{T}} = \text{sign}(\tau-\tau'). \tag{15.84}$$

One can then in principle follow the same sequence of steps as in the scalar QED case, to obtain an expression of the one-loop N-photon amplitude in spinor QED in terms of the propagators $G$, $\dot{G}$, $\ddot{G}$ and $S$. In fact, it was shown by Bern and Kosower that this expression can be obtained from the corresponding scalar QED amplitude by a simple substitution. Starting from the final scalar QED expression in terms of the polynomial $Q_N(\dot{G})$ (see eq. (15.82)), one should arrange each term of this polynomial as a product of cycles of the form

$$[1,2,3,\cdots c]_G \equiv \dot{G}(\tau_1,\tau_2)\dot{G}(\tau_2,\tau_3)\cdots \dot{G}(\tau_{c-1},\tau_c). \tag{15.85}$$

Then, the Bern–Kosower rule states that in order to obtain the analogous spinor QED amplitude, one should perform the following substitution on each such cycle:

$$[1,2,3,\cdots c]_G \quad \to \quad -2\big([1,2,3,\cdots c]_G - [1,2,3,\cdots c]_S\big), \tag{15.86}$$

where $[\cdots]_S$ is the same cyclic product made of the propagator $S$ instead of $\dot{G}$.

### 15.4.4 Example: QED Polarization Tensor

**Scalar QED:** As an illustration of the calculation of amplitudes in the worldline formalism, let us study the one-loop photon polarization tensor in QED, starting first from the simpler case of scalar QED. In d dimensions, the polarization tensor is related to the one-particle irreducible two-point function $\Gamma_2^{\mu_1\mu_2}(p_1,p_2)$ by

$$\Gamma_2^{\mu\nu}(p,q) \equiv (2\pi)^4 \delta(p+q)\, \Pi^{\mu\nu}(p). \tag{15.87}$$

From eq. (15.77), we obtain the following expression in scalar QED

$$\Pi^{\mu\nu}_{\text{scalar}}(p) = -e^2 \int_0^\infty \frac{dT}{T} e^{-m^2 T} \int_{\zeta(0)=\zeta(T)} [D\zeta(\tau)]\, e^{\int_0^T d\tau\, \tfrac{1}{4}\dot\zeta\dot\zeta}$$
$$\times \int_0^T d\tau_1 d\tau_2\, e^{ip\cdot\zeta(\tau_1)} e^{-ip\cdot\zeta(\tau_2)}\, \dot\zeta^\mu(\tau_1)\dot\zeta^\nu(\tau_2). \tag{15.88}$$

---

[4] Anti-periodic functions defined over the interval $[0,T]$ can be written as

$$\psi(\tau) \equiv \sum_{n\in\mathbb{Z}} \psi_n\, e^{2i\pi(n+\tfrac{1}{2})\frac{\tau}{T}}.$$

When restricted to these functions, the derivative operator $\partial_\tau$ has no zero mode and is thus invertible.

The path integration over $\zeta$ leads to

$$\int_{\zeta(0)=\zeta(T)} [D\zeta(\tau)] \, e^{\int_0^T d\tau \, \frac{1}{4}\zeta\ddot\zeta} \, e^{ip\cdot\zeta(\tau_1)} e^{-ip\cdot\zeta(\tau_2)} \, \dot\zeta^\mu(\tau_1)\dot\zeta^\nu(\tau_2)$$

$$= \frac{1}{(4\pi T)^{d/2}} \left[ g^{\mu\nu} \ddot G(\tau_1,\tau_2) - p^\mu p^\nu \dot G^2(\tau_1,\tau_2) \right] e^{-p^2 G(\tau_1,\tau_2)}$$

$$= \frac{1}{(4\pi T)^{d/2}} \left( g^{\mu\nu} p^2 - p^\mu p^\nu \right) \dot G^2(\tau_1,\tau_2) \, e^{-p^2 G(\tau_1,\tau_2)}, \tag{15.89}$$

where in the third line we have anticipated an integration by parts on $\tau_1$ for the term in $\ddot G$. We can already see that the polarization tensor is transverse. At this point, it is convenient to use rescaled variables $\tau_i \equiv T\vartheta_i$. Moreover, thanks to the translation invariance of the integrand in $\tau_{1,2}$ and to its T-periodicity, we are free to set $\vartheta_2 \equiv 0$. Having done that, the propagator and its derivative become simple functions of $\vartheta_1$:

$$G(T\vartheta_1, 0) = T\vartheta_1(1-\vartheta_1), \quad \dot G(T\vartheta_1, 0) = 1 - 2\vartheta_1. \tag{15.90}$$

(We have already dropped the constant term in $-T/6$ from the propagator, since it does not contribute to amplitudes thanks to momentum conservation.) At this point, the polarization tensor reads

$$\Pi^{\mu\nu}_{\text{scalar}}(p) = -\frac{e^2}{(4\pi)^{d/2}} \left( g^{\mu\nu} p^2 - p^\mu p^\nu \right) \int_0^1 d\vartheta_1 \, (1-2\vartheta_1)^2$$

$$\times \int_0^\infty \frac{dT}{T} \, T^{2-d/2} \, e^{-T(m^2 + p^2 \vartheta_1(1-\vartheta_1))}$$

$$= -\frac{e^2}{(4\pi)^{d/2}} \left( g^{\mu\nu} p^2 - p^\mu p^\nu \right) \int_0^1 d\vartheta_1 \, (1-2\vartheta_1)^2$$

$$\times \Gamma(2 - \tfrac{d}{2}) \left[ m^2 + p^2 \vartheta_1(1-\vartheta_1) \right]^{d/2-2}. \tag{15.91}$$

One may check that this expression is identical to the one we would have obtained from the two Feynman diagrams of Figure 15.3, after introducing Feynman parameters and performing the integration over the loop momentum.

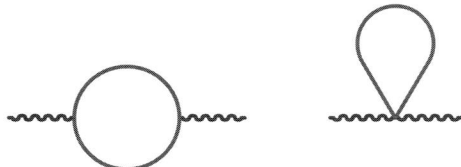

FIGURE 15.3: Feynman graphs contributing to the one-loop photon polarization tensor in scalar QED.

## 15.4 CALCULATION OF ONE-LOOP AMPLITUDES

**Spinor QED:** Let us now consider the same quantity in QED with a spin-1/2 fermion. Eq. (15.88) is replaced by

$$\Pi^{\mu\nu}_{\text{spin } 1/2}(p) = \frac{e^2}{2} \int_0^\infty \frac{dT}{T} e^{-m^2 T} \int_{\substack{\zeta(0)=\zeta(T) \\ \psi(0)=-\psi(T)}} [D\zeta(\tau)D\psi(\tau)] \, e^{\int_0^T d\tau \, (\frac{1}{4}\zeta\ddot{\zeta} - \frac{1}{2}\psi\cdot\dot\psi)}$$

$$\times \int_0^T d\tau_1 d\tau_2 \, e^{ip\cdot\zeta(\tau_1)} e^{-ip\cdot\zeta(\tau_2)}$$
$$\times \left(\dot\zeta^\mu(\tau_1) + 2i\,\psi^\mu(\tau_1)\,(\psi(\tau_1)\cdot p)\right)$$
$$\times \left(\dot\zeta^\nu(\tau_2) - 2i\,\psi^\nu(\tau_2)\,(\psi(\tau_2)\cdot p)\right). \tag{15.92}$$

The path integral for the term in $\dot\zeta^\mu\dot\zeta^\nu$ is the same as in eq. (15.89), but we should now multiply the result by[5]

$$\int_{\psi(0)=-\psi(T)} [D\psi(\tau)] \, e^{-\int_0^T d\tau \, \frac{1}{2}\psi\cdot\dot\psi} = 4. \tag{15.93}$$

For the terms involving the Grassmann variables, we have

$$4\int_{\psi(0)=-\psi(T)} [D\psi(\tau)] \, e^{-\int_0^T d\tau \, \frac{1}{2}\psi\cdot\dot\psi} \, \psi^\mu(\tau_1)\,(\psi(\tau_1)\cdot p)\psi^\nu(\tau_2)\,(\psi(\tau_2)\cdot p)$$
$$= -S^2(\tau_1, \tau_2)\left(g^{\mu\nu}p^2 - p^\mu p^\nu\right), \tag{15.94}$$

and this result should be multiplied by $(4\pi T)^{-d/2}$ to account for the integration over the variable $\zeta$. Thus, we see here an example of the Bern–Kosower substitution rule: the spin-1/2 loop can be obtained from the scalar loop by replacing $\dot{G}^2(\tau_1, \tau_2)$ by $\dot{G}^2(\tau_1, \tau_2) - S^2(\tau_1, \tau_2)$ and by multiplying by an overall factor $-2$ (this comes from a $-1/2$ due to the different prefactors in the scalar and spin-1/2 one-loop effective actions, times the factor 4 from eq. (15.93)). In terms of the variables $\vartheta_{1,2}$ and after setting $\vartheta_2 = 0$, we have simply $S(T\vartheta_1, 0) = 1$, and therefore the factor $(1 - 2\vartheta_1)^2$ from $\dot{G}^2$ becomes

$$(1 - 2\vartheta_1)^2 - 1 = -4\vartheta_1(1 - \vartheta_1). \tag{15.95}$$

Therefore, the worldline expression of the one-loop photon polarization tensor in spinor QED reads

$$\Pi^{\mu\nu}_{\text{spin } 1/2}(p) = \frac{8\,e^2}{(4\pi)^{d/2}}\left(g^{\mu\nu}p^2 - p^\mu p^\nu\right)\int_0^1 d\vartheta_1 \, \vartheta_1(1 - \vartheta_1)$$
$$\times \Gamma(2 - \tfrac{d}{2})\left[m^2 + p^2\vartheta_1(1 - \vartheta_1)\right]^{d/2-2}, \tag{15.96}$$

which agrees with the expression obtained from Feynman graphs (only the first topology in Figure 15.3, with the scalar loop replaced by a spinor loop, contributes in this case). Remark-

---

[5]This formula may be obtained by zeta function regularization. If we denote $\mathbf{A} \equiv -\partial_\tau$ (restricted to the subspace of anti-periodic functions) and $\lambda_n = -2i\pi(n + \tfrac{1}{2})$ its eigenvalues, the $\zeta$ function of this operator is $\zeta_A(s) \equiv \sum_{n\in\mathbb{Z}} \lambda_n^{-s}$. Since there are four variables $\psi_\mu$, the value of the path integral is $[\det \mathbf{A}]^2 = \exp(-2\zeta_A'(0))$. On the other hand, we have $\zeta_A(s) = (i\pi)^{-s}(1 + e^{i\pi s})(1 - 2^{-s})\zeta(s)$, where $\zeta(s)$ is Riemann's zeta function. This function can be expanded at small s, giving: $\zeta_A(s) = -\ln(2)\,s + \mathcal{O}(s^2)$.

ably, all the Dirac algebra usually involved in the calculation of fermion loops is completely avoided in the worldline formalism, since it is encapsulated into the Grassmann functional integration over $\psi^\mu$.

## Exercises

**15.1** Establish eq. (15.10). *Hint: Apply a Fourier transform to both sides of the equation.*

**\*15.2** Fill in the details of the derivation of eq. (15.26).

**15.3** Justify eq. (15.34).

**15.4** Check eqs. (15.41), (15.47), (15.48) and (15.51). *Hint: Calculate the overlap $\langle mn|\xi\rangle$ between the coherent state and the Fock states. Then, check that*

$$\int d\bar{\xi}d\xi\, e^{-\bar{\xi}\xi}\langle -\xi|n_1 n_2\rangle\langle m_2 m_1|\xi\rangle = \delta_{m_1 n_1}\delta_{m_2 n_2}.$$

**15.5** Derive the details of eq. (15.53).

**\*15.6** Start from the representation of eq. (15.57) for the one-loop effective action in a background electromagnetic field.

- Introduce the new variables $s \equiv m^2 T$, $u \equiv \tau/T$. Show that after this transformation, the integration over $s$ can be done in closed form, yielding a Bessel function.
- Assuming that $m^2 \int_0^1 du\, \left(\frac{dz}{du}\right)^2 \gg 1$, derive an approximate form of the worldline path integral.
- Evaluate the latter integral by a stationary phase approximation. Show that the stationary solutions correspond to $\left(\frac{dz}{du}\right)^2 = \text{const}$.
- Application: Consider the following time-independent and $z^3$-dependent electrical field in the $z^3$ direction:

$$E_3(z^3) \equiv \frac{E}{\cosh^2(kz^3)}, \quad A^4 = -i\frac{E}{k}\tanh(kz^3).$$

(We choose a specific gauge to obtain this gauge potential.) Derive the equations satisfied by the stationary solutions, and show that they can be parametrized as

$$z^3(u) = \frac{m}{eE}\frac{1}{\gamma}\text{arcsinh}\left(\frac{\gamma}{\sqrt{1-\gamma^2}}\sin(2\pi n\, u)\right)$$

$$z^4(u) = \frac{m}{eE}\frac{1}{\gamma\sqrt{1-\gamma^2}}\arcsin\left(\gamma\cos(2\pi n\, u)\right),$$

where $\gamma \equiv mk/(eE)$ and $n$ is an integer. Discuss the limits $\gamma \to 0$ and $\gamma \to \infty$, and conclude that spatial inhomogeneities tend to reduce particle production.

**\*15.7** Check the formulas (15.70) and (15.84).

**\*15.8** For $N = 2$, show explicitly that eq. (15.73) is equivalent to what one would obtain from Feynman diagrams.

**15.9** Set $T = 1$, and denote $J_n(\tau, \tau')$ the inverse of $\partial_\tau^n$ over the set of T-periodic functions of null average.

- Show that
$$J_n(\tau, \tau') = -\frac{1}{n!} \left(\text{sign}(\tau - \tau')\right)^n B_n(|\tau - \tau'|),$$

where $B_n(z)$ is the nth Bernouilli polynomial (use $B_n' = nB_{n-1}$).
- Show that the trace of this inverse is given by $\text{tr}\,(\partial_\tau^{-n}) = -B_n/n!$, where the $B_n$ are the Bernouilli numbers ($B_n \equiv B_n(z=0)$).
- Denoting $K_n$ the inverse of $\partial_\tau^n$ in the anti-periodic case, show that
$$K_n(\tau, \tau') = \frac{1}{2(n-1)!} \left(\text{sign}(\tau - \tau')\right)^n E_{n-1}(|\tau - \tau'|),$$

where $E_n(z)$ is the nth Euler polynomial (use $E_n' = nE_{n-1}$).

**\*15.10** (This exercise is a continuation of Exercise 15.9) Consider the QED worldline formalism, and add to the background field $\mathcal{A}^\mu$ another field $a^\mu$ that corresponds to a constant field strength $f^{\mu\nu} = \partial^\mu a^\nu - \partial^\nu a^\mu$.

- Show that, in the Fock–Schwinger gauge $x_\mu \mathcal{A}^\mu = 0$, it is possible to choose this additional gauge potential as $a^\mu = \frac{1}{2} x_\nu f^{\nu\mu}$.
- Explain why, when inserted into eq. (15.27) or (15.53), this additional field can be incorporated exactly in the worldline propagators. Show that these dressed propagators obey
$$\left[\partial_\tau^2 \delta_{\mu\nu} - 2ie\, f_{\mu\nu}\, \partial_\tau\right] \mathcal{G}_{\nu\rho}(\tau, \tau') = 2\,\delta_{\mu\rho}\delta(\tau-\tau'),$$
$$\left[\partial_\tau \delta_{\mu\nu} - 2ie\, f_{\mu\nu}\right] \mathcal{S}_{\nu\rho}(\tau, \tau') = 2\,\delta_{\mu\rho}\delta(\tau-\tau'),$$

respectively for the bosonic and fermionic cases.
- Check that these equations are solved by
$$\mathcal{G}(\tau, \tau') = \frac{T}{2\,Z^2}\left(\frac{Z}{\sin(Z)} e^{-iZ\dot{G}(\tau,\tau')} + iZ\,\dot{G}(\tau, \tau') - 1\right),$$
$$\mathcal{S}(\tau, \tau') = \frac{e^{-iZ\dot{G}(\tau,\tau')}}{\cos(Z)} S(\tau, \tau'),$$

where $G, S$ are the vacuum worldline propagators and $Z^{\mu\nu} \equiv eTf^{\mu\nu}$ (the above expressions should thus be understood as functions of matrices in what concerns the Lorentz structure of the propagators). *Hint:* Use the results of the previous exercise, and use
$$\frac{t\,e^{tz}}{e^t - 1} = \sum_{n=0}^\infty B_n(z) \frac{t^n}{n!}, \quad \frac{2\,e^{zt}}{e^t + 1} = \sum_{n=0}^\infty E_n(z) \frac{t^n}{n!}.$$

- Show that the path integral given in eq. (15.13) must be replaced by
$$\frac{1}{(4\pi T)^{d/2}} \left(\det \frac{\sin Z}{Z}\right)^{-1/2} \quad \text{(scalar QED)},$$
$$\frac{1}{(4\pi T)^{d/2}} \left(\det \frac{\tan Z}{Z}\right)^{-1/2} \quad \text{(spinor QED)}.$$

# CHAPTER 16

# Lattice Field Theory

We have seen in Chapter 9 that the running coupling in an SU(N) non-Abelian gauge theory decreases at large energy (provided the number of quark flavors is less than $11N/2$). The counterpart of asymptotic freedom is that the coupling increases toward lower energies, precluding the use of perturbation theory to study phenomena in this regime. Among such properties is that of color confinement, i.e., the fact that colored states cannot exist as asymptotic states. Instead, the quarks and gluons arrange themselves into color-neutral bound states, which can be mesons (e.g., pions, kaons) made of a quark and an anti-quark or baryons (e.g., protons, neutrons) made of three quarks.[1] A legitimate question would be to determine the mass spectrum of the asymptotic states of QCD from its Lagrangian.

Since the perturbative expansion is not applicable to this type of problem, one would like to be able to attack it via some *non-perturbative* approach. By non-perturbative, we mean a method by which observables would directly be obtained to all orders in the coupling constant, without any expansion. One such method, known as *lattice field theory*, consists in discretizing space-time in order to evaluate numerically the path integral. The continuous space-time is replaced by a discrete grid of points, the simplest arrangement being a hyper-cubic lattice such as the one shown in Figure 16.1. The distance between nearest-neighbor sites is called the lattice spacing, and is usually denoted $a$. The lattice spacing, being the smallest distance that exists in this setup, therefore provides a natural ultraviolet regularization. Indeed, on a lattice of spacing $a$, the largest conjugate momentum is of order $a^{-1}$.

---

[1] More exotic bound states made of four valence quarks (tetraquarks) or five valence quarks (pentaquarks) have also been speculated, but the experimental evidence for these states is so far not fully conclusive. Likewise, there may exist bound states without valence quarks, the *glueballs*.

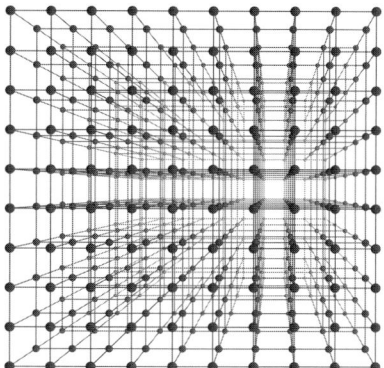

FIGURE 16.1: Discretization of Euclidean space-time on a hyper-cubic lattice (here shown in three dimensions).

Moreover, one usually uses periodic boundary conditions; if the lattice has $N$ spacings in all directions, then we have $\phi(x + N\widehat{\mu}) = \phi(x)$ for bosonic fields and $\phi(x + N\widehat{\mu}) = -\phi(x)$ for fermionic fields ($\widehat{\mu}$ is the displacement vector by one lattice spacing in the direction $\mu$ of space-time).

## 16.1 Discretization of Bosonic Actions

### 16.1.1 Scalar Field Theory

As an illustration of some of the issues involved in the discretization of a quantum field theory, let us consider a simple scalar field theory with a local interaction in $\phi^4$, whose action in continuous space-time is

$$S = \int d^4x \left\{ -\frac{1}{2}\phi(x)(\partial_\mu \partial^\mu + m^2)\phi(x) - \frac{\lambda}{4!}\phi^4(x) \right\}. \tag{16.1}$$

A natural choice is to replace the integral over space-time by a discrete sum over the sites of the lattice, weighted by the volume $a^4$ of the elementary cells of the lattice:

$$a^4 \sum_{x \in \text{lattice}} \underset{a \to 0}{\longrightarrow} \int d^4x. \tag{16.2}$$

Then we replace the continuous function $\phi(x)$ by a discrete set of real numbers that live on the lattice nodes. For simplicity, we keep denoting $\phi(x)$ the value of the field on the lattice site x. The discretization of the mass and interaction terms is trivial, but the discretization of the derivatives that appear in the D'Alembertian operator is not unique. Using only two nearest neighbors, one may define forward or backward finite differences,

$$\nabla_F^\mu f(x) \equiv \frac{f(x + \widehat{\mu}) - f(x)}{a}, \quad \nabla_B^\mu f(x) \equiv \frac{f(x) - f(x - \widehat{\mu})}{a}, \tag{16.3}$$

that both go to the continuum derivative $\partial^\mu f$ in the limit $a \to 0$. However, unlike the continuous derivative, $\nabla_F^\mu$ and $\nabla_B^\mu$ are not anti-adjoint. Instead, assuming periodic boundary conditions, we have

$$\sum_{x \in \text{lattice}} f(x) \left( \nabla_F^\mu g(x) \right) = - \sum_{x \in \text{lattice}} \left( \nabla_B^\mu f(x) \right) g(x). \tag{16.4}$$

In other words, $\nabla_F^{\mu\dagger} = -\nabla_B^\mu$. From this, we may construct a self-adjoint discrete second derivative as follows:

$$\nabla_B^\mu \nabla_F^\mu f(x) = \frac{f(x + \hat\mu) + f(x - \hat\mu) - 2 f(x)}{a^2} \xrightarrow[a \to 0]{} \partial^\mu \partial^\mu f(x). \tag{16.5}$$

(There is no summation on $\mu$ on the left-hand side.) Thus, a self-adjoint discretization of the scalar Lagrangian leads to

$$S_{\text{lattice}} = a^4 \sum_{x \in \text{lattice}} \left\{ -\frac{1}{2} \phi(x) (g_{\mu\nu} \nabla_B^\mu \nabla_F^\nu + m^2) \phi(x) - \frac{\lambda}{4!} \phi^4(x) \right\}. \tag{16.6}$$

Let us make a few remarks concerning the errors introduced by the discretization. First, the continuous space-time symmetries (translation and rotation invariance) of the underlying theory are now reduced to the subgroup of the discrete symmetries of a cubic lattice. They are recovered in the limit $a \to 0$. Another source of discrepancy between the continuum and discrete theories is the dispersion relation that relates the energy and momentum of an on-shell particle. In the continuum theory, this relation is of course

$$E^2 = \mathbf{p}^2 + m^2, \tag{16.7}$$

where $-\mathbf{p}^2$ is an eigenvalue of the Laplacian. In order to find its counterpart with the above discretization, we must determine the spectrum of the finite difference operator $\nabla_B^\mu \nabla_F^\mu$. On a lattice with $N$ sites and periodic boundary conditions, its eigenfunctions are given by

$$\phi_{\mathbf{k}}(x) \equiv \exp\left( 2i\pi \frac{k_1 x + k_2 y + k_3 z}{Na} \right) \quad \text{with } k_i \in \mathbb{Z}, \ -\tfrac{N}{2} \leq k_i \leq \tfrac{N}{2}. \tag{16.8}$$

The associated eigenvalue is

$$\lambda_{\mathbf{k}} \equiv \frac{2}{a^2} \sum_{i=1,2,3} \left( \cos \frac{2\pi k_i}{N} - 1 \right) = -\frac{4}{a^2} \sum_{i=1,2,3} \sin^2 \frac{\pi k_i}{N}. \tag{16.9}$$

Thus, the discrete analogue of the continuum $\mathbf{p}^2 + m^2$ is

$$m^2 + \frac{4}{a^2} \sum_{i=1,2,3} \sin^2 \frac{\pi k_i}{N}. \tag{16.10}$$

As long as $k_i \ll N$, this agrees quite well with the continuum dispersion relation (with the correspondence $p_i = 2\pi k_i/Na$), but the agreement is not good for larger values of $k_i$. This discrepancy is illustrated in Figure 16.2. This mismatch does not improve by increasing the number of lattice points: Only the center of the Brillouin zone has a dispersion relation that agrees with the continuum one. In order to mitigate this problem, one should choose the parameters of the lattice in such a way that the physically relevant scales correspond to values of $k_i$ for which the distortion of the dispersion curve is small (see the Exercise 16.1 for an improved discrete D'Alembertian with a reduced error).

## 16.1 DISCRETIZATION OF BOSONIC ACTIONS

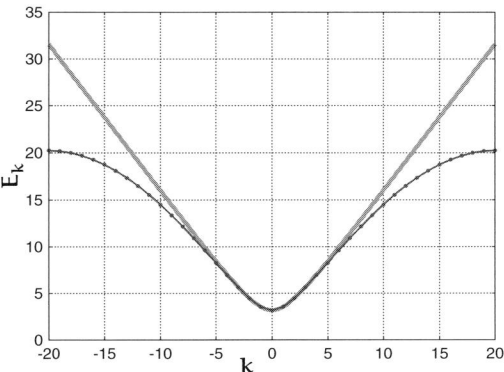

FIGURE 16.2: Discrepancy between the continuous (solid curve) and discrete (points) dispersion relations, on a one-dimensional lattice with $N = 40$.

### 16.1.2 Gluons and Wilson Action

Non-Abelian gauge theories pose an additional difficulty: Since the local gauge invariance plays a central role in their properties, any attempt at discretizing gauge fields should preserve this symmetry. It turns out that there exists a discretization of the Yang–Mills action that goes to the continuum action in the limit where $a \to 0$, and has an exact gauge invariance. The main ingredient in this construction is eq. (7.164), which relates the Wilson loop along a small square,

$$[\Box]_{x;\mu\nu} \equiv U_\nu^\dagger(x)\, U_\mu^\dagger(x+\hat{\nu})\, U_\nu(x+\hat{\mu})\, U_\mu(x), \tag{16.11}$$

to the squared field strength. These elementary lattice Wilson loops are called *plaquettes*. In the fundamental representation of $\mathfrak{su}(N)$, we have

$$\operatorname{tr}\left([\Box]_{x;\mu\nu}\right) = N - \frac{g^2 a^4}{4} F_a^{\mu\nu}(x) F_{\mu\nu}^a(x) + \mathcal{O}(a^6). \tag{16.12}$$

Note that, although the first two terms on the right-hand side are real-valued, the remainder (terms of order $a^6$ and beyond) may be complex. Therefore, it is convenient to take the real part of the trace of the Wilson loop in order to construct a real-valued discrete action. By summing this equation over all the lattice points $x$ and all the pairs of distinct directions $(\mu, \nu)$, we obtain

$$a^4 \sum_{x \in \text{lattice}} \left(-\frac{1}{4} F_a^{\mu\nu}(x) F_{\mu\nu}^a(x)\right)$$
$$= \underbrace{\frac{1}{g^2} \sum_{x \in \text{lattice}} \sum_{(\mu,\nu)} \left(\operatorname{tr}\left(\operatorname{Re}[\Box]_{x;\mu\nu}\right) - N\right)}_{\text{Wilson action, denoted } \frac{1}{g^2} S_W[U]} + \mathcal{O}(a^2). \tag{16.13}$$

Note that the error term of order $a^6$ becomes a term of order $a^2$ after summation over the lattice sites, since the number of sites grows like $a^{-4}$ if the volume is held fixed. Thus, the sum

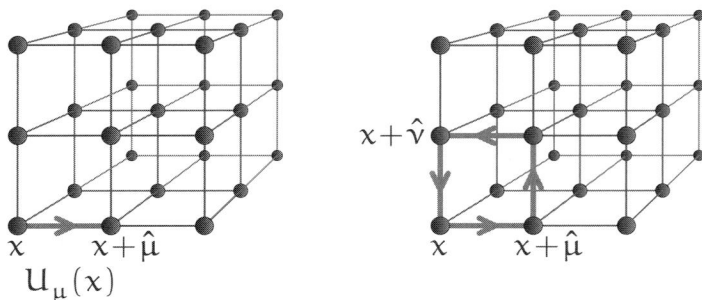

FIGURE 16.3: Left: link variable. Right: plaquette on an elementary square of the lattice.

of the traces of the Wilson loops over all the elementary plaquettes of the lattice provides a discretization of the Yang–Mills action. In this discrete formulation, the natural variables are not the gauge potentials $A^\mu(x)$ themselves, but the Wilson lines $U_\mu(x)$ that live on the edges of the lattice, called *link variables*. In this notation, $x$ is the starting point and $\mu$ the direction of the Wilson line, as illustrated in the left panel of Figure 16.3. The Wilson line oriented in the $-\hat\mu$ direction, i.e., starting at the point $x + \hat\mu$ and ending at the point $x$, is simply the Hermitian conjugate of $U_\mu(x)$. Under a local gauge transformation, the link variables are changed as follows:

$$U_\mu(x) \quad \to \quad \Omega^\dagger(x+\hat\mu)\, U_\mu(x)\, \Omega(x). \tag{16.14}$$

The plaquette variable, shown in the right panel of Figure 16.3, can then be obtained by multiplying four link variables, as indicated by eq. (16.11), and its trace is obviously invariant under the transformation of eq. (16.14).

At this stage, the discrete analogue of the path integral that gives the expectation value of a gauge invariant operator reads:

$$\langle \mathcal{O} \rangle = \int \prod_{x,\mu} dU_\mu(x)\; \mathcal{O}[U]\; \exp\Big\{\frac{i}{g^2} \sum_x \sum_{(\mu,\nu)} \Big(\mathrm{tr}\,\big(\mathrm{Re}\,[\square]_{x;\mu\nu}\big) - N\Big)\Big\}. \tag{16.15}$$

Since there exists a left and right invariant[2] group measure $dU_\mu(x)$, the left-hand side of this formula is gauge invariant. Moreover, it goes to the expectation value of the continuum theory in the limit of zero lattice spacing.

### 16.1.3 Monte-Carlo Sampling

Thanks to the discretization, the path integral of the original theory is replaced by an ordinary integral over each of the link variables $U_\mu(x)$, whose number is finite. A non-perturbative

---

[2]This means that
$$\int dU\; f[U] = \int dU\; f[\Omega U] = \int dU\; f[U\Omega].$$

Such a measure, known as the Haar measure, exists for compact Lie groups, like $SU(N)$.

answer could be obtained if one were able to evaluate these integrals numerically. However, because of the prefactor i inside the exponential in eq. (16.15), the integrand is a strongly oscillating function, whose numerical integration is practically impossible except on lattices with a very small number of sites. In order to be amenable to a numerical calculation, this integral must be transformed into a Euclidean one:

$$\langle \mathcal{O} \rangle_E = \int \prod_{x,\mu} dU_\mu(x) \, \mathcal{O}[U] \, \exp\left\{ \frac{N}{g^2} \sum_x \sum_{(\mu,\nu)} \left( N^{-1} \mathrm{tr} \left( \mathrm{Re} \, [\Box]_{x;\mu\nu} \right) - 1 \right) \right\}. \quad (16.16)$$

The exponential under the integral is now real-valued, and thus positive definite. Note that numerical quadratures (such as Simpson's rule) are not practical for this problem, given the huge number of dimensions of the integral to be evaluated. For instance, for the eight-dimensional Lie group SU(3), in four space-time dimensions, on a lattice with $N^4$ points, this dimension is $8 \times 4 \times N^4$. For $N = 32$, the path integral is thus transformed into a $2^{25}$-dimensional ($2^{25} \sim 3.10^7$) ordinary integral. Instead, one views the exponential of the Wilson action as a probability distribution (up to a normalization constant) for the link variables, which may be sampled by a Monte-Carlo algorithm (e.g., the *Metropolis–Hastings algorithm*) in order to estimate the integral.

In this approach, as long as one is evaluating the expectation value of gauge invariant observables, it is not necessary to fix the gauge in lattice QCD calculations. Gauge fixing is necessary when calculating non-gauge invariant quantities, such as propagators. The Landau gauge is the most commonly used, because the Landau gauge condition is realized at the extrema of a functional of the link variables. However, the comparison between gauge fixed lattice calculations and analytical calculations is very delicate because of the existence of Gribov copies (the problem stems from the fact that the two setups may not select the same Gribov copy – see Section 16.5).

Although considering the Euclidean path integral instead of the Minkowski one allows for a numerical evaluation by Monte-Carlo sampling, this leads to a serious limitation: Only quantities that can be expressed as a Euclidean expectation value are directly calculable. Others could in principle be reached by an analytic continuation from imaginary to real time, but this turns out to be practically impossible numerically. For instance, the masses of hadrons are accessible to lattice QCD calculations (see Section 16.3 for an example), while scattering amplitudes cannot be calculated by this method.

## 16.2 Lattice Fermions

### 16.2.1 Discretization of the Dirac Action

Consider now the Dirac action, whose expression in continuum space reads

$$S_D = \int d^4x \, \overline{\psi}(x) \left( i\gamma^\mu D_\mu - m \right) \psi(x). \quad (16.17)$$

In the discretization, we assign a spinor $\psi(x)$ to each site of the lattice. Under a gauge transformation $\Omega(x)$, these spinors transform in the same way as in the continuous theory:

$$\psi(x) \rightarrow \Omega^\dagger(x) \psi(x), \qquad \overline{\psi}(x) \rightarrow \overline{\psi}(x) \Omega(x). \quad (16.18)$$

The main difficulty in defining a discrete covariant derivative that transforms appropriately under a gauge transformation is that $\psi(x)$ and $\psi(x \pm \hat{\mu})$ transform differently when $\Omega(x)$ depends on space-time. This problem can be remedied by using a link variable between point x and its neighbors. Like with the ordinary derivatives, one may define forward and backward discrete covariant derivatives,

$$D_F^\mu \psi(x) \equiv \frac{U_\mu^\dagger(x)\psi(x+\hat{\mu}) - \psi(x)}{a}, \quad D_B^\mu \psi(x) \equiv \frac{\psi(x) - U_\mu(x-\hat{\mu})\psi(x-\hat{\mu})}{a}, \quad (16.19)$$

that both transform like a spinor at point x, and therefore are valid discretizations of a covariant derivative. However, neither of these two operators is anti-adjoint, and therefore they would not give a Hermitian Lagrangian density. This may be achieved by using instead $\frac{1}{2}(D_F^\mu + D_B^\mu)$, which corresponds to a symmetric forward–backward difference

$$\frac{1}{2}(D_F^\mu + D_B^\mu)\psi(x) = \frac{U_\mu^\dagger(x)\psi(x+\hat{\mu}) - U_\mu(x-\hat{\mu})\psi(x-\hat{\mu})}{2a}. \quad (16.20)$$

(This also has the benefit of reducing the discretization errors by one power of $a$.)

### 16.2.2 Fermion Doublers

Let us now study how the dispersion relation of fermions is modified by this discretization. This can easily be done in the vacuum, i.e., by setting all the link variables to the identity. In this case, the eigenfunctions of the operator $\frac{1}{2}(D_F^\mu + D_B^\mu)$ are

$$\psi_k(x) = e^{2i\pi \frac{(k_\mu + 1/2)x^\mu}{Na}}, \quad \text{with } k_\mu \in \mathbb{Z}, \ -\frac{N}{2} \leq k_\mu \leq \frac{N}{2}. \quad (16.21)$$

The action of the discrete Dirac operator on this plane wave reads

$$(i\slashed{D} - m)\psi_k = -(\lambda_k + m)\psi_k, \quad (16.22)$$

with

$$\lambda_k = \frac{1}{a}\sum_\mu \gamma^\mu \sin\left(\frac{2\pi(k_\mu + 1/2)}{N}\right). \quad (16.23)$$

In order to find the dispersion curve, we use $(\slashed{b} + c)^{-1} = (\slashed{b} - c)/(b^2 - c^2)$ to invert $\lambda_k + m$. This gives a dispersion relation of the form $E^2 = \hat{p}^2 + m^2$ with

$$\hat{p}^2 \equiv \frac{1}{a^2} \sum_{i=1,2,3} \sin^2\left(\frac{2\pi(k_i + 1/2)}{N}\right). \quad (16.24)$$

This dispersion relation is shown in Figure 16.4. As in the bosonic case, the discrete dispersion relation agrees with the continuous one only for small enough $k_i$. However, the discrepancy at large $k_i$ is now much more serious, because the discrete dispersion curve has another minimum at the edge of the Brillouin zone. This additional minimum indicates the existence of a second

## 16.2 LATTICE FERMIONS

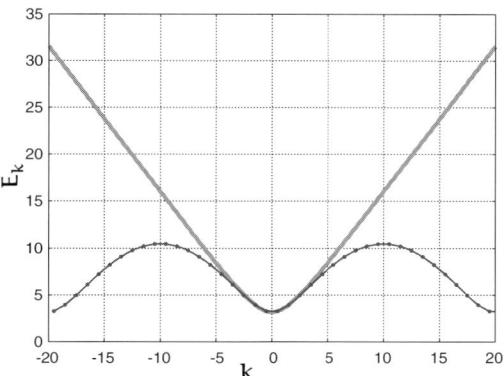

FIGURE 16.4: Discrepancy between the continuous (solid curve) and discrete (points) dispersion curves for fermions, on a one-dimensional lattice with $N = 40$.

propagating mode of mass m. This spurious mode is called a *fermion doubler*. In d dimensions, the number of these fermionic modes is $2^d$, while our goal was to have only one. This problem is quite serious, because it affects all quantities that depend on the number of quark flavors. In particular, this is the case of the running of the coupling constant, whenever quark loops are included.

### 16.2.3 Wilson Term

Various modifications of the discretized Dirac action have been proposed to remedy the problem of fermion doublers. One of these modifications, known as the *Wilson term*, consists in adding to the Lagrangian the following term:

$$-\frac{1}{2a} \overline{\psi}(x) \sum_{\mu} \left[ U_\mu^\dagger(x)\, \psi(x + \widehat{\mu}) + U_\mu(x - \widehat{\mu})\, \psi(x - \widehat{\mu}) - 2\psi(x) \right], \tag{16.25}$$

which is nothing but a D'Alembertian (or the opposite of a Laplacian in the Euclidean theory where the metric has only minus signs) constructed with covariant derivatives. The corresponding operator in the continuum theory is

$$\frac{a}{2} g_{\mu\nu} \overline{\psi} \left( D^\mu D^\nu \right) \psi. \tag{16.26}$$

Note that the denominator in eq. (16.25) has a single power of the lattice spacing a, hence the prefactor a in the previous equation. Therefore, this term goes to zero in the limit $a \to 0$, and it should have no effect in the continuum limit. In the absence of gauge field ($U_\mu(x) \equiv 1$), the functions of eq. (16.21) are still eigenfunctions after adding the Wilson term, but with modified eigenvalues,

$$\lambda_k + m \to \lambda_k + \mu_k + m, \quad \text{with } \mu_k = \frac{1}{a} \sum_{\mu} \left( \cos \frac{2\pi (k_\mu + 1/2)}{N} - 1 \right). \tag{16.27}$$

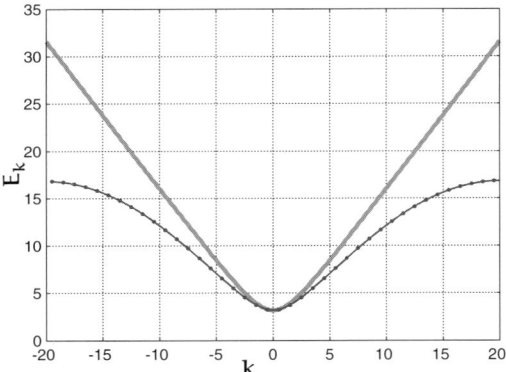

FIGURE 16.5: Discrepancy between the continuous (solid curve) and discrete (points) dispersion curves in the fermionic case, on a one-dimensional lattice with $N = 40$, after inclusion of the Wilson term.

Thus, the Wilson term does not modify the spectrum at small k, but lifts the spurious minimum that existed at the edge of the Brillouin zone, as shown in Figure 16.5. Roughly speaking, the Wilson term gives a mass of order $a^{-1}$ to the fermion doublers, making them decouple from the rest of the degrees of freedom when $a \to 0$. See Exercise 16.3 for a discussion of the errors introduced by the Wilson term, and of a way to mitigate them.

However, the Wilson term has an important drawback: There is no Dirac matrix $\gamma^\mu$ in eqs. (16.25) and (16.26) since the Lorentz indices are contracted directly between the two covariant derivatives. Therefore, the Wilson term – like an ordinary mass term – breaks explicitly the chiral symmetry of the Dirac Lagrangian in the case of massless fermions. The fermion doublers are in fact intimately related to chiral symmetry. Without the Wilson term, lattice QCD with massless quarks has an exact chiral symmetry unbroken by the lattice regularization, and therefore there cannot be a chiral anomaly. In fact, this absence of anomaly is precisely due to a cancellation of anomalies among the multiple copies (the doublers) of the fermion modes. This argument is completely general and not specific to the Wilson term: Any mechanism that lifts the degeneracy among the doublers will spoil the anomaly cancellation and thus break chiral symmetry (this result is known as the *Nielsen–Ninomiya theorem*).

### 16.2.4 Evaluation of the Fermion Path Integral

The path integral representation for fermions uses anti-commuting Grassmann variables. However, such variables are not representable as ordinary numbers in a numerical implementation. To circumvent this difficulty, one exploits the fact that the Dirac action is quadratic in the fermion fields (this remains true after adding the Wilson term to remove the fermion doublers). Therefore, the path integral over the fermion fields can be done exactly. In addition to the fermion fields contained in the action, there may be $\psi$s and $\overline{\psi}$s (in equal numbers) in the operator whose expectation value is being evaluated. The result of such a fermionic path integral is given by

$$\int \left[D\psi D\overline{\psi}\right] \left(\psi(x_1)\overline{\psi}(x_2)\right) e^{iS_D[\psi,\overline{\psi}]} = S(x_1, x_2) \times \det\left(i\gamma^\mu D_\mu - m\right), \tag{16.28}$$

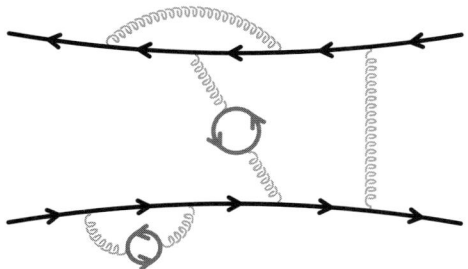

FIGURE 16.6: Illustration of the two types of quark contributions. Dark: quark propagators (i.e., inverse of the Dirac operator) that connect the ψs and $\bar{\psi}$s in the operator being evaluated. Light: quark loops coming from the determinant of the Dirac operator.

where $S(x_1, x_2)$ is the inverse of the Dirac operator $i\slashed{D} - m$ between the points $x_1$ and $x_2$. When there is more than one $\psi\bar{\psi}$ pair in the operator, one must sum over all the ways of connecting the $\psi$s and the $\bar{\psi}$s by the fermion propagators $S(x, y)$. The same can be done in the lattice formulation. In this case, the Dirac operator is simply a (very large) matrix that depends on the configuration of link variables. Therefore one needs the inverse of this matrix, and its determinant (see Exercise 16.2 for a discussion of its properties). The determinant is not computed by deterministic formulas, but may be obtained from the inverse of the Dirac operator by Monte-Carlo sampling, by writing

$$\left[\det\left(i\gamma^\mu D_\mu - m\right)\right]^2 = \int [D\phi D\phi^*]\, e^{-\int \phi^\dagger (i\gamma^\mu D_\mu - m)^{-2} \phi}, \tag{16.29}$$

where $\phi$ is an auxiliary bosonic field.

In eq. (16.28), the Dirac determinant provides closed quark loops, while the propagator $S(x_1, x_2)$ connects the external points of the operator under consideration. This observation, illustrated in Figure 16.6, clarifies the meaning of the *quenched approximation*, in which the determinant of the Dirac operator is replaced by 1. This approximation, motivated primarily by the computational difficulty of evaluating the Dirac determinant, was widely used in lattice QCD computations until advances in algorithms and computer hardware made it unnecessary. Note that, although quark loops are not included in the quenched approximation, gluon loops are present to their full extent. In contrast, lattice QCD calculations that include the Dirac determinant, and thus the effect of quark loops, are said to use *dynamical fermions*.

## 16.3 Hadron Mass Determination on the Lattice

Let us consider a hadronic state $|h\rangle$. Any operator $\mathcal{O}$ that carries the same quantum numbers as this hadron leads to a non-zero matrix element $\langle h|\mathcal{O}|0\rangle$. The vacuum expectation value of the product of two $\mathcal{O}$ s at different times 0 and T can be rewritten as follows:

$$\langle 0|\mathcal{O}^\dagger(0)\mathcal{O}(T)|0\rangle = \sum_n \langle 0|\mathcal{O}^\dagger(0)|\Psi_n\rangle\langle\Psi_n|\mathcal{O}(T)|0\rangle$$

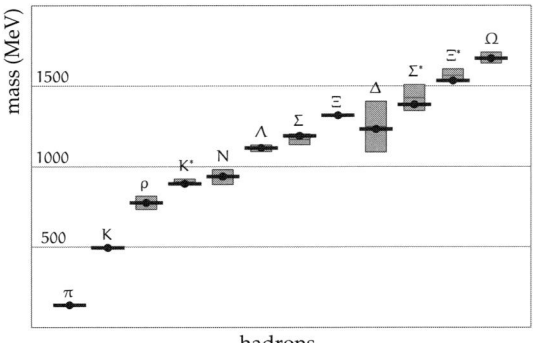

FIGURE 16.7: Hadron mass determination from lattice QCD. Gray boxes: predictions of lattice QCD and associated uncertainty (note: the experimental masses of the $\pi$, K and $\Xi$ were used for calibration and thus have no error). Black bars: experimental values. Data from Durr et al., Science 322 (2008) 1224.

$$\begin{aligned} &= \sum_n \langle 0|\mathcal{O}^\dagger(0)|\Psi_n\rangle\langle\Psi_n|\mathcal{O}(0)|0\rangle\, e^{-M_n T} \\ &= \sum_n \left|\langle\Psi_n|\mathcal{O}(0)|0\rangle\right|^2 e^{-M_n T}. \end{aligned} \qquad (16.30)$$

In the first equality, we have inserted a complete basis of eigenstates of the QCD Hamiltonian, and the second equality follows from the fact that $\langle\Psi_n|$ is an eigenstate of rest energy $M_n$ (there is no factor $i$ inside the exponential because of the Euclidean time used in lattice QCD). The sum in the last equality receives non-zero contributions from all the states $\Psi_n$ that possess the quantum numbers carried by the operator $\mathcal{O}$. However, taking the limit $T \to \infty$ selects the one among these eigenstates that has the smallest mass. This observation can be turned into a method to determine hadron masses in lattice QCD:

1. Choose an operator $\mathcal{O}$ that has the quantum numbers of the hadron h of interest. The choice of operator is not crucial, as long as the overlap $\langle h|\mathcal{O}|0\rangle$ is not zero. However, eq. (16.30) suggests that a better result, i.e., less noisy with limited statistics, may be obtained by trying to maximize this overlap.

2. Evaluate the vacuum expectation value of $\mathcal{O}^\dagger(0)\mathcal{O}(T)$ by Monte-Carlo sampling, as a function of T.

3. Fit the large T tail of this expectation value. The slope of the exponential gives the mass of the lightest hadron that possesses these quantum numbers.

The discretized QCD Lagrangian contains several dimensionful parameters: the lattice spacing $a$ and the quark masses $m_f$ (one for each quark flavor), whose values need to be fixed before novel predictions can be made. One must choose (at least) an equal number of physical quantities that are known experimentally. Their computed values depend on $a$, $m_f$, and one should adjust these parameters so that they match the experimental values. After this has been done, quantities computed in lattice QCD do not contain any free parameter anymore and are thus predictions. Figure 16.7 shows hadron masses calculated using lattice QCD.

## 16.4 Wilson Loops and Color Confinement

### 16.4.1 Strong Coupling Expansion

While perturbation theory is an expansion in powers of $g^2$, it is possible to use the lattice formulation of a Yang–Mills theory in order to perform an expansion in powers of the quantity $\beta \equiv g^{-2}$ that appears as a prefactor in the Wilson action. This is called a *strong coupling expansion*, since it becomes exact in the limit of infinite coupling.

This expansion produces integrals over the gauge group, such as

$$\int dU \, U_{i_1 j_1} \cdots U_{i_n j_n} \, U^\dagger_{k_1 l_1} \cdots U^\dagger_{k_m l_m}. \tag{16.31}$$

The simplest of these integrals are

$$\int dU = 1, \quad \int dU \, U_{ij} = 0. \tag{16.32}$$

(The first one is simply a choice of normalization[3] of the invariant group measure.) Because of the second of eqs. (16.32), a given link variable $U$ cannot appear isolated in the integrand. To obtain a non-zero integral, $U$ may be combined with the corresponding $U^\dagger$, since we have (see Exercise 16.5)

$$\int dU \, U_{ij} U^\dagger_{kl} = \frac{1}{N} \delta_{jk} \delta_{il}. \tag{16.33}$$

(The combination $U_{ij} U_{kl}$ also gives a non-zero integral in the special case of $SU(2)$, while it gives zero for $SU(N)$ with $N \geq 3$.) Note that the link variables on different edges of the lattice are independent variables, and there is a separate integral for each of them. Therefore, the integral of the Wilson loop defined on an elementary plaquette is zero,

$$\int \prod_{x,\mu} dU_\mu(x) \, \mathrm{tr} \Big( \underbrace{U^\dagger_\nu(x) \, U^\dagger_\mu(x+\hat{\nu}) \, U_\nu(x+\hat{\mu}) \, U_\mu(x)}_{[\Box]_{x;\mu\nu}} \Big) = 0. \tag{16.34}$$

In contrast, the integral of the trace of a plaquette times the trace of the conjugate plaquette is non-zero,

$$\int \prod_{x,\mu} dU_\mu(x) \, \left( \mathrm{tr} \, [\Box]_{x;\mu\nu} \right) \left( \mathrm{tr} \, [\Box]^\dagger_{x;\mu\nu} \right) = 1. \tag{16.35}$$

---

[3] For $SU(2)$, one may parameterize the group elements in the fundamental representation by $U = \theta_0 + 2i \, \theta_a \, t^a_f$ with $\theta_0^2 + \theta_1^2 + \theta_2^2 + \theta_3^2 = 1$, and the invariant group measure normalized according to eq. (16.32) is $dU = d\theta_1 d\theta_2 d\theta_3 / (\pi^2 \sqrt{1-\theta^2})$ (with $\theta_0 = \sqrt{1-\theta^2}$). By using this measure and the Fierz identity satisfied by the generators $t^a_f$, an explicit calculation leads easily to eq. (16.33).

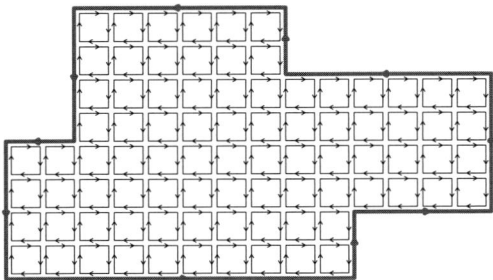

FIGURE 16.8: Tiling of a closed loop by elementary plaquettes.

Using these results, we can calculate to order $\beta$ the expectation value of the trace of a plaquette,

$$\langle \mathrm{tr}\,[\Box]_{x;\mu\nu}\rangle \equiv \frac{\int \prod_{x,\mu} dU_\mu(x) \left(\mathrm{tr}\,[\Box]_{x;\mu\nu}\right) \exp\left\{\beta N \sum_{y;\rho\sigma} \left(N^{-1}\mathrm{tr}\,\mathrm{Re}\,[\Box]_{y;\rho\sigma} - 1\right)\right\}}{\int \prod_{x,\mu} dU_\mu(x) \exp\left\{\beta N \sum_{y;\rho\sigma} \left(N^{-1}\mathrm{tr}\,\mathrm{Re}\,[\Box]_{y;\rho\sigma} - 1\right)\right\}}$$

$$= \frac{\beta}{2} + \mathcal{O}(\beta^2). \tag{16.36}$$

Consider now the trace of a more general Wilson loop along a path $\gamma$ (planar, to simplify the discussion). Each $U$ and $U^\dagger$ in the Wilson loop must be compensated by a link variable coming from the $\beta$ expansion of the exponential of the Wilson action. The lowest-order term in $\beta$ corresponds to a minimal tiling of the Wilson loop by elementary plaquettes, as illustrated in Figure 16.8. The corresponding contribution is

$$\langle \mathrm{tr}\, W_\gamma\rangle = \left(\frac{\beta}{2}\right)^{\mathrm{Area}\,(\gamma)} + \cdots, \tag{16.37}$$

where the dots are terms of higher order in $\beta$ (that can be constructed from non-minimal tilings of the contour $\gamma$, such that all the $U$s and $U^\dagger$s are still paired appropriately).

## 16.4.2 Heavy Quark Potential

Let us consider now a rectangular loop, with an extent R in the spatial direction 1 and an extent T in the Euclidean time direction 4. The previous result indicates that the expectation value of the trace of the corresponding Wilson loop has the following form:

$$\langle \mathrm{tr}\, W_\gamma\rangle \sim e^{-\sigma RT} + \cdots, \tag{16.38}$$

where $\sigma$ is a constant. Although it is gauge invariant, this expectation value is easier to interpret in the axial gauge $A_4 \equiv 0$. Indeed, in this gauge, the Wilson loop receives only contributions from gauge links along the spatial direction, as shown in Figure 16.9. Note that the

## 16.4 WILSON LOOPS AND COLOR CONFINEMENT

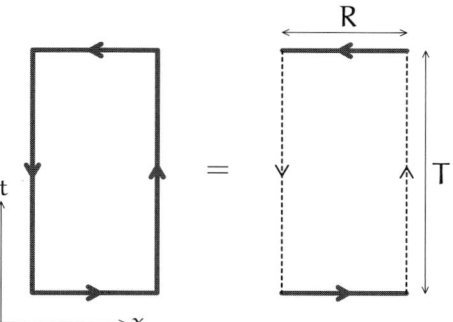

FIGURE 16.9: Rectangular Wilson loop in the $A_4 \equiv 0$ gauge.

remaining Wilson lines are precisely those that are needed to build a non-local gauge invariant operator with a quark at $x = R$ and an antiquark at $x = 0$,

$$\mathcal{O}_{q\bar{q}}(t) \equiv \overline{\psi}(t,0)\, W_{[0,R]}\, \psi(t,R)\,, \tag{16.39}$$

where $W_{[0,R]}$ is a (spatial) Wilson line going from $(t, R)$ to $(t, 0)$. Consider now the vacuum expectation value $\langle 0|\mathcal{O}^\dagger_{q\bar{q}}(0)\mathcal{O}_{q\bar{q}}(T)|0\rangle$. In this expectation value, the fermionic path integral produces two quark propagators that connect the $\psi$s to the $\overline{\psi}$s. However, in the limit of infinite quark mass, the quarks are static and their propagator is just a Wilson line in the temporal direction, which reduces to the identity in the $A_4 = 0$ gauge (represented by the dotted lines in Figure 16.9). Thus, we have

$$\langle \operatorname{tr} W_\gamma \rangle \propto \lim_{M \to \infty} \langle 0|\mathcal{O}^\dagger_{q\bar{q}}(0)\mathcal{O}_{q\bar{q}}(T)|0\rangle. \tag{16.40}$$

By inserting a complete basis of eigenstates of the Hamiltonian in the right-hand side of eq. (16.40) and by taking the limit $T \to \infty$, we find a result dominated by the quark–antiquark state of lowest energy $E_0$,

$$\lim_{T \to \infty} \langle 0|\mathcal{O}^\dagger_{q\bar{q}}(0)\mathcal{O}_{q\bar{q}}(T)|0\rangle = \left|\langle 0|\mathcal{O}^\dagger_{q\bar{q}}(0)|\Psi_0\rangle\right|^2 e^{-E_0 T}. \tag{16.41}$$

Moreover, in the limit of large mass, the energy $E_0$ of this state is dominated by the potential energy $V(R)$ between the quark and the antiquark (the quark and antiquark are non-relativistic, and their kinetic energy behaves as $\mathbf{P}^2/2M \to 0$),

$$\lim_{M,T \to \infty} \langle 0|\mathcal{O}^\dagger_{q\bar{q}}(0)\mathcal{O}_{q\bar{q}}(T)|0\rangle = \left|\langle 0|\mathcal{O}^\dagger_{q\bar{q}}(0)|\Psi_0\rangle\right|^2 e^{-V(R) T}. \tag{16.42}$$

By comparing this result with that of the strong coupling expansion, eq. (16.38), we conclude that $V(R) = \sigma R$. This linear potential indicates that the force between the quark and antiquark is constant at large distance, in sharp contrast with a Coulomb potential in electrodynamics. This is a consequence of the *color confinement* property of QCD.

## 16.5 Gauge Fixing on the Lattice

Until now, we have described lattice field theory restricted to the evaluation of the expectation value of gauge-invariant operators. One may legitimately argue that this is sufficient as far as the computation of physical observables is concerned. There are, however, some applications that involve gauge-dependent quantities, for which gauge fixing is necessary. This is the case, for instance, if one wishes to compare the behavior of propagators or vertex functions evaluated non-perturbatively on the lattice and in perturbation theory.

For practical reasons, it is convenient to consider gauge conditions that may be recast into the problem of finding the extrema of a functional. This is the reason why the Landau gauge, i.e., the strict covariant gauge $\partial_\mu A^\mu = 0$, is most often employed in these lattice studies. In the continuum theory, this condition is satisfied at the extrema of the following functional:

$$\mathcal{F}_{\text{Landau}}[A, \Omega] \equiv \frac{1}{2} \int d^4x \, A^a_{\Omega\mu}(x) A^{\mu a}_\Omega(x) , \tag{16.43}$$

where $A^\mu_\Omega$ is the gauge transform of the field configuration $A^\mu$ we aim at bringing to Landau gauge. Indeed, if we apply an infinitesimal gauge transformation to the field $A^\mu_\Omega$, the corresponding variation of $\mathcal{F}_{\text{Landau}}[A, \Omega]$ is

$$\delta\mathcal{F}_{\text{Landau}}[A, \Omega] = -\int d^4x \, (D^\Omega_{\mu ab}\theta_b) A^{\mu a}_\Omega = -\int d^4x \, (\partial_\mu\theta_a - gf^{cab}A^c_{\Omega\mu}\theta_b) A^{\mu a}_\Omega$$

$$= \int d^4x \, \theta_a (\partial_\mu A^{\mu a}_\Omega). \tag{16.44}$$

Therefore, if $A^\mu_\Omega$ realizes an extremum of the functional, then this variation must be zero for all possible $\theta_a(x)$, which means that $A^\mu_\Omega$ obeys the Landau gauge condition. The discrete analogue of the functional defined in eq. (16.43) reads (see Exercise 16.6)

$$\mathcal{F}_{\text{Landau}}[U, \Omega] \equiv -2\, a^2 \sum_x \sum_\mu \mathrm{Re\, tr}\left(\Omega(x) U_\mu(x) \Omega^\dagger(x+\widehat{\mu})\right). \tag{16.45}$$

Finding extrema of such a functional is a rather straightforward task, for instance using the steepest descent algorithm.

Due to the existence of *Gribov copies*, the gauge fixed field configuration is not defined uniquely by the gauge condition, which implies that this functional has more than one extremum corresponding to the various solutions of $\partial_\mu A^\mu_\Omega = 0$ along the same gauge orbit (see Figure 16.10). A natural criterion to decide which extrema to take into account could be to try to reproduce the perturbative Fadeev–Popov procedure. Let us recall here the starting point, which amounts to inserting under the path integral the left-hand side of the following equation:

$$Z_{\text{FP}} \equiv \int [D\Omega(x)] \, \delta[G^a(A^\mu_\Omega)] \det\left(\frac{\delta G^a}{\delta \Omega}\right)_{G^a(A^\mu_\Omega)=0} = 1 , \tag{16.46}$$

## 16.5 GAUGE FIXING ON THE LATTICE

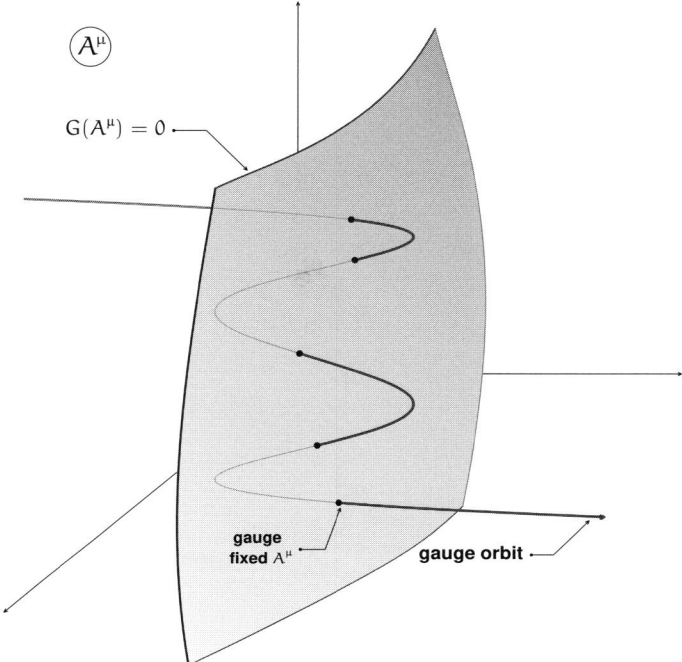

FIGURE 16.10: Gauge orbit that intersects multiple times the gauge fixing manifold.

where $G^a(A) = 0$ is the gauge fixing condition. However, two conditions must be met for this integral to be really equal to 1:

- $G^a(A_\Omega)$ has a unique zero along each gauge orbit; and
- the determinant is positive at this zero.

However, it was shown by Gribov that the unicity condition is generically not satisfied: The gauge condition has multiple solutions (now called Gribov copies). When these conditions are not satisfied, the inserted factor is not 1, but instead

$$Z_{\text{FP}} = \sum_{\text{zeroes } i} \det\left(\frac{\delta G^a}{\delta \Omega}\right)_i \left|\det\left(\frac{\delta G^a}{\delta \Omega}\right)_i\right|^{-1} = \sum_{\text{zeroes } i} \underbrace{\text{sign}\left(\det\left(\frac{\delta G^a}{\delta \Omega}\right)_i\right)}_{\equiv \text{sign}(i)}. \quad (16.47)$$

Thus, one may try to mimic the perturbative Fadeev–Popov procedure by the following definition of a gauge fixed operator on the lattice:

$$\mathcal{O}[A]\Big|_{\text{Landau}} \equiv \frac{\sum_{\text{extrema } i} \text{sign}(i)\, \mathcal{O}[A_{\Omega_i}]}{\sum_{\text{extrema } i} \text{sign}(i)}, \quad (16.48)$$

where the denominator follows from the requirement that gauge invariant operators should remain unaffected by the gauge fixing. However, it was shown by Neuberger that this definition is flawed, because the set of Gribov copies is such that both the numerator and the denominator are exactly zero. In order to see this, consider a gauge invariant observable $\mathcal{O}[U]$, and let us try to mimic closely the continuum BRST quantization, by introducing ghosts and antighosts, the BRST variation B of the antighost ($B \equiv Q_{BRST}\bar{\chi}$, $Q_{BRST}B = 0$) and a gauge fixing parameter $\xi$. By doing so, the expectation value of the observable would read

$$\langle \mathcal{O} \rangle = Z^{-1} \int [D(U,\bar{\chi},\chi,B)] \; \mathcal{O}[U] \; e^{-\frac{1}{g^2}S_W[U] - \frac{1}{2\xi}\sum_x BB + Q_{BRST}\sum_x \bar{\chi}G},$$

$$Z \equiv \int [D(U,\bar{\chi},\chi,B)] \; e^{-\frac{1}{g^2}S_W[U] - \frac{1}{2\xi}\sum_x BB + Q_{BRST}\sum_x \bar{\chi}G}. \tag{16.49}$$

(Here, we are assuming a completely generic gauge fixing function $G(U)$.) Note that with a compact gauge group and a finite lattice, all the integrals involved in this formula are finite. Consider now the following quantity:

$$\mathcal{F}_{\mathcal{O}}(t) \equiv \int [D(U,\bar{\chi},\chi,B)] \; \mathcal{O}[U] \; e^{-\frac{1}{g^2}S_W[U] - \frac{1}{2\xi}\sum_x BB + tQ_{BRST}\sum_x \bar{\chi}G}. \tag{16.50}$$

The numerator in the gauge fixed definition of $\langle \mathcal{O} \rangle$ is nothing but $\mathcal{F}_{\mathcal{O}}(1)$. The derivative of this function is given by

$$\frac{d\mathcal{F}_{\mathcal{O}}}{dt} = \int [D(U,\bar{\chi},\chi,B)] \left[ Q_{BRST} \sum_x \bar{\chi}G \right]$$

$$\times \underbrace{\mathcal{O}[U] \; e^{-\frac{1}{g^2}S_W[U] - \frac{1}{2\xi}\sum_x BB + tQ_{BRST}\sum_x \bar{\chi}G}}_{\text{BRST invariant}}$$

$$= \int [D(U,\bar{\chi},\chi,B)] \; Q_{BRST} \Big\{ \Big[\sum_x \bar{\chi}G\Big] \mathcal{O}[U]$$

$$\times e^{-\frac{1}{g^2}S_W[U] - \frac{1}{2\xi}\sum_x BB + tQ_{BRST}\sum_x \bar{\chi}G} \Big\} = 0. \tag{16.51}$$

In the last equality, we have used the fact that the integral of a total BRST variation is zero. Thus, we have

$$\mathcal{F}_{\mathcal{O}}(1) = \mathcal{F}_{\mathcal{O}}(0) = \int [D(U,\bar{\chi},\chi,B)] \; \mathcal{O}[U] \; e^{-\frac{1}{g^2}S_W[U] - \frac{1}{2\xi}\sum_x BB} = 0. \tag{16.52}$$

This time, the zero follows from the fact that the integrand does not depend on $\chi$ or $\bar{\chi}$, hence the integrals over the ghost and antighost are equal to zero. The same reasoning applies to the denominator Z in the gauge-fixed definition of $\langle \mathcal{O} \rangle$, hence we have an undefined ratio,[4]

$$\langle \mathcal{O} \rangle_{\text{gauge fixed}} = \frac{0}{0}. \tag{16.53}$$

If we interpret this result in the light of eq. (16.48), we see that these zeroes result from an even number of Gribov copies with alternating signs for the determinant of the Fadeev–Popov operator. One may view this issue as a fundamental obstruction for a non-perturbative

---

[4]The same conclusion holds if the operator $\mathcal{O}$ is not gauge invariant, but simply BRST invariant.

## 16.6 Lattice Hamiltonian

definition of gauge fixing by the Fadeev–Popov procedure. Because of this problem, the practical lattice implementation of the Landau gauge fixing is simply to pick one of the extrema of the functional (16.45), without any special selection rule. One should be aware of this procedure when comparing with perturbative results, since it is a priori not guaranteed that the solutions of the gauge condition used in the perturbative and in the non-perturbative calculations are the same.

## 16.6 Lattice Hamiltonian

Until now, we have discussed lattice field theory as a way to numerically implement the Euclidean path integral in non-Abelian gauge theories. In some applications (e.g., semi-classical approximations based on solving the classical Yang–Mills equations) it is useful to have a discrete formulation of the Hamiltonian. Since time plays a special role in the Hamiltonian formalism (in particular, the canonical conjugate of the gauge potential $A^0$ is zero), it is convenient to adopt the temporal gauge condition $A^0 = 0$. Note that this does not fix completely the gauge, since the system is still invariant under time-independent gauge transformations. With this choice, the conjugate momentum of $A^i$ is the corresponding chromo-electrical field $E^i \equiv F^{0i} = \partial^0 A^i$, and the Hamiltonian reads

$$\mathcal{H} = \int d^3x \left( \tfrac{1}{2} E_a^i E_a^i + \tfrac{1}{4} F_a^{ij} F_a^{ij} \right). \tag{16.54}$$

(In the Hamiltonian formalism, the electrical field should be treated as a variable independent of $A^i$.) The Hamilton's equations,

$$\partial_0 A_a^i(x) = E_a^i(x), \quad \partial_0 E_a^i(x) = -\frac{\delta \mathcal{H}}{\delta A_a^i(x)} = \left[ -\partial_j \delta_{ab} + ig(A_j^{adj})_{ab} \right] F_b^{ji}, \tag{16.55}$$

must be supplemented by Gauss' law, $\left(D_i^{adj}\right)_{ab} E_a^i = 0$ (before fixing the temporal gauge, this additional relationship is just the Yang–Mills equation $D_\mu^{adj} F^{\mu 0} = 0$; after setting $A^0 = 0$, it becomes a constraint between $A^i$ and $E^i$ at equal time). Note that Hamilton's equations are consistent with Gauss' law: If their initial condition satisfies Gauss' law, then it remains true at all times.

Let us now derive a discrete version of the Hamiltonian and Hamilton's equations. We already known how to discretize in a gauge invariant way the term $F^{ij}F^{ij}$ in terms of plaquettes. Since the electrical fields transform as $E^i(x) \to \Omega^\dagger(x) E^i(x) \Omega(x)$ under gauge transformations, it is natural to assign them to the nodes of the lattice. Thus, the discrete Hamiltonian is given by

$$\mathcal{H} = a^3 \sum_{x,i} \text{tr}\left( E^i(x) E^i(x) \right) + \frac{1}{g^2 a} \sum_{x,ij} \left( N - \text{tr}\left( \text{Re}\, [\square]_{x;ij} \right) \right). \tag{16.56}$$

To obtain the corresponding Hamilton's equations, we write the link variable as $U_i(x) \approx \exp(igaA_a^i(x)t^a)$. Therefore, we have

$$t^a \frac{\partial \, \text{tr}\, [\square]_{x;ij}}{\partial A_a^i(x)} = iga\, t^a\, \text{tr}\left( t^a [\square]_{x;ij} \right) = \frac{iga}{2}\left( [\square]_{x;ij} - \frac{1}{N} \text{tr}\, [\square]_{x;ij} \right). \tag{16.57}$$

(The second equality is obtained by using the $\mathfrak{su}(N)$ Fierz identity, assuming that we use the fundamental representation.) Then, we obtain the following equations of motion,

$$\partial_0 U_i(x) = iga\, E^i(x) U_i(x), \quad \partial_0 E^i = -\frac{ig}{4g} \sum_{j\neq i} \left\{ [\Box]_{x;ij} + [\Box]_{x;i-j} - \text{h.c.} - \frac{\text{traces}}{N} \right\}. \tag{16.58}$$

($[\Box]_{x;i-j}$ denotes the plaquette that starts at point $x$ by a segment in direction $\hat{\imath}$, followed by a segment in direction $-\hat{\jmath}$.) Note that in this formulation, the time is still a continuous variable (but the numerical resolution of the equations will of course call for some discretization of time as well).

## 16.7 Lattice Worldline Formalism

### 16.7.1 Discrete Analogue of the Heat Kernel

In Chapter 15 we exposed the worldline representation for a quantum field theory in a continuous space-time. However, a similar representation is also possible for propagators in a field theory defined on a discrete space-time. For the sake of simplicity, let us first consider a free scalar field theory, defined on a cubic lattice instead of a continuous space-time. As we saw earlier in this chapter, the second derivatives $\partial^\mu \partial^\mu$ that appear in the inverse propagator are replaced by centered finite differences, and with a Euclidean metric we have

$$(\Box + m^2)\,\phi(x) = m^2\,\phi(x) + \sum_\mu \frac{2\phi(x) - \phi(x+\hat{\mu}) - \phi(x-\hat{\mu})}{a^2}. \tag{16.59}$$

We could in principle introduce a continuous fictitious time $T$ as in eq. (15.3) in order to write a heat-kernel representation of this finite difference operator. But it turns out to be more convenient to introduce a discrete variable here as well, based on the identity

$$A^{-1} = \sum_{n=0}^\infty (1-A)^n. \tag{16.60}$$

### 16.7.2 Random Walks on a Cubic Lattice

This formula should be applied to a dimensionless operator, for instance $a^2(\Box+m^2)$ instead of $\Box+m^2$ itself. Since the discrete operator in eq. (16.59) contains a term $(m^2+2da^{-2})\phi(x)$ that acts as the identity, it is in fact more convenient to choose $A = (m^2+2da^{-2})^{-1}(\Box+m^2)$, in order to make $1-A$ simpler. Therefore, we write

$$(\Box+m^2)^{-1}\Big|_{\text{lattice}} = \frac{1}{m^2+2da^{-2}} \sum_{n=0}^\infty \left[\frac{\mathbb{D}}{2d+m^2 a^2}\right]^n, \tag{16.61}$$

## 16.7 LATTICE WORLDLINE FORMALISM

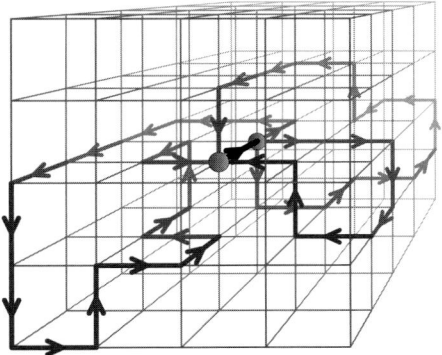

FIGURE 16.11: Worldlines on a cubic lattice. The points x and x' are materialized by the two little balls, and we have represented three different paths on the lattice connecting these two points.

where $\mathbb{D}$ is the discrete operator defined as follows:

$$\mathbb{D}\,\phi(x) \equiv \sum_{\mu} \Big(\phi(x+\widehat{\mu}) + \phi(x-\widehat{\mu})\Big). \tag{16.62}$$

(In d dimensions, there are 2d terms in the sum of the right-hand side.) The operator $\mathbb{D}/(2d + m^2 a^2)$ realizes one hop from a lattice site to one of its nearest neighbors, with a probability $(2d + m^2 a^2)^{-1}$ for a jump in any given direction. Raised to the power n, we get an operator that performs n successive jumps. The probability of a given sequence of n hops is $(2d + m^2 a^2)^{-n}$, and there are $(2d)^n$ such sequences, hence a total probability $(1 + m^2 a^2/2d)^{-n}$. This is equal to unity in a massless theory, but suppressed exponentially at large n with a non-zero mass (this observation is the discrete analogue of the fact that long worldlines are suppressed exponentially by a mass in the continuous case).

The propagator evaluated between the sites x and x' is proportional to the total probability to connect these two sites by a sequence of jumps, regardless of its length (because of the sum on n):

$$\left[(\square + m^2)^{-1}\right]_{xx'} \underset{\text{lattice}}{=} \frac{1}{m^2 + 2da^{-2}} \sum_{n=0}^{\infty} \frac{1}{(2d + m^2 a^2)^n} \sum_{\gamma \in \mathcal{P}_n(x,x')} 1, \tag{16.63}$$

where $\mathcal{P}_n(x, x')$ is the set of all paths of length n drawn on the edges of the lattice that connect x to x' (see Figure 16.11). Therefore, the second sum merely counts the number of such paths. This number has an upper bound of $(2d)^n$, which implies trivially the convergence of the sum on n in the massive case.

### 16.7.3 Scalar Electrodynamics

Let us now consider a complex scalar field in a background Abelian gauge field. The D'Alembertian is replaced by the square of the covariant derivative. In order to maintain an exact gauge symmetry in the discrete lattice formulation, the gauge field is represented by link variables $U_\mu(x)$ defined on each edge of the lattice, and the forward and backward discrete covariant derivatives read

$$D_F^\mu \phi(x) \equiv \frac{U_\mu^\dagger(x)\phi(x+\hat{\mu}) - \phi(x)}{a} \;,\quad D_B^\mu \phi(x) \equiv \frac{\phi(x) - U_\mu(x-\hat{\mu})\phi(x-\hat{\mu})}{a}. \quad (16.64)$$

In order to evaluate the scalar propagator in this gauge background, we can reproduce the derivation of the previous subsection, and the only change will arise in the definition of the operator $\mathbb{D}$, whose action now reads

$$\mathbb{D}\,\phi(x) \equiv \sum_\mu \Big( U_\mu^\dagger(x)\phi(x+\hat{\mu}) + U_\mu(x-\hat{\mu})\phi(x-\hat{\mu}) \Big). \quad (16.65)$$

Note that the right-hand side transforms like a scalar field at point x under a gauge transformation. The consequence of this modification is that the jumps performed by the operator $\mathbb{D}$ are now weighted by $U(1)$ phases corresponding to the link variables that appear in the right-hand side of eq. (16.65). The lattice worldline representation of the dressed scalar propagator is

$$\big[(D^2+m^2)^{-1}\big]_{xx'}\Big|_\text{lattice} = \frac{1}{m^2+2da^{-2}} \sum_{n=0}^\infty \frac{1}{(2d+m^2a^2)^n} \sum_{\gamma\in\mathcal{P}_n(x,x')} W_\gamma[U]\,, \quad (16.66)$$

where $W_\gamma[U]$ is the product of all the link variables collected along the path $\gamma$, which is nothing but the Wilson line defined on this path. This expression transforms as expected for a dressed scalar propagator, thanks to the properties of Wilson lines.

A representation similar to eq. (16.66) is also possible for the one-loop quantum effective action in the gauge background, which can be obtained by first noting that

$$\ln(A) = \ln(1-(1-A)) = -\sum_{n=1}^\infty \frac{1}{n}(1-A)^n. \quad (16.67)$$

The derivation mimics that of the propagator, and we finally obtain

$$\Gamma[U]\Big|_\text{lattice} = -\frac{a^4}{m^2+2da^{-2}} \sum_x \sum_{n=1}^\infty \frac{1}{2n(2d+m^2a^2)^{2n}} \sum_{\gamma\in\mathcal{P}_{2n}(x,x)} W_\gamma[U]. \quad (16.68)$$

Note that now the paths $\gamma$ involved in the sum are the closed paths starting and ending at point x, which is why only even values of the length are allowed.

## 16.7.4 Combinatorics of Loop Areas on a Planar Square Lattice

In two dimensions, the representation of eq. (16.66) may be used in order to study the properties of charged particles on an atomic lattice, under the influence of an external electromagnetic field. In particular, when this field is purely magnetic and transverse to the plane of the lattice (i.e., the field strength $F_{12}$ is non-zero), this model (known as the Hofstadter model) is related to the quantum Hall effect.

## 16.7 LATTICE WORLDLINE FORMALISM

This relationship may also be exploited in order to derive explicit formulas for the moments of the distribution of areas of random closed loops on a square lattice. For this application, it is interesting to consider a two-dimensional *anisotropic* lattice, with lattice spacings $a_{1,2}$ in the two directions. On this lattice, consider the propagator of a massless scalar at equal points in the presence of a transverse magnetic field. It is straightforward to generalize the previous derivation to the anisotropic case, and one obtains the following expression for the massless propagator:

$$[(D^2)^{-1}]_{xy} = \frac{a^2}{4} \sum_{n_{1,2}=0}^{\infty} \left(\frac{h_1}{4}\right)^{n_1} \left(\frac{h_2}{4}\right)^{n_2} \sum_{\gamma \in \mathcal{P}_{n_1,n_2}(x,y)} W_\gamma[U], \qquad (16.69)$$

where $\mathcal{P}_{n_1,n_2}(x,y)$ is the set of paths drawn on the edges of the lattice, that connect $x$ to $y$, and contain $n_1$ jumps in the first direction and $n_2$ jumps in the second direction. In this formula, we have also defined

$$\frac{2}{a^2} \equiv \frac{1}{a_1^2} + \frac{1}{a_2^2}, \qquad h_{1,2} \equiv \frac{a^2}{a_{1,2}^2} \qquad (h_1 + h_2 = 2). \qquad (16.70)$$

If we specialize to a closed path $x = y = 0$ and to a transverse magnetic field B, the paths $\gamma$ on the right-hand side are closed loops, and we have

$$W_{\text{closed } \gamma}[U] = e^{i\Phi \,\text{Area}(\gamma)}, \qquad (16.71)$$

where we denote $\Phi \equiv B a_1 a_2$ the magnetic flux through an elementary plaquette of the lattice, and where Area $(\gamma)$ is the area (in lattice units) enclosed by the loop $\gamma$. Note that this is an algebraic area whose sign depends on the orientation of the loop, and which accounts for the winding number of the loop. Expanding the exponential, we have

$$[(D^2)^{-1}]_{00} = \frac{a^2}{4} \sum_{n_{1,2}=0}^{\infty} \left(\frac{h_1}{4}\right)^{2n_1} \left(\frac{h_2}{4}\right)^{2n_2} \sum_{l=0}^{\infty} \frac{(i\Phi)^{2l}}{(2l)!} \sum_{\gamma \in \mathcal{P}_{2n_1,2n_2}(0,0)} (\text{Area}(\gamma))^{2l}. \qquad (16.72)$$

(Because the area is algebraic, the odd moments are all zero.) In this formula, we have made explicit the fact that a closed path must have an even number of hops in each direction. On the other hand, the propagator can be determined perturbatively order by order in $\Phi$, since this corresponds to a weak field expansion. This calculation requires that one chooses a gauge.[5] A convenient choice is provided by the following link variables:

$$U_1(x) \equiv 1, \qquad U_2(x) \equiv e^{i\Phi \, i_1}, \qquad (16.73)$$

---

[5] The product of the link variables along a closed loop does not depend on the choice of the gauge, as can be seen from the following gauge transformation formulas for the link variables

$$U_\mu(x) \to \Omega^*(x + \hat{\mu}) \, U_\mu(x) \, \Omega(x).$$

where $i_1$ is the integer that labels the first coordinate ($x \equiv \sum_\mu i_\mu \hat{\mu}$). By performing the expansion of $G(0,0)$ and identifying the coefficient of the term of order $\Phi^{2l} h_1^{2n_1} h_2^{2n_2}$ on both sides of eq. (16.72), one can show that

$$\sum_{\gamma \in \mathcal{P}_{2n_1,2n_2}(0,0)} (\text{Area}(\gamma))^{2l} = \frac{(2(n_1+n_2))!}{n_1!^2 n_2!^2} \mathcal{P}_{2l}(n_1,n_2), \tag{16.74}$$

where $\mathcal{P}_{2l}$ is a symmetric polynomial in $n_1, n_2$ of degree $2l$. Note that the combinatorial factor $(2(n_1+n_2))!/(n_1!^2 n_2!^2)$ is the number of loops in $\mathcal{P}_{2n_1,2n_2}(0,0)$. This expansion also provides a semi-explicit form of the polynomial $\mathcal{P}_{2l}$, and the evaluation of the first few terms gives

$$\mathcal{P}_0(n_1,n_2) = 1, \quad \mathcal{P}_2(n_1,n_2) = \frac{n_1 n_2}{3},$$
$$\mathcal{P}_4(n_1,n_2) = \frac{n_1 n_2 (7 n_1 n_2 - (n_1+n_2))}{15}. \tag{16.75}$$

For $l > 0$, all these polynomials satisfy two simple identities:

$$\mathcal{P}_{2l}(n_1,0) = \mathcal{P}_{2l}(0,n_2) = 0, \quad \mathcal{P}_{2l}(1,1) = \frac{1}{3}. \tag{16.76}$$

The first one is a consequence of the fact that if $n_1$ or $n_2$ is zero, then all the closed paths one can construct have a vanishing area. The second one follows from the fact that for $n_1 = n_2 = 1$, all the closed paths have area $-1, 0$ or $+1$, and therefore contribute equally to all the even moments.

## Exercises

**16.1** Check that with the scalar lattice action (eq. (16.6)), the error on the dispersion relation of free particles is of order $a^2$. Show that this dispersion relation can be made correct to (including) the order $a^2$, with an error now relegated to order $a^4$, by adding to the action an operator of the form

$$c\, a^2 \left(\nabla_B^\mu \nabla_F^\mu \phi\right)^2.$$

Calculate the dimensionless prefactor $c$ for this to happen.

**\*16.2** In Euclidean space-time, the Dirac matrix $\gamma^0$ is replaced by a matrix $\gamma^4$ such that $\gamma^4 \gamma^4 = -1$ and $\{\gamma^4, \gamma^{1,2,3}\} = 0$.
- What is an appropriate choice of $\gamma^4$?
- We want now to define a matrix $\gamma^5$ such that $(1 \pm \gamma^5)/2$ are projectors. Show that $\gamma^5 \equiv \gamma^1 \gamma^2 \gamma^3 \gamma^4$ is a good choice.
- Check that $\{\gamma^5, \gamma^{1,2,3,4}\} = 0$, $(\gamma^{1,2,3,4})^\dagger = \gamma^5 \gamma^{1,2,3,4} \gamma^5$ and $(\gamma^5)^\dagger = \gamma^5$.
- From this, conclude that the determinant of the lattice Dirac operator is real, and real positive for two degenerate flavors.

- The chemical potential μ associated with charge conservation modifies the Dirac operator as follows:

$$i\slashed{D} - m \to i\slashed{D} - m + \mu\gamma^0.$$

What can be said about its determinant?

**\*16.3** What is the discretization error of the first derivative in the Dirac operator? Explain why the Wilson term introduces an error of order $a^1$.

In the rest of this exercise we derive an extra term that reduces this error, while keeping the benefit of the Wilson term concerning the fermion doublers. We start from the continuum Dirac action given in eq. (16.17).

- Consider the following transformation of $\psi, \overline{\psi}$:

$$\psi \to \exp\left(ra(i\slashed{D} - m)\right)\psi, \quad \overline{\psi} \to \overline{\psi} \exp\left(r'a(i\slashed{D} - m)\right).$$

Justify heuristically that the Jacobian of this transformation deviates from 1 only at the order $a^2$ and beyond.
- How does the transformation change the Dirac Lagrangian at order $a$? Note that it contains a Wilson term, plus an extra term, called the Clover term.
- What is the error made if we include both the Wilson and the Clover terms, with their respective weights as found in the previous question?

**\*16.4** Consider Euclidean QCD in the continuum.
- Show that the expectation value $\langle \overline{\psi}(x)\psi(x)\rangle$ can be written as

$$\langle \overline{\psi}(x)\psi(x)\rangle = \lim_{V\to\infty} \frac{1}{ZV}\int [DA^\mu]\, e^{-S_{YM}[A]} \det(i\slashed{D} - m)\, \mathrm{tr}_V\left[\frac{1}{i\slashed{D} - m}\right],$$

where $V$ is a four-dimensional Euclidean volume to which the trace is restricted, and $Z$ is the same integral without the trace.
- Recalling the properties of the Euclidean Dirac operator, write the trace as follows:

$$\mathrm{tr}_V\left[\frac{1}{i\slashed{D} - m}\right] = \sum_n \frac{1}{i\lambda_n - m} = -\sum_n \frac{m}{\lambda_n^2 + m^2},$$

where the $\lambda_n$ are real numbers.
- Show that

$$\langle \overline{\psi}(x)\psi(x)\rangle = -\lim_{V\to\infty}\int d\lambda\, \frac{m\,\rho_V(\lambda)}{\lambda^2 + m^2},$$

with $\rho_V(\lambda) \equiv \frac{1}{ZV}\int [DA^\mu]\, e^{-S_{YM}[A]} \det(i\slashed{D} - m) \sum_n \delta(\lambda - \lambda_n).$

- Show that in the limits of infinite volume and zero mass, we have $\langle \overline{\psi}(x)\psi(x)\rangle = -\pi \lim_{V\to\infty} \rho_V(0)$. This identity is known as the *Banks–Casher relation*.

**16.5** Check that the SU(2) group measure given in footnote 3 is normalized according to eq. (16.32). Then, show that the integral of a single power of $U$ is zero, and check eq. (16.33).

**\*16.6** Show that the extrema of the functional defined in eq. (16.45) give configurations of the link variables that are the discrete analogue of gauge fields that fulfill the Landau gauge condition.

**16.7** Explain why (on a square planar lattice) the number of closed paths with $2n_1$ hops in the $\pm 1$ directions and $2n_2$ hops in the $\pm 2$ directions is $(2(n_1 + n_2))!/n_1!^2 n_2!^2$. Using eqs. (16.75), show that the first two coefficients in the $\Phi$-expansion of the equal-point propagator, $[D^{-2}]_{00} = c_0 + c_2 \Phi^2 + \cdots$, can be expressed as follows in terms of Bessel functions:

$$c_0 = \frac{a^2}{4} \int_0^\infty dt\, e^{-t}\, I_0\left(\tfrac{h_1 t}{2}\right) I_0\left(\tfrac{h_2 t}{2}\right), \quad c_2 = -\frac{a^2 h_1 h_2}{384} \int_0^\infty dt\, e^{-t}\, t^2\, I_1\left(\tfrac{h_1 t}{2}\right) I_1\left(\tfrac{h_2 t}{2}\right).$$

# CHAPTER 17

# Quantum Field Theory at Finite Temperature

Historically, the main area of developments and applications of quantum field theory (QFT) has been high-energy particle physics. This corresponds to situations in which the background is the vacuum, only perturbed by the presence of a few excitations whose interactions one aims at studying. Consequently, most of the QFT tools we have encountered so far are adequate for calculating transition amplitudes between pure states that contain only a few particles.

However, there are interesting physical problems that depart from this simple situation. For instance, in the early universe, particles are believed to have been in thermal equilibrium and formed a hot and dense plasma. The typical energy of a particle in this thermal bath is of the order of the temperature,[1] which implies that this hot environment may have an influence on all processes whose energy scale is comparable or lower. As a consequence, these problems contain some element of many body physics that was not present in applications of QFT to scattering reactions.

Another class of problems in which many body effects are important is condensed matter physics. When studied at some sufficiently large-distance scale, where the atomic discreteness is no longer important, these problems may be described in terms of (non-relativistic) quantum fields in which collective effects are usually important.

## 17.1 Canonical Thermal Ensemble

Usually, the system one would like to study is a little part of a much larger system (this is quite obvious in the case of the early universe, but is also generally true in condensed matter physics).

---

[1] In this chapter, we extend the natural system of units we have used so far to also set the Boltzmann constant $k_B = 1$. Therefore, the temperature has the dimension of an energy.

Thus, its energy and other conserved quantities are not fixed. Instead, they fluctuate due to exchanges with the surroundings, which play the role of a thermal reservoir. The appropriate statistical ensemble for describing this situation is the canonical ensemble, in which the system is described by the following density operator:

$$\rho \equiv \exp\left\{-\beta \mathcal{H}\right\}, \tag{17.1}$$

where $\beta = T^{-1}$ is the inverse temperature and $\mathcal{H}$ is the Hamiltonian. Given an operator $\mathcal{O}$, one is usually interested in calculating its expectation value in the above statistical ensemble:

$$\langle \mathcal{O} \rangle \equiv \frac{\text{tr}\,(\rho\,\mathcal{O})}{\text{tr}\,(\rho)}. \tag{17.2}$$

Let us span the Hilbert space by states $|n\rangle$ that are eigenstates of $\mathcal{H}$,

$$\mathcal{H}|n\rangle = E_n |n\rangle. \tag{17.3}$$

In terms of these states, the trace of $\rho\,\mathcal{O}$ can be represented as follows:

$$\text{tr}\,(\rho\,\mathcal{O}) = \sum_n e^{-\beta E_n} \langle n|\mathcal{O}|n\rangle. \tag{17.4}$$

From this representation, it is easy to see that the zero-temperature limit selects the state of lowest energy, i.e., the ground state of the Hamiltonian. Assuming that this state is non-degenerate, this corresponds to the vacuum expectation value,

$$\lim_{T\to 0} \text{tr}\,(\rho\,\mathcal{O}) = \langle 0|\mathcal{O}|0\rangle. \tag{17.5}$$

In this sense, eq. (17.2) should be viewed as an extension of the formalism we already know, rather than something entirely different. In this chapter, we discuss various aspects of these thermal averages, starting with the necessary extensions to the formalism in order to perform their perturbative calculation.

## 17.2 Finite-T Perturbation Theory

### 17.2.1 Naive Approach

The extension of ordinary perturbation theory to calculate expectation values such as (17.2) is usually called *quantum field theory at finite temperature*. A first approach for evaluating such an expectation value could be to use the representation of the trace provided by eq. (17.4), and a similar formula for the denominator, which would fall back to the perturbative rules we already know (since the temperature and chemical potential appear only in the form of numerical prefactors). Note, however, a peculiarity of the matrix elements that appear in eq. (17.4): $\langle n|$ and $|n\rangle$ are identical states since they come from a trace (they are both *in*-states, since $\rho$ defines the initial state of the system). This is a bit different from the transition amplitudes that enter in scattering cross-sections, where the matrix elements are evaluated between an *in*-state and an *out*-state. The perturbative rules to compute these in–in expectation values are provided by the Schwinger–Keldysh formalism introduced in Section 2.4.3.

A difficulty with this naive approach is that the number of states that contribute significantly to the sum in eq. (17.4) is large at high temperature, especially when the temperature is large compared to the masses of the fields (and even more so with massless particles like photons). In fact, it is possible to encapsulate the sum over the eigenstates $|n\rangle$ and the canonical weight of these states $\exp(-\beta E_n)$ directly into the Schwinger–Keldysh rules by a modest modification of its propagators.

## 17.2.2 Thermal Time Contour

To mimic closely the derivation of the Feynman rules at zero temperature, let us consider an observable made of the time-ordered product of elementary fields,

$$\mathcal{O} \equiv T\, \phi(x_1) \cdots \phi(x_n). \tag{17.6}$$

Each Heisenberg representation field $\phi$ can be related to the corresponding field in the interaction representation by

$$\phi(x) = U(t_i, x^0)\, \phi_{in}(x)\, U(x^0, t_i), \tag{17.7}$$

where $t_i$ is the time at which the system is prepared in equilibrium, and $U$ is the time evolution operator defined by

$$U(t_2, t_1) \equiv T\, \exp i \int_{t_1}^{t_2} d^4x\, \mathcal{L}_I(\phi_{in}(x)), \tag{17.8}$$

$\mathcal{L}_I$ is the interaction term in the Lagrangian. Thanks to eq. (17.7), we remove all the interactions from the field, and relegate them into the evolution operator where they can easily be Taylor expanded.

In the canonical ensemble at non-zero temperature, there is another source of dependence on the interactions, hidden in the Hamiltonian inside the density operator. Indeed, for the system to be in statistical equilibrium, the canonical density operator should be defined with the same Hamiltonian as the one that drives the time evolution, i.e., a Hamiltonian that also contains the interactions of the system.[2] If we decompose the full Hamiltonian as $\mathcal{H} \equiv \mathcal{H}_0 + \mathcal{H}_I$, we have (see Exercise 17.1)

$$e^{-\beta \mathcal{H}} = e^{-\beta \mathcal{H}_0}\, \underbrace{T\, \exp i \int_{t_i}^{t_i - i\beta} d^4x\, \mathcal{L}_I(\phi_{in}(x))}_{U(t_i - i\beta,\, t_i)}. \tag{17.9}$$

(This formula in fact does not depend on $t_i$.) It can be proven by noticing that right- and left-hand sides are equal for $\beta = 0$, and by checking that their derivatives with respect to $\beta$

---

[2] An alternative point of view is to decide that $\rho$ is the density operator of the system at $x^0 = -\infty$. There, we may turn off adiabatically the interactions, and therefore use only the free Hamiltonian inside $\rho$. In this section, we derive the formalism for an initial equilibrium state specified at a finite time $x^0 = t_i$.

are also equal (for this, we use the fact that the derivative of the time evolution with respect to its final time is known).

From the previous formulas, we can write

$$e^{-\beta\mathcal{H}}\, T\, \phi(x_1)\cdots\phi(x_n)$$
$$= e^{-\beta\mathcal{H}_0}\, P\, \phi_{in}(x_1)\cdots\phi_{in}(x_n)\, \exp i\int_{\mathcal{C}} dx^0 d^3x\, \mathcal{L}_I(\phi_{in}(x)), \qquad (17.10)$$

where the symbol P indicates a *path ordering*, and where the time integration contour is $\mathcal{C} = [t_i, +\infty] \cup [+\infty, t_i] \cup [t_i, t_i - i\beta]$,

$$\mathcal{C} = \begin{array}{c} t_i \\ \phantom{xx} \\ t_i - i\beta \end{array} \qquad (17.11)$$

In this contour, $t_i$ is the time at which the system is prepared in thermal equilibrium. As we shall see shortly, all observables are independent of this time, which physically means that a system in equilibrium has no memory of when it was put in equilibrium. Note also that in eq. (17.10), the times $x_1^0, \cdots, x_n^0$ are on the upper branch of the contour (but this constraint will be relaxed shortly).

### 17.2.3 Generating Functional

The time contour of eq. (17.11) is very similar to the contour of Figure 2.3, with the addition of a vertical part that captures the interactions hidden in the density operator. Since we had to extend the real-time axis into the contour $\mathcal{C}$, it is natural to extend also the observable of eq. (17.6) to allow the field operators to be located anywhere on $\mathcal{C}$, with a path ordering instead of a time ordering,

$$\mathcal{O} \equiv P\, \phi(x_1)\cdots\phi(x_n). \qquad (17.12)$$

The expectation values of these operators can be encapsulated in the following generating functional:

$$Z[j] \equiv \frac{\operatorname{tr}\left(\rho\, P\, \exp\left\{i\int_{\mathcal{C}} d^4x\, j(x)\phi(x)\right\}\right)}{\operatorname{tr}(\rho)}, \qquad (17.13)$$

where the fictitious source $j(x)$ also lives on the contour $\mathcal{C}$. In order to bring this generating functional to a useful form, we can follow very closely the derivation of Section 1.7.2, by first pulling out a factor that contains the interactions, and by rearranging the ordering of the free factor with two successive applications of the Baker–Campbell–Hausdorff formula. This leads to

$$Z[j] = \exp\left\{i\int_{\mathcal{C}} d^4x\, \mathcal{L}_I\left(\frac{\delta}{i\delta j(x)}\right)\right\}\, \exp\left\{-\frac{1}{2}\int_{\mathcal{C}} d^4x d^4y\, j(x)j(y)\, G^0(x,y)\right\}, \qquad (17.14)$$

where the free propagator $G^0(x,y)$, defined on the contour $\mathcal{C}$, is given by

$$G^0(x,y) \equiv \frac{\mathrm{tr}\left(e^{-\beta\mathcal{H}_0}\,P\,\phi_{\mathrm{in}}(x)\phi_{\mathrm{in}}(y)\right)}{\mathrm{tr}\left(e^{-\beta\mathcal{H}_0}\right)}. \qquad (17.15)$$

### 17.2.4 Expression for the Free Propagator

In order to calculate the free propagator, we need the free Hamiltonian expressed in terms of creation and annihilation operators,[3]

$$\mathcal{H}_0 = \int \frac{d^3\mathbf{p}}{(2\pi)^3 2E_\mathbf{p}}\, E_\mathbf{p}\, a^\dagger_{\mathbf{p},\mathrm{in}} a_{\mathbf{p},\mathrm{in}}, \qquad (17.16)$$

and the canonical commutation relation of the latter,

$$\left[a_{\mathbf{p},\mathrm{in}}, a^\dagger_{\mathbf{p}',\mathrm{in}}\right] = (2\pi)^3\, 2E_\mathbf{p}\, \delta(\mathbf{p}-\mathbf{p}'). \qquad (17.17)$$

From this, we get (see Exercise 17.2)

$$\begin{aligned}
\left[e^{-\beta\mathcal{H}_0}, a_{\mathbf{p},\mathrm{in}}\right] &= e^{-\beta\mathcal{H}_0}\left(1 - e^{-\beta E_\mathbf{p}}\right) a_{\mathbf{p},\mathrm{in}} \\
\mathrm{tr}\left(e^{-\beta\mathcal{H}_0} a_{\mathbf{p},\mathrm{in}}\right) &= 0 \\
\frac{\mathrm{tr}\left(e^{-\beta\mathcal{H}_0} a^\dagger_{\mathbf{p},\mathrm{in}} a_{\mathbf{p}',\mathrm{in}}\right)}{\mathrm{tr}\left(e^{-\beta\mathcal{H}_0}\right)} &= (2\pi)^3\, 2E_\mathbf{p}\, n_{_B}(E_\mathbf{p})\,\delta(\mathbf{p}-\mathbf{p}'),
\end{aligned} \qquad (17.18)$$

where $n_{_B}(E)$ is the Bose–Einstein distribution,

$$n_{_B}(E) \equiv \frac{1}{e^{\beta E}-1}. \qquad (17.19)$$

This leads to the following formula for the free propagator:

$$G^0(x,y) = \int \frac{d^3\mathbf{p}}{(2\pi)^3 2E_\mathbf{p}} \Big[\left(\theta_c(x^0-y^0) + n_{_B}(E_\mathbf{p})\right) e^{-ip\cdot(x-y)} \\
+ \left(\theta_c(y^0-x^0) + n_{_B}(E_\mathbf{p})\right) e^{+ip\cdot(x-y)}\Big], \qquad (17.20)$$

where $\theta_c$ generalizes the step function to the contour $\mathcal{C}$ (i.e., $\theta_c(x^0-y^0)$ is non-zero if $x^0$ is posterior to $y^0$ according to the contour ordering). This expression of the propagator generalizes to a non-zero temperature eq. (1.136) (the Bose–Einstein distribution goes to zero when $T \to 0$). Let us postpone a bit the calculation of the propagator in momentum space. For now, we just note the following rules for the perturbative expansion in coordinate space:

1. Draw all the graphs (with vertices corresponding to the interactions of the theory under consideration) that connect the $n$ points of the observable. Graphs containing disconnected subgraphs can be ignored. Each graph should be weighted by its symmetry factor.

---

[3] We can drop the zero point energy here. It would simply multiply the density operator by a constant factor, which would be canceled since all expectation values are normalized by the factor $1/\mathrm{tr}(\rho)$.

2. Each line of a graph brings a free propagator $G^0(x,y)$.
3. Each vertex brings a factor $-i\lambda$. The space-time coordinate of this vertex is integrated out, but the time integration runs over the contour $\mathcal{C}$.

Thus, the only differences with the zero-temperature Feynman rules are the explicit form of the free propagator, and the fact the time integrations are over the contour $\mathcal{C}$ instead of the real axis.

### 17.2.5 Kubo–Martin–Schwinger Symmetry

The canonical density operator $\exp(-\beta\mathcal{H})$ can be viewed as an evolution operator for an imaginary time shift, which implies the following formal identity:

$$e^{-\beta\mathcal{H}}\,\phi(x^0-i\beta,\mathbf{x})\,e^{\beta\mathcal{H}} = \phi(x^0,\mathbf{x}). \tag{17.21}$$

Let us now consider the following correlator:

$$\mathcal{G}(t_i,\cdots) \equiv \mathrm{tr}\left(e^{-\beta\mathcal{H}}\,P\,\phi(t_i,\mathbf{x})\cdots\right), \tag{17.22}$$

which contains a field whose time argument is the initial time $t_i$ (the other fields it contains need not be specified in this discussion). Since $t_i$ is the "smallest" time on the contour $\mathcal{C}$, the field operator that carries it is placed at the rightmost position by the path ordering. Thus, we have

$$\mathcal{G}(t_i,\cdots) = \mathrm{tr}\left(e^{-\beta\mathcal{H}}\,[P\,\cdots]\,\phi(t_i,\mathbf{x})\right), \tag{17.23}$$

where the path ordering now applies only to the remaining (unwritten) fields. Using the cyclic invariance of the trace and eq. (17.21), we then get

$$\begin{aligned}\mathcal{G}(t_i,\cdots) &= \mathrm{tr}\left(e^{-\beta\mathcal{H}}\,\phi(t_i-i\beta,\mathbf{x})\,[P\,\cdots]\right)\\ &= \mathrm{tr}\left(e^{-\beta\mathcal{H}}\,P\,\phi(t_i-i\beta,\mathbf{x})\cdots\right)\\ &= \mathcal{G}(t_i-i\beta,\cdots),\end{aligned} \tag{17.24}$$

where in the second line we have used the fact that $t_i - i\beta$ is the "latest" time on the contour $\mathcal{C}$ in order to put back the operator carrying it inside the path ordering. This equality is one of the forms of the *Kubo–Martin–Schwinger (KMS) symmetry*: All bosonic path-ordered correlators take identical values at the two endpoints of the contour $\mathcal{C}$. Note that, although we have singled out the first field in the correlator, this identity applies equally to all the fields.

The KMS symmetry is very closely tied to the fact that the system is in thermal equilibrium, since it is satisfied only when the density operator is the canonical equilibrium one. One of its consequences is that all the equilibrium correlation functions are independent of the initial time $t_i$. In order to prove this assertion, let us first note that the free propagator satisfies the KMS symmetry, and does not contain $t_i$ explicitly. A generic Feynman graph leads to time integrations that have the following structure:

$$\mathcal{G}(x_1,\cdots,x_n) = \int_\mathcal{C} dy_1^0\cdots dy_p^0\,\mathcal{F}(y_1^0,\cdots,y_p^0\,|\,x_1,\cdots,x_n). \tag{17.25}$$

(We assume that the integrals over the positions at every vertex have already been performed.) Since the free propagator does not depend on $t_i$, the derivative of the integral with respect to $t_i$ comes only from the endpoints of the integration contour, and we can write

$$\frac{\partial \mathcal{G}(x_1, \cdots, x_n)}{\partial t_i} = \sum_{i=1}^{p} \int_{\mathcal{C}} \prod_{j \neq i} dy_j^0 \left[ \mathcal{F}(\cdots, y_i^0 = t_i - i\beta, \cdots | x_1, \cdots, x_n) \right.$$
$$\left. - \mathcal{F}(\cdots, y_i^0 = t_i, \cdots | x_1, \cdots, x_n) \right] = 0. \quad (17.26)$$

The vanishing result follows from the fact that the bracket in the integrand is zero, since it is built from objects that obey the KMS symmetry. The independence with respect to $t_i$ merely reflects the fact that, in a system in thermal equilibrium, no measurement can tell at what time the system was prepared in equilibrium. From the analyticity properties of the integrand, the result of the integrations in eq. (17.25) is in fact invariant under all the deformations of the contour $\mathcal{C}$ that preserve the spacing $-i\beta$ between its endpoints.

## 17.2.6 Conserved Charges

Until now, we have considered only the simplest case of a boson field coupled to a thermal bath. Although energy is conserved, the system under consideration may exchange energy with the environment, which translates into the canonical density operator $\exp(-\beta \mathcal{H})$.

Let us now consider a Hermitian operator $\mathcal{Q}$ that commutes with the Hamiltonian, i.e., that corresponds to a conserved quantity. A field $\phi$ is said to carry a charge $q$ if it obeys the following commutation relation:

$$[\mathcal{Q}, \phi(x)] = -q\, \phi(x). \quad (17.27)$$

Note that $q$ is an eigenvalue of $\mathcal{Q}$ and is therefore real. Thus, only complex fields can have $q \neq 0$.

When there are additional conserved quantities such as $\mathcal{Q}$, their conservation constrains in a similar fashion how they may be exchanged with the heat bath. The canonical equilibrium ensemble must be generalized into the grand canonical ensemble, in which the density operator of the subsystem is given by

$$\rho \equiv \exp\left\{-\beta\left(\mathcal{H} + \mu \mathcal{Q}\right)\right\}, \quad (17.28)$$

where $\mu$ is the chemical potential associated to the charge $\mathcal{Q}$. Although we have introduced a single such charge, there could be any number of them, each accompanied by its own chemical potential. A first consequence of this generalization is that the KMS symmetry is modified by the conserved charge. Now it reads

$$\mathcal{G}(t_i, \cdots) = e^{\beta \mu q} \mathcal{G}(t_i - i\beta, \cdots), \quad (17.29)$$

where $q$ is the charge carried by the field on which the identity applies. Thus, the values of correlation functions at the endpoints are equal up to a twist factor that depends on the chemical potential.

The simplest field that can carry a non-zero charge is a complex scalar field. In the interaction picture, it can be decomposed as follows on a basis of creation and annihilation operators:

$$\phi_{in}(x) = \int \frac{d^3p}{(2\pi)^3 2E_p} \left[ a_{p,in} e^{-ip\cdot x} + b^\dagger_{p,in} e^{+ip\cdot x} \right]. \tag{17.30}$$

(This field requires two sets $\{a_{p,in}, b_{p,in}\}$ of such operators, because it describes a particle which is distinct from its anti-particle.) With this field, it is possible to construct a theory that has a global U(1) symmetry, corresponding to the conservation of the following charge:

$$Q \equiv \int \frac{d^3p}{(2\pi)^3 2E_p} \left\{ b^\dagger_{p,in} b_{p,in} - a^\dagger_{p,in} a_{p,in} \right\}. \tag{17.31}$$

It is then easy to obtain the grand-canonical averages,

$$\begin{aligned} \operatorname{tr}\left(e^{\beta(\mathcal{H}_0+\mu Q)} a^\dagger_{p,in} a_{p',in}\right) &= (2\pi)^3 \, 2E_p \, n_{\scriptscriptstyle B}(E_p - \mu q) \, \delta(\mathbf{p}-\mathbf{p}'), \\ \operatorname{tr}\left(e^{\beta(\mathcal{H}_0+\mu Q)} b^\dagger_{p,in} b_{p',in}\right) &= (2\pi)^3 \, 2E_p \, n_{\scriptscriptstyle B}(E_p + \mu q) \, \delta(\mathbf{p}-\mathbf{p}'), \end{aligned} \tag{17.32}$$

and finally obtain the free propagator for a complex scalar carrying the charge q:

$$\begin{aligned} G^0(x,y) = \int \frac{d^3p}{(2\pi)^3 2E_p} \Big[ & \left(\theta_c(x^0-y^0) + n_{\scriptscriptstyle B}(E_p-\mu q)\right) e^{-ip\cdot(x-y)} \\ &+ \left(\theta_c(y^0-x^0) + n_{\scriptscriptstyle B}(E_p+\mu q)\right) e^{+ip\cdot(x-y)} \Big]. \end{aligned} \tag{17.33}$$

### 17.2.7 Fermions

Consider now spin-1/2 fermions, whose interaction picture representation reads

$$\psi_{in}(x) = \sum_{s=\pm} \int \frac{d^3p}{(2\pi)^3 2E_p} \left\{ a^\dagger_{sp,in} v_s(p) e^{+ip\cdot x} + b_{sp,in} u_s(p) e^{+ip\cdot x} \right\}, \tag{17.34}$$

where the creation and annihilation operators obey canonical *anti-commutation* relations (see eq. (3.22)). Because they are anti-commuting fields, a minus sign appears in the derivation of the KMS identity:

$$\mathcal{G}(t_i, \cdots) = -e^{\beta\mu q} \, \mathcal{G}(t_i - i\beta, \cdots). \tag{17.35}$$

Moreover, eq. (17.32) is modified into

$$\begin{aligned} \operatorname{tr}\left(e^{\beta(\mathcal{H}_0+\mu Q)} a^\dagger_{p,in} a_{p',in}\right) &= (2\pi)^3 \, 2E_p \, n_{\scriptscriptstyle F}(E_p - \mu q) \, \delta(\mathbf{p}-\mathbf{p}'), \\ \operatorname{tr}\left(e^{\beta(\mathcal{H}_0+\mu Q)} b^\dagger_{p,in} b_{p',in}\right) &= (2\pi)^3 \, 2E_p \, n_{\scriptscriptstyle F}(E_p + \mu q) \, \delta(\mathbf{p}-\mathbf{p}'), \end{aligned} \tag{17.36}$$

where $n_{\scriptscriptstyle F}(E)$ is the Fermi–Dirac distribution,

$$n_{\scriptscriptstyle F}(E) \equiv \frac{1}{e^{\beta E}+1}, \tag{17.37}$$

## 17.2 FINITE-T PERTURBATION THEORY

and the free propagator reads

$$S^0(x,y) = \int \frac{d^3p}{(2\pi)^3 2E_p} \Big[ (\not{p}_+ + m)\big(\theta_c(x^0-y^0) - n_{_F}(E_p - \mu q)\big) e^{-ip\cdot(x-y)}$$
$$+ (\not{p}_- + m)\big(\theta_c(y^0-x^0) - n_{_F}(E_p + \mu q)\big) e^{+ip\cdot(x-y)} \Big], \tag{17.38}$$

with the notation $\not{p}_\pm \equiv \pm E_p \gamma^0 - \mathbf{p}\cdot\boldsymbol{\gamma}$.

### 17.2.8 Examples of Observables

**Thermodynamical quantities:** A central quantity that encapsulates the thermodynamical properties of a system is its partition function, defined in the canonical ensemble as

$$Z \equiv \mathrm{tr}\left(e^{-\beta\mathcal{H}}\right). \tag{17.39}$$

In perturbation theory, the logarithm of Z is obtained as the sum of all the connected vacuum graphs at finite temperature. For instance, for a scalar field, its perturbative expansion starts with the following diagrams:

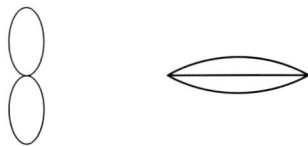

From Z, one may access various thermodynamical quantities as follows:

$$\begin{aligned}
\text{Energy:} & \quad E = -\frac{\partial Z}{\partial \beta}, \\
\text{Entropy:} & \quad S = \beta E + \ln(Z), \\
\text{Free energy:} & \quad F = E - TS = -\frac{1}{\beta}\ln(Z).
\end{aligned} \tag{17.40}$$

These quantities encode the bulk properties of the system, such as its equation of state or the existence of phase transitions.

**Production rates of weakly coupled particles:** In a system at high temperature, it is sometimes interesting to calculate the production rate of a given species of particles. First, note that this quantity is not interesting for particles that are in thermal equilibrium, since by definition they are produced and destroyed in equal amounts, so that their net production rate is zero. The real interest arises for weakly coupled particles that are not in thermal equilibrium with the bulk of the system. For instance, in a hot plasma of quarks and gluons interacting predominantly via the strong nuclear force, photons are also produced since quarks are electrically charged. However, since photons interact only electromagnetically, they may not be thermalized. This is the case for instance when the system size is small compared to the mean free path of photons (i.e., the average distance between two interactions of a photon), because in this situation the produced photons escape without re-interactions.

A pedestrian method for calculating a production rate is the following formula:

$$E_p \frac{dN_\gamma}{dt d^3x d^3p} \propto \int_{\substack{\text{unobserved} \\ \text{particles}}} \left| \begin{array}{c} E_p \\ \vdots \end{array} \right|^2 \begin{array}{c} n(E_1) \cdots n(E_n) \\ \times (1 \pm n(E_1')) \cdots (1 \pm n(E_p')) \end{array}, \qquad (17.41)$$

where the integration is over the invariant phase-space of the unobserved incoming and outgoing particles, weighted by the appropriate occupation factor ($n_B$ or $n_F$ for a particle in the initial state, and $1 + n_B$ or $1 - n_F$ for a particle in the final state). In this formula, the gray blob should be calculated with the finite-T Feynman rules.

The previous approach becomes rapidly cumbersome as the numbers of initial and final state particles increase. The bookkeeping may be simplified by using a finite-T generalization of the formula that relates the decay rate of a particle to the imaginary part of its self-energy:

$$E_p \frac{dN_\gamma}{dt d^3x d^3p} = \frac{1}{e^{\beta E_p} - 1} \operatorname{Im} \Pi_{++}(E_p, \mathbf{p}), \qquad (17.42)$$

where $\Pi_{++}$ is the photon self-energy at finite temperature (contracted with physical polarization vectors). Moreover, there exists a finite-T generalization of Cutkosky's rules (one major difference compared to zero temperature is that these rules cannot be phrased in terms of "cuts" – see the Exercise 17.3), in order to organize the perturbative calculation of the imaginary part that appears in the right-hand side.

**Transport coefficients:** Let us now discuss the case of *transport coefficients*. As their name suggest, these quantities characterize the ability of the system to carry certain (locally conserved) quantities. For instance, the electrical conductivity encodes the properties of the system with respect to the transport of electrical charges, the shear viscosity tells us about how the system reacts to a shear stress (this coefficient is related to the transport of momentum), etc. Note that in their simplest version, these quantities do not depend on frequency (in fact, they are the zero-frequency limit of a two-point function), and therefore they describe the response of the system to an infinitely slow perturbation. They can be generalized into frequency-dependent quantities that also contain information about the response to a dynamical disturbance.

The standard approach for evaluating transport coefficients is to use the *Green–Kubo formula* that relates the transport coefficient to the two-point correlation function of a current J that couples to the quantity of interest (electrical charge, momentum, etc.):

$$\begin{bmatrix} \text{transport} \\ \text{coefficient} \end{bmatrix} \sim \lim_{\omega \to 0} \frac{1}{\omega} \operatorname{Im} \int_0^{+\infty} dt d^3\mathbf{x}\, e^{-i\omega t} \left\langle [J(t, \mathbf{x}), J(0, \mathbf{0})] \right\rangle. \qquad (17.43)$$

The physical meaning of this formula is that the system is perturbed at the origin by a current J, and one measures the linear response by evaluating the same current at a generic point $(t, \mathbf{x})$. The transport coefficient is proportional to the Fourier transform of this correlation function at zero energy and momentum. Note that this formula contains the commutator of the two currents, since the two points must be causally connected (see the Exercise 14.4).

## 17.2.9 Matsubara Formalism

The perturbative rules that we have derived so far are expressed in coordinate space, which is usually not very appropriate for explicit calculations. The standard way of turning them into a set of rules in momentum space is to Fourier transform all the propagators, and to rely on the fact that the Fourier transform of a convolution product is the ordinary product of the Fourier transforms, i.e., symbolically:

$$\text{FT}\left(F * G\right) = \text{FT}\left(F\right) \times \text{FT}\left(G\right). \tag{17.44}$$

However, the main difficulty in doing this at finite temperature is that the time integration in the "convolution product" involves an integration over the complex-shaped contour $\mathcal{C}$, which makes it unclear whether we may use the above identity.

Two main solutions to this problem have been devised. The first one is the *imaginary time formalism*, also known as the *Matsubara formalism*, that we have already presented superficially in Section 5.7.2. The main motivation for this formulation is that the quantities that describe the thermodynamics of a system in thermal equilibrium are time independent. Therefore, one may exploit the freedom to deform the contour $\mathcal{C}$ in order to simplify it, as shown in the following diagram:

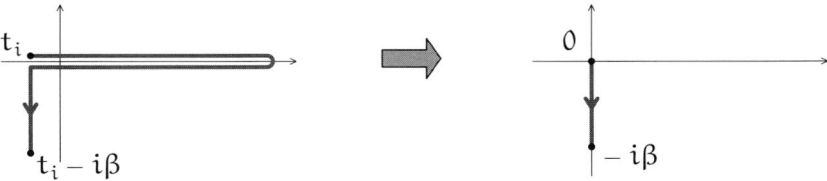

It is customary to denote $x^0 = -i\tau$, so that the variable $\tau$ is real and spans the range $[0, \beta)$ (the point $\tau = \beta$ should be removed – indeed, because of the KMS symmetry, it is redundant with the point $\tau = 0$). The imaginary time formalism corresponds to the Feynman rules derived earlier, specialized to this purely imaginary time contour. Note that one could in principle use this formalism in order to calculate time-dependent quantities. One would first obtain them as a function of imaginary times $\tau_1, \tau_2, \cdots$ and their dependence upon real times $x_1^0, x_2^0, \cdots$ may then be obtained by an analytic continuation.

From the KMS symmetry, we see that the propagator, and more generally the integrand of any Feynman diagram, is periodic (for bosons) in the variable $\tau$, with a period $\beta$. Therefore, one can go to Fourier space by decomposing the time dependence in the form of a Fourier series and by doing an ordinary Fourier transform in space,

$$G^0(\tau_x, \mathbf{x}, \tau_y, \mathbf{y}) \equiv T \sum_{n=-\infty}^{+\infty} \int \frac{d^3\mathbf{p}}{(2\pi)^3} \, e^{i\omega_n(\tau_x - \tau_y)} e^{-i\mathbf{p}\cdot(\mathbf{x}-\mathbf{y})} \, G^0(\omega_n, \mathbf{p}), \tag{17.45}$$

with $\omega_n \equiv 2\pi n T$. These discrete frequencies are called *Matsubara frequencies*. Note that for fermions, the propagator is anti-periodic with period $\beta$, and the discrete frequencies that appear in the Fourier series are $\omega_n = 2\pi(n + \frac{1}{2})T$. Moreover, if the line carries a conserved charge $q$, the Matsubara frequencies are shifted by $-i\mu q$, i.e., $\omega_n \to \omega_n - i\mu q$ ($\mu$ is the chemical potential associated to this conservation law). In the case of scalar fields, an explicit calculation gives the following free bosonic propagator in Fourier space:

$$G^0(\omega_n, \mathbf{p}) = \frac{1}{\omega_n^2 + \mathbf{p}^2 + m^2} \equiv \frac{1}{P^2 + m^2}. \quad (17.46)$$

(For the sake of brevity, we denote $P^2 \equiv \omega_n^2 + \mathbf{p}^2$.) Note that, up to a factor $-i$, this propagator is the usual free zero-temperature Feynman propagator in which one has substituted $p^0 \to i\omega_n$. Let us list here the Feynman rules for perturbative calculations in this formalism:

- Propagators:

$$G^0(\omega_n, \mathbf{p}) = \frac{1}{P^2 + m^2},$$

- Vertices: Each vertex brings a factor $\lambda$. Moreover, the sum of the $\omega_n$ and of the $\mathbf{p}$ that enter into each vertex is zero.
- Loops:

$$T \sum_{n \in \mathbb{Z}} \int \frac{d^3\mathbf{p}}{(2\pi)^3} \equiv \sumint_P.$$

(The right-hand side of this equation is a frequently used compact notation for the combination of discrete sums and integrals that appear in the Matsubara formalism. This notation includes a factor T that makes its dimension equal to four in four space-time dimensions.)

As an illustration of the use of this formalism, let us give two examples of vacuum graphs:

$$\bigcirc\!\bigcirc = \frac{\lambda}{8} \sumint_P \sumint_Q \frac{1}{(P^2 + m^2)(Q^2 + m^2)},$$

$$\bigcirc\!\!\!\bigcirc = \frac{g^2}{6} \sumint_P \sumint_Q \frac{1}{(P^2 + m^2)(Q^2 + m^2)((P+Q)^2 + m^2)}. \quad (17.47)$$

The Fourier space version of the Matsubara formalism is structurally very similar to the zero-temperature Feynman rules, which makes it quite appealing. There is one caveat, however: The continuous integrations over energies are now replaced by discrete sums, which are considerably harder to calculate. Let us expose here two general methods for evaluating these sums. The first one is based on the following representation of the propagator of eq. (17.46),

$$G^0(\omega_n, \mathbf{p}) = \frac{1}{2E_\mathbf{p}} \int_0^\beta d\tau \, e^{-i\omega_n \tau} \left[ (1 + n_{\text{B}}(E_\mathbf{p})) e^{-E_\mathbf{p}\tau} + n_{\text{B}}(E_\mathbf{p}) e^{E_\mathbf{p}\tau} \right], \quad (17.48)$$

where the integrand in the right-hand side is a mixed representation that depends on the momentum $\mathbf{p}$ and the imaginary time $\tau$. By replacing each propagator of a given graph by this formula, the discrete sums can be easily performed since they are all of the form

$$\sum_{n \in \mathbb{Z}} e^{i\omega_n \tau} = \beta \sum_{n \in \mathbb{Z}} \delta(\tau - n\beta). \quad (17.49)$$

## 17.2 FINITE-T PERTURBATION THEORY

(The left-hand side is obviously periodic in $\tau$ with period $\beta$, which is ensured in the right-hand side by the sum over infinitely many shifted copies of the delta function.) At this point, one has to integrate over the $\tau$s that have been introduced when replacing the propagators by (17.48), but these integrals are straightforward since the dependence on these times is in the form of delta functions and exponentials. Moreover, only a finite number of the delta functions that appear on the right-hand side of eq. (17.49) actually contribute, due to the constraint that each $\tau$ must be in the range $[0, \beta)$. As an illustration, consider the evaluation of the one-loop tadpole in a scalar theory with quartic coupling:

$$\begin{aligned}
\bigcirc &= \frac{\lambda}{2} \sum_P \frac{1}{P^2 + m^2} \\
&= \frac{\lambda}{2} \int \frac{d^3 p}{(2\pi)^3 2E_p} \int_0^\beta d\tau \sum_{n \in \mathbb{Z}} \delta(\tau - n\beta) \left[(1 + n_B(E_p))e^{-E_p \tau} + n_B(E_p)e^{E_p \tau}\right] \\
&= \frac{\lambda}{2} \int \frac{d^3 p}{(2\pi)^3 2E_p} \left[1 + 2n_B(E_p)\right] \\
&= \lambda \left[\frac{\Lambda^2}{16\pi^2} + \frac{T^2}{24} + \cdots\right],
\end{aligned} \qquad (17.50)$$

where $\Lambda$ is an ultraviolet cutoff that restricts the integration range $|\mathbf{p}| \leq \Lambda$ (the final formula assumes that $\Lambda \gg T$, and we have not written the terms that depend on the mass). The first term is the usual zero-temperature ultraviolet divergence, while the term coming from the Bose–Einstein distribution exists only at non-zero temperature. This second term is ultraviolet finite, thanks to the exponential decrease of the Bose–Einstein distribution at large energy. We can already note on this example that the ultraviolet divergences are identical to the zero-temperature ones. This is a general property: If the action has already been renormalized at zero temperature, there are no additional ultraviolet divergences at finite temperature. This is quite clear on physical grounds: Being at finite temperature means that one has a dense medium in which the average inter-particle distance is $T^{-1}$. However, in the ultraviolet limit, one probes distance scales that are much smaller than the inverse temperature, for which the effects of the surrounding medium are irrelevant.

An alternate method for evaluating the sums over the discrete Matsubara frequencies is to note that the function

$$\mathcal{P}(z) \equiv \frac{\beta}{e^{\beta z} - 1} \qquad (17.51)$$

has simple poles of residue 1 at all the $z = i\omega_n$. Therefore, we can write

$$\sum_{n \in \mathbb{Z}} f(i\omega_n) = \oint_\gamma \frac{dz}{2i\pi} \mathcal{P}(z) f(z), \qquad (17.52)$$

where $\gamma$ is an integration contour made of infinitesimal circles around each pole of $\mathcal{P}(z)$, as shown in the left part of Figure 17.1. The second step is to deform the contour $\gamma$, as shown in the middle of Figure 17.1. For this transformation to hold as is, with no extra term, the function $f(z)$ should not have any pole on the imaginary axis, which is usually the case. Finally, a second deformation brings the contour along the real axis. If the function $f(z)$ has poles,

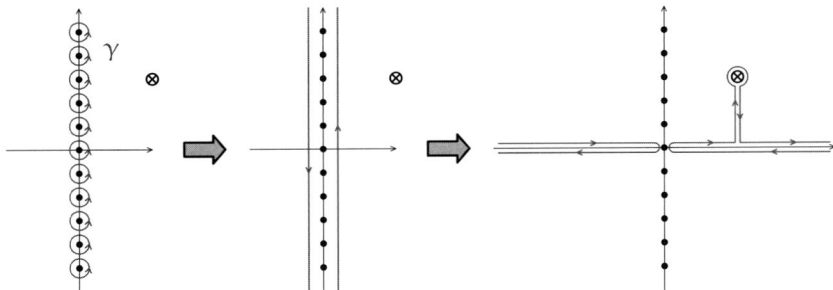

FIGURE 17.1: Successive deformations of the contour in order to calculate the discrete sums over Matsubara frequencies. The cross denotes a pole of the function $f(z)$, while the solid dots on the imaginary axis are the poles of $\mathcal{P}(z)$.

the new contour should wrap around these poles, which gives an additional contribution. Thus, after these transformations, the discrete sum over the Matsubara frequencies has been rewritten as a continuous integral along the real axis (and the weight $\mathcal{P}(z)$ becomes a Bose–Einstein distribution), plus some isolated contributions coming from poles of the summand.

## 17.2.10 Momentum Space Schwinger–Keldysh Formalism

The imaginary time formalism is particularly well suited to calculating the time-independent thermodynamical properties of a system at finite temperature. However, interesting dynamical information is also contained in time-dependent objects. In principle, one could first evaluate them in the Matsubara formalism in terms of imaginary times $\tau$ (or imaginary frequencies $i\omega_n$), and then perform an analytic continuation to real times or energies. Beyond two-point functions (i.e., for functions that depend on more than one energy, taking into account energy conservation), this analytic continuation is usually complicated and for this reason it is often desirable to be able to obtain the result directly in terms of real energies.

In fact, we may ignore[4] the vertical part of the contour $\mathcal{C}$. A heuristic justification is to let the initial time $t_i$ go to $-\infty$ and turn off adiabatically the interactions in this limit. Therefore, the canonical density operator becomes $\exp(-\beta\,\mathcal{H}_0)$ and there is no need for the vertical part of the time contour. Let us call $+$ and $-$ respectively the upper and lower horizontal branches of the contour. We may then break down the free propagator $G^0(x, y)$ into four propagators $G^0_{++}$, $G^0_{--}$, $G^0_{+-}$ and $G^0_{-+}$ depending on where $x^0, y^0$ are located, and Fourier transform each of them separately. For a scalar field, this gives

$$\begin{aligned}
G^0_{++}(p) &= \frac{i}{p^2 - m^2 + i\epsilon} + 2\pi\, n_{\scriptscriptstyle B}(E_{\mathbf{p}})\, \delta(p^2 - m^2), \\
G^0_{+-}(p) &= 2\pi\, (\theta(-p^0) + n_{\scriptscriptstyle B}(E_{\mathbf{p}}))\, \delta(p^2 - m^2), \\
G^0_{--}(p) &= \left[G^0_{++}(p)\right]^*, \qquad G^0_{-+}(p) = G^0_{+-}(-p).
\end{aligned} \qquad (17.53)$$

---

[4] A more careful treatment of the vertical part of the contour indicates that its effect it to replace the statistical distribution $n_{\scriptscriptstyle B}(E_{\mathbf{p}})$ by $n_{\scriptscriptstyle B}(|p^0|)$ in eq. (17.53). See the discussion after eq. (17.61).

## 17.2 FINITE-T PERTURBATION THEORY

Note that these propagators are very closely related to those of the Schwinger–Keldysh formalism at zero temperature (see eq. (2.76)), since we have

$$\text{for } \epsilon, \epsilon' = \pm, \qquad G^0_{\epsilon\epsilon'}(p) = \left[G^0_{\epsilon\epsilon'}(p)\right]_{T=0} + 2\pi\, n_{\text{B}}(E_{\mathbf{p}})\, \delta(p^2 - m^2). \tag{17.54}$$

The rules for the vertices and loops are identical to those of the Schwinger–Keldysh formalism at zero temperature, namely:

- one must assign types $+$ and $-$ to the vertices of a diagram in all the possible ways;
- each vertex of type $+$ brings a factor $-i\lambda$ and each type $-$ vertex a factor $+i\lambda$;
- a vertex of type $\epsilon$ and a vertex of type $\epsilon'$ are connected by the free propagator $G^0_{\epsilon\epsilon'}$;
- each loop momentum must be integrated with the measure $d^4p/(2\pi)^4$.

Since this formalism is a simple extension of the zero-temperature Schwinger–Keldysh formalism (the only difference being the propagators in eq. (17.54)), it also makes the connection with perturbation theory at zero temperature more transparent.

In the Matsubara formalism, the KMS symmetry is trivially encoded in the fact that all the objects depend only on the discrete frequencies $\omega_n$. In the Schwinger–Keldysh formalism, it is somewhat more obfuscated. A generic n-point function $\Gamma_{\epsilon_1\cdots\epsilon_n}(p_1\cdots p_n)$, amputated of its external legs, obeys the following two identities:

$$\sum_{\epsilon_1\cdots\epsilon_n=\pm} \Gamma_{\epsilon_1\cdots\epsilon_n}(k_1,\cdots,k_n) = 0,$$

$$\sum_{\epsilon_1\cdots\epsilon_n=\pm} \left[\prod_{\{i|\epsilon_i=-\}} e^{-\beta k_i^0}\right] \Gamma_{\epsilon_1\cdots\epsilon_n}(k_1,\cdots,k_n) = 0. \tag{17.55}$$

It is the second of these identities that encodes the KMS symmetry.

Finally, let us note for later use that the four propagators of eqs. (17.53) can be related to the zero-temperature Feynman propagator and its complex conjugate by the following formula:

$$\begin{pmatrix} G^0_{++} & G^0_{+-} \\ G^0_{-+} & G^0_{--} \end{pmatrix} = U \begin{pmatrix} G^0_F & 0 \\ 0 & G^{0\,*}_F \end{pmatrix} U \tag{17.56}$$

with

$$U(p) \equiv \begin{pmatrix} \sqrt{1+n_{\text{B}}} & \dfrac{\theta(-p^0)+n_{\text{B}}}{\sqrt{1+n_{\text{B}}}} \\ \dfrac{\theta(+p^0)+n_{\text{B}}}{\sqrt{1+n_{\text{B}}}} & \sqrt{1+n_{\text{B}}} \end{pmatrix} \tag{17.57}$$

and

$$G^0_F(p) \equiv \frac{i}{p^2 - m^2 + i0^+}. \tag{17.58}$$

**Resummation of a mass:** In eq. (17.57), we have voluntarily not written the argument of the Bose–Einstein distribution. Given its origin in the propagators listed in eqs. (17.53), this argument could be $E_{\mathbf{p}}$ or $|p^0|$. It turns out that the second possibility is the correct one.

In order to see this, consider a mass term, but instead of including the mass directly into the propagators (as in eq. (17.53)), let us start from massless propagators and resum the mass to all orders. In order to simplify this calculation, let us introduce the following compact notations:

$$\mathbb{G}_0 \equiv \begin{pmatrix} G^0_{++} & G^0_{+-} \\ G^0_{-+} & G^0_{--} \end{pmatrix}_{m=0}, \quad \mathbb{G}_m \equiv \begin{pmatrix} G^0_{++} & G^0_{+-} \\ G^0_{-+} & G^0_{--} \end{pmatrix}_m, \tag{17.59}$$

the $2 \times 2$ matrix of Schwinger–Keldysh propagators, without and with the mass, and

$$\mathbb{D}_0 \equiv \begin{pmatrix} G^0_F & 0 \\ 0 & G^{0*}_F \end{pmatrix}_{m=0}, \quad \mathbb{D}_m \equiv \begin{pmatrix} G^0_F & 0 \\ 0 & G^{0*}_F \end{pmatrix}_m \tag{17.60}$$

the corresponding diagonal matrices made of the Feynman propagator and its complex conjugate. The massive propagators obtained by explicitly summing the mass term are given by

$$\begin{aligned}
\mathbb{G}_m &= \mathbb{G}_0 \sum_{n=0}^{\infty} \left( -im^2 \sigma^3 \mathbb{G}_0 \right)^n \\
&= \mathbb{U} \mathbb{D}_0 \sum_{n=0}^{\infty} \left( -im^2 \mathbb{U} \sigma^3 \mathbb{U} \mathbb{D}_0 \right)^n \mathbb{U} \\
&= \mathbb{U} \mathbb{D}_0 \sum_{n=0}^{\infty} \left( -im^2 \sigma^3 \mathbb{D}_0 \right)^n \mathbb{U} = \mathbb{U} \mathbb{D}_m \mathbb{U}.
\end{aligned} \tag{17.61}$$

In the first line, the third Pauli matrix $\sigma^3$ provides the necessary signs for the vertices of types $+$ and $-$ in the Schwinger–Keldysh formalism. The third line uses the fact that $\mathbb{U}\sigma^3\mathbb{U} = \sigma^3$. In the final result, only the matrix $\mathbb{D}$ is affected by the mass, while the matrix $\mathbb{U}$ has remained unchanged. If we use the on-shell energy $E_p$ as the argument of $n_B$ in the matrix $\mathbb{U}$, then this argument is $|\mathbf{p}|$ since we started from a massless propagator. With this choice, the final result would be inconsistent, since the poles of the massive propagator are at $p^0 = \pm\sqrt{\mathbf{p}^2 + m^2}$ (since the matrix $\mathbb{D}_m$ in the middle now contains the mass), but the statistical information contained in the $\mathbb{U}$s would still be massless. In contrast, using $|p^0|$ as the argument of $n_B$ ensures that the energy inside $n_B$ follows the poles of the propagator, and correctly picks the change due to the mass. We also see that the (incorrect) prescription $n_B(E_p)$ is equivalent to neglecting the vertical path of the contour, since it amounts to keeping the interactions (here, the mass term, treated as an interaction) in the time evolution of the system but not in the density operator.

## 17.2.11 Retarded Basis

**Change of basis:** All the objects that appear in the Schwinger–Keldysh formalism carry indices that take the values $+$ or $-$. Variants of this formalism may be obtained by performing linear combinations of these two indices, akin to a change of basis. For any n-point function $G^{\{\epsilon_i\}}$ in the $\pm$ basis, we may define

$$G^{\{\chi_i\}}(k_1, \cdots, k_n) \equiv \sum_{\{\epsilon_i = \pm\}} G^{\{\epsilon_i\}}(k_1, \cdots, k_n) \prod_{i=1}^{n} U^{\chi_i \epsilon_i}(k_i), \tag{17.62}$$

## 17.2 FINITE-T PERTURBATION THEORY

where $U$ is an invertible "rotation" matrix. The new indices $X_i$ also take two values, which we may denote 1 and 2. For consistency, Feynman graphs amputated of their external lines must be related by

$$\Gamma^{\{X_i\}}(k_1,\cdots,k_n) \equiv \sum_{\{\epsilon_i=\pm\}} \Gamma^{\{\epsilon_i\}}(k_1,\cdots,k_n) \prod_{i=1}^{n} V^{X_i\epsilon_i}(k_i), \qquad (17.63)$$

where the matrix $V$ is defined by

$$V^{X\epsilon}(k) \equiv ((U^T)^{-1})^{X\epsilon}(-k). \qquad (17.64)$$

In particular, this formula gives the expression of the vertices in the new formalism. For instance, in a $\phi^4$ scalar theory, we have

$$-i\lambda^{ABCD}(k_1,\cdots,k_4) = -i\lambda \sum_{\epsilon=\pm} \epsilon \, V^{A\epsilon}(k_1) V^{B\epsilon}(k_2) V^{C\epsilon}(k_3) V^{D\epsilon}(k_4). \qquad (17.65)$$

Note that the new vertices may be momentum dependent if the rotation matrix is. Moreover, there could be up to $2^4$ non-zero quartic vertices, while there are only two in the original Schwinger–Keldysh formalism (but we will see shortly that these rotations may reduce the number of non-zero entries for the propagators, which is sometimes an advantage). The n-point functions in the new basis may be obtained directly in perturbation theory, in terms of Feynman diagrams made of the bare propagators and vertices of the new basis.

**Retarded-advanced formalism:** A convenient choice of rotation consists in exploiting the two relations satisfied by the Schwinger–Keldysh propagators,

$$G_{++}(p) + G_{--}(p) = G_{-+}(p) + G_{+-}(p),$$
$$G_{-+}(p) = e^{p^0/T} G_{+-}(p), \qquad (17.66)$$

in order to generate two zero entries in the new propagators. Note that the second of these identities is equivalent to KMS, and is therefore only valid in thermal equilibrium. For bosons, the matrix $U$ that achieves this is

$$U(k) = \frac{1}{a(-k^0)} \begin{pmatrix} a(k^0)a(-k^0) & -a(k^0)a(-k^0) \\ -n_B(-k^0) & -n_B(k^0) \end{pmatrix}, \qquad (17.67)$$

where $a(k^0)$ is an arbitrary non-vanishing function. A similar transformation exists for fermions. In all cases, it is such that the new propagators become

$$G^{XY}(k) = \begin{pmatrix} 0 & G_A(k) \\ G_R(k) & 0 \end{pmatrix}, \qquad (17.68)$$

where $G_{R,A}$ are the retarded and advanced propagators:

$$G_R(x,y) = G_{++}(x,y) - G_{+-}(x,y), \quad G_A(x,y) = G_{++}(x,y) - G_{-+}(x,y).$$

(See the Exercise 2.11.) In this formalism, it is customary to denote R and A the values taken by the indices X, Y (therefore, the term in the upper-right location is $G^{RA} \equiv G_A$, and the other non-zero term is $G^{AR} \equiv G_R$). The bare rotated propagators do not depend on temperature, which is now relegated into the vertices. A convenient choice is $a(k^0) = -n_B(k^0)$, which leads to the following vertices in a $\phi^4$ scalar theory,

$$\begin{aligned}
\lambda^{AAAA} &= \lambda^{RRRR} = 0, \\
\lambda^{ARRR} &= \lambda, \\
\lambda^{AARR}(k_1, \cdots, k_4) &= -\lambda \left[1 + n_B(k_1^0) + n_B(k_2^0)\right], \\
\lambda^{AAAR}(k_1, \cdots, k_4) &= \lambda \big[(1 + n_B(k_1^0))(1 + n_B(k_2^0))(1 + n_B(k_3^0)) \\
&\quad - n_B(k_1^0) n_B(k_2^0) n_B(k_3^0)\big].
\end{aligned} \qquad (17.69)$$

(The vertices we have not written explicitly are obtained by circular permutations.) The general expression of the vertices in the retarded–advanced formalism is

$$\lambda^{\{X_i\}} = \lambda \frac{\prod\limits_{i | X_i = A} n_B(-k_i^0)}{n_B\left(\sum\limits_{i | X_i = R} k_i^0\right)}. \qquad (17.70)$$

(For fermionic lines, we must replace $n_B$ by $-n_F$, and shift the argument by $-q\mu$ if the line carries a conserved charge.) This formalism, compared to the original Schwinger–Keldysh one, has a number of advantages:

- Thanks to eq. (17.70), the Bose–Einstein (or Fermi–Dirac in the case of fermions) functions are conveniently factorized in each Feynman graph.
- In this formalism, the two identities (17.55) satisfied by n-point functions take a particularly simple form,

$$\Gamma^{A\cdots A} = \Gamma^{R\cdots R} = 0, \qquad (17.71)$$

  which renders immediate the simplifications allowed by these identities.
- The retarded-advanced formalism has close connections to the Matsubara formalism, since every R/A n-point function can be obtained as a linear combination of the analytical continuations ($i\omega_n \to p^0 \pm i0^+$) of the corresponding function in the Matsubara formalism.

## 17.3 Large-Distance Effective Theories

### 17.3.1 Infrared Divergences

Quantum field theories with massless bosons at non-zero temperature suffer from pathologies in the infrared sector, due to the low-energy behavior of the Bose–Einstein distribution,

$$n_B(E) \underset{E \ll T}{\approx} \frac{T}{E} \gg 1. \qquad (17.72)$$

## 17.3 LARGE-DISTANCE EFFECTIVE THEORIES

As we shall see now, using a massless $\phi^4$ scalar field theory as a playground, this leads to loop contributions that exhibit soft divergences. The simplest graph that suffers from this problem is the two-loop graph

which has two nested tadpoles. Let us assume that the uppermost tadpole has already been combined with the corresponding one-loop ultraviolet counterterm, so that only the finite part remains, and denote $\mu^2$ the finite remainder. From eq. (17.50), its expression is given by

$$\mu^2 \equiv \frac{\lambda T^2}{24}. \tag{17.73}$$

(This is the exact result for the temperature-dependent part in a massless theory.) With this shorthand, we have

$$
\begin{aligned}
\text{\raisebox{-0.5em}{\includegraphics[height=2em]{tadpole}}} &= \frac{\lambda \mu^2}{2} \sum_P \frac{1}{(P^2)^2} \\
&= \frac{\lambda \mu^2}{2} \int \frac{d^3p}{(2\pi)^3} \underbrace{\frac{n_B(p)(1+n_B(p))}{4p^2} \left\{ \frac{2}{T} + \frac{e^{\beta p} - e^{-\beta p}}{p} \right\}}_{\underset{p \ll T}{\approx} \frac{T}{p^4}} \\
&= \frac{\lambda \mu^2}{4\pi^2} \frac{T}{\Lambda_{IR}} + \text{infrared finite terms}, \tag{17.74}
\end{aligned}
$$

where in the last line we have introduced an *infrared cutoff* $\Lambda_{IR}$ in order to prevent a divergence at the lower end of the integration range. A similar calculation would indicate an even worse infrared singularity in the following three-loop graph,

$$\text{\raisebox{-0.5em}{graph}} \sim \lambda T \mu \frac{\mu^3}{\Lambda_{IR}^3} + \text{infrared finite terms}, \tag{17.75}$$

and more generally for n insertions of the base tadpole on the main loop,

$$\text{\raisebox{-0.5em}{graph}} \sim \lambda T \mu \left(\frac{\mu}{\Lambda_{IR}}\right)^{2n-1} + \text{infrared finite terms}. \tag{17.76}$$

Unlike ultraviolet divergences that can, in a renormalizable theory, be disposed of systematically by a redefinition of the couplings in front of a few *local* operators in the Lagrangian, it is not possible to handle infrared divergences in this manner because they correspond to long-distance phenomena. Fortunately, there is a simple way out in the present case: The series of graphs that we have started evaluating are the first terms of a geometrical series, since the repeated insertions of a tadpole equal to $\mu^2$ (after subtraction of the appropriate counterterm) merely amounts to dressing by a mass $\mu^2$ an originally massless propagator. Namely, we have

$$= \frac{\lambda}{2} \int \frac{d^3\mathbf{p}}{(2\pi)^3 2\sqrt{\mathbf{p}^2 + \mu^2}} \left[1 + 2 n_{\mathrm{B}}(\sqrt{\mathbf{p}^2 + \mu^2})\right]. \tag{17.77}$$

(The thicker propagator indicates a massive scalar with mass $\mu^2$.) The procedure used here, which consists in summing an infinite subset of (individually divergent) perturbative contributions, is a simple form of *resummation*. We can readily see that it leads to an infrared finite sum, since now the quantity $\mu^2$ plays the role of a cutoff at small momentum.

Let us now estimate the contribution of the infrared sector to this integral. At weak coupling, we have $\mu \sim \sqrt{\lambda} T \ll T$. Therefore, for momenta $p \sim \mu$, we have

$$\lambda \int dp \, p^2 \, \frac{1 + 2 n_{\mathrm{B}}(\sqrt{\mathbf{p}^2 + \mu^2})}{\sqrt{\mathbf{p}^2 + \mu^2}} \sim \lambda \int \frac{dp \, p^2 \, T}{p^2 + \mu^2} \sim \lambda T \mu \sim \lambda^{3/2} T^2. \tag{17.78}$$

This contribution comes in addition to the ultraviolet divergence $\lambda \Lambda^2$ and the contribution $\lambda T^2$ that are both contained in the first diagram of the resummed series (these terms come from momenta of order $T$ or above). We observe here an unexpected feature; the appearance of half powers of the coupling constant $\lambda$. On the surface, this is quite surprising since the power counting indicates that one power of $\lambda$ should come with each loop. This oddity is in fact a consequence of the infrared behavior of the Bose–Einstein distribution, in $T/E$, combined with the fact that the $\mu$ introduced in the resummation is of order $\lambda^{1/2}$. Although the loop expansion generates a series which is analytic in $\lambda$, this property may be spoiled if some parameters in the integrands depend on $\lambda^{1/2}$.

## 17.3.2 Screened Perturbation Theory

The resummation of the finite part of the one-loop tadpole is sufficient to screen the infrared divergences in the graphs corresponding to a strict loop expansion. However, since such a resummation amounts to a reorganization of perturbation theory (here, by already including an infinite set of graphs in the propagator), it should be done in a careful way that avoids any double counting, and ensures that we are not modifying the original theory. This can be achieved by a method, called *screened perturbation theory*, that consists of adding and subtracting a mass term to the Lagrangian:

$$\mathcal{L} = \frac{1}{2}(\partial_\mu \phi)(\partial^\mu \phi) - \frac{\lambda}{4!}\phi^4 - \frac{1}{2}\mu^2 \phi^2 + \frac{1}{2}\mu^2 \phi^2. \tag{17.79}$$

This manipulation clearly ensures that nothing is changed in the original theory. The reorganization of perturbation theory allowed by this trick comes from treating the two mass terms on different footings: the term $-\frac{1}{2}\mu^2 \phi^2$ is treated non-perturbatively by including it directly

## 17.3 LARGE-DISTANCE EFFECTIVE THEORIES

in the definition of the free propagator, while the term $+\frac{1}{2}\mu^2\phi^2$ is treated order by order, as a finite counterterm.

In this reorganization, the value of $\mu^2$ has so far been left unspecified, and it could a priori be chosen arbitrarily. A general rule governing this choice is to include in $\mu^2$ as much as possible of the large contributions coming from loop corrections to the propagator. The one-loop contribution in $\lambda T^2$ is an obvious candidate for including in $\mu^2$, since for momenta $p^2 \lesssim \lambda T^2$ this is indeed a large correction to the denominator of the propagator. At small coupling $\lambda \ll 1$, this is the dominant one. However, when the coupling increases, the propagator may receive additional large corrections from higher-order loop corrections, and an improved resummation scheme could include these additional corrections.

A further improvement, sometimes considered in some applications, is to let $\mu^2$ free and to use some reasonable condition to choose an "optimal" value. For instance, this condition may be the minimization of the one-loop correction, which in a sense would indicate that the resummation has shifted most of the relevant physics into the free propagator. For instance, one may try to achieve

$$0 = \bigcirc + \text{counterterms} = \frac{\lambda}{2} \int^{\Lambda} \frac{d^3p}{(2\pi)^3} \frac{1 + 2n_{\text{B}}(\sqrt{p^2 + \mu^2})}{2\sqrt{p^2 + \mu^2}} - \frac{\lambda \Lambda^2}{16\pi^2} - \mu^2, \quad (17.80)$$

where the two subtractions are respectively the ultraviolet counterterm and the finite counterterm necessary in order not to overcount the mass $\mu^2$. This equation, which provides an implicit definition of the mass $\mu^2$, is called a *gap equation*.[5] Because this equation is nonlinear in $\mu^2$, its solution contains all orders in $\lambda$, but at small $\lambda$ it is dominated by the one-loop result $\mu^2 = \lambda T^2/24$.

We show an application of this method to the calculation of the free energy F in Figure 17.2. In this figure, the results obtained at one loop and two loops in screened perturbation theory are compared to the first two orders ($\lambda$ and $\lambda^{3/2}$) of the ordinary perturbative expansion. First, we can see that the latter is quite unstable except at low coupling: The two subsequent orders differ substantially, and even the sign of the correction due to the interactions flips. In contrast, screened perturbation theory leads to a remarkably stable result, with very small changes when going from one loop to two loops. To a large extent, this success is due to the non-trivial coupling dependence of the mass $\mu^2$, acquired by solving the gap equation (17.80) (screened perturbation theory with only the one-loop mass, would be better than strict perturbation theory, but would encounter some difficulties at large coupling).

### 17.3.3 Symmetry Restoration at High Temperature

The thermal correction to the mass, $\mu^2 = \lambda T^2/24$, also explains why symmetries that may be spontaneously broken at low temperature are generically restored at high temperature. Let us consider, for instance, a scalar theory whose potential at zero temperature is

$$V(\phi) = -\frac{m_0^2}{2}\phi^2 + \frac{\lambda}{4!}\phi^4. \quad (17.81)$$

---

[5]The terminology comes from the fact that the solution of this equation usually shifts the energy of a particle, generating a "gap" in the spectrum, and thus requiring a non-zero energy to create such a particle.

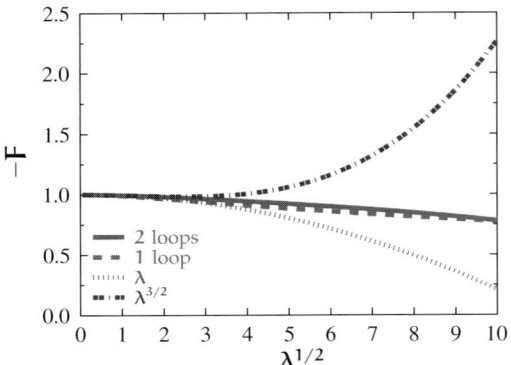

FIGURE 17.2: Free energy at non-zero temperature in the $\phi^4$ scalar field theory (normalized to the free energy of the non-interacting theory). The horizontal axis is the coupling strength $\lambda^{1/2}$. Curves "$\lambda$" and "$\lambda^{3/2}$" give the contributions of orders $\lambda$ and $\lambda^{3/2}$ in the original perturbative expansion. Curves "1 loop" and "2 loops" give the contributions of screened perturbation theory at one loop and two loops, with the mass $\mu^2$ determined as the exact solution of the gap equation. Figure from Karsch *et al.*, Phys. Lett. B401 (1997) 69.

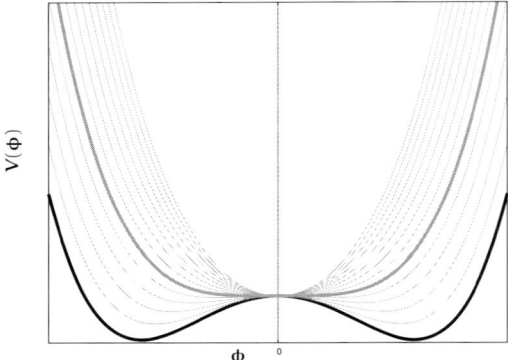

FIGURE 17.3: Evolution of a scalar potential with increasing temperature. Thick dark curve: potential with degenerate non-trivial minima at low temperature, leading to spontaneous symmetry breaking. Thick light curve: potential at the critical temperature.

Because of the sign of the mass term, this potential has two degenerate minima. The true vacuum of this theory is at a non-zero value of $\phi$, and the discrete symmetry $\phi \to -\phi$ is thus spontaneously broken. When the temperature increases, the thermal fluctuations generate a positive correction to the square of the mass, proportional to $\lambda T^2$. Eventually $m^2$ becomes positive, i.e., the potential has a unique minimum at $\phi = 0$, and the symmetry is restored (see Figure 17.3). The critical temperature, that separates the low temperature broken phase and the high temperature symmetric phase, is the point at which $m^2 = 0$.

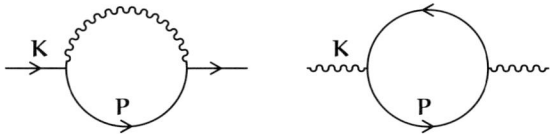

FIGURE 17.4: Fermion and photon self-energies at one loop.

### 17.3.4 Hard Thermal Loops

The scalar $\phi^4$ theory considered in the previous subsection is rather special because the one-loop tadpole diagram that gives the thermal correction to the mass is momentum independent, and because it is calculable analytically in a massless theory. In gauge theories at finite temperature, there are also important thermal corrections to the propagator of fermions and gauge bosons, but their structure is much richer. As we shall see now, the calculation of the corresponding one-loop self-energies requires an approximation based on the assumption that the loop momentum is of order of T while the external momentum is much smaller, $p \ll T$. The resulting thermal contributions are known as *hard thermal loops* (HTLs).

**Photon hard thermal loop:** A simple example of an HTL is that of a photon in QED, for which the only graph is shown on the right of Figure 17.4. In the Matsubara formalism, the expression of the photon polarization tensor is

$$\Pi^{\mu\nu}(K) = e^2 \sum_P \frac{\text{tr}\,(\gamma^\mu (\slashed{P} - \slashed{K})\gamma^\nu \slashed{P})}{P^2 (P-K)^2}. \tag{17.82}$$

(We neglect the fermion mass in this expression.)

Let us pause a moment in order to discuss the possible form of this tensor. In QED, $\Pi^{\mu\nu}(K)$ must be symmetric under the exchange of the Lorentz indices, and must obey the following Ward–Takahashi identity:

$$K_\mu \Pi^{\mu\nu}(K) = K_\nu \Pi^{\mu\nu}(K) = 0. \tag{17.83}$$

In the vacuum, this relation is sufficient to fully constrain the tensorial structure of $\Pi^{\mu\nu}$, up to an overall function of $K^2$. In the presence of a surrounding thermal bath, the situation is more complicated: Besides the metric tensor $g^{\mu\nu}$ and the 4-momentum $K^\mu$, this tensor may also depend on the 4-velocity $U^\mu$ of the thermal bath (with respect to the observer). Let us first introduce

$$V^\mu \equiv K^2 U^\mu - (K \cdot U) K^\mu. \tag{17.84}$$

Then, one may check that the Ward–Takahashi identity is satisfied by two symmetric tensors,

$$P_T^{\mu\nu} \equiv g^{\mu\nu} - \frac{K^\mu K^\nu}{K^2} - \frac{V^\mu V^\nu}{V^2}, \quad P_L^{\mu\nu} \equiv \frac{V^\mu V^\nu}{V^2}. \tag{17.85}$$

Besides being transverse to $K_\mu$, these two tensors satisfy

$$P_{T\,\nu}^{\mu} P_T^{\nu\rho} = P_T^{\mu\rho}, \quad P_{L\,\nu}^{\mu} P_L^{\nu\rho} = P_L^{\mu\rho}, \quad P_{T\,\nu}^{\mu} P_L^{\nu\rho} = 0,$$
$$P_{T\,\mu}^{\mu} = 2, \quad P_{L\,\mu}^{\mu} = 1, \tag{17.86}$$

which means that they are mutually orthogonal projectors (the values of their traces indicate that $P_T^{\mu\nu}$ encodes two degrees of freedom, while $P_L^{\mu\nu}$ contains only one). Moreover, in the rest frame of the thermal bath, we have $U^\mu = \delta^{\mu 0}$, and the first of these tensors reads

$$P_T^{00} = P_T^{i0} = P_T^{0i} = 0, \quad P_T^{ij} = \delta^{ij} - \frac{k^i k^j}{k^2}. \tag{17.87}$$

Therefore, $P_T^{\mu\nu}$ is a projector orthogonal to the 3-momentum **k**.

In terms of these projectors, the most general photon polarization tensor is of the form

$$\Pi^{\mu\nu}(K) = P_T^{\mu\nu} \Pi_T(K) + P_L^{\mu\nu} \Pi_L(K). \tag{17.88}$$

Note that in the presence of a heat bath, the functions $\Pi_{T,L}(K)$ may depend on the four components of $K^\mu$ separately (in the vacuum, the corresponding function would depend only on the Lorentz invariant $K^2$). This complication is due to the fact that the thermal bath imposes a preferred frame, which breaks Lorentz invariance. If the photon self-energy is resummed on the propagator, one obtains the following dressed propagator in a generic covariant gauge:

$$-D^{\mu\nu}(K) = \frac{P_T^{\mu\nu}}{K^2 + \Pi_T(K)} + \frac{P_L^{\mu\nu}}{K^2 + \Pi_L(K)} + \xi \frac{K^\mu K^\nu}{K^2}, \tag{17.89}$$

thanks to the orthogonality properties of the projectors (the gauge-dependent term in the propagator is not affected by the resummation). The two functions $\Pi_{T,L}(K)$ may be obtained from $\Pi^\mu{}_\mu$ and $\Pi^{00}$ by using

$$\Pi^\mu{}_\mu = 2\Pi_T + \Pi_L \quad \Pi^{00} = -\frac{k^2}{K^2} \Pi_L. \tag{17.90}$$

The fully traced polarization tensor, $\Pi^\mu{}_\mu$, is the easiest to evaluate:

$$\Pi^\mu{}_\mu(K) = e^2 \sum_P \frac{\text{tr}(\gamma^\mu(\slashed{P}-\slashed{K})\gamma_\mu \slashed{P})}{P^2(P-K)^2} = 4e^2 \sum_P \left\{ \frac{K^2}{P^2(P-K)^2} - \frac{2}{P^2} \right\}. \tag{17.91}$$

The *hard thermal loop approximation* consists in assuming that the external momentum K is much smaller than the temperature, which controls the typical loop momentum. In this approximation, we have

$$\Pi^\mu{}_\mu(K) \underset{\text{HTL}}{=} -8e^2 \underbrace{\sum_P \frac{1}{P^2}}_{-\frac{T^2}{24}} = \frac{e^2 T^2}{3}. \tag{17.92}$$

The sum-integral in this expression has a very simple tadpole structure, but note that the Matsubara frequencies are the fermionic ones, hence the result $-T^2/24$ for its thermal

## 17.3 LARGE-DISTANCE EFFECTIVE THEORIES

contribution (instead of $T^2/12$ in the bosonic case – see Exercise 17.10). The 00 component is a bit more complicated (see the Exercise 17.10):

$$\Pi^{00}(K) = e^2 \sum_P \frac{\text{tr}(\gamma^0(\slashed{P}-\slashed{K})\gamma^0 \slashed{P})}{P^2(P-K)^2} \underset{\text{HTL}}{=} e^2 \sum_P \left\{ \frac{8P^0 P^0}{P^2(P-K)^2} - \frac{4}{P^2} \right\}$$

$$\underset{\text{HTL}}{=} \frac{e^2 T^2}{3} \left[1 - \frac{k^0}{2k} \ln\left(\frac{k^0+k}{k^0-k}\right)\right]. \tag{17.93}$$

In the second equality, we have dropped non-HTL terms, and in the second line we have analytically continued the discrete Matsubara frequency to a real energy, $K^0 \to ik^0$. Therefore, the transverse and longitudinal self-energies of the photon in the HTL approximation read

$$\Pi_T(K) = \frac{e^2 T^2}{6} \frac{k^0}{k} \left[ \frac{k^0}{k} + \frac{1}{2}\left(1 - \frac{k_0^2}{k^2}\right) \ln\left(\frac{k^0+k}{k^0-k}\right) \right]$$

$$\Pi_L(K) = \frac{e^2 T^2}{3} \left(1 - \frac{k_0^2}{k^2}\right) \left[1 - \frac{k^0}{2k} \ln\left(\frac{k^0+k}{k^0-k}\right)\right]. \tag{17.94}$$

**Electron hard thermal loop:** A similar approximation can be used for the fermions in QED. Due to the breaking of Lorentz invariance caused by the thermal bath, the self-energy may be decomposed as

$$\Sigma(K) \equiv \alpha(K)\gamma^0 + \beta(K)\hat{\mathbf{p}} \cdot \boldsymbol{\gamma}, \tag{17.95}$$

where $\hat{\mathbf{p}} \equiv \mathbf{p}/|\mathbf{p}|$. Using the same method as above, one finds

$$\text{tr}(\slashed{K}\Sigma(K)) = 4(K^0\alpha(K) + k\beta(K)) \underset{\text{HTL}}{=} \frac{e^2 T^2}{2},$$

$$\text{tr}(\gamma^0\Sigma(K)) = 4\alpha(K) \underset{\text{HTL}}{=} \frac{e^2 T^2}{4k} \ln\left(\frac{k^0+k}{k^0-k}\right). \tag{17.96}$$

Moreover, the HTL approximation leads to a fermion self-energy that does not depend on the gauge chosen for the photon propagator. After summation of this self-energy to all orders, the fermion propagator becomes

$$S(K) = \frac{\gamma^0 + \hat{\mathbf{k}} \cdot \boldsymbol{\gamma}}{2(k^0 - k - \Sigma_+(K))} + \frac{\gamma^0 - \hat{\mathbf{k}} \cdot \boldsymbol{\gamma}}{2(k^0 + k + \Sigma_-(K))}, \tag{17.97}$$

where $\Sigma_\pm(K) \equiv \beta(K) \pm \alpha(K)$.

**Non-Abelian gauge theories:** In the case of a non-Abelian gauge theory such as QCD, the fermion self-energy is given by the same graph as in QED, the only change being the substitution $e^2 \to g^2 t_f^a t_f^a = g^2(N^2-1)/(2N)$ (assuming fermions in the fundamental representation of $\mathfrak{su}(N)$). Interestingly, although it is given by four graphs (see the second line in Figure 17.5), the gluon self-energy in the HTL approximation has the same form as the photon one, modulo the change $e^2 \to g^2(N + N_f/2)$ with $N_f$ the number of quark flavors. In addition, there are HTLs specific to non-Abelian gauge theories, both in the n-gluon function, and in the function with a quark–antiquark pair and $n-2$ gluons, as shown in Figure 17.5.

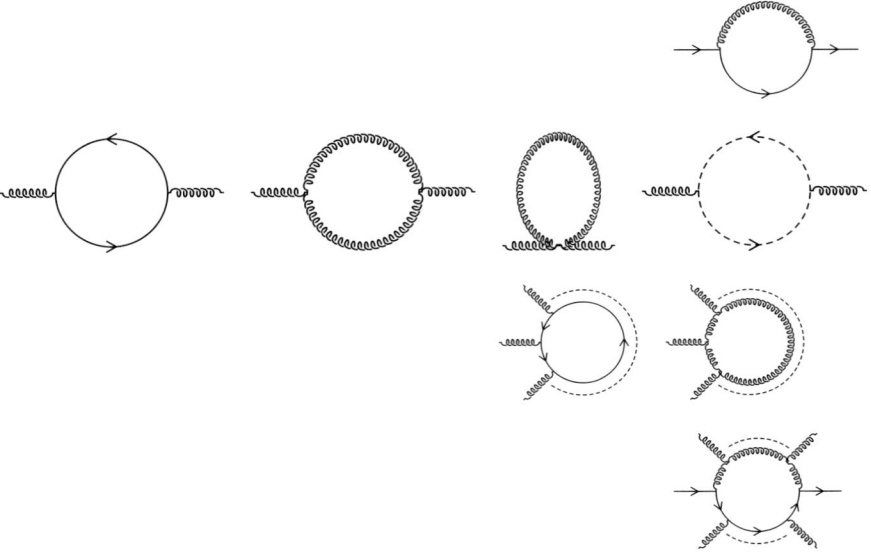

FIGURE 17.5: List of hard thermal loops in QCD.

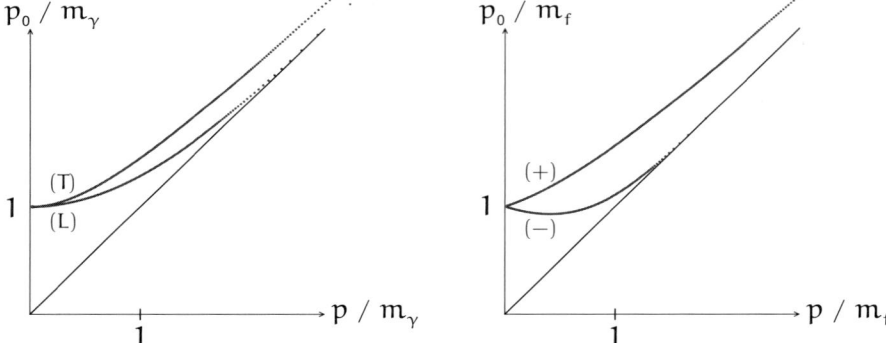

FIGURE 17.6: Photon (left) and quark (right) dispersion relations.

**Quasi-particles:** One of the effects of the summation of HTLs on the boson and fermion propagators is to shift their poles, i.e., to modify their dispersion relations (such modified excitations are called *quasi-particles*). This can be seen from the fact that the self-energies $\Pi_{T,L}$ and $\Sigma_\pm$ do not vanish on the original mass-shell, at $k^0 = \pm k$.

This modification is due to the multiple interactions of a particle with those of the surrounding bath,[6] which tend to make it heavier than it would be in the vacuum.

In the case of gauge bosons, the dispersion relations of the transverse and longitudinal modes become distinct (except at zero momentum), as shown in the left part of Figure 17.6. Another peculiarity of this change of the dispersion curves is that it does not correspond

---

[6]The same happens to electrons in a crystal, due to their interactions with phonons.

to a constant mass, but rather to a momentum-dependent one (in the figure, $m_\gamma^2 \equiv e^2T^2/9$ denotes the mass of long wavelength excitations). The residues of the poles are calculated in Exercise 17.9. For $k \gg T$, the longitudinal pole is exponentially close to the light-cone and becomes indistinguishable from the gauge dependent term. This is indeed expected, since the longitudinal mode is unphysical in the vacuum. Thus, the longitudinal mode is a purely collective phenomenon that exists only in the presence of a dense medium. In contrast, the residue of the transverse pole goes to 1 at large momentum, and one thus recovers in this limit the vacuum gauge boson propagator. Furthermore, the self-energies in eq. (17.94) are purely real above the light-cone (they can have an imaginary part only when the argument of the logarithm is negative), which implies that the shifted poles remain on the real axis. In other words, in the HTL approximation, the gauge boson excitations remain infinitely long-lived.

In the case of fermions, there are also two distinct modes, denoted $(+)$ and $(-)$, that merge at zero momentum and $k_0^2 = m_f^2 \equiv e^2T^2/8$. The $+$ mode is the analogue of the zero-temperature fermion, modified by the surrounding thermal bath (the residue of this pole goes to 1 when $k \gg T$). In contrast, the $-$ mode is a purely collective mode (this mode decouples at low temperature). Like for bosons, these fermionic modes have an infinite lifetime in the HTL approximation.

**Debye screening:** The HTL correction to the gauge boson propagator also encodes interesting phenomena in the space-like region. In particular, by taking the zero-frequency limit of the photon self-energy, and then its zero-momentum limit (in this order), one can determine how the Coulomb potential of a static electrical charge is modified at long distance. Simply recall that the Coulomb potential is given by the Fourier transform of the longitudinal[7] term in the propagator,

$$A^0(r) \sim \int \frac{d^3k}{(2\pi)^3} \frac{e^{i\mathbf{k}\cdot\mathbf{r}}}{k^2 + \Pi_L(0, \mathbf{k})}. \tag{17.98}$$

At large distance, we need the small $k$ behavior of $\Pi_L(0, \mathbf{k})$, which is given by

$$\lim_{k \to 0} \Pi_L(k_0 = 0, \mathbf{k}) = \underbrace{\frac{e^2 T^2}{3}}_{m_D^2}. \tag{17.99}$$

(The mass $m_D$ is called the Debye mass.) The Fourier transform then gives the following Coulomb potential at long distance:

$$A^0(r) \sim \frac{e^{-m_D r}}{r}, \tag{17.100}$$

which is exponentially attenuated compared to the vacuum Coulomb potential of a point-like charge. The inverse of the Debye mass characterizes the typical distance beyond which this screening is sizeable. Physically, this phenomenon is due to the fact that the test charge polarizes the charged medium surrounding it, by attracting in its vicinity charges of the opposite sign. Because of this, a distant observer sees an effective charge which is much small than the bare charge visible at short distance (see Figure 17.7).

---

[7]The transverse projector does not couple to the electromagnetic current of a static charge, e.g., an infinitely heavy charged particle. Indeed, $\bar{u}_r(\mathbf{p})\gamma^\mu u_s(\mathbf{p}+\mathbf{k}) \approx 2m\delta_{rs}\delta^{\mu 0}$ if $p^\mu = (m, \mathbf{0})$ and $\mathbf{k} \to \mathbf{0}$.

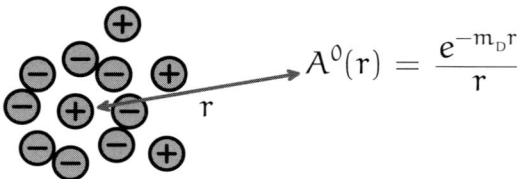

FIGURE 17.7: Debye screening in QED.

**Landau damping:** The last collective phenomenon included in HTLs is *Landau damping*, which manifests itself in the fact that the HTL self-energies have an imaginary part in the space-like region. This imaginary part indicates that a wave propagating in such a dense medium is attenuated over distance scales of order $(eT)^{-1}$. In the case of photons, the microscopic mechanism of this damping is the absorption of a photon by the surrounding electrical charges, by a process such as $e^-\gamma \to e^-$.

**Sum rules:** Propagators resummed with HTL self-energies should be used in processes involving soft momenta of order $eT$ or below, in order to capture the main collective effects. However, given the explicit form of the self-energies (recall for instance eq. (17.94)), the use of these dressed propagators complicates significantly the calculations in which they appear. Nevertheless, some integrals in which these propagators enter can be evaluated in closed form by exploiting their analytical properties.

Let us return to Minkowski space-time, and consider the retarded resummed propagators, defined as

$$\Delta^R_{T,L}(k_0, \mathbf{k}) \equiv \frac{i}{k_0^2 - k^2 - \Pi_{T,L}(k_0, \mathbf{k}) + ik_0 0^+}. \qquad (17.101)$$

This propagator admits the following *spectral representation* (see Exercise 17.7),

$$\frac{i}{k_0^2 - k^2 - \Pi_{T,L}(k_0, \mathbf{k}) + ik_0 0^+} = \int_{-\infty}^{+\infty} \frac{d\omega}{2\pi} \, \omega \, \rho_{T,L}(\omega, \mathbf{k}) \, \frac{i}{k_0^2 - \omega^2 + ik_0 0^+}, \qquad (17.102)$$

where the spectral function $\rho_{T,L}$ is defined by

$$\rho_{T,L}(k_0, \mathbf{k}) \equiv 2\,\text{Re}\,\Delta^R_{T,L}(k_0, \mathbf{k}). \qquad (17.103)$$

From eq. (17.102), we may derive other useful integrals that contain the spectral functions $\rho_{T,L}$. The starting point is to take the imaginary part of eq. (17.102), by denoting $\omega \equiv kx$ and $k_0 \equiv ky$, which gives the following identity

$$\int_{-\infty}^{+\infty} \frac{dx}{2\pi} \, x \, \rho_{T,L}(kx, \mathbf{k}) \, P\left[\frac{1}{y^2 - x^2}\right]$$

$$= \frac{k^2(y^2 - 1) - \text{Re}\,\Pi_{T,L}(ky, \mathbf{k})}{(k^2(y^2 - 1) - \text{Re}\,\Pi_{T,L}(ky, \mathbf{k}))^2 + (\text{Im}\,\Pi_{T,L}(ky, \mathbf{k}))^2}. \qquad (17.104)$$

## 17.3 LARGE-DISTANCE EFFECTIVE THEORIES

Various interesting integrals can then be obtained by taking special values of y. With $y = 0$, we obtain

$$\int_{-\infty}^{+\infty} \frac{dx}{2\pi} \frac{\rho_T(kx, \mathbf{k})}{x} = \frac{1}{k^2}, \quad \int_{-\infty}^{+\infty} \frac{dx}{2\pi} \frac{\rho_L(kx, \mathbf{k})}{x} = \frac{1}{k^2 + m_D^2}, \quad (17.105)$$

while $y = +\infty$ leads to

$$\int_{-\infty}^{+\infty} \frac{dx}{2\pi} x \rho_{T,L}(kx, \mathbf{k}) = \frac{1}{k^2}. \quad (17.106)$$

Let us also mention another exact integral involving the HTL photon self-energies (see Exercise 17.11):

$$\int_0^1 \frac{dx}{x} \frac{2\,\text{Im}\,\Pi(x)}{(k^2 + \text{Re}\,\Pi(x))^2 + (\text{Im}\,\Pi(x))^2} = \pi \left[ \frac{1}{k^2 + \text{Re}\,\Pi(\infty)} - \frac{1}{k^2 + \text{Re}\,\Pi(0)} \right], \quad (17.107)$$

where $\Pi(x)$ is any of $\Pi_{T,L}$ (the bosonic HTL self-energies depend only on the ratio $x = k^0/k$). The values at $x = \infty$ and $x = 0$ of these self-energies that appear in the right-hand side are easily determined from eq. (17.94). This integral appears for instance in the scattering cross-section of a hard particle off another particle of the thermal bath, by exchange of a soft photon (the momentum of this photon is space-like, hence $|x| \leq 1$).

**Relevant physical scales:** When discussing the physics of a weakly coupled system of particles at high temperature T (much larger than the masses), it is useful to have in mind the following hierarchy of length scales (illustrated in Figure 17.8):

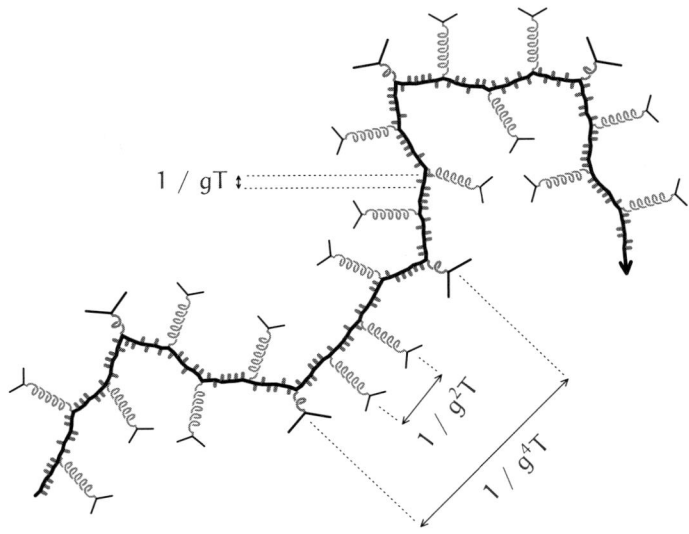

FIGURE 17.8: Relevant distance scales in a relativistic plasma at high temperature.

- $\ell = T^{-1}$. T is the typical momentum of a particle in this system, and its inverse is the typical separation between two neighboring particles. At shorter-distance scales, a particle behaves exactly as if it were in the vacuum. This is why ultraviolet renormalization at non-zero temperature can be done with the zero-temperature counterterms.
- $\ell = (gT)^{-1}$ is the typical distance over which a particle "feels" modifications of its dispersion relation. Besides the appearance of a thermal gap in the spectrum of gauge bosons and matter fields, the HTL self-energies also encode Debye screening and Landau damping.
- $\ell = (g^2 T)^{-1}$ is the mean distance between scatterings with a soft color exchange. These are forward scatterings, since the momentum transfer (of order gT, the scale of the infrared cutoff provided by the dressing of the gluon propagator) is much smaller than the momentum of the incoming particles (typically T). A gross way to obtain this scale is by estimating the corresponding scattering rate,

$$\Gamma_{\substack{\text{soft}\\\text{collisions}}} = \left| \vcenter{\hbox{\scriptsize [diagram]}} \uparrow p_\perp \right|^2 \sim g^4 T^3 \int_{gT \lesssim p_\perp} \frac{d^2 p_\perp}{p_\perp^4} \sim g^2 T, \qquad (17.108)$$

where $p_\perp$ is the momentum transfer transverse to the momentum of the incoming particles. Although these scatterings do not lead to an appreciable transport of momentum, they reshuffle the color of the particles and hence contribute to the color conductivity.
- $\ell = (g^4 T)^{-1}$ is the mean distance between scatterings with a momentum transfer of order T, i.e., those that scatter particles at large angles. Estimating this scale is done as above, but with a lower limit of order T for the momentum transfer,

$$\Gamma_{\substack{\text{hard}\\\text{collisions}}} = \left| \vcenter{\hbox{\scriptsize [diagram]}} \uparrow p_\perp \right|^2 \sim g^4 T^3 \int_{T \lesssim p_\perp} \frac{d^2 p_\perp}{p_\perp^4} \sim g^4 T. \qquad (17.109)$$

This scale is usually called the *mean free path*. This is the relevant scale for all transport phenomena that require significant momentum exchanges – for instance, the viscosity. Beyond this scale is the realm of collective effects such as sound waves (on these scales, it is more appropriate to describe the system as a fluid rather than in terms of elementary particle excitations).

**Perturbative and non-perturbative modes:** Although it is in principle possible to study any phenomenon at finite temperature in terms of the bare Lagrangian, this becomes increasingly difficult at large distance because of non-trivial collective effects. In order to circumvent this difficulty, various resummation schemes and effective descriptions have been devised, one of which is the resummation of HTLs discussed above.

Our goal here is not to give a detailed account of these various techniques, but to provide general principles regarding what can and cannot be treated perturbatively, focusing on gauge bosons. Let us first recall that a mode is perturbative if its kinetic energy dominates its interaction energy. For a mode of momentum k, the kinetic energy of a gauge field can be estimated as

$$K \sim \langle (\partial A)^2 \rangle \sim k^2 \langle A^2 \rangle. \qquad (17.110)$$

## 17.4 OUT-OF-EQUILIBRIUM SYSTEMS

For the interaction energy, we have

$$I \sim g^2 \langle A^4 \rangle \sim g^2 \langle A^2 \rangle^2. \qquad (17.111)$$

(The second part of the equation is of course not exact, but it gives the correct order of magnitude.) Thus, a mode of momentum k is perturbative if $k^2 \gg g^2 \langle A^2 \rangle$. When discussing the order of magnitude of $\langle A^2 \rangle$, it is useful to distinguish the contribution of the various momentum scales by defining

$$\langle A^2 \rangle_{\kappa^*} \sim \int^{\kappa^*} \frac{d^3 \mathbf{p}}{E_\mathbf{p}} \, n_{\text{B}}(E_\mathbf{p}), \qquad (17.112)$$

the contribution of all the thermal modes up to the scale $\kappa^*$. From these considerations, we can now distinguish three types of modes:

- **Hard modes:** $k \sim T$. For these modes, we have $\langle A^2 \rangle_T \sim T^2$, and $K \gg I$. They are therefore fully perturbative at weak coupling g.
- **Soft modes:** $k \sim gT$. For these modes, $k^2 \sim g^2 \langle A^2 \rangle_T$, which implies that the soft modes interact strongly with the hard modes. However, we also have $\langle A^2 \rangle_{gT} \sim gT^2$, so that $k^2 \gg g^2 \langle A^2 \rangle_{gT}$. Thus, the soft modes interact perturbatively among themselves. Consequently, it is possible to describe perturbatively the soft modes, provided one has performed first a resummation of the contribution of the hard modes. Screened perturbation theory and hard thermal loops are a realization of this idea.
- **Ultrasoft modes:** $k \sim g^2 T$. For these modes, we have $\langle A^2 \rangle_{g^2 T} \sim g^2 T^2$, so that $k^2 \sim g^2 \langle A^2 \rangle_{g^2 T}$. Therefore, the ultrasoft modes interact non-perturbatively among themselves, and there is no way to treat them in a perturbative approach. A non-perturbative approach, such as lattice field theory, is necessary for this.

## 17.4 Out-of-Equilibrium Systems

Until now, we have discussed only systems in equilibrium, whose initial state is described by the canonical density operator $\rho \equiv \exp(-\beta \mathcal{H})$. However, many interesting questions could also be asked for a system which is not initially in thermal equilibrium, the prime one being to describe its relaxation toward equilibrium. In this section, we discuss a few aspects of the QFT treatment of out-of-equilibrium systems.

### 17.4.1 Pathologies of the Naive Approach

First, let us note that the Matsubara formalism does not seem prone to a simple out-of-equilibrium generalization, since the KMS symmetry (that encodes into the correlation functions the fact that the system is in equilibrium) is in a sense hardwired into the discrete Matsubara frequencies.

The Schwinger–Keldysh formalism appears to be a more adequate starting point for such a generalization. Let us first discuss a simple extension that does not work, because the reasons for its failure will teach us a useful lesson. Since in eq. (17.54) the only reference to the statistical state of the system is contained in the Bose–Einstein distribution $n_{_B}(E_{\mathbf{p}})$, we may try to replace it by an arbitrary distribution $f(\mathbf{p})$ that describes the particle distribution in an out-of-equilibrium system:[8]

$$\forall \epsilon, \epsilon' = \pm, \qquad G^0_{\epsilon\epsilon'}(p) = \left[G^0_{\epsilon\epsilon'}(p)\right]_{T=0} + 2\pi f(\mathbf{p})\,\delta(p^2 - m^2). \qquad (17.113)$$

Consider now the insertion of a self-energy $\Sigma$ on the bare propagator:

$$\mathord{-\!\!-\!\!\bigcirc\!\!\!\Sigma\!\!-\!\!-} \; = \; \sum_{\epsilon,\epsilon'=\pm} G^0_{+\epsilon}(p)\,\Sigma_{\epsilon\epsilon'}(p)\,G^0_{\epsilon'+}(p). \qquad (17.114)$$

Such an expression is delicate to expand, because it involves products of distributions that are ill-defined, such as $\delta^2(p^2 - m^2)$. Let us first determine which of these products are well defined and which are not. For this, let us write

$$\left[iP\left(\tfrac{1}{z}\right) + \pi\delta(z)\right]^2 = \pi^2\delta^2(z) - \left[P\left(\tfrac{1}{z}\right)\right]^2 + 2i\pi\delta(z)P\left(\tfrac{1}{z}\right)$$

$$= \left[\frac{i}{z+i0^+}\right]^2 = -i\frac{d}{dz}\left[\frac{i}{z+i0^+}\right]$$

$$= -i\frac{d}{dz}\left[iP\left(\tfrac{1}{z}\right) + \pi\delta(z)\right] = \left[P\left(\tfrac{1}{z}\right)\right]' - i\pi\delta'(z). \qquad (17.115)$$

From this exercise, we obtain the following two identities:

$$\pi^2\delta^2(z) - \left[P\left(\tfrac{1}{z}\right)\right]^2 = \left[P\left(\tfrac{1}{z}\right)\right]'$$

$$2\delta(z)P\left(\tfrac{1}{z}\right) = -\delta'(z). \qquad (17.116)$$

Since the derivative of a distribution is well defined, this indicates that certain products (or combinations of products) of delta functions and principal values are well defined, but not all of them (for instance, the product $\delta^2(z)$ makes no sense).

Returning now to eq. (17.114) and expanding the propagators, we see that it contains terms that are ill-defined:

$$\mathord{-\!\!-\!\!\bigcirc\!\!\!\Sigma\!\!-\!\!-} \; = \; \left[\begin{array}{c}\text{well-defined}\\\text{distributions}\end{array}\right] + \pi^2\delta^2(p^2-m^2)\left[(1+f(\mathbf{p}))\Sigma_{+-} - f(\mathbf{p})\Sigma_{-+}\right], \qquad (17.117)$$

where we have used the first of eqs. (17.55) in order to simplify the combination of self-energies that appear in the square bracket. Note that the square bracket vanishes *in equilibrium* thanks to the KMS symmetry. We are thus facing a very peculiar pathology that exists only out of equilibrium.

---

[8] Note, however, that this would not encompass the most general initial states, only those for which the initial correlations are only two-point correlations.

## 17.4 OUT-OF-EQUILIBRIUM SYSTEMS

We may learn a bit more about this issue by formally resumming the self-energy $\Sigma$ on the propagator. Let us introduce the following notations:

$$\mathbb{G}^0 \equiv \begin{pmatrix} G^0_{++} & G^0_{+-} \\ G^0_{-+} & G^0_{--} \end{pmatrix}, \quad \mathbb{D} \equiv \begin{pmatrix} G^0_F & 0 \\ 0 & G^{0*}_F \end{pmatrix}, \quad \mathbb{S} \equiv \begin{pmatrix} \Sigma_{++} & \Sigma_{+-} \\ \Sigma_{-+} & \Sigma_{--} \end{pmatrix}, \quad (17.118)$$

and consider the resummed propagator defined by

$$\mathbb{G} \equiv \sum_{n=0}^{\infty} \left[\mathbb{G}^0(-i\mathbb{S})\right]^n \mathbb{G}^0. \quad (17.119)$$

A straightforward calculation shows that

$$\mathbb{G} = \mathbb{U} \begin{pmatrix} G_F & G_F \widetilde{\Sigma} G^*_F \\ 0 & G^*_F \end{pmatrix} \mathbb{U} \quad (17.120)$$

where $\mathbb{U}$ is the matrix defined in eq. (17.57), but with $f(\mathbf{p})$ instead of the Bose–Einstein distribution, and where we have used the following notations:

$$G_F(p) \equiv \frac{i}{p^2 - m^2 - \Sigma_F + i0^+},$$
$$\Sigma_F \equiv \Sigma_{++} + \Sigma_{+-},$$
$$\widetilde{\Sigma} \equiv \frac{1}{1 + f(\mathbf{p})}\left[(1 + f(\mathbf{p}))\Sigma_{+-} - f(\mathbf{p})\Sigma_{-+}\right]. \quad (17.121)$$

Note that the Feynman propagator and its complex conjugate have mirror poles on each side of the real energy axis. If the self-energy $\Sigma_F$ has no imaginary part, then these poles "pinch" the real axis and lead to a singularity in the product $G_F G^*_F$ (this is in fact a pathology of the same nature as the product $\delta^2$ in eq. (17.117)). By performing explicitly the multiplication with the matrix $\mathbb{U}$, we obtain the resummed propagator in the following form:

$$G_{\epsilon\epsilon'}(p) = \left[G^0_{\epsilon\epsilon'}(p)\right]_{T=0} + 2\pi f(\mathbf{p})\delta(p^2 - m^2)$$
$$+ \left[(1 + f(\mathbf{p}))\Sigma_{+-} - f(\mathbf{p})\Sigma_{-+}\right] G_F(p) G^*_F(p). \quad (17.122)$$

Since it does not depend on the indices $\epsilon\epsilon'$, the pathological term (on the second line) appears on the same footing as the second term, which contains the distribution $f(\mathbf{p})$. Thus, the lesson of this calculation is that one may consider hiding this pathology into a redefinition of the distribution $f(\mathbf{p})$. However, the naive formalism that we have tried to use so far is too rigid for doing this consistently, and must be amended in a number of ways:

- The initial time $t_i$ should not be taken to $-\infty$, as is done when using the Schwinger–Keldysh formalism in momentum space. Indeed, this is the time at which the system was prepared in an out-of-equilibrium state. If it were equal to $-\infty$, the system would have had an infinite amount of time for relaxing to equilibrium at the finite time where a measurement is performed. Note that observables will in general depend on the initial time $t_i$, in contrast with what happens in equilibrium.

- The Schwinger–Keldysh formalism in momentum space assumes that the system is invariant by translation, in particular in the time direction. This is clearly not the case when the system starts out-of-equilibrium, since it is expected to evolve toward equilibrium. Thus, one should stick to the formalism in coordinate space.

### 17.4.2 Kadanoff–Baym Equations

The Kadanoff–Baym equations, which we shall derive now, may be viewed as a kind of *quantum kinetic equation*. These equations are exact, but contain a self-energy that must be truncated to a manageable number of diagrams in order to be usable in practical applications. In the next subsection, we will show how the traditional kinetic equations can be derived from the Kadanoff–Baym equations.

The starting point is the Dyson equation, written in coordinate space, that expresses the resummation of a self-energy on the propagator:

$$G(x,y) = G^0(x,y) + \int_{\mathcal{C}} d^4u\, d^4v\, G^0(x,u)\big(-i\Sigma(u,v)\big)G(v,y),$$
$$G(x,y) = G^0(x,y) + \int_{\mathcal{C}} d^4u\, d^4v\, G(x,u)\big(-i\Sigma(u,v)\big)G^0(v,y), \quad (17.123)$$

where $G^0$ is the free propagator and $G$ is the resummed one. Note that the time integrations run over the Schwinger–Keldysh contour $\mathcal{C}$. Here, we have written the equation in two ways, depending on whether the self-energy is inserted on the right or on the left of the bare propagator (in the end, the resulting propagator $G$ is the same in both cases). Next, we apply the operator $\Box_x + m^2$ on the first equation and $\Box_y + m^2$ on the second equation. This eliminates the bare propagators, and we obtain

$$(\Box_x + m^2)G(x,y) = -i\delta_c(x-y) - \int_{\mathcal{C}} d^4v\, \Sigma(x,v)\, G(v,y),$$
$$(\Box_y + m^2)G(x,y) = -i\delta_c(x-y) - \int_{\mathcal{C}} d^4v\, G(x,v)\, \Sigma(v,y), \quad (17.124)$$

where $\delta_c(x-y)$ is the generalization of the delta function to the contour $\mathcal{C}$. This is one of the forms of the Kadanoff–Baym equations.

### 17.4.3 From QFT to Kinetic Theory

Kinetic theory is an approximation of the underlying dynamics in terms of a space-time-dependent distribution of particles $f(x,\mathbf{p})$. One may note right away that this is necessarily an approximate description, because it is not possible to define simultaneously the position and momentum of a particle.

In the Kadanoff–Baym equations (17.124), the dressed propagator $G$ and the self-energy $\Sigma$ are in general not invariant under translations, precisely because the system is out of equilibrium. Therefore, one may not Fourier transform them in the usual way. Instead, one uses a *Wigner transform*, defined as follows:

$$F(X, p) \equiv \int d^4 s \, e^{ip \cdot s} \, F\left(X + \frac{s}{2}, X - \frac{s}{2}\right), \quad (17.125)$$

where $F(x, y)$ is a generic two-point function (we use the same symbol for its Wigner transform, since the arguments are sufficient to distinguish them). In other words, the Wigner transform is an usual Fourier transform with respect to the separation $s \equiv x - y$, and the result still depends on the mid-point $X \equiv (x + y)/2$. Note that in eq. (17.125), the time integration is over the real axis, not over the contour $\mathcal{C}$. Wigner transforms do not share with the Fourier transform their properties with respect to convolution. Given two two-point functions F and G, let us define

$$H(x, y) \equiv \int d^4 z \, F(x, z) \, G(z, y). \quad (17.126)$$

The Wigner transform of H is given by (see Exercise 17.12)

$$H(X, p) = F(X, p) \exp\left\{\frac{i}{2} [\overleftarrow{\partial}_x \overrightarrow{\partial}_p - \overrightarrow{\partial}_x \overleftarrow{\partial}_p]\right\} G(X, p), \quad (17.127)$$

where the arrows indicate on which side the corresponding derivative acts. The right-hand side of this formula reduces to the ordinary product of the transforms when there is no X dependence, i.e., when the functions F and G are translation invariant. The first correction to the translation invariant case is proportional to the Poisson bracket of the functions F and G,

$$H(X, p) = F(X, p) G(X, p) + \frac{i}{2} \{F(X, p), G(X, p)\} + \cdots. \quad (17.128)$$

The derivatives with respect to $x$ and $y$ that appear in the Kadanoff–Baym equations can be written in terms of derivatives with respect to X and s,

$$\partial_x = \frac{1}{2}\partial_X + \partial_s, \quad \partial_y = \frac{1}{2}\partial_X - \partial_s$$
$$\Box_x = \frac{1}{4}\Box_X + \partial_X \cdot \partial_s + \Box_s, \quad \Box_y = \frac{1}{4}\Box_X - \partial_X \cdot \partial_s + \Box_s. \quad (17.129)$$

In these operators, the Wigner transform just amounts to a substitution

$$\partial_s \to -ip, \quad \Box_s \to -p^2. \quad (17.130)$$

In order to go from the Kadanoff–Baym equations to kinetic equations, two approximations are necessary:

1. **Gradient approximation:** $p \sim \partial_s \gg \partial_X$. The derivatives with respect to the mid-point X characterize the space and time scales over which the properties of the system (e.g., its particle distribution) change significantly. This approximation therefore means that these scales, that characterize the off-equilibriumness of the system, should be much larger than the De Broglie wavelength of the particles. Another way to state this approximation is that the mean free path in the system should be much larger than the wavelength of the particles, which amounts to a certain diluteness of the system. Using this approximation

in the two Kadanoff–Baym equations (17.124), taking their difference, and breaking it down into its $++$, $--$, $+-$ and $-+$ components, one obtains

$$-2ip \cdot \partial_X \left(G_{+-}(X,p) - G_{-+}(X,p)\right) = 0,$$
$$-2ip \cdot \partial_X \left(G_{+-}(X,p) + G_{-+}(X,p)\right) = 2\left[G_{-+}\Sigma_{+-} - G_{+-}\Sigma_{-+}\right]. \tag{17.131}$$

2. **Quasi-particle approximation:** This approximation amounts to assuming that the dressed propagators $G_{\epsilon\epsilon'}$ can be written in terms of a local particle distribution $f(X,\mathbf{p})$ as in eqs. (17.113). This is equivalent to

$$G_{-+}(X,p) = (1 + f(X,\mathbf{p}))\,\rho(X,p),$$
$$G_{+-}(X,p) = f(X,\mathbf{p})\,\rho(X,p), \tag{17.132}$$

where $\rho(X,p) \equiv G_{-+}(X,p) - G_{+-}(X,p)$. This would be exact for non-interacting, infinitely long-lived particles. In the presence of interactions, the approximation is justified when the time between two collisions of a particle is large compared to its wavelength.

Using eqs. (17.131) and (17.132), we obtain an equation for $f(X,\mathbf{p})$, which is nothing but a *Boltzmann equation*,

$$\left[\partial_t + \mathbf{v_p} \cdot \boldsymbol{\nabla}_\mathbf{x}\right] f(X,\mathbf{p}) = \underbrace{\frac{i}{2E_\mathbf{p}}\left[(1 + f(X,\mathbf{p}))\Sigma_{+-} - f(X,\mathbf{p})\Sigma_{-+}\right]}_{\mathbb{C}_\mathbf{p}[f;X]}, \tag{17.133}$$

where $\mathbf{v_p} \equiv \mathbf{p}/E_\mathbf{p}$ is the velocity vector for particles of momentum $\mathbf{p}$. Note that the Boltzmann equation is spatially local since all the objects it contains are evaluated at the coordinate $X$, but its right-hand side is non-local in momentum. The right-hand side, $\mathbb{C}_\mathbf{p}[f;X]$, is called the *collision term*. The combination $\partial_t + \mathbf{v_p} \cdot \boldsymbol{\nabla}_\mathbf{x}$ that appears in the left-hand side is called the transport derivative. It is zero on any function whose $t$ and $\mathbf{x}$ dependence arise only in the combination $\mathbf{x} - \mathbf{v_p} t$ (this is the case for a distribution of non-interacting particles that move at the constant velocity $\mathbf{v_p}$ prescribed by their momentum).

In order to obtain an explicit expression of the collision term, it is necessary to truncate the self-energies to a certain order (usually, the lowest order that gives a non-zero result) in the loop expansion. In a scalar theory with a $\phi^4$ interaction, the self-energies should be evaluated at two loops:

$$\Sigma = \;\begin{array}{c}\longrightarrow\!\bigcirc\!\longrightarrow\end{array}. \tag{17.134}$$

Using the Feynman rules of the Schwinger–Keldysh formalism, this diagram leads to the following collision term:

$$\mathbb{C}_\mathbf{p}[f;X] = \frac{\lambda^2}{4E_\mathbf{p}} \int \frac{d^3\mathbf{p}_1}{(2\pi)^3 2E_1} \frac{d^3\mathbf{p}_2}{(2\pi)^3 2E_2} \frac{d^3\mathbf{p}_3}{(2\pi)^3 2E_3} (2\pi)^4 \delta(p+p_3-p_1-p_2)$$
$$\times \Big[ f(X,\mathbf{p}_1)f(X,\mathbf{p}_2)(1+f(X,\mathbf{p}_3))(1+f(X,\mathbf{p})) $$
$$- f(X,\mathbf{p}_3)f(X,\mathbf{p})(1+f(X,\mathbf{p}_1))(1+f(X,\mathbf{p}_2))\Big]. \tag{17.135}$$

The expression describes the rate of change of the particle distribution, under the effect of two-body elastic collisions. It is the difference between a production rate (coming from the term in which the particle of momentum $\mathbf{p}$ is produced, and thus weighted by a factor $1 + f(X, \mathbf{p})$) and a destruction rate (from the term in which the particle of momentum $\mathbf{p}$ is destroyed, and has a weight $f(x, \mathbf{p})$).

To close this section, let us mention an additional term that arises when the self-energy contains a local part, i.e., a term proportional to a delta function in space-time:

$$\Sigma(u, v) = \Phi(u)\delta_c(u - v) + \Pi(u, v). \tag{17.136}$$

When such a local term is present, the difference of the two Kadanoff–Baym equations contains $\Phi(y)G(x, y) - \Phi(x)G(x, y)$, whose Wigner transform at lowest order in the gradient approximation is

$$i\left(\partial_X \Phi(X)\right) \cdot \left(\partial_p G(X, p)\right). \tag{17.137}$$

This extra term leads to a somewhat modified Boltzmann equation,

$$\left[\partial_t + \mathbf{v_p} \cdot \nabla_x\right] f + \underline{\frac{1}{2E_p} \partial_X \Phi \cdot \partial_p f} = \frac{i}{2E_p}\left[(1 + f)\Pi_{+-} - f\Pi_{-+}\right]. \tag{17.138}$$

In the new term (underlined), one may interpret $\partial_X \Phi$ as a mean force field acting on the particles. Under the action of this force, the particles accelerate, which implies a change of their momentum. The left-hand side of the above equation thus describes the change of the distribution of particles under the effect of this mean field, in the absence of any collisions (which are described by the right-hand side).

## Exercises

**17.1** Prove eq. (17.9).

**\*17.2** Check eq. (17.18). Generalize the last equation by deriving the analogue of Wick's theorem for thermal expectation values of products of creation and annihilation operators.

**17.3** Reproduce the derivation of Cutkosky's cutting rules at finite temperature. At which point does the temperature introduce a structural difference? What is the physical origin of the Bose–Einstein prefactor in the right-hand side of eq. (17.42)?

**\*17.4** Derive the Green–Kubo formula that gives the electrical conductivity of a plasma of charged particles and photons.

**17.5** Derive *Abel-Plana's summation formula*,

$$\sum_{n=0}^{\infty} f(n) = \tfrac{1}{2} f(0) + \int_0^{\infty} dx\, f(x) + i \int_0^{\infty} dt\, \frac{f(it) - f(-it)}{e^{2\pi t} - 1}.$$

**\*17.6** Prove the following identity:

$$\int_0^{+\infty} dx \, \frac{x^n}{(e^x + \epsilon)^{p+1}} = (-\epsilon)^{p+1} \frac{n!}{p!} \sum_{i=0}^{p} \alpha_{p,i} \left[ 2^{i-n-1}(1+\epsilon) - \epsilon \right] \zeta(n+1-i),$$

where $\epsilon = \pm 1$ and the coefficients $\alpha_{p,i}$ are defined by $(x-1)(x-2)\cdots(x-p) \equiv \alpha_{p,0} + \alpha_{p,1} x + \cdots + \alpha_{p,p} x^p$.

**17.7** Consider the n-point path-ordered field correlation function, $G(x_1, \cdots, x_n)$.
- Using the definition of the path ordering, write this function as a sum of $n!$ terms corresponding to all the possible orderings.
- Use the KMS relations to show that this representation can be reduced to a sum of $(n-1)!$ independent terms that can be written as follows:

$$G(\{x_i^0, \mathbf{x}_i\}) = \int \left[ \prod_{i=1}^{n} \frac{d\omega_i}{2\pi} e^{-i\omega_i x_i^0} \right] \sum_{\sigma \in \mathfrak{S}_n / \mathbb{Z}_n} g_\sigma(\{\omega_i, \mathbf{x}_i\})$$

$$\times \sum_{k=1}^{n} \left[ \prod_{i=1}^{k} e^{-\beta \omega_{\sigma(i)}} \right] \theta_c(x^0_{\sigma\tau^k(1)} - x^0_{\sigma\tau^k(2)}) \cdots \theta_c(x^0_{\sigma\tau^k(n-1)} - x^0_{\sigma\tau^k(n)}),$$

where $\tau$ is the cyclic permutation $1 \to 2 \to \cdots \to n \to 1$.
- Specialize the above representation to the case $n = 2$, and massage it to obtain the spectral representation of eq. (17.102).

**\*17.8** Show that in the calculation of an all-plus correlation function, the $-$ sector decouples completely in the limit $T \to 0^+$. *Hint: Use the KMS relations in the Schwinger–Keldysh formalism to prove that*

$$\lim_{T \to 0^+} \Gamma_{\{\epsilon_i\}}(\{k_i\}) \propto \theta \left( \sum_{i | \epsilon_i = -} k_i^0 \right).$$

**17.9** Consider the transverse and longitudinal parts of the HTL-dressed photon propagator, $(K^2 - \Pi_{T,L}(K))^{-1}$ (the explicit form of the self-energies are given in eq. (17.94)). Show that the residues at the quasi-particle poles are given by

$$Z_T = \frac{2\omega_T^2(\omega_T^2 - k^2)}{\omega_T^2 m_\beta^2 - (\omega_T^2 - k^2)^2}, \quad Z_L = \frac{2\omega_L^2}{m_\beta^2 + k^2 - \omega_L^2},$$

where $\omega_{T,L}$ are the respective values of $k^0$ at the pole, and $m_\beta^2 \equiv e^2 T^2/3$.

**17.10** Perform explicitly the sum-integrals in order to obtain eq. (17.93).

**17.11** Derive eq. (17.107). *Hint: Write the spectral representation of a fictitious propagator obtained by resumming $(1-x^2)\Pi(x)$. Assume that $\Pi(x)$ obeys $\mathrm{Im}\,\Pi(x=0) = 0$, $\mathrm{Im}\,\Pi(x) = 0$ if $x \geq 1$ and $\mathrm{Re}\,\Pi(x) \geq 0$ if $x \geq 1$, like the photon or gluon HTLs.*

**\*17.12** Prove eq. (17.127).

**\*17.13** Consider SU(N) Yang–Mills theory at finite T, in the imaginary time formalism.

# EXERCISES

- The *center* of the gauge group is the subgroup made of the elements that commute with the entire group. In the case of $SU(N)$, show that the center is $\mathbb{Z}_N \equiv \{e^{2i\pi k/N}\}_{0 \leq k < N}$, i.e., the set of Nth roots of unity (multiplied by the identity matrix).
- At finite temperature, legitimate gauge transformations should be single-valued in imaginary time, i.e., be $\beta$-periodic, $\Omega(\beta, \mathbf{x}) = \Omega(0, \mathbf{x})$. Consider now a more general transformation, called a *center transformation*, defined as a gauge transformation with a non-periodic $\Omega$. Instead, $\Omega(\tau, \mathbf{x})$ is assumed to obey $\Omega(\beta, \mathbf{x}) = \xi \Omega(0, \mathbf{x})$ with $\xi \in \mathbb{Z}_N$. Show that center transformations preserve the periodicity of the gauge field.
- A *Polyakov loop* is the trace of a Wilson loop that wraps around the imaginary time direction, at fixed position $\mathbf{x}$,

$$L(\mathbf{x}) \equiv N^{-1} \operatorname{tr} \left\langle P \exp \int_0^\beta d\tau\, A^0(\tau, \mathbf{x}) \right\rangle.$$

  Show that $L(\mathbf{x})$ is invariant under the usual (periodic) gauge transformations, but not invariant under the center transformations. What does this imply for the center transformations?
- Explain why $L(\mathbf{x})$ is related to the free energy $F_q$ of an infinitely massive quark added to the system at point $\mathbf{x}$ by $|L(\mathbf{x})| = e^{-\beta F_q}$. From this interpretation, conclude that $L(\mathbf{x})$ should be zero in the confining phase of Yang–Mills theory, and that one may view deconfinement as a breaking of the center symmetry.
- Why is the center symmetry explicitly broken if dynamical quarks in the fundamental representation are added to the theory? What about matter fields in the adjoint representation?

# CHAPTER 18

# Strong Fields and Semi-Classical Methods

Except for Chapter 13, where we discussed extended classical field configurations such as instantons, most of our discussion of quantum field theory has been centered on an expansion about the vacuum, i.e., on situations involving a system with few particles. This is also a regime in which the fields are in a certain sense[1] small. The connection between the field amplitude and the density of particles may be grasped by writing the Lehmann–Symanzik–Zimmermann (LSZ) reduction formula that gives the expectation value of the number operator (for simplicity, imagine a system whose initial state is the vacuum and that evolves away from it under the influence of an external source). Mimicking the derivation of Section 1.5.3, one gets (see Exercise 18.1)

$$\langle 0_{in} | a^\dagger_{\mathbf{p},out} a_{\mathbf{p},out} | 0_{in} \rangle = \frac{1}{Z} \int d^4x d^4y \, e^{-i p \cdot (x-y)} (\Box_x + m^2)(\Box_y + m^2)$$
$$\times \langle 0_{in} | \phi(x) \phi(y) | 0_{in} \rangle$$
$$\langle 0_{in} | \phi(x) \phi(y) | 0_{in} \rangle = \int [D\phi_\pm(z)] \, \phi_-(x) \phi_+(y) \, e^{i(S[\phi_+] - S[\phi_-])}, \tag{18.1}$$

where in the second line we have sketched the path integral representation of the matrix element that appears in the reduction formula. Note that, since there is no time ordering in this matrix element, the Schwinger–Keldysh formalism must be used here (hence the fields $\phi_+$ and $\phi_-$ in the path integral). These formulas illustrate the direct relationship between large particle occupation numbers (the left-hand side of the first equation), and large fields in

---

[1] When we talk of small or large fields, we are referring either to the magnitude of their expectation value, or to the commuting field that appears in a path integral (it does not make sense to apply these qualifiers to the field operator itself).

a path integral (the integrand in the right-hand side of the second equation). In this chapter, we discuss several situations leading to strong fields, and derive some tools for dealing with them.

## 18.1 Situations Involving Strong Fields

### 18.1.1 Power Counting

There is an implicit assumption of weak fields in the perturbative machinery that we have studied so far, which is best viewed in the path integral formalism. For instance, in the second line of eq. (18.1), the perturbative expansion amounts to writing $S = S_0 + S_{int}$, and to expanding the exponential in powers of $S_{int}$. In a scalar field theory with a quartic coupling, the interaction part of the action reads

$$S_{int}[\phi] = -\frac{\lambda}{4!} \int d^4x\, \phi^4(x), \tag{18.2}$$

while the free action (that we keep inside the exponential) is given by

$$S_0[\phi] = \frac{1}{2} \int d^4x\, \left[(\partial_\mu \phi)(\partial^\mu \phi) - m^2 \phi^2\right]. \tag{18.3}$$

The common justification of the perturbative expansion is that, when the coupling constant $\lambda$ is small, we have $S_{int} \ll S_0$. However, since $S_0[\phi]$ is quadratic in the field while $S_{int}[\phi]$ contains higher powers of $\phi$, this inequality may not be true if the field is large, even at weak coupling. In order to make this statement more precise, we must account for the fact that the field has mass dimension 1. Let us denote by Q the typical momentum scale in the problem under consideration (for simplicity we assume that there is only one), and then we write

$$\phi(x) \sim \vartheta\, Q, \tag{18.4}$$

where $\vartheta$ is a dimensionless number that encodes the order of magnitude of the field. Naive dimensional analysis tells us that

$$(\partial_\mu \phi)(\partial^\mu \phi) \sim \vartheta^2\, Q^4, \quad \lambda \phi^4 \sim \lambda \vartheta^4\, Q^4. \tag{18.5}$$

For the interaction term to be small compared to the kinetic term, we must have

$$\lambda \vartheta^2 \ll 1, \tag{18.6}$$

which is slightly different from the usual criterion of small $\lambda$, since this condition depends on the field magnitude via $\vartheta$. The purpose of this chapter is to explore situations of weak coupling (i.e., $\lambda \ll 1$) where the inequality (18.6) is not satisfied because of strong fields. We call this the *strong field regime* of quantum field theory. We will discuss two main situations in which strong fields may occur:

- the initial state is a highly occupied state, such as a coherent state;
- the initial state is the vacuum, but the system is driven by a strong external source.

As we shall see, since the coupling constant is assumed to be small, there is nevertheless a loop expansion, but each loop order (including the tree-level approximation) is non-perturbative in a sense that we will clarify in the rest of the chapter.

## 18.1.2 Expectation Values in a Coherent State

In Section 2.4.3, we presented the Schwinger–Keldysh formalism, which allows the evaluation of expectation values of an observable in the in-vacuum state, $\langle 0_{in}|\mathcal{O}|0_{in}\rangle$. In the previous chapter, we generalized this technique to expectation values in a thermal state, i.e., a mixed state whose density matrix is the canonical equilibrium one, $\rho \equiv \exp(-\beta\,\mathcal{H})$. Another generalization, which we shall study in this section, is to consider an expectation value in a (non-interacting) coherent state, which may be defined from the perturbative in-vacuum as follows:

$$|\chi_{in}\rangle \equiv \mathcal{N}_\chi \, \exp\left\{\int \frac{d^3k}{(2\pi)^3 2E_k}\,\chi(\mathbf{k})\,a^\dagger_{\mathbf{k},in}\right\}|0_{in}\rangle, \tag{18.7}$$

where $\chi(\mathbf{k})$ is a function of 3-momentum and $\mathcal{N}_\chi$ a normalization constant adjusted so that $\langle \chi_{in}|\chi_{in}\rangle = 1$. From the canonical commutation relation

$$[a_{\mathbf{p},in},\,a^\dagger_{\mathbf{q},in}] = (2\pi)^3\,2E_\mathbf{p}\,\delta(\mathbf{p}-\mathbf{q}), \tag{18.8}$$

it is easy to check the following identities

$$a_{\mathbf{p},in}\,|\chi_{in}\rangle = \chi(\mathbf{p})\,|\chi_{in}\rangle,$$
$$|\mathcal{N}_\chi|^2 = \exp\left\{-\int \frac{d^3k}{(2\pi)^3 2E_k}|\chi(\mathbf{k})|^2\right\}. \tag{18.9}$$

The first equation tells us that $|\chi_{in}\rangle$ is an eigenstate of annihilation operators, which is another definition of coherent states, and the second one provides the value of the normalization constant. The occupation number in the initial state is closely related to the function $\chi(\mathbf{k})$. Indeed, we have

$$\langle \chi_{in}|a^\dagger_{\mathbf{p},in}a_{\mathbf{p},in}|\chi_{in}\rangle = |\chi(\mathbf{p})|^2\,. \tag{18.10}$$

In other words, the density of particles in the mode of momentum $\mathbf{p}$ is the squared modulus of the function $\chi(\mathbf{p})$. A large $\chi$ thus corresponds to a highly occupied initial state (at the opposite, $\chi(\mathbf{p}) \equiv 0$ corresponds to the vacuum).

Consider now the generating functional for the extension of the Schwinger–Keldysh formalism in this coherent state,

$$Z_\chi[j] \equiv \langle\chi_{in}|P\,\exp i\int_{\mathcal{C}} d^4x\,j(x)\phi(x)|\chi_{in}\rangle$$
$$= \langle\chi_{in}|P\,\exp i\int_{\mathcal{C}} d^4x\,\big[\mathcal{L}_{int}(\phi_{in}(x)) + j(x)\phi_{in}(x)\big]|\chi_{in}\rangle, \tag{18.11}$$

where $j(x)$ is a fictitious source that lives on the closed-time contour $\mathcal{C}$ introduced in Figure 2.3. As usual, the first step is to factor out the interactions as follows:

$$Z_\chi[j] = \exp\left\{i\int_{\mathcal{C}} d^4x\,\mathcal{L}_{int}\!\left(\frac{\delta}{i\delta j(x)}\right)\right\}\underbrace{\langle\chi_{in}|P\,\exp i\int_{\mathcal{C}} d^4x\,j(x)\phi_{in}(x)|\chi_{in}\rangle}_{Z_{\chi 0}[j]}. \tag{18.12}$$

## 18.1 SITUATIONS INVOLVING STRONG FIELDS

A first application of the Baker–Campbell–Hausdorff formula enables one to remove the path ordering, which gives

$$Z_{\chi 0}[j] = \langle \chi_{\text{in}} | \exp i \int_{\mathcal{C}} d^4x \, j(x)\phi_{\text{in}}(x) | \chi_{\text{in}} \rangle$$
$$\times \exp\left\{ -\frac{1}{2} \int_{\mathcal{C}} d^4x d^4y \, j(x)j(y) \, \theta_c(x^0 - y^0) \, [\phi_{\text{in}}(x), \phi_{\text{in}}(y)] \right\}, \qquad (18.13)$$

where $\theta_c(x^0 - y^0)$ generalizes the step function to the ordered contour $\mathcal{C}$. Note that the factor on the second line is a commuting number and thus can be removed from the expectation value. A second application of the Baker–Campbell–Hausdorff formula allows to us normal-order the first factor. Decomposing the in-field,

$$\phi_{\text{in}}(x) \equiv \underbrace{\int \frac{d^3k}{(2\pi)^3 2E_k} \, a_{k,\text{in}} \, e^{-ik\cdot x}}_{\phi_{\text{in}}^{(-)}(x)} + \underbrace{\int \frac{d^3k}{(2\pi)^3 2E_k} \, a_{k,\text{in}}^{\dagger} \, e^{+ik\cdot x}}_{\phi_{\text{in}}^{(+)}(x)}, \qquad (18.14)$$

we obtain the following expression for the free generating functional:

$$Z_{\chi 0}[j] = \langle \chi_{\text{in}} | \exp\left\{ i \int_{\mathcal{C}} d^4x \, j(x)\phi_{\text{in}}^{(+)}(x) \right\} \exp\left\{ i \int_{\mathcal{C}} d^4y \, j(y)\phi_{\text{in}}^{(-)}(y) \right\} | \chi_{\text{in}} \rangle$$
$$\times \exp\left\{ +\frac{1}{2} \int_{\mathcal{C}} d^4x d^4y \, j(x)j(y) \, [\phi_{\text{in}}^{(+)}(x), \phi_{\text{in}}^{(-)}(y)] \right\}$$
$$\times \exp\left\{ -\frac{1}{2} \int_{\mathcal{C}} d^4x d^4y \, j(x)j(y) \, \theta_c(x^0 - y^0) \, [\phi_{\text{in}}(x), \phi_{\text{in}}(y)] \right\}. \qquad (18.15)$$

The factor of the first line can be evaluated by using the fact that the coherent state is an eigenstate of annihilation operators,

$$\langle \chi_{\text{in}} | \exp\left\{ i \int_{\mathcal{C}} d^4x \, j(x)\phi_{\text{in}}^{(+)}(x) \right\} \exp\left\{ i \int_{\mathcal{C}} d^4y \, j(y)\phi_{\text{in}}^{(-)}(y) \right\} | \chi_{\text{in}} \rangle$$
$$= \exp\left\{ i \int_{\mathcal{C}} d^4x \, j(x) \underbrace{\int \frac{d^3k}{(2\pi)^3 2E_k} \left( \chi(k) e^{-ik\cdot x} + \chi^*(k) e^{+ik\cdot x} \right)}_{\Phi_\chi(x)} \right\}. \qquad (18.16)$$

We denote by $\Phi_\chi(x)$ the field obtained by substituting the creation and annihilation operators of the in-field by $\chi^*(k)$ and $\chi(k)$ respectively. Note that this is no longer an operator, but a (real-valued) commuting classical field. Moreover, because it is a linear superposition of plane waves, this field is a free field,

$$(\Box_x + m^2) \, \Phi_\chi(x) = 0. \qquad (18.17)$$

The second and third factors of eq. (18.15) are commuting numbers, provided we do not attempt to disassemble the commutators. Using the decomposition of the in-field in terms of creation and annihilation operators, and the canonical commutation relation of the latter, we obtain

$$\theta_c(x^0 - y^0) \left[\phi_{in}(x), \phi_{in}(y)\right] - \left[\phi_{in}^{(+)}(x), \phi_{in}^{(-)}(y)\right]$$
$$= \theta_c(x^0 - y^0) \int \frac{d^3k}{(2\pi)^3 2E_k} e^{-ik\cdot(x-y)}$$
$$+ \underbrace{\theta_c(y^0 - x^0) \int \frac{d^3k}{(2\pi)^3 2E_k} e^{+ik\cdot(x-y)}}_{G_c^0(x,y)}, \quad (18.18)$$

which is nothing but the usual free path-ordered propagator $G_c^0(x, y)$. Collecting all the factors, the generating functional for path-ordered Green's functions in the Schwinger–Keldysh formalism with an initial coherent state reads

$$Z_\chi[j] = \exp\left\{i \int_e d^4x \, \mathcal{L}_{int}\left(\frac{\delta}{i\delta j(x)}\right)\right\} \underline{\exp\left\{i \int_e d^4x \, j(x) \, \Phi_\chi(x)\right\}}$$
$$\times \exp\left\{-\frac{1}{2} \int_e d^4x d^4y \, j(x) j(y) \, G_c^0(x, y)\right\}. \quad (18.19)$$

It differs from the corresponding functional with the perturbative vacuum[2] as the initial state only by the second factor, which we have underlined. This generating functional is also equal to[3]

$$Z_\chi[j] = \exp\left\{i \int_e d^4x \, j(x) \, \Phi_\chi(x)\right\} \exp\left\{i \int_e d^4x \, \mathcal{L}_{int}\left(\Phi_\chi(x) + \frac{\delta}{i\delta j(x)}\right)\right\}$$
$$\times \exp\left\{-\frac{1}{2} \int_e d^4x d^4y \, j(x) j(y) \, G_c^0(x, y)\right\}. \quad (18.20)$$

The first factor has the effect of shifting the fields by $\Phi_\chi(x)$. The simplest way to see this is to write

$$\phi \equiv \Phi_\chi + \zeta. \quad (18.21)$$

In eq. (18.11), this leads to

$$Z_\chi[j] = \exp\left\{i \int_e d^4x \, j(x) \, \Phi_\chi(x)\right\} \langle \chi_{in} | P \exp i \int_e d^4x \, j(x) \, \zeta(x) | \chi_{in} \rangle, \quad (18.22)$$

where the second factor on the right-hand side is the generating functional for correlators of $\zeta$. Comparing with eq. (18.20), we see that the generating functional for $\zeta$ is identical to the vacuum one, except that the argument $\phi$ of the interaction Lagrangian is replaced by $\Phi_\chi + \zeta$,

$$\mathcal{L}_{int}(\phi) \quad \to \quad \mathcal{L}_{int}(\Phi_\chi + \zeta). \quad (18.23)$$

---

[2]The vacuum initial state corresponds to the function $\chi(k) \equiv 0$, i.e., to $\Phi_\chi(x) = 0$.
[3]In this transformation, we use the functional analogue of (see Exercise 18.2)

$$F(\partial_x) e^{\alpha x} G(x) = e^{\alpha x} F(\alpha + \partial_x) G(x).$$

FIGURE 18.1: Vertices that appear in the perturbative expansion for the calculation of expectation values with a coherent initial state. The circled cross denotes the field $\Phi_\chi$.

In other words, the field $\zeta$ appears to be coupled to a background field $\Phi_\chi$. For instance, for a $\phi^4$ interaction term, we have

$$\mathcal{L}_{\text{int}}(\Phi_\chi + \zeta) \equiv -U(\Phi_\chi + \zeta) = -\lambda \left\{ \frac{\zeta^4}{4!} + \frac{\zeta^3 \Phi_\chi}{8} + \frac{\zeta^2 \Phi_\chi^2}{4} + \frac{\zeta \Phi_\chi^3}{8} + \frac{\Phi_\chi^4}{4!} \right\}. \tag{18.24}$$

The first term in $\zeta^4$ gives the usual four-leg vertex in the Feynman rules, and the following terms describe the interactions of $\zeta$ with the background field $\Phi_\chi$. The last term plays no role since it does not contain the quantum field $\zeta$. Except for the appearance of these new vertices that involve a background field (see Figure 18.1), the Feynman rules are the same as in the Schwinger–Keldysh formalism for a vacuum initial state, with $+$ and $-$ vertices, and bare propagators $G^0_{++}$, $G^0_{+-}$, $G^0_{-+}$ and $G^0_{--}$ to connect them. In summary, replacing the vacuum initial state by a coherent state amounts to extending the usual Schwinger–Keldysh formalism with a background field $\Phi_\chi$.

As in eq. (18.4), let us assume for the purpose of power counting that

$$\Phi_\chi \sim \vartheta Q, \tag{18.25}$$

and consider a connected graph $\mathcal{G}$ made of $n_{\text{E}}$ external lines, $n_{\text{I}}$ internal lines, $n_1$ vertices $\zeta \Phi_\chi^3$, $n_2$ vertices $\zeta^2 \Phi_\chi^2$, $n_3$ vertices $\zeta^3 \Phi_\chi$, $n_4$ vertices $\zeta^4$ and $n_{\text{L}}$ loops. These parameters are related by the following two identities:

$$\begin{aligned} n_{\text{E}} + 2n_{\text{I}} &= 4n_4 + 3n_3 + 2n_2 + n_1, \\ n_{\text{L}} &= n_{\text{I}} - (n_1 + n_2 + n_3 + n_4) + 1. \end{aligned} \tag{18.26}$$

Then, the order in $\lambda$ and $\vartheta$ of this graph is given by

$$\begin{aligned} \mathcal{G} &\sim \lambda^{n_1+n_2+n_3+n_4} \vartheta^{3n_1+2n_2+n_3} \\ &\sim \lambda^{n_{\text{L}}-1+n_{\text{E}}/2} \left(\sqrt{\lambda}\vartheta\right)^{3n_1+2n_2+n_3}. \end{aligned} \tag{18.27}$$

The first factor is nothing but the usual order in $\lambda$ of a connected graph with $n_{\text{E}}$ external lines and $n_{\text{L}}$ loops. The second factor counts the number of insertions ($3n_1 + 2n_2 + n_3$) of the background field $\Phi_\chi$. Interestingly, it involves only the combination $\sqrt{\lambda}\vartheta$, which appears also in the inequality (18.6) that delineates the strong field regime. From eq. (18.27), we can draw the following conclusions:

- When $\lambda\vartheta^2 \ll 1$, i.e., in the weak field regime, we can make a double perturbative expansion in $\lambda$ and in $\vartheta$ (i.e., in the occupation of the initial coherent state). Leading order results correspond to tree diagrams with zero (or the minimal number necessary for the observable under consideration to be non-zero) insertions of the background field.

- When $\lambda\vartheta^2 \gtrsim 1$, i.e., in the strong field regime, the expansion in powers of $\lambda$ is still possible if $\lambda \ll 1$ (and is organized by the number of loops in the graphs). But the expansion in powers of the background field becomes illegitimate, and one should instead treat $\Phi_\chi$ to all orders. As we shall see now, this leads to important modifications in the calculation of observables in the strong field regime.

Note that for a system prepared in a coherent initial state, it is the function $\chi(\mathbf{k})$ defining the coherent state that determines whether we are in the weak or strong field regime.

In order to illustrate the changes to the perturbative expansion in the strong field regime, let us consider a very simple observable, the expectation value of the field operator,

$$\Phi(x) \equiv \langle \chi_{\text{in}} | \phi(x) | \chi_{\text{in}} \rangle = \Phi_\chi(x) + \langle \chi_{\text{in}} | \zeta(x) | \chi_{\text{in}} \rangle. \tag{18.28}$$

The beginning of the diagrammatic representation of $\Phi(x)$ at tree level reads

$$\Phi(x)\Big|_{\text{tree}} = -\bullet + \cdots + \cdots + \cdots + \cdots + \cdots \tag{18.29}$$

In fact, at tree level, $\Phi(x)$ is the sum of all the tree diagrams (weighted by the appropriate symmetry factor) whose root is the point $x$ and whose leaves are the coherent field $\Phi_\chi$. This infinite set of trees is generated recursively by the following integral representation:

$$\Phi(x) = \Phi_\chi(x) + i \int d^4y \, \underbrace{\left[G^0_{++}(x,y) - G^0_{+-}(x,y)\right]}_{G^0_R(x,y)} \underbrace{\left(-\frac{\lambda}{6}\Phi^3(y)\right)}_{-U'(\Phi(y))}. \tag{18.30}$$

Interestingly, after one has summed over the $+$ and $-$ indices carried by the vertices, the propagators $G^0_{++}$ and $G^0_{+-}$ of the Schwinger–Keldysh diagrammatic rules always appear via their difference, which is nothing but the bare retarded propagator:

$$G^0_{++}(x,y) - G^0_{+-}(x,y) = G^0_{-+}(x,y) - G^0_{--}(x,y) = G^0_R(x,y). \tag{18.31}$$

Since this propagator obeys

$$\left(\Box_x + m^2\right) G^0_R(x,y) = -i\delta(x-y), \quad G^0_R(x,y) = 0 \quad \text{if } x^0 < y^0, \tag{18.32}$$

the expectation value $\Phi(x)$ at tree level satisfies

$$\left(\Box_x + m^2\right) \Phi(x) + U'(\Phi(x)) = 0,$$
$$\lim_{x^0 \to -\infty} \Phi(x) = \Phi_\chi(x). \tag{18.33}$$

In other words, at tree level, the field expectation value obeys the classical field equation of motion, with the boundary value $\Phi_\chi(x)$ at the initial time. The nonlinearity of this equation of motion is crucial in the strong field regime, and all the terms of the series (18.29) have the same magnitude when $\lambda\vartheta^2 \sim 1$. Nevertheless, the representation of this series as the solution of the classical field equation of motion with a retarded boundary condition is very useful, since it turns the problem of summing an infinite series of Feynman graphs into the much simpler (at least conceptually) problem of solving a partial differential equation.

This result for the expectation value of $\phi(x)$ generalizes to the expectation value of any observable built from the field operator: at tree level, its expectation value is obtained by replacing the operator $\phi(x)$ by the commuting classical field $\Phi(x)$ inside the observable,

$$\langle \chi_{\text{in}} | \mathcal{O}(\phi(x)) | \chi_{\text{in}} \rangle \underset{\text{tree level}}{=} \mathcal{O}(\Phi(x)). \tag{18.34}$$

We will defer the study of loop corrections to these expectation values until Section 18.2, because this discussion will be common with the case of quantum field theories coupled to a strong external source, which we shall discuss first.

### 18.1.3 Quantum Field Theory with External Sources

Let us now consider a second way to reach the large field regime. This time, the initial state of the system is the vacuum, but the field is coupled to an external source that drives the system away from the ground state. When the external source is large, the field expectation value will eventually become large itself, and the system will be in the strong field regime. Let us consider a scalar field theory with quartic interaction coupled to a source J, whose Lagrangian is

$$\mathcal{L} \equiv \frac{1}{2}(\partial_\mu \phi)(\partial^\mu \phi) - \frac{1}{2} m^2 \phi^2 - \underbrace{\frac{\lambda}{4!} \phi^4}_{U(\phi)} + J\phi. \tag{18.35}$$

Although we consider here the example of a $\phi^4$ interaction term, we will often write the equations for a generic potential $U(\phi)$, and sometimes diagrammatic illustrations will be given for a cubic interaction for simplicity. These more general interaction terms scale as $\lambda^{-1+n/2} Q^{4-n} \phi^n$, where Q is an object of mass dimension 1. The Feynman rules for this theory are the usual ones, with the addition of a special rule for the external current J. In momentum space, a source J attached to the end of a propagator of momentum p contributes a factor $i\tilde{J}(p)$ (where $\tilde{J}$ is the Fourier transform of J).

The source $J(x)$ is a function of space-time, fixed once for all. As we shall see shortly, the strong field regime corresponds to large sources $J \sim \lambda^{-1/2}$ – we shall call this situation the *strong source, or dense, regime*. In contrast, the situation where the external source J is small is called the *weak source, or dilute, regime*.

Consider a simply connected diagram (see Figure 18.2) with $n_E$ external legs, $n_I$ internal lines, $n_L$ independent loops, $n_J$ sources and $n_3$ cubic vertices, $n_4$ quartic vertices, etc. These parameters are not all independent. First, the number of propagator endpoints should match the available sites to which they can be attached. This leads to a first identity,

$$n_E + 2n_I = n_J + 3n_3 + 4n_4 + 5n_5 + \cdots \tag{18.36}$$

A second identity expresses the number of independent loops in terms of the other parameters:

$$n_L = n_I - (n_3 + n_4 + n_5 + \cdots) - n_J + 1. \tag{18.37}$$

Thanks to these two relations, the order of a diagram $\mathcal{G}$ can be written as

$$\mathcal{G} \sim J^{n_J} \lambda^{\frac{1}{2}n_3 + n_4 + \frac{3}{2}n_5 + \cdots} = \lambda^{n_L - 1 + n_E/2} \left(\sqrt{\lambda} J\right)^{n_J}. \tag{18.38}$$

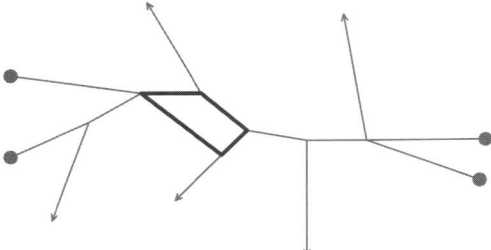

FIGURE 18.2: Generic connected graph in the strong source regime. In this example, $n_E = 5$, $n_I = 11$, $n_J = 4$, $n_L = 1$, $n_3 = 5$ and $n_4 = 2$.

This formula is very similar to eq. (18.27). First, it does not depend on the number of vertices and on the number of internal lines; only the number of external legs, the number of loops and the number of sources appear in the result. The strong source regime is the regime in which the factor $\sqrt{\lambda}\, J$ is not small, and it is therefore not legitimate to expand in powers of J. In this case, the order of a diagram does not depend on its number of sources, and an infinite number of diagrams – with fixed $n_E$ and $n_L$ but arbitrary $n_J$ – contribute at each order.

## 18.2 Observables at Leading and Next-to-Leading Orders

**Leading order:** Let us consider an observable $\mathcal{O}(\phi)$, possibly non-local but with fields only at the same time $t_f$ (the results extend trivially to the case of space-like separations – for time-like separations, see Exercise 18.3), at leading order in the strong field regime. As we have seen in the previous sections, this can be achieved by the presence of strong external sources, or by starting from a highly occupied coherent state. In both cases, the calculation of expectation values is done with the Schwinger–Keldysh formalism. Note that since the field operators in the observable are taken at equal times, they commute and the result does not depend on the + or − assignments for those fields. But it is crucial to sum over all the ± indices in the internal vertices of the graphs.

At leading order in $\lambda$, its expectation value is obtained by simply replacing the field operator $\phi$ by the solution $\Phi$ of the classical equations of motion,

$$\left\langle \mathcal{O}(\phi) \right\rangle_{\text{LO}} = \mathcal{O}(\Phi), \tag{18.39}$$

with

$$(\Box_x + m^2)\Phi + U'(\Phi) = J,$$
$$\lim_{x^0 \to t_i} \Phi(x) = \Phi_\chi(x). \tag{18.40}$$

(We have combined in a single description the two situations, with an external source J and starting from a non-trivial coherent state $\left|\chi_{\text{in}}\right\rangle$.) Recall that the internal sums over the ± indices of the Schwinger–Keldysh formalism lead to retarded boundary conditions, by virtue of eq. (18.31).

## 18.2 OBSERVABLES AT LEADING AND NEXT-TO-LEADING ORDERS

FIGURE 18.3: The two contributions to observables at NLO in the strong field regime.

**Next-to-leading order:** For such an observable, the corresponding next-to-leading order correction can be formally written as follows:

$$\langle \mathcal{O}(\phi) \rangle_{\text{NLO}} = \int_{t_f} d^3x \, \delta\Phi(x) \, \frac{\delta \mathcal{O}(\Phi)}{\delta \Phi(x)} + \frac{1}{2} \int_{t_f} d^3x d^3y \, \mathcal{G}(x,y) \, \frac{\delta^2 \mathcal{O}(\Phi)}{\delta \Phi(x) \delta \Phi(y)}, \quad (18.41)$$

where $\delta\Phi$ is the one-loop correction to the classical field $\Phi$, and $\mathcal{G}(x,y)$ is the propagator dressed by the background field $\Phi$. The two contributions of eq. (18.41) are illustrated in Figure 18.3. Since the field operators in the observable $\mathcal{O}(\phi)$ are all separated by space-like intervals, it is not necessary to indicate the $\pm$ indices in $\delta\Phi$ and $\mathcal{G}$, and we have in fact

$$\delta\Phi_+(x) = \delta\Phi_-(x),$$
$$\mathcal{G}_{++}(x,y) = \mathcal{G}_{--}(x,y) = \mathcal{G}_{-+}(x,y) = \mathcal{G}_{+-}(x,y) \quad \text{if } (x-y)^2 < 0. \quad (18.42)$$

Let us start with $\delta\Phi_\pm$. The propagators in the diagram on the left of Figure 18.3 are the Schwinger–Keldysh propagators in the presence of a background field $\Phi$, i.e., the propagators $\mathcal{G}_{\epsilon\epsilon'}$. For a generic interaction potential, we can write $\delta\Phi_\pm(x)$ as follows:

$$\delta\Phi_\epsilon(x) = -\frac{i}{2} \sum_{\epsilon'=\pm} \int d^4z \, \epsilon' \, \mathcal{G}_{\epsilon\epsilon'}(x,z) \, U'''(\Phi(z)) \, \mathcal{G}_{\epsilon'\epsilon'}(z,z). \quad (18.43)$$

In this formula, the $1/2$ is a symmetry factor, the factor $\epsilon'$ in the integrand takes into account the fact that vertices of type $-$ have an opposite sign in the Schwinger–Keldysh formalism, and the factor $-i\,U'''(\Phi(z))$ is the general form of the trivalent vertex in the presence of an external field (for an arbitrary interaction potential U).

Thus, we have reduced the calculation to that of the two-point functions $\mathcal{G}_{\pm\pm}$. These four propagators are defined recursively by the following equations:

$$\mathcal{G}_{\epsilon\epsilon'}(x,y) = G^0_{\epsilon\epsilon'}(x,y) - i \sum_{\eta=\pm} \eta \int d^4z \, G^0_{\epsilon\eta}(x,z) \, U''(\Phi(z)) \, \mathcal{G}_{\eta\epsilon'}(z,y). \quad (18.44)$$

Here, $-i\,U''(\Phi(z))$ is the general form for the insertion of a background field on a propagator in a theory with potential U. From these equations, we obtain:

$$[\Box_x + m^2 + U''(\Phi(x))] \, \mathcal{G}_{+-}(x,y) = [\Box_y + m^2 + U''(\Phi(y))] \, \mathcal{G}_{+-}(x,y) = 0,$$
$$[\Box_x + m^2 + U''(\Phi(x))] \, \mathcal{G}_{-+}(x,y) = [\Box_y + m^2 + U''(\Phi(y))] \, \mathcal{G}_{-+}(x,y) = 0. \quad (18.45)$$

In addition to these equations of motion, these propagators must become equal to their free counterparts $G^0_{+-}$ and $G^0_{-+}$ when $x^0, y^0 \to -\infty$. From the definition of the various components of the Schwinger–Keldysh propagators, $\mathcal{G}_{++}$ and $\mathcal{G}_{--}$ are given in terms of $\mathcal{G}_{+-}$ and $\mathcal{G}_{-+}$ by the following expressions,

$$\mathcal{G}_{++}(x,y) = \theta(x^0 - y^0)\,\mathcal{G}_{-+}(x,y) + \theta(y^0 - x^0)\,\mathcal{G}_{+-}(x,y),$$
$$\mathcal{G}_{--}(x,y) = \theta(x^0 - y^0)\,\mathcal{G}_{+-}(x,y) + \theta(y^0 - x^0)\,\mathcal{G}_{-+}(x,y). \tag{18.46}$$

The above conditions determine $\mathcal{G}_{+-}$ and $\mathcal{G}_{-+}$ uniquely. In order to find these propagators, let us recall the following representation of their bare counterparts:

$$G^0_{\pm\mp}(x,y) = \int \frac{d^3\mathbf{p}}{(2\pi)^3 2E_\mathbf{p}}\, a_{\mp\mathbf{p}}(x) a_{\pm\mathbf{p}}(y), \tag{18.47}$$

where

$$(\Box_x + m^2)\, a_{\pm\mathbf{p}}(x) = 0, \qquad \lim_{x^0 \to -\infty} a_{\pm\mathbf{p}}(x) = e^{\mp i p\cdot x}. \tag{18.48}$$

(Obviously, the functions $a_{\pm\mathbf{p}}(x)$ are plane waves in the entire space-time.) It is trivial to generalize this representation of the off-diagonal propagators to the case of a non-zero background field, by writing

$$\mathcal{G}_{\pm\mp}(x,y) = \int \frac{d^3\mathbf{p}}{(2\pi)^3 2E_\mathbf{p}}\, a_{\mp\mathbf{p}}(x) a_{\pm\mathbf{p}}(y), \tag{18.49}$$

with

$$\left[\Box_x + m^2 + \underline{U''(\Phi(x))}\right] a_{\pm\mathbf{p}}(x) = 0, \qquad \lim_{x^0 \to -\infty} a_{\pm\mathbf{p}}(x) = e^{\mp i p\cdot x}. \tag{18.50}$$

(The only difference due to the background field is the term in $U''$, underlined in the equation of motion. It leads to a distortion of the functions $a_{\pm\mathbf{p}}$, and they are no longer plane waves.) By construction, these expressions of $\mathcal{G}_{+-}$ and $\mathcal{G}_{-+}$ obey the appropriate equations of motion, and go to the correct limit in the remote past. The functions $a_{\pm\mathbf{p}}(x)$ are sometimes called *mode functions*. They provide a complete basis for the linear space of solutions of eq. (18.50), i.e., the space of linearized perturbations to the classical solution of the field equation of motion.

**Relationship between LO and NLO:** At this point, we have all the building blocks in order to obtain the expectation value of $\mathcal{O}(\phi)$ at NLO. One can go further and obtain a formal relationship between the LO and NLO results. A key observation (we postpone its justification to the next Section, 18.3.2) for this is that the functions $a_{\pm\mathbf{k}}$ that appear in the dressed propagators $\mathcal{G}_{\pm\mp}$ can be obtained from the classical field $\Phi$ as follows:

$$a_{\pm\mathbf{k}}(x) = \mathbb{T}_{\pm\mathbf{k}}\, \Phi(x), \tag{18.51}$$

where the operator $\mathbb{T}_{\pm\mathbf{k}}$ is defined by

$$\mathbb{T}_{\pm\mathbf{k}}\cdots \equiv \int_{u^0=-\infty} d^3\mathbf{u}\, e^{\mp i\mathbf{k}\cdot\mathbf{u}} \left[\frac{\delta}{\delta\Phi_{\text{ini}}(u)} \mp iE_\mathbf{k}\frac{\delta}{\delta(\partial^0\Phi_{\text{ini}}(u))}\right]\cdots\bigg|_{\Phi_{\text{ini}}\equiv\Phi_x}. \tag{18.52}$$

## 18.2 OBSERVABLES AT LEADING AND NEXT-TO-LEADING ORDERS

In words, the operator $\mathbb{T}_{\pm\mathbf{k}}$ in eq. (18.51) differentiates the classical field $\Phi$ with respect to its initial condition $\Phi_{\text{ini}}$, and replaces it by the initial condition of $a_{\pm\mathbf{k}}$. Since $a_{\pm\mathbf{k}}$ is a *linear* perturbation to $\Phi$, this indeed gives the correct result. Thus, the propagator $\mathcal{G}_{+-}(x,y)$ that enters in the NLO result can be written as:

$$\mathcal{G}_{+-}(x,y) = \int \frac{d^3\mathbf{k}}{(2\pi)^3 2E_\mathbf{k}} \left[\mathbb{T}_{-\mathbf{k}}\,\Phi(x)\right]\left[\mathbb{T}_{+\mathbf{k}}\,\Phi(y)\right]. \tag{18.53}$$

In the rest of the NLO calculation, we only need this propagator for a space-like separation between $x$ and $y$, which implies that $\mathcal{G}_{+-}(x,y) = \mathcal{G}_{-+}(x,y)$. In this case, we can symmetrize the expression of the propagator as follows,

$$\mathcal{G}_{+-}(x,y) = \frac{1}{2}\int \frac{d^3\mathbf{k}}{(2\pi)^3 2E_\mathbf{k}} \left\{\left[\mathbb{T}_{-\mathbf{k}}\,\Phi(x)\right]\left[\mathbb{T}_{+\mathbf{k}}\,\Phi(y)\right] \right. \\ \left. + \left[\mathbb{T}_{+\mathbf{k}}\,\Phi(x)\right]\left[\mathbb{T}_{-\mathbf{k}}\,\Phi(y)\right]\right\}. \tag{18.54}$$

As we shall see now, a similar expression can be obtained for $\delta\Phi_\pm$. Let us start from eq. (18.43). Since the propagators $\mathcal{G}_{++}$ and $\mathcal{G}_{--}$ are equal when the two endpoints are evaluated at equal times, we have

$$\delta\Phi_\epsilon(x) = -\frac{i}{2}\int \frac{d^3\mathbf{k}}{(2\pi)^3 2E_\mathbf{k}} d^4z \underbrace{\left[\mathcal{G}_{\epsilon+}(x,z) - \mathcal{G}_{\epsilon-}(x,z)\right]}_{\mathcal{G}_R(x,z)} \\ \times U'''(\Phi(z))\,a_{-\mathbf{k}}(z)a_{+\mathbf{k}}(z), \tag{18.55}$$

where $\mathcal{G}_R$ is the retarded propagator in the presence of the background field $\Phi$. By writing more explicitly the interactions with the background field,

$$\delta\Phi_\epsilon(x) = -i\int d^4y\; G_R^0(x,y)\left[U''(\Phi(y))\,\delta\Phi_\epsilon(y) \right. \\ \left. + \frac{1}{2}U'''(\Phi(y))\int \frac{d^3\mathbf{k}}{(2\pi)^3 2E_\mathbf{k}}\,a_{-\mathbf{k}}(y)a_{+\mathbf{k}}(y)\right] \tag{18.56}$$

(with $G_R^0$ the bare retarded propagator), one may prove that

$$\delta\Phi_\epsilon(x) = \frac{1}{2}\int \frac{d^3\mathbf{k}}{(2\pi)^3 2E_\mathbf{k}}\,\mathbb{T}_{+\mathbf{k}}\mathbb{T}_{-\mathbf{k}}\,\Phi(x). \tag{18.57}$$

By inserting this expression, as well as eq. (18.54), in eq. (18.41), we can write the NLO expectation value as follows:

$$\langle\mathcal{O}\rangle_{\text{NLO}} = \left[\frac{1}{2}\int \frac{d^3\mathbf{k}}{(2\pi)^3 2E_\mathbf{k}}\,\mathbb{T}_{+\mathbf{k}}\mathbb{T}_{-\mathbf{k}}\right]\langle\mathcal{O}\rangle_{\text{LO}}. \tag{18.58}$$

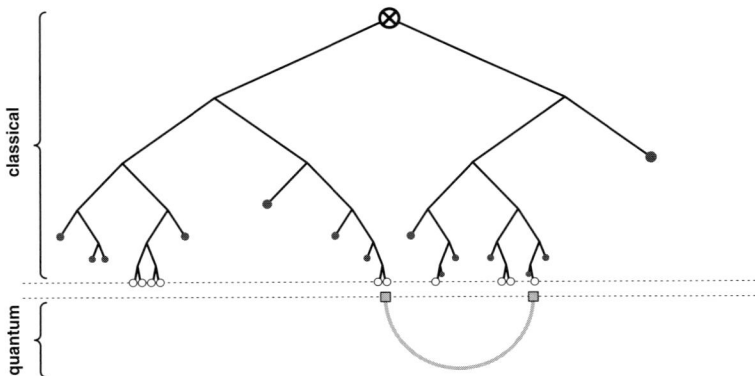

FIGURE 18.4: Illustration of eq. (18.58). The open squares represent the operator $\mathbb{T}_k\mathbb{T}_{-k}$. Its action is to remove two instances of the initial classical field (the open circles), and to connect them with the light-colored link to form a loop.

This central result is illustrated in Figure 18.4 (see Exercise 18.4 for a generalization). Some remarks should be made about this formula:

1. In this formula, the LO observable that appears on the right-hand side must be considered as a functional of the initial classical field.
2. The LO and NLO observables cannot be obtained in closed analytical form, because they contain the classical field $\Phi$ – retarded solution of a nonlinear partial differential equation that cannot be solved analytically in general. Nevertheless, eq. (18.58) is an exact relationship between the two.

**Why is the NLO "nearly classical"?:** In a sense, eq. (18.58) indicates that observables at NLO in the strong field regime are almost classical, since they can be obtained from the LO result (that depends only on the classical field $\Phi$) by acting with the operators $\mathbb{T}_{\pm k}$ (i.e., derivatives with respect to the initial value of the classical field). This is in fact not specific to the strong field regime nor to quantum field theory, but is a general property of quantum mechanics. To see this, consider a generic quantum system of Hamiltonian H and density operator $\rho_t$. The latter evolves according to the Liouville-von Neumann equation:

$$i\hbar \frac{\partial \rho_t}{\partial t} = [H, \rho_t]. \tag{18.59}$$

The next step is to introduce the Wigner transform of the density operator:

$$W_t(\mathbf{x}, \mathbf{p}) \equiv \int d\mathbf{s} \, e^{i\mathbf{p}\cdot\mathbf{s}} \langle \mathbf{x} + \frac{\mathbf{s}}{2} | \rho_t | \mathbf{x} - \frac{\mathbf{s}}{2} \rangle. \tag{18.60}$$

The Wigner transform of an operator is a Fourier transform of the matrix elements of the operator in the position basis with respect to the difference of coordinates. The function

$W_t(\mathbf{x}, \mathbf{p})$ may be viewed in a loose sense[4] as a probability distribution in the classical phase-space of the system ($\mathbf{x}$ and $\mathbf{p}$ are classical variables, not operators). Note that the Wigner transform of the Hamiltonian operator H is the classical Hamiltonian $\mathcal{H}$ (see Exercise 5.1). One may show that the Liouville-von Neumann equation is equivalent to (see Exercise 18.5)

$$\frac{\partial W_t}{\partial \tau} = \mathcal{H}(\mathbf{x}, \mathbf{p}) \, \frac{2}{i\hbar} \sin\left(\frac{i\hbar}{2}\left(\overleftarrow{\partial}_p \overrightarrow{\partial}_x - \overleftarrow{\partial}_x \overrightarrow{\partial}_p\right)\right) W_t(\mathbf{x}, \mathbf{p}) \tag{18.61}$$

$$= \underbrace{\{\mathcal{H}, W_t\}}_{\text{Poisson bracket}} + \mathcal{O}(\hbar^2) \tag{18.62}$$

The first line is an exact equation, known as the *Moyal-Groenewold equation*. In the second line, we have performed an expansion in powers of $\hbar$, and one can readily see that the order zero in $\hbar$ is nothing but the classical Liouville equation (it thus describes a system whose time evolution is classical). The first quantum correction to the time evolution arises only at the order $\hbar^2$. Therefore, at the order $\hbar$ (i.e., NLO in the language of quantum field theory), the time evolution of the system remains purely classical. This does not mean that there are no quantum corrections of order $\hbar$, but that these corrections can only come from the initial state of the system (since a quantum system cannot have well-defined $\mathbf{x}$ and $\mathbf{p}$ at the same time, the Wigner distribution $W_t(\mathbf{x}, \mathbf{p})$ must have a support whose area in each pair of canonically conjugate variables $(x_i, p_i)$ must be at least of order $\hbar$). The effect of the operator in $\mathbb{T}_{+\mathbf{k}}\mathbb{T}_{-\mathbf{k}}$ that acts on the LO in eq. (18.58) is precisely to restore this quantum width of the initial state.

## 18.3 Green's Formulas

Eq. (18.51), which formally relates a small field perturbation to the background field on top of which it propagates, plays a crucial role in discussing many questions related to strong fields. The proof of this formula relies on Green's formulas, which we discuss in this section.

### 18.3.1 Green's Formula for a Retarded Classical Scalar Field

Consider the following partial differential equation,[5]

$$(\Box_x + m^2)\Phi(x) + U'(\Phi(x)) = J(x). \tag{18.63}$$

Since this equation contains time derivatives up to second order, it is necessary to specify the initial value of $\Phi$ itself as well as that of its first time derivative. Let us assume that we know these values on the surface $x^0 = 0$. We wish to obtain a formula for $\Phi(x)$ at a time $x^0 > 0$ in

---

[4]$W_t$ is not a bona fide probability distribution, because it is not positive definite in general. But the regions of phase-space where it is negative are small, typically of order $\hbar$. After being integrated either over $\mathbf{x}$ or over $\mathbf{p}$, it becomes a genuine probability distribution for the expectation values of $\mathbf{p}$ or $\mathbf{x}$, respectively.

[5]This equation is the classical equation of motion in the scalar field theory of Lagrangian

$$\mathcal{L} \equiv \frac{1}{2}(\partial_\mu \phi)(\partial^\mu \phi) - \frac{1}{2}m^2 \phi^2 - U(\phi) + J\phi.$$

terms of this initial data. In order to do this, we must introduce the retarded Green's function of the operator $\Box_x + m^2$, defined by

$$(\Box_x + m^2) G_R^0(x,y) = -i\delta(x-y),$$
$$G_R^0(x,y) = 0 \quad \text{if } x^0 < y^0. \tag{18.64}$$

(The superscript 0 is a reminder of the fact that this is a *free* Green's function that does not depend on the interaction potential $U(\Phi)$.) Note that $G_R^0(x,y)$ obeys the same equation if acted upon with $\Box_y + m^2$ instead. From the equations obeyed by $\Phi$ and by $G_R^0$, we obtain

$$G_R^0(x,y)(\overrightarrow{\Box}_y + m^2)\Phi(y) = G_R^0(x,y)\big[J(y) - U'(\Phi(y))\big],$$
$$G_R^0(x,y)(\overleftarrow{\Box}_y + m^2)\Phi(y) = -i\delta(x-y)\Phi(y), \tag{18.65}$$

where the arrows on the d'Alembertian operators indicate on which side they act. By integrating these equations over $y$ in the domain $y^0 > 0$, and by subtracting them, we get the following relation:

$$\Phi(x) = i \int_{y^0 > 0} d^4y\, G_R^0(x,y)\Big[(\overleftarrow{\Box}_y - \overrightarrow{\Box}_y)\Phi(y) + J(y) - U'(\Phi(y))\Big]. \tag{18.66}$$

The last step is to show that the term that involves the difference between the two d'Alembertian operators is in fact a boundary term that depends only on the initial conditions. Note first the following identity:

$$A(\overleftarrow{\Box} - \overrightarrow{\Box})B = \partial^\mu A(\overleftarrow{\partial}_\mu - \overrightarrow{\partial}_\mu)B, \tag{18.67}$$

where the leftmost $\partial^\mu$ acts on everything on its right. In other words, the left-hand side is a total derivative, and its integral over $d^4y$ can be rewritten as a surface integral thanks to the Stokes theorem. The integration domain defined by $y^0 > 0$ has three boundaries:

1. $y^0 = +\infty$: This boundary at infinite time does not contribute, since the retarded propagator $G_R^0(x,y)$ vanishes if $y^0 > x^0$.
2. $y^0 = 0$: This boundary gives a non-zero contribution that depends only on the initial conditions for the field $\Phi$.
3. Boundary at spatial infinity: This boundary does not contribute if we assume that the field vanishes when $|\mathbf{x}| \to \infty$, or for a finite volume with periodic boundary conditions in the spatial directions.

Therefore, we obtain

$$\Phi(x) = i \int_{y^0 > 0} d^4y\, G_R^0(x,y)\Big[J(y) - U'(\Phi(y))\Big]$$
$$+ i \int_{y^0 = 0} d^3\mathbf{y}\, G_R^0(x,y)(\overrightarrow{\partial}_{y^0} - \overleftarrow{\partial}_{y^0})\Phi_{\text{ini}}(y). \tag{18.68}$$

## 18.3 GREEN'S FORMULAS

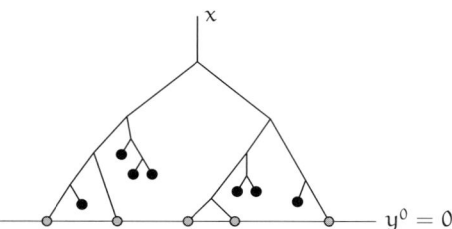

FIGURE 18.5: Typical contribution to $\Phi(x)$ in the diagrammatic representation of eq. (18.68), in the case of cubic interactions. The solid dots represent the sources J, the open circles represent the initial value of the field or field derivatives on the surface $y^0 = 0$. The lines are retarded propagators $G_R^0$.

In this *Green's formula*, the first term on the right-hand side provides the dependence on the source J, and on the interactions, while the second term tells us how $\Phi(x)$ depends on the initial value $\Phi_{\rm ini}(y)$ and on that of its first time derivative.

Except in the trivial case where the potential $U(\Phi)$ is zero, eq. (18.68) does not provide an explicit result for $\Phi(x)$, since the right-hand side depends on $\Phi(y)$ at points above the initial surface. Despite this limitation, this is a very useful tool for performing formal manipulations involving retarded solutions of eq. (18.63). To end this section, let us mention a diagrammatic interpretation of eq. (18.68), illustrated in Figure 18.5. One can expand the right-hand side of eq. (18.68) in powers of the interactions. The starting point is the zeroth order approximation, obtained by setting the potential to $U = 0$. Then, one proceeds recursively in order to obtain the higher orders in $U$. The outcome of this expansion is an infinite series of terms that have a tree structure. The root of this tree is the point $x$ where the field is evaluated, and its leaves are either sources J (if there are any above the surface $y^0 = 0$) or the initial data on the surface $y^0 = 0$. In particular, if the source $J(x)$ vanishes at $y^0 > 0$, then all the J dependence of the classical field is implicitly hidden in the $\Phi_{\rm ini}(y)$ that appears in the boundary term.

**Extension to a generic initial surface:** In eq. (18.68), the initial conditions for the field $\Phi$ have been set on the surface of constant time $y^0 = 0$. However, there are many situations in which this initial data is known on a different initial surface. Let us consider a generic surface $\Sigma$, on which the field $\Phi_{\rm ini}$ and its derivatives are known. As before, we wish to obtain a formula that expresses $\Phi(x)$ at some point $x$ above $\Sigma$ in terms of these initial conditions on $\Sigma$.

Most of the derivation is identical to the case of a constant time initial surface, with all the integrals over the domain $y^0 > 0$ replaced by integrals over the domain $\Omega$ located above $\Sigma$. The only significant change occurs when we apply the Stokes theorem in order to transform the four-dimensional integral of a total derivative into an integral over the boundary of $\Omega$. As in the previous case, the boundaries at positive infinite time, and at infinity in the spatial directions, do not contribute, and we have only a contribution from the surface $\Sigma$. The Stokes theorem can then be written as

$$\int_\Omega d^4y\, \partial_\mu F^\mu(y) = -\int_\Sigma d^3S_y\, n_\mu F^\mu(y), \qquad (18.69)$$

where $d^3S_y$ is the measure on the surface $\Sigma$, and $n^\mu$ is a 4-vector normal to the surface $\Sigma$ at the point $y$, pointing above the surface $\Sigma$. In the important case where the initial surface

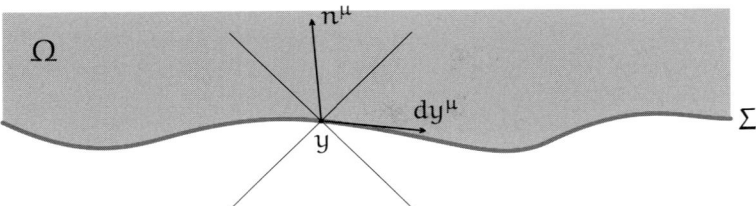

FIGURE 18.6: Illustration of eq. (18.70).

is invariant by translation in the transverse directions, the proper normalization for $n^\mu$ and $d^3 S_y$ can be obtained as follows. Parametrize an arbitrary displacement $dy^\mu$ on the surface $\Sigma$ about the point $y$ as $dy^\mu = (\beta dy^3, dy^1, dy^2, dy^3)$, where $\beta$ is the local slope of the surface $\Sigma$ in the $(y^3, y^0)$ plane. Then, we have

$$\begin{aligned} n_\mu dy^\mu &= 0, \\ n_\mu n^\mu &= 1, \quad n^0 > 0, \\ d^3 S_y &= \sqrt{1-\beta^2}\, dy^1 dy^2 dy^3. \end{aligned} \quad (18.70)$$

The second and third conditions require that $\beta < 1$ in order to make sense (see Exercise 18.7 for the case of a light-like surface). This implies that the surface $\Sigma$ must be locally space-like. Physically, this means that a signal emitted from a point of the surface $\Sigma$ cannot reach the surface again in the future. The relations (18.70) are illustrated in Figure 18.6. Note that the orthogonality defined by $n_\mu dy^\mu = 0$ does not correspond to the Euclidean concept of orthogonality.

Thanks to eq. (18.69), it is possible to write the Green's formula for an arbitrary initial surface $\Sigma$ as

$$\Phi(x) = i \int_\Omega d^4y\, G_R^0(x,y) \Big[J(y) - U'(\Phi(y))\Big]$$
$$+ i \int_\Sigma d^3 S_y\, G_R^0(x,y)(n \cdot \overrightarrow{\partial}_y - n \cdot \overleftarrow{\partial}_y) \Phi_{\text{ini}}(y). \quad (18.71)$$

For an arbitrary surface $\Sigma$, the second term on the right-hand side of this formula tells us explicitly what information about $\Phi$ we must provide on the initial surface in order to determine it uniquely above the surface: At every point $y \in \Sigma$, one must specify the values of the field $\Phi_{\text{ini}}(y)$ and of its *normal derivative* $n \cdot \partial_y \Phi_{\text{ini}}(y)$.

### 18.3.2 Green's Formula for Small Field Perturbations

Consider now a small perturbation $a(x)$ to the classical field, and assume that $a(x) \ll \Phi(x)$. Therefore, one can linearize the equation of motion of $a(x)$, and we get

$$\Big[\Box_x + m^2 + U''(\Phi(x))\Big] a(x) = 0. \quad (18.72)$$

## 18.3 GREEN'S FORMULAS

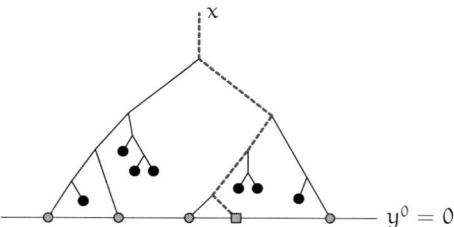

FIGURE 18.7: Typical contribution to $a(x)$ in the diagrammatic representation of eq. (18.73), in the case of cubic interactions. The solid dots represent the sources J, the open circles represent the initial data for $\Phi(y)$ on the surface $y^0 = 0$, and the open square the initial data for $a(y)$. The lines are retarded propagators $G_R^0$. The dashed line is the retarded propagator of the perturbation in the background $\Phi$, i.e., an inverse of the operator $\Box + m^2 + U''(\Phi)$.

Treating the term $U''(\Phi(x))a(x)$ as an interaction, we can easily derive a Green's formula that expresses the field perturbation $a(x)$ in terms of its initial conditions on a surface $\Sigma$:

$$a(x) = i \int_\Omega d^4y \, G_R^0(x,y) \Big[ -U''(\Phi(y)) a(y) \Big]$$
$$+ i \int_\Sigma d^3S_y \, G_R^0(x,y)(n \cdot \vec{\partial}_y - n \cdot \overleftarrow{\partial}_y) a_{\rm ini}(y). \qquad (18.73)$$

Eq. (18.73) is illustrated in Figure 18.7. Every diagram contributing to $a(x)$ has exactly one instance of the initial value $a_{\rm ini}(y)$ (represented by an open square in the figure) on the initial surface. Indeed, it is easy to see from eq. (18.73) that $a(x)$ depends linearly on its value on the initial surface. This is a consequence of the fact that the equation of motion for a small perturbation is linear.

By comparing Figures 18.5 and 18.7, one sees that they differ only by the fact that one instance of the field $\Phi_{\rm ini}(y)$ has been replaced by the small perturbation $a_{\rm ini}(y)$ on the initial surface. Therefore, we expect a linear relationship between $a(x)$ and $\Phi(x)$, of the form

$$a(x) = \mathbb{T}_a \, \Phi(x), \qquad (18.74)$$

where $\mathbb{T}_a$ is a linear operator that substitutes one power of $\Phi_{\rm ini}(y)$ by $a_{\rm ini}(y)$ on $\Sigma$ (i.e., an operator that involves first derivatives with respect to the initial conditions on $\Sigma$). It is easy to prove this relation by using eqs. (18.71) and (18.73). In order to do so and at the same time determine the form of the operator $\mathbb{T}_a$, let us apply $\mathbb{T}_a$ to the Green's formula that gives $\Phi(x)$. We get[6]

$$\mathbb{T}_a \Phi(x) = i \int_\Omega d^4y \, G_R^0(x,y) \Big[ -U''(\Phi(y)) \, \mathbb{T}_a \Phi(y) \Big]$$
$$+ i \mathbb{T}_a \int_\Sigma d^3S_y \, G_R^0(x,y)(n \cdot \vec{\partial}_y - n \cdot \overleftarrow{\partial}_y) \Phi_{\rm ini}(y). \qquad (18.75)$$

---

[6] Since $\mathbb{T}_a$ acts only on the initial fields on $\Sigma$, we have $\mathbb{T}_a J = 0$.

If the boundary term in this formula can be made identical to that of eq. (18.73), then this equation will be identical to the Green's formula for $a(x)$ and we will have proven the announced relationship between $a(x)$ and $\Phi(x)$. This is the case if the operator $\mathbb{T}_a$ is chosen as

$$\mathbb{T}_a \equiv \int_\Sigma d^3 S_y \left[ a_{ini}(y) \frac{\delta}{\delta \Phi_{ini}(y)} + (n \cdot \partial a_{ini}(y)) \frac{\delta}{\delta(n \cdot \partial \Phi_{ini}(y))} \right], \tag{18.76}$$

which is nothing but the operator that substitutes $a_{ini}(y)$ to $\Phi_{ini}(y)$ on the initial surface $\Sigma$, as announced (this definition generalizes eq. (18.52) to a generic initial surface and to generic initial conditions for the perturbation).

Note that $\mathbb{T}_a$ is an operator that performs an infinitesimal translation (by an amount $a(y)$) to the initial condition of the classical field. By exponentiation, it may be promoted into an operator that performs a finite shift of the initial condition. In particular, if we denote by $\Phi[\Phi_{ini}]$ the classical field whose initial value on $\Sigma$ is $\Phi_{ini}$, then we have

$$e^{\mathbb{T}_a} \Phi[\Phi_{ini}] = \Phi[\Phi_{ini} + a_{ini}]. \tag{18.77}$$

### 18.3.3 Schwinger–Keldysh Formalism

It is often useful to obtain Green's formulas for fields in the Schwinger–Keldysh formalism. In this case, one has a pair of fields $\Phi_\pm(x)$ that are both solutions of the classical equation of motion:

$$(\Box_x + m^2)\Phi_\pm(x) + U'(\Phi_\pm(x)) = J(x). \tag{18.78}$$

(For simplicity we take the same source $J(x)$ for the two fields, but this limitation is easily circumvented if necessary.) Since the Schwinger–Keldysh propagator $G^0_{++}$ is also a Green's function of the operator $\Box_x + m^2$, we can reproduce the previous derivation of the Green's formula, which leads to[7]

$$\Phi_+(x) = i \int d^4y \, G^0_{++}(x,y) \left[ J(y) - U'(\Phi_+(y)) \right]$$
$$+ i \int d^3y \left[ G^0_{++}(x,y)(\overleftarrow{\partial}_{y^0} - \overrightarrow{\partial}_{y^0})\Phi_+(y) \right]_{y^0=-\infty}^{y^0=+\infty}, \tag{18.79}$$

where we used the notation $[f(y^0)]_{y^0=a}^{y^0=b} \equiv f(b) - f(a)$. The only difference with the Green's formula derived with retarded propagators is the boundary term: Since $G^0_{++}(x,y)$ does not vanish when $y^0 > x^0$, there is also a non-zero contribution from the boundary at $y^0 = +\infty$. Then, by using the fact that $(\Box_y + m^2)G^0_{+-}(x,y) = 0$, we obtain in a similar way

$$0 = i \int d^4y \, G^0_{+-}(x,y) \left[ J(y) - U'(\Phi_-(y)) \right]$$
$$+ i \int d^3y \left[ G^0_{+-}(x,y)(\overleftarrow{\partial}_{y^0} - \overrightarrow{\partial}_{y^0})\Phi_-(y) \right]_{y^0=-\infty}^{y^0=+\infty}. \tag{18.80}$$

---

[7] Here also, the prefactors $i$ follow from our convention for the propagators of the Schwinger–Keldysh formalism (see eq. (2.76)).

## 18.3 GREEN'S FORMULAS

Subtracting this equation from eq. (18.79), we obtain

$$
\begin{aligned}
\Phi_+(x) &= i\int d^4y\, G^0_{++}(x,y)\Big[J(y)-U'(\Phi_+(y))\Big] - G^0_{+-}(x,y)\Big[J(y)-U'(\Phi_-(y))\Big] \\
&\quad - i\int d^3\mathbf{y}\, \Big[G^0_{++}(x,y)\overleftrightarrow{\partial}_{y^0} \Phi_+(y) - G^0_{+-}(x,y)\overleftrightarrow{\partial}_{y^0} \Phi_-(y)\Big]_{y^0=-\infty}^{y^0=+\infty},
\end{aligned}
\quad (18.81)
$$

where $A\overleftrightarrow{\partial}_{y^0} B \equiv A(\overrightarrow{\partial}_{y^0} - \overleftarrow{\partial}_{y^0})B$. Similarly, we obtain the following expression for $\Phi_-(x)$:

$$
\begin{aligned}
\Phi_-(x) &= i\int d^4y\, G^0_{-+}(x,y)\Big[J(y)-U'(\Phi_+(y))\Big] - G^0_{--}(x,y)\Big[J(y)-U'(\Phi_-(y))\Big] \\
&\quad - i\int d^3\mathbf{y}\, \Big[G^0_{-+}(x,y)\overleftrightarrow{\partial}_{y^0} \Phi_+(y) - G^0_{--}(x,y)\overleftrightarrow{\partial}_{y^0} \Phi_-(y)\Big]_{y^0=-\infty}^{y^0=+\infty}.
\end{aligned}
\quad (18.82)
$$

At this point, these formulas are rather formal, and it is not clear why we have gone through the trouble of subtracting the quantity given by eq. (18.80), since it is identically zero. This will become transparent in the next section, where we show that these formulas enable one to sum series of tree diagrams encountered in the Schwinger–Keldysh formalism.

Note also that the only property of the propagators $G^0_{-+}$ and $G^0_{+-}$ that we have used in this derivation is the fact that they are annihilated by $\square + m^2$. Therefore, eqs. (18.81) and (18.82) remain valid if we replace these propagators by any other pair of propagators sharing the same property. For instance, one can replace the propagators $G^0_{+-}$ and $G^0_{-+}$ of eq. (2.76) by the following objects:

$$
\begin{aligned}
\overline{G}^0_{+-}(x,y) &= \int \frac{d^4p}{(2\pi)^4}\, e^{-ip\cdot(x-y)}\, u(\mathbf{p})\, 2\pi\theta(-p^0)\delta(p^2), \\
\overline{G}^0_{-+}(x,y) &= \int \frac{d^4p}{(2\pi)^4}\, e^{-ip\cdot(x-y)}\, v(\mathbf{p})\, 2\pi\theta(+p^0)\delta(p^2),
\end{aligned}
\quad (18.83)
$$

where $u(\mathbf{p})$ and $v(\mathbf{p})$ are some arbitrary functions of the momentum $\mathbf{p}$, without altering any of the formulas in this section. We will make use of this freedom in the next section.

### 18.3.4  Summing Tree Diagrams Using Green's Formulas

Many problems involving strong fields require that one sums infinite series of tree diagrams. These sums of diagrams can in general be expressed in terms of solutions of the classical equations of motion. However, in order to determine them uniquely, one must know the boundary conditions obeyed by these classical solutions. The strategy in order to obtain them is to write the sum of tree diagrams as a recursive integral equation. Then, by comparing this integral equation with a Green's formula such as eq. (18.68), one can read off the boundary conditions easily.

**Sum of retarded trees:** Let us first illustrate this in the simplest case, where one must sum all the tree diagrams built with retarded propagators, and whose leaves are a source $J(x)$. Let us call $\Phi(x)$ the sum of all such tree diagrams. Given the recursive structure of such trees, one can write immediately:

$$\Phi(x) = i \int d^4y \; G_R^0(x,y) \Big[ J(y) - U'(\Phi(y)) \Big], \tag{18.84}$$

where the integration over $d^4y$ is extended to the entire[8] space-time. Therefore, we see that this formula is identical to the Green's formula (18.68), with the initial surface at $y^0 = -\infty$ instead of $y^0 = 0$, and where the boundary term is identically zero. This means that the sum of these tree diagrams is a retarded solution of the classical equation of motion with a null boundary condition in the remote past:

$$(\Box_x + m^2)\Phi(x) + U'(\Phi(x)) = J(x),$$
$$\lim_{y^0 \to -\infty} \Phi(y^0, \mathbf{y}) = 0, \quad \lim_{y^0 \to -\infty} \partial_0 \Phi(y^0, \mathbf{y}) = 0. \tag{18.85}$$

**Sum of trees in the Schwinger–Keldysh formalism:** Consider now a more complicated example, in which one must sum tree diagrams in the Schwinger–Keldysh formalism. Now, each vertex is carrying an index $\epsilon = \pm$. For simplicity, we assume that the $+$ and $-$ sources are identical, so that we still have a single source $J(x)$. Because this is necessary in certain applications, we are going to use the modified propagators $\overline{G}_{+-}^0$ and $\overline{G}_{-+}^0$ defined in eqs. (18.83), instead of the propagators $G_{+-}^0$ and $G_{-+}^0$ defined in eqs. (2.76) (the propagators $G_{++}^0$ and $G_{--}^0$ are kept unchanged). In addition to summing over all the possible trees, we sum over all the combinations of $\pm$ indices at every internal vertex. First, the sum of these trees can be written in the form of two coupled integral equations (there are now two fields $\Phi_\pm(x)$ depending on the index carried by the root of the tree):

$$\Phi_+(x) = i \int d^4y \; G_{++}^0(x,y) \Big[ J(y) - U'(\Phi_+(y)) \Big]$$
$$\quad - i \int d^4y \; \overline{G}_{+-}^0(x,y) \Big[ J(y) - U'(\Phi_-(y)) \Big],$$
$$\Phi_-(x) = i \int d^4y \; \overline{G}_{-+}^0(x,y) \Big[ J(y) - U'(\Phi_+(y)) \Big]$$
$$\quad - i \int d^4y \; G_{--}^0(x,y) \Big[ J(y) - U'(\Phi_-(y)) \Big]. \tag{18.86}$$

At this point, we recognize that the right-hand side of these equations is identical to the first term on the right-hand side of eqs. (18.81) and (18.82). From this observation, we conclude that $\Phi_+(x)$ and $\Phi_-(x)$ are solutions of the classical equation of motion,

$$(\Box_x + m^2)\Phi_\pm(x) + U'(\Phi_\pm(x)) = J(x), \tag{18.87}$$

---

[8] In the Feynman rules the integration at each vertex is extended to the full space-time $\mathbb{R}^4$.

## 18.3 GREEN'S FORMULAS

and that they obey the following boundary conditions:

$$\int d^3y \left[ G^0_{++}(x,y) \overset{\leftrightarrow}{\partial}_{y^0} \Phi_+(y) - \overline{G}^0_{+-}(x,y) \overset{\leftrightarrow}{\partial}_{y^0} \Phi_-(y) \right]_{y^0=-\infty}^{y^0=+\infty} = 0,$$

$$\int d^3y \left[ \overline{G}^0_{-+}(x,y) \overset{\leftrightarrow}{\partial}_{y^0} \Phi_+(y) - G^0_{--}(x,y) \overset{\leftrightarrow}{\partial}_{y^0} \Phi_-(y) \right]_{y^0=-\infty}^{y^0=+\infty} = 0. \quad (18.88)$$

We have now coupled boundary conditions for the fields $\Phi_+$ and $\Phi_-$, which involve the value of the fields both at $y^0 = -\infty$ and at $y^0 = +\infty$. In addition, these boundary conditions are non-local in coordinate space, since they involve integrals over $d^3y$ on the surfaces $y^0 = \pm\infty$. However, they can be simplified considerably if one uses the Fourier representations for the propagators:

$$G^0_{++}(x,y) = \int \frac{d^3p}{(2\pi)^3 2E_p} \left[ \theta(x^0-y^0) e^{-ip\cdot(x-y)} + \theta(y^0-x^0) e^{+ip\cdot(x-y)} \right],$$

$$G^0_{--}(x,y) = \int \frac{d^3p}{(2\pi)^3 2E_p} \left[ \theta(x^0-y^0) e^{+ip\cdot(x-y)} + \theta(y^0-x^0) e^{-ip\cdot(x-y)} \right],$$

$$\overline{G}^0_{+-}(x,y) = \int \frac{d^3p}{(2\pi)^3 2E_p} u(p) e^{+ip\cdot(x-y)},$$

$$\overline{G}^0_{-+}(x,y) = \int \frac{d^3p}{(2\pi)^3 2E_p} v(p) e^{-ip\cdot(x-y)}. \quad (18.89)$$

Compared to the expressions of these propagators in eqs. (2.76) and (18.83), we have performed explicitly the integration over $p^0$ in order to obtain these formulas. Thus, whenever $p^0$ appears in these expressions, it should be replaced by the positive on-shell value $p^0 = +E_p$. The other ingredient we need in order to simplify the boundary conditions is a Fourier representation for the fields $\Phi_\pm(y)$,

$$\Phi_\epsilon(y) \equiv \int \frac{d^3p}{(2\pi)^3 2E_p} \left[ f^{(+)}_\epsilon(y^0, p) e^{-ip\cdot y} + f^{(-)}_\epsilon(y^0, p) e^{+ip\cdot y} \right]. \quad (18.90)$$

The superscripts $(\pm)$ on the Fourier coefficients serve to distinguish the positive and negative frequency modes. Note that because the fields $\Phi_\pm(y)$ are not free fields, these Fourier coefficients are time-dependent. In practice, one may assume that the interactions are switched off at $y^0 = \pm\infty$, so that the coefficients $f^{(\pm)}_\epsilon(y^0, p)$ tend to constants when $y^0 \to \pm\infty$. However, these limiting values are different at $y^0 = +\infty$ and at $y^0 = -\infty$, and we must keep the $y^0$ argument to distinguish them. Using the identity

$$\int d^3y \, e^{i\epsilon p\cdot(x-y)} \overset{\leftrightarrow}{\partial}_{y^0} e^{i\epsilon' p'\cdot y} = i\delta_{\epsilon\epsilon'} e^{i\epsilon p\cdot x} (2\pi)^3 2E_p \delta(p-p') \quad (18.91)$$

(valid for $\epsilon, \epsilon' = \pm$), it is easy to rewrite the boundary conditions (18.88) as a set of separate conditions for each Fourier mode $p$,

$$f^{(+)}_+(-\infty, p) = f^{(-)}_-(-\infty, p) = 0,$$
$$f^{(-)}_+(+\infty, p) = u(p) f^{(-)}_-(+\infty, p),$$
$$f^{(+)}_-(+\infty, p) = v(p) f^{(+)}_+(+\infty, p). \quad (18.92)$$

The boundary conditions have a very compact expression in terms of the Fourier coefficients of the fields $\Phi_\pm$. At $y^0 = -\infty$, $\Phi_+(y)$ has no positive frequency modes and $\Phi_-(y)$ has no negative frequency modes. At $y^0 = +\infty$, the negative frequency modes of $\Phi_+(y)$ and $\Phi_-(y)$ are proportional (with a proportionality relation that involves the function $u(\mathbf{p})$). A similar relation that involves the function $v(\mathbf{p})$ holds between their positive frequency modes at $y^0 = +\infty$. Eq. (18.92), together with the equations of motion (18.87), determine uniquely the fields $\Phi_\pm(x)$ and therefore provide the solution to our original problem of summing tree diagrams in the Schwinger–Keldysh formalism. One should, however, keep in mind that this solution is somewhat academic, because it is in general extremely difficult to solve a nonlinear field equation of motion with boundary conditions specified both at $y^0 = -\infty$ and $y^0 = +\infty$.

Let us also mention that these boundary conditions become considerably simpler in the case where $u(\mathbf{p}) = v(\mathbf{p}) \equiv 1$. Indeed, from the second and third of eqs. (18.92), we see that the fields $\Phi_\pm(y)$ have identical Fourier coefficients at $y^0 = +\infty$. Therefore, the two fields must be equal in the limit $y^0 \to +\infty$. Then, by solving their equation of motion backwards in time, one sees trivially that they are equal at all times (since they obey identical equations of motion):

$$\text{if } u(\mathbf{p}) = v(\mathbf{p}) \equiv 1, \quad \Phi_+(x) = \Phi_-(x), \quad \text{for all } x \in \mathbb{R}^4. \tag{18.93}$$

Finally, the first part of eq. (18.92) tells us that

$$\text{if } u(\mathbf{p}) = v(\mathbf{p}) \equiv 1, \quad \lim_{x^0 \to -\infty} \Phi_\pm(x) = 0. \tag{18.94}$$

To summarize, when $u(\mathbf{p}) = v(\mathbf{p}) \equiv 1$, the two fields $\Phi_\pm(x)$ are equal to the retarded field that vanishes when $x^0 \to -\infty$. This result could in fact have been obtained by a much more elementary argument. Indeed, when $u(\mathbf{p}) = v(\mathbf{p}) \equiv 1$, the summation over the $\pm$ indices at the vertices of tree diagrams always leads to the following combinations of propagators:

$$G^0_{++} - G^0_{+-} = G^0_{-+} - G^0_{--} = G^0_R. \tag{18.95}$$

In other words, summing over these indices amounts to replacing all the propagators in a given tree by retarded propagators, and one is thus led to the problem discussed in the first paragraph of Section 18.3.4.

## 18.4 Mode Functions

### 18.4.1 Propagators in a Background Field

We have introduced in eq. (18.50) a set of small perturbations on top of a background field $\Phi$, as a way to express propagators dressed by this background. We shall discuss further properties of these functions in this section. However, before we do so, let us propose an alternative derivation of the dressed propagators. In eq. (18.49), this problem was solved by writing the equations of motion obeyed by the various propagators, as well as their boundary conditions, and by exhibiting an expression that fulfills both. The method proposed here simply amounts to performing explicitly the resummation of the background field insertions. As one will see, this approach is arguably more tedious, but is in a sense much more elementary.

## 18.4 MODE FUNCTIONS

The starting point is eq. (18.44), which performs the resummation of the background field. In this form, the equation is fairly complicated to solve because the four components of the Schwinger–Keldysh propagator get mixed already after the first insertion of the background field. However, there is a simple way to simplify these equations. It is based on the observation that the four propagators are not independent, but satisfy a linear relation,

$$\begin{aligned} G^0_{++} + G^0_{--} &= G^0_{+-} + G^0_{-+}, \\ \mathcal{G}_{++} + \mathcal{G}_{--} &= \mathcal{G}_{+-} + \mathcal{G}_{-+}, \end{aligned} \quad (18.96)$$

which follows immediately from their definition as path-ordered products of two fields, and from the identity $\theta(x) + \theta(-x) = 1$. It is possible to exploit this relation as follows: Perform a rotation on the matrix made of the four propagators so that one component of the rotated matrix becomes zero. Having a zero in the matrix of propagators makes the resummation of the background field considerably simpler. Therefore, let us define

$$\mathbb{G}_{\alpha\beta} \equiv \sum_{\epsilon,\epsilon'=\pm} \Omega_{\alpha\epsilon}\Omega_{\beta\epsilon'}\mathcal{G}_{\epsilon\epsilon'}. \quad (18.97)$$

(The same rotation is applied to the free propagators.) The choice of the matrix $\Omega_{\alpha\epsilon}$ that gives a zero component in $\mathbb{G}_{\alpha\beta}$ is not unique, but the following one is convenient:

$$\Omega_{\alpha\epsilon} \equiv \begin{pmatrix} 1 & -1 \\ 1/2 & 1/2 \end{pmatrix}. \quad (18.98)$$

The rotated propagators read

$$\mathbb{G}^0_{\alpha\beta} = \begin{pmatrix} 0 & G^0_A \\ G^0_R & G^0_S \end{pmatrix}, \quad \mathbb{G}_{\alpha\beta} = \begin{pmatrix} 0 & \mathcal{G}_A \\ \mathcal{G}_R & \mathcal{G}_S \end{pmatrix}, \quad (18.99)$$

where we have introduced

$$\begin{aligned} G^0_R &= G^0_{++} - G^0_{+-}, & \mathcal{G}_R &= \mathcal{G}_{++} - \mathcal{G}_{+-}, \\ G^0_A &= G^0_{++} - G^0_{-+}, & \mathcal{G}_A &= \mathcal{G}_{++} - \mathcal{G}_{-+}, \\ G^0_S &= G^0_{++} + G^0_{--}, & \mathcal{G}_S &= \mathcal{G}_{++} + \mathcal{G}_{--}. \end{aligned} \quad (18.100)$$

(The subscripts R, A and S stand respectively for *retarded, advanced* and *symmetric*.) After having performed this rotation, eq. (18.44) is transformed into

$$\mathbb{G}_{\alpha\beta}(x,y) = \mathbb{G}^0_{\alpha\beta}(x,y) - i\sum_{\delta,\gamma}\int d^4z\, \mathbb{G}^0_{\alpha\delta}(x,z)\, U''(\Phi(z))\, \sigma^1_{\delta\gamma}\, \mathbb{G}_{\gamma\beta}(z,y), \quad (18.101)$$

where $\sigma^1$ is the first Pauli matrix,

$$\sigma^1 \equiv \begin{pmatrix} 0 & 1 \\ 1 & 0 \end{pmatrix}. \quad (18.102)$$

In order to make the notations simpler, let us introduce the following shorthand:

$$\left[\mathbb{A} \circ \mathbb{B}\right]_{\alpha\beta}(x,y) \equiv -i \sum_{\delta,\gamma} \int d^4z\, \mathbb{A}_{\alpha\delta}(x,z)\, U''(\Phi(z))\, \sigma^1_{\delta\gamma}\, \mathbb{B}_{\gamma\beta}(z,y). \tag{18.103}$$

With this notation, eq. (18.101) takes a very compact form,

$$\mathbb{G} = \mathbb{G}^0 + \mathbb{G}^0 \circ \mathbb{G}, \tag{18.104}$$

and its solution is

$$\mathbb{G} = \sum_{n=0}^{\infty} \left[\mathbb{G}^0\right]^{\circ n}, \quad \text{with } \mathbb{A}^{\circ n} \equiv \underbrace{\mathbb{A} \circ \cdots \circ \mathbb{A}}_{n \text{ times}}. \tag{18.105}$$

What makes the calculation of this infinite sum easy after the rotation we have performed is the fact that the elementary object $\mathbb{G}^0 \sigma^1$ is the sum of a diagonal and a nilpotent matrix:

$$\mathbb{G}^0 \sigma^1 = \mathbb{D} + \mathbb{N}, \quad \mathbb{D} \equiv \begin{pmatrix} G^0_A & 0 \\ 0 & G^0_R \end{pmatrix}, \quad \mathbb{N} \equiv \begin{pmatrix} 0 & 0 \\ G^0_S & 0 \end{pmatrix}. \tag{18.106}$$

One has $\mathbb{N}^2 = 0$, which simplifies a lot the calculation of the $n$th power of $\mathbb{G}^0 \sigma^1$. From this observation, it is easy to obtain

$$\left[\mathbb{G}^0\right]^{\circ(n+1)} = \begin{pmatrix} 0 & \left[G^0_A\right]^{\star(n+1)} \\ \left[G^0_R\right]^{\star(n+1)} & \sum_{i=0}^{n} \left[G^0_R\right]^{\star i} \star G^0_S \star \left[G^0_A\right]^{\star(n-i)} \end{pmatrix}, \tag{18.107}$$

with the notation

$$[A \star B](x,y) \equiv -i \int d^4z\, A(x,z)\, U''(\Phi(z))\, B(z,y), \tag{18.108}$$

and an obvious definition for the $\star$-exponentiation. The summation of the off-diagonal components of eq. (18.107) is trivial since these terms do not mix. Moreover, the resummed $\mathcal{G}_S$ propagator has a simple expression in terms of the resummed retarded and advanced propagators. These results can be summarized by

$$\mathcal{G}_R = \sum_{n=0}^{\infty} \left[G^0_R\right]^{\star n}, \quad \mathcal{G}_A = \sum_{n=0}^{\infty} \left[G^0_A\right]^{\star n},$$
$$\mathcal{G}_S = \mathcal{G}_R\, (G^0_R)^{-1}\, G^0_S\, (G^0_A)^{-1}\, \mathcal{G}_A. \tag{18.109}$$

At this stage, we know all the components of the resummed propagator in the rotated basis. In order to obtain them in the original basis, we just have to invert the rotation of eq. (18.97), which gives

$$\mathcal{G}_{-+} = \mathcal{G}_R\, (G^0_R)^{-1}\, G^0_{-+}\, (G^0_A)^{-1}\, \mathcal{G}_A,$$
$$\mathcal{G}_{+-} = \mathcal{G}_R\, (G^0_R)^{-1}\, G^0_{+-}\, (G^0_A)^{-1}\, \mathcal{G}_A. \tag{18.110}$$

It is easy to check that these equations are equivalent to eq. (18.49) (see Exercise 18.9).

## 18.4.2 Basis of Retarded Small Perturbations

Since small perturbations obey a linear equation of motion, it is always possible to write them as a linear superposition of small perturbations that obey retarded boundary conditions. Let us use as a basis the functions $a_{\pm k}(x)$, defined by the equation of motion

$$\left[\Box_x + m^2 + U''(\Phi(x))\right] a_{\pm k}(x) = 0, \qquad (18.111)$$

and the retarded boundary condition

$$\lim_{x^0 \to -\infty} a_{\pm k}(x) = e^{\mp i k \cdot x}. \qquad (18.112)$$

(These functions are called *mode functions*.) Note that for a real potential $U(\Phi)$, $a_{\pm k}(x)$ are mutual complex conjugates. Any solution of the equation of motion for small perturbations can be written as

$$a(x) = \int \frac{d^3 k}{(2\pi)^3 2 E_k} \left[\alpha_+^k a_{+k}(x) + \alpha_-^k a_{-k}(x)\right], \qquad (18.113)$$

where the $\alpha_\pm^k$ are constant coefficients that depend on the boundary conditions (the boundary conditions in general lead to a set of linear equations for the coefficients).

## 18.4.3 Completeness Relations

The set of small perturbations $\{a_{+k}(x), a_{-k}(x)\}$ obey some useful relations that are a consequence of unitarity. Consider first two generic solutions $a_1(x)$ and $a_2(x)$ of the equation of small perturbations. In order to make the notations more compact in the rest of this section, it is useful to introduce the following notations:

$$|a) \equiv \begin{pmatrix} a(x) \\ \dot{a}(x) \end{pmatrix}, \quad (a| \equiv \begin{pmatrix} a^*(x) & \dot{a}^*(x) \end{pmatrix} \sigma_2, \qquad (18.114)$$

where the dot denotes a time derivative and $\sigma_2$ is the second Pauli matrix. Thanks to the fact that the background potential $U''(\Phi(x))$ is real, one can construct from $a_1$ and $a_2$ an inner product which is an invariant of the time evolution of the two perturbations. This quantity is reminiscent of the Wronskian for two solutions of a second-order ordinary differential equation, and it is defined as follows (see Exercises 18.10 and 18.11 for extensions):

$$(a_1|a_2) \equiv i \int d^3 x \left[\dot{a}_1^*(x) a_2(x) - a_1^*(x) \dot{a}_2(x)\right]. \qquad (18.115)$$

Although $(a_1|a_2)$ could in principle depend on time (since one integrates only over space in its definition), it is immediate to verify that

$$\frac{\partial}{\partial x^0} (a_1|a_2) = 0. \qquad (18.116)$$

Since it is a constant in time, one can compute this inner product from the value of the field perturbations in the remote past. This is particularly handy when the perturbations under consideration are specified by retarded boundary conditions, as is the case for the mode functions $a_{\pm k}(x)$. One finds

$$\begin{aligned}(a_{+k}|a_{+l}) &= (2\pi)^3 2E_k\, \delta(\mathbf{k}-\mathbf{l}),\\ (a_{-k}|a_{-l}) &= -(2\pi)^3 2E_k\, \delta(\mathbf{k}-\mathbf{l}),\\ (a_{+k}|a_{-l}) &= (a_{-k}|a_{+l}) = 0.\end{aligned} \qquad (18.117)$$

Consider now a generic solution $a(x)$ of eq. (18.111). Since the $a_{\pm k}$ form a basis of the linear space of solutions, one can write $a(x)$ as a linear superposition,

$$|a) = \int \frac{d^3 k}{(2\pi)^3 2k} \left[\alpha_+^k |a_{+k}) + \alpha_-^k |a_{-k})\right], \qquad (18.118)$$

where the coefficients $\alpha_\pm^k$ do not depend on time or space. By using the orthogonality relations obeyed by the vectors $|a_{\pm k})$, one gets easily

$$\alpha_+^k = (a_{+k}|a), \quad \alpha_-^k = -(a_{-k}|a). \qquad (18.119)$$

By inserting these relations back into eq. (18.118), and by using the fact that it is valid for any small perturbation $a(x)$ solution of eq. (18.111), we obtain the following identity:

$$\int \frac{d^3 k}{(2\pi)^3 2k} \left[|a_k)(a_k| - |a_{-k})(a_{-k}|\right] = 1. \qquad (18.120)$$

This identity, valid at all times, is a manifestation of the fact that when the background field is real, the time evolution preserves the completeness of the set of states $|a_{\pm k})$.

## 18.5 Multi-Point Correlation Functions at Tree Level

### 18.5.1 Generating Functional for Local Measurements

**Definition:** In Section 18.2, we have studied a generic observable at leading and next-to-leading orders in the coupling $\lambda$, and we have established a general functional relationship that relates the leading and next-to-leading orders. In a sense, this relationship reflects the fact that the first $\hbar$ correction in a quantum theory is not fully quantum: At this order only the initial state contains quantum effects, but the time evolution of the system is still classical.

Let us consider now an observable involving multiple points $x_1, \cdots, x_n$, corresponding to $n$ measurements. For simplicity, we assume that the points $x_i$ where the measurements are performed lie on the same surface of constant time $x^0 = t_f$, but the final results are valid for any locally space-like surface (this ensures that there is no causal relation between the points $x_i$, and also that the ordering between the operators in the correlator does not matter). In this case, the leading order is a completely disconnected contribution made of $n$ separate factors, that does not contain any correlation between the $n$ measurements. However, the physically interesting information often lies in the correlation between these measurements,

$$\mathcal{C}_{\{1\cdots n\}} \equiv \langle \mathcal{O}(x_1)\cdots \mathcal{O}(x_n)\rangle_c, \qquad (18.121)$$

## 18.5 MULTI-POINT CORRELATION FUNCTIONS AT TREE LEVEL

where the subscript c indicates that we retain only the connected part of the correlator. From the generic power counting arguments developed in the previous sections, these connected correlators are all of order $\lambda^{-1}$ in the strong field regime. It is also important to realize that the connected part of these correlators is subleading compared to the disconnected parts, since

$$\langle \mathcal{O}(x_1)\cdots\mathcal{O}(x_n)\rangle = \underbrace{\langle \mathcal{O}(x_1)\rangle \cdots \langle \mathcal{O}(x_n)\rangle}_{\lambda^{-n}} + \underbrace{\sum_{i<j}\langle \mathcal{O}(x_i)\mathcal{O}(x_j)\rangle_c \prod_{k\neq i,j}\langle \mathcal{O}(x_k)\rangle}_{\lambda^{1-n}}$$
$$+\cdots+\underbrace{\langle \mathcal{O}(x_1)\cdots\mathcal{O}(x_n)\rangle_c}_{\mathcal{C}_{\{1\cdots n\}}\sim\lambda^{-1}}. \qquad (18.122)$$

We see in this formula that, in the strong field regime, the fully connected part of an n-point correlator is suppressed by $\lambda^{n-1}$ compared to the trivial disconnected term. Thus, even at tree level, the correlated part of an n-point function is not a leading-order quantity, but arises only at order $n-1$ in the expansion in powers of $\lambda$.

To facilitate the bookkeeping, one can encapsulate all the correlation functions (18.121) into a generating functional defined as follows:[9]

$$\mathcal{F}[z(\mathbf{x})] \equiv \langle 0_{in}| \exp \int_{t_f} d^3\mathbf{x}\, z(\mathbf{x})\, \mathcal{O}(\phi(x))|0_{in}\rangle, \qquad (18.123)$$

where the argument of the field in $\mathcal{O}$ is $x \equiv (t_f, \mathbf{x})$. From this generating functional, the correlation functions are obtained by differentiating with respect to $z(\mathbf{x}_i)$ and by setting $z \equiv 0$ afterwards. In order to remove the uncorrelated part of the n-point function, we should differentiate the logarithm of $\mathcal{F}$, i.e.,

$$\mathcal{C}_{\{1\cdots n\}} = \left.\frac{\delta^n \ln \mathcal{F}}{\delta z(\mathbf{x}_1)\cdots \delta z(\mathbf{x}_n)}\right|_{z\equiv 0} \qquad (18.124)$$

The observable $\mathcal{O}(\phi(x))$ is made of the field in the Heisenberg picture, $\phi(x)$, which can be related to the field $\phi_{in}(x)$ of the interaction picture by $\phi(x) = U(-\infty, x^0)\,\phi_{in}(x)\,U(x^0, -\infty)$, where the evolution operator $U(t_1, t_2)$ is expressed in terms of the interactions by the following formula:

$$U(t_2, t_1) = T \exp i \int_{t_1}^{t_2} d^4x\, \mathcal{L}_{int}(\phi_{in}(x)). \qquad (18.125)$$

We can therefore rewrite the generating functional solely in terms of the interaction picture field $\phi_{in}$,

$$\mathcal{F}[z(\mathbf{x})] = \langle 0_{in}|P \exp \int d^3\mathbf{x}\, \Big\{ z(\mathbf{x})\, \mathcal{O}(\phi_{in}(t_f, \mathbf{x})) + i\int_{\mathcal{C}} dx^0\, \mathcal{L}_{int}(\phi_{in}(x)) \Big\}|0_{in}\rangle, \qquad (18.126)$$

---

[9]This is easily generalized to the case where the initial state is a coherent state instead of the vacuum, simply by changing the state in which the expectation value of the exponential is evaluated.

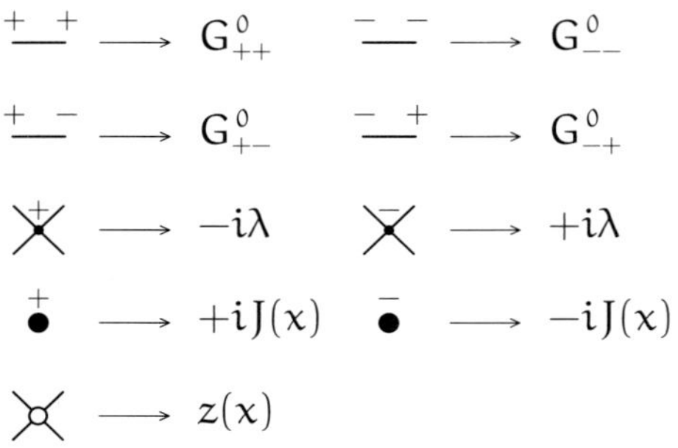

$$\underset{+\ +}{\underline{\phantom{xxx}}} \longrightarrow G^0_{++} \qquad \underset{-\ -}{\underline{\phantom{xxx}}} \longrightarrow G^0_{--}$$

$$\underset{+\ -}{\underline{\phantom{xxx}}} \longrightarrow G^0_{+-} \qquad \underset{-\ +}{\underline{\phantom{xxx}}} \longrightarrow G^0_{-+}$$

$$\overset{+}{\times} \longrightarrow -i\lambda \qquad \overset{-}{\times} \longrightarrow +i\lambda$$

$$\overset{+}{\bullet} \longrightarrow +iJ(x) \qquad \overset{-}{\bullet} \longrightarrow -iJ(x)$$

$$\times \longrightarrow z(x)$$

FIGURE 18.8: Rules for the extended Schwinger–Keldysh formalism that gives the generating functional. The Feynman rules shown here for the self-interactions correspond to a $\lambda\phi^4/4!$ interaction term. In this illustration, we have assumed that the observable is quartic in the field when drawing the corresponding vertex (proportional to $z(x)$).

where P denotes the path ordering on the Schwinger–Keldysh time contour $\mathcal{C}$. We denote by $\phi^+_{\text{in}}$ the (interaction picture) field that lives on the upper branch and by $\phi^-_{\text{in}}$ the field on the lower branch The operator $\mathcal{O}(\phi_{\text{in}}(x))$ lives at the final time of this contour, and could be viewed as made of fields of type $+$ or of type $-$ (the two choices lead to the same results).

**Expression in the Schwinger–Keldysh formalism:** Since the initial state is the vacuum, the generating functional defined in eq. (18.126) can be represented diagrammatically as the sum of all the vacuum-to-vacuum graphs (i.e., graphs without external legs) in the Schwinger–Keldysh formalism (see Section 2.4.3), extended by an extra vertex that corresponds to the insertions of the observable $\mathcal{O}$. The additional vertex exists only on the final surface, at time $t_f$. It is accompanied by a factor $z(x)$ and has as many legs as there are fields in $\mathcal{O}(\phi)$. There is only one kind of this vertex (we can decide to call it $+$ or $-$ without affecting anything). We recapitulate the Feynman rules in Figure 18.8. In the case of the vacuum initial state, we recall that the propagators have the following explicit expressions:

$$G^0_{-+}(x,y) = \int \frac{d^3k}{(2\pi)^3 2E_k} e^{-ik\cdot(x-y)}, \quad G^0_{+-}(x,y) = \int \frac{d^3k}{(2\pi)^3 2E_k} e^{ik\cdot(x-y)},$$
$$G^0_{++}(x,y) = \theta(x^0 - y^0) G^0_{-+}(x,y) + \theta(y^0 - x^0) G^0_{+-}(x,y),$$
$$G^0_{--}(x,y) = \theta(x^0 - y^0) G^0_{+-}(x,y) + \theta(y^0 - x^0) G^0_{-+}(x,y). \tag{18.127}$$

Note that when we set $z \equiv 0$, these diagrammatic rules fall back to the pure Schwinger–Keldysh formalism, for which all the connected vacuum-to-vacuum graphs are zero. This implies that

$$\mathcal{F}[z \equiv 0] = 1, \tag{18.128}$$

in accordance with the fact that this should be $\langle 0_{\text{in}} | 0_{\text{in}} \rangle = 1$.

**Retarded–advanced representation:** In order to clarify which approximations are legitimate in the strong field regime, it is useful to use a different basis of fields by introducing

$$\phi_2 \equiv \tfrac{1}{2}(\phi_+ + \phi_-), \qquad \phi_1 \equiv \phi_+ - \phi_-. \tag{18.129}$$

The half-sum $\phi_2$ in a sense captures the classical content (plus some quantum corrections), while the difference $\phi_1$ is purely quantum (because it represents the different histories of the fields in the amplitude and in the complex conjugated amplitude). To see how the Feynman rules are modified in terms of these new fields, let us start from

$$\phi_\alpha = \sum_{\epsilon=\pm} \Omega_{\alpha\epsilon} \phi_\epsilon \qquad (\alpha = 1, 2), \tag{18.130}$$

with $\Omega_{\alpha\epsilon}$ the matrix defined in eq. (18.98). The new propagators obtained after this rotation were calculated earlier:

$$\begin{aligned}
G^0_{21} &= G^0_{++} - G^0_{+-}, \\
G^0_{12} &= G^0_{++} - G^0_{-+}, \\
G^0_{22} &= \tfrac{1}{2} \left[ G^0_{+-} + G^0_{-+} \right], \\
G^0_{11} &= 0.
\end{aligned} \tag{18.131}$$

Note that $G^0_{21}$ is the bare retarded propagator, while $G^0_{12}$ is the bare advanced propagator. The vertices in the new formalism (here written for a quartic interaction) are given by

$$\Lambda_{\alpha\beta\gamma\delta} \equiv -i\lambda \left[ \Omega^{-1}_{+\alpha}\Omega^{-1}_{+\beta}\Omega^{-1}_{+\gamma}\Omega^{-1}_{+\delta} - \Omega^{-1}_{-\alpha}\Omega^{-1}_{-\beta}\Omega^{-1}_{-\gamma}\Omega^{-1}_{-\delta} \right], \tag{18.132}$$

where

$$\Omega^{-1}_{\epsilon\alpha} = \begin{pmatrix} 1/2 & 1 \\ -1/2 & 1 \end{pmatrix} \qquad (\Omega_{\alpha\epsilon}\Omega^{-1}_{\epsilon\beta} = \delta_{\alpha\beta}). \tag{18.133}$$

More explicitly, we have

$$\begin{aligned}
\Lambda_{1111} &= \Lambda_{1122} = \Lambda_{2222} = 0 \\
\Lambda_{1222} &= -i\lambda, \quad \Lambda_{1112} = -i\lambda/4.
\end{aligned} \tag{18.134}$$

(The vertices not listed explicitly here are obtained by permutations.) Finally, the rules for an external source in the retarded-advanced basis are

$$J_1 = J, \quad J_2 = 0. \tag{18.135}$$

Note also that the observable depends only on the field $\phi_2$, i.e., $\mathcal{O} = \mathcal{O}(\phi_2)$. Indeed, the fields $\phi_+$ and $\phi_-$ represent the field in the amplitude and in the conjugated amplitude, and their difference should vanish when a measurement is performed.

## 18.5.2 First Derivative at Tree Level

**First derivative of $\ln \mathcal{F}$:** Differentiating the generating functional with respect to $z(x)$ amounts to exhibiting a vertex $\mathcal{O}$ at the point $x$ at the final time (as opposed to weighting this vertex by $z(x)$ and integrating over $x$). Furthermore, by considering the logarithm of the generating functional rather than $\mathcal{F}$ itself, we have only diagrams that are connected to point $x$, as shown in this representation:

$$\frac{\delta \ln \mathcal{F}}{\delta z(x)} = x \!\!\!\bigcirc\!\!\!\!\blacksquare , \qquad (18.136)$$

where the gray blob is a sum of graphs constructed with the Feynman rules of Figure 18.8, or their analogue in the retarded-advanced formulation. Therefore, these graphs still depend implicitly on $z$. Note that this blob does not have to be connected.

**Tree-level expression:** Without further specifying the content of the blob, eq. (18.136) is valid to all orders, both in $z$ and in $\lambda$. At lowest order in $\lambda$ (tree level), a considerable simplification happens because the blob must be a product of disconnected subgraphs, one for each line attached to the vertex $\mathcal{O}(\varphi(x))$:

$$\left.\frac{\delta \ln \mathcal{F}}{\delta z(x)}\right|_{\text{tree}} = x \!\!\!\bigcirc\!\!\!\!\bigcirc , \qquad (18.137)$$

where now each of the light-colored blobs is a *connected tree* one-point diagram. In the retarded-advanced formalism, there are two of these one-point functions, which we denote $\varphi_1$ and $\varphi_2$. At tree level, they can be defined recursively by the following pair of coupled integral equations:

$$\varphi_1(x) = i\int_\Omega d^4y \; G^0_{12}(x,y) \frac{\partial L_{\text{int}}(\varphi_1,\varphi_2)}{\partial \varphi_2(y)}$$
$$+ \int_{t_f} d^3y \; G^0_{12}(x,y) \, z(y) \, \mathcal{O}'(\varphi_2(y)),$$

$$\varphi_2(x) = i\int_\Omega d^4y \left\{ G^0_{21}(x,y) \frac{\partial L_{\text{int}}(\varphi_1,\varphi_2)}{\partial \varphi_1(y)} + G^0_{22}(x,y) \frac{\partial L_{\text{int}}(\varphi_1,\varphi_2)}{\partial \varphi_2(y)} \right\}$$
$$+ \int_{t_f} d^3y \; G^0_{22}(x,y) \, z(y) \, \mathcal{O}'(\varphi_2(y)). \qquad (18.138)$$

In these equations, $\mathcal{O}'$ is the derivative of the observable with respect to the field, $\Omega$ is the space-time domain comprised between the initial and final times, and we denote

$$L_{\text{int}}(\varphi_1,\varphi_2) \equiv \mathcal{L}_{\text{int}}(\varphi_2 + \tfrac{1}{2}\varphi_1) - \mathcal{L}_{\text{int}}(\varphi_2 - \tfrac{1}{2}\varphi_1). \qquad (18.139)$$

For an interaction Lagrangian $-\frac{\lambda}{4!}\varphi^4 + J\varphi$, this difference reads

$$L_{\text{int}}(\varphi_1,\varphi_2) = -\frac{\lambda}{6}\varphi_2^3\varphi_1 - \frac{\lambda}{4!}\varphi_1^3\varphi_2 + J\varphi_1. \qquad (18.140)$$

In terms of these fields, we have

$$\left.\frac{\delta \ln \mathcal{F}}{\delta z(x)}\right|_{\text{tree}} = \mathcal{O}(\varphi_2(x)), \qquad (18.141)$$

## 18.5 MULTI-POINT CORRELATION FUNCTIONS AT TREE LEVEL

i.e., simply the observable $\mathcal{O}$ evaluated on the field $\phi_2(x)$ (but this field depends on $z$ to all orders, via the boundary terms in eqs. (18.138)).

**Classical equations of motion:** Using the fact that $G^0_{12}$ and $G^0_{21}$ are Green's functions of $\Box+m^2$ obeying the following equations,

$$(\Box_x + m^2)\, G^0_{12}(x,y) = (\Box_x + m^2)\, G^0_{21}(x,y) = -i\delta(x-y), \tag{18.142}$$

while $G^0_{22}$ vanishes when acted upon by this operator,

$$(\Box_x + m^2)\, G^0_{22}(x,y) = 0, \tag{18.143}$$

we see that $\phi_1$ and $\phi_2$ obey the following classical field equations of motion,

$$(\Box_x + m^2)\, \phi_1(x) = \frac{\partial L_{int}(\phi_1,\phi_2)}{\partial \phi_2(x)},$$

$$(\Box_x + m^2)\, \phi_2(x) = \frac{\partial L_{int}(\phi_1,\phi_2)}{\partial \phi_1(x)}. \tag{18.144}$$

Note that here the point $x$ is located in the "bulk" $\Omega$; this is why the observable does not enter in these equations of motion. In fact, the observable enters only in the boundary conditions satisfied by these fields on the hypersurface at $t_f$. For later reference, let us also rewrite these equations of motion in the specific case of a scalar field theory with a $\lambda\phi^4/4!$ interaction term and an external source $J$,

$$\left[\Box_x + m^2 + \tfrac{\lambda}{2}\phi_2^2\right]\phi_1 + \frac{\lambda}{4!}\phi_1^3 = 0,$$

$$(\Box_x + m^2)\,\phi_2 + \frac{\lambda}{6}\phi_2^3 + \frac{\lambda}{8}\phi_1^2\phi_2 = J. \tag{18.145}$$

**Boundary conditions:** The equations of motion (18.144) are easier to handle than the integral equations (18.138), but they must be supplemented with boundary conditions in order to uniquely define the solutions. The standard procedure for deriving the boundary conditions is to consider the combination $G^0_{12}(x,y)\,(\Box_y + m^2)\,\phi_1(y)$, and let the operator $\Box_y + m^2$ act alternatively on the right and on the left:

$$G^0_{12}(x,y)\,(\vec{\Box}_y + m^2)\,\phi_1(y) = G^0_{12}(x,y)\,\frac{\partial L_{int}(\phi_1,\phi_2)}{\partial\phi_2(y)}$$

$$G^0_{12}(x,y)\,(\overleftarrow{\Box}_y + m^2)\,\phi_1(y) = -i\delta(x-y)\,\phi_1(y). \tag{18.146}$$

By subtracting these equations and integrating over $y \in \Omega$, we obtain

$$\phi_1(x) = i\int_\Omega d^4y\, G^0_{12}(x,y)\,\frac{\partial L_{int}(\phi_1,\phi_2)}{\partial\phi_2(y)} - i\int_\Omega d^4y\, G^0_{12}(x,y)\,\overleftrightarrow{\Box}_y\,\phi_1(y). \tag{18.147}$$

The second term of the right-hand side is a total derivative that can be rewritten as a surface integral extended to the boundary of the domain $\Omega$. With reasonable assumptions on the spatial localization of the source $J(x)$ that drives the field, we may disregard the contribution from the boundary at spatial infinity. The remaining boundaries are at the initial time $t_i$ and final time $t_f$,

$$\phi_1(x) = i \int_\Omega d^4y \, G^0_{12}(x,y) \frac{\partial L_{int}(\phi_1,\phi_2)}{\partial \phi_2(y)} - i \int d^3y \left[ G^0_{12}(x,y) \stackrel{\leftrightarrow}{\partial}_{y_0} \phi_1(y) \right]^{t_f}. \tag{18.148}$$

Note that the boundary term vanishes at the initial time $t_i$, because $G^0_{12}$ is the advanced propagator. Likewise, we obtain the following equation for $\phi_2$:

$$\phi_2(x) = i \int_\Omega d^4y \left\{ G^0_{21}(x,y) \frac{\partial L_{int}(\phi_1,\phi_2)}{\partial \phi_1(y)} + G^0_{22}(x,y) \frac{\partial L_{int}(\phi_1,\phi_2)}{\partial \phi_2(y)} \right\}$$
$$- i \int d^3y \left[ G^0_{21}(x,y) \stackrel{\leftrightarrow}{\partial}_{y_0} \phi_2(y) + G^0_{22}(x,y) \stackrel{\leftrightarrow}{\partial}_{y_0} \phi_1(y) \right]^{t_f}_{t_i}. \tag{18.149}$$

The boundary conditions at $t_i$ and $t_f$ are obtained by comparing eqs. (18.138), (18.148) and (18.149). At the final time $t_f$, the boundary condition is

$$\phi_1(t_f, \mathbf{x}) = 0, \qquad \partial_0 \phi_1(t_f, \mathbf{x}) = i z(\mathbf{x}) \mathcal{O}'(\phi_2(t_f, \mathbf{x})). \tag{18.150}$$

At the initial time $t_i$, we must have

$$\int_{y^0=t_i} d^3y \left[ G^0_{21}(x,y) \stackrel{\leftrightarrow}{\partial}_{y_0} \phi_2(y) + G^0_{22}(x,y) \stackrel{\leftrightarrow}{\partial}_{y_0} \phi_1(y) \right] = 0. \tag{18.151}$$

Some simple manipulations lead to the following equivalent form:

$$\int_{y^0=t_i} d^3y \, G^0_{-+}(x,y) \stackrel{\leftrightarrow}{\partial}_{y_0} (\phi_2(y) + \tfrac{1}{2}\phi_1(y))$$
$$= \int_{y^0=t_i} d^3y \, G^0_{+-}(x,y) \stackrel{\leftrightarrow}{\partial}_{y_0} (\phi_2(y) - \tfrac{1}{2}\phi_1(y)) = 0. \tag{18.152}$$

From the explicit form of the propagators $G^0_{+-}$ and $G^0_{-+}$ (see eq. (18.127)), we see that, at the initial time, the combination $\phi_2 + \tfrac{1}{2}\phi_1$ has no positive frequency components and the combination $\phi_2 - \tfrac{1}{2}\phi_1$ has no negative frequency components. An equivalent way to state this boundary condition is in terms of the Fourier coefficients of the fields $\phi_{1,2}$. Let us decompose them at time $t_i$ as follows:

$$\phi_{1,2}(t_i, \mathbf{x}) \equiv \int \frac{d^3k}{(2\pi)^3 2E_k} \left\{ \widetilde{\phi}^{(+)}_{1,2}(\mathbf{k}) e^{-i k \cdot x} + \widetilde{\phi}^{(-)}_{1,2}(\mathbf{k}) e^{+i k \cdot x} \right\}. \tag{18.153}$$

In terms of the coefficients introduced in this decomposition, the boundary conditions at the initial time read

$$\widetilde{\phi}^{(+)}_2(\mathbf{k}) = -\frac{1}{2} \widetilde{\phi}^{(+)}_1(\mathbf{k}), \qquad \widetilde{\phi}^{(-)}_2(\mathbf{k}) = \frac{1}{2} \widetilde{\phi}^{(-)}_1(\mathbf{k}). \tag{18.154}$$

### 18.5.3 Correlations in the Quasi-Classical Regime

**Quasi-classical approximation:** It is in principle possible to solve order by order in the function $z(x)$ the equations of motion (18.144) (or (18.145) for a quartic interaction term) with the boundary conditions (18.150) and (18.154). At order 0 in $z$, one easily recovers the result of eq. (18.34) for the one-point function, which states that the expectation value of an observable at leading order is given by the solution $\Phi$ of the classical field equation of motion,

$$(\Box_x + m^2)\, \Phi(x) + U'(\Phi(x)) = J(x), \qquad (18.155)$$

with a boundary condition at the initial time that depends on the coherent state in which the system is initialized (or $\Phi_{\text{ini}} \equiv 0$ when the initial state is the vacuum). However, this expansion becomes increasingly cumbersome beyond this simple result. Instead of pursuing this very complicated expansion in powers of $z$, we present an approximation that allows for an all-orders solution of eqs. (18.144), (18.150) and (18.154). Here, we give only a very sketchy motivation for this approximation, and a lengthier discussion of its validity will be provided later in this section (after we have derived expressions for the fields $\phi_1$ and $\phi_2$).

Let us first recall that the fields $\phi_+$ and $\phi_-$ represent, respectively, the space-time evolution of the field in amplitudes and in conjugate amplitudes. The fact that they are distinct leads to interferences when squaring amplitudes, a quantum effect controlled by $\hbar$. Consequently, in the strong field regime, we expect the difference $\phi_1 \equiv \phi_+ - \phi_-$ to be small compared to $\phi_\pm$ themselves, i.e.,

$$\phi_1 \ll \phi_2. \qquad (18.156)$$

In this situation, which we call the *quasi-classical approximation*, we can approximate the equations of motion (18.144) by keeping only the lowest order in $\phi_1$. This amounts to keeping only the terms linear in $\phi_1$ in eq. (18.139) (in the case of a $\phi^4$ theory, it means dropping the $\phi_1^3 \phi_2$ term in eq. (18.140)). After this approximation, they become

$$\begin{aligned}
\left[\Box + m^2 - \mathcal{L}''_{\text{int}}(\phi_2)\right] \phi_1 &= 0, \\
(\Box + m^2)\, \phi_2 - \mathcal{L}'_{\text{int}}(\phi_2) &= 0,
\end{aligned} \qquad (18.157)$$

while the boundary conditions are still given by (18.150) and (18.152). The simpler problem one must now solve is illustrated in Figure 18.9. The field $\phi_1$ obeys a linear equation of motion (dressed by the field $\phi_2$, although this aspect is not visible in the figure), with an advanced boundary condition that depends on $\phi_2$. In parallel, the field $\phi_2$ obeys the nonlinear classical field equation of motion, with a retarded boundary condition that depends on $\phi_1$. As we shall show, this tightly constrained problem admits a formal solution, valid to all orders in the function $z$, in the form of an implicit functional equation for the first derivative of $\ln \mathcal{F}[z]$.

**Formal solution:** In order to solve the equation of motion for $\phi_1$, we use the *mode functions* $a_{\pm \mathbf{k}}(x)$ defined as follows:

$$\begin{aligned}
\left[\Box_x + m^2 - \mathcal{L}''_{\text{int}}(\phi_2(x))\right] a_{\pm \mathbf{k}}(x) &= 0, \\
\lim_{x^0 \to t_i} a_{\pm \mathbf{k}}(x) &= e^{\mp i k \cdot x}.
\end{aligned} \qquad (18.158)$$

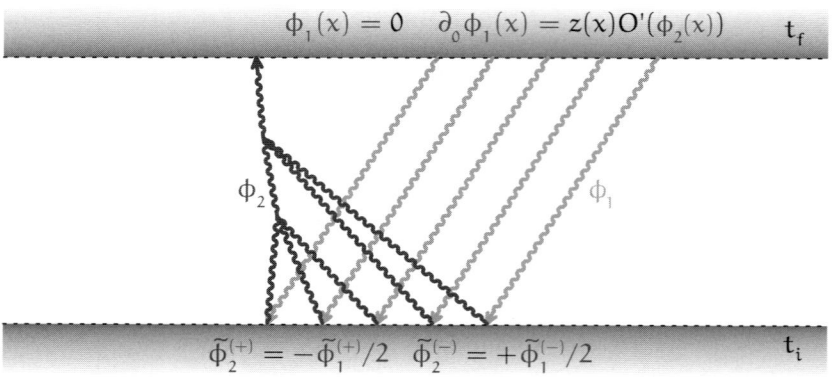

FIGURE 18.9: Relationship between the fields $\phi_1$ and $\phi_2$ in the quasi-classical approximation.

They form a basis of the linear space of solutions of the equation obeyed by $\phi_1$, and therefore we may express $\phi_1$ as a linear superposition of the mode functions. At this point, we need a slightly more explicit form of eq. (18.120),

$$\int \frac{d^3 k}{(2\pi)^3 2E_k} \left\{ \begin{pmatrix} a_{+k}(x)\dot{a}_{-k}(y) & a_{-k}(x)a_{+k}(y) \\ \dot{a}_{+k}(x)\dot{a}_{-k}(y) & \dot{a}_{-k}(x)a_{+k}(y) \end{pmatrix} - (+k \leftrightarrow -k) \right\}$$

$$\underset{x^0=y^0}{=} i\delta(x-y) \begin{pmatrix} 1 & 0 \\ 0 & 1 \end{pmatrix}. \tag{18.159}$$

Thanks to these identities, it is easy to check that the field $\phi_1$ that obeys the required equation of motion and boundary conditions is given by

$$\phi_1(x) = \int \frac{d^3 k\, d^3 u}{(2\pi)^3 2E_k} \left\{ a_{-k}(x)a_{+k}(t_f, u) - a_{+k}(x)a_{-k}(t_f, u) \right\} z(u)\, \mathcal{O}'(\phi_2(t_f, u)). \tag{18.160}$$

The above equation formally defines $\phi_1(x)$ in the bulk, $x \in \Omega$, in terms of the field $\phi_2$ at the final time. Besides the explicit factor $z(u)$, the right-hand side contains also an implicit $z$ dependence (to all orders in $z$) in the field $\phi_2(t_f, u)$ and in the mode functions $a_{\pm k}$ (since they evolve on top of the background $\phi_2$).

Then, using the boundary condition at the initial time, we obtain the following expression for the field $\phi_2$ at $t_i$:

$$\phi_2(t_i, y) = \int \frac{d^3 k\, d^3 u}{(2\pi)^3 4E_k} \left\{ e^{+ik\cdot y}\, a_{+k}(t_f, u) + e^{-ik\cdot y}\, a_{-k}(t_f, u) \right\} z(u)\, \mathcal{O}'(\phi_2(t_f, u)). \tag{18.161}$$

This can be expressed in a more convenient way with eq. (18.51). In terms of the operators $\mathbb{T}_{\pm k}$, we may rewrite $\phi_2$ at the initial time as follows:

$$\phi_2(t_i, y) = \int \frac{d^3 k\, d^3 u}{(2\pi)^3 4E_k}\, z(u)\, \mathcal{O}(\phi_2(t_f, u)) \left\{ \overleftarrow{\mathbb{T}}_{+k}\, e^{+ik\cdot y} + \overleftarrow{\mathbb{T}}_{-k}\, e^{-ik\cdot y} \right\}, \tag{18.162}$$

## 18.5 MULTI-POINT CORRELATION FUNCTIONS AT TREE LEVEL

where the arrows indicate on which side the $\mathbb{T}_{\pm k}$ operators act. This expression gives the initial condition for the second of eqs. (18.157), in the form of a linear superposition of plane waves $\exp(\pm i k \cdot y)$. The next step is to note that the field $\phi_2(x)$ that satisfies this equation of motion, and has the initial condition $\phi_2(t_i, y)$, is formally given by

$$\phi_2(x) = \exp\left\{\int d^3 y \left[\phi_2(t_i, y) \frac{\delta}{\delta \Phi_{\text{ini}}(t_i, y)} + (\partial^0 \phi_2(t_i, y)) \frac{\delta}{\delta(\partial^0 \Phi_{\text{ini}}(t_i, y))}\right]\right\} \Phi(x). \tag{18.163}$$

This formula follows from the fact that the derivative with respect to the initial field is the generator for shifts of the initial condition of $\Phi$; its exponential is therefore the corresponding translation operator (see eq. (18.77)). The same formula applies also to any function of the field, e.g. $\mathcal{O}(\phi_2)$. Substituting $\phi_2(t_i, y)$ by eq. (18.162) inside the exponential, this leads to

$$\mathcal{O}(\phi_2(x)) = \exp\left\{\frac{1}{2}\int \frac{d^3 k}{(2\pi)^3 2E_k} \int d^3 u\, z(u)\, \mathcal{O}(\phi_2(t_f, u)) \right.$$
$$\left. \times \left[\overleftarrow{\mathbb{T}}_{+k}\overrightarrow{\mathbb{T}}_{-k} + \overleftarrow{\mathbb{T}}_{-k}\overrightarrow{\mathbb{T}}_{+k}\right]\right\} \mathcal{O}(\Phi(x))\bigg|_{\Phi_{\text{ini}} \equiv 0}$$

$$= \exp\left\{\int d^3 u\, z(u)\, \mathcal{O}(\phi_2(t_f, u)) \otimes \right\} \mathcal{O}(\Phi(x))\bigg|_{\Phi_{\text{ini}} \equiv 0}, \tag{18.164}$$

where the $\otimes$ operation is defined by

$$A \otimes B \equiv \frac{1}{2}\int \frac{d^3 k}{(2\pi)^3 2E_k}\, A\left[\overleftarrow{\mathbb{T}}_{+k}\overrightarrow{\mathbb{T}}_{-k} + \overleftarrow{\mathbb{T}}_{-k}\overrightarrow{\mathbb{T}}_{+k}\right] B. \tag{18.165}$$

Setting $x^0 = t_f$ and denoting $\mathcal{D}[x_1; z] \equiv \mathcal{O}(\phi_2(t_f, x_1))$ the first derivative of $\ln \mathcal{F}$, we see that it obeys the following recursive formula:

$$\mathcal{D}[x_1; z] = \exp\left\{\int d^3 u\, z(u)\, \mathcal{D}[u; z] \otimes \right\} \mathcal{O}(\Phi(t_f, x_1))\bigg|_{\Phi_{\text{ini}} \equiv 0}. \tag{18.166}$$

**Realization of the quasi-classical approximation:** Let us now return to the condition $\phi_1 \ll \phi_2$, which was used in the derivation of eq. (18.166), in order to see a posteriori when it is satisfied. To that effect, we can use eq. (18.160) for $\phi_1$. For $\phi_2$, the initial condition at $t_i$ is given by eq. (18.162). For the sake of this discussion, it is sufficient to use a linearized solution for $\phi_2$ in the bulk, which reads

$$\phi_2(x)\bigg|_{\text{lin}} = \frac{1}{2}\int \frac{d^3 k}{(2\pi)^3 2E_k}\int d^3 u\, \{a_{-k}(x)a_{+k}(t_f, u) + a_{+k}(x)a_{-k}(t_f, u)\}$$
$$\times z(u)\, \mathcal{O}'(\phi_2(t_f, u)). \tag{18.167}$$

First, a comparison between eqs. (18.160) and (18.167) indicates that $\phi_1$ and $\phi_2$ have the same order in the coupling constant $\lambda$, since they are made of the same building blocks (the only difference is the sign between the two terms of the integrand, and an irrelevant overall factor $\tfrac{1}{2}$).

However, a hierarchy between $\phi_1$ and $\phi_2$ arises dynamically when the classical solutions of the field equation of motion (18.155) are unstable. Such instabilities are fairly generic in several quantum field theories; in particular the scalar field theory with a $\phi^4$ coupling that we are using as an example is known to have a parametric resonance (see Exercise 18.12 – in Yang–Mills theory, the classical solutions are also subject to instabilities). Since the mode functions $a_{\pm k}$ are linearized perturbations on top of the classical field $\phi_2$, an instability of the classical solution $\phi_2$ is equivalent to the fact that some of the mode functions grow exponentially with time, as $\exp(\mu(x^0 - t_i))$ (where $\mu$ is the Lyapunov exponent). Thus, since eq. (18.167) is bilinear in the mode functions, we expect that

$$\phi_2(x)\Big|_{\text{lin}} \sim e^{\mu(x^0 + t_f - 2t_i)}. \tag{18.168}$$

Estimating the magnitude of $\phi_1$ requires more care. Indeed, from eq. (18.159), antisymmetric combinations of the mode functions at equal times remain of order 1 even if individual mode functions grow exponentially with time. Thus, at the final time, we have

$$\phi_1(t_f, x) \sim 1 \quad \text{and} \quad \frac{\phi_2(t_f, x)}{\phi_1(t_f, x)} \sim e^{2\mu(t_f - t_i)} \gg 1, \tag{18.169}$$

for sufficiently large $t_f - t_i$.

In order to estimate the ratio $\phi_2/\phi_1$ at intermediate times, one may use the following reasoning. The antisymmetric combination of mode functions that enters in eq. (18.160) is the advanced propagator $G_A$ in the background $\phi_2$. This advanced propagator may also be expressed in terms of a different set of mode functions $b_{\pm k}$ defined to be plane waves at the final time $t_f$,

$$\left[\Box_x + m^2 - \mathcal{L}''_{\text{int}}(\phi_2(x))\right] b_{\pm k}(x) = 0, \quad \lim_{x^0 \to t_f} b_{\pm k}(x) = e^{\mp i k \cdot x}. \tag{18.170}$$

In terms of these alternate mode functions, we also have

$$\phi_1(x) = \int \frac{d^3 k}{(2\pi)^3 2 E_k} \int d^3 u \left\{ b_{-k}(x) b_{+k}(t_f, u) - b_{+k}(x) b_{-k}(t_f, u) \right\}$$
$$\times z(u)\, \mathcal{O}'(\phi_2(t_f, u)). \tag{18.171}$$

In the presence of instabilities, these backward-evolving mode functions grow when $x^0$ decreases away from $t_f$, as $\exp(\mu(t_f - x^0))$ (in this sketchy argument, the Lyapunov exponent $\mu$ is assumed to be the same for the forward and backward mode functions). This implies

$$\phi_1(x) \sim e^{\mu(t_f - x^0)}, \tag{18.172}$$

and the following magnitude for the ratio $\phi_2/\phi_1$ at intermediate times:

$$\frac{\phi_2(x)}{\phi_1(x)} \sim \frac{e^{\mu(x^0 + t_f - 2t_i)}}{e^{\mu(t_f - x^0)}} \sim e^{2\mu(x^0 - t_i)}. \tag{18.173}$$

Thus, with instabilities and non-zero Lyapunov exponents, the quasi-classical approximation is generically satisfied thanks to the exponential growth of perturbations over the background.

## 18.5 MULTI-POINT CORRELATION FUNCTIONS AT TREE LEVEL

**Expansion of eq. (18.166) in powers of $z$:** Although eq. (18.166) cannot be solved explicitly, it is fairly easy to obtain a diagrammatic representation of its solution. For this, let us introduce the following graphical notations:

$$\text{(i)} \equiv \mathcal{O}(\Phi(t_f, \mathbf{x}_i)),$$

$$\bullet \equiv \int d^3\mathbf{u}\, z(\mathbf{u})\, \mathcal{O}(\Phi(t_f, \mathbf{u})),$$

$$\text{(A)} \longleftrightarrow \text{(B)} \equiv A \otimes B,$$

in terms of which the functional equation obeyed by $\mathcal{D}[\mathbf{x}_1; z]$ reads

$$\mathcal{D}[\mathbf{x}_1; z] = \exp\left\{\left(\int d^3\mathbf{u}\, z(\mathbf{u})\, \mathcal{D}[\mathbf{u}; z]\right) \longleftrightarrow \right\} \text{(1)} \bigg|_{\Phi_{\text{ini}} \equiv 0}.$$

At the order 0 in $z$, we just need to set $z \equiv 0$ inside the exponential, to obtain

$$\mathcal{D}^{(0)}[\mathbf{x}_1; z] = \text{(1)}. \tag{18.174}$$

Then, we proceed recursively. We insert the zeroth order result in the exponential, and expand to order 1 in $z$, leading to the following result at order 1:

$$\mathcal{D}^{(1)}[\mathbf{x}_1; z] = \bullet\!\!\longleftrightarrow\!\!\text{(1)}. \tag{18.175}$$

The next two iterations give

$$\mathcal{D}^{(2)}[\mathbf{x}_1; z] = \bullet\!\!\longleftrightarrow\!\!\bullet\!\!\longleftrightarrow\!\!\text{(1)} + \frac{1}{2!}\,\bullet\!\!\longleftrightarrow\!\!\text{(1)}\!\!\longleftrightarrow\!\!\bullet \tag{18.176}$$

and

$$\mathcal{D}^{(3)}[\mathbf{x}_1; z] = \bullet\!\!\longleftrightarrow\!\!\bullet\!\!\longleftrightarrow\!\!\bullet\!\!\longleftrightarrow\!\!\text{(1)} + \bullet\!\!\longleftrightarrow\!\!\bullet\!\!\longleftrightarrow\!\!\text{(1)}\!\!\longleftrightarrow\!\!\bullet$$

$$+ \frac{1}{2!}\,\text{(tree)}\!\!\longleftrightarrow\!\!\text{(1)} + \frac{1}{3!}\,\text{(tree)}\!\!\longleftrightarrow\!\!\text{(1)}\!\!\longleftrightarrow\!\!\bullet. \tag{18.177}$$

These examples generalize to all orders in $z$: The functional $\mathcal{D}[\mathbf{x}_1; z]$ can be represented as the sum of all the rooted trees (the root being the node carrying the fixed point $\mathbf{x}_1$) weighted by the corresponding symmetry factor $1/\mathcal{S}(T)$,

$$\frac{\delta \ln \mathcal{F}[z]}{\delta z(z_1)} = \mathcal{D}[\mathbf{x}_1; z] = \sum_{\substack{\text{rooted} \\ \text{trees } T}} \frac{1}{\mathcal{S}(T)}\,\text{(tree diagram)}. \tag{18.178}$$

FIGURE 18.10: Causal structure of the three-point correlation function in the quasi-classical regime.

**Correlation functions:** The n-point correlation function is obtained by differentiating this expression $n-1$ times, with respect to $z(x_2), \cdots, z(x_n)$, and by setting $z \equiv 0$ afterwards. This selects all the trees with n distinct labeled nodes[10] (including the node at $x_1$). Moreover, since derivatives commute, these successive differentiations eliminate the symmetry factors, leading to

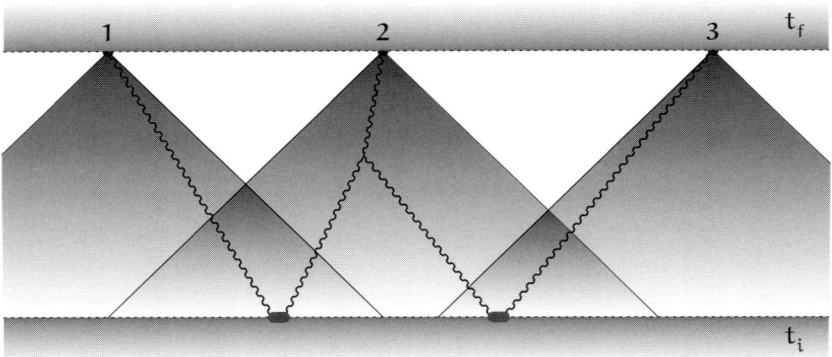

$$\left.\frac{\delta \ln \mathcal{F}[z]}{\delta z(z_1) \cdots \delta z(x_n)}\right|_{z\equiv 0} = \mathcal{C}_{\{1 \cdots n\}} = \sum_{\substack{\text{trees with n} \\ \text{labeled nodes}}} \qquad . \qquad (18.179)$$

The number of trees contributing to this sum is equal to $n^{n-2}$ (*Cayley's formula*) – see Exercise 18.13. Equation (18.179) tells us that, at tree level in the quasi-classical regime, all the n-point correlation functions are entirely determined by the functional dependence of the solution of the classical field equation of motion with respect to its initial condition. Moreover, this formula provides a way to construct explicitly the correlation functions in terms of functional derivatives with respect to the initial field.

In the quasi-classical approximation, the final state correlations are entirely due to quantum fluctuations in the initial state, which are encoded in the function $G_{22}^0(x,y)$. If the initial state is the vacuum, it reads

$$G_{22}^0(x,y) = \int \frac{d^3\mathbf{k}}{(2\pi)^2 2E_k} \, e^{i\mathbf{k}\cdot(x-y)}. \qquad (18.180)$$

The support of this function is dominated by distances $|x-y|$ smaller than the Compton wavelength $m^{-1}$. Thus, in the tree representation of eq. (18.179), a link between the points $x_i$ and $x_j$ is non-zero, provided that the past light-cones of summits $x_i$ and $x_j$ overlap at the initial time (or at least approach each other within distances $\lesssim m^{-1}$), as illustrated in Figure 18.10. A more thorough analysis would indicate that eq. (18.179) is exact at tree level

---

[10]Thus, permuting nodes yields a different tree, and the symmetry factors are now equal to 1.

for the one-point and two-point correlations, but is incomplete (even at tree level) beyond two points (see Exercise 18.14). The corrections to this formula are nevertheless suppressed if the condition $\phi_1 \ll \phi_2$ holds.

## Exercises

**18.1** Justify the reduction formula in the first part of eq. (18.1).

**18.2** Prove the formula quoted in footnote 3.

**\*18.3** What is the main change in the evaluation of non-local observables containing field operators with time-like separations? (One may consider the retarded propagator $G_R(x,y) \equiv \theta(x^0 - y^0)\langle[\phi(x), \phi(y)]\rangle$ as a toy example of this type of observable.) Try to generalize as much as possible of the Section 18.2 to this more general type of observable.

**\*18.4** In eq. (18.58), the operators $\mathbb{T}_{\pm k}$ are defined as derivatives with respect to the initial fields at the time $x^0 = -\infty$ (see eq. (18.52)). Derive a generalization of eq. (18.58) in which the derivatives would be with respect to an initial field $\Phi_{\text{ini}}$ on an arbitrary space-like initial surface (this is useful for initial value problems where the initial time is not $-\infty$).

**18.5** Recall Weyl's quantization prescription introduced in Exercise 5.1. Consider two functions f, g over the classical phase-space and define their Moyal bracket by

$$[f, g]_\star \equiv \frac{2}{i} f \sin\left(\frac{i}{2}(\overleftarrow{\partial}_p \overrightarrow{\partial}_x - \overleftarrow{\partial}_x \overrightarrow{\partial}_p)\right) g.$$

Show that the Weyl mapping of $[f, g]_\star$ is the commutator $[F, G]$ of the respective Weyl mappings F, G of f and g.

**18.6** By methods similar to those used in Section 18.2, show that the spectrum of produced fermions in a gauge background field is given at tree level by

$$2E_p \frac{dN_f}{d^3p} = \frac{1}{(2\pi)^3} \sum_{\sigma, s = \pm} \int \frac{d^3k}{(2\pi)^3 2E_k} \lim_{x^0 \to +\infty} \left| \int d^3x \, e^{ip \cdot x} u_\sigma^\dagger(p) \psi_{ks}(x) \right|^2,$$

where $\psi_{ks}(x)$ is a spinor defined by

$$(i\slashed{D} - m)\psi_{ks}(x) = 0, \quad \lim_{x^0 \to -\infty} \psi_{ks}(x) = v_s(k) e^{ik \cdot x}.$$

- Why does this formula assume that the background field is zero when $x^0 \to +\infty$?
- How should the formula be modified to cope with situations in which the background field is a non-zero pure gauge at $x^0 \to +\infty$? What happens if we use this formula if the background field at infinite time is not a pure gauge?
- If space is discretized as a lattice with $N^3$ nodes, and the Dirac evolution is solved by doing $N_t$ discrete timesteps, what is the computational cost for calculating this spectrum?
- Show that the spectrum can be obtained by

$$2E_p \frac{dN_f}{d^3p} = \frac{1}{(2\pi)^3} \sum_{\substack{\sigma = \pm \\ \text{confs } \alpha}} \lim_{x^0 \to +\infty} \left| \int d^3x \, e^{ip \cdot x} u_\sigma^\dagger(p) \psi_\alpha(x) \right|^2,$$

where $\psi_\alpha(x)$ is a spinor that obeys the same Dirac equation and has an initial condition of the form

$$\lim_{x^0 \to -\infty} \psi_\alpha(x) = \sum_{s=\pm} \int \frac{d^2 k}{(2\pi)^3 2 E_k} c_{ks}(\alpha) v_s(k) e^{ik\cdot x}$$

with random Gaussian distributed coefficients $c_{ks}(\alpha)$. What should be the distribution of these random coefficients to recover the correct result (in the limit of infinite sampling statistics)? What is the computational cost of this approach if $N_\alpha$ configurations $\alpha$ are used?

**18.7** Derive the Green's formula for a classical scalar field whose initial condition is set on the time-like surface $t = z$. *Hint: Use light-cone coordinates to simplify the notations.*

**\*18.8** Consider a locally space-like (possibly light-like) surface $\Sigma$, and two points $x, y$, with $y$ below $\Sigma$ and $x$ above $\Sigma$.
- Show that the scalar retarded propagator from $y$ to $x$ can be split as follows:

$$G_R(x, y) = i \int_\Sigma d^3 S_z\, G_R(x, y)(n \cdot \overleftrightarrow{\partial}_z) G_R(z, y),$$

where $n^\mu$ is the unit vector normal to the surface $\Sigma$ at point $z$, pointing upwards.
- Is this formula true for time-ordered propagators?
- Derive a similar formula for the fermionic retarded propagator.

**18.9** Explain why eqs. (18.49) and (18.110) are equivalent.

**\*18.10** Generalize the inner product defined in eq. (18.115) to a generic space-like surface $\Sigma$, in such a way that it does not depend on $\Sigma$ (this property is the generalization of the time independence of eq. (18.115)).

**\*18.11** Write the analogue of eq. (18.111) for the mode functions in Yang–Mills theory. Try to define an inner product analogous to eq. (18.115). Is it conserved in general? If one evaluates this inner product on surfaces of constant $x^0$, what is the most convenient choice of gauge?

**18.12** In the massless scalar theory with $\phi^4$ interaction, consider a spatially homogeneous background $\Phi$.

- Show that $\Phi$ is a periodic function of time that can be written in terms of an elliptic cosine function. How does the period of the solution depend on the amplitude of the oscillations?
- Consider now the mode function of momentum $k$ on top of this background. What is its equation of motion? What is the dimension of the linear space of solutions?
- The *monodromy matrix* is the matrix that transforms a basis of solutions of the previous equation from $t$ to $t + T$ where $T$ is the period of the background. Derive a criterion of instability of the mode functions based on the value of the trace of this matrix.

**18.13** Denote

$$w_n \equiv \sum_{\substack{\text{rooted}\\ \text{trees } T_{n+1}}} \frac{1}{S(T_{n+1})}$$

the sum of the symmetry factors over all the $(n+1)$-nodes rooted trees (i.e., trees with one labeled node and $n$ unlabeled nodes).

- Why do we have $w_n = (n+1)^{n-1}/n!$? (Assume Cayley's formula.)
- Define the function $w(z) \equiv \sum_{n \geq 0} w_n z^n$. Explain why $w(z) = \exp(zw(z))$.
- Solve iteratively the above functional equation to recover the values of the first $w_n$s.

**18.14** Up to $n = 2$, the semi-classical formula (18.179) actually gives the exact tree-level answer. Consider the three-point correlation function $\langle 0_{in} | \phi(x_1) \phi(x_2) \phi(x_3) | 0_{in} \rangle$, and determine the tree-level contributions not included in the semi-classical approximation. What is their physical interpretation? Check a posteriori that they are negligible in a system subject to instabilities.

**\*18.15** Consider the Boltzmann equation (17.133)–(17.135) for a scalar theory with $\phi^4$ interaction.

- Rewrite the collision term (right-hand side of eq. (17.133)) in terms of the self-energies in the retarded-advanced basis.
- Calculate the collision term in the semi-classical approximation, $\phi_1 \ll \phi_2$, where the vertex $\Gamma_{1112}$ can be neglected. What are the fixed points of the resulting Boltzmann equation?
- In the previous collision term, substitute $f + \frac{1}{2} \to f$. What are the new fixed points, and what is their interpretation? *Hint: Note that the low-energy expansion of the Bose–Einstein distribution is $n_B(E) = \frac{T}{E} - \frac{1}{2} + \cdots$.*

# CHAPTER 19

# From Trees to Loops

The past 25 years have witnessed spectacular progress, which we presented in Chapter 14, of the tools available for calculating physical on-shell amplitudes at tree level: color decomposition, spinor helicity formalism and on-shell recursion. However, especially in quantum chromodynamics, quantitative calculations usually require going beyond tree level. As in the case of tree-level amplitudes, the techniques developed for the evaluation of loop corrections have followed the same trend, which is to avoid as much as possible to refer to the underlying Feynman diagrams. Nowadays, the main approach for loop calculations is to find ways of breaking them down to on-shell trees, in order to reuse tree-level results. At one-loop, although analytical results remain difficult to obtain, the available tools have evolved to a point where automated computer codes exist that provide numerical results for most experimentally relevant one-loop amplitudes. The main purpose of this chapter is to expose the set of techniques that make this possible. The chapter focuses mostly on the one-loop case, at the exception of the last section where we discuss some recently proposed ideas to extend the one-loop tools to the multi-loop case. Note that we describe only methods that are generic enough to be applicable to Yang–Mills theory and QCD. Another important stream of new ideas, that we shall not present here, aim at gaining a deeper – geometrical – understanding of amplitudes in theories with more symmetry, such as supersymmetric extensions of Yang–Mills theory. Let us end this introduction by a word of caution: by its very nature, this area is a rather fast-moving target, and one should view this chapter as a (certainly incomplete) snapshot of the situation at this point in time.

## 19.1 Dualities Between Loops and Trees

In order to benefit from the developments in the calculation of tree on-shell amplitudes, the general strategy for computing loop amplitudes is to try to relate them to tree-level ones.

## 19.1 DUALITIES BETWEEN LOOPS AND TREES

Before a thorough discussion of this approach, let us first discuss an old observation by Feynman, who noted that a loop can always be rewritten as a sum of terms that have at least one on-shell propagator. We will also present some more recent attempts to write one-loop amplitudes as a sum of trees that are directly inspired by this observation. As we shall see once more in this section, the one-loop corrections are somewhat special because they do not yet exhibit quantum physics in its full glory (as we have seen in the previous chapter, the time evolution of any system at order $\hbar^1$ is still classical).

### 19.1.1 Feynman Tree Theorem

Feynman's tree theorem stems from the observation that if all the time-ordered propagators around a loop were replaced by advanced[1] propagators, then the integration over the loop momentum would give zero. This can be seen in several ways. In coordinate space, the loop momentum integration is replaced by a sequence of space-time integrations at each vertex around the loop,

$$\int \prod_{i=1}^{n} d^4 x_i \underbrace{\left[ G_A^0(x_1, x_2) G_A^0(x_2, x_3) \cdots G_A^0(x_n, x_1) \right]}_{=0} \times \begin{bmatrix} \text{external} \\ \text{propagators} \end{bmatrix} = 0. \tag{19.1}$$

The product of advanced propagators around the loop is zero, because each of them is proportional to a step function in time, as in $G_A^0(x_1, x_2) \propto \theta(x_2^0 - x_1^0)$, the product of which has no support for propagators that form a loop. This vanishing result can also be seen directly in momentum space, by recalling that an advanced propagator has its two poles above the real energy axis, unlike the Feynman propagator that has one above and one below. Because of this, the integration over the loop energy is over an integrand with no poles below the real axis. By turning the energy integration into a contour integral (the added contour at infinity does not contribute) that we close in the lower half-plane, we see that the integral is zero.

The next step is to recall a relation analogous to (18.95), $G_A^0 = G_{++}^0 - G_{-+}^0$, where $G_{++}^0$ is the Feynman propagator (the notation comes from the Schwinger–Keldysh formalism) and the propagator $G_{-+}^0$ has the following Fourier representation:

$$G_{-+}^0(x, y) = \int \frac{d^4 k}{(2\pi)^4} e^{-ik\cdot(x-y)} \underbrace{2\pi \theta(k^0) \delta(k^2)}_{G_{-+}^0(k)}. \tag{19.2}$$

Therefore, the fact that the product of the advanced propagators around the loop is zero can be symbolically (one should not take the nth power in the next equation literally, since the various propagators in fact carry different arguments) recast into

$$0 = \left[ G_{++}^0 - G_{-+}^0 \right]^n. \tag{19.3}$$

---

[1] Retarded propagators would work too, but then one would get on-shell propagators with negative energy.

By expanding the product and isolating on one side of the equation the term that contains only time-ordered propagators (this is the combination of propagators in the loop of the original amplitude), we get

$$\left[G_{++}^0\right]^n = \sum_{0\leq p\leq n-1} \left[G_{++}^0\right]^p G_{-+}^0 \left[G_{++}^0\right]^{n-p-1}$$
$$- \sum_{0\leq p+q\leq n-2} \left[G_{++}^0\right]^p G_{-+}^0 \left[G_{++}^0\right]^q G_{-+}^0 \left[G_{++}^0\right]^{n-p-q-2}$$
$$+ \sum_{0\leq p+q+r\leq n-3} \left[G_{++}^0\right]^p G_{-+}^0 \left[G_{++}^0\right]^q G_{-+}^0 \left[G_{++}^0\right]^r G_{-+}^0 \left[G_{++}^0\right]^{n-p-q-r-3}$$
$$+ \cdots \tag{19.4}$$

Given the explicit form of $G_{-+}^0$ given in eq. (19.2), we see that the right-hand side contains terms with one on-shell propagator (first line), two on-shell propagators (second line), etc. (up to a term where they are all on-shell). In total, there are $2^n - 1$ terms in the right-hand side of this equation. Each of these on-shell propagators effectively reduces by one unit the dimensionality of the loop integration. Therefore, in four dimensions, terms with more than four on-shell propagators have in general no support. All the factors interspersed between these on-shell propagators, with the external lines that are attached to them, are on-shell tree amplitudes for which there already exists powerful calculation tools.

## 19.1.2 Tree–Loop Duality for Time-Ordered One-Loop Diagrams

Variants of Feynman's tree theorem have been looked for in order to mitigate the proliferation of terms that are encountered in the right-hand side of eq. (19.4). In fact, there exists a decomposition of one-loop time-ordered diagrams as a sum of terms in which only one loop propagator is put on-shell, at the expense of a modification of the $i0^+$ prescriptions in the propagators of these tree diagrams. Unlike Feynman's approach in which the time-ordered loop propagators are written as the sum of the advanced propagators and an on-shell piece, we start from the integral over the loop momentum, and perform the integration over $l^0$ by extending the real axis by a semi-circle at infinity in the lower half of the complex plane (we assume that there are enough denominators to ensure that the integral on the semi-circle vanishes when its radius becomes infinite). Each time-ordered loop propagator has exactly one pole in the domain enclosed by this contour, and we can write (see Figure 19.1 for the labeling of the momenta around the loop),[2]

---

[2] In this discussion, we consider a scalar loop. In a gauge theory, the momentum dependence introduced in the numerators by gauge couplings is polynomial and does not interfere with the theorem of residues. Regarding the gauge boson propagators, the conclusion is unchanged in Feynman gauge. In other covariant gauges, one would have to handle possible double poles. In axial gauges $n \cdot A = 0$, it has been shown by Catani et al. that the tree–loop duality holds unmodified provided that $n^2 \leq 0$.

## 19.1 DUALITIES BETWEEN LOOPS AND TREES

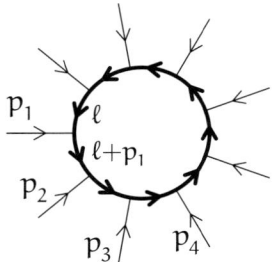

FIGURE 19.1: Loop momentum labeling.

$$\text{Loop} = \int \frac{d^d\ell}{(2\pi)^d} \prod_{i=1}^{m} G^0_{++}(\ell + q_i)$$

$$= \int \frac{d^{d-1}\ell}{(2\pi)^{d-1}} \sum_{i=1}^{m} \frac{1}{2|\ell + q_i|} \left[ \prod_{j \neq i} G^0_{++}(\ell + q_j) \right]_{\substack{(\ell+q_i)^2+i0^+ = 0 \\ \ell^0 + q_i^0 = |\ell+q_i| - i0^+}}, \quad (19.5)$$

where $q_i \equiv \sum_{j=1}^{i} p_j$ (and $q_0 \equiv 0$). To evaluate the denominators of the remaining propagators, we use

$$(\ell + q_j)^2 + i0^+ = (\ell + q_i + q_j - q_i)^2 + i0^+$$
$$= \underbrace{(\ell + q_i)^2 + i0^+}_{0} + 2(\ell + q_i) \cdot (q_j - q_i) + (q_i - q_j)^2$$
$$= \left[(\ell + q_j)^2\right]_{\ell^0 + q_i^0 = |\ell + q_i|} - i\eta \cdot (q_j - q_i) 0^+. \quad (19.6)$$

In the last line, we have used the fact that only the sign of the prefactor in front of $i0^+$ matters in order to replace $q_j^0 - q_i^0$ by $\eta \cdot (q_j - q_i)$, where $\eta^\mu \equiv (1, \mathbf{0})$ is a time-like future directed 4-vector.[3] By a simple relabeling of the loop momentum, the result of the $\ell^0$ integration takes the form

$$\text{Loop} = \int \frac{d^{d-1}\mathbf{k}}{(2\pi)^{d-1} 2|\mathbf{k}|} \sum_{i=1}^{m} \left[ \prod_{j \neq i} G_D(k + q_j - q_i) \right]_{k^0 = |\mathbf{k}|}, \quad (19.7)$$

where $G_D$ is a "dual" propagator defined by

$$G_D(k + q_j - q_i) \equiv \frac{i}{(k + q_j - q_i)^2 - i\eta \cdot (q_j - q_i) 0^+}. \quad (19.8)$$

The quantity in the square brackets on the right-hand side of eq. (19.7) is a forward tree on-shell diagram, with some propagators that have an unusual $i0^+$ prescription that depends on the signs of the differences $q_j^0 - q_i^0$ (note that this depends only on the energies external to the loop). It is this modifications that allows a seemingly "one-cut" expression to provide all the multi-cut terms contained in Feynman's tree theorem.

---

[3] A different vector $\eta^\mu$ may appear here if one chooses to integrate over a different variable instead of $\ell^0$.

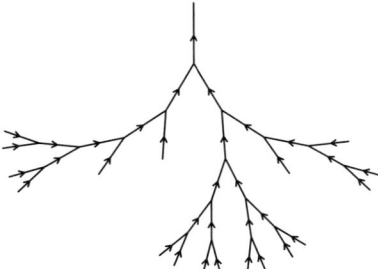

FIGURE 19.2: Example of a tree-level diagram contributing to the response function. The arrows indicate the flow of time on retarded propagators.

### 19.1.3 Response Functions and Forward Tree Amplitudes

Extensions of eq. (19.7) have been derived for two-loop and three-loop diagrams, but they do not have the nice simplicity encountered at one loop. In order to gain some physical insight on why such a tree–loop duality is natural at one loop, and what the obstructions are beyond one loop, it is useful to consider *response functions* rather than time-ordered Green's functions. The simplest context in which they occur is that of a quantum field theory coupled to an external source, as considered in the previous chapter. The n-point response function encodes the evolution of the field at order $n-1$ in the external sources. It can be obtained perturbatively by a sum of diagrams in the Schwinger–Keldysh formalism that are "oriented" by the fact that we have singled out one line corresponding to the measured field. At tree level, the sum over all the $\pm$ indices of the Schwinger–Keldysh formalism simply amounts to replacing the time-ordered propagators by retarded propagators that point toward the measured field, as shown in Figure 19.2. The perturbative rules for the calculation of a response function are more transparent when expressed in the retarded-advanced representation introduced in Section 18.5. The line at which a field is measured is of type 2, while the $n-1$ endpoints where sources are attached are of type 1. At tree level, the only possibility is that all propagators are of type 21 (i.e., retarded), and all vertices have a single index 1 (they are all of type 122 in the example in Figure 19.2 that has only trivalent vertices).

At one loop (see Figure 19.3 for an example of contribution), the situation is a bit more complicated, because the direction of the flow of time inside the loop is not fixed. The index assignment of the legs attached to the loop is always $122\cdots 2$, as shown in Figure 19.3. First, none of the vertices around the loop can be of type 111. This result is in fact general for all one-loop graphs, no matter how many legs are attached to the loop. By considering all the possible assignments for the indices carried by the propagators around the loop, one finds that the loop can be written as a sum of terms where all propagators are retarded, except for one propagator of type 22 the we represent with a cross,

$$\tag{19.9}$$

19.1 DUALITIES BETWEEN LOOPS AND TREES 551

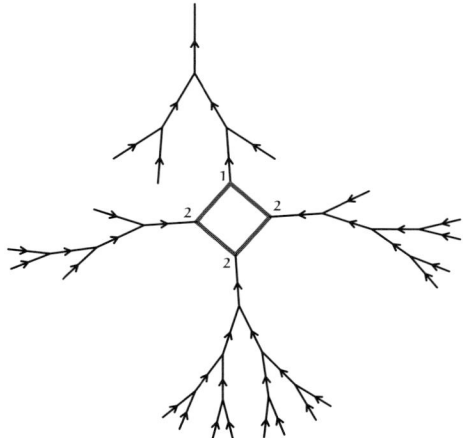

FIGURE 19.3: Example of a one-loop diagram contributing to the response function. We have indicated the retarded-advanced indices assignments of the legs attached to the loop.

(The dotted lines could be arbitrary trees attached to the loop.) Recall that[4]

$$G^0_{22}(k) = \pi \delta(k^2). \tag{19.10}$$

Since the support of the crossed propagator is restricted to on-shell momenta, we may view each of these terms as a tree amplitude integrated over the phase-space of an on-shell particle. For instance,

$$\cdots \diamond \cdots = \int \frac{d^4k}{(2\pi)^4} \, \pi \delta(k^2) \left[ \begin{array}{c} k \quad\quad k \\ \longrightarrow \quad \longleftarrow \end{array} \right]. \tag{19.11}$$

If the left-hand side is a loop with $m$ external on-shell legs, the integrand on the right-hand side is a forward (because the same momentum $k$ enters and leaves the graph) on-shell tree amplitude with $m + 2$ points, with the special feature that its internal propagators are all retarded propagators. Given eq. (19.9), the complete one-loop correction to the response function is the sum of $m$ ($m = 4$ in the example) such forward amplitudes. The physical picture that emerges from this representation is that the quantum vacuum is filled by on-shell particles with a flat occupation number equal to $1/2$ (see footnote 4), and these quantum fluctuations provide all the necessary corrections at one loop. Note that eqs. (19.9) and (19.11) are merely a perturbative – in the external source – version of eq. (18.58), which we have

---

[4] We give here the expression of $G^0_{22}$ (for real scalar fields) in the perturbative vacuum. More generally, the 22 propagator is the one that encodes the initial conditions of the system. For instance, at finite temperature, the prefactor $\pi$ is replaced by $2\pi(\frac{1}{2} + f_0(\mathbf{k}))$, where $f_0(\mathbf{k})$ is the (bosonic) particle distribution in the thermal bath. Clearly, the $\frac{1}{2}$ in $\frac{1}{2} + f_0(\mathbf{k})$ can be interpreted as an occupation number, which is nothing but the zero point occupation of all momentum modes.

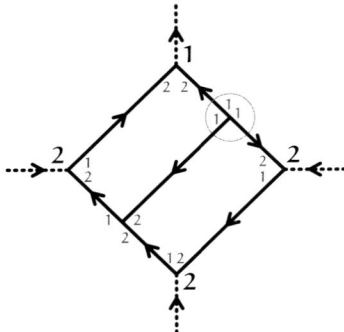

FIGURE 19.4: Example of a two-loop contribution to the response function that involves the 111 vertex (indicated by a circle), and cannot be expressed as a tree.

derived in the previous chapter in a situation in which quantum fields are coupled to a strong external source (see also Figure 18.4).

For generic external momenta, the internal propagators of a tree diagram are not on-shell, and therefore their $i0^+$ prescription (time-ordered, retarded, etc.) does not matter. This means that for these non-exceptional configurations of external momenta, the tree amplitude with inner retarded propagators is equal to the time-ordered one and can thus be calculated by the techniques exposed in Chapter 14. But the integration over k in eq. (19.11) leads to a serious complication, since for certain values of $k^\mu$, some internal propagators can reach their mass-shell. When this happens, the difference between retarded and time-ordered propagators cannot be ignored, since we have

$$G_F^0(k) = i\, P\left(\frac{1}{p^2}\right) + \pi \delta(p^2)\,, \quad G_R^0(k) = i\, P\left(\frac{1}{p^2}\right) + \pi\, \text{sign}\,(p^0)\, \delta(p^2), \qquad (19.12)$$

and the two propagators differ precisely by the sign that weights the delta function.

Even though the practical applications of this approach are contingent on being able to analytically continue graphs between retarded and time-ordered boundary conditions, it provides another illustration of the fact that one-loop diagrams can generically be broken into trees. This discussion also shows that one loop is rather special because the vertex 111 never appears. This peculiarity ceases at two loops, as one can see in the example Figure 19.4. This diagram, which contains a vertex of type 111, is non-zero (one may check that it does not contain any closed retarded loop). Moreover, all its internal propagators are of type 12, i.e., retarded, which means that none of them is on-shell. Thus, this contribution cannot readily be rewritten as a forward tree graph, and thus does not appear to be a correction due to vacuum quantum fluctuations. This is consistent with the observation made in the previous chapter that, starting at the order $\hbar^2$, quantum effects also alter the time evolution of a system.[5]

---

[5]The naive generalization of eq. (19.11) to two loops, in which one would simply cut open two lines in order to transform the graph into a tree graph, is incorrect because it misses these quantum corrections to the dynamical evolution of the system. It has been noted by Caron-Huot that two-loop corrections can formally be written as the phase-space integral of on-shell forward tree diagrams (with some special weights that depend on the energy of the particles that are forward scattered). But these tree graphs lack a clear physical interpretation in general, because they may involve ghosts or longitudinal polarizations.

## 19.2 Reduction of One-Loop Amplitudes

### 19.2.1 Color Decomposition

As in the case of tree gluon amplitudes, it is useful to first separate the color degrees of freedom from the kinematical factors. This is done by rewriting all the structure constants $f^{abc}$ contained in the three- and four-gluon vertices (as well as the ghost-gluon vertex in covariant gauges) as traces of three generators in the fundamental representation of $\mathfrak{su}(N)$, and then by performing the color contractions among these generators by means of the Fierz identity. For illustration purposes, let us consider the example of the following four-point one-loop graphs:

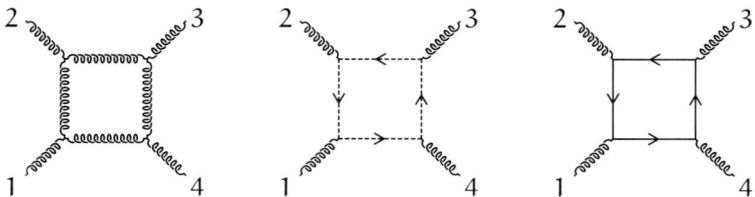

Consider first the gluon loop, which may be written as

$$\mathcal{M}_4(1234) = \left(\mathrm{i}f^{da_1 a}\right)\left(\mathrm{i}f^{a a_2 b}\right)\left(\mathrm{i}f^{b a_3 c}\right)\left(\mathrm{i}f^{c a_4 d}\right) \mathcal{A}_4(1234), \tag{19.13}$$

where $\mathcal{A}_4(1234)$ denotes all the kinematical factors of this graph. Using the usual techniques for eliminating the structure constants, we obtain

$$\begin{aligned}\mathcal{M}_4(1234) =\ & 2^4\, \mathcal{A}_4(1234)\, \mathrm{tr}\left(t_f^d t_f^{a_1} t_f^a - t_f^a t_f^{a_1} t_f^d\right)\mathrm{tr}\left(t_f^a t_f^{a_2} t_f^b - t_f^b t_f^{a_2} t_f^a\right)\\ & \times\, \mathrm{tr}\left(t_f^b t_f^{a_3} t_f^c - t_f^c t_f^{a_3} t_f^b\right)\mathrm{tr}\left(t_f^c t_f^{a_4} t_f^d - t_f^d t_f^{a_4} t_f^c\right)\\ =\ & \mathcal{A}_4(1234)\, \Big\{ N\, \mathrm{tr}\left(t_f^{a_1} t_f^{a_2} t_f^{a_3} t_f^{a_4}\right) + N\, \mathrm{tr}\left(t_f^{a_1} t_f^{a_4} t_f^{a_3} t_f^{a_2}\right)\\ & + 2\, \mathrm{tr}\left(t_f^{a_1} t_f^{a_2}\right)\mathrm{tr}\left(t_f^{a_3} t_f^{a_4}\right) + 2\, \mathrm{tr}\left(t_f^{a_1} t_f^{a_3}\right)\mathrm{tr}\left(t_f^{a_2} t_f^{a_4}\right)\\ & + 2\, \mathrm{tr}\left(t_f^{a_1} t_f^{a_4}\right)\mathrm{tr}\left(t_f^{a_2} t_f^{a_3}\right)\Big\}. \end{aligned} \tag{19.14}$$

At one loop, this color decomposition leads to single-trace terms weighted by the number of colors (these terms are sometimes called leading color terms, since they dominate the behavior of the amplitude at large N), and double-trace terms without a prefactor N. This structure is completely general for all graphs with one gluon loop, but the number of matrices in each trace of the double-trace terms may vary (in the above example, we have ignored terms with the trace of a single matrix due to the tracelessness of the generators of $\mathfrak{su}(N)$). The ghost loop has the same color factors but a different kinematical one, and the quark loop only involves the single trace $\mathrm{tr}\left(t_f^{a_1} t_f^{a_2} t_f^{a_3} t_f^{a_4}\right)$ (without a prefactor N).

This color decomposition in terms of single-trace terms and double-trace terms is true for all one-loop gluon amplitudes, and we may write the following general decomposition:

$$\mathcal{M}_n^{1-\text{loop}}(1\cdots n) = N \sum_{\sigma \in \mathfrak{S}_n/Z_n} \mathcal{A}_{n;0}(\sigma_1 \cdots \sigma_n) \,\text{tr}\left(t_f^{a_{\sigma_1}} \cdots t_f^{a_{\sigma_n}}\right)$$

$$+ \sum_{c=2}^{\lfloor \frac{n}{2} \rfloor} \sum_{\sigma \in \mathfrak{S}_n/Z_{c;n-c}} \mathcal{A}_{n;c}(\sigma_1 \cdots \sigma_n) \,\text{tr}\left(t_f^{a_{\sigma_1}} \cdots t_f^{a_{\sigma_c}}\right) \text{tr}\left(t_f^{a_{\sigma_{c+1}}} \cdots t_f^{a_{\sigma_n}}\right), \quad (19.15)$$

where $\lfloor \frac{n}{2} \rfloor$ is the integer value of $n/2$ and $Z_{c;n-c}$ is the product of cyclic permutations of $c$ elements by cyclic permutations of $n - c$ elements. Therefore, the second sum is over all permutations of $n$ elements that do not leave the two traces invariant. There is no term with $c = 1$ because the generators of $\mathfrak{su}(N)$ are traceless. Loop particles in the fundamental representation, such as quarks, contribute only to the single-trace terms. The leading-color partial amplitudes $\mathcal{A}_{n;0}$ for a given cyclic ordering are obtained as the sum of the corresponding color-ordered one-loop planar diagrams. This is not true for the partial amplitudes that appear as coefficients of the double traces. As a counter-example, note for instance in eq. (19.14) how a partial amplitude with cyclic ordering (1234) appears as the coefficient of $\text{tr}\,(t^{a_1}t^{a_3})\text{tr}\,(t^{a_2}t^{a_4})$. But the partial amplitudes $\mathcal{A}_{n;c}$ with $c \geq 2$ can be expressed as sums of permutations of the $\mathcal{A}_{n;0}$, thereby implying that it is sufficient to evaluate the leading-color partial amplitudes.

### 19.2.2 Tensor Reduction

Once the color structure of an amplitude has been disentangled from the kinematical information, we are left with the evaluation of the primitive partial amplitudes $\mathcal{A}_{n;0}$. At one loop, the typical terms produced by the color-ordered Feynman rules have the structure of a loop with "trees" attached to it (these trees may be either the polarization vector of an on-shell external leg of the amplitude directly glued to the loop, or genuine trees with one off-shell leg attached to the loop). If the amplitude has a total of $n$ external legs, the loop itself has $m \leq n$ external points. Let us call $p_1, \cdots, p_m$ the momenta incoming via these points. These momenta may be off-shell, and their sum is zero by momentum conservation. Generically, the loop integral is a linear combination of integral of the following type:[6]

$$L_m[q_1 \cdots q_{m-1}|N(\ell)] \equiv \int \frac{d^d\ell}{(2\pi)^d} \frac{N(\ell)}{\ell^2(\ell + q_1)^2(\ell + q_2)^2 \cdots (\ell + q_{m-1})^2}. \quad (19.16)$$

By a proper choice of the loop momentum $\ell$, only $m-1$ external momenta appear explicitly in the denominators (with the labeling chosen in Figure 19.1, we have $q_i \equiv p_1 + \cdots + p_i$, and we define $q_0 \equiv 0$). The $\ell$ dependence in the numerator $N(\ell)$ arises from momentum-dependent vertices such as the three-gluon vertex in QCD, implying that $N(\ell)$ has a polynomial dependence on $\ell$. In renormalizable theories, each vertex provides at most one power of momentum, and the degree of $N(\ell)$ is at most $m$. Note that $N(\ell)$ also depends on the momenta external to the loop and on polarization vectors, but we do not write this explicitly here. When the

---

[6]Masses can be included in the denominators in order to account for massive loop particles, but they augment the algebraic complexity of the subsequent manipulations. For this reason, we consider only the massless case here, having gluons and ghosts in mind.

## 19.2 REDUCTION OF ONE-LOOP AMPLITUDES

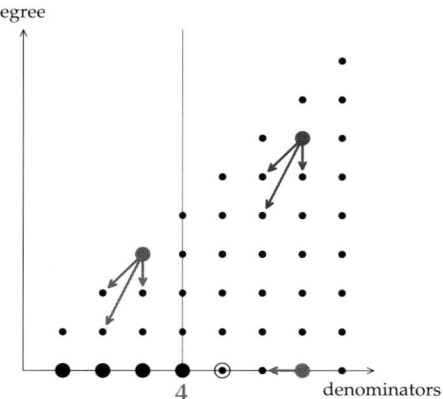

FIGURE 19.5: Pattern of reductions of one-loop tensor integrals. The arrows indicate the change in degree and number of denominators in one step of the algorithms described in this section. The thick blobs represent the master scalar integrals in four dimensions. The open circle is the additional pentagon scalar integral necessary in dimensional reduction.

numerator $N(\ell)$ is equal to 1, the loop integral is said to be scalar. *Tensor reduction* is a set of manipulations by which any loop integral such as eq. (19.16) can be reduced to a linear combination of scalar integrals (see Figure 19.5 for the possible integral reductions).

For the purpose of regulating the integral, it is useful to consider the loop momentum $\ell^\mu$ to be extended to $d = 4 - 2\epsilon$ dimensions, while the external momenta remain four-dimensional. With this prescription, the scalar products that involve only external momenta, such as $q_i \cdot q_j$, are independent of $\epsilon$. This is also the case of those that mix the loop momentum and an external momentum, like $\ell \cdot q_i$. The only $\epsilon$-dependent scalar product is $\ell^2$.

**Reduction of terms with $m \geq 5$:** A factor $\ell^2$ in the numerator can be canceled against the corresponding denominator,

$$\begin{aligned}
L_m[q_1 \cdots q_{m-1}|\ell^2 N(\ell)] &= \int \frac{d^d\ell}{(2\pi)^d} \frac{\ell^2 N(\ell)}{\ell^2(\ell+q_1)^2(\ell+q_2)^2 \cdots (\ell+q_{m-1})^2} \\
&= \int \frac{d^d\ell}{(2\pi)^d} \frac{N(\ell-q_1)}{\ell^2(\ell+q_2-q_1)^2 \cdots (\ell+q_{m-1}-q_1)^2}.
\end{aligned} \quad (19.17)$$

The change of variable $\ell + q_1 \to \ell$ in the second line brings the integrand back to the canonical form of eq. (19.16). After this cancellation, the number of denominators has decreased by one, and the degree in $\ell$ of the numerator by two.

By repeating the above step as many times as needed, one arrives at loop integrals with numerators that are polynomials in $P \cdot \ell$ where the $P^\mu$ are some combinations of the other vectors of the problem. For those integrals for which $m \geq 5$, the denominators contain generically four linearly independent vectors $q_i$ that we may use as a basis to decompose $P$. For instance,

$$P^\mu = \alpha_1 q_1^\mu + \alpha_2 q_2^\mu + \alpha_3 q_3^\mu + \alpha_4 q_4^\mu. \quad (19.18)$$

In order to determine the coefficients, we contract this equation with $q_1, \cdots, q_4$, which gives a $4 \times 4$ linear system,

$$\underbrace{(q_i \cdot q_j)}_{\Delta_{ij}} \alpha_j = P \cdot q_i, \tag{19.19}$$

which can be inverted[7] as $\alpha_i = [\Delta^{-1}]_{ij}(P \cdot q_j)$ (this gives coefficients that are rational functions of momenta and polarization vectors). Note that these coefficients involve only four-dimensional quantities. Once the coefficients $\alpha_i$ have been determined, we write

$$P \cdot \ell = \sum_{i=1}^{4} \alpha_i \, q_i \cdot \ell = \tfrac{1}{2} \sum_{i=1}^{4} \alpha_i \left((\ell + q_i)^2 - \ell^2 - q_i^2\right), \tag{19.20}$$

and we eliminate the $\ell$ dependence of the numerator by canceling either $(\ell + q_i)^2$ or $\ell^2$ with the corresponding denominator. This procedure should be repeated until all the $\ell$ dependence of the numerator has been removed, or there are fewer than five denominators. Note that during one of these steps, an integral with $m$ denominators and a numerator of degree $d$ is transformed into a linear combination of integrals with $m - 1$ denominators (and a degree $d - 1$ for the numerator) and integrals with $m$ denominators and a numerator of degree $d - 1$. In any case, the condition that the numerator has a degree at most equal to the number of denominators is preserved at every step.

**Passarino–Veltman reduction for $m \leq 4$:** When there are four or fewer denominators, there are not enough independent momenta in the denominators to form a basis in four dimensions, and we have to use a different method for removing the $\ell$ dependence of the numerator. This can be done by exhausting all the possibilities for the tensorial form of the integral. Let us illustrate this explicitly in the case of numerators of degree 1 and 2 (see Exercises 19.1 and 19.2 for simple explicit examples – but the procedure can be carried out explicitly for degrees 3 and 4 as well).

Consider first the integral $L_m[q_1 \cdots q_{m-1}|\ell^\mu]$. Since the result of the integral is a vector, it must be a linear combination of $q_i$:

$$L_m[q_1 \cdots q_{m-1}|\ell^\mu] \equiv \sum_{j=1}^{m-1} c_{m,j} \, q_j^\mu, \tag{19.21}$$

where the $c_{m,j}$ are scalar coefficients. By contracting this equation with $q_i^\mu$, we get

$$\sum_{j=1}^{m-1} c_{m,j} \, (q_j \cdot q_i) = \int \frac{d^d \ell}{(2\pi)^d} \, \frac{q_i \cdot \ell}{\ell^2 (\ell + q_1)^2 (\ell + q_2)^2 \cdots (\ell + q_{m-1})^2}$$

$$= \frac{1}{2} \int \frac{d^d \ell}{(2\pi)^d} \, \frac{(\ell + q_i)^2 - \ell^2 - q_i^2}{\ell^2 (\ell + q_1)^2 (\ell + q_2)^2 \cdots (\ell + q_{m-1})^2}. \tag{19.22}$$

---

[7]The matrix $\Delta_{ij}$, known as a *Gram matrix*, is invertible if the $q_i$ are linearly independent.

## 19.2 REDUCTION OF ONE-LOOP AMPLITUDES

The first two terms in the numerator each cancel against one of the denominators, producing scalar integrals with $m - 1$ denominators (we may have to translate the integration variable to bring the integrand back to the canonical form), and the last term gives a scalar integral with $m$ denominators. If we introduce the Gram matrix $\Delta_{ij} \equiv q_i \cdot q_j$, the coefficients $c_{m,j}$ are obtained by multiplying the right-hand side of the previous equation by the inverse of this matrix:

$$c_{m,j} = \frac{1}{2} \sum_{i=1}^{m-1} [\Delta^{-1}]_{ji} \int \frac{d^d\ell}{(2\pi)^d} \frac{(\ell + q_i)^2 - \ell^2 - q_i^2}{\ell^2(\ell + q_1)^2(\ell + q_2)^2 \cdots (\ell + q_{m-1})^2}. \tag{19.23}$$

The only impediment to this strategy could be that $\Delta$ is non-invertible. But since $m \leq 4$, this can happen only for exceptional configurations of the external momenta.

Consider now the second-degree case, i.e., the integral $L_m[q_1 \cdots q_{m-1} | \ell^\mu \ell^\nu]$. We may use the following decomposition:

$$L_m[q_1 \cdots q_{m-1} | \ell^\mu \ell^\nu] \equiv c_{m,00} g^{\mu\nu} + \sum_{i,j=1}^{m-1} c_{m,ij} q_i^\mu q_j^\nu. \tag{19.24}$$

Since the coefficients $c_{m,ij}$ are symmetric in $ij$, there are $\frac{1}{2}m(m-1)$ of them. Contracting with $g_{\mu\nu}$ (in d dimensions, note that $g_{\mu\nu} g^{\mu\nu} = d$), we obtain a first constraint

$$d\, c_{m,00} + \sum_{i,j=1}^{m-1} c_{m,ij} \Delta_{ij} = \int \frac{d^d\ell}{(2\pi)^d} \frac{1}{(\ell + q_1)^2(\ell + q_2)^2 \cdots (\ell + q_{m-1})^2}, \tag{19.25}$$

where the right-hand side is a scalar integral with $m - 1$ denominators. Another set of $\frac{1}{2}m(m-1)$ relations may be obtained via the contraction with $q_k^\mu q_l^\nu$:

$$\Delta_{kl} c_{m,00} + \sum_{i,j=1}^{m-1} c_{m,ij} \Delta_{ik}\Delta_{jl} = \int \frac{d^d\ell}{(2\pi)^d} \frac{(q_k \cdot \ell)(q_l \cdot \ell)}{\ell^2(\ell + q_1)^2(\ell + q_2)^2 \cdots (\ell + q_{m-1})^2}. \tag{19.26}$$

The next step is to reduce the degree in $\ell^\mu$ in the numerator of the integrands in the right-hand side of these equations. For this, we apply

$$q_k \cdot \ell = \tfrac{1}{2}\big[(\ell + q_k)^2 - \ell^2 - q_k^2\big]. \tag{19.27}$$

(We keep the second factor $q_l \cdot \ell$ unchanged.) This produces a number of integrals that have only one power of $\ell^\mu$ in the numerator, which we have already learned to treat when we considered numerators of degree 1. We shall not carry out the cases of degrees 3 and 4 here. They follow a similar strategy but are considerably more challenging given the large size of the tensor basis on which the result may be decomposed.

### 19.2.3 Van Neerven–Vermaseren Reduction of Scalar Integrals

A systematic application of the tensor reduction described above leads to purely scalar loop integrals, with a number of denominators that can be as large as the number of external legs on the loop. It turns out that in dimension four, all these scalar integrals may be rewritten as linear combinations of scalar integrals with four denominators (boxes), three denominators (triangles), two denominators (bubbles), and a single denominator (tadpoles[8]). Obviously, in order to prove this result, it is sufficient to consider the integrals with more than four denominators, and to show that they can be reduced to integrals with at most four denominators. Therefore, let us consider a generic scalar integral, with $m > 4$ denominators:

$$L_m[q_1 \cdots q_{m-1}|1] \equiv \int \frac{d^d\ell}{(2\pi)^d} \frac{1}{\ell^2(\ell+q_1)^2(\ell+q_2)^2 \cdots (\ell+q_{m-1})^2}. \quad (19.28)$$

**Six points and more:** Since it is much easier, let us start with the case $m \geq 6$, i.e., integrals with at least five momenta in the denominator. With the first four of these vectors, we may define

$$v_1^\alpha \equiv Q^{-1} \epsilon^{\alpha\nu\rho\sigma} q_{2\nu} q_{3\rho} q_{4\sigma}, \quad v_2^\alpha \equiv Q^{-1} \epsilon^{\mu\alpha\rho\sigma} q_{1\mu} q_{3\rho} q_{4\sigma},$$
$$v_3^\alpha \equiv Q^{-1} \epsilon^{\mu\nu\alpha\sigma} q_{1\mu} q_{2\nu} q_{4\sigma}, \quad v_4^\alpha \equiv Q^{-1} \epsilon^{\mu\nu\rho\alpha} q_{1\mu} q_{2\nu} q_{3\rho}, \quad (19.29)$$

with $Q \equiv \epsilon^{\mu\nu\rho\sigma} q_{1\mu} q_{2\nu} q_{3\rho} q_{4\sigma}$. These vectors can be used as a basis of the four-dimensional space spanned by $q_{1,2,3,4}$. If the loop momentum $\ell$ is defined in $d = 4 - 2\epsilon$ dimensions, we may write it as

$$\ell^\alpha = \sum_{i=1}^4 (\ell \cdot q_i) v_i^\alpha + \sum_\epsilon (\ell \cdot n_\epsilon) n_\epsilon^\alpha, \quad (19.30)$$

where $\{n_\epsilon^\mu\}$ is a symbolic notation for the orthonormal basis that spans the $d-4$ additional dimensions of the loop momentum. The coefficients follow from

$$q_i \cdot v_j = \delta_{ij}, \quad v_i \cdot n_\epsilon = 0, \quad q_i \cdot n_\epsilon = 0, \quad n_\epsilon \cdot n_{\epsilon'} = \delta_{\epsilon\epsilon'}. \quad (19.31)$$

A convenient property of the basis $\{v_i^\mu\}$ is that the coordinate of a vector $\ell$ on $v_i$ is its projection along $q_i$, which simplifies analytical expressions.[9] Contracting then eq. (19.30) with $q_5$

---

[8] In a massless theory, the tadpoles vanish in dimensional regularization.

[9] A similar construction, known as the *van Neerven–Vermaseren basis*, exists when there are fewer than four vectors $q_i$. From $m < 4$ momenta $q_i$, one may still define $m$ 4-vectors $v_i$ such that $q_i \cdot v_j = \delta_{ij}$, that form a basis of the *physical space* of dimension $m < 4$. These vectors should be supplemented by $4 - m$ vectors $n_r^\mu$, chosen to satisfy $v_i \cdot n_r = 0$, $n_r \cdot n_s = \delta_{rs}$, in order to span $4 - m$ dimensions transverse to the physical space. As in eq. (19.30), the $d-4 = -2\epsilon$ additional dimensions introduced in dimensional regularization are represented by vectors $n_\epsilon^\mu$, so that we have:

$$\ell^\alpha = \sum_{i=1}^m (\ell \cdot q_i) v_i^\alpha + \sum_{r=m+1}^4 (\ell \cdot n_r) n_r^\mu + \sum_\epsilon (\ell \cdot n_\epsilon) n_\epsilon^\mu. \quad (19.32)$$

Formal manipulations of the $n_\epsilon^\mu$ use the fact that $\sum_\epsilon n_\epsilon^\mu n_\epsilon^\nu$ is the metric tensor on the $d-4$ extra dimensions, and that its contraction with the full metric tensor is $\sum_\epsilon g_{\mu\nu} n_\epsilon^\mu n_\epsilon^\nu = d - 4$.

## 19.2 REDUCTION OF ONE-LOOP AMPLITUDES

provides a linear relationship between the $\ell \cdot q_i$ for $i = 1, \cdots, 5$, which can be rewritten as a relationship among denominators:

$$\left[\sum_{i=1}^{4}(q_5 \cdot v_i) - 1\right]\ell^2 - \sum_{i=1}^{4}(q_5 \cdot v_i)(\ell + q_i)^2 + (\ell + q_5)^2 = q_5^2 - \sum_{i=1}^{4}(q_5 \cdot v_i)\, q_i^2. \quad (19.33)$$

By dividing this identity by the m denominators, and integrating over $\ell$, we conclude that the m-point function can be written as a sum of six $(m-1)$-point functions. But it was crucial for obtaining eq. (19.33) to have at least five vectors $q_i$, i.e., to start from a function with at least six denominators.

**Five points:** We can apply recursively the previous reduction until we reach scalar integrals with five denominators. At this point, a different strategy is necessary because there is no longer $q_5$. In this case, a crucial observation is that

$$L_4\left[q_1 \cdots q_3 \big| l_\alpha\right] = \sum_{i=1}^{3} \alpha_i\, q_{i\alpha}, \quad (19.34)$$

which implies that

$$L_4\left[q_1 \cdots q_3 \big| \epsilon^{\mu\nu\rho\alpha} q_{1\mu} q_{2\nu} q_{3\rho} l_\alpha\right] = 0. \quad (19.35)$$

By contracting eq. (19.30) with $\ell$, we obtain

$$\ell^2 = \tfrac{1}{2}\sum_{i=1}^{4}(\ell \cdot v_i)\big((\ell + q_i)^2 - \ell^2 - q_i^2\big) + \sum_{\epsilon}(\ell \cdot n_\epsilon)^2. \quad (19.36)$$

Note that the last term in the right-hand side exists only in dimensional regularization, when $d \neq 4$. Let us now divide this equation by the five denominators of the pentagon integral, and integrate over the loop momentum. By our observation (19.35), we have $L_5\left[q_1 \cdots q_4 \big| (\ell \cdot v_i)(\ell + q_i)^2\right] = 0$. Consider now the terms of the right-hand side whose numerator is in $(\ell \cdot v_i)\ell^2$. Their sum can be rearranged as follows:

$$L_5\left[q_1 \cdots q_4 \Big| \sum_{i=1}^{4}(\ell \cdot v_i)\ell^2\right] = \int \frac{d^d\ell}{(2\pi)^d} \frac{\sum_{i=1}^{4}(\ell - q_1)\cdot v_i}{\ell^2 \prod_{i=2}^{4}(\ell + q_i - q_1)^2}$$

$$= \int \frac{d^d\ell}{(2\pi)^d} \frac{Q^{-1}\epsilon^{\alpha\nu\rho\sigma}\ell_\alpha(q_2 - q_1)_\nu(q_3 - q_1)_\rho(q_4 - q_1)_\sigma - 1}{\ell^2 \prod_{i=2}^{4}(\ell + q_i - q_1)^2}$$

$$= -L_5\left[q_1 \cdots q_4 \big| \ell^2\right]. \quad (19.37)$$

The first equality is merely a change of variables $\ell \to \ell - q_1$. The second equality follows from an algebraic reorganization of the numerator. The last line results from the lemma

(19.35) and the change of variables $\ell \to \ell + q_1$. Therefore, at this stage, we have shown that

$$L_5[q_1 \cdots q_4|\ell^2 + \sum_{i=1}^{4} q_i^2(\ell \cdot v_i)] = L_5[q_1 \cdots q_4|2\sum_{\epsilon}(\ell \cdot n_\epsilon)^2]. \tag{19.38}$$

Note now that

$$(\ell \cdot v_i) = \sum_{1 \le j \le 4}(\ell \cdot q_j)(v_i \cdot v_j), \tag{19.39}$$

which implies

$$\sum_{i=1}^{4} q_i^2(\ell \cdot v_i) = \sum_{j=1}^{4}(\ell \cdot q_j) v_j \cdot \underbrace{\sum_{i=1}^{4} q_i^2 v_i}_{\equiv w} = \tfrac{1}{2}\sum_{j=1}^{4}\left[(\ell + q_j)^2 - \ell^2\right](w \cdot v_j) - \tfrac{w^2}{2}. \tag{19.40}$$

Combining eq. (19.38) with the previous identity and canceling some denominators gives

$$\tfrac{w^2}{2} L_5[q_1 \cdots q_4|1] = \underline{-L_5[q_1 \cdots q_4|2\sum_{\epsilon}(\ell \cdot n_\epsilon)^2]} + \tfrac{1}{2}\sum_{j=1}^{4}(w \cdot v_j) L_4[q_{i \ne j}|1]$$

$$+ \left(1 - \tfrac{1}{2}\sum_{j=1}^{4} w \cdot v_j\right) L_4[q_2 - q_1, q_3 - q_1, q_4 - q_1|1]. \tag{19.41}$$

In other words, the pentagon scalar integral can be written as a sum of five scalar box integrals, plus a term (underlined) in $(\ell \cdot n_\epsilon)^2$ that exists only if one extends the loop momentum to $d \ne 4$. This term is a correction of order $\epsilon$ relative to the box integrals. Thus, if we need the full $\epsilon$ dependence of the pentagon integral, we could alternatively consider the scalar pentagon as an independent master integral on its own. Alternatively, we may use instead

$$\int \frac{d^d\ell}{(2\pi)^d} \frac{\sum_{\epsilon}(\ell \cdot n_\epsilon)^2}{\ell^2(\ell+q_1)^2(\ell+q_2)^2(\ell+q_3)^2(\ell+q_4)^2} \tag{19.42}$$

as a master integral. Even though this is not a scalar integral, it offers the advantage of vanishing when $d \to 4$. Note that, except for this last reduction from scalar pentagons to scalar boxes, all the other reductions are valid even in $d \ne 4$ dimensions (but some may produce $\epsilon$-dependent coefficients).

### 19.2.4 General Form of One-Loop Amplitudes

In the previous subsections, we have seen that by a systematic reduction of the tensor structures and of the number of denominators in the resulting scalar integrals, we arrive at an universal decomposition of all one-loop color-ordered amplitudes as a sum of tadpole (if

## 19.2 REDUCTION OF ONE-LOOP AMPLITUDES

there are internal masses), bubble, triangle, box and pentagon (assuming that dimensional regularization is used) scalar integrals:

$$
\begin{aligned}
\mathcal{A} = {} & A \left[\vcenter{\hbox{⬭}}\right]_{i_1} + B \left[\vcenter{\hbox{⬭}}_{K_1}^{K_2}\right] + C \left[\vcenter{\hbox{△}}_{K_1,K_3}^{K_2}\right] \\
& + D \left[\vcenter{\hbox{□}}_{K_4,K_3}^{K_1,K_2}\right] + E \left[\vcenter{\hbox{⬠}}_{K_1,K_5}^{K_2,K_3,K_4}\right] \\
= {} & \sum_{i_1} A_{i_1} L_{i_1}^{(d)} + \sum_{[i_1|i_2]} B_{i_1 i_2} L_{i_1 i_2}^{(d)} + \sum_{[i_1|i_3]} C_{i_1 i_2 i_3} L_{i_1 i_2 i_3}^{(d)} \\
& + \sum_{[i_1|i_4]} D_{i_1 i_2 i_3 i_4} L_{i_1 i_2 i_3 i_4}^{(d)} + \sum_{[i_1|i_5]} E_{i_1 i_2 i_3 i_4 i_5} L_{i_1 i_2 i_3 i_4 i_5}^{(d)},
\end{aligned}
\tag{19.43}
$$

where $A, B, C, D, E$ are the coefficients of the one-, two-, three-, four- and five-point scalar master integrals. In the more explicit second equality, the labeling of the sums should be interpreted as $[i_1|i_p] \equiv 0 \leq i_1 < i_2 < \cdots < i_p \leq n-1$, and $L_{i_1 \cdots i_p}^{(d)}$ is a notation for the d-dimensional scalar integral with a given set of p denominators $(\ell + q_{i_1})^2, \cdots, (\ell + q_{i_p})^2$,

$$
L_{i_1 \cdots i_p}^{(d)} \equiv \int \frac{d^d \ell}{(2\pi)^d} \frac{1}{(\ell + q_{i_1})^2 \cdots (\ell + q_{i_p})^2}. \tag{19.44}
$$

Since all the one-loop scalar integrals that appear in eq. (19.43) form a finite set of (known) integrals, the problem of calculating one-loop amplitudes amounts to that of determining the coefficients in the decomposition. The naive way would be to apply the sequence of reductions we have just described to every Feynman diagram that contributes to the amplitude $\mathcal{A}$, but this is precisely what we would like to avoid. Indeed, this is completely impractical because it suffers from a huge combinatorial barrier due to the proliferation of one-loop graphs as one increases the number of external legs and to the rapidly increasing complexity of the recursive reductions that must be applied on each of them. Nevertheless, the mere existence of the decomposition (19.43) is a very profound result, which has important practical applications provided one finds other methods to evaluate the coefficients.

**Rational term in four dimensions:** In $d = 4 - 2\epsilon$, the coefficients themselves may contain a (regular) dependence on $\epsilon$. If the integrals they happen to multiply have a pole $\epsilon^{-1}$, their product contains an extra finite term in the limit $\epsilon \to 0$, obtained as the product of the residue of the pole by the coefficient of the term linear in $\epsilon$ in the coefficient. Generically, the product of a coefficient by a scalar master integral can be arranged as follows:

$$
C \, L^{(4-2\epsilon)} = \Big(C_0 + C_1 \epsilon + \cdots\Big)\Big(\frac{L_{-1}}{\epsilon} + L_0 + \cdots\Big)
$$

$$= \underbrace{C_0\left(\frac{L_{-1}}{\epsilon} + L_0\right)}_{[C]_{d=4}\left[L^{(4-2\epsilon)}\right] + \mathcal{O}(\epsilon)} + \underbrace{C_1 L_{-1}}_{\text{extra term}} + \mathcal{O}(\epsilon). \quad (19.45)$$

The first term is what one would obtain naively by evaluating the coefficients with only four-dimensional momenta (since they are finite rational functions), and the master loop integral in d dimensions to regularize its divergences. Therefore, an alternative to eq. (19.43) is a decomposition with coefficients evaluated in four dimensions, without pentagons, and with an additional term,

$$\mathcal{A} = \sum_{[i_1|i_1]} [A_{i_1}]_{d=4} L_{i_1}^{(4-2\epsilon)} + \sum_{[i_1|i_2]} [B_{i_1 i_2}]_{d=4} L_{i_1 i_2}^{(4-2\epsilon)} + \sum_{[i_1|i_3]} [C_{i_1 i_2 i_3}]_{d=4} L_{i_1 i_2 i_3}^{(4-2\epsilon)}$$
$$+ \sum_{[i_1|i_4]} [D_{i_1 i_2 i_3 i_4}]_{d=4} L_{i_1 i_2 i_3 i_4}^{(4-2\epsilon)} + \mathcal{R} + \mathcal{O}(\epsilon). \quad (19.46)$$

(See Exercise 19.3 for an example.) Note that the extra term $\mathcal{R}$ is rational in momenta, since by Weinberg's theorem the ultraviolet divergences are polynomial. The only possible singularities of this term are poles in the kinematical invariants, in contrast with the first four terms that may have branch cuts. This remark is very important when discussing how to obtain the coefficients from four-dimensional unitarity cuts.

### 19.2.5 Ossola-Papadopoulos-Pittau Integrand Reduction

The decompositions (19.43) and (19.46) are valid for one-loop amplitudes in which the integral over the loop momentum has already been performed. Indeed, some steps of the reduction rely on the fact that certain integrals over $\ell$ vanish. However, it would also be very useful to have a similar decomposition at the integrand level. One advantage is that the integrand is a rational function of the momenta, and thus has simpler singularities (only poles). Another benefit of working at the integrand level is that, instead of matching the branch cut singularities of an amplitude with those of master integrals, one just needs to fit the coefficients of a polynomial, which is a much simpler problem (the actual expressions of the master integrals are not needed until the very last step). Intuitively, we may expect that a similar decomposition of the integrand exists, but with the presence of additional terms that vanish only once integrated over the loop momentum. Such a reduction of the integrand $\mathcal{I}(\ell)$ of a one-loop n-point amplitude was obtained by Ossola, Papadopoulos and Pittau (OPP), and reads as follows:

$$\mathcal{I}(\ell) = \prod_{i=0}^{n-1} \frac{1}{(\ell+q_i)^2} \Bigg\{ \sum_{i_1} \widetilde{A}_{i_1}(\ell) \prod_{j \neq i_1} (\ell+q_j)^2 + \sum_{[i_1|i_2]} \widetilde{B}_{i_1 i_2}(\ell) \prod_{j \notin \{i_1,2\}} (\ell+q_j)^2$$
$$+ \sum_{[i_1|i_3]} \widetilde{C}_{i_1 i_2 i_3}(\ell) \prod_{j \notin \{i_1,2,3\}} (\ell+q_j)^2 + \sum_{[i_1|i_4]} \widetilde{D}_{i_1 i_2 i_3 i_4}(\ell) \prod_{j \notin \{i_1,2,3,4\}} (\ell+q_j)^2$$
$$+ \sum_{[i_1|i_5]} \widetilde{E}_{i_1 i_2 i_3 i_4 i_5}(\ell) \prod_{j \notin \{i_1,\cdots,5\}} (\ell+q_j)^2 \Bigg\}. \quad (19.47)$$

## 19.2 REDUCTION OF ONE-LOOP AMPLITUDES

In each term, the product in the numerator precisely removes all denominators except those that correspond to the list $i_1, \cdots, i_p$. The coefficients are $\ell$-dependent, but give back the coefficients of eq. (19.43) after this expression has been integrated over $\ell$. By a reduction procedure somewhat similar to the one described earlier, one finds the following generic form for these coefficients (see Exercises 19.5 and 1936 for a derivation of the first two coefficients):

$$\widetilde{E}_{01234}(\ell) = \mathbf{E}^{(0)}_{01234},$$

$$\widetilde{D}_{0123}(\ell) = \mathbf{D}^{(0)}_{0123} + \mathbf{D}^{(1)}_{0123}(\ell \cdot n_4) + \mathbf{D}^{(2)}_{0123} \sum_{\epsilon}(\ell \cdot n_\epsilon)^2$$
$$+ \mathbf{D}^{(3)}_{0123}(\ell \cdot n_4) \sum_{\epsilon}(\ell \cdot n_\epsilon)^2 + \mathbf{D}^{(4)}_{0123} \sum_{\epsilon,\epsilon'}(\ell \cdot n_\epsilon)^2 (\ell \cdot n_{\epsilon'})^2,$$

$$\widetilde{C}_{012}(\ell) = \mathbf{C}^{(0)}_{012} + \mathbf{C}^{(1)}_{012}(\ell \cdot n_3) + \mathbf{C}^{(2)}_{012}(\ell \cdot n_4) + \mathbf{C}^{(3)}_{012}((\ell \cdot n_3)^2 - (\ell \cdot n_4)^2)$$
$$+ \mathbf{C}^{(4)}_{012}(\ell \cdot n_3)(\ell \cdot n_4) + \mathbf{C}^{(5)}_{012}(\ell \cdot n_3)^3 + \mathbf{C}^{(6)}_{012}(\ell \cdot n_4)^3$$
$$+ \mathbf{C}^{(7)}_{012} \sum_{\epsilon}(\ell \cdot n_\epsilon)^2 + \mathbf{C}^{(8)}_{012}(\ell \cdot n_3) \sum_{\epsilon}(\ell \cdot n_\epsilon)^2 + \mathbf{C}^{(9)}_{012}(\ell \cdot n_4) \sum_{\epsilon}(\ell \cdot n_\epsilon)^2,$$

$$\widetilde{B}_{01}(\ell) = \mathbf{B}^{(0)}_{01} + \mathbf{B}^{(1)}_{01}(\ell \cdot n_2) + \mathbf{B}^{(2)}_{01}(\ell \cdot n_3) + \mathbf{B}^{(3)}_{01}(\ell \cdot n_4)$$
$$+ \mathbf{B}^{(4)}_{01}((\ell \cdot n_2)^2 - (\ell \cdot n_4)^2) + \mathbf{B}^{(5)}_{01}((\ell \cdot n_3)^2 - (\ell \cdot n_4)^2)$$
$$+ \mathbf{B}^{(6)}_{01}(\ell \cdot n_2)(\ell \cdot n_3) + \mathbf{B}^{(7)}_{01}(\ell \cdot n_3)(\ell \cdot n_4) + \mathbf{B}^{(8)}_{01}(\ell \cdot n_2)(\ell \cdot n_4)$$
$$+ \mathbf{B}^{(9)}_{01} \sum_{\epsilon}(\ell \cdot n_\epsilon)^2,$$

$$\widetilde{A}_0(\ell) = \mathbf{A}^{(0)}_0 + \mathbf{A}^{(1)}_0(\ell \cdot n_1) + \mathbf{A}^{(2)}_0(\ell \cdot n_2) + \mathbf{A}^{(3)}_0(\ell \cdot n_3) + \mathbf{A}^{(4)}_0(\ell \cdot n_4). \qquad (19.48)$$

The other coefficients are given by identical formulas, with the appropriate mapping $[01234] \to [i_1 i_2 i_3 i_4 i_5]$. The coefficients of the five-denominator terms are independent of the loop momentum. For the other terms, the maximal degree in $\ell$ is controlled by the highest rank tensor that may appear in the numerator of the corresponding function (in Yang–Mills theory, the maximal rank is p for a loop with p denominators). In this parametrization, the $n_r$ form the basis of the space transverse to the $q_i$ that appear in a given function, and $n_\epsilon$ is the extension needed for the loop momentum to be in $d = 4 - 2\epsilon$ dimensions.

From the decomposition of eq. (19.32), we can rewrite the denominators as

$$(\ell + q_i)^2 = \ell^2 + 2\ell \cdot q_i + q_i^2$$
$$= q_i^2 + \sum_{j,k}(\nu_j \cdot \nu_k)(\ell \cdot q_j)(\ell \cdot q_k) + 2\sum_j(\ell \cdot q_j)(q_i \cdot \nu_j)$$
$$+ \underline{\sum_r(\ell \cdot n_r)^2 + \sum_\epsilon(\ell \cdot n_\epsilon)^2}, \qquad (19.49)$$

which implies that they are isotropic in the transverse space (underlined terms). Therefore, any term in the numerator which is odd in one of the coordinates of the transverse space gives zero upon (angular) integration over the transverse space. In eq. (19.48), this integration eliminates all terms except the constants ($\mathbf{E}^{(0)}_{01234}$, $\mathbf{D}^{(0)}_{0123}, \cdots$) and the terms in $(\ell \cdot n_\epsilon)^2$ or $(\ell \cdot n_\epsilon)^2 (\ell \cdot n_{\epsilon'})^2$. The latter terms, which do not exist in four dimensions because their numerators are of order $\epsilon$, lead to a finite remainder if the corresponding integrals have a pole $\epsilon^{-1}$ when evaluated in $4 - 2\epsilon$ dimensions. By a systematic inspection of these terms and calculation of their integrals, we find that the rational term $\mathcal{R}$ can be related to the coefficients of the OPP decomposition as follows (see Exercise 19.7):

$$\mathcal{R} = \frac{i}{(4\pi)^2}\left[ -\frac{1}{6}\sum_{[i_1|i_4]} D^{(4)}_{i_1 i_2 i_3 i_4} - \frac{1}{2}\sum_{[i_1|i_3]} C^{(7)}_{i_1 i_2 i_3} + \frac{1}{6}\sum_{[i_1|i_2]} B^{(9)}_{i_1 i_2}(q_{i_1} - q_{i_2})^2 \right]. \quad (19.50)$$

As with eqs. (19.43) or (19.46), the OPP decomposition (19.47) is most useful if one can determine the coefficients without resorting to the underlying Feynman diagrams. The advantage of having a decomposition at the level of the integrand is that if we were able to calculate the integrand $\mathcal{I}(\ell)$ for a large enough number of values of the loop momentum $\ell^\mu$, then eq. (19.47) may be viewed as a linear system of equations for the coefficients and inverted (for this, it is important that we know a priori the $\ell$ dependence of these coefficients, via eq. (19.48)). Note that only the coefficients that survive upon integration over the loop momentum are really necessary, since we are interested in the integrated amplitude in four dimensions (the others are sometimes called *spurious* terms). As we shall see later, there are special choices of $\ell^\mu$, such that a certain subset of denominators vanish, for which the calculation of the integrand simplifies because we can reuse results from tree-level amplitudes.

## 19.3 One-Loop Amplitudes from Unitarity Cuts

Since the scalar master integrals that enter in the decomposition are all tabulated, the main task in fully determining a one-loop amplitude is to obtain the coefficients. In this section, we explain how to derive the coefficients from the OPP decomposition of the integrand. The general idea is to evaluate both sides of eq. (19.47) for a sufficient number of values of the loop momentum $\ell^\mu$, in order to obtain a large enough system of linear equations for all the coefficients that appear in eqs. (19.48). Of course, the evaluation of the right-hand side is trivial at any $\ell$, since the OPP decomposition tells us its functional form (up to unknown coefficients). The problem lies in the evaluation of the left-hand side, since it seems on the surface that this is as complicated as summing over all the Feynman diagrams that contribute to the amplitude under consideration.

The trick is to choose certain special values of the loop momentum for which this evaluation is simplified. As we have seen in the derivation of the BCFW on-shell recursion, on-shell internal propagators are particularly nice because they correspond physically to infinitely long-lived intermediate states, which effectively factorizes the graph as the product of simpler graphs that are both on-shell. In the case of tree amplitudes, putting a single internal propagator on-shell (which corresponds to studying the poles of the amplitude) is enough to split the amplitude into two smaller on-shell amplitudes. As we shall see, for one-loop amplitudes, it is also convenient to choose values of the loop momentum for which several denominators in the loop vanish, which correspond to branch cut singularities of the amplitude rather than poles. At these special values of the loop momentum, the loop can be split into a product of on-shell *tree amplitudes*, which are much simpler to evaluate. Moreover, by varying the set of propagators that are on-shell, we will obtain enough relations to determine the complete amplitude.

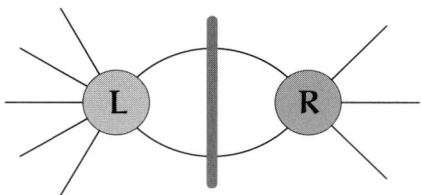

FIGURE 19.6: Generic structure of Cutkosky's unitarity cuts for a one-loop diagram. The left and right blobs are connected tree diagrams.

### 19.3.1 Reminder on Cutkosky's Cutting Rules

We have in fact already encountered a basic version of this procedure, in the discussion of perturbative unitarity in Section 2.4. There, we derived a set of diagrammatic rules, known as Cutkosky's rules, that allow the determination of the imaginary part of Feynman diagrams. These rules amount to drawing in all possible ways a line (the "cut") that divides the graph in two halves. For each cut, the propagators intercepted by the cut are on-shell, and the Feynman rules on the left side of the cut are the complex conjugate of the normal rules. In a two-point function $\Sigma$ that depends on a single external momentum p, the imaginary part can also be viewed as a branch cut discontinuity of $\Sigma(s)$, where $s = p^2$ is the Lorentz invariant upon which it depends. Indeed, for $s \in \mathbb{R}$ and negative, $\Sigma(s)$ is analytic and real. Therefore, all its Taylor coefficients are real, and we have $\Sigma^*(s) = \Sigma(s^*)$ in the entire complex plane of the variable s. Specializing this relation to $s + i0^+$, we have

$$\Sigma(s + i0^+) = \Sigma^*(s - i0^+). \tag{19.51}$$

Therefore, if $\Sigma(s)$ is purely real, then it is continuous across the real axis at this s. Conversely, the presence of a non-zero imaginary part indicates a branch cut discontinuity,

$$\Sigma(s + i0^+) - \Sigma(s - i0^+) = 2 i \operatorname{Im} \Sigma(s + i0^+). \tag{19.52}$$

In the case of n-point functions (see Figure 19.6), which are multivariate functions of all the possible Lorentz invariants constructed with the external momenta, each unitarity cut contributes to the discontinuity across the real axis in one of these variables.[10]

### 19.3.2 Generalized Unitarity Cuts

Unfortunately, cuts such as the ones that appear in Cutkosky's rules make it difficult to disambiguate between the many coefficients of eq. (19.48), because at one loop only two propagators go on-shell simultaneously. The form of the numerator on the right-hand side of eq. (19.47) suggests that a much stronger discriminating power would be achieved if one could make more than two denominators vanish at the same time, a procedure that may be viewed as a

---

[10] In the case of color-ordered partial amplitudes, only the invariants made of adjacent momenta, of the form $s_{ij} \equiv \left(\sum_{l=i}^{j} p_l\right)^2$, can appear in these cuts.

generalization of the usual unitary cuts. In order to see this more clearly, let us parametrize the integrand as a numerator $\mathcal{N}(\ell)$ multiplied by loop propagators,

$$\mathcal{I}(\ell) \equiv \mathcal{N}(\ell) \prod_{i=0}^{n-1} \frac{1}{(\ell+q_i)^2}. \tag{19.53}$$

Factoring out the denominators, the OPP decomposition reads

$$\mathcal{N}(\ell) = \sum_{i_1} \widetilde{A}_{i_1}(\ell) \underbrace{\prod_{j \neq i_1}(\ell+q_j)^2}_{n-1 \text{ factors}} + \sum_{[i_1|i_2]} \widetilde{B}_{i_1 i_2}(\ell) \underbrace{\prod_{j \notin \{i_{1,2}\}}(\ell+q_j)^2}_{n-2 \text{ factors}}$$

$$+ \sum_{[i_1|i_3]} \widetilde{C}_{i_1 i_2 i_3}(\ell) \underbrace{\prod_{j \notin \{i_{1,2,3}\}}(\ell+q_j)^2}_{n-3 \text{ factors}} + \sum_{[i_1|i_4]} \widetilde{D}_{i_1 i_2 i_3 i_4}(\ell) \underbrace{\prod_{j \notin \{i_{1,2,3,4}\}}(\ell+q_j)^2}_{n-4 \text{ factors}}$$

$$+ \sum_{[i_1|i_5]} \widetilde{E}_{i_1 i_2 i_3 i_4 i_5}(\ell) \underbrace{\prod_{j \notin \{i_1,\cdots,5\}}(\ell+q_j)^2}_{n-5 \text{ factors}}, \tag{19.54}$$

where we have indicated how many factors appear in each product. Before proceeding with the determination of the coefficients, let us make a few preliminary remarks. Recall that many terms in the $\ell$ dependence of the coefficients integrate to zero because they are odd in the coordinates of the transverse space. In the end, we need only the constant coefficients, and a few of the coefficients of the terms that depend on $\ell \cdot n_\epsilon$ because they end up in the rational term. Moreover, the coefficient of the pentagon seems superfluous in four dimensions, since we know that the scalar pentagon becomes a sum of five scalar boxes when $d \to 4$. Nevertheless, all these seemingly useless coefficients are necessary in the intermediate steps.[11] Furthermore, because the terms that contain $\ell \cdot n_\epsilon$ only exist if $d \neq 4$, it is necessary to perform these manipulations for an arbitrary dimension.

**Pentagon coefficients:** Consider first a loop momentum $\ell_*^\mu$ for which five denominators vanish simultaneously:

$$(\ell_* + q_i)^2 = 0 \quad \text{for } i \in \{i_1, \cdots, i_5\}. \tag{19.55}$$

Note that this is a set of five constraints on $\ell^\mu$, which in general admits a solution only if space-time is at least five dimensional. Let us assume for the time being that the number of dimensions is a free parameter, chosen to be $\geq 5$. For such an $\ell^\mu$, the first four terms of the right-hand side of eq. (19.54) vanish. Indeed, they all have at least $n-4$ factors, which implies that they contain at least one of the five vanishing denominators. Regarding the pentagon terms, most of them are zero, except for the one for which the absent factors are precisely the

---

[11]If one is only interested in the non-rational terms (also called the cut-constructible part) of the amplitude, there are purely four-dimensional algorithms that avoid the calculation of many of the coefficients.

denominators we have chosen to make vanish. Therefore, for any $\ell_*^\mu$ that satisfies eq. (19.55), we have

$$\widetilde{E}_{i_1 i_2 i_3 i_4 i_5}(\ell) = E^{(0)}_{i_1 i_2 i_3 i_4 i_5} = \mathcal{N}(\ell_*) \prod_{i \notin \{i_1, \cdots, 5\}} \frac{1}{(\ell_* + q_i)^2}. \tag{19.56}$$

Since the pentagon coefficient is independent of the loop momentum, a single $\ell_*^\mu$ is sufficient to determine it (in fact, eqs. (19.61) have a unique solution). By varying in eq. (19.55) the set of five denominators that vanish, we can determine successively all the pentagon coefficients. In order to evaluate the left-hand side in eq. (19.56), we use

$$\frac{\mathcal{N}(\ell_*)}{\prod_{i \notin \{i_1, \cdots, 5\}} (\ell_* + q_i)^2} = \phantom{xxx} = \mathcal{A}_1^{\text{tree}} \mathcal{A}_2^{\text{tree}} \mathcal{A}_3^{\text{tree}} \mathcal{A}_4^{\text{tree}} \mathcal{A}_5^{\text{tree}}. \tag{19.57}$$

The second equality is the statement of the fact that this amplitude in which five loop momenta are put on-shell (and the corresponding propagators removed from the equation) factorizes into the product of five on-shell tree amplitudes. This factorization is closely related to unitarity, and shares the same properties as the usual Cutkosky's rules regarding the cancellation between ghosts and the unphysical gluon polarizations. Thus, on each cut, we need only to sum over physical degrees of freedom (gluon transverse polarizations, or fermions).

The only caveat of this method, on which we will expand a bit later, is that it requires evaluating these amplitudes in a dimension larger than four for $\ell_*$ to exist. Consider for instance the case of $\widetilde{E}_{01234}$, in order to keep the notations simple. First, note that out of the five vectors $q_{0,\cdots,4}$, only four are independent, which we may choose as $\overline{q}_i \equiv q_i - q_0$. Then, we decompose $\overline{\ell}^\mu \equiv \ell^\mu + q_0^\mu$ on the van Neerven–Vermaseren basis as follows:

$$\overline{\ell}^\mu = \underbrace{-\tfrac{1}{2} \sum_{i=1}^{4} v_i^\mu \overline{q}_i^2}_{\overline{\ell}_\parallel^\mu} + \underbrace{\sum_\epsilon (\overline{\ell} \cdot n_\epsilon) n_\epsilon^\mu}_{\overline{\ell}_\perp^\mu}. \tag{19.58}$$

The condition $(\ell + q_0)^2 = \overline{\ell}^2 = 0$ relates the norms of the longitudinal and transverse components by $\overline{\ell}_\parallel^2 + \overline{\ell}_\perp^2 = 0$. Then, the specific form of $\overline{\ell}_\parallel^\mu$ in eq. (19.58) ensures that the other constraints are satisfied:

$$(\ell + q_i)^2 = (\overline{\ell} + \overline{q}_i)^2 = 2\overline{\ell} \cdot \overline{q}_i + \overline{q}_i^2$$
$$= -\sum_{j=1}^{4} \underbrace{(v_j \cdot \overline{q}_i)}_{\delta_{ij}} \overline{q}_j^2 + \overline{q}_i^2 = 0. \tag{19.59}$$

Except for the orientation of $\bar{\ell}_\perp^\mu$ in the $d-4$ extra dimensions (but the integrand does not depend on that), the momentum $\ell_*^\mu$ is completely fixed.[12]

**Box coefficients:** Once all the pentagon coefficients $\mathsf{E}^{(0)}_{i_1 i_2 i_3 i_4 i_5}$ have been determined, we rewrite eq. (19.54) as

$$\underbrace{\mathcal{N}(\ell) - \sum_{[i_1|i_5]} \mathsf{E}^{(0)}_{i_1 i_2 i_3 i_4 i_5} \prod_{j \notin \{i_1,\cdots,5\}} (\ell+q_j)^2}_{\mathcal{N}_5(\ell),\ \text{known at this point}} = \sum_{[i_1|i_4]} \widetilde{\mathsf{D}}_{i_1 i_2 i_3 i_4}(\ell) \underbrace{\prod_{j \notin \{i_1,2,3,4\}} (\ell+q_j)^2}_{n-4\ \text{factors}} + \cdots, \tag{19.60}$$

and now we choose values of the loop momentum $\ell_*^\mu$ that put four propagators on-shell,

$$(\ell_* + q_i)^2 = 0 \quad \text{for } i \in \{i_1, \cdots, i_4\}. \tag{19.61}$$

In the right-hand side of eq. (19.60), such a loop momentum kills all the triangles, bubbles and tadpoles, and among the boxes it selects precisely the one for which these four denominators are absent. This leads to

$$\widetilde{\mathsf{D}}_{i_1 i_2 i_3 i_4}(\ell_*) = \left[\mathcal{N}(\ell_*) - \mathcal{N}_5(\ell_*)\right] \prod_{i \notin \{i_1,\cdots,4\}} \frac{1}{(\ell_* + q_i)^2}$$

$$= \underbrace{\vcenter{\hbox{[box diagram with legs $\ell_*+q_{i_2}$, $\ell_*+q_{i_1}$, $\ell_*+q_{i_3}$, $\ell_*+q_{i_4}$]}}}_{\mathcal{A}_1^{\text{tree}} \mathcal{A}_2^{\text{tree}} \mathcal{A}_3^{\text{tree}} \mathcal{A}_4^{\text{tree}}} - \frac{\mathcal{N}_5(\ell_*)}{\prod_{i \notin \{i_1,\cdots,4\}}(\ell_* + q_i)^2}. \tag{19.62}$$

In the second line, we have rewritten the amplitude with four on-shell loop momenta as a product of four tree on-shell amplitudes. The second term, containing $\mathcal{N}_5$, is known since the pentagon coefficients have been obtained at the previous step.

Compared to the pentagon case, we need now to determine a polynomial $\widetilde{\mathsf{D}}_{i_1 i_2 i_3 i_4}(\ell)$ instead of a constant. But from the second part of eq. (19.48), we know that it can be parametrized by five constant coefficients. Therefore, we need to find five independent momenta $\ell_*^\mu$ that fulfill the constraints of eq. (19.61). Consider the case of $\widetilde{\mathsf{D}}_{0123}(\ell)$. By

---

[12]The only extra freedom obtained by going to $d \neq 4$ dimensions is the parameter $\sum_\epsilon (\bar{\ell} \cdot n_\epsilon)^2$. Therefore, no matter how large is the dimension $d$, at most five on-shell conditions can be satisfied simultaneously.

## 19.3 ONE-LOOP AMPLITUDES FROM UNITARITY CUTS

defining $\bar{q}_i \equiv q_i - q_0$, $\bar{\ell}^\mu \equiv \ell^\mu + q_0^\mu$ and decomposing it on the van Neerven–Vermaseren basis,[13]

$$\bar{\ell}^\mu = \underbrace{-\frac{1}{2}\sum_{i=1}^{3} v_i^\mu \bar{q}_i^2}_{\bar{\ell}_\parallel^\mu} + \underbrace{\bar{\ell}_\perp \left(x_4 n_4^\mu + \sum_\epsilon x_\epsilon n_\epsilon^\mu\right)}_{\bar{\ell}_\perp^\mu}, \tag{19.63}$$

with $x_4^2 + \sum_\epsilon x_\epsilon^2 = 1$, we ensure that the four cut conditions are satisfied. In order to determine the five constants that enter in the parametrization of $\widetilde{D}_{i_1 i_2 i_3 i_4}(\ell)$, we need five different vectors $\ell_*^\mu$, which can be obtained by varying the values of $x_4$ and $x_\epsilon$. This procedure gives a $5 \times 5$ linear system whose solution provides the coefficients.[14]

**Triangle coefficients:** The strategy used to obtain the box coefficients can be reproduced for the triangular ones. Now, we rewrite eq. (19.54) as

$$\underbrace{N(\ell) - N_5(\ell) - \sum_{[i_1|i_4]} D_{i_1 i_2 i_3 i_4 i_5}(\ell) \prod_{j \notin \{i_1,\ldots,4\}} (\ell+q_j)^2}_{N_4(\ell),\text{ known at this point}} = \sum_{[i_1|i_3]} \widetilde{C}_{i_1 i_2 i_3}(\ell) \underbrace{\prod_{j \notin \{i_1,2,3\}} (\ell+q_j)^2}_{n-3\text{ factors}} + \cdots, \tag{19.64}$$

and we consider loop momenta $\ell_*^\mu$ that put three propagators on-shell,

$$(\ell_* + q_i)^2 = 0 \quad \text{for } i \in \{i_1, i_2, i_3\}. \tag{19.65}$$

In the right-hand side of eq. (19.64), only the term that contains $\widetilde{C}_{i_1 i_2 i_3}(\ell_*)$ contributes, and we have

$$\widetilde{C}_{i_1 i_2 i_3}(\ell_*) = \left[N(\ell_*) - N_4(\ell_*) - N_5(\ell_*)\right] \prod_{i \notin \{i_1,\ldots,3\}} \frac{1}{(\ell_* + q_i)^2}.$$

$$= \underbrace{\begin{array}{c}\text{[diagram with } \ell_*+q_{i_2}, \ell_*+q_{i_1}, \ell_*+q_{i_3}\text{]}\end{array}}_{\mathcal{A}_1^{\text{tree}} \mathcal{A}_2^{\text{tree}} \mathcal{A}_3^{\text{tree}}} - \frac{N_4(\ell_*) + N_5(\ell_*)}{\prod_{i \notin \{i_1,2,3\}}(\ell_* + q_i)^2}. \tag{19.66}$$

---

[13] By setting $x_\epsilon = 0$, $x_4 = 1$, and defining $\bar{\ell}_\pm^\mu = \bar{\ell}_\parallel^\mu \pm \bar{\ell}_\perp n_4^\mu$, one can obtain directly the coefficient $D_{0123}^{(0)}$ from the half-sum of four-particle cuts evaluated at these two solutions. This is the first step in four-dimensional approaches (since all the scalar pentagons can be rewritten as sums of scalar boxes in four dimensions). There are also (more complicated) ways of getting the non-spurious triangle, bubble and tadpole coefficients without evaluating the spurious ones. Moreover, four-dimensional methods to obtain the rational term also exist.

[14] There exist clever choices of $x_{4,\epsilon}$ for which the system becomes particularly sparse and can be solved readily. These improved sampling methods are even more important in the case of the triangle and bubble coefficients, in order to avoid potential numerical instabilities when solving a $10 \times 10$ linear system.

We see here a practical illustration of a comment made earlier: Even though most of the coefficients in $D_{i_1 i_2 i_3 i_4 i_5}(\ell)$ will not enter in the final answer, it was nevertheless necessary to determine all of them at the previous step because they play a role in the subtraction on the left-hand side of eq. (19.64). Without having calculated them, we would not be able to determine the triangle coefficients. There are ten constants in the parametrization of a triangle coefficient such as $\widetilde{C}_{012}(\ell)$ and consequently we need ten independent values of $\ell_*$. Mimicking the procedure used before, we define $\bar{\ell} \equiv \ell + q_0$, $\bar{q}_i \equiv q_i - q_0$ and parametrize $\bar{\ell}$ by

$$\bar{\ell}^\mu = \underbrace{-\tfrac{1}{2}\sum_{i=1}^{2} v_i^\mu \, \bar{q}_i^2}_{\bar{\ell}_\parallel^\mu} + \underbrace{\bar{\ell}_\perp \left(x_3\, n_3^\mu + x_4\, n_4^\mu + \sum_\epsilon x_\epsilon\, n_\epsilon^\mu\right)}_{\bar{\ell}_\perp^\mu}, \tag{19.67}$$

with $\bar{\ell}_\parallel^2 + \bar{\ell}_\perp^2 = 0$ and $x_3^2 + x_4^2 + \sum_\epsilon x_\epsilon^2 = 1$. The needed values of $\ell_*^\mu$ are obtained by varying $x_{3,4,\epsilon}$, which gives a $10 \times 10$ linear system for the ten unknown coefficients in $\widetilde{C}_{012}(\ell)$.

**Bubble coefficients:** Let us be quick with the determination of bubble coefficients, since the procedure is exactly the same. For every $\ell_*^\mu$ that puts on-shell two loop momenta,

$$(\ell_* + q_i)^2 = 0 \quad \text{for } i \in \{i_1, i_2\}, \tag{19.68}$$

we obtain

$$\widetilde{B}_{i_1 i_2}(\ell_*) = \underbrace{\begin{array}{c}\ell_*+q_{i_1}\\ \text{[diagram]}\\ \ell_*+q_{i_2}\end{array}}_{\mathcal{A}_1^{\text{tree}}\mathcal{A}_2^{\text{tree}}} - \frac{\mathcal{N}_3(\ell_*) + \mathcal{N}_4(\ell_*) + \mathcal{N}_5(\ell_*)}{\prod_{i \notin \{i_{1,2}\}}(\ell_* + q_i)^2}. \tag{19.69}$$

A bubble coefficient like $\widetilde{B}_{01}(\ell)$ contains ten constants, which can be determined by taking ten independent values of $\ell_*$, parametrized in the van Neerven–Vermaseren basis by

$$\bar{\ell}^\mu = \underbrace{-\tfrac{1}{2} v_1^\mu \, \bar{q}_1^2}_{\bar{\ell}_\parallel^\mu} + \underbrace{\bar{\ell}_\perp \left(x_2\, n_2^\mu + x_3\, n_3^\mu + x_4\, n_4^\mu + \sum_\epsilon x_\epsilon\, n_\epsilon^\mu\right)}_{\bar{\ell}_\perp^\mu}, \tag{19.70}$$

with $\bar{\ell}_\parallel^2 + \bar{\ell}_\perp^2 = 0$ and $x_2^2 + x_3^2 + x_4^2 + \sum_\epsilon x_\epsilon^2 = 1$. If there are no massive particles in the loop, we do not need to consider the tadpoles.

### 19.3.3 Higher Dimensional Unitarity Cuts

It should be rather clear from the previous discussion that, despite being a systematic method for determining the coefficients in the OPP decomposition, it leads to very heavy algebra and is therefore more suitable for a numerical implementation. On the other hand, we have stressed that in order to be able to obtain all the coefficients, we need to work in d > 4, and in view of a numerical implementation d should be an integer, meaning at least d = 5. Moreover, in order to hold a d-dimensional loop momentum $\ell^\mu$, the dimension $d_s$ of space-time should be $d_s \geq d$. A larger space-time dimension implies an increase of the number of polarization degrees of freedom for gluons, since there are $d_s - 2$ physical polarizations for spin-1 particles in $d_s$ dimensions, which modifies the tensor dependence in the numerator of gluon propagators. Note that, if we consider d and $d_s$ as independent dimensions, then the dimension d of $\ell^\mu$ does not appear explicitly in the integrand of the one-loop amplitudes. There is an implicit reference to the dimension d in the factors $\sum_\epsilon (\ell \cdot n_\epsilon)^2$ that enter in the OPP decomposition (but not in the constant coefficients), which becomes explicit after we perform the integral $d^d \ell$ since this produces combinations like $g_{\mu\nu} \sum_\epsilon n_\epsilon^\mu n_\epsilon^\nu = d - 4$.

At the end of the calculation, the dimension $d_s$ of space-time should be set to 4 ("four dimensional helicity" (FDH) scheme) or $4 - 2\epsilon$ ("t'Hooft-Veltman" (tHV) scheme). It seems a priori quite challenging to obtain any of the limits $d_s \to 4$ or $d_s \to 4 - 2\epsilon$ from computations with an integer $d_s > 4$. Fortunately, we know beforehand what $d_s$ dependence to expect. Indeed, since all the external momenta and external polarization vectors remain four-dimensional, the $d_s$ dependence of a gluon loop[15] can only arise via traces of the metric tensor in $d_s$ dimensions in the numerator of the integrand. At one-loop, at most one such trace can be produced, implying that the numerator $\mathcal{N}(\ell)$ is linear in $d_s$:

$$\mathcal{N}_{(d_s)}(\ell) = \mathcal{N}_1(\ell) + d_s \mathcal{N}_2(\ell). \qquad (19.71)$$

Therefore, it is sufficient to determine the numerator for two (possibly integer) dimensions $d_s$ ($d_1$ and $d_2$) in order to obtain its full $d_s$ dependence:

$$\mathcal{N}_{(d_s)}(\ell) = \frac{d_2 - d_s}{d_2 - d_1} \mathcal{N}_{(d_1)}(\ell) + \frac{d_s - d_1}{d_2 - d_1} \mathcal{N}_{(d_2)}(\ell). \qquad (19.72)$$

For instance, assuming that we have obtained the numerator for $d_s = 5, 6$, the above two limits are then given by

$$\mathcal{N}_{\text{FDH}}(\ell) = 2\mathcal{N}_{(5)}(\ell) - \mathcal{N}_{(6)}(\ell),$$
$$\mathcal{N}_{\text{tHV}}(\ell) = \mathcal{N}_{\text{FDH}}(\ell) + 2\epsilon \left( \mathcal{N}_{(5)}(\ell) - \mathcal{N}_{(6)}(\ell) \right). \qquad (19.73)$$

The same separation can be performed for a fermion loop, but we must choose two even dimensions $d_{1,2}$.

The next step is to define polarization vectors for gluons that live in $d_s \neq 4$ dimensions, in order to perform the polarization sums in the products of on-shell tree amplitudes that appear in the coefficients of the OPP reduction. Let us assume that the momentum of the gluons on the loop is d = 5 dimensional. We can parametrize the loop momentum as $\ell^\mu \equiv \ell_\parallel^\mu + x_5 n_5^\mu$, where $\ell_\parallel^\mu$ lives in ordinary four-dimensional space-time and $n_5^\mu$ is the unit vector in the fifth dimension (therefore, $\ell_\parallel \cdot n_5 = 0$). For $d_s = 5$, we need $d_s - 2 = 3$ polarization vectors $\epsilon_\lambda^\mu$,

---

[15] For a quark loop, there is an additional overall $d_s$-dependent factor, tr $(\mathbf{1}) = 2^{d_s/2}$. It comes from the fact that the Dirac matrices must be enlarged since there are $2^{(d_s-2)/2}$ spin-1/2 eigenstates in (even) $d_s$ dimensions.

which satisfy $\ell \cdot \epsilon_\lambda = 0$. Therefore, we may choose them in the three-dimensional subspace that is orthogonal both to $\ell_\parallel^\mu$ and to $n_5^\mu$. This choice leads to the following polarization sums:

$$\sum_{\lambda=1}^{3} \epsilon_\lambda^\mu(\ell) \epsilon_\lambda^\nu(\ell) = \begin{cases} -g^{\mu\nu} + \dfrac{\ell_\parallel^\mu \ell_\parallel^\nu}{\ell_\parallel^2} & \text{if } \mu, \nu \leq 4, \\ 0 & \text{otherwise.} \end{cases} \quad (19.74)$$

In the case where $d = 5$ and $d_s = 6$, we need an additional polarization vector, $\epsilon_4^\mu$. Since $\ell^\mu$ still lives in the first five dimensions, a valid choice for this extra polarization is $\epsilon_4^\mu = n_6^\mu$ (this ensures that $\ell \cdot \epsilon_4 = 0$). Now the polarization sum becomes

$$\sum_{\lambda=1}^{4} \epsilon_\lambda^\mu(\ell) \epsilon_\lambda^\nu(\ell) = \begin{cases} -g^{\mu\nu} + \dfrac{\ell_\parallel^\mu \ell_\parallel^\nu}{\ell_\parallel^2} & \text{if } \mu, \nu \leq 4, \\ n_6^\mu n_6^\nu & \text{if } \mu, \nu = 6, \\ 0 & \text{otherwise.} \end{cases} \quad (19.75)$$

It turns out that the calculations with $d_s = 5$ and $d_s = 6$ differ in a rather simple way. Since the extra polarization vector in $d_s = 6$ is taken to be the unit vector $n_6^\mu$ of the sixth dimension, its contraction with any other vector of the problem is zero since all momenta are at most five-dimensional. The only non-zero contractions of $n_6^\mu$ are with itself or with the metric tensor (but in that case the second index of the metric tensor must also be contracted into another $n_6$). Therefore, if one of the cuts around the loop carries the polarization $n_6^\mu$, then all the cuts must carry that polarization. In other words, the polarization $n_6^\mu$ does not mix with the others in the loop. Therefore, the difference $\mathcal{N}_{(6)} - \mathcal{N}_{(5)}$ is given by a loop carrying only $n_6$-polarized gluons. Moreover, when contracted with two polarizations $n_6^\mu$, the three-gluon and four-gluon vertices become scalar-like:

$$\left[ \begin{array}{c} \mu \\ k \\ q \\ \rho \quad p \quad \nu \end{array} \right] n_6^\nu n_6^\rho = g(p-q)_\mu, \quad \left[ \begin{array}{c} \mu \quad \nu \\ \sigma \quad \rho \end{array} \right] n_6^\nu n_6^\rho = i g^2 g_{\mu\sigma}. \quad (19.76)$$

(These are indeed the Feynman rules for a scalar minimally coupled to gauge fields.)

### 19.3.4 Berends–Giele Recursion

As we have seen, the calculation of the coefficients of the OPP decomposition can be broken down into that of on-shell tree amplitudes. However, these amplitudes are a bit more general than those we considered in Chapter 14 since some of their momenta are in $d = 5$ dimensions, and some of their external polarizations are defined in $d_s = 5, 6$ dimensions. It turns out the spinor-helicity formalism we have introduced in this chapter is very much tied to four dimensions, and difficult to generalize to these higher dimensions. An alternative is a recursion formula for tree amplitudes developed by Berends and Giele, which applies to color-ordered tree amplitudes with one off-shell leg not contracted with a polarization vector (all the remaining legs are on-shell and contracted with a physical polarization). With n on-shell legs and one off-shell, let us denote this object by

$$J^\mu(1,\cdots,n) = \text{[diagram]}, \qquad (19.77)$$

where each gluon terminated by a cross indicates an on-shell line contracted with a polarization vector. Berends and Giele's recursion formula is obtained as follows: Starting from the off-shell leg, the first encountered vertex may either be a three-gluon or a four-gluon vertex, and the objects attached to the other legs of this vertex are $J^\mu$s with fewer on-shell legs. Diagrammatically, this recursion reads

$$\text{[diagram]} = \sum_{1\le k<n} \text{[diagram]} + \sum_{1\le k<l<n} \text{[diagram]}. \qquad (19.78)$$

We see that the fact that the ordering of the external gluons is fixed plays a crucial role in limiting the number of terms that may appear in the right-hand side. The starting point of the recursion is $J^\mu(1)$, i.e., a single polarization vector $\epsilon^\mu(\mathbf{p}_1)$. Although it is less effective than BCFW recursion for deriving compact analytical results, the numerical implementation of this recursion formula is straightforward and it is thus quite suitable for the numerical determination of the OPP coefficients (in particular, this recursion can be used with $d, d_s > 4$).

## 19.4 The Frontier: Multi-Loop Amplitudes

### 19.4.1 Generalities

In the previous sections of this chapter, we presented a series of techniques for reducing and organizing the calculation of any one-loop amplitude in a gauge theory like QCD. At the end of the day, these tools break down the one-loop computation into a series of on-shell tree-level calculations. Even if the process is too heavy to be carried out analytically in most cases, it can be automated in numerical codes, and we can thus consider the one-loop case as a "solved problem." Given the success of this approach at one loop, a natural question is whether it may be extended at two loops and higher orders. An important step toward repeating the one-loop treatment is to bring the integrand[16] to a universal form with a reduced number of denominators. In this section, we present some recent progress in formalizing the problem of

---

[16]Of course, reductions at the level of the integrand require many more coefficients than reductions in terms of master integrals, because of the presence of numerous spurious terms that integrate to zero in the end. On the other hand, identifying linear dependences among integrands is easier because it is a purely algebraic question.

integrand reduction in terms of algebraic geometry. Although this is not yet a fully complete framework for multi-loop calculations (especially with many external legs), it brings some appreciable order into this important question, as well as the possibility of benefiting from algorithmic developments in computational algebraic geometry.

Let us consider an L-loop n-point amplitude $\mathcal{A}_{n,L}$ in a renormalizable theory such as Yang–Mills theory in four dimensions, and denote $\ell \equiv (\ell_1, \ell_2, \cdots, \ell_L)$ the loop momenta. When needed, the components of this vector will be arranged as follows:

$$\ell \equiv (\underbrace{\ell_1, \cdots \ell_d}_{\ell_1^\mu}, \underbrace{\ell_{d+1}, \cdots, \ell_{2d}}_{\ell_2^\mu}, \cdots, \underbrace{\ell_{(L-1)d+1}, \cdots, \ell_{Ld}}_{\ell_L^\mu}). \qquad (19.79)$$

Generically, a contribution to this amplitude with m denominators may be written as

$$\mathcal{A}_{n,L} \sim \int \prod_{i=1}^{L} \frac{d^d \ell_i}{(2\pi)^d} \frac{\mathcal{N}(\ell)}{d_1(\ell) \cdots d_m(\ell)}, \qquad (19.80)$$

where the denominators $d_i(\ell)$ are defined in the massless case as

$$d_i(\ell) \equiv \Big(\sum_i \epsilon_i \ell_i + q_i\Big)^2, \qquad (19.81)$$

where the $q_i$ are sums of external momenta, and $\epsilon_i \in \{-1, 0, +1\}$. The numerator $\mathcal{N}(\ell)$ results from the Lorentz structures carried by the vertices and the numerators of propagators. It also depends on the external momenta and polarization vectors, but we do not need to make this dependence explicit. Both the numerator and the denominators are multivariate polynomials in the various components of the loop momenta. The reduction of such an integrand encompasses various sub-questions:

1. Detect when a given numerator contains powers of some of the denominators, and thus leads to integrands with fewer denominators. This step requires a multivariate generalization of the usual univariate polynomial division (Euclid algorithm).
2. What is the most general parametrization of an irreducible numerator, given a certain set of denominators (and assuming that the theory is renormalizable, which limits the possible degree of the numerator)?
3. Can we obtain the coefficients in this parametrization by considering the multi-cut configurations of the loop momenta that make the corresponding denominators vanish?

## 19.4.2 Numerator Reduction

Let us denote $\mathbb{C}[\ell]$ the ring of multivariate polynomials in the components of the loop momenta, with complex coefficients. Then, consider the set of linear combinations of the denominators $d_i(\ell)$ with polynomial coefficients,

$$\mathcal{I}_m \equiv \Big\{\sum_{i=1}^m q_i(\ell) d_i(\ell) \Big| q_i \in \mathbb{C}[\ell]\Big\}. \qquad (19.82)$$

This set is called the *ideal generated by the* $d_i$. Its relevance for the present problem is twofold. First, any element of $\mathcal{I}_m$ divided by the denominators $d_1 \cdots d_m$ simplifies into a sum of terms with one fewer denominator. Second, if $\ell_*$ is a common zero of all the $d_s(\ell)$, then it is also a zero of all the elements of the ideal. Given the ideal $\mathcal{I}_m$, a very important object is its *Gröbner basis* $\mathbb{G} = \{g_1, \cdots, q_p\}$, defined as a set of polynomials[17] $g_k$ that generates $\mathcal{I}_m$ and such that the leading term of any element of $\mathcal{I}_m$ is divisible by the leading term of one of the $g_j$. The Gröbner basis has several important properties:

1. It is unique, given the ideal and an ordering of the monomials, and there are efficient algorithms to compute it (e.g., the Buchberger algorithm).
2. $f \in \mathcal{I}_m$ if its division by $\mathbb{G}$ (see eq. (19.83)) has a null remainder. Therefore, the Gröbner basis may be used for tests of ideal membership.
3. The set of polynomial equations $d_1 = \cdots = d_m = 0$ has no solution if and only if $1 \in \mathbb{G}$, i.e., the ideal is identical to the whole ring. This result is known as the *weak*[18] Hilbert's *"nullstellensatz"* (theorem of zeroes).

Given the Gröbner basis, a polynomial $\mathcal{N}(\ell)$ may be decomposed as follows:

$$\mathcal{N}(\ell) = \underbrace{r(\ell)}_{\text{remainder}} + \sum_{j=1}^{p} q_j(\ell) g_j(\ell) = r(\ell) + \underbrace{\sum_{i=1}^{m} p_i(\ell) d_i(\ell)}_{\text{reducible part}}. \tag{19.83}$$

This decomposition is a multivariate generalization of the division of polynomials. Equation (19.83) is sometimes called the *division of* $\mathcal{N}$ *by* $\mathbb{G}$. The remainder belongs to the quotient ring $\mathbb{C}[\ell]/\mathcal{I}_m$, while the second term is an element of the ideal, i.e., a linear combination of the denominators.

Clearly, a numerator $\mathcal{N}(\ell)$ is reducible if its division by the elements of the Gröbner basis has a null remainder, which also means that $\mathcal{N} \in \mathcal{I}_m$. Consider now the set of polynomial equations $d_1 = \cdots = d_m = 0$, and assume that it has no solution. By property **3** above, the Gröbner basis contains 1, which means that any polynomial (in particular, $\mathcal{N}(\ell)$) belongs to the ideal. Thus, *a numerator $\mathcal{N}(\ell)$ is always reducible on a set of denominators that do not have a common zero. Conversely, all irreducible terms in the integrand are such that their denominators have at least one common zero.* This result provides a deeper explanation of why scalar pentagons are reducible to scalar boxes in four dimensions: The five-cut equations have no solution if the loop momentum depends on four components only. In contrast, in dimensional regularization, the loop momentum provides one more variable (the square of the component transverse to the physical space), which generically leads to the existence of solutions.

From the decomposition of eq. (19.83), the integrand can be rewritten as

$$\frac{\mathcal{N}(\ell)}{d_1(\ell) \cdots d_m(\ell)} = \frac{r(\ell)}{d_1(\ell) \cdots d_m(\ell)} + \sum_{i=1}^{m} \frac{p_i(\ell)}{\prod_{j \neq i} d_j(\ell)}. \tag{19.84}$$

---

[17] In general, the size p of the *Gröbner basis* may differ from the number m of denominators.

[18] The strong version states that if a polynomial $p(\ell)$ vanishes at all the common roots of the $d_i$, then there is a positive integer r such that $p^r \in \mathcal{I}_m$. Thus, $p(\ell)$ does not need to be itself in the ideal, but some of its powers must.

The first term in a genuine irreducible m-denominator integrand, while each term in the sum has only $m - 1$ denominators. This process can be repeated for each of the terms with $m - 1$ denominators (now, the division must be performed with the Gröbner basis corresponding to the remaining denominators), until no denominator is left. If the quotient of the last division is non-zero, we are left with a pure polynomial in $\ell$, which does not have any cut singularity and is therefore not cut-constructible (but for such terms to occur, the degree of the numerator must be larger than the number of denominators, which does not happen in a renormalizable theory).

**Reconstruction from cuts:** From this definition, an obvious method could be to sum all the Feynman diagrams in order to obtain $N(\ell)$. Once the numerator and denominator are known, one would calculate the Gröbner basis and perform the explicit polynomial division to achieve the reduction. However, as in the one-loop case, summing over Feynman diagrams is not the best approach except in the simplest cases. Instead, it would be preferable to determine the remainder by evaluating the amplitude at the common roots of the denominators, since this breaks down the integrand into on-shell trees.

Note that this determination is impossible if the numerator contains a polynomial $p(\ell)$ that vanishes at these roots but does not belong to the ideal $\mathcal{I}_m$. Indeed, since $p \notin \mathcal{I}_m$, this polynomial is not removed by the multivariate polynomial division and stays in the remainder, and yet is undetectable by the generalized unitarity method because it vanishes on the cut. From the strong form of Hilbert's nullstellensatz, if a polynomial $p(\ell)$ vanishes at the common zeroes of the denominators, then we have

$$p \in \sqrt{\mathcal{I}_m}, \quad \text{with} \quad \sqrt{\mathcal{I}_m} \equiv \{q \in \mathbb{C}[\ell] \mid q^r \in \mathcal{I}_m, \, r > 0\}. \tag{19.85}$$

In order to avoid this problem, the ideal generated by the denominators should satisfy $\mathcal{I}_m = \sqrt{\mathcal{I}_m}$ (such an ideal is said to be *radical*). That this property is true should be checked when extracting coefficients from cuts.

In addition, this approach requires knowing a priori the most general allowed structure for the residual numerator (given the constraints posed by renormalizability, that limits the highest possible degree in each loop momentum), and one needs information on the set of common roots of the denominators (dimensionality of the set of roots, number of branches).

### 19.4.3 Maximal Cuts

As in the one-loop case, the unitarity cut approach starts with the terms that have the largest number of denominators, and proceeds iteratively to terms with fewer denominators. Assuming that there are enough external legs, the first step consists in putting 4L (in $d = 4$ dimensions) propagators on-shell. This configuration, called a *maximal cut*, is the simplest. Intuitively, since each loop momentum $\ell_i$ has four components, there are as many on-shell conditions as components in the loop momenta, and we expect that the solutions of the cut conditions form a discrete (zero-dimensional) algebraic set. By extension, the corresponding ideal $\mathcal{I}_m$ is also said to be zero-dimensional. Let us call $n_s$ the number of independent solutions, i.e.,

$$d_1(\ell_*^{(s)}) = \cdots = d_{4L}(\ell_*^{(s)}) = 0, \quad \text{for } s = 1, \cdots, n_s. \tag{19.86}$$

## 19.4 THE FRONTIER: MULTI-LOOP AMPLITUDES

We may assume that all the roots $\ell_*^{(s)}$ have multiplicity 1 and have distinct first coordinates,[19] i.e., $\ell_{*1}^{(s)} \neq \ell_{*1}^{(s')}$ if $s \neq s'$ (this is the generic situation for non-exceptional configurations of the external momenta). Then, define the polynomials

$$P_i(\ell) \equiv \ell_i - \sum_{s=1}^{n_s} \ell_{*i}^{(s)} \prod_{t \neq s} \frac{(\ell_1 - \ell_{*1}^{(t)})}{(\ell_{*1}^{(s)} - \ell_{*1}^{(t)})}. \tag{19.87}$$

Note that $P_{i \neq 1}$ depends only on $\ell_1, \ell_i$ (and is of degree 1 in $\ell_i$), and vanishes at all the $\ell_*^{(s)}$. In terms of these polynomials, the Gröbner basis (in lexicographical order) of the ideal generated by the on-shell denominators is

$$\mathbb{G} = \left\{ \prod_{s=1}^{n_s} (\ell_1 - \ell_{*1}^{(s)}), P_2(\ell), P_3(\ell), \cdots, P_{4L}(\ell) \right\}. \tag{19.88}$$

The remainder of the division of any polynomial by $\mathbb{G}$ is a univariate polynomial in $\ell_1$ of degree $< n_s$. Thus, *the residue of a maximal cut that has $n_s$ solutions depends polynomially on a single component of the loop momenta, and contains exactly $n_s$ parameters. Therefore, all the coefficients in this residue can be obtained from an $n_s \times n_s$ linear system by evaluating the numerator at these $n_s$ solutions.* The number $n_s$ of solutions can be obtained as follows. Given $m$ denominators $d_i$ and their ideal $\mathfrak{I}_m$, their *initial ideal* $\mathrm{In}\,(\mathfrak{I}_m)$ is the ideal generated by the leading[20] terms of the elements of $\mathfrak{I}_m$. A monomial is then called *standard* if it is not contained in the initial ideal. When finite, the number of solutions (counted with their multiplicities) is equal to the number of standard monomials. In four dimensions, this result generalizes to all loop orders the second part of eq. (19.48), whose right-hand side contains exactly two terms ($D_{0123}^{(0)} + D_{0123}^{(1)}(\ell \cdot n_4)$) if one excludes those that are due to dimensional regularization. Indeed, we have seen that the four-cut equations in four dimensions have two solutions.[21]

Dimensional regularization amounts to adding to each loop momentum extra coordinates that are transverse to the four-dimensional physical space. Because these additional coordinates do not couple to any physical object (external particle momentum or polarization vector), integrands depend on them only through the scalar products of the transverse components of the loop momenta, and the precise number of transverse dimensions does not matter. With L independent loop momenta, there are $\frac{1}{2}L(L+1)$ of these scalar products, which brings the total number[22] of free parameters in the loop momenta to $4L + \frac{1}{2}L(L+1) =$

---

[19] Zero-dimensional ideals whose roots all have multiplicity 1 are radical. Therefore, in this case, there is no ambiguity when extracting the residue from the cut.

[20] The initial ideal $\mathrm{In}\,(\mathfrak{I}_m)$ is also generated by the leading terms of the elements of the Gröbner basis.

[21] The initial ideal generated by the four-cut equations contains the monomials $\{\ell_0^2, \ell_1, \ell_2, \ell_3\}$. There are two standard monomials, 1 and $\ell_0$, and therefore two solutions.

[22] At one loop, there are five parameters, which is consistent with the fact that it was sufficient to use an embedding dimension $d = 5$ when extracting the coefficients of the OPP integrand decomposition. At two loops, the loop momenta have 11 components. Thus, the minimal embedding dimension is now $d = 6$. Loop momenta in six dimensions can be parametrized with an extension of the spinor-helicity formalism (since $d = 6$ is even). With $d_s$ space-time dimensions, the numerators of purely gluonic amplitudes are polynomials of degree L in $d_s$. When $d_s > d$, the contribution of the extra gluon polarizations is equivalent to that of scalars.

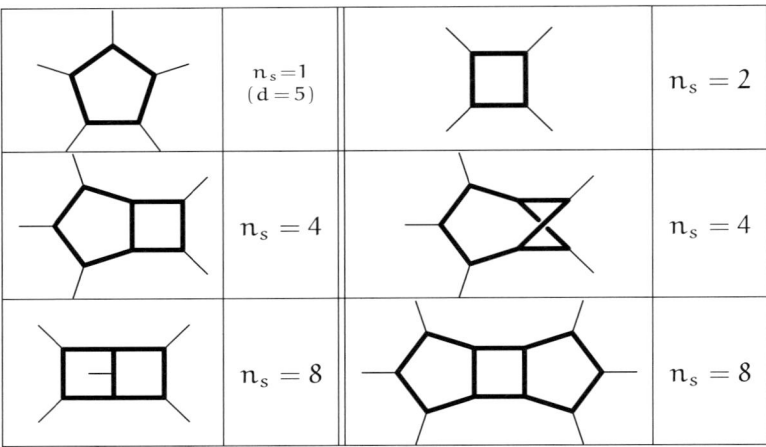

FIGURE 19.7: Number of solutions (or equivalently, number of parameters in the residue) for several types of maximal cuts (in d = 5 for the pentagon, and in d = 4 for all the others). Results from Mastrolia et al., Phys. Lett. B718 (2012) 173.

$\frac{1}{2}L(L+9)$. The *embedding dimension* is the smallest integer dimension equal to or larger than $\frac{1}{2}(L+9)$. Maximal cuts in dimensional regularization thus correspond to putting $\frac{1}{2}L(L+9)$ denominators on-shell, instead of 4L. Except for this difference, the reasoning developed in this section still applies. For L = 1, this result, combined with the fact that the five-cut equations have a single solution,[23] allows one to recover the fact that the pentagon terms have a unique coefficient $E^{(0)}_{01234}$. In Figure 19.7 we give a few examples of maximal cut configurations in four dimensions at two and three loops, and the corresponding number of solutions of the cut equations.

### 19.4.4 Non-Maximal Cuts

Let us now consider the case in which the number m of denominators $(d_1, \cdots, d_m)$ is smaller than 4L (in four dimensions). We assume that all the terms whose denominators form a strict superset of $\{d_1, \cdots, d_m\}$ have already been reduced. This situation is more complicated, but insights about the structure of the corresponding residue can also be gained from computational algebraic geometry.

**Independent scalar products:** In the case of maximal cuts, we managed to obtain explicitly the Gröbner basis, from which we concluded that the residue depends on a single variable. A more mundane way to state this simplification is that the 4L loop variables obey 4L − 1 linear relationships *when the maximal cut conditions are satisfied*, allowing elimination of all

---

[23] When using the number of standard monomials to count the solutions in d = 4 − 2ϵ at one loop, we should use the fact that the integrand depends only on $x_\epsilon \equiv (\ell \cdot n_\epsilon)^2$, rather than $\ell \cdot n_\epsilon$ itself. With this in mind, the initial ideal generated by the five-cut equations contains the monomials $\{\ell_0, \ell_1, \ell_2, \ell_3, x_\epsilon\}$, and the only standard monomial is 1.

but one. A simple explicit example of this phenomenon is the four cut conditions for the one-loop box:

$$\ell^2 = 0, \quad (\ell + q_1)^2 = 0, \quad (\ell + q_2)^2 = 0, \quad (\ell + q_3)^2 = 0. \tag{19.89}$$

When they are satisfied, we have

$$\ell \cdot q_1 \underset{\text{cut}}{=} -\frac{q_1^2}{2}, \quad \ell \cdot q_2 \underset{\text{cut}}{=} -\frac{q_2^2}{2}, \quad \ell \cdot q_3 \underset{\text{cut}}{=} -\frac{q_3^2}{2}. \tag{19.90}$$

Thus, although the components of the loop momentum are independent variables in general, they obey three linear relationships on the cut. We can use this linear system to eliminate three components of the loop momentum, and in four dimensions only one remains.

Such linear relations also exist for non-maximal cuts, and they can be obtained systematically from the Gröbner basis of the ideal generated by the on-shell denominators. In dimensional regularization, the numerator and denominators of any integrand are polynomials in the following variables:

$$\ell_i^\mu \equiv \underbrace{\ell_{i\parallel}^\mu}_{d=4} + \underbrace{\ell_{i\perp}^\mu}_{-2\epsilon}, \quad \mathbf{L} \equiv \underbrace{(\{\ell_i \cdot e_j\})}_{4L}, \quad \underbrace{\mu_{ij} \equiv \ell_{i\perp} \cdot \ell_{j\perp}}_{L(L+1)/2}, \tag{19.91}$$

where $e_{1,2,3,4}$ is a basis for the four dimensions in which live physical momenta and polarization vectors. Linear relations among these variables that are valid on the cut can be written as follows:

$$\sum_l \alpha_{kl} L_l + \beta_k = \underbrace{\sum_{i=1}^m q_{ki}(\mathbf{L}) \, d_i(\mathbf{L})}_{\in \mathcal{I}_m}, \tag{19.92}$$

where the $\alpha_{kl}, \beta_k$ are constants (the index $k \in [1, p]$ enumerates these linear relationships). By Gaussian elimination, these linear relations can be arranged in such a way that the coefficient matrix $\alpha_{kl}$ is triangular. Eq. (19.92) implies that the ideal $\mathcal{I}_m$ contains $p$ linearly independent polynomials of degree 1, whose leading monomials are $L_N, \cdots, L_{N-p+1}$. Therefore, the Gröbner basis $\mathcal{G}$ in graded lexicographical order[24] contains $p$ polynomials of degree 1 whose leading terms are respectively $L_N, \cdots, L_{N-p+1}$. In other words, the calculation of the Gröbner basis with this ordering tells us which variables are linearly dependent on the others on the cut, simply by inspecting the degree 1 polynomials it contains and their leading terms. The residue can only depend on the remaining variables, $L_1, \cdots, L_{N-p}$, called the irreducible scalar products.

**Set of cut solutions:** In the case of maximal cuts, we found that the number of discrete solutions coincides with the number of parameters to be determined in the residues. When

---

[24]This ordering of monomials, also called the degree lexicographical order (*deglex*), is obtained by comparing first the total degree, and in the case of a tie to use lexicographical order. If the comparison of monomials is denoted $\prec$, for two indeterminates $L_{1,2}$ this ordering reads $1 \prec L_1 \prec L_2 \prec L_1^2 \prec L_1 L_2 \prec L_2^2 \prec L_1^3 \prec L_1^2 L_2 \prec \cdots$.

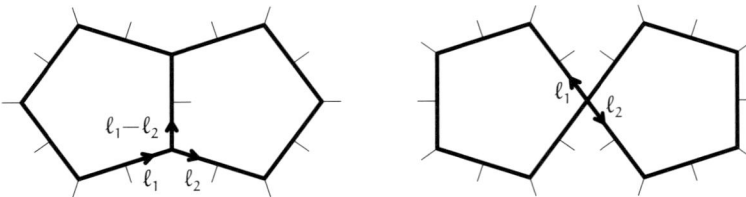

FIGURE 19.8: Two-loop skeleton topologies. Left: topology with overlapping loops. Right: "butterfly" topology with non-overlapping loops.

$m < 4L$ propagators $\{d_1, \cdots, d_m\}$ are cut, the naive count of the equations versus indeterminates suggests that the solutions span a $4L - m$ dimensional variety $Z(\{d_i\})$. This is certainly a minimum, but the actual dimension could be larger if the cut equations are not all independent. Because $Z(\{d_i\})$ is also the set of common roots of all the polynomials in $\mathcal{I}_m$, its dimension can be related to a purely algebraic concept,

$$\dim \left( Z(\{d_i\}) \right) = \text{Krull dimension of } \mathbb{C}[\ell]/\mathcal{I}_m, \qquad (19.93)$$

where the *Krull dimension* of the quotient ring can be computed from the Gröbner basis.[25] In the special case where the quotient is itself a polynomial ring $\mathbb{C}[\ell_1, \cdots, \ell_k]$, the dimension is k.

Algebraic geometry also tells us the number of branches of solutions. Let us denote $Z(\mathcal{I})$ the set of common zeroes of all polynomials in an ideal $\mathcal{I}$. Then, there is a correspondence between the fact that $Z(\mathcal{I})$ is an union of smaller irreducible algebraic varieties (one for each branch of solution), $Z(\mathcal{I}) = Z(\mathcal{I}_1) \cup Z(\mathcal{I}_2) \cup \cdots \cup Z(\mathcal{I}_q)$, and the *primary decomposition* of the ideal, $\mathcal{I} = \mathcal{I}_1 \cap \mathcal{I}_2 \cap \cdots \cap \mathcal{I}_q$, where the $\mathcal{I}_i$ obey the following:

- $\mathcal{I}_i$ is a primary ideal (i.e., such that $AB \in \mathcal{I}_i$ implies $A \in \mathcal{I}_i$ or $B^r \in \mathcal{I}_i$ with $r > 0$);
- the intersection is not redundant, i.e., removing any of the $\mathcal{I}_i$ changes the intersection (in other words, none of the $\mathcal{I}_i$ contain the intersection of the others); and
- the $\sqrt{\mathcal{I}_i}$ are all distinct.

Here also, there are algorithms based on Gröbner bases in order to perform explicitly this decomposition.

**Two-loop example:** Let us make this discussion more concrete with the example of two-loop topologies, starting with the case shown in the left diagram of Figure 19.8, where the two loops overlap. The two loop momenta $\ell_{1,2}^\mu$ can be parametrized in terms of eight scalar products $\ell_i \cdot e_j$ (with $\{e_j\}_{1 \leq j \leq 4}$ a basis of the physical four-dimensional space), and three scalar products $\mu_{11}, \mu_{12}, \mu_{22}$ among the transverse components introduced when $d \neq 4$, i.e., a total of 11 loop parameters. Let us denote $m$ the number of cut propagators. For this topology, three of the cut conditions are quadratic in the loop momenta,

$$\ell_1^2 = \ell_2^2 = (\ell_1 - \ell_2)^2 = 0, \qquad (19.94)$$

---

[25] Given the Gröbner basis in graded lexicographical order, construct the set $\{m_i\}$ of its leading monomials. The Krull dimension is the maximal size of a subset $\mathcal{S}$ of the variables such that none of the $m_i$ depends only on the variables in $\mathcal{S}$.

while the remaining m − 3 conditions have one of the following three forms:

$$2\ell_1 \cdot q + q^2 = 0, \quad 2\ell_2 \cdot q + q^2 = 0, \quad 2(\ell_1 - \ell_2) \cdot q + q^2 = 0, \quad (19.95)$$

with q a combination of external momenta or polarization vectors. These m − 3 linear equations can be used to eliminate an equal number of the linear scalar products. Let us denote collectively $\mathbf{L} \equiv (L_1, \cdots, L_{11-m})$ the unconstrained variables among the set $\{\ell_i \cdot e_j\}$. The remaining three cut equations express the $\mu_{ij}$ in terms of the independent $L_i$,

$$\mu_{ij} = p_{ij}(\mathbf{L}), \quad (19.96)$$

where the $p_{ij}$ are polynomials. The residue associated to this cut belongs to the quotient ring $\mathbb{C}[\mathbf{L}, \mu_{11}, \mu_{22}, \mu_{12}]/\mathcal{I}_m$. Thank to eqs. (19.96), this quotient is in fact isomorphic to $\mathbb{C}[\mathbf{L}]$. Since this is an integral domain (i.e., such that $AB = 0$ implies either $A = 0$ or $B = 0$), the ideal $\mathcal{I}_m$ is a *prime* ideal (i.e., an ideal such that $AB \in \mathcal{I}_m$ implies either $A \in \mathcal{I}_m$ or $B \in \mathcal{I}_m$), and therefore a radical ideal. Moreover, since $\mathcal{I}_m$ is prime, its primary decomposition is reduced to a single factor, and therefore the cut solution has only one branch. The dimension of the set of roots is $11 - m$, equal to the number of variables on which the residue depends.

Consider now the "butterfly" topology (the graph on the right of Figure 19.8). Now, only two of the cut conditions are quadratic, $\ell_1^2 = \ell_2^2 = 0$, and the remaining m − 2 cut conditions become linear constraints among the $\ell_i \cdot e_j$. The independent ones are $\mathbf{L} \equiv (L_1, \cdots, L_{10-m})$, and the $\mu_{ij}$ obey two equations,

$$\mu_{11} = p_{11}(\mathbf{L}), \quad \mu_{22} = p_{22}(\mathbf{L}). \quad (19.97)$$

($\mu_{12}$ is not constrained at all, and should be added to the list of independent variables upon which the residue depends.) The residue belongs to $\mathbb{C}[\mathbf{L}, \mu_{11}, \mu_{22}, \mu_{12}]/\mathcal{I}_m$, which is now isomorphic to $\mathbb{C}[\mathbf{L}, \mu_{12}]$. The conclusions about the radicality of $\mathcal{I}_m$ and the existence of a single branch of cut solutions still hold, and its dimension is also $11 - m$.

## Exercises

**19.1** Perform the reduction of the following rank-1 triangle integral:

$$T^\mu(1,2) \equiv \int \frac{d^d\ell}{(2\pi)^d} \frac{\ell^\mu}{\ell^2(\ell+q_1)^2(\ell+q_2)^2}.$$

**\*19.2** Do the same as Exercise 19.1 with the following rank-2 triangle integral:

$$T^{\mu\nu}(1,2) \equiv \int \frac{d^d\ell}{(2\pi)^d} \frac{\ell^\mu \ell^\nu}{\ell^2(\ell+q_1)^2(\ell+q_2)^2}.$$

**\*19.3** Consider massless quantum electrodynamics in 1 + 1 dimensions (also known as the Schwinger model), and the one-loop photon polarization tensor,

$$\Pi^{\mu\nu}(q) \equiv -e^2 \int \frac{d^d\ell}{(2\pi)^d} \frac{\text{tr}\left(\gamma^\mu \slashed{\ell} \gamma^\nu (\slashed{q}+\slashed{\ell})\right)}{\ell^2(\ell+q)^2}.$$

- Keeping in mind that the loop momentum $\ell$ is d-dimensional, while the external Lorentz indices $\mu, \nu$ and the momentum $q$ are strictly two-dimensional, use the OPP decomposition to reduce the integral to

$$\Pi^{\mu\nu}(q) = -4e^2 \int \frac{d^d\ell}{(2\pi)^d} \frac{1}{\ell^2(\ell+q)^2} \left[\ell_\perp^\mu \ell_\perp^\nu + \frac{q^2}{4}\left(g^{\mu\nu} - \frac{q^\mu q^\nu}{q^2}\right)\right],$$

where the loop momentum has been parametrized as follows:

$$\ell^\mu \equiv \underbrace{\frac{(\ell \cdot q)q^\mu}{q^2} + \ell_\perp^\mu}_{\text{2 dim}} + \underbrace{\sum_\epsilon (\ell \cdot n_\epsilon) n_\epsilon^\mu}_{d-2 \text{ dim}}.$$

- Show that the angular integration over the orientation of $\ell_\perp$ simply amounts to the substitution

$$\ell_\perp^\mu \ell_\perp^\nu \to -\left(\sum_\epsilon (\ell \cdot n_\epsilon)^2 + \frac{q^2}{4}\right)\left(g^{\mu\nu} - \frac{q^\mu q^\nu}{q^2}\right).$$

- Show that the photon polarization tensor is entirely given by a rational term, and calculate it. What happens to the pole of the photon propagator after this correction is resummed?

**19.4** Show that in a generic renormalizable field theory in four dimensions, the rank of the numerator of a one-loop integral is at most equal to the number of denominators.

**\*19.5** Consider the integrand of a rank-5 five-point function,

$$\mathcal{J} \equiv \frac{(\ell \cdot k_1) \cdots (\ell \cdot k_5)}{d_0 d_1 d_2 d_3 d_4},$$

with $d_i \equiv (\ell + q_i)^2$ ($q_0 = 0$) and $k_{1,2,3,4,5}$ purely four-dimensional vectors.

- Using the van Neerven–Vermaseren parametrization of the loop momentum,

$$\ell^\mu \equiv \sum_{i=1}^4 (\ell \cdot q_i) v_i^\mu + \sum_\epsilon (\ell \cdot n_\epsilon) n_\epsilon^\mu,$$

show that the numerator of the integrand can be written as

$$\text{Numerator} = \frac{1}{2}\left[\prod_{i=1}^4 \ell \cdot k_i\right] \sum_{j=1}^4 (v_j \cdot k_5)(d_j - d_0 - q_j^2),$$

which will lead to a sum of rank-4 four-point and five-point functions.
- Repeat in order to show that the coefficient of the five-denominator term in the OPP reduction is independent of $\ell$.
- What is the highest-rank integrand left at the end of the process?

**\*19.6** This is a sequel to the previous exercise. Consider now the integrand of a rank-4 four-point function,

$$\mathcal{J} \equiv \frac{(\ell \cdot k_1) \cdots (\ell \cdot k_4)}{d_0 d_1 d_2 d_3}.$$

- Write down the van Neerven–Vermaseren parametrization of the loop momentum relevant to this case (denote $n_4^\mu$ the additional vector introduced to make a basis of the four-dimensional physical space).
- Eliminate the factor $\ell \cdot k_4$ from the numerator, and show that the terms with four denominators in the integrand have the following structure:

$$\frac{\left[\prod_{i=1}^{3} \ell \cdot k_i\right]}{d_0 d_1 d_2 d_3} \left\{1 \oplus (\ell \cdot n_4)\right\}.$$

- Apply the same method on $\ell \cdot k_3$ to reach

$$\frac{\left[\prod_{i=1}^{2} \ell \cdot k_i\right]}{d_0 d_1 d_2 d_3} \left\{1 \oplus (\ell \cdot n_4) \oplus (\ell \cdot n_4)^2\right\}.$$

- Show that $(\ell \cdot n_4)^2 = -\sum_\epsilon (\ell \cdot n_\epsilon)^2 + \text{(constants)} + \text{(terms in } d_{0,1,2,3}\text{)}$.
- Repeat and show that the coefficient of the four-denominator term is of the form

$$D^{(0)} + D^{(1)}(\ell \cdot n_4) + D^{(2)} \sum_\epsilon (\ell \cdot n_\epsilon)^2$$
$$+ D^{(3)}(\ell \cdot n_4) \sum_\epsilon (\ell \cdot n_\epsilon)^2 + D^{(4)} \sum_{\epsilon,\epsilon'} (\ell \cdot n_\epsilon)^2 (\ell \cdot n_{\epsilon'})^2.$$

**19.7** From eq. (19.48), determine the terms that will be in "$\epsilon/\epsilon$" after performing the loop integration. Evaluate explicitly their integral in order to check the formula (19.50) for the rational term.

**19.8** Recall that $Z(\mathcal{I})$ is the set of the common zeroes of the polynomials in the ideal $\mathcal{I}$. Given two ideals $\mathcal{I}, \mathcal{J}$, their product is defined by

$$\mathcal{I}\mathcal{J} \equiv \left\{\sum_i f_i g_i | f_i \in \mathcal{I}, g_i \in \mathcal{J}\right\}.$$

(In words, it is the ideal generated by products of elements of $\mathcal{I}$ and $\mathcal{J}$.) Show that

a. $\mathcal{I}\mathcal{J} \subset \mathcal{I} \cap \mathcal{J}$,
b. $\sqrt{\mathcal{I}\mathcal{J}} = \sqrt{\mathcal{I} \cap \mathcal{J}} = \sqrt{\mathcal{I}} \cap \sqrt{\mathcal{J}}$,
c. $Z(\mathcal{I}) = Z(\sqrt{\mathcal{I}})$,
d. $Z(\mathcal{I} \cup \mathcal{J}) = Z(\mathcal{I}) \cap Z(\mathcal{J})$,
e. $Z(\mathcal{I} \cap \mathcal{J}) = Z(\mathcal{I}\mathcal{J}) = Z(\mathcal{I}) \cup Z(\mathcal{J})$.

# Further Reading

Álvarez-Gaumé, L. and Ginsparg, P. H. (1985), The structure of gauge and gravitational anomalies. *Annals Phys* 161: 423–490.

Banks, T. (2014), *Modern Quantum Field Theory* (Cambridge University Press).

Bilal, A. (2008), *Lectures on Anomalies* (arXiv:0802.0634).

Blaizot, J.-P. and Iancu, E. (2001), The quark gluon plasma: collective dynamics and hard thermal loops. *Phys Rept* 359: 355–528.

Bogolyubov, N. N. and Shirkov, D. V. (1983), *Quantum Fields* (Benjamin Cummings).

Cheung, C. (2018), *TASI Lectures on Scattering Amplitudes* (arXiv:1708.03872).

Coleman, S. (1985), *Aspects of Symmetry* (Cambridge University Press).

DeGrand, T. and Detar, C. E. (2006), *Lattice Methods for Quantum Chromodynamics* (World Scientific).

Dokshitzer, Y. L., Khoze, V. A., Mueller, A. H. and Troyan, S. I. (1991), *Basics of Perturbative QCD* (Editions Frontieres).

Elvang, H. and Huang, Y.-T. (2015), *Scattering Amplitudes in Gauge Theory and Gravity* (Cambridge University Press).

Itzykson, C. and Zuber, J. B. (1980), *Quantum Field Theory* (McGraw-Hill).

Kapusta, J. I. and Gale, C. (2011), *Finite-Temperature Field Theory: Principles and Applications* (Cambridge University Press).

Kovchegov, Y. V. and Levin, E. (2012), *Quantum Chromodynamics at High Energy* (Cambridge University Press).

Lancaster, T. and Blundell, S. J. (2014), *Quantum Field Theory for the Gifted Amateur* (Oxford University Press).

Le Bellac, M. (2011), *Thermal Field Theory* (Cambridge University Press).

Peskin, M. E. and Schroeder, D. V. (1995), *An Introduction to Quantum Field Theory* (Addison-Wesley).

Pich, A. (1998), *Effective Field Theory*. Course (arXiv:hep-ph/9806303).

Preskill, J. (1984), Magnetic monopoles. *Ann Rev Nucl Part Sci* 34: 461–530.

Ryder, L. H. (1996), *Quantum Field Theory* (Cambridge University Press).

Schäfer, T. and Shuryak, E. V. (1998), Instantons in QCD. *Rev Mod Phys* 70: 323–426.

Schubert, C. (2001), Perturbative quantum field theory in the string inspired formalism. *Phys Rept* 325: 73–234.

Schwartz, M. D. (2014), *Quantum Field Theory and the Standard Model* (Cambridge University Press).

Srednicki, M. (2007), *Quantum Field Theory* (Cambridge University Press).

Sterman, G. F. (1993), *An Introduction to Quantum Field Theory* (Cambridge University Press).

Weinberg, S. (2005), *The Quantum Theory of Fields. Vol. 1: Foundations* (Cambridge University Press).

Weinberg, S. (2013), *The Quantum Theory of Fields. Vol. 2: Modern Applications* (Cambridge University Press).

Weinberg, S. (2013), *The Quantum Theory of Fields. Vol. 3: Supersymmetry* (Cambridge University Press).

Zee, A. (2003), *Quantum Field Theory in a Nutshell* (Princeton University Press).

Zinn-Justin, J. (2002), *Quantum Field Theory and Critical Phenomena* (Oxford University Press).

# Index

1PI
    diagram, 132
    effective action, 129, 236, 421
2PI effective action, 138

Abel-Plana formula, 501
adjoint mapping, 177
adjoint representation, *see* representation, adjoint
Aharonov–Bohm effect, 347
anomalous dimension, 253, 256, 261, 267
anomaly, 312
    axial, *see* chiral anomaly
    from functional measure, 162
    function, 164, 357
    scale, 334
anti-commutation relations, 67
asymptotic freedom, 250, 272, 440
asymptotic safety, 272
Atiyah–Singer theorem, 168, 200, 201, 370
axial current, 166, 314, 319
axial symmetry, *see* chiral symmetry

background field method, 245
background field propagator, 526
Baker–Campbell–Hausdorff formula, 28, 117, 177, 421, 468, 507
Banks–Casher relation, 463
baryon
    current, 368
    number, 368
BCFW recursion, 388, 394, 401, 404, 408, 416
Berends–Giele recursion, 406, 572
Berezin integral, *see* Grassmann

Bern–Carrasco–Johansson, *see* color-kinematics duality
Bern–Kosower rule, 435
beta function, 253, 254, 339
Bianchi identity, 188
Bogoliubov inequality, 113, 115
Bogomol'nyi inequality, 344, 356
Boltzmann equation, 500, 545
Bose–Einstein distribution, 469, 477, 478
box, *see* master integral
BPHZ renormalization, 51
brownian motion, 420, 422
BRST
    charge, 230
    cohomology, 230
    current, 230
    symmetry, 227, 238, 331, 455, 456
bubble, *see* master integral
Buchberger algorithm, 575

Cabbibo–Kobayashi–Maskawa matrix, 192, 284
Cachazo–Svrcek–Witten rules, 406, 412
Callan–Symanzik equation, 252, 256, 257, 259, 307
canonical quantization
    fermions, 66
    photons (Coulomb gauge), 72
    scalars, 11
Cartan–Maurer invariant, 359, 363, 369
Casimir operator, 189, 210
Cayley's formula, 542
center symmetry, 503
charge conjugation, 79, 94, 287, 309
charge renormalization, 81
chemical potential, 471

chiral anomaly, 164, 166, 316, 322, 323, 448
chiral gauge theory, 190, 326
chiral Lagrangian, 307, 368
chiral symmetry, 109, 163, 190, 282, 300, 313, 448
chiral transformation, 199
Christoffel symbol, 324
classical electrodynamics, 70
classical field
    boundary conditions, 517, 525, 526, 535
    retarded, 510, 512, 517, 524, 526
clover term, 463
coherent state, 114, 506
    fermionic, 426
Coleman's theorem, 105
Coleman–Weinberg potential, 135
collision term, 500
color glass condensate, 295, 298
color ordering, 373, 553
    Feynman rules, 376
color-kinematics duality, 376, 400, 417
commutation relations
    canonical, 11
composite operator, 256
conformal transformation, 341
connected graph, 38, 62, 511
contact term, 161
contour ordering, *see* path ordering
cosmological constant, 282
counterterm, 50, 83, 233, 241, 254, 263, 316, 339, 483
covariant derivative, 75, 165, 181, 244, 293, 324, 352, 391, 429, 446
critical
    point, 270, 272
    surface, 272
cross-section, 17
current conservation, 13, 71, 81, 230, 261, 312
    covariant, 188
curvature tensor, 175, 305, 308, 326, 398
Cutkosky, *see* cutting rules
cutting rules, 54, 474, 565
    generalized, 564
    QED, 86

Debye
    mass, 491
    screening, 491, 494
decay rate, 21
degenerate vacua, 95, 107, 112
density operator, 465, 471, 495, 516
derivative expansion, 294
Derrick theorem, 367
dilatation current, 335, 340
dimensional regularization, 42, 45, 315, 558, 578, 579
Dirac
    equation, 65, 79
    Lagrangian, 65, 445
    matrices, 64, 93
    operator, 69, 164, 168, 198, 324, 424, 446
    spinor, 64
domain wall, 343
doubler, *see* Lattice doubler
Duhamel's formula, 178
dynamical fermion, 449
Dyson equation, 143

Eikonal approximation, 205, 391
electroweak theory, 262, 284
embedding dimension, 577
energy–momentum tensor, 336

Fadeev–Popov
    ghost, 215, 456
    method, 215, 454
Fermi theory, 261, 282
Fermi–Dirac distribution, 472
Feynman
    diagram, 36
    parametrization, 43, 314, 432
Feynman propagator, 30
    fermion, 69
Feynman rules, 37
    QED, 76
    Yang–Mills, 213, 219
Feynman tree theorem, 547
field strength, 70, 182, 323, 333, 368, 443
Fierz identity, 188, 265, 374, 382, 553
fine structure constant, 1

Fock state, 17, 426
form factor, 284
functional
    derivative, 27
    determinant, 127
    Fourier transform, 124
fundamental representation, *see* representation, fundamental
Furry's theorem, 83, 94

Galilean boost, 7
gap equation, 485
gauge fixing
    axial gauge, 71, 220
    background field, 246, 391
    Coulomb gauge, 71
    covariant gauge, 158, 214, 218
    Feynman gauge, 159, 213
    Fock–Schwinger gauge, 310, 439
    lattice, 454
    Lorenz gauge, 71, 158, 445, 454, 464
gauge invariance
    Abelian, 70
    non-Abelian, 181
gauge transformation, 70, 75, 181, 186
    Abelian, 70, 424
Gaussian fixed point, 272
generalized unitarity cut, 565, 570
    maximal cut, 576
generating functional, 531
    fermions, 69
    of connected graphs, 38
    of time-ordered products, 27, 468, 506
    photons, 74
    QED, 76
Georgi–Glashow model, 348
ghost field, *see* Fadeev–Popov, ghost
Gluon saturation, 296
Goldstone
    boson, 105, 112, 300, 305, 366
    theorem, 95, 105, 192
gradient approximation, 499
Grassmann
    algebra, 151

    complex variable, 155
    delta function, 171
    derivative, 151
    determinant, 154
    Fourier transform, 171
    function, 151
    Gaussian integral, 153, 154, 427
    integral, 151, 427
    Jacobian, 153
    variable, 151, 426, 448
gravitational amplitudes, 398
Green's formula, 517, 520, 522, 524
Green–Kubo formula, 474
Gribov copy, 214, 445, 454
Gröbner basis, 575–577, 579

Haar measure, 444, 451
Hamilton's equations, 12
hard thermal loop, 487, 488, 494
heat kernel, 418, 458
heavy quark
    potential, 452
    symmetry, 292
hedgehog field, 349
Heisenberg model, 111, 369
Higgs mechanism, 192
Hilbert's nullstellensatz, *see* theorem of zeroes
Hilbert–Einstein action, 398
Hofstadter model, 460
homogeneous space, 304
homotopy group, 170, 200, 201, 349, 352, 357, 367

ideal, 575, 576
    initial, 577
    primary decomposition, 580
    quotient ring, 580
    radical, 576
imaginary time formalism, *see* Matsubara, formalism
infrared divergence, 88, 482
instanton, 355, 363
interaction representation, 15
irreducible representation, *see* representation, irreducible
irrelevant operator, 272, 280
Ising model, 268

Jacobi identity, 177, 228, 244, 416
JIMWLK equation, 300

Kadanoff blocking, 268
Kadanoff–Baym equations, 498
kaon decay, 262
Kawai–Lewellen–Tye relations, 400
Killing form, 361
kinetic equation, *see* Boltzmann equation
kinetic theory, 498
Klein–Gordon equation, 13, 14, 507
Kleiss–Kuijf relations, 376, 416
Krull dimension, 580
Kubo–Martin–Schwinger symmetry, 145, 470, 479, 496

Landau damping, 492, 494
Landau gauge, *see* gauge fixing, Lorenz
largest time equation, 54
Lattice action, 442
lattice action, 443
lattice doubler, 447
Legendre transform, 131, 139, 149, 239
Lie algebra, 174
  compact, 185
  simple, 184
Lie bracket, 174, 177
Lie group, 173, 354
light-cone
  coordinates, 206, 297
  quantization, 11, 34
  wave function, 209
linear sigma model, 107, 368
link variable, 444
Liouville equation, 517
Liouville–von Neumann equation, 516
little group, 6
  scaling, 384, 399, 415
loop integral reduction
  Passarino–Veltman, 556
  van Neerven–Vermaseren, 557
loop integrand reduction, 562, 573
Lorentz
  group, 3, 4, 379
  transformation, 3
LSZ reduction formula, 22
  fermions, 68
  photons, 73
Lyapunov exponent, 540

Majorana fermion, 309
Mandelstam variables, 77
marginal operator, 272, 280
mass matrix, 97
master integral, 560, 566, 568–570
Matsubara
  formalism, 147, 475
  frequency, 147, 475
maximal cut, *see* generalized unitarity cut
maximally symmetric space, 304
Maxwell's equations, 70, 79
McLerran–Venugopalan model, 297
mean free path, 494
Mermin–Wagner's theorem, 112
MHV, *see* scattering amplitude, MHV
Milne coordinates, 310
minimal momentum shift, 389
mode function, 514, 526, 540
  completeness, 529
monomial
  order, 575, 577, 579
  standard, 577
monopole
  charge quantization, 347, 352
  Dirac, 346
  non-Abelian, 348
Moyal–Groenewold equation, 517

Nambu–Goldstone, *see* Goldstone
natural units, 9
Newton's constant, 398
Nielsen–Ninomiya theorem, 448
Noether's theorem, 13, 161, 312, 335
nonlinear sigma model, 301, 366
normal-ordered exponential, 29
Nullstellensatz, *see* theorem of zeroes

occupation number, 9
operator product expansion, 258, 262, 280

optical theorem, 53
Ossola–Papadopoulos–Pittau, *see* loop integrand reduction

Parke–Taylor formula, 394, 397, 406, 409, 416
partition function, 473
Parton model, 296
Passarino–Veltman, *see* loop integral reduction
path integral
    classical limit, 120
    ground state projection, 123
    quantum mechanics, 116
    scalar field, 125
    statistical mechanics, 144
    time-ordered product, 120
path ordering, 59, 468
Pauli statistics, 67
pentagon, *see* master integral
pion decay, 167
Plaquette, 204, 451
plaquette, 443
Poincaré group, 5
polarization tensor, 85, 435
polarization vector, 73, 382, 399, 571
polynomial division, 574
power counting, 40, 48, 280, 317, 386, 432, 505, 509, 511
    gluon saturation, 298
    QED, 82
    Yang–Mills, 234
principal value, 31
pure gauge field, 182, 203

quantum anomaly, *see* anomaly
quantum effective action, *see* 1PI
quantum electrodynamics, 63, 424, 434
    scalar, 423, 428, 433, 459
quark confinement, 452
quasi-classical approximation, 537, 539
quasi-particle, 490
    approximation, 500
quenched approximation, 449

rational term, *see* scattering amplitude
relevant operator, 272, 280, 281
renormalizability, 49, 281, 303

QED, 83
    Yang–Mills, 241
renormalization condition, 51, 85
renormalization group, 253, 267, 268
    functional, 273
    Wilson, 270
representation, 179
    adjoint, 179
    fundamental, 179
    irreducible, 179
    singlet, 179
retarded propagator, 480, 510, 515, 518, 524, 527, 532, 547
Ricci flow, 308
running coupling, 250, 255, 339, 440
    QCD, 247
$R_\xi$ gauge, *see* gauge fixing, covariant

scale anomaly, *see* anomaly, 339
scale invariance, 334
scattering amplitude, 17, 372, 546, 560, 573
    MHV, 386, 394, 406
    next-to-MHV, 407
    rational term, 561, 562, 564
Schouten identity, 381, 383, 407, 412
Schwinger
    mechanism, 427
    model, 312
Schwinger–Dyson equations, 160
Schwinger–Keldysh
    formalism, 58, 467, 478, 480, 498, 504, 512, 513, 522, 524, 532, 550
    propagators, 56, 479, 513, 514, 547
    retarded basis, 527
seesaw mechanism, 288
simple Lie algebra, *see* Lie algebra
singlet representation, *see* representation, singlet
skyrmion, 366
S-matrix, 17, 53, 205, 222, 291
special conformal transformation, 3, 341
spectral function, 25, 34, 492
spin wave, 112
spin-flavor, *see* heavy quark, symmetry
spin-statistics theorem, 67, 217

spinor-helicity formalism, 379
    polarization vector, 382
        three-point amplitudes, 382, 399
spontaneous symmetry breaking, 95, 190, 282, 300, 345, 349, 366
spurious term, 564
stabilizer subgroup, 192, 300
Standard Model, 186, 192, 278, 282, 285
    anomaly cancellation, 326
stereographic projection, 367
Stokes theorem, 197, 321, 348, 358, 518
strong coupling expansion, 451
strong CP problem, 199
strong field, 363, 504, 509, 512
structure constant, 177, 183, 348
Sudakov factor, 92
superficial degree of divergence, 48
symmetry factor, 14, 38, 61, 76, 149, 225, 541

tetrad formalism, 325
theorem of zeroes, 575
thermal
    contour, 467, 470, 475
    ensemble, 465, 473
    mass, 483
    propagator, 469, 472, 473, 476
    sum rules, 492, 502
    symmetry restoration, 485
$\theta$ term, 194, 357, 363, 364
't Hooft anomaly matching, 333
time evolution operator, 15, 467
time ordered
    exponential, 15, 28
    product, 24
    propagator, *see* Feynman
transport coefficient, 474, 494
tree–loop duality, 548
triangle, *see* master integral
triviality, 272
Trotter formula, 174

ultraviolet
    divergences, 44
    regularization, 45
unitarity, 53
    QED, 86
    Yang–Mills, 223, 232
universality class, 272

vacuum graphs, 37
van Neerven–Vermaseren, *see* loop integral reduction
van Neerven–Vermaseren basis, 557, 567, 569
Vierbein, *see* tetrad formalism

ward identity, 81, 85, 313, 326, 487
    non-Abelian, 221, 224, 392
Weierstrass transform, 148, 420
Weinberg
    operator, 287
    theorem, 48
Wess–Zumino conditions, 329
Weyl
    mapping, 147
    representation, 64, 264
    spinor, 380
Wick
    rotation, 42, 165
    theorem, 148
Wigner transform, 148, 498, 516
Wilson
    action, 443
    coefficient, 259, 263
    line, 202, 429, 444, 460
    loop, 204, 443, 451
    term, 447, 463
worldline representation, 418, 421, 423, 427, 431, 432, 458

Yang–Mills
    energy–momentum tensor, 338
    equation, 187
    Lagrangian, 181, 212, 443
    theory, 181, 372
Yukawa coupling, 191, 287, 288

$\zeta$-function regularization, 128, 437
Zinn–Justin equation, 238